Astrophysics with Radioactive Isotopes

Astrophysics and Space Science Library

More information about this series at http://www.springer.com/series/5664

Roland Diehl • Dieter H. Hartmann • Nikos Prantzos
Editors

Astrophysics with Radioactive Isotopes

Second Edition

Springer

Editors
Roland Diehl
MPI für Extraterrestrische Physik
Garching, Germany

Dieter H. Hartmann
Dept. of Physics & Astronomy, Kinard Lab
Clemson University
Clemson
South Carolina, USA

Nikos Prantzos
Institut d'Astrophysique de Paris
Paris, France

Originally published with the title: Astronomy with Radioactivities

ISSN 0067-0057 ISSN 2214-7985 (electronic)
Astrophysics and Space Science Library
ISBN 978-3-030-06331-3 ISBN 978-3-319-91929-4 (eBook)
https://doi.org/10.1007/978-3-319-91929-4

Cover illustration: The gamma-ray emission from ^{26}Al radioactivity, superimposed onto an optical image showing the stars of the milky way. On the righthand side a schematic illustration of a ^{26}Al nucleus is shown. Credits: MPE Garching, Germany

This Springer imprint is published by the registered company Springer Nature Switzerland AG
The registered company address is: Gewerbestrasse 11, 6330 Cham, Switzerland

Preface to the Second Edition

After the success of the first edition of this book, we were stimulated, but also challenged, towards updating what had seemed to be a useful compilation of basic knowledge for this inter-disciplinary field. While some of our contributors were heavily engaged in new projects that demanded their full energy, others had opportunities to include such a project into their professional calendars, and actually did this with great enthusiasm.

A change in the title reflects the new emphasis to discuss the astro-physical considerations, more than merely presenting them as background for the astronomical achievements only. As a result, several chapters were re-written, and improved in details, as the new, in some cases dramatic, insights of the past 8 years were woven in.

This book is structured into four parts: (I) an introduction, (II) a treatment of sources of radioactivity, (III) a presentation of observed cosmic radioactivities, and a final part (IV) addressing characteristic elements of research with cosmic radioisotopes. In the book's Introduction part, the first chapter presents radioactivity in general with a view on adjacent disciplines of astronomy, and is followed by Chap. 2 providing an account of the history of astrophysical studies with radioactive isotopes. The latter was written by D.D. Clayton, one of the founders of the field, who also contributed a fundamental textbook on stellar astrophysics and many stimulating ideas that determined the course of this field, in particular the need for a concerted observational effort to understand nucleosynthesis from both measurements of electromagnetic radiation and meteoritic material. We enjoyed Donald D. Clayton's enthusiasm, which delivered an update that illuminates many astrophysics ideas within their historical framework. This is followed by Part II on the specific sources of cosmic radioactivity. In Chaps. 3–5, stars with their hydrostatic interiors, then the more-massive stars and their core-collapse supernova explosions, and last binary-star interactions including thermonuclear supernova explosions are described with their astrophysical ingredients and their links to radioactivities, summarising our current theoretical models and understanding of each of the cosmic sources of radioactivities, and the account of their key observations. It was a remarkable experience for the second edition of the book to see how these 8 years of research led to significant updates, starting from stellar structure and the lessons from solar neutrinos and pre-solar grain data, then re-evaluating cosmic explosions with the new role that appeared from detailed data from supernovae such as Cas A and SN2014J, and a spectacular merger of two neutron stars with an event named GW170817. Then in Part III we turn our perspective more towards the observational side and present the prominent locations of radioactivities as they are observed: Chap. 6 addresses the solar system, for which new views have emerged in recent years that make it appear possibly less special than thought a decade ago, and Chap. 7 more broadly then presents the different diffuse radioactivities seen in interstellar gas. Here, long-lived radioactivities from ^{26}Al and ^{60}Fe established an important astronomical window of their own in these recent years of emerging multi-messenger studies, and new spectroscopy of positron annihilation is now capable of discriminating among components of different origins or physical environments. The book is completed by chapters of Part IV, which present the most important "tools" that are characteristic of the field, that is, modelling astrophysical sources, studying nuclear reactions, building and using astronomical instruments, describing the compositional evolution of cosmic gas, and a commented list of branching-point isotopes (in Chaps. 8–12). With respect to the first Edition, we updated these "tools" chapters so that they reflect the current status. Continued advances can be demonstrated, for nuclear physics and reaction rates as well as for astronomical instrumentation. Such progress often occurs in shadows of the big headlines, but these are the steps that are needed to sustain scientific progress in complex, inter-disciplinary fields. We added a new chapter on chemical evolution among these "tools"; this completes our treatment of the cycle of matter, which starts at the microscopic level of atomic nuclei, using their properties to make changes to the compositional mix of atomic nuclei throughout the evolution of our great universe, thus providing the elements that enable also organic life. We conclude with some views on future perspectives. An Appendix holds a timeline of milestones for the field, and a guide through the basic terminology of *astrophysics with radioactive isotopes*.

Beyond the general guidance of the first two book chapters subsequent chapters can be read in any sequence suitable to the interests and curiosities of the reader, and we made a special attempt to provide cross references throughout the book. Enjoy the ride!

Garching, Germany
26 July 2018

Roland Diehl

Preface to the First Edition

About a 100 years ago, it was found that nuclear reactions were the source of the energy which makes the stars shine. It became clear that stars, supernovae, and other cosmic sites of nuclear fusion reactions are the agents that drive the evolution from the primordial element abundances, which consist of mainly hydrogen and helium, into the rich variety of more than 100 chemical elements and more than 3000 known isotopes. As part of this *cosmic nucleosynthesis*, unstable isotopes are created. These lead to unique astronomical signatures about the cosmic nucleosynthesis process and its environments, which need to be captured and unravelled to teach us about the astrophysics of cosmic nucleosynthesis and related processes.

It is the goal of this book to introduce this specific kind of *astronomy*, i.e. the acquisition and reading of measurements on unstable isotopes in different parts of the universe, and to relate other astronomical and nuclear-physics measurements of very different kinds and corresponding theories to the *astrophysical studies on nuclear-physics aspects of cosmic objects and evolutionary processes*. This field is part of what is also often called *"nuclear astrophysics"*. Pursuing this goal, we describe the stars of different types, stellar explosions that often characterise the ends of stellar evolution, and other explosions triggered by mass transfers and instabilities in binary stars. We also address nuclear reactions and transport processes in interstellar space, in the contexts of diffuse radioactivities, of cosmic rays, and of chemical evolution. A special chapter is dedicated to the solar system, where we have our best material samples of one star that formed 4.6 Gigayears ago with its planets and our Earth. We add to this book a description of key tools which astrophysicists employ in those particular studies.

This book has its roots in a series of small-scale, bi-annual workshops, which started in 1996 in Clemson, South Carolina (USA), as *"The Radioactive Galaxy"*, and was later continued mostly at the Max Planck Society's "Ringberg" castle in the Bavarian alps (Germany) under the title of *Astronomy with Radioactivities*. During these cross-disciplinary workshops, the variety of themes of nuclear astrophysics were discussed among experimenters in meteoritics, astronomy with spectral lines from gamma-rays to radio wavelengths, cosmic rays, and theorists of stellar evolution and explosions, of cosmic chemical evolution, and of nuclear reaction experts. These fruitful interactions among scientists are reflected in this book. We are grateful to the team of authors who took their time to write down their lessons, thoughts, and puzzles, in this inter-disciplinary spirit of trying to understand how nature made the atomic nuclei and elements that we are composed of.

Garching, Germany Roland Diehl
Clemson, CA, USA Dieter H. Hartmann
Paris, France Nikos Prantzos
25 March 2010

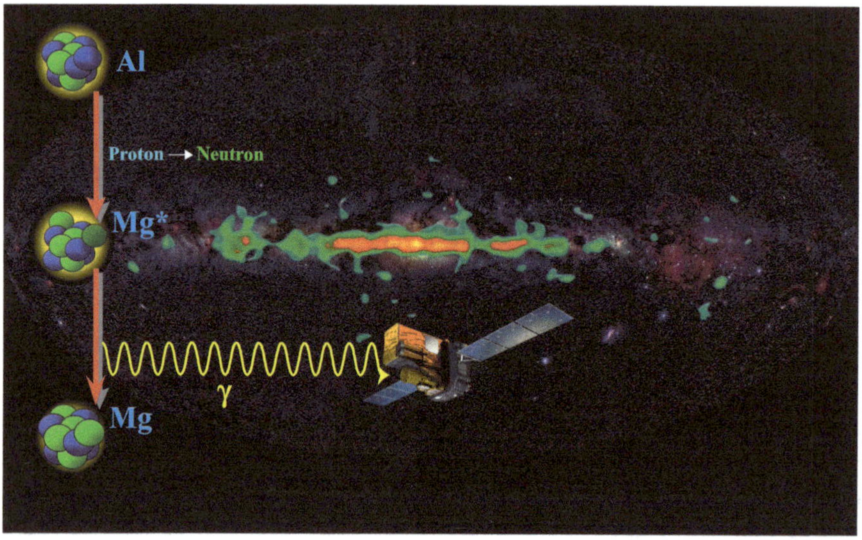

Acknowledgements

Many colleagues have contributed to this field, yet are not explicitly co-authors of this book. The choice of book authors grew out of opportunities around 2008, being aware that there were many more who could (and probably would) have joined into this adventure. All book authors are grateful for stimulating discussions and for collaborative work and publications shared with many dear friends and colleagues. We explicitly would like to mention here Sachiko Amari, Peter von Ballmoos, John Beacom, Peter Biermann, Bob Binns, Andreas Burkert, Roger Chevalier, John Danziger, Mounib El Eid, Bruce Elmegreen, Brian Fields, Claes Fransson, Roberto Gallino, Neil Gehrels, Matthieu Gounelle, Alexander Heger, Wolfgang Hillebrandt, Rob Hoffman, Peter Hoppe, Christian Illiadis, Anatoli Iyudin, Thomas Janka, Franz Käppeler, Jürgen Knödlseder, Gunther Korschinek, Karl-Ludwig Kratz, Pavel Kroupa, Karlheinz Langanke, John Lattanzio, Bruno Leibundgut, Mark Leising, Marco Limongi, Günther W. Lugmair, Gabriel Martinez-Pinedo, Bradley Meyer, Georges Meynet, Peter Milne, Yuko Mochizuki, Thierry Montmerle, Nami Mowlavi, Ewald Müller, Ken'ichi Nomoto, Uli Ott, Etienne Parizot, Volker Schönfelder, David Smith, Andrew Strong, Vincent Tatischeff, Lih-Sin The, James Truran, Jacco Vink, Stan Woosley, Hans Zinnecker, and the late Ernst Zinner.

For the second edition, special acknowledgements go to Albino Perego for contributing to Chap. 4, and all the chapter authors for their enthusiasm to review and update their earlier work. Thank you all.

Garching, Germany
26 July 2018

Roland Diehl

Contents

Part V Epilogue

Part I
The Role of Radioactivities in Astrophysics

In the following two chapters we introduce basic astrophysical considerations related to radioactive isotopes and their decay. A general introduction to the terminology of radioactivity and its physical meaning is followed by a personal account from one of the fathers of this field. The latter segment presents the development of the science of cosmic radioactivity and discusses the fundamental questions that were raised from the early years to those challenging its present day practitioners.

Chapter 1
Astrophysics with Radioactive Isotopes

Roland Diehl

Radioactivity was discovered as a by-product of searching for elements with suitable chemical properties. The efforts to understand its characteristics led to the development of nuclear physics, understanding that unstable configurations of nucleons transform into stable end products through radioactive decay. In the universe, nuclear reactions create new nuclei under the energetic circumstances characterising cosmic nucleosynthesis sites, such as the cores of stars and supernova explosions. Observing the radioactive decays of unstable nuclei, which are by-products of such cosmic nucleosynthesis, is a special discipline of astronomy. Understanding these special cosmic sites, their environments, their dynamics, and their physical processes, is the research goal of the *Astrophysics with Radioactivities* that makes the subject of this book. We address the history, the candidate sites of nucleosynthesis, the different observational opportunities, and the tools of this field of astrophysics.

1.1 Origin of Radioactivity

The nineteenth century spawned various efforts to bring order into the elements encountered in nature. Among the most important was an inventory of the *elements* assembled by the Russian chemist Dimitri Mendeleyev in 1869, which grouped elements according to their chemical properties, their *valences*, as derived from the compounds they were able to form, at the same time sorting the elements by atomic weight. The genius of Mendeleyev lay in his confidence in these sorting principles, which enforce gaps in his table for expected but then unknown elements, and Mendeleyev was able to predict the physical and chemical properties of such

R. Diehl (✉)
Max Planck Institut für extraterrestrische Physik, Garching, Germany
e-mail: rod@mpe.mpg.de

R. Diehl et al. (eds.), *Astrophysics with Radioactive Isotopes*, Astrophysics and Space Science Library 453, https://doi.org/10.1007/978-3-319-91929-4_1

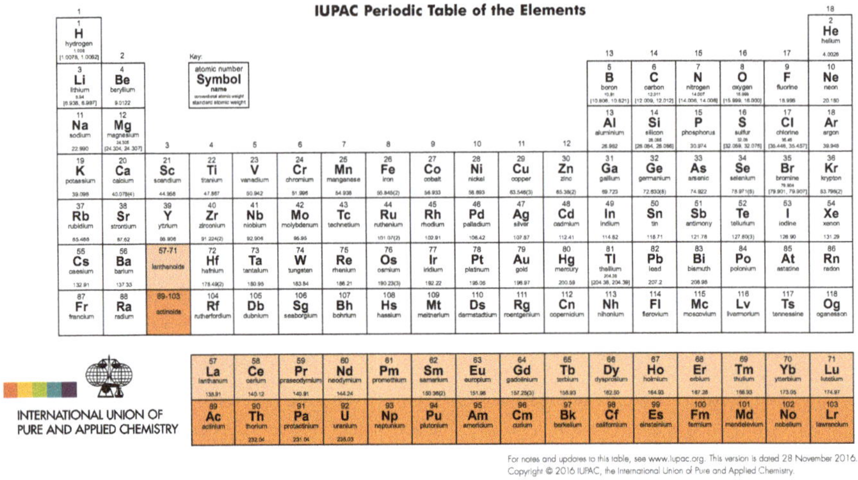

Fig. 1.1 The periodic table of elements, grouping chemical elements according to their chemical-reaction properties and their atomic weight, after Mendeleyev (1869), in its 2016 version (https://IUPAC.org)

elements-to-be-found. The tabular arrangement invented by Mendeleyev (Fig. 1.1) still is in use today, and is being populated at the high-mass end by the great experiments in heavy-ion collider laboratories to create the short-lived elements predicted to exist. The second half of the nineteenth century thus saw scientists being all-excited about chemistry and the fascinating discoveries one could make using Mendeleyev's sorting principles. Note that this was some 30 years before sub-atomic particles and the atom were discovered. Today the existence of 118 elements is firmly established,[1] the latest additions no. 113–118 all discovered in year 2016, which reflects the concerted experimental efforts.

In the late nineteenth century, scientists also were excited about new types of penetrating radiation. Conrad Röntgen's discovery in 1895 of *X-rays* as a type of electromagnetic radiation is important for understanding the conditions under which Antoine Henri Becquerel discovered radioactivity in 1896. Becquerel also was engaged in chemical experiments, in his research on phosphorescence exploiting the chemistry of photographic-plate materials. At the time, Becquerel had prepared some plates treated with uranium-carrying minerals, but did not get around to make the planned experiment. When he found the plates in their dark storage some time later, he accidentally processed them, and was surprised to find an image of a coin which happened to have been stored with the plates. Excited about X-rays, he believed he had found yet another type of radiation. Within a few years, Becquerel with Marie and Pierre Curie and others recognised that the origin of the observed radiation were elemental transformations of the uranium minerals: The physical

[1] IUPAC, the international union of chemistry, coordinates definitions, groupings, and naming; see www.IUPAC.org.

process of *radioactivity* had been found! The revolutionary aspect of elements being able to spontaneously change their nature became masked at the beginning of the twentieth century, when sub-atomic particles and the atom were discovered. But well before atomic and quantum physics began to unfold, the physics of *weak interactions* had already been discovered in its form of *radioactivity*.

The different characteristics of different chemical elements and the systematics of Mendeleyev's periodic table were soon understood from the atomic structure of a compact and positively charged nucleus and a number of electrons orbiting the nucleus and neutralising the charge of the atom. Bohr's atomic model led to the dramatic developments of quantum mechanics and spectroscopy of atomic shell transitions. But already in 1920, Ernest Rutherford proposed that an electrically neutral particle of similar mass as the hydrogen nucleus (proton) was to be part of the compact atomic nucleus. It took more than two decades to verify by experiment the existence of this 'neutron', by James Chadwick in 1932. The atomic nucleus, too, was seen as a quantum mechanical system composed of a multitude of particles bound by the strong nuclear force. This latter characteristic is common to 'hadrons', i.e. the electrically charged proton and the neutron, the latter being slightly more massive.[2] Neutrons remained a mystery for so long, as they are unstable and decay with a mean life of 880 s from the weak interaction into a proton, an electron, and an anti-neutrino. This is the origin of radioactivity.

The chemical and physical characteristics of an element are dominated by their electron configuration, hence by the number of charges contained in the atomic electron cloud, which again is dictated by the charge of the atomic nucleus, the number of protons. The number of neutrons included in the nucleus are important as they change the mass of the atom, however the electron configuration and hence the properties are hardly affected. Therefore, we distinguish *isotopes* of each particular chemical element, which are different in the number of neutrons included in the nucleus, but carry the same charge of the nucleus. For example, we know of three stable isotopes of oxygen as found in nature, ^{16}O, ^{17}O, and ^{18}O. There are more possible nucleus configurations of oxygen with its eight protons, ranging from ^{13}O as the lightest and ^{24}O as the most massive known isotope.

An *isotope* is defined by the number of its two types of nucleons,[3] *protons* (the number of protons defines the charge number Z) and *neutrons* (the sum of the numbers of protons and neutrons defines the mass number A), written as ^{A}X for an element 'X'. Note that some isotopes may exist in different nuclear quantum states which have significant stability by themselves, so that transitions between

---

[2]The mass difference is (Patrignani and Particle Data Group 2016) 1.293332 MeV = 939.565413 − 938.272081 MeV for the mass of neutron and proton, respectively. One may think of the proton as the lowest-energy configuration of a hadron, that is the target of matter in a higher state, such as the combined proton-electron particle, more massive than the proton by the electron mass plus some binding energy of the quark constituents of hadrons.

[3]The sub-atomic particles in the nucleus are composed of three quarks, and also called *baryons*. Together with the two-quark particles called *mesons*, they form the particles called *hadrons*, which obey the strong nuclear force.

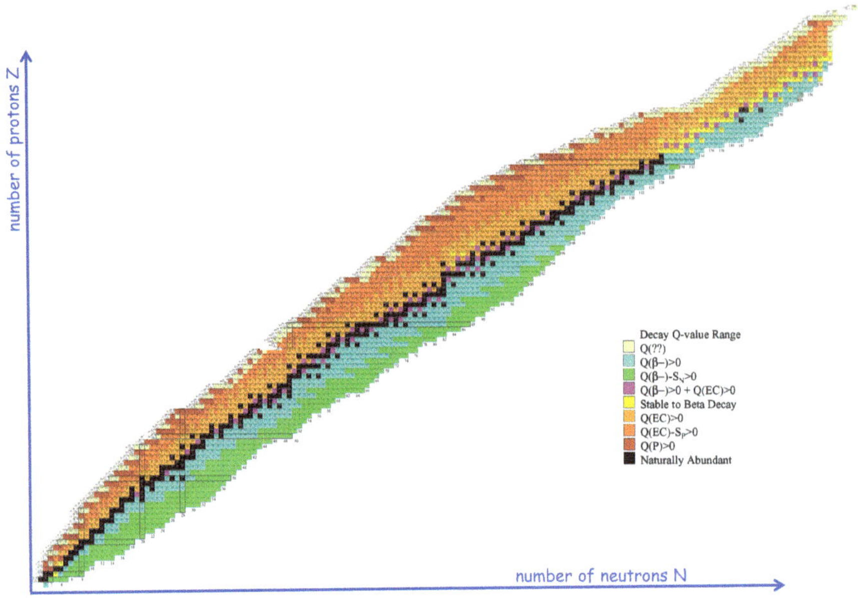

Fig. 1.2 The table of isotopes, showing nuclei in a chart of neutron number (abscissa) versus proton number (ordinate). The stable elements are marked in black. All other isotopes are unstable, or radioactive, and will decay until a stable nucleus is obtained

these configurations may liberate the binding energy differences; such states of the same isotope are called *isomers*. The landscape of isotopes is illustrated in Fig. 1.2, with black symbols as the naturally-existing stable isotopes, and coloured symbols for unstable isotopes.

Unstable isotopes, once produced, will be *radioactive*, i.e. they will transmute to other isotopes through nuclear interactions, until at the end of such a decay chain a stable isotope is produced. Weak interactions will mediate transitions between protons and neutrons and lead to neutrino emission, involvements of atomic-shell electrons will result in X-rays from atomic-shell transitions after electron capture and internal-conversion transitions, and γ-rays will be emitted in electromagnetic transitions between excitation levels of a nucleus.

The production of non-natural isotopes and thus the generation of man-made radioactivity led to the Nobel Prize in Chemistry being awarded to Jean Frédéric Joliot-Curie and his wife Iréne in 1935—the second Nobel Prize awarded for the subject of radioactivity after the 1903 award jointly to Pierre Curie, Marie Skłodowska Curie, and Henri Becquerel, also in the field of Chemistry. At the time of writing, element 118 called oganesson (Og) is the most massive superheavy element which has been synthesised and found to exist at least for short time intervals, although more massive elements may exist in an island of stability beyond. Depending on the astrophysical objective, radioactive isotopes may be called *short-lived*, or *long-lived*, depending on how the radioactive lifetime compares to astrophysical time scales of interest. Examples are the utilisation of ^{26}Al and

^{60}Fe (τ ~My) diagnostics of the early solar system (*short-lived*, Chap. 6) or of nucleosynthesis source types (*long-lived*, Chaps. 3–5).

Which radioactive decays are to be expected? What are stable configurations of nucleons inside the nuclei involved in a production and decay reaction chain? The answer to this involves an understanding of the nuclear forces and reactions, and the structure of nuclei. This is an area of current research, characterised by combinations of empirical modeling, with some capability of *ab initio* physical descriptions, and far from being fully understood.

Nevertheless, a few general ideas appear well established. One of these is recognising a system's trend towards minimising its total energy, and inspecting herein the concept of *nuclear binding energy*. It can be summarised in the expression for nuclear masses (Weizsäcker 1935):

$$m(Z, A) = Zm_p + (A - Z)m_n - BE \tag{1.1}$$

with

$$BE = a_{volume}A - a_{surface}A^{2/3} - a_{coulomb}\frac{Z^2}{A^{1/3}} - a_{asymmetry}\frac{(a-2Z)^2}{4A} - \frac{\delta}{A^{1/2}} \tag{1.2}$$

The total *binding energy* (BE) is used as a key parameter for a system of nucleons, and nucleons may thus adopt bound states of lower energy than the sum of the free nucleons, towards a global minimum of system energy. Thus, in a thermal mixture of nucleons, bound nuclei will be formed, and their abundance depends on their composition and shape, and on the overall system temperature, defining how the totally-available phase space of internal and kinetic energy states is populated. The nucleonic systems would thus have local maxima of binding energy from (1) the *odd-even* effect described by the last term, which results in odd-nucleon nuclei being less favored that even-nucleon nuclei, and (2) a general excess of neutrons would be favored by the asymmetry term, which results in heavier nuclei being relatively more neutron rich.

The other concept makes use of *entropy*, recognising the relation of this thermodynamic variable to the over-all state of a complex multi-particle and multi-state system. A change in entropy corresponds to a change in the micro-states available to the system. For an infinitesimal change in entropy, we have

$$Tds = -\sum_i \mu_i dY_i \tag{1.3}$$

where Y_i are the fractional abundances by number of a species i, e.g. i= ^{12}C, or ^{4}He, or protons ^{1}H, and μ is the thermodynamic potential[4] of species i.

[4]This is often called *chemical potential*, and describes the energy that is held as internal energy in species i, which could potentially be liberated when binding energy per nucleon would change as nucleons would be transferred to different species $j, k, l \ldots$.

Hence, for our application, if the isotopic composition of a nucleonic mixture changes, its entropy will also change. Or, conversely, the entropy, normalised by the number of baryons in the system, will be a characteristic for the composition:

$$Y_i \propto \frac{S}{n_b} = s \tag{1.4}$$

with the interpretation of entropy related to the (logarithm of) the number Γ of micro-states available:

$$S = k_b \cdot ln\Gamma \tag{1.5}$$

This thermodynamic view allows to calculate *equilibrium* compositions, as they depend on the temperature and on the entropy per baryon. With

$$\frac{S}{n_b} \propto \frac{n_\gamma}{n_b} \tag{1.6}$$

the photon to baryon ratio also serves as a measure of the entropy per baryon. This consideration of thermodynamic equilibrium can be carried through to write down the *nuclear Saha equation* for the composition for an isotope with mass A and charge Z:

$$Y_i = Y(Z_i, A_i) = G(Z_i, A_i)[\zeta(3)^{A_i-1}\pi^{(1-A_i)/2}2^{(3A_i-5)/2} \cdot$$
$$A_i^{3/2}(k_B T/m_N c^2)^{3(A_i-1)/2}\Phi^{1-A_i}Y_p^{Z_i}Y_n^{A_i-Z_i}exp[BE(Z_i, A_i)/k_B T] \tag{1.7}$$

Herein, $G(Z_i, A_i)$ is the nuclear partition function giving the number of micro-states for the particular isotope, $\zeta(3)$ is the Riemann function of argument 3, and we find again the *binding energy* BE and also the *thermal energy* $k_B T$. Φ is defined as ratio of photon number to baryon number, and is proportional to the entropy per baryon, thus including the phase space for the plasma constituents. This equation links the proton and neutron abundances to the abundances of all other isotopes, with the characteristic isotope properties of mass m_N, mass and charge numbers A, Z, and internal micro-states G, using the different forms of energy (rest mass, thermal, and binding), as well as the characteristic entropy.

Illustrative examples of how entropy helps to characterise isotopic mixtures are: For high temperatures and entropies, a composition with many nuclei, such as rich in α nuclei would be preferred (e.g. near the big bang in the early universe), while at lower entropy values characteristic for stellar cores a composition of fewer components favouring tightly-bound nucleons in Fe nuclei would be preferred (e.g. in supernova explosions).

With such knowledge about nuclear structure in hand, we can look at the possible configurations that may exist: Those with a minimum of total energy will be *stable*, all others *unstable* or *radioactive*. Figure 1.2 shows the table of isotopes, encoded as stable (black) and unstable isotopes, the latter decaying by β^--decay (blue) and β^+-decay (orange). This is an illustration of the general patterns among the available nuclear configurations. The *ragged* structure signifies that there are systematic variations of nuclear stability with nucleon number, some nucleonic numbers allowing for a greater variety of stable configurations of higher binding energy. These are, in particular, *magic numbers* of protons and neutrons of 2, 8, 20, 28, 50, and 82. We now know approximately 3100 such *isotopes* making up the 118 now-known chemical elements, but only 286 of these isotopes are considered stable. The (7th) edition of the Karlsruher Nuklidkarte (2007) (Pfennig et al. 2007) lists 2962 experimentally-observed isotopes and 652 isomers, its first edition (1958) included 1297 known isotopes of 102 then-known elements. Theoretical models of atomic nuclei, on the other hand, provide estimates of what might still be open to discovery, in terms of isotopes that might exist but either were not produced in the nearby universe or are too shortlived to be observed. Recent models predict existence of over 9000 nuclei (Erler et al. 2012; Xia et al. 2018).

It is the subject of this book to explain in detail the astrophysical implications of this characteristic process of nuclear rearrangements, and what can be learned from measurements of the messengers of radioactive decays. But first we describe the phenomenon of radioactivity in more detail.

1.2 Processes of Radioactivity

The number of decays at each time should be proportional to the number of currently-existing radioisotopes:

$$\frac{dN}{dt} = -\lambda \cdot N \tag{1.8}$$

Here N is the number of isotopes, and the *radioactive-decay constant* λ is the characteristic of a particular radioactive species.

Therefore, in an ensemble consisting of a large number of identical and unstable isotopes, their number remaining after radioactive decay declines exponentially with time:

$$N = N_0 \cdot exp\frac{-t}{\tau} \tag{1.9}$$

The decay time τ is the inverse of the radioactive-decay constant, and τ characterises the time after which the number of isotopes is reduced by decay to $1/e$ of the original number. Correspondingly, the radioactive half-life $T_{1/2}$, is defined as the time after which the number of isotopes is reduced by decay to $1/2$ of the original amount, with

$$T_{1/2} = \frac{\tau}{ln(2)} \tag{1.10}$$

The above exponential decay law is a consequence of a surprisingly simple physical property: The probability per unit time for a single radioactive nucleus to decay is independent of the age of that nucleus. Unlike our common-sense experience with living things, decay does not become more likely as the nucleus ages. Radioactive decay is a nuclear transition from one set of nucleons constituting a nucleus to a different and energetically-favored set with the same number of nucleons. Different types of interactions can mediate such a transition (see below). In *β-decays* it is mediated by the *weak transition* of a neutron into a proton and vice versa,[5] or more generally, nucleons of one type into the other type[6]:

$$n \longrightarrow p + e^- + \overline{\nu_e} \tag{1.11}$$

$$p \longrightarrow n + e^+ + \nu_e \tag{1.12}$$

If such a process occurs inside an atomic nucleus, the quantum state of the nucleus is altered. Depending on the variety of configurations in which this new state may be realized (i.e. the *phase space* available to the decaying nucleus), this change may be more or less likely, in nature's attempt to minimize the total energy of a composite system of nucleons. The decay probability λ per unit time for a single radioactive nucleus is therefore a property which is specific to each particular type of isotope. It can be estimated by Fermi's *Golden Rule* formula though time-dependent perturbation theory (e.g. Messiah 1962). When schematically simplified to convey the main ingredients, the decay probability is:

$$\lambda = \frac{4\pi^2}{h} V_{fi}^2 \, \rho(W) \tag{1.13}$$

where $\rho(W)$ is the number of final states having suitable energy W. The detailed theoretical description involves an integral over the final kinematic states, suppressed here for simplicity. The matrix element V_{fi} is the result of the transition-causing potential between initial and final states.

In the general laboratory situation, radioactive decay involves a transition from the ground state of the parent nucleus to the daughter nucleus in an excited state. But in cosmic environments, nuclei may be part of hot plasma, and temperatures exceeding millions of degrees lead to population of excited states of nuclei. Thus, quantum mechanical transition rules may allow and even prefer other initial and final states, and the nuclear reactions involving a radioactive decay become more complex. Excess binding energy will be transferred to the end products, which

[5]The mass of the neutron exceeds that of the proton by 1.2933 MeV, making the proton the most stable baryon.

[6]In a broader sense, nuclear physics may be considered to be similar to chemistry: elementary building blocks are rearranged to form different species, with macroscopically-emerging properties such as, e.g., characteristic and well-defined energies released in such transitions.

are the daughter nucleus plus emitted (or absorbed, in the case of electron capture transitions) leptons (electrons, positrons, neutrinos) and γ-ray photons.

The occupancy of nuclear states is mediated by the *thermal* excitation spectrum of the *Boltzmann distribution* of particles, populating states at different energies according to:

$$\frac{dN}{dE} = G_j \cdot e^{-\frac{E}{k_B T}} \tag{1.14}$$

Here k_B is Boltzmann's constant, T the temperature of the particle population, E the energy, and G_j the statistical weight factor of all different possible states j which correspond to a specific energy E.[7] In natural environments, particles will populate different states as temperature dictates. Transition rates among states thus will depend on temperature. Inside stars, and more so in explosive environments, temperatures can reach ranges which are typical for nuclear energy-level differences. Therefore, in cosmic sites, radioactive decay time scales may be significantly different from what we measure in terrestrial laboratories on *cold* samples (see Sect. 1.2 for more detail).

Also the atomic-shell environment of a nucleus may modify radioactive decay, if a decay involves *capture or emission of an electron* to transform a proton into a neutron, or vice versa. Electron capture decays are inhibited in fully-ionized plasma, due to the non-availability of electrons. Also β^--decays are affected, as the phase space for electrons close to the nucleus is influenced by the population of electron states in the atomic shell.

After Becquerel's discovery of radioactivity in 1896, Rutherford and others found out in the early twentieth century that there were different types of radioactive decay (Rutherford 1903). They called them *α decay, β decay* and *γ decay*, terms which are still used today. It was soon understood that they are different types of interactions, all causing the same, spontaneous, and time-independent decay of an unstable nucleus into another and more stable nucleus.

Alpha Decay This describes the ejection of a ^{4}He nucleus from the parent radioactive nucleus upon decay. ^{4}He nuclei have since been known also as *alpha particles* for that reason. This decay is intrinsically fast, as it is caused by the *strong* nuclear interaction quickly clustering the nucleus into an alpha particle and the daughter nucleus. Since α-nuclei are tightly-bound, they have been found as sub-structures even within nuclei. In the cases of nuclei much heavier than Fe, a nucleus thus consisting of many nucleons and embedded α clusters can find a preferred state for its number of nucleons by separation of such an α cluster, liberating the binding-energy difference.[8] In such heavy nuclei, Coulomb repulsion helps to overcome the potential barrier which is set up by the strong nuclear force, and decay can occur

[7] States may differ in their quantum numbers, such as spin, or orbital-momenta projections; if they obtain the same energy E, they are called *degenerate*.

[8] The binding energy *per nucleon* is maximized for nucleons bound as a Fe nucleus.

through emission of an α particle. The α particle *tunnels*, with some calculable probability, through the potential barrier, towards an overall more stable and less-energetic assembly of the nucleons.

An example of α decay is ${}_{88}Ra^{226} \Rightarrow {}_{86}Rn^{222} + {}_{2}He^{4}$, which is one step in the decay series starting from ^{238}U. The daughter nucleus , ${}_{86}Rn^{222}$, has charge $Z - 2$, where Z is the original charge of the radioactive nucleus (Z=88 in this example), because the α particle carried away two charge units from the original radioactive nucleus. Such decay frequently leads to an excited state of the daughter nucleus. Kinetic energy E_α for the α particle is made available from the nuclear binding energy liberation expressed by the *Q-value* of the reaction if the mass of the radioactive nucleus exceeds the sum of the masses of the daughter nucleus and of the helium nucleus[9]:

$$Q_\alpha = [M({}_{88}\mathrm{Ra}^{226}) - \mathrm{M}({}_{86}\mathrm{Rn}^{222}) - \mathrm{M}({}_{2}\mathrm{He}^{4})]\mathrm{c}^2 \tag{1.15}$$

The range of the α particle (its stopping length) is about 2.7 cm in standard air (for an α particle with E_α of 4 MeV), and it will produce about 2×10^5 ionizations before being stopped. Even in a molecular cloud, though its range would be perhaps 10^{14} times larger, the α particle would not escape from the cloud. Within small solids (dust grains), the trapping of radioactive energy from α decay provides a source of heat which may result in characteristic melting signatures.[10]

Beta Decay This is the most-peculiar radioactive decay type, as it is caused by the nuclear *weak interaction* which converts neutrons into protons and vice versa. The neutrino ν carries energy and momentum to balance the dynamic quantities, as Pauli famously proposed in 1930 (Pauli did not publish this conjecture until 1961 in a letter he wrote to colleagues). The ν was given its name by Fermi, and was discovered experimentally in 1932 by James Chadwick, i.e. *after* Wolfgang Pauli had predicted its existence. Neutrinos from the Sun have been discovered to *oscillate* between flavors. β decays are being studied in great detail by modern physics experiments, to understand the nature and mass of the ν. Understanding β decay challenges our mind, as it involves several such unfamiliar concepts and particles.

There are three types[11] of β-decay:

$$^{A}_{Z}X_N \longrightarrow {}^{A}_{Z-1}X_{N+1} + e^+ + \nu_e \tag{1.16}$$

$$^{A}_{Z}X_N \longrightarrow {}^{A}_{Z+1}X_{N-1} + e^- + \overline{\nu_e} \tag{1.17}$$

---

[9]These masses may be either nuclear masses or atomic masses, the electron number is conserved, and their binding energies are negligible, in comparison.

[10]Within an FeNi meteorite, e.g., an α particle from radioactivity has a range of only $\sim$10 µm.

[11]We ignore here two additional β decays which are possible from ν and $\overline{\nu}$ captures, due to their small probabilities.

$$ {}^{A}_{Z}X_{N} + e^{-} \longrightarrow {}^{A}_{Z-1}X_{N+1} + \nu_{e} \quad (1.18)$$

In addition to Eq. (1.11) (β^{-} *decay*), these are the conversion of a proton into a neutron (β^{+} *decay*), and *electron capture*. The weak interaction itself involves two different aspects with intrinsic and different strength, the vector and axial-vector couplings. The V_{fi}^{2} term in Eq. (1.13) thus is composed of two terms. These result in *Fermi* and *Gamow-Teller transitions*, respectively (see Langanke and Martínez-Pinedo 2003, for a review of weak-interaction physics in nuclear astrophysics).

An example of β decay is ${}^{13}_{7}$N $\longrightarrow$ ${}^{13}_{6}$C + e^{+} + ν, having mean lifetime τ near 10 min. The kinetic energy Q of the two leptons, as well as the created electron's mass, must be provided by the radioactive nucleus having greater mass than the sum of the masses of the daughter nucleus and of an electron (neglecting the comparatively-small neutrino mass).

$$Q_{\beta} = [M({}^{13}_{7}\mathrm{N}) - \mathrm{M}({}^{13}_{6}\mathrm{C}) - \mathrm{m_e}]\mathrm{c}^{2} \quad (1.19)$$

where these masses are nuclear masses, not atomic masses. A small fraction of the energy release Q_{β} appears as the recoil kinetic energy of the daughter nucleus, but the remainder appears as the kinetic energy of electron and of neutrino.

Capture of an electron is a *two-particle* reaction, the bound atomic electron e^{-} or a free electron in hot plasma being required for this type of β decay. Therefore, depending on availability of the electron, electron-capture β decay lifetimes can be very different for different environments. In the laboratory case, electron capture usually involves the 1s electrons of the atomic structure surrounding the radioactive nucleus, because those present their largest density at the nucleus.

In many cases the electron capture competes with e^{+} + ν emission. In above example, ^{13}N can decay not only by emitting e^{+} + ν, but also by capturing an electron: ${}^{13}_{7}$N + e^{-} $\longrightarrow$${}^{13}_{6}$C + ν. In this case the capture of a 1s electron happens to be much slower than the rate of e^{+} emission. But cases exist for which the mass excess is not large enough to provide for the creation of the e^{+} mass for emission, so that only electron capture remains to the unstable nucleus to decay. Another relevant example is the decay of ^{7}Be. Its mass excess over the daughter nucleus ^{7}Li is only 0.351 MeV. This excess is insufficient to provide for creation of the rest mass of an emitted e^{+}, which is 0.511 MeV. Therefore, the ^{7}Be nucleus is stable against e^{+} + ν emission. However, electron capture adds 0.511 MeV of rest-mass energy to the mass of the ^{7}Be nucleus, giving a total 0.862 MeV of energy above the mass of the ^{7}Li nucleus. Therefore, the e^{-} capture process (above) emits a monoenergetic neutrino having E_{ν}= 0.862 MeV.[12]

The situation for electron capture processes differs significantly in the interiors of stars and supernovae: Nuclei are ionized in plasma at such high temperature. The capture lifetime of ^{7}Be, for example, which is 53 days against 1s electron capture

[12]This neutrino line has just recently been detected by the Borexino collaboration arriving from the center of the Sun (Arpesella et al. 2008).

in the laboratory, is lengthened to about 4 months at the solar center (see theory by Bahcall 1964; Takahashi and Yokoi 1983), where the free electron density is less at the nucleus.

The range of the β particle (its stopping length) in normal terrestrial materials is small, being a charged particle which undergoes Coulomb scattering. An MeV electron has a range of several meters in standard air, during which it loses energy by ionisations and inelastic scattering. In tenuous cosmic plasma such as in supernova remnants, or in interstellar gas, such collisions, however, become rare, and may be unimportant compared to electromagnetic interactions of the magnetic field (*collisionless plasma*). Energy deposit or escape is a major issue in intermediate cases, such as expanding envelopes of stellar explosions, supernovae (positrons from ^{56}Co and ^{44}Ti) and novae (many β^+ decays such as ^{13}N) (see Chaps. 4, 5, and 7 for a discussion of the various astrophysical implications). Even in small solids and dust grains, energy deposition from ^{26}Al β-decay, for example, injects 0.355 W kg^{-1} of heat. This is sufficient to result in melting signatures, which have been used to study condensation sequences of solids in the early solar system (see Chap. 6).

Gamma Decay In γ decay the radioactive transition to a different and more stable nucleus is mediated by the *electromagnetic interaction*. A nucleus relaxes from its excited configuration of the nucleons to a lower-lying state of the same nucleons. This is intrinsically a fast process; typical lifetimes for excited states of an atomic nucleus are 10^{-9} s. We denote such electromagnetic transitions of an excited nucleus *radioactive γ-decay* when the decay, and, time of the excited nucleus is considerably longer and that nucleus thus may be considered a temporarily-stable configuration of its own, a *metastable* nucleus.

How is stability, or instability, of a nuclear-excited state effected? In electromagnetic transitions

$$A^{\star} \longrightarrow A^{g.s.} + \gamma \tag{1.20}$$

the spin (angular momentum) is a conserved quantity of the system. The spin of a nuclear state is a property of the nucleus as a whole, and reflects how the states of protons and neutrons are distributed over the quantum-mechanically allowed *shells* or nucleon wave functions (as expressed in the *shell model* view of an atomic nucleus). The photon (γ *quantum*) emitted (Eq. (1.20)) will thus have a *multipolarity* resulting from the spin differences of initial and final states of the nucleus. Dipole radiation is most common and has multipolarity 1, emitted when initial and final state have angular momentum difference $\Delta l = 1$. Quadrupole radiation (multipolarity 2, from $\Delta l = 2$) is ~6 orders of magnitude more difficult to obtain, and likewise, higher multipolarity transitions are becoming less likely by the similar probability decreases (the *Weisskopf estimates* (see Weisskopf 1951)). This explains why some excited states in atomic nuclei are much more long-lived (*meta-stable*) than others; their transitions to the ground state are also considered as *radioactivity*, and called *γ decay*.

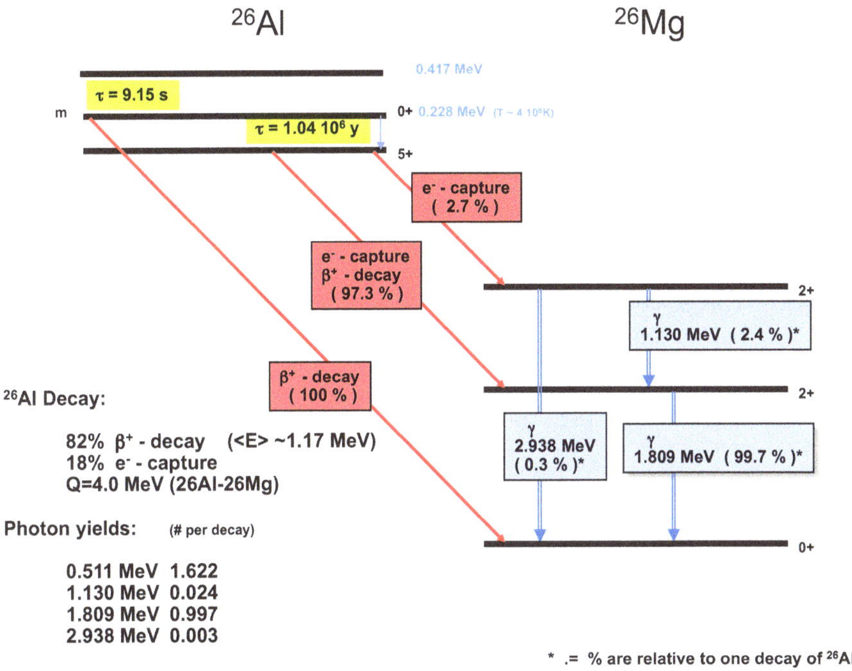

Fig. 1.3 ^{26}Al decay. The ^{26}Al nucleus ground state has a long radioactive lifetime, due to the large spin difference of its state to lower-lying states of the daughter nucleus ^{26}Mg. An isomeric excited state of ^{26}Al exists at 228 keV excitation energy. If thermally excited, ^{26}Al may decay through this state. Secondary products, lifetime, and radioactive energy available for deposits and observation depend on the environment

The range of a γ-ray (its stopping length) is typically about 5–10 g cm^{-2} in passing through matter of all types. Hence, except for dense stars and their explosions, radioactive energy from γ decay is of astronomical implication only.[13]

An illustrative example of radioactive decay is the ^{26}Al nucleus. Its decay scheme is illustrated in Fig. 1.3. The ground state of ^{26}Al is a 5+ state. Lower-lying states of the neighboring isotope ^{26}Mg have states 2+ and 0+, so that a rather large change of angular momentum Δl must be carried by radioactive-decay secondaries. This explains the large β-decay lifetime of ^{26}Al of τ ~1.04 10^6 years. In the level scheme of ^{26}Al, excited states exist at energies 228, 417, and 1058 keV.

[13]Gamma-rays from nuclear transitions following ^{56}Ni decay (though this is a β decay by itself) inject radioactive energy through γ-rays from such nuclear transitions into the supernova envelope, where it is absorbed in scattering collisions and thermalized. This heats the envelope such that thermal and optically bright supernova light is created. Deposition of γ-rays from nuclear transitions are the engines which make supernovae to be bright light sources out to the distant universe, used in cosmological studies (Leibundgut 2000) to, e.g., support evidence for *dark energy*.

The 0+ and 3+ states of these next excited states are more favorable for decay due to their smaller angular momentum differences to the ^{26}Mg states, although $\Delta l = 0$ would not be *allowed* for the 228 keV state to decay to ^{26}Mg's ground state. This explains its relatively long lifetime of 9.15 s, and it is a *metastable* state of ^{26}Al. If thermally excited, which would occur in nucleosynthesis sites exceeding a few 10^8 K, ^{26}Al may decay through this state without γ-ray emission as $^{26}\mathrm{Al}^{\mathrm{g.s.}} + \gamma \longrightarrow ^{26}\mathrm{Al}^{\mathrm{m}} \longrightarrow ^{26}\mathrm{Mg} + \mathrm{e}^+$, while the ground state decay is predominantly a β^+ *decay* through excited ^{26}Mg states and thus including γ-ray emission. Secondary products, lifetime, and radioactive energy available for deposits and observation depend on the environment.

1.3 Radioactivity and Cosmic Nucleosynthesis

Nuclear reactions in cosmic sites re-arrange the basic constituents of atomic nuclei (neutrons and protons) among the different allowed configurations. Throughout cosmic evolution, such reactions occur in various sites with different characteristic environmental properties. Each reaction environment leads to rearrangements of the relative abundances of cosmic nuclei. The cumulative process is called *cosmic chemical evolution*.[14]

The *cosmic abundance* of a specific isotope is expressed in different ways, depending on the purpose. Counting the atoms of isotope i per unit volume, one obtains n_i, the number density of atoms of species i (atoms cm^{-3}). The interest of cosmic evolution and nucleosynthesis lies in the fractional abundances of species i related to the total, and how it is altered by cosmic nuclear reactions. Observers count a species i and relate it to the abundance of a reference species. For astronomers this is hydrogen. Hydrogen is the most abundant element throughout the universe, and easily observed through its characteristic atomic transitions in spectroscopic astronomical measurements. Using the definition of Avogadro's constant A_{Av} as the number of atoms which make up A grams of species i (i.e., one mole), we can obtain abundances *by mass*; $A_{Av} = 6.02214 \times 10^{23}$ atoms mole^{-1}. The mass contained in a particular species S results from scaling its abundance N_S by its atomic weight A.

We can get a global measure for cosmic evolution of the composition of matter by tracing how much of the total mass is contained in hydrogen, helium, and the remainder of elements called *metals*,[15] calling these quantities X for hydrogen abundance, Y for helium abundance, and Z for the cumulative abundance of all

[14]We point out that there is no chemistry involved; the term refers to changes in abundances of the chemical elements, which are important for our daily-life experiences. But these are a result of the more-fundamental changes in abundances of isotopes mediated by cosmic nuclear reactions.

[15]This nomenclature may be misleading, it is used by convenience among astrophysicists. Only a part of these elements are actually metals.

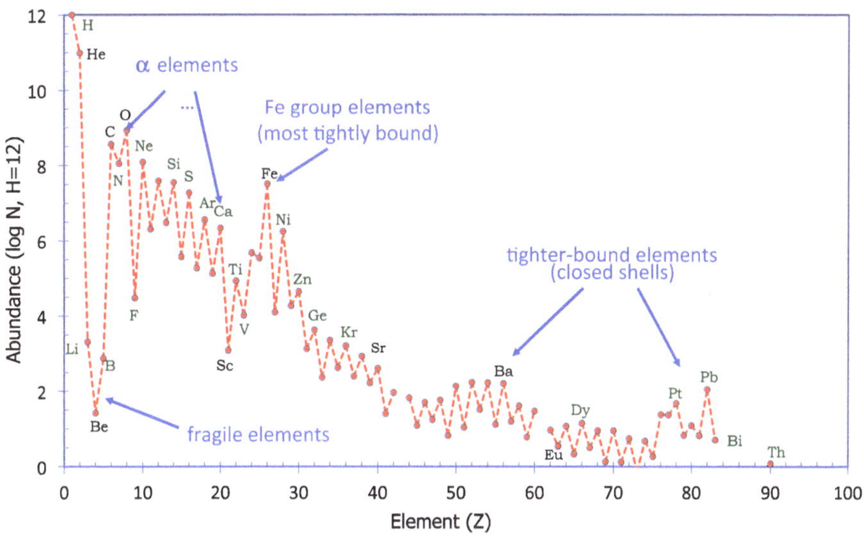

Fig. 1.4 The abundance of elements in the present-day nearby universe. Abundances (by number) are shown in a logarithmic scale, and span 12 orders of magnitude. The interplay of nuclear properties (several are indicated in the graph) with environmental conditions in cosmic nucleosynthesis sites has created this complex abundance pattern during the course of cosmic history

nuclei heavier than helium. We call these *mass fractions* of hydrogen X, helium Y, and metals Z, with $X + Y + Z = 1$. The metallicity Z is a key parameter used to characterise the evolution of elemental and isotopic composition of cosmic matter. The astronomical abundance scale is set from most-abundant cosmic element Hydrogen to $log(X_H) = 12$ (Fig. 1.4), but mineralogists and meteoriticians use Si as their reference element and set $log(X_{Si}) = 6$.

We often relate abundances also to our best-known reference, the solar system, denoting *solar-system* values by the $\odot$ symbol. Abundances of a species S are then expressed in *bracket notation*[16] as

$$[\frac{S}{H}] \equiv log(\frac{X_S}{X_H})_\star - log(\frac{X_S}{X_H})_\odot \qquad (1.21)$$

Depending on observational method and precision, our astronomical data are *metallicity*, elemental *enrichments* with respect to solar abundances, or isotopic abundances. Relations to nuclear reactions are therefore often indirect. Understanding the nuclear processing of matter in the universe is a formidable challenge, often listed as one of the *big questions* of science.

[16]Deviations from the standard may be small, so that $[\frac{S_1}{S_2}]$ may be expressed in δ units (parts per mil), or ϵ units (parts in 10^4), or ppm and ppb; $\delta(^{29}Si/^{28}Si)$ thus denotes excess of the ^{29}Si/^{28}Si isotopic ratio above solar values in units of 0.1%.

Big Bang Nucleosynthesis (BBN) about 13.8 Gyrs ago left behind a primordial composition where hydrogen (protons) and helium were the most-abundant species; the total amount of nuclei heavier than He (the *metals*) was less than 10^{-9} (by number, relative to hydrogen (Steigman 2007; Cyburt et al. 2016)). Today, the total mass fraction of metals in *solar abundances* is $Z = 0.0134$ (Asplund et al. 2009), compared to a hydrogen mass fraction of[17] $X = 0.7381$. This growth of metal abundances by about seven orders of magnitude is the effect of cosmic nucleosynthesis. Nuclear reactions in stars, supernovae, novae, and other places where nuclear reactions may occur, all contribute. But it also is essential that the nuclear-reaction products inside those cosmic objects will eventually be made available to observable cosmic gas and solids, and thus to later-generation stars such as our solar system born 4.6 Gyrs ago. This book will also discuss our observational potential for cosmic isotopes, and we address the constraints and biases which limit our ability to draw far reaching conclusions.

The growth of isotopic and elemental abundances from cosmic nucleosynthesis does not occur homogeneously. Rather, the cosmic abundances observed today span a dynamic range of twelve orders of magnitude between abundant hydrogen and rare heavy elements (Fig. 1.4). Moreover, the elemental abundance pattern already illustrates clearly the prominent effects of nuclear structure (see Fig. 1.4): Iron elements are among the most-tightly bound nuclei, and locally elements with even numbers of nucleons are more tightly bound than elements with odd numbers of nuclei. The Helium nucleus (*α-particle*) also is more tightly bound than its neighbours in the chart of nuclei, hence all elements which are multiples of α's are more abundant than their neighbours.

Towards the heavier elements beyond the Fe group, abundances drop by about five orders of magnitude again, signifying a substantially-different production process than the mix of charged-particle nuclear reactions that produced the lighter elements: *neutron capture* on Fe *seed nuclei*. The two abundance peaks seen for heavier elements are the results of different environments for cosmic neutron capture reactions (the *r-process* and *s-process*), both determined by neutron capture probabilities having local extrema near *magic numbers*. The different peaks arise from the particular locations at which the processes' reaction path encounters these *magic nuclei*, as neutron captures proceed much faster (slower) than beta decays in the r process (s process).

The subjects of cosmic nucleosynthesis research are complex and diverse, and cover the astrophysics of stars, stellar explosions, nuclear reactions on surfaces of compact stars and in interstellar space. For each of the potential nuclear-reaction sites, we need to understand first how nuclear reactions proceed under the local conditions, and then how material may be ejected into interstellar space from such a source. None of the nucleosynthesis sites is currently understood to a level of

[17]This implies a *metallicity* of solar matter of 1.4%. Our local reference for cosmic material composition seems to be remarkably universal. Earlier than ~2005, the commonly-used value for solar metallicity had been 2%.

detail which would be sufficient to formulate a physical description, sit back and consider cosmic nucleosynthesis *understood*. For example, one might assume we know our Sun as the nearest star in most detail; but solar neutrino measurements have been a puzzle only alleviated in recent years with the revolutionary adoption of non-zero masses for neutrinos, which allow for flavour oscillations; and even then, the abundances of the solar photosphere, revised by almost a factor two based on three-dimensional models of the solar photosphere (Asplund et al. 2009), created surprising tension with measurements of helioseismology and the vibrational behaviour reflected herein, and the physical descriptions in our currently-best solar model are under scrutiny (Vinyoles et al. 2017). As another example, there are two types of supernova explosions. Core-collapse supernovae are the presumed outcome of the final gravitational collapse of a massive star once its nuclear fuel is exhausted, and thermonuclear supernovae were thought to originate from detonation of degenerate stars once they exceed a critical threshold for nuclear burning of Carbon, the Chandrasekhar mass limit. The gravitational collapse can not easily be reverted into an explosion, and even the help of neutrinos from the newly-forming neutron star in the center appears only marginally sufficient, so that many massive stars that were thought to explode may collapse to black holes (Janka et al. 2016). And the thermonuclear supernova variety appears to require white dwarf collisions as triggering events in some well-constrained cases, while in other cases luminosities deviate by orders of magnitude from the expectation from a Chandrasekhar-mass white dwarf and its nuclear-burning demise that once was thought to be a cosmic standard candle (Hillebrandt et al. 2013). For neither of these supernovae, a *physical* model is available, which would allow us to calculate and predict the outcome (energy and nuclear ashes) under given, realistic, initial conditions (see Chaps. 4 and 5). Much research remains to be done in cosmic nucleosynthesis.

One may consider measurements of cosmic material in all forms to provide a wealth of data, which now has been exploited to understand cosmic nucleosynthesis. Note, however, that cosmic material as observed has gone through a long and ill-determined journey. We need to understand the trajectory in time and space of the progenitors of our observed cosmic-material sample if we want to interpret it in terms of cosmic nucleosynthesis. This is a formidable task, necessary for distant cosmic objects, but here averaging assumptions help to simplify studies. For more nearby cosmic objects where detailed data are obtained, astrophysical models quickly become very complex, and also need simplifying assumptions to operate for what they are needed. It is one of the objectives of cosmic nucleosynthesis studies to contribute to proper models for processes in such evolution, which are sufficiently isolated to allow their separate treatment. Nevertheless, carrying out *well-defined experiments* for a source of cosmic nucleosynthesis remains a challenge, due to this complex flow of cosmic matter (see Chaps. 6–8).

The special role of radioactivity in such studies is contributed by the intrinsic decay of such material after it has been produced in cosmic sites. This brings in a new aspect, the clock of the radioactive decay. Technical applications widely known are based on ^{14}C with its half life of 5700 years, while astrophysical applications extend this too much longer half lives up to Gyrs (^{235}U has a decay time of 10^9 years). Changes in isotopic abundances with time will occur at such natural and isotope-specific rates, and will leave their imprints in observable isotopic abundance records. For example, the observation of unstable technetium in stellar atmospheres of AGB stars was undisputable proof of synthesis of this element inside the same star, because the evolutionary time of the star exceeds the radioactive lifetime of technetium. Another example, observing radioactive decay γ-ray lines from short-lived Ni isotopes from a supernova is clear proof of its synthesis in such explosions; measuring its abundance through γ-ray brightness is a direct *calibration* of processes in the supernova interior. A last example, solar-system meteorites show enrichments in daughter products of characteristic radioactive decays, such as ^{26}Al and ^{53}Mn; the fact that these radioactive elements were still alive at the time those solids formed sets important constraints to the time interval between the latest nucleosynthesis event near the forming Sun and the actual condensation of solid bodies in the interstellar gas accumulating to form the young solar system. This book will discuss these examples in detail, and illustrate the contributions of radioactivity studies to the subject of cosmic nucleosynthesis.

1.4 Observing Radioactive Isotopes in the Universe

Astronomy has expanded beyond the narrow optical band into *new astronomies* in the past decades. By now, we are familiar with telescopes measuring radio and sub-mm through infrared emission towards the long wavelength end, and ultraviolet, X-ray, and γ-ray emission towards the short wavelength end (see Fig. 1.5). The physical origins of radiation are different in different bands. Thermal radiation dominates emission from cosmic objects in the middle region of the electromagnetic spectrum, from a few 10 K cold molecular clouds at radio wavelengths through dust and stars up to hot interstellar gas radiating X-rays. Non-thermal emission is characteristic for the wavelength extremes, both at radio and γ-ray energies. Characteristic spectral lines originate from atomic shell electrons over most of the spectrum; nuclear lines are visible only in roughly two decades of the spectrum at 0.1–10 MeV. Few exceptional lines arise at high energy from annihilations of positrons and pions. Cosmic *elements* can be observed in a wide astronomical range. *Isotopes*, however, are observed almost exclusively through ~MeV γ-rays (see Fig. 1.5). Note that nucleosynthesis reactions occur among isotopes, so that this is the prime[18] information of interest when we wish to investigate cosmic nucleosynthesis environment properties.

[18]Other astronomical windows may also be significantly influenced by biases from other astrophysical and astrochemical processes; an example is the observation of molecular isotopes of CO, where chemical reactions as well as dust formation can lead to significant alterations of the abundance of specific molecular species.

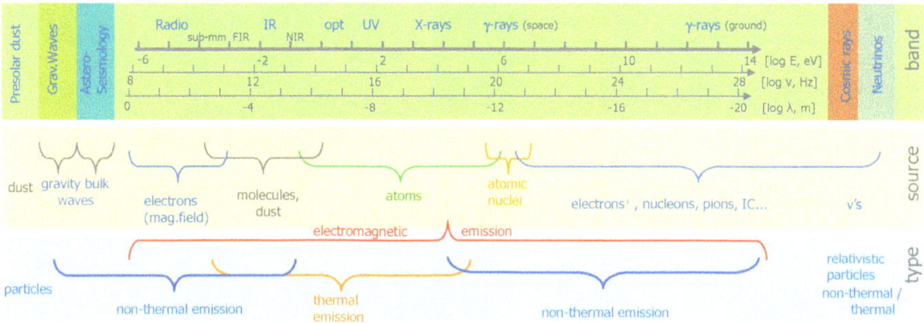

Fig. 1.5 The electromagnetic spectrum of candidate astronomical measurements ranges across more than 20 orders of magnitude. Not all are easily accessible. Information categories of thermal and non-thermal, and of molecular, atomic, nuclear, and elementary-particle physics origins of cosmic radiation extends over different parts of this broad spectrum. Nuclear physics is directly accessible in a small band (0.1–10 MeV) only. Non-electromagnetic astronomical messengers are indicated at both ends of the electromagnetic spectrum

Only few elements such as technetium (Tc) do not have any stable isotope; therefore, elemental photospheric absorption and emission line spectroscopy, the backbone of astronomical studies of cosmic nucleosynthesis, have very limited application in astronomy with radioactivities. This is about to change currently, as spectroscopic devices in the optical and radio/sub-mm regimes advance spectral resolutions. Observational studies of cosmic radioactivities are best performed by techniques which intrinsically obtain isotopic information. These are:

- Modern spectrographs on large ground-based telescopes reach R=20,000, sufficient to resolve fine structure lines and isotopic features in molecules (see Fig. 1.6). Radio spectroscopy with CO isotopes has been successfully applied since the 1990s, and has been used mainly to track the CO molecule at different columns densities, while sub-mm lines from molecules have been demonstrated to observe specific isotopes within molecules such as ^{36}ArN and ^{26}AlF (Schilke et al. 2014; Kamiński et al. 2018).
- Precision mass spectroscopy in terrestrial laboratories, which has been combined with sophisticated radiochemistry to extract meteoritic components originating from outside the solar system
- Spectroscopy of characteristic γ-ray lines emitted upon radioactive decay in cosmic environments

The two latter *astronomical disciplines* have a relatively young history. They encounter some limitations due to their basic methods of how astronomical information is obtained, which we therefore discuss in somewhat more detail:

- Precision mass spectrometry of meteorites for astronomy with radioactivity began about 1960 with a new discovery of now extinct radioactivity within the young solar system. From heating of samples of bulk meteorite material, the presence of a surprising excess ^{129}Xe had been puzzling. Through a variety of different chemical processing, this could be tracked to trapped gas enclosures

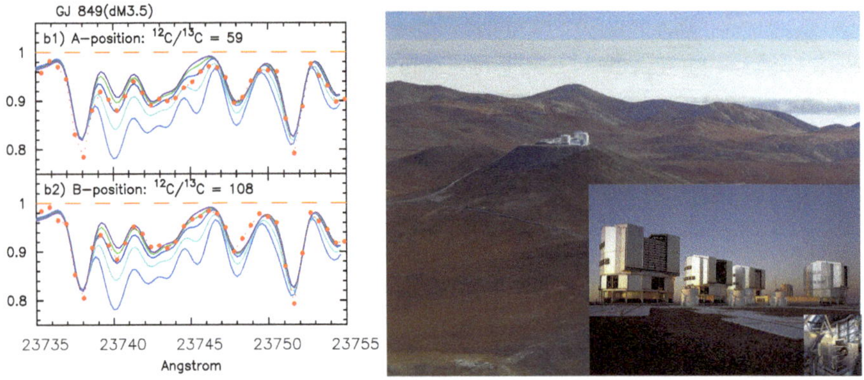

Fig. 1.6 *Left:* Example of an isotopic measurement in a stellar atmosphere. Shown is an absorption-line spectrum of a cool star with a present-generation optical telescope, here the Subaru telescope on Hawaii with its IR spectrograph at a resolution of 20,000. Molecular lines from the CO molecule isotopologues show isotopic shifts, which can be recognised as changes in line shapes, as resulting from the isotopic abundance ratio. Here the carbon isotopic ratio is determined for the stellar atmosphere of a M dwarf star, comparing the measurement (red dots) with expectations for different ratios $^{12}C/^{13}C$ (from Tsuji 2016). *Right:* The Very Large Telescope (VLT) on Mount Paranal in Chile, with four telescopes (lower right), is one of the modern optical instruments. Equipped with high-resolution spectrographs such as FLAMES (insert lower right), absorption-line spectroscopy of stars provides elemental abundances in stellar atmospheres, even in nearby galaxies (Figure ESO)

in rather refractory components, which must have been enriched in ^{129}I at the time of formation of this meteorite. From mineralogical arguments, this component could be associated with the early solar system epoch about 4.6 Gy ago (Reynolds 1960). This was the first evidence that the matter from which the solar system formed contained radioactive nuclei whose half-lives are too short to be able to survive from that time until today (^{129}I decays to ^{129}Xe within 1.7×10^7 years). Another component could be identified from most-refractory Carbon-rich material, and was tentatively identified with dust grains of pre-solar origins. Isotopic anomalies found in such *extra-solar* inclusions, e.g. for C and O isotopes, range over four orders of magnitude for such *star dust* grains as shown in Fig. 1.7 (Zinner 1998), while isotopic-composition variations among bulk meteoritic-material samples are a few percent at most. These mass spectroscopy measurements are characterised by an amazing sensitivity and precision, clearly resolving isotopes and counting single atoms at ppb levels to determine isotopic ratios of such rare species with high accuracy, and nowadays even for specific, single dust grains. They provide an *astronomy*

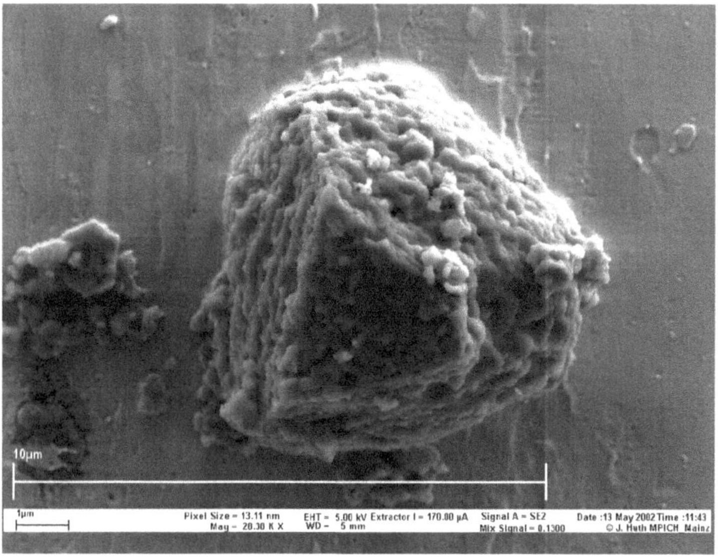

Fig. 1.7 Meteoritic inclusions such as this SiC grain are recognised as dust formed near a cosmic nucleosynthesis source outside the solar system, from their large isotopic anomalies, which cannot be explained by interstellar nor solar-system processing but are reminiscent of cosmic nucleosynthesis sites. Having condensed in the envelope of a source of new isotopes, laboratory mass spectroscopy can reveal isotopic composition for many elements, thus providing a remote probe of one cosmic nucleosynthesis source

in terrestrial laboratories (see Chap. 11 for instrumental and experimental aspects), and constitute now an established part of astrophysical research (see Clayton and Nittler 2004, for a review, and e.g. Amari et al. 2014; Zinner 2014).

Table 1.1 lists the radioactive isotopes used for studies of pre-solar grains (Groopman et al. 2015). Studies of pre-solar dust grain compositions have led to the distinctions of grain origins from AGB stars, from supernovae, and from novae, all of which are copious producers of dust particles. Formation of stardust occurs in circumstellar environments where temperatures are cool enough (e.g. Cherchneff and Sarangi 2017, for a recent review of the open issues). On their journey through the interstellar medium, heating and partial or complete destruction may occur from starlight or even shocks from supernovae (Zhukovska et al. 2016). Also a variety chemical and physical reactions may reprocess dust grains (Dauphas and Schauble 2016). Thus, the journey from the stardust source up to inclusion in meteorites which found their way to Earth remains subject to theoretical modelling and much residual uncertainty (Jones 2009). Nevertheless, cosmic dust particles are independent astrophysical messengers, and complement studies based on electromagnetic

Table 1.1 Radioactivities in presolar grains, sorted by ascending radioactive mean lifetime (from Groopman et al. 2015)

Isotope chain	Lifetime	Presolar grain type	Source
$^{49}V \longrightarrow {}^{49}Ti$	330 days	SiC, Graphite	SNe
$^{22}Na \longrightarrow {}^{22}Ne$	2.6 years	Graphite	SNe
$^{44}Ti \longrightarrow {}^{44}Ca$	60 years	SiC, Graphite, Hibonite	SNe
$^{32}Si \longrightarrow {}^{32}S$	153 years	SiC	SNe, post-AGB stars
$^{41}Ca \longrightarrow {}^{41}K$	1.02×10^5 years	SiC, Graphite, Hibonite	SNe, RGB, and AGB stars
$^{99}Tc \longrightarrow {}^{99}Ru$	2.11×10^5 years	SiC	AGB stars
$^{26}Al \longrightarrow {}^{26}Mg$	7.17×10^5 years	SiC, Graphite, Corundum, Spinel, Hibonite, Silicate	SNe, RGB, and AGB stars
$^{93}Zr \longrightarrow {}^{93}Nb$	1.61×10^6 years	SiC	AGB stars

radiation in important ways. Grain composition and morphology from the stardust laboratory measurements are combined with astronomical results such as characteristic spectral lines (e.g. from water ice, or a prominent feature associated with silicate dust), and interpreted through (uncertain) theories of cosmic dust formation and transport (Zinner 1998; Cherchneff 2016). Experimental difficulties and limitations arise from sample preparation through a variety of complex chemical methods, and by the extraction techniques evaporising material from the dust grain surfaces for subsequent mass spectrometry (see Chap. 10).

- Characteristic γ-ray lines from cosmic sources were not known until the 1960s, when spaceflight and its investigations of the near-earth space radiation environment had stimulated measurements of γ-rays. The discovery of a cosmic γ-ray line feature near 0.5 MeV from the direction towards the center of our Galaxy in 1972 (Johnson et al. 1972) stimulated balloon and satellite experiments for cosmic γ-ray line spectroscopy. Radioactive isotopes are ejected into the surroundings of their nucleosynthesis sources, and become observable through their gamma-ray line emission once having left dense production sites where not even gamma-rays may escape. Nuclear gamma-rays can penetrate material layers of integrated thickness of a few grams cm^{-2}. A typical interstellar cloud would have $\sim$0.1 g cm^{-2}, SNIa envelopes are transparent to gamma-rays after 30–100 days, depending on explosion dynamics. Depending on radioactive lifetime, gamma-ray lines measure isotopes which originate from single sources (the short-lived isotopes) or up to thousands of sources as accumulated in interstellar space over the radioactive lifetime of long-lived isotopes (see Table 1.2). Decays of the isotopes ^{26}Al, ^{60}Fe, ^{44}Ti, ^{57}Ni, and

Table 1.2 Radioactivities with gamma-ray line emission, sorted by ascending radioactive mean lifetime (updated from Diehl et al. 2006)

Decay chain	Lifetime [years]	γ-Ray Energy [keV] (branching ratio [%])	Site (detections)	Process type
$^{7}Be \rightarrow ^{7}Li$	0.21	478 (100)	Novae	Explosive H burning
$^{56}Ni \longrightarrow ^{56}Co \longrightarrow ^{56}Fe$	0.31	847 (100), 1238 (68) 2598 (17), 1771 (15) and 511 from e^{+}	SNe (SN1987A, (SN1991T, SN2014J)	NSE burning
$^{57}Co \longrightarrow ^{57}Fe$	1.1	122 (86), 136 (11)	SNe (SN1987A	NSE burning
$^{22}Na \longrightarrow ^{22}Ne$	3.8	1275 (100) and 511 from e^{+}	Novae	Explos. H burning
$^{44}Ti \longrightarrow ^{44}Sc \longrightarrow ^{44}Ca$	89	68 (95), 78 (96) 1156 (100) and 511 from e^{+}	SNe (Cas A, SN1987A)	NSE α freeze-out
$^{26}Al \longrightarrow ^{26}Mg$	1.04×10^{6}	1809 (100) and 511 from e^{+}	ccSNe, WR Novae, AGB (Galaxy) (Cygnus;Sco-Cen; Orion; Vela)	H burning (ν-proc.)
$^{60}Fe \longrightarrow ^{60}Co \longrightarrow ^{60}Ni$	3.8×10^{6}	1173 (100), 1332 (100) 59 (2)	SNe (Galaxy)	He,C shell burning
$e^{+} \longrightarrow Ps,.. \longrightarrow \gamma\gamma(\gamma)$	$\sim 10^{7}$	2·511 ($\sim$100), cont <510	Radioactivities Pulsars, μQSOs, …(Galactic bulge; disk)	β^{+} decay rel. plasma

^{56}Ni in distant cosmic sites are an established fact (see, e.g., Fig. 1.8), and astrophysical studies make use of such measurements. The downsides of those experiments is the rather poor resolution by astronomy standards (on the order of degrees), and the sensitivity limitations due to large instrumental backgrounds, which effectively only shows the few brightest sources of cosmic γ-rays until now (see Diehl et al. 2006; Diehl 2013, for a discussion of achievements and limitations).

Despite their youth and limitations, both methods to address cosmic radioactivities share a rather direct access to isotopic information, unlike other fields of astronomy. Isotopic abundance studies in the nuclear energy window will be complemented for specific targets and isotopes from the new opportunities in optical and radio/sub-mm spectroscopy. From a combination of all available astronomical methods, the study of cosmic nucleosynthesis will continue to advance towards a

Fig. 1.8 Example of a γ-ray line measurement: The characteristic line from ^{26}Al decay at 1808.63 keV appears Doppler-shifted from large scale galactic rotation, as it is viewed towards different galactic longitudes (left; from Kretschmer et al. 2013). This measurement was performed with the SPI spectrometer on INTEGRAL, an example of a present-generation space-borne γ-ray telescope. The INTEGRAL satellite (artist view picture, ESA) has two main telescopes; the spectrometer SPI, one of them, is shown at the lower-right schematically with its 19-detector Ge camera and the tungsten mask for imaging by casting a shadow onto the camera. Space-based instruments of this kind are required to directly measure the characteristic γ-ray lines from the decay of unstable isotopes near sites of current-epoch cosmic element formation

truly astrophysical decomposition of the processes and their interplays. This book describes where and how specific astronomical messages from cosmic radioactivity help to complement these studies.

References

Amari S, Zinner E, Gallino R (2014) Geochim Cosmochim Acta 133:479

Arpesella C, Back HO, Balata M et al (2008) Phys Rev Lett 101:091302

Asplund M, Grevesse N, Sauval AJ, Scott P (2009) Annu Rev Astron Astrophys 47:481

Bahcall JN (1964) Astrophys J 139:318

Cherchneff I (2016) IAU focus meeting, vol 29, p 166

Cherchneff I, Sarangi A (2017) In: Miroshnichenko A, Zharikov S, Korčáková D, Wolf M (eds) The B[e] phenomenon: forty years of studies. Astronomical Society of the Pacific conference series, vol 508, p 57

Clayton DD, Nittler LR (2004) Annu Rev Astron Astrophys 42:39

Cyburt et al. (2016) Rev Mod Phys 88:015004

Dauphas N, Schauble EA (2016) Annu Rev Earth Planet Sci 44:709

Diehl R (2013) Rep Prog Phys 76:026301

Diehl R, Prantzos N, von Ballmoos P (2006) Nucl Phys A 777:70

Erler J, Birge N, Kortelainen M et al (2012) Nature 486:509

Groopman E, Zinner E, Amari S et al (2015) Astrophys J 809:31

Hillebrandt W, Kromer M, Röpke FK, Ruiter AJ (2013) Front Phys 8:116–143. https://doi.org/10.1007/s11467-013-0303-2
Janka H-T, Melson T, Summa A (2016) Annu Rev Nucl Part Sci 66:341–375. https://doi.org/10.1146/annurev-nucl-102115-044747
Johnson WN III, Harnden FR Jr, Haymes RC (1972) Astrophys J 172:L1+
Jones A (2009) In: Pagani L, Gerin M (eds) EAS publications series, vol 34, pp 107–118
Kamiński T, Tylenda R, Menten K et al (2018) Nature Astron. https://doi.org/10.1038/s41550-018-0541-x
Kretschmer K, Diehl R, Krause M et al (2013) Astron Astrophys 559:A99
Langanke K, Martínez-Pinedo G (2003) Rev Mod Phys 75:819
Leibundgut B (2000) Astron Astrophys Rev 10:179
Messiah A (1961–1962) Quantum mechanics. North-Holland Publication, Amsterdam
Patrignani C, Particle Data Group (2016) Chin Phys C 40:100001
Pfennig G, Klewe-Nebenius H, Seelmann-Eggebert W (2007) Karlsruher nuklidkarte - chart of nuclides, 7th edn. Haberbeck, Lage. ISBN 3-92-1879-18-3
Reynolds JH (1960) Phys Rev Lett 4:8
Rutherford E (1903) Proc Phys Soc Lond 18:595
Schilke P, Neufeld DA, Müller HSP et al (2014) Astron Astrophys 566:A29
Steigman G (2007) Annu Rev Nucl Part Sci 57:463
Takahashi K, Yokoi K (1983) Nucl Phys A 404:578
Tsuji T (2016) Publ Astron Soc Jpn 68:84
Vinyoles et al. (2017) Astrophys J 835:202
Weisskopf VF (1951) Phys Rev 83:1073
Weizsäcker CFV (1935) Z Phys 96:431
Xia XW, Lim Y, Zhao PW, Liang HZ, Qu XY, Chen Y, Liu H, Zhang LF, Zhang SQ, Kim Y, Meng J (2018) Atom Data Nucl Data Tables 121:1–215. https://doi.org/10.1016/j.adt.2017.09.001
Zhukovska S, Dobbs C, Jenkins EB, Klessen RS (2016) Astrophys J 831:147
Zinner E (1998) Annu Rev Earth Planet Sci 26:147
Zinner E (2014) In: Davis AM (ed) Meteorites and cosmochemical processes. Elsevier, Amsterdam, pp 181–213

Chapter 2
The Role of Radioactive Isotopes in Astrophysics

Donald D. Clayton

Astronomy with radioactivity can be described as using the quantity of radioactivity within cosmic samples to infer their physical circumstances and history. The quantity of radioactive nuclei is inferred or measured by a variety of techniques reviewed here and in subsequent chapters. We also review the major disciplines of such knowledge in astronomy in the attempt to introduce the reader to its principles. Because the sciences of nucleosynthesis and of radioactivity have been so intertwined, brief histories of each are first given. Descriptions of recent discoveries in these disciplines are relegated to subsequent chapters where they can receive fuller treatment. The scientific disciplines of astronomy using radioactive nuclei are primarily these: (1) *nuclear cosmo-chronology*, the age of the oldest elements based the remaining natural abundances of live radioactive nuclei; (2) nuclear chronology of *extinct radioactive nuclei*, whose previous presence is measured by overabundances of their unsupported daughter nuclei; (3) spectroscopic abundances of uranium and thorium in old dwarf stars measuring the ages of those stars; (4) *cosmo-radiogenic chronologies*, using daughter nuclei of long-lived nuclei to measure the duration of galactic nucleosynthesis; (5) radioactive abundances within well-mixed interstellar gas calculated by models of *galactic chemical evolution*; (6) predictions and discovery of nuclear gamma-ray lines emitted by radioactive nuclei from discreet galactic supernovae and by diffuse galactic radioactivity; (7) discovery of ^{26}Al, the first detected gamma-ray line emitted by radioactive nuclei in interstellar galactic gas; (8) *radiogenic thermal luminosity* in explosive stars, the luminous radioactive power in supernovae, in novae, and in neutron-star mergers; (9) *extinct radioactivity*, evidence for the final pre-solar nucleosynthesis events that added to the solar-system cloud; (10) *Stardust*, solid dust grains carrying extinct radioactivity that condensed within pre-solar stars; (11) radioactivity-enabled condensation chemistry of dust within expanding supernova interiors.

D. D. Clayton (✉)
Clemson University, Clemson, SC, USA

R. Diehl et al. (eds.), *Astrophysics with Radioactive Isotopes*, Astrophysics and Space Science Library 453, https://doi.org/10.1007/978-3-319-91929-4_2

2.1 History of Nucleosynthesis and Radioactivity

Radioactive nuclei can not be infinitely old as they would now be gone. This point had first been made by Lord Rutherford, who concluded in 1929 (Rutherford 1929) that the elements had been created in the Sun 100 Myr ago and had somehow got from Sun to Earth! Amusing astrophysics, but his argument showed profound appreciation of the conundrum of live radioactivity on earth; namely, the nuclei must be created. By the middle of the twentieth century, the consequences had progressed no further. The question, "Why aren't they all gone?" went largely unspoken. It took much longer to realise that the galaxy was a prolific creator of radioactivity and that it could be directly observable.

Radioactivity had been discovered by Henri Becquerel in 1896 when he showed that the element uranium emitted radiation that would develop images of the uranium salts on an adjacent photographic plate. Marie Curie coined the term *radioactivity* to describe this phenomenon after she demonstrated that other elements possessed the same spontaneous property. Rutherford separated the rays into three types—α, β, and γ—by their physical properties. These are described in Chap. 1.

In mid-twentieth century, astronomers discovered metal-poor stars in increasing numbers, especially after the introduction of the CCD chip, which greatly increased the sensitivity of spectroscopic observations of stars. The abundances relative to hydrogen in old low-mass stars reveals the initial composition of the gas from which those stars formed. Observed metallicities ranged over a continuum of values from $(\mathrm{Fe/H})/(\mathrm{Fe/H})_{\odot} = 10^{-4}$ to 1 and even greater, so the synthesis of metals must have occurred after the Big Bang. Today we know that creation of the chemical elements occurs in stars by synthesis of them from initial H and He. By the same token, natural radioactivity must be the product of nucleosynthesis in stars.

2.1.1 Two Very Different Pioneers

There could hardly be two more different scientists than the two that made the first significant steps toward nucleosynthesis in stars of the elements heavier than carbon. Both were Europeans disrupted in careers by World War II. The originator of nucleosynthesis in stars arose from non-privileged Yorkshire mill villages. The pioneer in glimpsing how nuclear magic numbers can be observed in isotopic abundances arose from a distinguished Viennese family of geologists.

2.1.1.1 Sir Fred Hoyle (1946, 1954)

In a flash of astrophysical brilliance, Fred Hoyle created the theory of nucleosynthesis in stars with two monumental papers (Hoyle 1946, 1954). The first (Hoyle 1946) demonstrated that stars will naturally evolve to central temperatures 4×10^9 K and greater, and that nuclear statistical equilibrium (NSE) at high temperature T creates an abundance peak at ^{56}Fe that Hoyle interpreted as the explanation of that observed abundance peak. This explains why a graph of the abundances of chemical elements in the sun as a function of atomic weight declines almost monotonically from H to Sc (A = 1–45), whereafter it increases sharply to a broad abundance peak centered on A = 56 (Fe). Readers unfamiliar with the meaning and issues of abundances in the universe can find an eloquent essay in the introductory chapter of Clayton (2003).

Hoyle's second paper (Hoyle 1954) described how the elements from carbon to nickel are synthesized during the advanced evolution of massive stars. This is called *primary nucleosynthesis* because the elements are fused from the initial H and He of the massive stars. Hoyle described how the *ashes* of each thermonuclear burning epoch became, upon contraction and heating, the fuel for the next group of elements synthesized. When its core collapses, most of the mass of the overlying new elements is returned to the interstellar medium. Such nucleosynthesis creates new heavy nuclei and is responsible for increasing the metallicity of the interstellar medium as it ages.

Because the origin of our chemical elements is one of the grand theories of science, Clayton (2007) went to some lengths to create what he called *Hoyle's equation* as determined from careful reading of his 1954 paper (Hoyle 1954). Hoyle's words and calculations pointed clearly to ideas of nucleosynthesis in stars that he was advancing for the first time and that are more sweeping than detail-oriented sequels. Hoyle's discussion was phrased in terms of the mass Δm_{new} of new primary isotopes that are being ejected from massive stars. His basic approach to stellar nucleosynthesis is

$$\frac{dm_{new}}{dt} = H_{nucl} \tag{2.1}$$

where

$$H_{nucl} = B(t')\mathbf{Ev}(t' - t)\Sigma_k \Delta m_k \tag{2.2}$$

is the *Hoyle nucleosynthesis rate*. B(t') is the stellar birthrate of stars having total mass such that they evolve to end their lives at time t, $\mathbf{Ev}(t', t)$ is an operator (rather than a number) that expresses the nuclear and stellar evolution during its lifetime from t' to t, and Σm_k is the mass of isotope k ejected at time t. Then a sum over all pre-solar birthdates t' selects the appropriate stellar masses. Stellar evolution was only dimly perceived in 1953 when Hoyle wrote this paper. The structure of red giants was its current literature frontier, and Hoyle was perhaps the world's leading expert on its ideas, enabling him to discern the more fruitful subsequent evolution that he advanced in this 1954 paper.

Hoyle subtitled this paper *The synthesis of elements from Carbon to Nickel*. These are primary isotopes ejected from massive stars that progress through a series of core evolutions that Hoyle laid out for the first time. Hoyle explained that gravitational contraction causes temperature increases after each nuclear fuel is consumed, and he described the nuclear burning during each advanced core evolution. Because those massive stars all evolve almost instantaneously in comparison with galactic timescale, Hoyle takes $B_{M>}(t)$ to be the birthrate of all such massive stars at time t, and it clearly equals their death rate at the same time if the numbers of stars are to change only slowly. The subscript $M >$ characterizes stars too massive to become white dwarfs. For those large-mass stars Hoyle (1954) predicted that collapse of the final central evolved core is inevitable. For those massive stars Hoyle's equation expresses the rate of ejection of new primary isotopes from C to Ni as

$$\frac{dm(C - Ni)}{dt} = B_{M>}(t)\mathbf{Ev}_{nucl}\Sigma_k \Delta m_k \tag{2.3}$$

Ev is an operator (rather than a number) that expresses the nuclear and stellar evolution during the stellar lifetime. It replaces the initial composition of the star by the composition it has attained at the time when its core collapses. Hoyle attributed the mass and identity k of new primary isotopes ejected per massive star to the following successive core burning phases: ^{12}C and ^{16}O from He burning; ^{20}Ne, ^{23}Na, and ^{24}Mg from C burning; additional ^{16}O and ^{24}Mg from Ne burning; ^{28}Si and ^{32}S from O burning; ^{32}S, ^{36}Ar and ^{40}Ca from photo-alpha reactions on ^{32}S and heavier alpha nuclei during later heating of O-burned matter by the inevitable contraction; and finally ^{52}Cr, ^{56}Fe, ^{60}Ni from subsequent nuclear statistical equilibrium. Hoyle also correctly stated that neutrino emission governs the collapse timescale when core temperature exceeds $T = 3 \times 10^9$ K.

Hoyle's equation expresses a breathtakingly modern view of the metallicity-increasing nucleosynthesis during galactic history. Hoyle missed only the complete photo-nuclear rearrangement during Si burning and the n/p ratio within the NSE. But his equation, given above, remains correct today. Like Schroedinger's equation, for which much work is required to determine the appropriate Hamiltonian operator used within it, so Hoyle's equation involves much work to determine the Hoyle nuclear evolution H_{nucl}.

Hoyle (1954) also distinguished the idea of *secondary nuclei*, those whose created abundance derives from initial seed concentrations of primary nuclei that had been created in previous massive stars, that seeded the interstellar matter with primary elements. These seed nuclei are required in order that secondary nuclei be produced from them in stars. He emphasized especially ^{14}N, ^{18}O, ^{19}F and ^{22}Ne in that regard, each of which depends on initial primary C and O nuclei within the initial composition of later-generation massive stars. Their yields of secondary nuclei do not obey Hoyle's equation but instead are proportional to the initial metallicity of each star. Hoyle's paper also first noted that ^{22}Ne would be a source of free neutrons; indeed, it is today their major source in burning shells of massive stars, although that insight is usually attributed to later emphasis by others.

It is unfortunate that Hoyle did not put to paper the equation he envisioned and described. Had he done so, clearer scientific visibility of his unparalleled achievement would have followed more easily. Research in nucleosynthesis has followed his approach during five decades of countless numerical models evaluating Hoyle's equation. Nonetheless, the astrophysical world began to inappropriately cite a paper to appear, with Hoyle as coauthor, in 1957 for the general origin of nucleosynthesis in stars rather than Hoyle's earlier and original works. Hoyle's great achievement lay somewhat obscured until modern times, at least in terms of citations of his research works (see Sect. 2.1.2.1).

2.1.1.2 Hans Suess, Nuclear Shell Model, and Abundances

Hans Suess was born in Vienna in 1909 to a distinguished family of geologists. His chemical training and focus on abundances of the elements in meteorites enabled him to establish *Suess's rules* for the abundances of the chemical elements and their isotopes. The elemental abundances were not well known after World War II, but testing of a theory of nucleosynthesis needed that data base. In the late 1940s Suess began to argue that certain regularities of the abundances had to reflect nuclear properties of their isotopes. He used those systematics of the isotopes to coauthor with Otto Haxel and Hans Jensen a discovery paper (Haxel et al. 1949) for the magic numbers of nuclear shell structure. Suess had discerned the magic numbers from those isotopes which are very abundant relative to their neighbours. These are hugely important for nucleosynthesis, e.g., from having $N = 82$ neutrons for abundant ^{138}Ba. The magic numbers are favoured numbers of identical nucleons, either protons or neutrons, in the sense that clusters of those numbers of identical nucleons have larger than normal binding energy. They are 2, 8, 14, 20, 28, 50, 82 and 126, and they result from the combined effect of a deep spherical potential binding the nucleons together, coupled with a very strong spin-orbit energy that moderates energy gaps between differing nuclear shells. For that paper Jensen shared the 1963 Nobel Prize in Physics with Maria Mayer for their independent theoretical work on how the spin-orbit force establishes the magic numbers. Suess, working in Hamburg and Heidelberg, had seen the evidence of magic numbers in the abundance regularities shown by nuclei having a magic structure of neutrons or protons. For example, the abundances of the isotopes of barium implicated neutron irradiation in their nucleosynthesis. Suess amazingly divined that the very high abundance of ^{56}Fe had to reflect the doubly-magic properties of its ^{56}Ni isobar having N=28 for both protons and neutrons. That correct assertion was not accepted for a decade because Hoyle's papers (and also B^{2}FH) had maintained that that abundance peak was established within a nuclear equilibrium having excess of neutrons relative to protons, so that the abundance of Fe isotopes could reflect their own nuclear properties rather than those of Ni. This error persisted for a decade, and had many dead-end astrophysical consequences. These have been intensively reviewed (Clayton 1999), who also reviews Suess's pioneering papers from the 1940s.

The magic numbers were also to play a pivotal role in the theories for nucleosynthesis of the elements heavier than the Fe peak. Some of the earliest testaments to the correctness of the theoretical ideas lay in their interplay with the magic numbers. Suess himself called for two neutron-addition processes needed to account for magic-number abundances within the heavy elements. After moving to the United States in 1950, Suess coauthored with Harold Urey an immensely influential tabulation (Suess and Urey 1956) of the abundances of the elements, largely from chemical analyses of meteorites rather than from astronomical observations. That review paper by Suess and Urey became a cornerstone of empirical evidence for nucleosynthesis in stars.

2.1.2 The Second Decade

Hoyle's two papers were not enough to establish the theory of nucleosynthesis in stars. Contributions from other pioneers came together to launch the full theory and to win the acceptance that it enjoys today.

2.1.2.1 William A. Fowler and Burbidge, Burbidge, Fowler and Hoyle (B^2FH)

William A. Fowler had met Fred Hoyle in 1953 when Hoyle famously predicted the existence and the energy of the 0^+ second excited state of ^{12}C. In that first application to nuclear physics of what has come to be called *the anthropic principle*, Hoyle had argued that if such a nuclear state of ^{12}C did not exist, neither would we! Fowler said that that prediction was what "really hooked me on nucleosynthesis". Fowler met Hoyle again during his 1955 sabbatical leave in Cambridge UK. Fowler was at this time already the leader of the world effort (which Caltech championed) to determine the rates at which nuclear reactions would occur in stars. This involved the now familiar technique of measuring nuclear interactions at MeV energies with Van de Graff accelerator beams of charged ions, and extrapolating measured data downward in energy to the Gamow-peak energy within a thermal distribution (Clayton 1968, Ch. 4). Fowler's zest for nuclear astrophysics was boundless, and he soon had the entire Kellogg Radiation Laboratory (including this writer) at work on thermonuclear reactions rates between positive ions. Fowler was awarded a share of the 1983 Nobel Prize in Physics for that pioneering program of research. His efforts had been directed primarily to the question of thermonuclear power in stars; but Fowler became intensely attracted also to the larger question of the origin of the elements. In Cambridge in 1955 he met Geoffrey and Margaret Burbidge, and these three began a project with Fred Hoyle to write a survey paper on the issues of nucleosynthesis in stars. That paper was written at Caltech in 1956 and published the next year (Burbidge et al. 1957). Soon called simply B^2FH, it became one of the most celebrated papers in astrophysics.

The authors of B^2FH contributed creatively and energetically to formulating the *neutron-capture processes* for synthesizing the elements heavier than nickel. Ascribing crucial roles to the magic neutron numbers $N = 50$, 82 and 126 they described environmental situations in stars within which selected isotopes would be abundant, fleshing out ideas that Suess had envisioned earlier for two neutron-capture processes. Suess had, however, been unable to formulate these clearly enough to win contemporary acclaim. The slow neutron-capture process, named *s process* by B^2FH, envisioned 100-to-1000 years between neutron captures, so that radioactive isotopes would generally beta decay before capturing another neutron, keeping the capture path trailing along the valley of beta stability. The rapid-neutron-capture process, named *r process* by B^2FH, envisioned neutron densities so large that neutron captures occur in tens of milliseconds, faster than beta decays, with the result that the capture path moved into the realm of radioactive neutron-rich nuclei, being halted only when no additional neutrons could be stably added owing to their diminishing separation energies. These created *waiting points* at which the capture flow would halt and wait until beta decay occurs (see also Seeger et al. 1965; Clayton 1968). Defining the *s* process and the *r* process was the high point of the B^2FH contributions to nucleosynthesis theory. Their Appendix included an inspiring table of all heavy isotopes in which each was characterized as being either *s*-process only, mostly s process, comparable *s* process and *r* process, mostly *r* process and *r*-process only. This can be regarded as the next important step in nucleosynthesis theory.

An important aspect of astronomy with radioactivity lies in the competition between beta decay and neutron capture that ensues when neutron capture by stable isotopes create isotopes that can undergo beta decay. Most do so quickly in comparison with the time required to capture another neutron, but some key branch points are slow to decay. B^2FH had inventively shown that such competition at branch points could, when compared with the actual solar abundances, reveal the time scale and neutron density for the operation of the *s* process. Thermally populated excited states of radioactive nuclei often increase the effective decay rate, a delicate point in that aspect of astronomy with radioactivity.

Two cautions about B^2FH must be made in order to not overly eulogize what they achieved. Firstly, both the *s* process and the *r* process are, as described by them, secondary processes of nucleosynthesis. *Secondary nucleosynthesis* refers to the synthesis of new heavy nuclei from other existing heavy nuclei. It does not increase the galactic metallicity, a goal that Hoyle's founding papers had achieved spectacularly as the primary goal of the astronomy of nucleosynthesis. The B^2FH neutron-capture processes instead change one existing heavy nucleus into another. So those processes did not contribute to increasing metal abundances in the galaxy. Today it is known that, contrary to their description, the *r* process is actually primary, because the collapsed supernova core synthesizes the seed nuclei that rapidly capture its free neutrons. The details of this are still not understood, however. Astronomical observations of old metal-poor stars confirm that the *r* process indeed began earlier than did the *s* process. Secondly, the B^2FH descriptions of the *s* process and the *r* process could not immediately be used for astrophysics calculations because they were time-independent formulations. Both the *s* process and the *r* process were described by the static condition $dN/dt = 0$. The B^2FH descriptions were thus of stationary abundances that could exist within appropriate environments for these processes. This enabled a rough but clear correlation between the nuclear systematics of their process abundances and the solar isotopic abundances. But it did not enable calculation of the temporal growth of these abundances.

Over the next two decades B^2FH nonetheless became the default citation for workers wanting a reference to the general theory of nucleosynthesis in stars, vastly eclipsing the rate of citations to Hoyle's previous papers. One key to its success was its citation of over 100 stars showing nucleosynthesis effects in their spectra and of more than 100 astronomical research papers concerning those stars. This drew the astronomical community into the scientific culture of nucleosynthesis (which was then new to astronomers) as Hoyle's papers could not. Astronomers generally cited B^2FH rather than Hoyle, with the result that Hoyle's papers slipped into relative obscurity.

Fowler himself coauthored several other important works for nucleosynthesis, primarily with Fred Hoyle on supernovae and with Donald Clayton on nucleosynthesis by neutron capture chains; however, his most important and essential role in history was the empirical thermonuclear reaction rates determined in his laboratory. For this he shared the 1983 Nobel Prize in physics. Probably Fowler's most significant subsequent work with Hoyle concerned the puzzling nature of the supernova phenomenon and on nucleosynthesis within them (Hoyle and Fowler 1960). They divided supernovae into Types I and II based on theory, rather than on the presence of H absorption lines in the spectra. Type I was assigned to low-mass progenitor stars that evolved to white-dwarf stars of degenerate carbon, but then ultimately explode in an exothermic thermonuclear display that results in most of the iron in the universe. Electron-degenerate matter, supported by degeneracy pressure, is violently unstable to thermonuclear runaway. The Type II supernovae, on the other hand, occurs in massive stars whose cores are too massive to form a white dwarf structure. In these, continuing nuclear burning eventually exhausts the nuclear energy supply, and so the core must collapse. Fred Hoyle had predicted this inevitable collapse in 1954; but their 1960 paper carried the physical picture further. Today the world of astronomy uses this Hoyle-Fowler classification based on the physics of the device. It was a landmark in astrophysics (see Woosley 1999).

2.1.2.2 A.G.W. Cameron, a Parallel Force

The year 1957 also saw the emergence of A.G.W. Cameron as one of the pioneers of nucleosynthesis. The Atomic Energy of Canada Ltd laboratory at Chalk River published in bound-mimeograph form a series of lectures on nucleosynthesis that Cameron delivered at Purdue University in March/April 1957 (Cameron 1957). These lectures covered in an independent way essentially the same material as in B^2FH. Cameron had constructed his treatments by working alone, first as a new faculty member at Iowa State University and then at Chalk River, with only Hoyle's two papers to guide him as well as his training in nuclear physics from University of Saskatchewan. Owing to his independent treatment, Cameron's Lecture Notes became a valuable source of new ideas in nucleosynthesis. In his emphasis on nucleosynthesis within the separate shells in massive stars, Cameron's approach followed Hoyle's equation and thereby enlarged the ideas in Hoyle (1954) more effectively than did B^2FH. For the next two decades Cameron stressed in many publications with research students—ones that he had recruited during a guest lecture course at Yale University—the vast nucleosynthesis changes that occur in a massive star at its time of explosion. His papers establish him as the first great disciple of Hoyle's equation. Cameron's later Yale Lecture Notes (1963), written by his Yale students W.D. Arnett, C.J. Hansen and J.W. Truran, were much improved over Cameron (1957). They probably should have been published, but were not and are therefore not generally available; however, they were a strong influence on history, especially on Cameron's students and on Clayton and his students. Partly owing to their unavailability, Clayton published his own textbook on *principles of stellar evolution and nucleosynthesis* (Clayton 1968).

Cameron began his nucleosynthesis research, however, with the sources of free neutrons in stars that could be responsible for the observed radioactive Technetium observed by Paul Merrill in stellar atmospheres. Because all isotopes of the element Tc are radioactive, the presence of its absorption lines in stellar spectra argued that it had been created within the star during roughly its last lifetime. In 1955 Cameron proposed that the $^{13}C(\alpha, n)^{16}O$ reaction would liberate the extra bound neutron in the ^{13}C nucleus and that the liberated neutron could be captured to create heavier isotopes (Cameron 1955). Later Cameron (1959) calculated with the same motivation the number of neutrons liberated during carbon thermonuclear reactions following the exhaustion of He. The carbon fusion reactions were part of the sequence of thermonuclear stages within evolving massive stars (as Hoyle (1954) had first described).

In his large subsequent body of work, Cameron established himself as a true polymath. He introduced the speeding up of beta decay rate by thermal population of the excited states of a nucleus, many of which decay more rapidly than the ground state. During 1955–1956 Cameron introduced numerical computation on the first vacuum-tube computers into nucleosynthesis problems, and remained thereafter on the cutting edge of nucleosynthesis computation. Cameron and his students repeatedly blazed new paths by programming nuclear reaction networks onto the latest and newest computers. He became an expert on planetary sciences and an important advisor to NASA. In particular, Cameron is forever famous for his work on the origin of the moon as the result of a Mars-like planetary collision with the young earth, a theory of origin that now seems beyond doubt.

2.1.2.3 Donald D. Clayton and Time-Dependent Heavy Element Nucleosynthesis

Beginning as research student at Caltech with Fowler, Donald Clayton began constructing a time-dependent formulation of the *s* process in 1957, the year of B^2FH publication and of Cameron's Chalk River lecture notes. Clayton's discard of the assumption $dN/dt = 0$, an assumption requiring a constant σN curve for s-process abundances, altered profoundly the direction of s-process research by focusing on how efficiently seed nuclei could be converted to heavy s-process nuclei. B^2FH had not addressed that question. Clayton showed, as B^2FH had surmised, that the iron abundance peak must provide the seed nuclei being transmuted into the large overabundances of barium in stars whose spectra showed Ba/Fe some 20–50 times the solar ratio. More surprisingly his results also showed that, as cumulative neutron fluence increases, none of the sequential abundance distributions that are generated resemble the solar abundances (Clayton et al. 1961). The solar s-process abundances were required to be a superposition of differing numbers of Fe-seed nuclei (per Fe nucleus) exposed to differing integrated fluxes of free neutrons. The number exposed must be increasingly smaller for increasingly larger neutron irradiations. Galactic history or history within s-process stars is required to bring that superposition about. Therein lay new astrophysics. The solar s-process abundances were shown to not be simply a smoothly declining σN curve, as B^2FH had speculated, but a superposition of exposures generating narrow regions of atomic weight near the neutron-magic numbers where the assumption $dN/dt = 0$ is severely violated. Two decades of improved measurements and consequent fitting to solar abundances (Käppeler et al. 1982) would be required before advances in stellar evolution would be able to describe the exponential-like fluence distribution that was required. It was a sophisticated interplay between He-shell pulses and cyclically ignited H burning at the base of the envelope of AGB stars (see Chap. 3).

From the time of these first solutions of the neutron-irradiation superpositions resulting in the s-process abundances, new phenomenological aspects of heavy element nucleosynthesis were possible. The theory-based fit yielded all s-process abundances with meaningful accuracy. These allowed Clayton and Fowler (1961) to publish the first decomposition of heavy-element abundances into their s-process and r-process parts. B^2FH had suggested the dominant processes for each isotope in their spectacular appendix; but a quantitative decomposition became possible for the first time. Their initial effort has been redone at least a dozen times as new neutron-capture-cross-section data appeared, most notably first by Seeger et al. (1965) and later by a new measurement program in Karlsruhe (Käppeler et al. 1982). This s-r decomposition applied to astronomical spectroscopy of old stars has routinely produced meaningful new knowledge. Observations of old metal-poor dwarf stars indicated that the r-process abundances began to grow earlier, as star formation first began, than did the s-process abundances (Truran 1981; Burris et al. 2000). That result demonstrated that the r process is a primary nucleosynthesis process, rather than secondary as B^2FH had stated. This requires the r-process to occur within violent and explosive environments that include abundant existence of free nucleons, such as core-collapse supernovae. Quantitative s-r decomposition also inspired unforeseen new techniques for radioactivity-based cosmochronology.

Impressed by the new astrophysics lurking in time-dependence, Clayton advocated time-dependent formulation of the r process as well, again jointly with Fowler and with P.A. Seeger, Fowler's research student in Kellogg Lab. For the r process as for the s process, B^2FH had described only a time-independent steady flow that showed that the neutron-rich heavy isotopes did indeed have abundant progenitors in suitable time-independent settings; but they had not been able to address whether the entire r-process abundances can be synthesized at one set of conditions. Computers at that time (1963) were not capable of handling a full r-process network. Seeger et al. (1965) showed that the full mass range cannot be produced together (unless new seed nuclei are injected during the process). B^2FH had creatively defined the key nuclear physics relationships of the r process but were mute on its dynamics. The time-dependent formulation by Seeger et al. (1965) became a prototype for r process astrophysics. It showed the r process to also be a superposition of differing irradiation histories of seed nuclei. Even after four decades of subsequent computations, the nature of the r process superposition remains a frontier puzzle. It is noteworthy for astronomy with radioactivity that the entire r process reactions occur within the realm of radioactive nuclei. Only after rapid expansive cooling can that neutron-rich radioactive abundance distribution undergo a series of beta decays changing each isobar identity until resting at the most-neutron-rich stable isobar (isobar is an isotope having the same atomic weight). A major research goal of nuclear astrophysics today is better laboratory definition of the parameters defining accurately the properties of the neutron-rich radioactive nuclei.

Understanding the time dependence during silicon burning (silicon photoerosion) was the big scientific challenge of the mid-1960s. Clayton introduced nuclear quasiequilibrium as a physical concept (Bodansky et al. 1968a,b) to clarify how silicon transmutes to an iron abundance peak. Quasiequilibrium explained the only big gap in Hoyle's 1954 theory of primary nucleosynthesis, replacing the ill-formulated α process of B^2FH. A temporal sequence of quasiequilibrium states facilitated the calculation of the set Δm_k for k = 28–62 to be inserted in Hoyle's equation. The sequence of quasiequilibria again involved relaxing the assumption $dN/dt = 0$. The quasiequilibrium concept was powerful and new, and enriched many subsequent aspects of nucleosynthesis reaction networks (the r process, explosive oxygen burning, the α-rich freeze out, the origin of ^{48}Ca, and others). Just as all nuclear reactions proceed at the same rate as their inverses in full nuclear equilibrium (NSE), during quasiequilibrium one refractory nucleus violates that equilibrium assumption by changing abundance only slowly, while all others maintain equilibrium with it. During silicon burning the ^{28}Si nucleus is the slowly changing, refractory post to which the quasiequilibrium distribution is attached (Bodansky et al. 1968a,b). The most abundant isotopes between A = 44 and 62 are, in this quasiequilibrium sequence, created as radioactive progenitors rather than as stable isobars, with important consequences for emerging astronomies of radioactivity. The most shocking consequence was that supernovae are profoundly radioactive, a realisation that dominates study of supernovae today.

2.1.2.4 The Sequel

The decade 1956–1966 following Hoyle's pioneering two papers had witnessed profound enlargements of the theory of nucleosynthesis in stars. The years following the publication of B^2FH had been marked by vast improvement and reformulation of its influential processes. Calculable time-dependent descriptions of heavy-element nucleosynthesis processes refocused attention from simple correlations between nuclear properties and abundances to the astrophysical histories and stellar evolution that bring them into existence. The sometimes heard statement by astronomers that not much happened after B^2FH reflects lack of awareness of these historic changes. It may be of interest to note that the beginnings of nucleosynthesis theory was an international innovation. Of the pioneers named, only Fowler and Clayton were born Americans; Hoyle, E.M. Burbidge and G.R. Burbidge were English; Suess was Austrian; and Cameron was Canadian.

Innovations continued and accelerated during the next decade 1967–1977. These will not be reviewed here except to say that the evaluation of Hoyle's equation through numerical computation of the evolution of massive stars yielded repeated insights into the interplay between stellar evolution and nucleosynthesis. The B^2FH neutron-capture processes revealed layers of complexity associated with the stars and with time dependences. The innovative center of this research moved away from Caltech, initially to Yale University and to Rice University, where Cameron and Clayton respectively founded schools evaluating Hoyle's equation. New leaders developed within those schools included especially W. David Arnett from Yale (subsequently also Rice) and Stanford E. Woosley from Rice (subsequently U.C. Santa Cruz). Each has many important publications on the evolution of massive stars and the nucleosynthesis in its shells (Arnett 1996; Woosley and Weaver 1995) (see Chap. 4). A European role in this research also experienced rebirth during that decade, especially in Darmstadt and in München. It is also the decade 1967–1977 that sees the emergence of several new observational aspects of astronomy with radioactivity. A description of those developments follows.

2.1.3 New Astronomy with Radioactivity

The existence of natural radioactivity clearly holds implications for the origins of atomic nuclei. The chemical elements could not have always existed if the radioactive nuclei were created along with the stable nuclei. In this way naturally occurring radioactivity is intimately related to nucleosynthesis of the chemical elements. It was evident from the structure of the suggested processes of nucleosynthesis that radioactive nuclei played a large role in each of them. The key role of radioactive isotopes during nucleosynthesis and during thermonuclear power in stars was the first astronomy with radioactivity. Presumably the radioactive nuclei would be ejected from stars along with the new stable nuclei unless they decayed within the stars prior to ejection. Radioactivity plays a large role in the H-burning reactions, the PP chains and the CNO cycle (see Clayton 1968) responsible for the stellar power capable of keeping the stars from cooling. Hans Bethe was the 1968 Nobel Prize awardee for discovering these H-burning cycles in stars just prior to World War II.

So it was evident in mid-twentieth century that radioactive nuclei carry significant issues for astronomy. In the 1950s the idea arose of directly viewing the radioactivity at the solar center by detecting its neutrinos emitted from the sun's center. Raymond Davis Jr. would win a Nobel Prize for spearheading that grandiose effort.

What made astronomy with radioactivity so exciting scientifically was the discovery in the 1960s and 1970s of altogether new ways of observing radioactivity in astronomy. These aspects of astronomy with radioactivity generated new interdisciplinary connections to nucleosynthesis. All science needs observations to provide an empirical base, astronomy just as surely as laboratory science. To be sure, the scientific method happens differently in astronomy than in laboratory science. In astronomy there are no experiments that can change the initial conditions as a test of theories. No experiments can be planned to refute a hypothesis. One has instead only observations of natural events. In many cases simple observations can be sufficient to refute a hypothesis. It is fortunate that nature provides so many natural events, so that in many cases contrasting separate events constitutes a type of experimentation, giving a spectrum of observations in which conditions differ in ways that must be inferred but are nonetheless real. The experiments are natural and performed by nature herself, however, rather than by scientists.

New types of observations of the occurrence of natural radioactivity galvanised astronomy by providing new kinds of astronomical data. These first historical advances are reviewed in what follows, as previews to the more detailed chapters that discuss them. They are: cosmochronology, the age of the elements; gamma-ray-line astronomy of radioactivity; radioactive power for light curves of supernovae and of novae; extinct radioactivity measured by excess abundances of daughter nuclei in solids formed in the early solar system; extinct radioactivity measured by excess abundances of daughter nuclei in solids condensed while dust particles condense as gas leaves a site of stellar nucleosynthesis. What follows is a brief description of how these applications were discovered or anticipated.

2.1.4 Interpreting Exponential Decay

The familiar exponential law of radioactive decay follows from Eq. (1.1) from Chap. 1 for an ensemble of N radioactive nuclei from the constancy of decay probability for a single nucleus. Given N such nuclei the expected number of decays per unit time in the ensemble is given by the product of the number N of nuclei and the decay probability per unit time for a single nucleus. Accordingly

$$\frac{dN}{dt} = N\lambda = N/\tau \tag{2.4}$$

Because λ, and therefore τ, is a constant for that nuclear species when it exists outside of stars, this equation has a well-known integral form for the number of remaining radioactive nuclei

$$N(t) = N_0 \, e^{-t/\tau} \tag{2.5}$$

where N_0 is the initial number in the ensemble (at $t = 0$). Similarly, if a number N_0 is observed now at $t = t_0$, the number that existed at an earlier time t_1 would have been

$$N(t_1) = N_0 \, e^{(t_0 - t_1)/\tau} \tag{2.6}$$

provided that new nuclei have not been added to the ensemble during that time interval. In the same spirit, if N_1 were created at earlier time t_1 and N_2 were added at a subsequent time t_2, the number N_0 that would exist today at t_0 is

$$N_0 = N_1 \, e^{-(t_0 - t_1)/\tau} + N_2 \, e^{-(t_0 - t_2)/\tau}. \tag{2.7}$$

Generalizing slightly, let $B_\odot(t)$ represent the fractional age distribution of the primary stable solar nuclei at the time of solar formation at $t = t_\odot$. Then $dB_\odot(t)/dt$ is the number of stable solar nuclei that had been born per unit time at time t. It is the age distribution that a radioactive species within an ensemble would have *if it were not decaying*. The age distribution for radioactive parents may be thought of as the age distribution of stable nuclei that were synthesized along with the radioactive nuclei. A mnemonic for the symbol $B_\odot$ is the *birthrate* of those solar system nuclei. $dB_\odot/dt$ is the rate at which the stable nuclei and the primary parent radioactive nuclei were added to the total sample destined for the solar system rather than the rate at which they were produced by nucleosynthesis. In drawing this subtle distinction, Clayton (1988) demonstrated the error of the common practice of equating the age distribution of solar nuclei with the rate of galactic nucleosynthesis. The age distribution of solar-system nuclei is vastly different from the galactic nucleosynthesis rate because the evolution of the ISM is involved. Even if the ISM is instantaneously mixed, many of the old nuclei became trapped within stars, so that the age distribution in the ISM becomes biased toward more recent nucleosynthesis.

Given this definition of $dB_\odot/dt$ as the age distribution of solar-system nuclei, and given that their addition to the solar sample began at time t_1 and continued until $t_\odot$, the total number of any stable primary nucleosynthesis product would have been,

$$N_{total} = \int_{t_1}^{t_\odot} \frac{dB_\odot(t')}{dt'} \, dt' \tag{2.8}$$

Then in analogy with Eq. (2.7), if those nuclei are instead radioactive, the number surviving until solar formation at time $t_\odot$ would be

$$N_\odot = \int_{t_1}^{t_\odot} \frac{dB_\odot}{dt'} \, e^{-(t_\odot - t')/\tau} dt' \tag{2.9}$$

If the total duration $t_0 - t_1$ of production is much greater than the mean lifetime τ, and if $dB_\odot$/dt varies only slowly, Eq. (2.9) reduces approximately to

$$N_\odot = \left(\frac{dB_\odot}{dt}\right)_{t_\odot} \tau \tag{2.10}$$

This useful formula estimates the number of remaining radioactive nuclei to be equal to the number created during the last mean lifetime of that radioactive species. If $dB_\odot/dt$ has jagged temporal structure near $t_\odot$, however, a more complicated evaluation would be required. Equation (2.10) is very useful as a first approximation to the numbers of radioactive nuclei within the ISM during continuous nucleosynthesis in stars if the birthrate $B(t)$ is redefined as the birthrate spectrum of those stable ISM nuclei residing in the ISM rather than in stars.

These relationships expressing properties of exponential decay are central to understanding both extinct radioactivity in the early solar system and the numbers of radioactive nuclei within astronomical objects.

2.2 Disciplines of Astronomy with Radioactivity

2.2.1 Nuclear Cosmochronology

Today it is self evident that the existence of radioactive nuclei implies that they were created at some estimable moment in the past. Were that not the case, they would have long ago have disappeared. But the full implications were not evident to those engaged in the exciting day-to-day goals of understanding natural radioactivity. Nonetheless, in 1929 Ernest Rutherford wrote what may be the first paper on astronomy with radioactivity (Rutherford 1929). Today we use the term nuclear cosmochronology to mean the attempt to use natural abundances of radioactive nuclei to compute the age of an astronomical object or of the elements themselves. Determining the age of the chemical elements by assuming them to have the same age as the radioactive nuclei became a goal that has attracted many, most notably William A. Fowler and this writer.

2.2.1.1 Uranium and Thorium on Earth

When he began to think on these things, Ernest Rutherford concluded from the ratios of ^{235}U/^{238}U as measured in his laboratory that uranium was created somehow within the sun and transported to earth. Accepting Jeans's estimate that the age of the sun was an exuberant 7×10^{12} years, he wrote in a prophetic paper addressing the puzzle (Rutherford 1929):

> ..it is clear that the uranium isotopes which we observe on earth must have been forming in the sun at a late period of its history, namely, about 4×10^9 years ago. If the uranium could only be formed under special conditions in the early history of the sun, the actino-uranium on account of its average shorter life would have practically disappeared long ago. We may thus conclude, I think with some confidence, that the processes of production of elements like uranium were certainly taking place in the sun 4×10^9 years ago and probably still continue today.

Corrected modern numbers for those that Rutherford used are the observed abundance ratio $^{235}U/^{238}U = 7.25 \times 10^{-3}$ and the mean lifetimes against alpha decay, $\tau(^{235}U) = 1.029 \times 10^9$ years and $\tau(^{238}U) = 6.506 \times 10^9$ years. Since each isotopic abundance has been exponentially decaying during the age A_E of the earth, their abundance ratio when earth formed would have been $(^{235}U/^{238}U)_0 = 7.25 \times 10^{-3}$ $(e^{A_E/\tau(^{235}U)}/e^{A_E/\tau(^{238}U)})$.

The age of the earth has been reliably measured using the fact that these two isotopes of U come to rest, after a series of alpha decays, as different isotopes of Pb, namely ^{207}Pb and ^{206}Pb respectively. Using the measured earth age, $A_E = 4.57 \times 10^9$ years, yields the initial U ratio on earth to have been $(^{235}U/^{238}U)_O = 0.31$. These facts are beyond doubt.

The implication of great consequence stems from the expectation that the r process that is responsible for the nucleosynthesis of both isotopes, should make more ^{235}U than ^{238}U. ^{235}U has six non-fissioning progenitors whereas ^{238}U has but three. An r-process progenitor is a non-fissioning transuranic nucleus that after a series of relatively fast radioactive decays comes to rest at one of these long-lived isotopes of U. For ^{235}U those nuclei are $A = 235, 239, 243, 247, 251$ and 255-totaling six r-process progenitors-whereas for ^{238}U they are $A = 238$, 242, 246 and 35% of 250-totaling 3.35 progenitors. Taking into account the empirical evidence that production by the r-process favors even-A nuclei by a slight 20% over adjacent odd-A nuclei, one expects ^{235}U production to exceed that of ^{238}U by a factor near $P(^{235}U)/P(^{238}U) = 1.79$. This argument reveals that the abundance ratio $^{235}U/^{238}U$ has declined from near 1.79 at production to 0.31 when the earth formed. This decline takes considerable time, showing that U isotopes were synthesized during pre-earth astrophysical history.

This first calculation of nuclear cosmochronology reveals the nature of the problem, but also its uncertainties. Is the production ratio $P(^{235}U)/P(^{238}U) = 1.79$ correctly estimated? Were the U isotopes synthesized in one single pre-solar event, in which case it occurred about 6.6 billion years ago, 2 billion years prior to formation of the earth. Or was their production rate distributed in pre-solar time? If the age distribution of the solar system's r-process nuclei is flat between the time of first production and the earth's formation, the beginning of nucleosynthesis would fall near 13 billion years ago. The true age distribution can hardly be known with any assurance, revealing the severe limitation of this single pair for constraining the time of the beginning of nucleosynthesis in our galaxy.

The technique was extended to the ratio $^{232}Th/^{238}U$ by Fowler and Hoyle (1960); and Fowler returned to it many times in later years (e.g. Fowler 1971); see also the textbook by Clayton (1968). Relevant numbers used are the observed abundance ratio ^{232}Th/^{238}U = 4.0 and the mean lifetimes against alpha decay, $\tau(^{232}\text{Th}) = 20.04 \times 10^9$ years and $\tau(^{238}\text{U}) = 6.506 \times 10^9$ years. The abundance ratio is much more uncertain than it was for the pair ^{235}U/^{238}U because Th and U are different chemical elements. Any two elements having different fractionation chemistry pose a tough problem when seeking their ratio in the initial solar system (or in the sun). Since each isotopic abundance has been exponentially decaying during the age A_E of the earth, their abundance ratio when earth formed would have been $(^{232}\text{Th}/^{238}\text{U})_0 = 2.5$ instead of 4.0. And the production ratio in r-process events was inferred from the same counting arguments of odd and even progenitors of ^{232}Th and ^{238}U to be $P(^{232}\text{Th})/P(^{238}\text{U}) = 1.73$.

A problem is that this Th/U pair does not give transparently concordant numbers with the U pair. Uncertainties in r-process production ratios, in the relative abundances of Th and U, and in the arbitrary parameterizations of galactic chemical evolution that have been used each conspire to yield possible solutions in which nucleosynthesis began anywhere from 2 to 10 Gyr prior to solar birth. Fowler strove repeatedly to circumvent these uncertainties, trying to extract the true answer (for there obviously is a true answer!); but in fact, the data are not adequate and the astrophysical model of the galaxy used is not adequate for the task. Data from other chronological species would be needed, along with a more sophisticated appreciation of galactic chemical abundance evolution.

On a more positive note, Fowler's papers (Fowler and Hoyle 1960; Fowler et al. 1960; Fowler 1971, and several others) inspired many others to tackle this fascinating aspect of astronomy with radioactivity.

2.2.1.2 Cosmochronology with Extinct Radioactive Abundances

In 1960s a new discovery, excess trapped ^{129}Xe gas in meteorites (Reynolds 1960), provided the first evidence that the matter from which the solar system formed contained radioactive nuclei whose halflives are too short to be able to survive from that time until today. These are called *extinct radioactivity*. Discovery of extinct radioactivity utilized the buildup of the daughter abundance from a radioactive decay as a measure of how abundant that radioactive parent was in the initial solar system. What Reynolds (1960) showed was that the excess ^{129}Xe gas in meteorites had resulted from ^{129}I decay and that the initial ^{129}I abundance was about 10^{-4} of the initial ^{127}I abundance. He realized that this datum allowed an estimate of when the iodine isotopes had been created. By assuming that at the time of their nucleosynthesis the initial ^{129}I abundance had been equal to that of stable initial ^{127}I, Reynolds argued from the mean lifetime $\tau(^{129}\text{I}) = 23.5 \times 10^6$ years (23.5 My) that iodine had been created only about 300 My prior to solar system birth. This was the first new cosmochronology technique since Rutherford and brought Reynolds great fame. Unfortunately, Reynolds' conclusion was no more believable than Rutherford's had been, because his astrophysical model was also quite unrealistic. He mistakenly assumed that all of the iodine abundance in the solar system had been created at the same time, and therefore used Eq. (2.5) to estimate that time. Because 10^{-4} requires a decay period equal to about 13 mean lifetimes, he had concluded that the age of the elements was $13\tau(^{129}\mathrm{I}) = 300$ My. Fowler et al. (1960) countered that it was more realistic to assume that the synthesis of iodine isotopes was spread out by multiple r-process events, occurring between the time t_1 of a first nucleosynthesis event until the time t_0 of the last such event prior to solar birth. That concept, called *continuous nucleosynthesis*, required use of Eq. (2.10) limit of Eq. (2.9) instead of Eq. (2.5). That calculation gave the initial nucleosynthesis epoch to be $104\tau(^{129}\mathrm{I}) = 230$ Gy—very old indeed, unrealistically old. This picture was then modified to allow for an interval of sequestered interstellar gas, probably within a molecular cloud, of about 100 My during which no new nuclei were added to the solar mix—a so-called *free decay interval* in which radioactive abundances would follow Eq. (2.5). Such an interval would require that 100 My before the solar birth, when that interstellar gas withdrew from new nucleosynthesis, the interstellar abundance ratio would have been $N(^{129}\mathrm{I})/N(^{127}\mathrm{I}) = 2.8 \times 10^{-3}$ rather than 10^{-4}. Equation (2.10) then suggests that nucleosynthesis began 8 Gy prior to solar birth, a reasonable number, obtained, unfortunately, by construction. But that construction had nonetheless established a new paradigm for radioactive abundances in the early solar system. The upshot for the problem of radioactive chronology is that the abundance of now extinct radioactivity in the early solar system can say little about when nucleosynthesis first began. But it provides other equally interesting issues for the astronomy of radioactivity.

Radioactive ^{129}I is of special historical significance as the first extinct radioactive nucleus to be discovered that can be regarded as part of the average radioactivity in the galactic ISM at the time of solar birth. If so it is in that sense typical of what is expected in the ISM. Other extinct radioactivities may not be capable of such an interpretation but must instead be interpreted as produced by a special event associated with the birth of the solar system. Other subsequently discovered extinct nuclei that can be regarded in the first category of galactic survivors are ^{53}Mn, ^{107}Pd, ^{182}Hf, ^{146}Sm and ^{244}Pu.

2.2.1.3 Birthrate Function for Primary Solar Abundances

The history of nuclear cosmochronology reveals widespread conceptual confusion between the rate of galactic nucleosynthesis and the age spectrum of solar system nuclei. It is the latter that enters into cosmochronology through Eq. (2.9). So endemic is this confusion between these two fundamental concepts that it is important to define in this section the birthrate spectrum for primary abundances in the solar system.

Let $N_\odot$ be the solar abundance of a primary nucleosynthesis product. One may think of it either as the total number of those atoms in the solar system, which consists of awkwardly large numbers, or as the total number of those atoms in the solar system normalized to a defined abundance for a specified nucleus that sets the abundance scale. Two such normalized scales are very common: one used more in astronomy defines the solar abundance of hydrogen to be $N_H = 10^{12}$; another, used more in solar abundances derived from meteorites, defines the solar abundance of silicon as $N_{Si} = 10^6$. This book will primarily use the latter definition because most studies of nucleosynthesis in the literature use the scale Si = 10^6 (where the chemical symbol is often used to represent the abundance of that element). Let $B_\odot(t)$ be the cumulative fraction, as a function of time, of stable solar primary nuclei that already existed at time t. Letting for convenience the starting time $t = 0$ be the time when nucleosynthesis first began, the function $B_\odot(t)$ starts at 0 at $t = 0$ and rises to $B_\odot = 1$ at $t = t_0$, the time when the solar system formed. Note carefully, $B_\odot(t)$ is not the cumulative birth fraction of that nucleus in galactic nucleosynthesis, but rather the cumulative birth fraction of those primary nuclei that actually entered the solar system inventory. Normalizing this definition of $B_\odot(t)$ as a function that rises to unity at solar birth, the product

$$N_\odot \cdot B_\odot(t) = N(t) \tag{2.11}$$

where $N_\odot$ is the solar abundance of that stable species, $N(t)$ is the number of those solar-system nuclei that already existed at time t. It is equally clear that $dB_\odot(t)/dt$ is the fractional birthrate per unit time of primary solar nuclei. It is the age distribution of solar nuclei, and redefined as here to rise to unity it is approximately the same function for all primary stable nuclei.

A mental experiment is needed to set this birthrate function clearly in mind before considering why it differs so from the nucleosynthesis rate. The reader is warned that the research literature suffers endemic confusion over that distinction, requiring its careful definition. For any primary nucleus within solar abundances, *paint each solar-system atom of it red!* This cannot of course be done in reality, but the thought experiment helps understanding of the birthrate spectrum. On the scale Si = 10^6 one has by construction 10^6 *red* Si atoms in the solar system. Then imagine watching a film run backwards in time of these 10^6 *red* Si atoms. Back in the interstellar medium they are mixed with a much larger number of *unpainted* Si atoms. As the film runs to earlier times, these 10^6 *red* Si atoms have greatly differing histories. Some, after being created in a supernova, have later been inside a star, survived and come back out. But for each atom, its world line reaches a time when that atom first came into existence. Each of the 10^6 red solar Si atoms is labeled by that birthdate, born from distinct supernovae at differing times. Then a graph is constructed giving the total number that existed at time t. This graph starts at 1 atom when the first solar Si atom is created and reaches 10^6 by the time the solar system forms, when all 10^6 have been created. If the curve is then normalized by dividing by 10^6, one has the birthrate spectrum $B_\odot(t)$ of solar primary nuclei, which rises to unity at the time the solar system forms. Equation (2.11) depicts this situation, where $N(t)$ rises to 10^6 at solar birth, but $B_\odot(t)$ rises to unity at that time. Although different primary species have vastly differing abundances, for each the function $B_\odot(t)$ is approximately the same function because the abundances of all primary nucleosynthesis products rise approximately (but not exactly) together.

Equation (2.11) is the solution of the differential equation

$$dN/dt = dB_{\odot}(t)/dt \text{ for stable nuclei} \quad (2.12)$$

and

$$dN/dt = dB_{\odot}(t)/dt - \lambda N \text{ for radioactive nuclei} \quad (2.13)$$

One frequently sees these equations written for galactic nucleosynthesis, with dB/dt identified as the galactic nucleosynthesis rate and $N(t)$ as the ISM abundance. This is a fundamental error. Consider why.

The galactic nucleosynthesis rate for Si atoms was very large in the young galaxy, when the rate of formation of massive stars and supernovae was large. But those new Si atoms in the interstellar medium become mostly locked up in the interiors of subsequent low-mass stars, and are not available for incorporation into the solar system. This occurs because it is the fate of most interstellar gas to be locked up in low-mass stars. The initially gaseous galaxy is today about 10% gas, with 90% trapped in low-mass stars. The origin of the Si atoms that entered solar matter is therefore biased towards supernovae that occur relatively shortly before solar birth, and biased against those that were created long earlier. This important conceptual distinction was demonstrated quantitatively by Clayton (1988), who stressed the concept of the age spectrum of solar nuclei and who constructed analytic models of galactic chemical evolution that enable a full disclosure of such issues. As an example, the simplest of all realistic models supports a star formation rate and supernova rate that declines exponentially as the interstellar gas is exponentially consumed; but the age spectrum of solar nuclei in that model is constant—equal numbers from equal times, despite the strong bias of nucleosynthesis of Si toward early galactic times. The student can study Clayton (1988) for many related issues for cosmochronology. Some of these issues involving galactic chemical evolution will be addressed later.

2.2.1.4 Uranium and Thorium in Old Dwarf Stars

An altogether new technique in radioactive chronology became possible as CCD detectors enabled astronomers to measure line strengths for much weaker lines than had been previously possible. Butcher (1987) advanced the first such argument when he was able to accurately measure the abundance of thorium in old dwarf stars. The measured abundances enabled him to argue that the oldest stars are no older than 10 Gy. If they were, Butcher argued, the Th, which has resided in the old stars since their births, would by now have decayed to a smaller abundance than it is observed to have. Because the dwarf stars observed appear to be among the early stars formed in our galaxy, the argument concludes that the galaxy age is negligibly greater than 10 Gy. It will be clear that since the ^{232}Th mean life is a very long 20 Gy, it can have decayed from its initial abundance by only $e^{-0.5} = 0.6$ during a 10 Gy life of a dwarf star. Therefore the method requires not only an accurate measurement of the Th abundance but also an accurate estimate of its initial abundance when the star formed. It is the reasoning required to obtain that initial abundance that is a controversial aspect of this method. Realize also that the Th in an old star is decaying freely, so that its abundance declines faster than in the ISM where new nucleosynthesis of Th continuously replenishes it.

The method advocated measures the ratio of Th/Nd in the star and observes that the measured ratio is smaller than that seen in younger stars. It assumes that the initial Th/Nd ratio in the star would be the same as the ratio seen in young stars, since both Th and Nd are products of the *r* process and may be hoped to have a constant production ratio there. It was then counter argued (Clayton 1987, 1988) that a constant production ratio for Th/Nd is not to be expected because almost half of the solar Nd abundance has been created in the *s* process, and, furthermore, the *r* process is primary whereas the *s* process is secondary. There exists ample and exciting evidence from the observed *r*-process pattern of abundances in extremely metal-deficient old stars (Truran 1981; Gilroy et al. 1988; Beers and Christlieb 2005) that the *r* process nucleosynthesis began prior to the beginning of the *s* process. That seemed to support the initial skepticism about Butcher's technique. But astronomers hurried to point out that if one omits the very metal-deficient old stars, the ratio of *r*-to-*s* abundances in stars having more than 10% of solar abundances is observed to be a near constant. Clayton (1988) revisited that larger puzzle of parallel growth of *s* and *r* nuclei, noting that the abundance evidence suggests that the *s* process is primary despite its building upon iron seed nuclei; moreover, he presented a nucleosynthesis argument that showed how the *s* process could in fact resemble primary nucleosynthesis even though it is secondary! See section 3.1 of Clayton (1988) for that argument, which has proven to be of high significance for nucleosynthesis theory.

The lesson to be taken from this exciting new technique and its controversy is this: radioactive nuclear cosmochronology is vitally dependent upon a correct picture of the chemical evolution of the galaxy. As such it has become less the province of nuclear physics and more the province of astronomy. Tinsley and Clayton both made that point in timely and influential ways. Only when many essential details of the history of our galaxy and of the history of nucleosynthesis within it have been settled can these techniques of radioactive chronology yield a reliable answer for the age of the chemical elements. Nuclear cosmochronology is truly an astronomy with radioactivity.

2.2.1.5 Cosmoradiogenic Chronologies

In the early 1960s a different approach to radioactive chronology became possible. It utilizes the buildup of the daughter abundances of radioactive decay during the history of interstellar matter as a measure of how long that decay had been occurring and, therefore, how long ago the production of radioactivity began. One can imagine the abundance of the stable daughter of a radioactive decay as a bucket into which all interstellar decay of the radioactive parent has been collected. It integrates past decay rather than focusing on how much radioactivity remains. Such daughter-isotope buildup during the history of the earth was already known as *radiogenic abundance*. It had been applied to the ages of earth and of meteorites, samples in which it could be expected that the other initial isotopic compositions were well known-namely, the solar abundances. But radiogenic abundance collected during the history of interstellar matter, what Clayton (1964) called *cosmoradiogenic abundance*, was not seen as possible data for determining the age of the elements themselves because the interstellar abundances are increased by the processes of nucleosynthesis as well as by any radioactive decay for specific isotopes. Thus the cosmoradiogenic abundance could not be easily disentangled from the direct nucleosynthesis abundance.

What cut through that impasse was a credible *quantitative theory* of heavy-element nucleosynthesis. Two distinct neutron-capture processes, the s process and the r process, had been responsible for the creation of all but the very lightest isotopes of the elements heavier than about $Z = 32$. The disentanglement required two things: firstly, an accurate s-process theory and secondly, a parent-radioactivity abundance that could be produced only by the r process and a daughter that could be produced only by the s process. The radiogenic daughter of such a radioactive r-process isotope is called a shielded isotope, because it is shielded from r-process production by the radioactive parent.

Clayton (1964) made the key first step by noticing that the solar abundance of ^{187}Os is about twice as great as it is expected to be from s-process nucleosynthesis, and that it cannot be synthesized by the r process because neutron-rich matter at $A = 187$ will, upon decaying toward stability, arrest at ^{187}Re. The decay cannot reach ^{187}Os, which is shielded by ^{187}Re. The r process production flows into ^{187}Re. The abundance of ^{187}Os can then be thought of as having two parts; a part produced by the s process and a part produced by the decay of very-long-lived ^{187}Re. The quantitative s-process theory (Clayton et al. 1961; Käppeler et al. 1982) reliably accounted for about half of the ^{187}Os abundance, so the other half had to be the result of the beta decay of ^{187}Re during ISM residence:

$$^{187}\mathrm{Re} \longrightarrow {}^{187}\mathrm{Os} + e^- + \nu \qquad t_{1/2} = 43\ \mathrm{Gyr}$$

That the half-life exceeds the galaxy's age is useful, because only a modest fraction of the ^{187}Re can therefore have decayed during pre-solar history and most of it therefore still exists; but because ^{187}Re is six times more abundant than the s-process amount of ^{187}Os, a decay of only 1/6th of the ^{187}Re has doubled the s-process ^{187}Os abundance. The persistence of live ^{187}Re is even necessary for the method, because it allows one to equate the total amount of ^{187}Re nucleosynthesis that was destined for solar incorporation to the sum of the quantity of ^{187}Re remaining at solar birth and the quantity of ^{187}Re that had already decayed prior to solar birth. The latter is the difference between the total solar abundance of daughter ^{187}Os and the quantity produced by the s process. That is,

$$^{187}\mathrm{Os} = ^{187}\mathrm{Os_s} + ^{187}\mathrm{Os_c}$$

where $^{187}\mathrm{Os}_c$ is the cosmoradiogenic part of the ^{187}Os abundance owing to cosmic ^{187}Re decay. And the s-process part can be obtained from $^{187}\mathrm{Os}_s$ = ^{186}Os ($\sigma(186)/\sigma(187)$), where σ is the appropriate neutron-capture cross section during the operation of the s process. This last relationship is valid because the entirety of ^{186}Os is from the s process because it is shielded from r-process nucleosynthesis.

The discovery of the Re-Os clock was the key that Clayton (1964) used in presenting solutions to three cosmoradiogenic chronologies. These are the beta decay of ^{187}Re to ^{187}Os described above, the similar beta decay of ^{87}Rb to ^{87}Sr, and the α-decay chains by which ^{235,238}U decay to ^{207,206}Pb. Each had been known in the study of meteorites, but by this work they joined studies of the age of the elements themselves.

Cosmoradiogenic lead is the more interesting of the other two chronologies, in part because it couples cosmoradiogenic Pb to the older chronology based on ^{235}U/^{238}U, and in part because the Pb isotopes have three contributions to their abundances:

$$^{206}\mathrm{Pb} = ^{206}\mathrm{Pb}_s + ^{206}\mathrm{Pb}_r + ^{206}\mathrm{Pb}_c$$

$$^{207}\mathrm{Pb} = ^{207}\mathrm{Pb}_s + ^{207}\mathrm{Pb}_r + ^{207}\mathrm{Pb}_c$$

where $^{206}\mathrm{Pb}_c$ is the cosmoradiogenic part from ^{238}U decay, and where $^{207}\mathrm{Pb}_c$ is the cosmoradiogenic part from ^{235}U decay. Suffice it to be said here that the s-process part and the r-process part can both be estimated, and from them the chronological solutions can be displayed. Readers can turn to Clayton (1988) for these solutions.

The cosmoradiogenic chronologies seem to indicate an older galaxy than do the direct-remainder chronologies (Clayton 1988). But they also are compromised by unique and interesting uncertainties. For the ^{187}Re cosmoradiogenic chronology, the main uncertainty is how greatly its decay rate is speeded by its incorporation into stars, where ionization increases its beta decay rate. Yokoi et al. (1983) evaluated this effect within an ambitious galactic chemical evolution model that enabled them to take into account the fraction of ^{187}Re that is incorporated into stars and ejected again without nuclear processing and how much time such interstellar ^{187}Re spends inside of stars. Their results can be reproduced by a 40% increase of the neutral ^{187}Re decay rate. For the Pb cosmoradiogenic chronology, the main uncertainty is an especially interesting set of nucleosynthesis problems associated with both the s-process part and the r-process part of their abundances. Uncertainty about the Pb/U elemental abundance ratio also suggests some caution.

Certainly it can be said that these chronologies present intricate problems in the astronomy with radioactivity.

2.2.1.6 Radioactivity and Galactic Chemical Evolution

In astronomy, the abundances of radioactive species are interpreted through the lens of galactic abundance evolution, frequently called galactic chemical evolution (GCE). Interpretations of nuclear cosmochronology or of initial solar abundances of extinct radioactivities depend on the mean expected abundances in the ISM when solar birth occurred. Galactic Chemical Evolution traditionally concerns the chemical composition of the mean ISM. It features continuous nucleosynthesis between a starting time and the time when solar formation occurred. During that lengthy galactic period radioactive abundances undergo decay and are incorporated into new stars, but they are also replenished by the injection of freshly synthesized radioactivity and are diluted by low-metallicity matter falling onto the disk. Competition among these terms renders simple galactic chemical evolution a showplace for the behavior of radioactivity. One differential equation describing that abundance of those specific galactic atoms (the *red atoms*; see Sect. 2.2.1.3) that will later be incorporated into the solar system can be thought of as the solution of

$$\frac{dN}{dt} = \frac{dB_{\odot}(t)}{dt} - \frac{N(t)}{\tau} \tag{2.14}$$

where $\tau = 1/\lambda$ is the mean radioactive lifetime of the nucleus in question, N(t) is its time-dependent abundance within those (red) atoms that are destined for inclusion in the solar system, and $B_{\odot}(t)$ is the birthrate spectrum of those solar nuclei, defined such that $\frac{dB_{\odot}(t)}{dt}dt$ is the number of stable solar nuclei that were born between times t and t+dt. The birthrate spectrum of solar nuclei is unknown, however, so Eq. (2.14) can not be solved. Despite this limitation, workers persisted in attempting to fix the time of the beginning of nucleosynthesis from Eq. (2.9), which students can show does solve the differential equation (2.14) above. Many tried this approach by assuming an easily integrable form for $\frac{dB_{\odot}(t)}{dt}$, most often $exp(-Gt)$, where G is taken (erroneously) to be an unknown positive number. That assumption proved seductive to many, because the rate of galactic nucleosynthesis is believed to decrease smoothly with time, so maybe $\frac{dB_{\odot}(t)}{dt}$ can be assumed to decline as well. Importantly, however, $\frac{dB_{\odot}(t)}{dt}$ differs greatly from the galactic production rate and actually grows with time. The reason for this surprise requires understanding.

The birthrate spectrum imagines that the nuclei destined for the solar system could be tagged at birth (imagine red paint) and ignores all other ISM nuclei. Because $B_{\odot}(t)$ is unknown, however, astrophysicists instead calculate the evolution of the interstellar abundances. Their mean values in a well mixed ISM at the time of solar birth provide the expected solar abundances. The conceptual difference between $B_{\odot}(t)$ and the galactic nucleosynthesis rate has confused many unwary researchers. Many published papers have used their chosen form for $\frac{dB_{\odot}(t)}{dt}$ and integrated Eq. (2.9) in the hope of fixing the beginning time of galactic nucleosynthesis (the lower limit) by comparing those calculated results with known initial radioactive abundances in the solar system. A common form because Eq. (2.9) is then easily integrated is $\frac{dB_{\odot}(t)}{dt} = exp(-Gt)$. In retrospect, however, such poor choices for the form of $\frac{dB_{\odot}(t)}{dt}$ render their conclusions invalid.

How does GCE clarify the problem of galactic radioactivity? The mass M_G of ISM gas is reduced owing to the rate $\psi(t)$ of its incorporation into new stars, but it is increased by the rate $E(t)$ at which mass is ejected from old stars and by the rate $f(t)$ at which new mass falls onto the galactic disk:

$$\frac{dM_G}{dt} = -\psi(t) + E(t) + f(t) \tag{2.15}$$

If one takes the ejecta rate $E(t)$ from the spectrum of newly born stars to be a fixed return fraction R of the rate $\psi(t)$ at which mass joins new stars, and if one assumes linear models in which the rate of star formation is proportional to the mass of gas M_G , the equations governing the interstellar composition of both stable and radioactive nuclei can be solved analytically within families of choices for the infall rate $f(t)$, as Clayton (1988, 1985) has shown. These analytic solutions for ISM abundances do indeed clarify radioactive abundances. For that purpose Clayton writes

$$\frac{dM_G}{dt} = -\psi(t)\,(1 - R)\, + f(t) = -\omega M_G + f(t) \tag{2.16}$$

where $\omega M_G(t) = (1 - R)\psi(t)$ is valid for linear models. The constant ω is the rate of consumption of gas by star formation when compensated by gaseous return from stars. These are called *linear models* because the star formation rate $\psi(t)$ is taken to be proportional to the mass $M_G(t)$ of interstellar gas. That linear assumption is not strictly true; but it is plausible in taking the star formation rate to increase as the mass of gas increases, and to decline as it declines. Furthermore, it is supported by observations of star formation rates in spiral galaxies. From their observations, Gao and Solomon (2004) state, "The global star formation rate is linearly proportional to the mass of dense molecular gas in spiral galaxies." The purpose of models of galactic chemical evolution insofar as galactic radioactivity is concerned is to understand the mean expectation for radioactivity in the ISM. The families of linear analytic models are constructed for that purpose.

Coupled with Eq. (2.16) is an equation for the rate of increase of the concentration Z of each interstellar nucleosynthesis product. The concentration is defined as the mass of species Z in the ISM divided by the total mass M_G of ISM gas and dust. The concentration Z, rather than total numbers of atoms, is the quantity traditionally used in chemical evolution studies because it is concentration that astronomers measure. The mass $m_Z = ZM_G$ of interstellar species Z is governed by

$$\frac{dm_Z}{dt} = -Z\psi(t) + Z_E\, E(t) + Z_f\, f(t) \tag{2.17}$$

where Z_E and Z_f are, respectively, the concentration of Z in the spectrum of stellar ejecta, where it is large, and in the infalling gas where it is small. The metallicity in infalling gas Z_f may probably be neglected with good accuracy.

Many workers have shown that it is a good approximation for analytic understanding to assume that the ejecta $E(t)$ is returned at once from the entire spectrum of newly born stars despite the dependence of stellar lifetime on stellar mass. That assumption is called the *instantaneous recycling approximation*, and it is reasonable except late in the life of a galaxy when the gas mass and the star-formation rate have both become very small. But for galaxies in early and middle lifetime, Eq. (2.17) can then be written after some straightforward algebra as

$$\frac{dZ}{dt} = y\,\omega - \frac{Zf(t)}{M_G(t)} - \lambda\,Z \tag{2.18}$$

where $m_Z = ZM_G$, Z being the concentration in the ISM gas (taken to be well mixed), and y is the *yield* of element Z, defined as the mass m_Z of new Z ejected from the entire spectrum of newly born stars divided by the mass of stellar remnants left behind by that entire spectrum of stars. The yield y for primary nucleosynthesis products may be taken as a constant despite having small variations in full numerical models. The product $y\omega$ in Eq. (2.18) can be called the *galactic nucleosynthesis rate*, which, through ω, depends explicitly on the star formation rate ($\omega M_G = (1 - R)\psi(t)$). Equation (2.18) for the ISM concentration differs from Eq. (2.14) for the history of solar nuclei by the existence of the second term of Eq. (2.18). That term reflects the loss of interstellar metal concentration Z when ISM gas containing Z collapses into new stars and is simultaneously diluted by metal-poor infall.

Comparison of Eq. (2.18) with Eq. (2.14) shows clearly why it is an error to integrate Eq. (2.14) thinking that the solar-nuclei birthrate spectrum can be mentally equated with the spectral rate of galactic nucleosynthesis. Equation (2.18) similarly has the galactic nucleosynthesis rate as the first term and the radioactive decay rate as the last term; but the astrophysical Eq. (2.18) contains the middle term involving both the galactic infall rate and the mass of ISM gas, both of which are time dependent. Integrations of Eq. (2.14) by assuming $B_\odot(t)$ ignore the effects of the ISM and can be correct only if the assumed form $dB_\odot(t)/dt$ actually resembles the number of solar nuclei born per unit time rather than the galactic nucleosynthesis rate. As the birthdate of solar nuclei is usually what is being sought in nuclear cosmochronology, one cannot get the answer from Eq. (2.14) without first knowing and inserting the answer. For that reason nuclear cosmochronology must instead be investigated within the context of galactic chemical evolution. These points are central to the subject of astronomy with radioactivity within solar-system nuclei.

Clayton (1988) showed one way forward. One can integrate Eq. (2.18) analytically with the aid of flexible families of functions that enable analytic integration. This can be accomplished by form-fitting the ratio $f(t)/M_G(t)$ to an integrable family of functions $d\theta/dt$. Such form fitting is much more flexible than it would at first seem to be, because most physically plausible time-dependent behavior can be approximated by a specific form fitting choice. One such useful choice has been called the *Clayton Standard Model.* It takes

$$\frac{d\theta}{dt} = f(t)/M_G(t) = \frac{k}{(t + \Delta)} \tag{2.19}$$

and its explicit functions are given in Appendix A.1.1 of this book.

For the astronomy of radioactivity, the great merit of this approach is exact analytic solutions for $Z(t)$ for both stable and radioactive isotopes while simultaneously yielding exact functional representations of the mass $M_G(t)$ of interstellar gas and of the infall rate $f(t)$. Understanding these analytic models greatly aids understanding the behavior of radioactivity within more general numerical models of galactic chemical evolution. The numerical approach is to place the evolution of the galaxy on a computer, taking into account the evolutionary lifetime of each star formed and the specific nucleosynthesis products to be ejected from each star (e.g. Timmes et al. 1995). Although this approach is undoubtedly correct, it obscures theoretical understanding that can be seen more easily within analytic models. Furthermore, surveys of nuclear cosmochronology can more easily be carried out within analytic models.

All well mixed models of galactic chemical evolution can at best yield only an average expectation for the ISM. The true ISM is inhomogeneous in space and the nucleosynthesis rate is sporadic in time rather than maintaining its steady average. Nonetheless, well mixed models, both analytic and numerical, are important in laying out the results that would be true for a rapidly mixed ISM and a smooth rate of galactic nucleosynthesis. One interprets the observations of radioactive abundances against the backdrop of that expectation.

Another result of great importance for short-lived galactic radioactivity is best illustrated within Clayton's standard model. For short-lived radioactivity, i.e. whenever λ is large, the mean concentration in the galactic ISM is

$$Z_\lambda = y\,\omega \left(\lambda + \frac{k}{(t + \Delta)}\right)^{-1} \tag{2.20}$$

From this equation and one for a stable isotope the abundance ratio of a short-lived radioactive nucleus to that of a stable primary isotope in the ISM is larger by the factor $(k + 1)$ than one might estimate without taking galactic chemical evolution into account. As an example, consider the interstellar ratio of radioactive ^{26}Al to stable ^{27}Al. The formula

$$\frac{Z(^{26}\mathrm{Al})}{Z(^{27}\mathrm{Al})} = \frac{y(26)}{y(27)}\,(k + 1)\,\frac{\tau_{26}}{t_\odot} \tag{2.21}$$

can be derived from such a ratio (see Clayton et al. 1993, equations 8 and 9). This result is larger by the factor $(k+1)$ than an estimate using only Eq. (2.10), which had traditionally been used in oversimplified discussions of the steady-state amount of short-lived interstellar radioactivity. That oversimplification can be found in almost all published papers on short-lived interstellar radioactivity. The extra factor $(k+1)$, which is an effect of infall of low-metallicity gas, became important when a large mass of interstellar ^{26}Al was detected by gamma-ray astronomy (see Chap. 7).

2.2.2 *Gamma-Ray Lines from Galactic Radioactivity*

One thinks of at least three reasons that the idea of astronomically detecting galactic radioactivity did not occur until the 1960s. In the first place, MeV range gamma rays do not penetrate to the ground, but must be detected above the atmosphere. Secondly, detecting MeV gamma rays and measuring their energies is a quite difficult technology, especially so when the background of cosmic-ray induced events above the atmosphere is so large. Thirdly, even a back-of-the-envelope estimate of expected rate of arrival from a stellar source is discouragingly small.

2.2.2.1 The Rice University Program

In 1964 a new aspect of astronomy with radioactivity arose. Robert C. Haymes was hired by Rice University, and he spoke there with Donald Clayton about the possibility of sources of galactic radioactivity that he might seek with an active anticoincidence collimation for a NaI detector flown beneath a high-altitude balloon. Burbidge et al. (1957) had speculated (incorrectly) that the exponential 55-night decline of the luminosity of many Type Ia supernovae was the optical manifestation of the decline of the spontaneous-fission radioactive decay of ^{254}Cf in the ejecta of the supernova, which would quickly have become cold without some heating mechanism that had to decline with a roughly 2-month half-life. The ^{254}Cf nucleus would be synthesized by the r process (assumed to occur in Type Ia supernovae) along with uranium and thorium. The large kinetic energy of its spontaneous-fission fragments would be converted to optical emission by being degraded by atomic collisions within the ejected gas. This was called *the californium hypothesis*. But B^{2}FH had said nothing about gamma-ray lines. The novel excitement at Rice University was the new idea that concepts of nucleosynthesis could be tested directly if the associated gamma-ray lines from the r-process radioactivity could be detected on earth coming from supernova remnants. Haymes estimated rather optimistically that his detector could resolve lines having flux at earth greater than about $10^{-4}\,\mathrm{cm}^{-2}\,\mathrm{s}^{-1}$. The first scientific paper written with that goal was soon published (Clayton and Craddock 1965). It evaluated the full spectrum of radioactivity by an r process normalized to the yield proposed by the Cf hypothesis. They found several promising lines. For example, the strongest from a 900-year old Crab Nebula would be ^{249}Cf, presenting $10^{-4}\,\mathrm{cm}^{-2}\,\mathrm{s}^{-1}$ gamma-ray lines having energy 0.39 MeV. It is no accident that the ^{249}Cf half-life (351 years) is of the same order as the age of the Crab Nebula. Within a broad range of halflives in an ensemble of nuclei, the one giving the largest rate of decay has mean life comparable to the age of the ensemble. It can be confirmed that in a remnant of age T the rate of decay per initial nucleus is $\lambda e^{-\lambda T}$, which is easily shown to be maximal for that nucleus whose decay rate $\lambda = 1/\tau = 1/T$. A photograph (Fig. 2.1) shows Haymes with his gondola and Clayton from those Rice years.

Realistic expectations were not so sanguine, however. The r-process yield required by the Cf hypothesis appeared excessively large (Clayton and Craddock 1965). The r-process nuclei in solar abundances would be overproduced by a factor 100 if all Type Ia supernovae produced that yield. Despite that reservation, an exciting chord had been struck. A new astronomy of radioactivity appeared possible, one having significant scientific payoff, and Haymes's gamma-ray telescope began a program of balloon flights. The program did not discover r-process radioactivity, but it did discover apparent positron-annihilation radiation from the center of our galaxy (Johnson et al. 1972) and hard X-rays from several sources. The observational program at Rice University also trained two of the principal investigators (G.J. Fishman and J.D. Kurfess) and two of the project scientists (W.N. Johnson and C. Meegan) on NASA's later *Compton Gamma Ray Observatory*, which went into orbit in 1991. And Clayton began a NASA-sponsored research program for laying out theoretical expectations for nucleosynthesis produced radioactivity and

Fig. 2.1 Robert C. Haymes and Donald D. Clayton with Haymes' gondola for his balloon-borne gamma ray telescope at Rice University in 1973. Haymes pioneered gamma-ray astronomy in the MeV region and Clayton developed gamma-ray-line targets for this goal and for nucleosynthesis. This program at Rice University trained two Principal Investigators and two Project Scientists for later experiments on NASA's Compton Gamma Ray Observatory

its associated nuclear gamma-ray lines in astronomy. Gamma-ray-line astronomy was in that sense launched by the Rice University program.

At least two other groups followed the Rice University lead and became and remained leaders for the subsequent decades. They constructed differing telescope techniques from that of Haymes NaI scintillator. One was the JPL research group—A.S. Jacobson, J.C. Ling, W.A. Wheaton, and W.A. Mahoney. Using a cluster of four cooled high-purity germanium detectors on NASA's *HEAO* 3, which was launched in 1979, they discovered the first galactic gamma-ray line from radioactive decay. A group at MPI für Extraterrestrische Physik in Munich, V. Schönfelder, P. von Ballmoos and R. Diehl, developed a Compton-scattering telescope capable of good energy resolution and a greater angular view of the sky. Their active results from balloon-borne launches in the 1980s presaged the splendid results of their instrument on NASA's Compton Gamma Ray Observatory. Jacobson's JPL group was not so lucky, as their experiment was removed from *CGRO* as a cost saving. Other groups joined the effort to create this new astronomy with radioactivity.

2.2.2.2 Earliest Predictions and Detections of Gamma-Ray Lines

Inspired by the Rice University balloon program, a series of studies of the nucleosynthesis of radioactivity was undertaken. It may seem surprising in retrospect that this could not have been done by a routine search of the chart of nuclides in conjunction with nucleosynthesis theory. But the data base necessary did not exist. Four decades ago, nucleosynthesis theory was undergoing rapid expansion and clarification, so that it was a series of new insights that laid out each new prospect as it was discovered. Unquestionably the most important of these developments occurred in 1967 when a Caltech group (Bodansky et al. 1968a,b) and a Yale group (Truran et al. 1967) showed with differing research techniques that the *e* process of Hoyle (1946) and Burbidge et al. (1957) for the nucleosynthesis of the iron abundance peak was incorrectly applied. Instead of producing iron isotopes as themselves within a neutron-rich nuclear-statistical equilibrium (NSE), as Hoyle had suggested, they were produced as isobars of radioactive nickel that decayed to iron isobars only after nucleosynthesis and ejection from supernovae. Owing to the increasing strength of the Coulomb energy in nuclei, the most stable nucleus for $4n$ nuclei shifts from $Z = N$ below $A = 40$ to $N = Z + 4$ for $A = 44$–60. But insufficient neutron excess in the explosive supernova plasma required those abundant isobars to assemble as unstable $Z = N$ nuclei. This brought into view the astronomical significance of several important radioactive nuclei. A historical study, *Radiogenic Iron*, was written later to lay out the many significant astrophysical issues that hinged on this new understanding (Clayton 1999).

It is of historical interest to consider the sequence of subsequent predictions of target radioactivity for astronomy. Clayton et al. (1969) jumped on the newly discovered radioactivity in the revised equilibrium processes (preceding paragraph). Easily the most important nucleus was ^{56}Ni, having $Z = N$, whose beta decays to ^{56}Co and thereafter to ^{56}Fe, having $N = Z + 4$, are accompanied by numerous gamma ray lines that might be resolved by spectroscopic gamma-ray detectors. G.J. Fishman was a research student in R.C. Haymes' Rice University balloon program when he and Clayton began calculating this spectrum of lines and their time dependences. Their motivation was to test the fresh new view of explosive nucleosynthesis that had led to ^{56}Ni nucleosynthesis. S.A. Colgate joined this effort because he had independently begun to investigate whether the energy deposition from the radioactivity could explain the exponentially declining Type Ia supernova light curves (Colgate et al. 1969). Burbidge et al. (1957) had attributed those long lasting light curves to heating by spontaneous fission fragments from freshly produced ^{254}Cf. As the simplest model (Clayton et al. 1969) evaluated fluxes at earth from the rapid expansion of a solar mass of concrete, much of which had been explosively processed to ^{56}Ni in the way anticipated by Bodansky et al. (1968a,b). The lifetime of ^{56}Co was seen as favorable for sufficient expansion for the gamma rays to escape. They neglected a massive envelope, as might overlie Type II cores, realizing that such an envelope would seriously limit the escape of the ^{56}Ni gamma rays. Clayton et al. (1969) was a very important paper, not only because it was correct in its prediction of the strongest gamma-ray nuclear source, but also because of its galvanizing effect upon experimental teams and on NASA planning. It later was included in the AAS Centennial Volume of seminal papers of the twentieth century. NASA Headquarters suggested funding of a research program at Rice to lay out additional prospects for this new astronomy of radioactivity. The NASA grant at Rice was entitled *Prospects for Nuclear Gamma-Ray Astronomy*. But the detection of those ^{56}Co gamma rays did not become possible until SN1987A occurred. Detection of the gamma rays from the even more shortlived ^{56}Ni only succeeded with SN2014J, a nearby supernova of type Ia that was observed early with ESA's INTEGRAL satellite (Diehl et al. 2014).

Clayton et al. (1969) also first suggested that unknown galactic supernovae (unseen owing to optical obscuration) may be discovered in gamma-ray lines. The lines from radioactive ^{44}Ti, with 60-year half-life, ought to be detectable from several supernovae that have exploded during the past two centuries, considering that about three per century seem to occur on average within the Milky Way. This exciting idea has great implications for galactic nucleosynthesis and astrophysics; but its hope has been frustrated for astrophysical reasons that are not yet understood (The et al. 2006). But, at least from one source, the Cas A supernova remnant, ^{44}Ti gamma rays had been detected (Iyudin et al. 1994) with COMPTEL on NASA's Compton Gamma Ray Observatory. Only much later, the NuSTAR hard X-ray telescope made an important image of Cas A (Grefenstette et al. 2014) in the low-energy lines from ^{44}Sc de-excitation, and was capable to also detect ^{44}Ti lines from SN1987A (Boggs et al. 2015).

It was quickly realized (Clayton and Silk 1969) that the entire universe might be filled with detectable ^{56}Co gamma rays from the past history of supernova explosions throughout the universe. This work suggested that observable gamma rays owing to the collective effects of many supernovae rather than specific young supernovae might be targeted. Detecting the universe in this way remains a goal of the astronomy of radioactivity.

Two years later it was proposed (Clayton 1971) that the collective effects of galactic supernovae might allow one to observe gamma-ray lines from long-lived radioactive nuclei whose emission would be too weak from individual supernovae. That first work in that regard focused on ^{60}Fe, whose 2.6 Myr half-life[1] makes emission from an individual supernova too slow to be observable for the modest number of ^{60}Fe nuclei produced within a single supernova. The ^{60}Fe nucleus emits a 59 keV gamma ray upon decay, and its daughter ^{60}Co emits gamma-ray lines of 1.17 and 1.33 MeV. Reasoning that during its long mean lifetime some 50,000 supernovae occur in the Milky Way, their collective effect should be observable. This reasoning applied equally well 13 years later to the first interstellar radioactivity to be detected, that of ^{26}Al.

Surprisingly in retrospect, several years passed before Clayton (1974) realized that when ^{57}Co, the daughter of 36-h ^{57}Ni, decays, it also emits favorable gamma-ray lines and that its longer 272-day half-life would cause it to radiate these after the ^{56}Co was almost extinct. This made ^{57}Co much more significant than had been appreciated. Clayton (1974) drew attention to the significance of ^{57}Co and to several new ideas for astronomy with radioactivity. Firstly, given an appropriate galactic supernova one might measure the material thickness overlying the radioactive cobalt by the distinct times at which differing gamma-ray lines reach maximum flux. This happens because the structural opacity for the 2.60 MeV gamma ray is only about half that for the 0.84 MeV gamma ray (both from ^{56}Co decay), so that the 2.60 MeV gamma ray should peak earlier despite the equal rates of emission. Additionally, the 0.12 MeV gamma ray from the slower ^{57}Co decay suffers even more absorption and so peaks later yet in time. These time delays have not yet been measurable because a time-dependent gamma line flux requires a bright supernova which would have to occur within our own galaxy. Clayton (1974) also warned that hydrodynamic *instabilities could allow the central material to burst through in streams* and thereby appear earlier than anticipated. Early escape of lines from ^{56}Co was later detected in SN1987A, which exploded in the Magellanic Clouds, and from which gamma-ray lines from both ^{56}Co and ^{57}Co were first detected (see Chap. 3). Earliest of the recorded ^{56}Co photons were by the gamma ray spectrometer aboard NASA's *Solar Maximum Mission*, which happened to be in orbit when SN1987A exploded and whose sun-pointing spectrometer was reinterpreted as a gamma ray telescope (Leising and Share 1990). Several balloon-borne gamma-ray spectrometers were flown and also detected ^{56}Co lines. When ^{57}Co was detected in SN1987A by Kurfess et al. (1992) with the *Oriented Scintillation Spectrometer Experiment* on NASA's *Compton Gamma Ray Observatory* during summer 1991, 4 years (about six halflives of ^{57}Co) had elapsed since the explosion, so that the ^{57}Co abundance had decayed to only about 1.5% of its initial amount. Tension surrounded the hope that it would still be detectable. It was, with 4σ significance, implying that the initial abundance ratio ^{57}Ni/^{56}Ni was about twice the ratio ^{57}Fe/^{56}Fe measured in terrestrial iron.

[1] At that time, the half-life of ^{60}Fe was best-known as 1.5 My.

At almost the same time as the ^{57}Co prediction, Clayton and Hoyle (1974) proposed that gamma rays may be detectable from the more common nova explosions owing to the radioactivity created by the nova outburst. These involve the positron-annihilation line from the hot-CNO burning that powers the nova outburst (Starrfield et al. 1972). These 511 keV photons would have to be detected very quickly, within roughly 10^3 s after the outburst because of the short halflives of CNO radioactive nuclei. They also pointed out the prospect of detecting a 1.274 MeV gamma ray following ^{22}Na decay. Because of its longer half-life (2.6 years), ^{22}Na remains alive for the roughly 10^6 s required for the nova ejecta to become transparent to the gamma ray. These goals have not succeeded yet, but they remain a realistic hope of measuring the thermonuclear power of the nova. The model of the nova makes it a remarkable laboratory for thermonuclear explosions. Detection could confirm the model or rule it out. Some years later the possibility arose that the 478 keV gamma-ray line following the decay of radioactive ^{7}Be to ^{7}Li might also be detectable from novae (Clayton 1981). Detectability requires the nova envelope accreted from the companion star to be enriched by an order of magnitude in ^{3}He, but that is perhaps to be expected. If so, the nuclear reaction $^3\mathrm{He} + {}^4\mathrm{He} \Rightarrow {}^7\mathrm{Be} + \gamma$ creates the radioactive nucleus in the thermonuclear flash. One positive aspect is that the ^{7}Be half-life is large enough for the envelope to become transparent. The likelihood of ^{7}Be production in novae has been enhanced by observations of singly ionized ^{7}Be absorption line at 313 nm in Nova Delphini 2013 (Tajitsu et al. 2015). This observation solidifies novae as a likely galactic nucleosynthesis source of ^{7}Li.

The radioactive ^{22}Na nucleus is also produced in supernovae, although the complicated details of its nucleosynthesis therein were not well understood when Clayton (1975a) proposed its detectability within supernovae. It might be observable for a decade, allowing ample time to become transparent to the ^{22}Na gamma-ray line.

Almost the last good prospect to be predicted turned out to be the first actually observed! That was the radioactive ^{26}Al nucleus (Ramaty and Lingenfelter 1977). Like radioactive ^{60}Fe, the ^{26}Al nucleus is long-lived and thus decays too slowly to be detectable from individual explosions. It is the cumulative yield of many events over the past Myr or so that was discovered. This is the topic of the next section.

The predictions whose history has been recounted here placed tantalizing targets before the community of experimental physicists. These assembled teams to design, build and fly gamma-ray spectrometers capable of detecting these lines. NASA entertained a grants program pursuing these goals, which create a new wavelength range for astronomy. Moreover, the will to fund the *Compton Gamma Ray Observatory* was strengthened by these specific hopes. That hope was fueled also by the surprising discovery of a gamma-ray line from ^{26}Al nuclei in the interstellar medium.

2.2.2.3 The Surprise ^{26}Al

The first detection of an interstellar radioactive nucleus by its gamma-ray line emission came as a surprise. The history of this radioactive isotope in astrophysics reveals that correct predictions can be made on the basis of inadequate reasoning, that scientists may be blindsided by their own excessive faith in their own pictures and beliefs, and that experimental discovery is the arbiter. The possibility of detecting ^{26}Al nuclei by observing the 1.809 MeV gamma-ray line emitted following its decay to ^{26}Mg was suggested by Ramaty and Lingenfelter (1977) and independently by Arnett (1977). Their interesting papers were not quantitative predictions because the grounds for their suggestions were not convincing. They suggested wrongly that interstellar ^{26}Al nuclei would be detectable if the production ratio in massive stars is $P(^{26}\mathrm{Al})/P(^{26}\mathrm{Mg}) = 10^{-3}$. However, that production ratio, which was expected owing to carbon burning in massive stars, was inadequate for detectability. The argument showing that that production ratio was unobservable (Clayton 1984) was submitted for publication only after interstellar ^{26}Al was discovered (Mahoney et al. 1982, 1984). The predicted 1.809 MeV flux was too small by a wide margin to be detected by contemporary gamma-ray spectrometers. Something was wrong.

Using the gamma-ray spectrometer on NASA's third *High Energy Astrophysical Observatory* (HEAO 3) Mahoney et al. (1982, 1984) reported a measured flux of 1.809 gamma rays that required about 3 $M_\odot$ of interstellar ^{26}Al nuclei. That quantity was far larger than the predicted interstellar mass. Nonetheless, both Ramaty and Lingenfelter (1977) and Arnett (1977) had urged, on general grounds and on intuitive arguments, that the 1.809 MeV interstellar gamma-ray line be sought. It was indeed observable, the first to be detected.

What were the conflicts that this discovery illuminated? Using the mean lifetime ($\tau = 1.04 \times 10^6$ years) of ^{26}Al nuclei, the time-average interstellar abundance ratio for aluminum isotopes was traditionally estimated to be $^{26}\mathrm{Al}/^{27}\mathrm{Al} = P(^{26}\mathrm{Al})/P(^{27}\mathrm{Al})\ \tau/10^{10}$ years $= 10^{-7}$ if one uses $P(^{26}\mathrm{Al})/P(^{27}\mathrm{Al}) = 10^{-3}$. Since total ISM mass is about $10^{10}\,M_\odot$, it contains about $5.8 \times 10^5\,M_\odot$ of stable ^{27}Al. Multiplying by the above isotope ratio, the ISM would carry about 0.06 $M_\odot$ of ^{26}Al nuclei, woefully inadequate for the observed 3 $M_\odot$ of interstellar ^{26}Al (Mahoney et al. 1982, 1984). So although Ramaty and Lingenfelter (1977) and Arnett (1977) suggested that ^{26}Al nuclei might provide a suitable ISM radioactivity to seek, their reasoning did not convincingly justify that hope. The discovery of so much interstellar ^{26}Al was surprising and meant that some assumptions were quite in error.

Confirming detections of the ^{26}Al radioactivity were quickly made from balloon-borne gamma-ray spectrometers (von Ballmoos et al. 1987; MacCallum et al. 1987), and the total mass was later measured accurately by the gamma ray spectrometer aboard NASA's *Solar Maximum Mission*, whose sun-pointing spectrometer had been reinterpreted as a gamma ray telescope (Share et al. 1985). The existence of 2–3 $M_\odot$, of ^{26}Al was then beyond doubt. The 1991 launch of NASA's *Compton Gamma Ray Observatory* with its imaging COMPTEL Compton telescope produced the most detailed data about the spatial distribution of interstellar ^{26}Al (Diehl et al. 1995) (see detailed discussion in Chap. 7).

Stepping back to the history of the mid 1970s, a related issue for the astronomy of ^{26}Al radioactivity had arisen about 1 year prior to Ramaty and Lingenfelter's 1977 suggestion; namely, it was discovered (Gray and Compston 1974; Lee et al. 1977) that the molecular cloud from which the sun had formed apparently carried within it the large isotopic ratio $^{26}\text{Al}/^{27}\text{Al} = 5 \times 10^{-5}$ at the time the planetary system began to form. Had that ratio been applicable to the ISM as a whole, it would have corresponded to 29 $M_\odot$ of radioactive ^{26}Al nuclei, very much more than was discovered in 1982 by Mahoney et al. The seeming impossibility of producing such a large quantity by nucleosynthesis prompted the idea (Cameron and Truran 1977; Wasserburg and Papanastassiou 1982) that a supernova within the molecular cloud from which the sun was born had injected the solar ^{26}Al radioactive nuclei into the collapsing solar portion of the cloud just prior to the solar birth. In that case the solar cloud was quite atypical of the ISM at large. The initial solar ^{26}Al nuclei in meteorites became the standard bearer for a class of extinct radioactive nuclei that required special local production related to solar birth. But owing to the huge consequent ratio in the forming sun, it seemed to bear no transparent relationship to the existence of several solar masses of interstellar ^{26}Al nuclei, which necessarily represents nucleosynthesis of ^{26}Al nuclei by many thousands of supernovae spread out in time over 1–2 Myr prior to today. This discovery from meteorites brought the idea of interstellar ^{26}Al nuclei to the attention of astrophysicists.

What new ideas brought the expected 0.06 $M_\odot$ of ^{26}Al nuclei in the ISM into line with the observed 2–3 $M_\odot$ of ^{26}Al nuclei? Some suggested that novae (Clayton 1984; Woosley and Weaver 1980) or AGB stars, rather than supernovae, were the source of the ^{26}Al nuclei; but that possibility was ruled out by observations made later with the Compton Telescope (COMPTEL) following the launch of *Compton Gamma Ray Observatory*. Those observations (Diehl et al. 1995) showed spatial concentrations of ^{26}Al nuclei in ISM regions where star formation is currently active. Spatial correlation of ^{26}Al nuclei with massive stars was clear. Attention therefore returned to supernovae sources with the realization (Arnould et al. 1980) that hot hydrogen burning shells of massive stars constitutes a significant extra source of ^{26}Al. The small production ratio in carbon burning ($P(^{26}\text{Al})/P(^{27}\text{Al}) = 10^{-3}$) must be augmented by ^{26}Al-rich convective shells in massive stars. Those calculations became a major industry with numerical models of pre-supernova evolution. Because $^{26}\text{Al}/^{27}\text{Al}$ ratios of order 0.1 exist in these shells, their contribution calculated by using time-dependent numerical models of pre-supernova evolution would be needed. Weaver and Woosley (1993) calculated that the average production ratio from a standard spectrum of massive stars was near $\text{P}(^{26}\text{Al})/\text{P}(^{27}\text{Al}) = 0.006$, six times larger than the earlier estimates. Secondly, the estimated ratio of mean interstellar $^{26}\text{Al}/^{27}\text{Al}$ must be increased because most of the stable ^{27}Al is locked up inside old stars whereas the live ^{26}Al nuclei will still be overwhelmingly in the gaseous ISM. This reasoning augments the expected ratio by the factor $(\text{k} + 1)$ derived from the standard analytic models of chemical evolution of the mean galaxy (see Eq. (2.21)). Clayton et al. (1993) presented that argument for the expected interstellar (^{26}Al/^{27}Al) abundance ratio. Because k = 2–4 seems likely from other astrophysical arguments, this (k + 1) correction also increases the interstellar ratio by another factor 3–5. These two effects combined amounted to an increase of a factor 20–30 in the initial expectations, bringing the expected value from 0.06 $M_\odot$ to 1–2 $M_\odot$ of ^{26}Al nuclei in the ISM. The original conflict with global theory was largely resolved.

Moreover, the largest observed flux from ^{26}Al nuclei concentrated in ISM regions where star formation is currently active. This directly challenged the theoretical simplification that adopts a star-formation rate that is constant in time and spread smoothly through the spatial ISM. Models of chemical evolution of the galaxy use that simplification in order to be calculable. But the observations of 1.809 MeV gamma rays painted a picture of star formation occurring sporadically whenever and wherever large regions of ISM become vulnerable to prolific star formation. The massive ^{26}Al emission regions stunned and surprised many experts (including this writer). The fluxes from these are moderated by the inverse square of their respective distances, further complicating discussions of the mass of interstellar ^{26}Al. The history of ^{26}Al nuclei in the ISM provides a textbook example of scientific progress into a new discipline. Experts disagree in their predictions; correct predictions often rely on intuition as much as on scientific justification; surprises often attend exploratory experimental surveys; and scientists from many disciplines amplify the relevant ideas and measurements. Astronomy with radioactivity is such a science, and its first discovery of an interstellar gamma-ray line emitter, the surprise of ^{26}Al, was bathed in confusion.

2.2.3 Radiogenic Luminosity

A big problem lay in the path of trying to understand how explosive objects in astronomy could remain bright. Their luminosity was expected to fall rapidly as the objects expand. Suppose a rapidly expanding object consists of an ideal non-degenerate gas and that the expansion is too rapid for gain or loss of heat; that is, the expansion is adiabatic. The internal heat of such an object is rapidly lost to the mechanical work of the expansion. For adiabatic expansion $TV^{\gamma-1}$ = constant during the expansion, where γ is the ratio of specific heats at constant pressure c_P to the specific heat at constant volume c_V. For a perfect monatomic gas $\gamma = c_P/c_V = 5/3$. Thus $TV^{2/3}$ = constant for such an expanding hot object. If the expanding object can escape gravitational binding, the radius of an exploding object may be approximately $R = vt$, so that volume V is proportional to t^3. Thus one would expect that Tt^2 is constant during the expansion. The problem is then that if luminosity L is proportional to R^2T^4, as in a black body, one expects L to be proportional to t^{-6}. However, many supernovae that are bright after 1 week are still bright after 3 weeks instead of dimmer by the large expected factor near 3^{-6}. Similar problems exist for novae expansions. The problem, then, lay in discerning what source of energy could keep the expanding objects hot. The answer is radioactivity. The radiogenic luminosity of exploding objects became one of the major aspects of astronomy with radioactivity.

2.2.3.1 Radiogenic Luminosity in Supernovae

It was excellent luck that SN 1987A exploded nearby in the Magellanic Cloud in 1987. It became the most observed explosive astronomical event of all time. Astronomers recorded its emissions in every possible wavelength band, which proved necessary in exposing how radiogenic luminosity worked in that kind of supernova. The observations showed that after July 1987 the total power output of SN 1987A declined exponentially for about 2 years, accurately tracking the 77.2-day half-life of ^{56}Co. This confirmed that the energy of the positrons emitted and of the subsequent gamma-ray emission following each ^{56}Co decay were efficiently converted to optical and infrared luminosity. This old idea for declining supernova light curves was first treated for the identification of ^{56}Co as the relevant radioactivity by Colgate et al. (1969). The observed value of the total luminosity of SN 1987A required 0.075 $M_{\odot}$ of ^{56}Ni to have been synthesized within the core matter that escaped the central neutron star. With each scattering of a gamma ray it loses roughly half of its energy to the recoil electron, which is quickly degraded into heat. Once its energy has been scattered below 5 keV the large photoelectric opacity converts its entire remaining energy to electron energy, which is also degraded into heat.

Supernovae come in two main types, with additional structural subtypes (see Chaps. 4 and 5). Type Ia and Type II differ in how compact they are, how much radioactivity is produced, how that radioactivity is distributed, and how much envelope overlies the radioactivity. The nucleosynthesis of ^{56}Ni is easily the largest and most important source of heating power for the radiating gas. Such issues impact the way in which radioactivity provides luminosity for the remnant. To truly judge the degree of scientific understanding of radiogenic luminosity requires astronomers to infer, to the best of their ability, the structure of the exploding object. The luminosity being observed has complicated relationship to the location of the radioactivity. In SN1987A itself, for example, the pre-supernova star was blue, implying smaller size than the common red pre-supernova stars. The amount of overlying matter was modest owing to a large extent of mass loss during the prior evolution of the star. These characteristics influenced the relative importance of the radiogenic luminosity to the luminosity caused by the shock heated envelope.

Although ^{56}Ni provides the largest and most important source of heating power, other radioactive nuclei also play very important roles. The ^{56}Co daughter of ^{56}Ni is actually more observable in its effect on radioactive luminosity. This is primarily because its 77.2-day half-life allows it to remain alive after the 6.08-day parent ^{56}Ni has decayed. Therefore, ^{56}Co delivers heating power at a later time when the expanding remnant is much larger and has suffered more adiabatic cooling. In these circumstances, quantitative astronomy with radioactivity requires careful evaluation of how, when and where the energy released by radioactivity is deposited in the expanding gas (The et al. 1990).

The nucleosynthesis of ^{57}Ni also plays a role in radiogenic luminosity because it decays rather quickly to 272-day ^{57}Co. Owing to its longer half-life, ^{57}Co is still providing radioactive power when the ^{56}Co has decayed to negligible abundance. This transition of dominant radioactive power occurs after about 2 years. When *Compton Gamma Ray Observatory* (Kurfess et al. 1992) measured with OSSE the actual mass of ^{57}Co in SN1987A, it corresponded to twice the solar ratio ^{57}Fe/^{56}Fe. This surprised many who had inferred its abundance to be five times the solar ratio, a large isotopic ratio that had been deduced on the basis of the bolometric luminosity exceeding the instantaneous power from ^{56}Co decay (Suntzeff et al. 1992; Dwek et al. 1992). The radiogenic luminosity truly deriving from ^{57}Co decay was demonstrated by the OSSE measurement to have been overestimated by a factor near 2.5. That discovery team (Clayton et al. 1992) then advanced a new theoretical aspect of radiogenic luminosity, what they called *delayed power*. They presented a model showing that when the remnant gets sufficiently dilute owing to its expansion, the rate of electronic recombination cannot keep up with the rate of ionization that had been produced somewhat earlier by residual amounts of ^{56}Co radioactivity. Because of the high degree of ionization, the radiogenic luminosity from ^{56}Co radioactivity begins to exceed the instantaneous rate of energy deposition from ^{56}Co decay. In effect, the radiogenic luminosity reflects the rate of ^{56}Co decay from a somewhat earlier time than the time of observation. This observed upturn in luminosity had been attributed to ^{57}Co instead of delayed power from ^{56}Co decay. Such time-dependence became a new aspect of astronomy with radioactivity.

A more general aspect of radiogenic emission lies in any astronomical investigation in which the gas possesses a higher degree of energy excitation than could be expected in the absence of radioactivity. For example, the hydrogenic lines of He^{+} in SN 1987A were interpreted as a consequence of the high-energy photons escaping from the supernova interior. Hard X rays emanating from the expanding supernova 1987A were another clear example of Compton-scattered radioactivity gamma rays escaping from the interior; and only radioactivity can produce so many hard X rays. The Compton electrons dissociate the abundant CO molecules (Clayton et al. 1999), so the free C atoms represent abnormal excitation caused by radioactivity within the gas. Another example might be the detectable presence of doubly ionized species, an ionization state that would make no sense in a purely thermal setting. All such radiogenic possibilities are aspects of radiogenic luminosity, and as such a part of astronomy with radioactivity.

A similar transition between identities of powering radioactive nuclei occurs when the decay power of 60-year ^{44}Ti exceeds that of ^{57}Co. And no doubt there exists a similar problem, namely, does the rate of radiant emission of ^{57}Co power keep up with the rate of ^{57}Co decay power. The radioactive ^{44}Ti is created in almost the same location as are the Ni isotopes. This occurs in the explosive burning of oxygen and silicon (Bodansky et al. 1968a,b; Woosley et al. 1973) and reflects the extent to which ^{28}Si has been decomposed into the silicon-burning quasiequilibrium. But radiogenic luminosity from this ^{44}Ti nucleus depends even more strongly on how its emissions are converted to luminosity and the extent of the time lag between radioactive decay and the creation of observed photons. Its contributions to delayed radiogenic luminosity require careful assessment of those issues.

Such considerations are significant in the scientific understanding of astronomy with radioactivity. The luminosity of a young supernova remnant is one of the easiest observations of radioactivity in astronomy, because it does not require direct detection of the radioactivity but only the associated increase of the photospheric luminosity. But its interpretation requires fairly accurate description of radiation transport and the structure of the exploding object.

2.2.3.2 Radiogenic Luminosity in Novae

The brightening of a nova explosion and the associated ejection of matter from its expanded envelope are also issues that are dependent upon the deposition of radioactive heat to the envelope (Starrfield et al. 1972) (see Chap. 5). Those authors showed that a successful nova model needs to mix carbon from the surface of the underlying white dwarf into the hydrogen-rich accreted envelope in order to have sufficient radioactivity produced by the thermonuclear flash that triggers the nova event. That burning is the hot CN cycle. They showed that radioactive decay keeps the envelope hot while it expands to larger radiating surface area and therefore increased luminosity. They also showed that without the *radioactive afterburner*, mass ejection would not occur. Truran (1982) gives more nuclear-physics details of the nova explosions.

2.2.3.3 Radiogenic Luminosity in Neutron-Star Mergers

A new manifestation of radiogenic luminosity was discovered in 2017 after the LIGO and Virgo gravitational-wave observatories discovered a merger of two neutron stars (Abbott et al. 2017) (see also Chap. 4). The localization enabled Arcavi et al. (2017), Pian et al. (2017), Smartt et al. (2017) to study the optical counterpart, finding spectroscopic evidence of r-process material thrown off by the merging neutron stars. The bulk of this material seems to consist of two types: hot blue masses of highly radioactive r-process matter of lower-mass-range nuclei $A<140$ and cooler red masses of higher mass-number r-process nuclei $A>140$ rich in lanthanides. When released from the huge pressure of the neutron star, these ejecta expand and radiate detected optical light for about a week. Such duration of luminosity would not be possible without heating by internal radioactive decay, which is provided by r-process nuclei near their waiting points. Two distinct mass regions ($A<140$ and $A>140$) for the r-process yields have been known since the first time dependent calculations of the r-process (Seeger et al. 1965). Because of these spectroscopic features it has been argued that such nucleosynthesis in the Milky Way has been primarily ejecta from neutron-star mergers rather than from supernovae

(Kasen et al. 2017). These results offer a new possibility for clarifying six decades of uncertainty over the site of origin of r-process nuclei (see Chap. 4 for a discussion of the candidate sites of supernovae and neutron star mergers for the r-process). The relevance to the present section is that it is radiogenic luminosity that maintains the visibility of these r-process-rich ejecta.

2.2.4 *Extinct Radioactivity and Immediate Pre-solar Nucleosynthesis*

2.2.4.1 Xenology Revisited

John Reynolds had discovered in 1959 that extinct radioactive ^{129}I appeared to have existed initially in meteorites. The observed ratio to stable I is near $^{129}\mathrm{I}/^{127}\mathrm{I} = 10^{-4}$. This stood for decades as a nearly unique example. Then in the mid-1970s new discoveries and ideas greatly enlarged the context of astronomy with radioactivity. Using Eq. (2.21) for the number of radioactive nuclei born during the last mean lifetime ($\tau = 23.5 \times 10^6$ years) of ^{129}I nuclei, the mean interstellar abundance ratio for iodine isotopes is expected (see Eq. (2.21)) to be

$$^{129}\mathrm{I}/^{127}\mathrm{I} = P(^{129}\mathrm{I})/P(^{127}\mathrm{I})(\mathrm{k}+1)\tau_{129}/\mathrm{t}_{\odot} = 0.013 - 0.022$$

using $P(^{129}\mathrm{I})/P(^{127}\mathrm{I}) = 1.5$, $t_{\odot} = 8$ Gy for the pre-solar duration of nucleosynthesis, and $k = 2$–4 within the standard model of radioactivity in galactic chemical evolution. This result is valid because both ^{129}I and ^{127}I are primary-nucleosynthesis products via the r process. The measured ratio is very much smaller than this expected ratio. The interpretation was that the solar molecular cloud sat dormant, with no new nucleosynthesis input, for 4–6 mean lifetimes of ^{129}I to allow decay to reduce its activity to the measured level. As plausible as this *waiting period* seemed to be for the next two decades, it was not correct.

Reynolds' laboratory also discovered that that extinct radioactive ^{244}Pu appeared to have existed initially in meteorites. It was measured by an anomalous pattern of several Xe isotopes that was consistent with the spontaneous-fission spectrum from ^{244}Pu. Hudson et al. (1989) presented more modern data setting its abundance relative to that of ^{232}Th as $^{244}\mathrm{Pu}/^{232}\mathrm{Th} = 3 \times 10^{-3}$. The many studies of these first two extinct radioactivities came to be called *xenology* because of their reliance on excess isotopic abundances of Xe isotopes in meteorites. With a production ratio $P(^{244}\mathrm{Pu}/^{232}\mathrm{Th}) = 0.5$ and mean lifetime $\tau_{244} = 115 \times 10^6$ years, the expected interstellar ratio for $^{244}\mathrm{Pu}/^{232}\mathrm{Th}$ would be 0.02–0.03—larger by tenfold than the solar amount. But in this case a sequestering of the solar cloud to allow ^{129}I nuclei to decay would not sufficiently reduce the ^{244}Pu abundance. Trying to resolve the xenology puzzles occupied many meteoriticists for two decades. Then three new developments altered xenology studies.

Firstly, existence of such ^{129}Xe-rich and fission-Xe-rich interstellar dust was predicted (Clayton 1975b) by the argument that dust containing evidence of extinct radioactivity should condense within supernovae and appear in the early solar system. This was an exciting new idea in the astronomy of radioactivity. However, it appears that chemical ways of subsequently forming meteoritic minerals from dust without loss of the carried xenon isotopes is implausible; so this new idea was not supported for isotopes of Xe in meteorites. But the idea of such supernova dust had been firmly implanted among the concepts of astronomy with radioactivity; moreover, the future may yet find evidence of fossil xenon. Several other extinct radioactive nuclei were predicted to exist within supernova dust (Clayton 1975b,c), and many of these are now demonstrated to have existed in meteorites.

Secondly, the ^{26}Al extinct radioactivity was discovered in meteorites. But its mean lifetime is too short ($\tau_{26} = 1.05 \times 10^6$ years) for any to survive the duration of sequestering without fresh nucleosynthesis input that xenology had suggested. This led to the second big change. The seeming impossibility of maintaining the large measured quantity of interstellar ^{26}Al nuclei by nucleosynthesis prompted the idea that a supernova within the molecular cloud from which the sun was born had injected the solar ^{26}Al radioactive nuclei into the solar collapse just prior to the solar birth. Such a unique association would render the solar cloud atypical of the ISM at large (Cameron and Truran 1977; Wasserburg and Papanastassiou 1982). The substantial solar ^{26}Al abundance initially in meteorites became the standard bearer for a class of extinct radioactive nuclei that required special local production. Cameron and Truran (1977) suggested that far from being a coincidence, the supernova producing the solar ^{26}Al nuclei also triggered the collapse of the solar molecular cloud by the overpressure its shock wave brought to bear on that cloud. That model for live short-lived radioactivity in the solar cloud came to be referred to as *the supernova trigger*. But Cameron and Truran went too far. They supposed that the xenology-producing radioactivities (^{129}I and ^{244}Pu) were also injected by that supernova trigger. They ignored that the ISM already was expected to contain a hundred times too much ^{129}Xe for Reynolds' measurement, not too little, and ten times too much ^{244}Pu. So injecting those radioactivities seems to not be correct either. Nor does the idea of fossil ^{26}Mg nuclei from Al-rich interstellar dust rather than live ^{26}Al in meteorites seem correct. But the puzzles of extinct radioactivity had come more sharply in focus, and these new ideas were instrumental for astronomy with radioactivity.

Thirdly, additional extinct radioactive nuclei were discovered. So diverse were their lifetimes that the idea of exponential decay from a starting abundance became increasingly untenable. Then another new idea of great importance for astronomy with radioactivity appeared. Clayton (1983) pointed out that the concept of a homogeneous ISM was a faulty expectation. He argued that supernova ejecta enter a hot phase of an ISM in which matter cycles on average among three phases, but that the sun was born from the cold molecular cloud phase. This distinction greatly modifies the expectation of exponential decay in the ISM and of extinct radioactivities in molecular clouds. Huss et al. (2009) provide a modern list of the many extinct radioactivity for which evidence exists (see Fig. 2.2).

The revisiting of xenology and these three new developments left extinct radioactivity research changed conceptually.

2.2.4.2 Neon-E

A different noble gas, neon, provided an early but poorly understood evidence of stardust in the meteorites. As with xenology, the $^{20}Ne/^{22}Ne$ isotopic ratio can be studied by incremental heating of a meteoritic sample in a mass spectrometer. David Black discovered that in the Ivuna meteorite the measured isotopic ratio $^{20}Ne/^{22}Ne$ decreased precipitously when the sample heating reached 1000 °C, dropping to a ratio near three, much smaller than the common ratio near ten. Variation by a factor three in a trapped noble-gas isotopic component was unprecedented. This suggested that some unknown mineral releases its trapped neon gas near 1000 °C, and that mineral contains ^{22}Ne-rich neon. Black (1972) suggested that this small $^{20}Ne/^{22}Ne$ isotopic ratio was so bizarre that it can not be accounted for by conventional means within an initially homogeneous, gaseous solar system. He suggested that interstellar grains that had formed somewhere where the $^{20}Ne/^{22}Ne$ isotopic ratio was smaller than three had survived the origin of the solar system and imprinted the 1000 °C temperature outgassing with its isotopic signature. He called this exotic neon component *Ne-E*. This was a radical far-reaching conclusion, perhaps the first of its kind based on sound analyzed data rather than on pure speculation.

But where might this dust have formed? Arnould and Beelen (1974) remarked that neon gas ejected from explosive He-burning shells of massive stars could resemble Ne-E, but they offered no suggestion for forming carriers of Ne-E there or of otherwise getting such gas into the early solar system. Clayton (1975c) offered a more concrete proposal, arguing that supernova dust would condense before the ejecta could mix with circumstellar matter, roughly within the first year, and that ^{22}Na (2.6 years) produced by the explosion would condense as the element sodium and decay to daughter ^{22}Ne only after the grains had been grown. In that manner nearly isotopically pure ^{22}Ne could be carried within supernova dust into the forming solar system. This was an early prediction of the supernova-dust phenomenon in meteorites. Clayton and Hoyle (1976) argued that nova dust provided another possible source for Ne-E. Either radioactive venue would amount to a new technique for astronomy with radioactivity.

The leading alternative to radiogenic ^{22}Ne is condensation of dust in a ^{22}Ne-rich red-giant stellar atmosphere. This can indeed occur, because during He burning the ^{14}N residue of the prior CN cycle can be converted to ^{22}Ne by two successive radiative alpha-particle captures (Arnould and Norgaard 1978). The question arises whether this Ne is sufficiently ^{22}Ne-rich. And that required much more study of meteoritic specimens. But the possibility was experimentally strengthened by the discovery (Srinivasan and Anders 1978) that at least one Ne-E component was associated with *s*-process xenon from red-giant atmospheres.

These were exciting developments in isotopic astronomy and for astronomy with radioactivity. This fresh new set of ideas intensified interrelationships among interstellar stardust, the origin of the solar system, and meteorites.

2.2.4.3 ^{26}Al: Fossil or Injected Fuel

During the mid 1970s it was discovered (Gray and Compston 1974; Lee et al. 1977) that very refractory Al-rich minerals that apparently were among the first solids to form in the early solar system contained elevated isotopic ratios ^{26}Mg/^{27}Mg. Furthermore, the number of excess ^{26}Mg atoms within a mineral was shown to be proportional to the number of Al atoms. That correlation could be reproduced (Lee et al. 1977) if the Mg had been initially isotopically normal and the initial Al contained a small component of radioactive ^{26}Al, whose later decay produced the excess ^{26}Mg atoms. The simplest way for this to have occurred is that the molecular cloud from which the sun had formed carried within it the large isotopic ratio $^{26}\mathrm{Al}/^{27}\mathrm{Al} = 5 \times 10^{-5}$ at the time the planetary system began to form. The reason for calling that ratio large is that it far exceeded the ratio that had been anticipated to exist within the ISM. Had that ratio been applicable to the ISM as a whole, it would have corresponded to 29 $M_{\odot}$ of radioactive ^{26}Al nuclei—much more than was discovered in 1982 by Mahoney et al by their detection of the 1.809 MeV gamma ray line that is emitted following its decay.

The immediate issue was whether these radioactive ^{26}Al nuclei were actually alive in the early solar system or whether ^{26}Al was alive only when Al-rich dust formed in earlier galactic supernovae. If the latter option were true, the excess ^{26}Mg nuclei were fossils of radioactive ^{26}Al decay within interstellar Al-rich dust. Existence of such ^{26}Mg-rich and Al-rich interstellar dust had been predicted (Clayton 1975b,c) in the first works to propose that dust containing evidence of extinct radioactivity should condense within supernova ejecta. This was an exciting new idea in the astronomy of radioactivity. Clayton (1977a,b) presented a solid-carrier model for excess ^{26}Mg nuclei in Al-rich minerals based on the fossil picture. In the first case, that of live ^{26}Al nuclei during the formation of the minerals, some explanation was required for why so much radioactive ^{26}Al should have been present in the solar matter when it could not have been a general property of the ISM. In the second case, fossil excess ^{26}Mg nuclei, why did interstellar Al dust containing abundant radioactive ^{26}Al nuclei form within supernova ejecta and how did it participate in the chemical growth of the Al-rich minerals found in meteorites. In the first case, the heat of the radioactive decay of radioactive ^{26}Al nuclei within those planetesimals that had been the parent bodies for the meteorites would have been sufficient to melt large parent bodies, allowing them to differentiate chemically (as the earth has done). In that case ^{26}Al also played a role as fuel for melting of the meteorite parent bodies. In the second case, the excess ^{26}Mg nuclei represented a fossil of interstellar decay. Tension between these two possibilities was reflected in the phrase *"Fossil or Fuel"* within the title of the paper by Lee et al, who argued in favor of the live ^{26}Al nuclei in the solar gas. This debate raged for a few years but lost steam as chemical arguments for the growth of the mineral phases seemed increasingly likely to require live ^{26}Al nuclei.

The struggle over the interpretation very short-lived extinct radioactivity was heightened later by the discovery of now-extinct ^{41}Ca in meteorites. Its abundance relative to that of ^{41}K is only $^{41}\mathrm{Ca}/^{41}\mathrm{K} = 1.5 \times 10^{-8}$ (Srinivasan et al. 1996); however, even that small abundance looms large because its mean lifetime is but $\tau_{41} = 0.144 \times 10^6$ years. Surely none can survive galactic nucleosynthesis to the time the solar system formed. So the supernova trigger injection would be needed to account for it as well as ^{26}Al. But fossil effects are also definitely possible. Huge ^{41}K excesses in Ca-rich supernova dust had been predicted two decades earlier (Clayton 1975c).

The history of documentation of extinct radioactive nuclei that had been alive in the early solar system was led for decades by G.J. Wasserburg and his laboratory at Caltech (jokingly called "the lunatic asylum" owing to their research on lunar samples). His laboratory was of such renown for its measurements using traditional mass spectrometry with extreme sensitivity that Wasserburg was chosen recipient of the 1986 Crafoord Prize for geosciences. Similar mass spectrometric research of great consequence for this problem was conducted by Günter Lugmair in his laboratory at UCSD.

2.2.4.4 Extinct Radioactivity and Mixing of Interstellar Phases

Xenology soon gave way to a host of newly discovered extinct radioactive nuclei. ^{129}I and ^{244}Pu could seen to be members of a larger class of now extinct parents. A very special class contains those nuclei having mean lifetimes long enough that some of that radioactive abundance might survive from galactic nucleosynthesis but short enough that their survival live to the time of the solar system depends on the timing of interstellar mixing. That class of nuclei also includes ^{129}I and ^{244}Pu. Their mean lifetimes are, in ascending order: $\tau(^{60}\mathrm{Fe}) = 2.16$[2] $\times 10^6$ years; $\tau(^{53}\mathrm{Mn}) = 5.34 \times 10^6$ years; $\tau(^{107}\mathrm{Pd}) = 9.38 \times 10^6$ years; $\tau(^{182}\mathrm{Hf}) = 13.8 \times 10^6$ years; $\tau(^{129}\mathrm{I}) = 23.5 \times 10^6$ years; $\tau(^{244}\mathrm{Pu}) = 115 \times 10^6$ years; and $\tau(^{146}\mathrm{Sm}) = 149 \times 10^6$ years. The relative abundances that they possess and the three-isotope correlation plots by which they are measured are discussed in Chap. 6. What became abundantly clear is that simple exponential decay from a set of starting abundances is not consistent with the data. Something more complex is at work.

Four Layers of Interpretation The new data and new creative ideas led to a picture of cosmic radioactivity that is interpreted on four layers of complexity. Like the proverbial *Russian doll*, one opens each layer only to reveal a more complex version inside. What are these four layers?

[2]The mean life of ^{60}Fe is now known as 3.8 My.

1. *Mean galactic nucleosynthesis and the well-mixed ISM:* Using Eq. (2.21) for the mean interstellar ratio of a radioactive abundance to that of a reference isotope gives that mean abundance ratio to be

$$N(^{A}Z)/N(^{A*}Z) = P(A^{Z})/P(^{A*}Z)\,(k+1)\,\tau_{Z}/t_{\odot} \tag{2.22}$$

   If the half-life of species ^{A}Z is long enough that the ISM can be considered to be well-mixed, it may be interpretable by the mean expectation of Eq. (2.22). An example might be the uranium isotope ratio, since both have halflives of 10^9 years or greater. The ISM probably mixes well during 10^9 years.
2. *Mean galactic nucleosynthesis and the ISM phases:* If the lifetime τ_Z is of order 10^8 years or less, the abundances of that radioactive nucleus may differ in the different phases of the ISM. The three-phase mixing model (Clayton 1983) that describes this in the mean is described below. The idea is that supernova ejecta appear initially in the hot ISM, and considerable mean time is required for it to work its way via mass exchange between phases into the molecular clouds where the sun might have formed. That delay reduces radioactive abundances, depending upon their lifetimes. This idea regards each phase of the ISM as having the mean concentration of radioactive nuclei everywhere in that phase; but the three phases have differing mean concentrations of the radioactivity, with the molecular-cloud abundance being the least. This picture divides the mean ratio Eq. (2.21) into differing steady-state ratios in each of the three phases. An example might be the iodine isotope ratio, since its half-life of 17 My is shorter than the expected phase change times (perhaps 50 Ma) for the ISM phases. Thus each phase will in the mean have different $^{129}I/^{127}I$ ratios.
3. *Mean galactic nucleosynthesis and deviations from mean ISM mixing times:* All galactic samples of a given ISM phase will surely not be identical. Differences occur because nucleosynthesis does not follow the galactic mean rate at all locations, or because the interphase mixing times may vary from one place to another. Separate portions of a phase will scatter about that mean for that phase, perhaps by sizeable amounts. Astronomical observations of very old, very metal-poor stars reveal this phenomenon clearly (Burris et al. 2000). The iodine isotopes can illustrate how this works. Consider the molecular cloud from which the sun formed. It may have existed in a portion of the galaxy whose hot phase received less than the average amount of new nucleosynthesis during the past 24 My (the ^{129}I mean lifetime). Thus the ^{129}I concentration locally would be less than its galactic mean, whereas stable ^{127}I would have the same concentration everywhere. Furthermore, the time delay for the hot phase to be admixed into the solar molecular cloud may locally have exceeded the galactic mean delay, in which case the ^{129}I concentration in the solar molecular cloud would have decayed to a smaller value than typical of that phase. This layer of interpretation deals with specific local galactic workings that differ from the galactic mean.
4. *Nucleosynthesis from a single neighboring supernova:* The solar concentration of a short-lived radioactive nucleus may have been the result of a single galactic supernova. The shorter the half-life, the more likely it becomes that a neighboring

dying star is needed to create it. In such a case the concept of mean galactic nucleosynthesis rate has no relevance. Short-lived ^{26}Al is historically the prime example of this situation. Variations on the supernova injection model then appear: When did that supernova occur? With what efficiency did its ejecta admix into the solar molecular cloud? How long was required for that mixing to occur? Such questions are so specific that they are frustrating. Almost any answer seems possible, depending upon hydrodynamic modeling for its credibility.

Finally, some radioactive nuclei may have multiple causes for their solar abundances. Each layer of interpretation may contribute. Take ^{244}Pu as a likely example ($\tau(^{244}\mathrm{Pu}) = 115$ Ma). A portion of the solar ^{244}Pu abundance may survive at the level expected from mean galactic nucleosynthesis (layer 1). Layer 2 may have reduced that somewhat, depending on the mean interphase mixing times. Layer 3 describes a non-average ^{244}Pu solar concentration owing to atypical (non-mean) local nucleosynthesis rate and mixing parameters. Finally, a portion of solar ^{244}Pu abundance may have resulted from a single nearby supernova at the time of solar birth. It will be clear that scientific understanding must seek the best picture for simultaneously fitting all of the radioactivities. Each individual abundance will be inadequate by itself to determine the physics of its solar presence. But a model that fits them all is a serious contender for the truth. To advance more into this topic, consider the mean interphase mixing model.

Mean Mixing Model for ISM Radioactivity Astrophysics provides this novel example of the layer-2 mathematics of radioactive decay. The distinct ISM phases have vastly differing temperatures and densities. These phases mix with one another on timescales that are longer than the shortest of the extinct radioactivities but shorter than the longest. As a consequence, the concentration X of a radioactive nucleus will differ from phase to phase, with the differences greater for shorter half-life. Solar-system samples revealing the presence of once-live radioactive nuclei are obtained from meteorites and other planetary objects formed in the solar accretion disk. Measurements of solar samples therefore are of the concentrations in the cold molecular cloud (one of the phases) from which the sun was born. The key point is that radioactive concentrations in that molecular-cloud phase should be smaller than in the mean ISM, which includes all of its phases. This consequence of the mixing times among the phases, taking into account their levels of radioactivity, was devised by Clayton (1983).

A Simplified Two-Phase Model To focus on mathematical aspects of radioactivity rather than on aspects of astrophysics, consider first a simplified model that illustrates the essence of the problem. Suppose that the ISM consists of but two phases. Stars are born in phase 1, but the freshly synthesized radioactivity is ejected from supernovae into phase 2. Let those two phases have equal mass M and exchange matter with the other phase on a timescale T. The mass exchange rate is then $(dM/dt)_{exch} = M/T$. Matter from phase 1 joins phase 2 at that rate, and conversely matter from phase 2 joins phase 1 at that same rate. The masses of each phase then remain constant and equal.

In reality, spatial inhomogeneity also exists, depending on where in phase 2 the fresh radioactivity is created and how long is required for its homogenization. Roughly 30,000 supernovae occur per My in the Milky Way, and they have similar nucleosynthesis yields, so an almost homogeneous injection phase occurs faster than the output of a single supernova can be spread uniformly. Even so, concentration differences and isotopic differences will exist spatially in that injection phase. The best that can be done without a specific calculation defined for a specific configuration is to calculate the *average concentration* in the injection phase. To this end consider the injection phase to be uniformly mixed at all times (Layer 2). With that assumption the average difference in radioactive concentration between the phases can be calculated with a simple mathematical model. The answer can not be assumed to have applied exactly to the unique case of solar formation. Nonetheless, such a calculation reveals what is anticipated on average, without inhomogeneous effects.

Returning to the simplified two-phase model for those average concentrations, the mixing rate between them is $(dM/dt)_{exch} = M/T$. Then with X_1 being the concentration (grams of X per gram of ISM) of a radioactive nucleus in phase 1, the star-forming phase, its total radioactive mass MX_1 changes as follows: $d/dt(MX_1) = MdX_1/dt + X_1 dM/dt = MdX_1/dt$ because the second equality is a consequence of the unchanging mass M of each phase. How shall one evaluate dX_1/dt? It changes owing to two effects: first, the radioactive decay of X_1; second, the mass exchange with phase 2 having concentration X_2. One writes for those two terms $MdX_1/dt = -MX_1/\tau + (X_2 - X_1)M/T$ where the first term is the radioactive decay rate with mean lifetime τ. But the left-hand side must also vanish because $dX_1/dt = 0$ in the steady state. This reasoning yields at once $X_1 = \tau/(T + \tau)X_2$ This factor expressing the ratio of the concentrations can differ significantly from unity. Suppose the mixing time $T = 00$ Ma. Then for ^{129}I, X_1 = (23.5Ma/123.5Ma)$X_2 = 0.19X_2$. The ^{129}I concentration in star-forming clouds is then fivefold smaller than in the injection phase. This reduction is even greater for a nucleus having smaller mean lifetime. But for very long lifetimes, $\tau/(T + \tau)$ approaches unity, both phases having the same concentration.

Both X_2 and X_1 can be individually evaluated from the requirement that their average (since the two phases have the same mass) must equal the mean expected ISM concentration for that nucleus (Eqs. (2.20) and (2.21)). Chapter 6 includes a table of the known extinct radioactive nuclei and their abundances in the early solar system.

The significance for differing ratios of extinct nuclei between the two phases is that those differences do not reflect the simple expectation of exponential decay. The ratio $\tau/(T + \tau)$ is very far from exponential in dependence on lifetime τ. Clayton (1983) introduced that effect when comparing the solar abundance levels of differing radioactive nuclei. This simplified two-phase mean model illustrates that idea.

Clayton Three-Phase Model The initial (Clayton 1983) introduction of this physical idea for interpreting relative abundances of extinct radioactive nuclei in this manner actually suggested a three-phase ISM rather than only two. Those masses and phases were:

1. M_1 is the molecular-cloud mass in which stars form. It exchanges mass with M_2.
2. M_2 is the mass of large HI complexes that can not be disrupted by the supernova shock waves that frequently traverse the ISM. Phase 2 exchanges mass with M_1 with lifetime T_1 and also with M_3 with lifetime T_2.
3. M_3 is the mass average HI clouds that are sufficiently small to be disrupted by passing supernova shock waves and which therefore are part of this warm neutral phase. Phase 3 exchanges mass with phase 2 with lifetime T_2.

It is not clear that this three-phase model is more realistic than the simpler two-phase model; but the need for fresh radioactivity to mix from its injection phase through a second phase before admixing with molecular clouds does amplify the distinction between radioactive lifetimes. Future progress with this multiphase idea, which definitely is significant for the extinct radioactivity problem, will require more modern astrophysical work on the differing phases, their masses, and how effectively they exchange mass with one another. Finally, it may be that time-dependence must be introduced. The growth of each phase, rather than static mean masses for each, needs evaluation. Over long times the global masses of each phase will be static; but their cyclic growth and disruption may be an essential part of the problem.

An insightful paper in regard to these issues has recently appeared (Huss et al. 2009). Those authors present current data for the observed abundances of each extinct radioactivity. They also review the sites of nucleosynthesis of each nucleus. Then they evaluate the mean abundance expected within the standard model of galactic chemical evolution moderated by the interphase mixing described above. Figure 2.2 (from Huss et al. 2009) displays the expectation as a function of the mean decay lifetime for a wide range of interphase-mixing timescales and compares those results to those for two differing free-decay intervals (the solid curves). Note that interphase mixing fits more extinct radioactivities than does either free-decay interval; but the four short-lived isotopes are too abundant for any galactic scheme, apparently demanding injection from an adjacent supernova. Keep in mind that what is being examined by these mathematical procedures is a fundamental property of interstellar radioactivity, couched for convenience and transparency in terms of realistic analytic representations of the true galactic mixing processes.

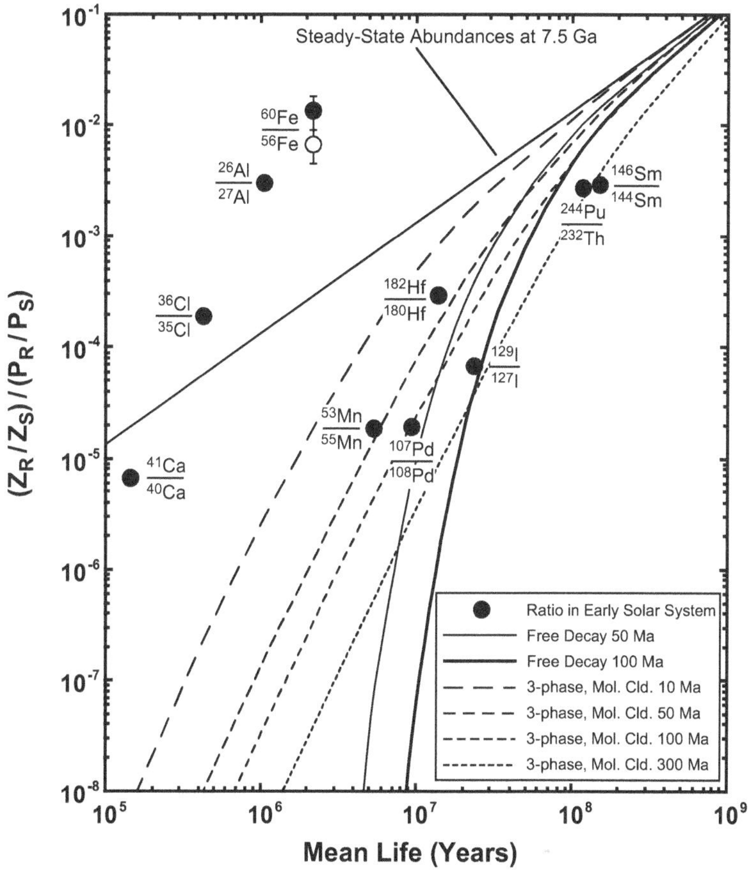

Fig. 2.2 Calculated surviving levels of the extinct radioactive nuclei in the solar cloud at the time of solar birth are expressed as a ratio to a neighboring stable isotope of the same element. Galactic steady state modified by the continuous interphase mixing is shown for several different mixing times (from Huss et al. 2009)

2.2.5 *Stardust: Radioactivity in Solid Samples of Pre-solar Stars*

A pre-solar stardust grain is composed only of atoms from a single donor star. The word *stardust* applies to only those grains. It is used as a scientific word rather than a poetic word. It may be hard to accept that the isotopic composition of elements in an interstellar grain attests to it being a solid chunk of a single star. An average ISM grain has suffered a lengthy residence time within the interstellar medium before being incorporated into a new stellar accretion disk. Do not high energy interstellar ions implant within the grains as well as sputter them? Can interstellar chemistry add more atoms to mineralized grains? Would not grain ejection from a protostellar disk after chemically adding atoms to it within the hot dense disk structure also produce mineralized grains? Many questions flood the skeptical mind. Such doubts are reasonable considering the incomplete knowledge of these and other processes.

Stardust grains are not average interstellar grains. Those grains themselves provide the best answers to the doubts expressed above. Their well ordered crystal structures suggest thermal condensation at high temperatures, as would be expected within hot but slowly cooling gases leaving a star. Annealed crystals would not, in contrast, be expected from accreted interstellar atoms. The dramatic isotopic ratios measured within individual stardust grains, obtained from nearly homogeneous grains having large numbers of atoms (10^6 to 10^{12}, say), strongly suggests that the condensation was from a gas having the extremely non-solar isotopic composition of the stardust grain. The SiC grains, one of the most abundant of all pre-solar stardust grains, were assembled from C atoms having observed isotopic ratios in the grains between $^{12}C/^{13}C = 3–5000$, whereas hot chemistry in an accretion disk would shift ratios toward interstellar norms, $^{12}C/^{13}C = 89$ in the case of the solar accretion disk. Similar evidence appears in isotopic compositions of N and Si, to name only the most studied of the elements. Those ratios attest to an extreme condensation environment wherein isotopic composition differs significantly from that found in the ISM. These isotopic studies of individual stardust grains have been made using SIMS (secondary-ion mass spectrometry) in terrestrial laboratories (Bernatowicz and Zinner 1997).

What followed from the discovery of stardust was nothing less than a revolution in astronomy. Solid pieces of stars are now characterized routinely in terrestrial laboratories, including isotopic analysis of many elements with a precision not attainable at telescopes. Distinguishing isotopes in stellar spectra is very difficult. Three scientists dominated the early development of stardust research experimentally. Robert M. Walker committed the lunar-sample laboratory at Washington University to development of new technology capable of better laboratory study of primitive solar system samples. Owing to the limited numbers of atoms within a stardust grain, high efficiencies were needed both for liberation of atoms and for their counting. He foresaw the capability of the sputtering ion probe, and recruited Ernst Zinner to devote a decade to improving and studying that technique. Their laboratory at Washington University (St. Louis) was then equipped to study the newly isolated meteoritic grains of silicon carbide, which they documented to have condensed in AGB stars, from their measured C and N isotopic ratios. Zinner and his students led and dominated the new astronomy that blossomed from their isotopic analysis. An engaging history of this has been written in tribute to Walker and Zinner by McKeegan (2007). Thirdly, Edward Anders inspired his associates at University of Chicago to the detective-story pursuit that isolated stardust from the bulk meteorite rock. They termed their approach *burning down the haystack to find the needle*. It tracked the chemical carriers within the meteorites of isotopically anomalous neon and xenon through a sequential dissolving the bulk of the rock in strong acids (Amari et al. 1994) and finding that the undissolved residue became increasingly anomalous. The carriers of anomalous noble gas were, fortunately, carbonaceous, and therefore did not dissolve in acid. From these residues they were able to extract the individual stardust grains. This too was a profound adventure in scientific exploration. Finally it may be noted that predictions of the existence of isotopically anomalous stardust had existed for a decade prior to their experimental discovery.

Within oxide grains, Nittler et al. (1997) showed that stunningly abnormal families of correlated O isotopes rule out any growth within mixed ISM. Equally stunning was the almost pure isolated s-process isotopic compositions that were predicted for noble xenon in red giant stardust (Clayton and Ward 1978) and discovered (Srinivasan and Anders 1978) in bulk carbon-rich residues of acid dissolution of meteorites. These Xe_s-rich residues are dominated by collections of SiC grains. By contrast, interstellar bombardment would implant normal Xe isotopic composition. Neon isotopes also revealed an exciting story. Individual mainstream SiC grains have been shown to carry almost pure s-process Mo, a quite reactive trace element rather than a noble gas, showing that the Mo-containing condensing gases had not mixed with the ISM prior to condensation. Isotopic evidence of stardust abounds. By careful consideration of the entirety of the known properties of stardust grains, researchers become confident that they are indeed solid samples of stars that died before the sun was born.

2.2.5.1 Stardust Predicted

With the realization that computer models of massive-star supernovae generate shells of abundant intermediate-mass elements, Hoyle and Wickramasinghe (1970) argued that the adiabatic expansion and cooling of these newly synthesized nuclei within the supernova interior should be accompanied by condensation of a potentially large amount of silicate dust. The ejected gas cools to below 2000 K after only several months when the density is still quite high. Their goal had been to account for the interstellar dust mass of silicates and of graphite. It did not occur to them that such grains could be identified by their unusual isotopic compositions. Nor did they suggest that they might be found within meteorites. They simply tried to explain the observed existence of dust in astronomy.

Clayton (1975b,c, 1978) advanced exciting observational considerations on the grains' condensed radioactive nuclei. He suggested that grains from the supernova-condensation process would be identifiable by their extreme isotopic signatures, inherited from the isotopic compositions of those supernova shells. Radioactivity is prominent within supernova shells. Short-lived radioactivity was proposed to decay within each grain during its interstellar residence. The ISM should contain everywhere numerous interstellar grains from the very large number of pre-solar supernovae. He also proposed a chemical memory model (Clayton 1977a,b, 1978, 1982) in which supernova grains were incorporated into solids accreting in the solar accretion disk during later growth of larger solar-system solids. Many of those solids were incorporated into meteorites. That picture suggested potential explanations for several isotopic anomalies that had been discovered in meteorite solids; but it was rejected initially by meteoriticists, who focussed their doubts on the potential explanations of xenon anomalies from extinct radioactivity. Clayton and Wickramasinghe (1976), Clayton (1979), and Kozasa et al. (1989) demonstrated that a sufficient number of collisions of supernova-interior atoms will occur with any new grain there during expansion of the supernova interiors, to grow grains greater than one micrometer in radius. None of these works was so optimistic as to predict that individual supernova grains would be found intact within meteorites; but they did establish the ideas by which those discoveries could be recognized. Radioactivity lay at the heart of identifying supernova grains.

In a paper outlining a system for the different types of precondensed matter in the early solar system, the supernova condensates were named SUNOCONs and the term STARDUST was restricted to hot thermal condensation during mass loss from other stars (Clayton 1978), primarily red giants. NEBCON was suggested for grain mantles grown by nebular sticking of atoms and molecules to preexisting dust. These names have not found the favor of usage, despite describing a theoretical picture of what to expect in the early solar system. Only the term stardust is commonly used, and applied to all types of high-temperature thermal condensates from stars. The title *Precondensed matter: key to the early solar system* of Clayton (1978) explicitly contradicted that epoch's favored picture in which the initial solar system was initially hot and totally gaseous and condensed its solids from that solar gas.

The predictions of isotopically extreme stardust were, after some initial controversy, wonderfully confirmed by the discovery of supernova stardust and the later documentation of chemical memory within large solids, earning broad acceptance.

2.2.5.2 Stardust Discovered: Fossil Extinct Radioactivity

Secondary-ion mass spectrometry (SIMS) of ions sputtered from isolated single meteoritic grains by a focused ion beam identified stardust in the late 1980s. A history may be read in Anders and Zinner (1993). First to be discovered (Bernatowicz et al. 1987; Zinner et al. 1989) was silicon carbide (SiC), with isotopic patterns that identified it as thermal condensate in matter flowing away from the photospheres of asymptotic-giant-branch red giants. Clayton and Ward (1978) added the proof that such stardust would contain pure s-process xenon, which it does (Srinivasan and Anders 1978). The sensational identification of stardust is now beyond doubt, buttressed by a huge number of experimental and theoretical studies. These carbide grains can condense in AGB star atmospheres and winds only after the composition changes caused by the third dredge-up of carbon, which eventually cause carbon to be more abundant than oxygen in the envelope. As long as O remains more abundant than C, it oxidizes (combusts) all C into CO molecules. Those SiC grains condensed after the carbon-star transition contained $^{26}Al/^{27}Al$ near 10^{-3} based on the number of excess ^{26}Mg atoms they contain. When supernova SiC *X* grains were later identified as a subfamily of SiC stardust, many revealed $^{26}Al/^{27}Al$ near 0.1 when they condensed within the expanding supernova interior.

Such measurements of extinct ^{26}Al in stardust immediately highlighted a new aspect of astronomy with radioactivity.

Oxide stardust too was identified and classified into isotopic families a few years later (Nittler et al. 1997). Its most common pattern is ^{17}O-richness owing to secondary production in the parent star of ^{17}O from initial ^{16}O during hydrogen burning . At about the same time supernova stardust was identified unambiguously (Hoppe et al. 1996; Nittler et al. 1996) using the prediction (Clayton 1975c) that excess ^{44}Ca would exist in the Ti-bearing phases within grains owing to the condensation of radioactive ^{44}Ti after the explosion. Because the elemental ratio Ti/Ca is for crystal-lattice reasons larger than solar in SiC grains, the supernova SiC grains possess very large isotopic excess at ^{44}Ca. These were found in the family of SiC grains called *X grains*, which were already suspected of being supernova grains on the basis of deficiencies in the heavy isotopes of Si and C and on excesses of ^{15}N (Alexander et al. 1990; Amari et al. 1992).

The extremely large isotopic excesses of ^{26}Mg and ^{44}Ca in these SiC X grains were the most dramatic discoveries of fossil extinct radioactivity in stardust. They demonstrated that radioactive ^{26}Al and ^{44}Ti had been quite abundant when the grains condensed during the supernova expansion. Predictions had also suggested that fossil abundances in interstellar grains might contribute to excesses of the daughter isotope in solids grown later from interstellar grains (*chemical memory*). But Lee et al. (1977) emphasized grounds for believing that the excess ^{26}Mg within solids made in the solar system reflected instead live ^{26}Al at the time the solids were assembled from pre-solar dust and fused thermally along with hot condensing solar-system atoms. Both fossil and live ^{26}Al now appear to have existed. These discoveries dramatically spotlighted rich new roles for the astronomy of radioactivity in early solar system chemistry.

Many other fossil extinct radioactive nuclei have been added to the observed list. In supernova stardust these include excess ^{22}Ne owing to ^{22}Na decay, excess ^{41}K owing to ^{41}Ca decay, and excess ^{49}Ti owing to ^{49}V decay within the grains (see Clayton and Nittler 2004). All had been predicted to exist in supernova dust, but their discovery surprised and delighted isotopic chemists and astronomers alike.

Following the exciting discovery of pre-solar stardust, its study has emerged as a new area of astronomy. In just 23 years its existence has evolved from a bewildering new discovery into several new techniques for measuring isotopic abundance ratios with high precision in stars. This has been especially important for astronomy with radioactivity, because in stardust the level of extinct radioactivity reflects its abundance during nucleosynthesis in stars, whereas in solar system samples it reflects the level of its survival. Precisely measured isotopic ratios for four-to-eight distinct chemical elements endows each gemlike refractory mineral with significance for stellar structure and evolution, and for the chemical evolution of the Milky Way during the epoch 7–5 Gyr ago, and for new insight into nucleosynthesis (Clayton and Nittler 2004). This rich harvest is compromised only by the fact that the donor stars can not be seen, because they died before our solar system began. Their stardust bears no label, save that of their measurable properties. The nature and evolutionary state of the donor stars must be ascertained from the detailed properties, primarily isotopic, of each grain and of the way each grain fits into the spectrum of the thousands of stardust grains that have been analyzed to date. The rapidly growing numbers of analyzed grains allows evolutionary trends within their distinct families to be identified.[3]

2.2.5.3 Radioactivity and Chemistry of Condensation in Supernovae

An unanticipated aspect of astronomy with radioactivity lay in the role of radioactivity in the chemistry of the condensation process. The supernova interior offers a unique laboratory for condensation physics. It guarantees that chemistry begins with gaseous atoms, with no trace of previous molecules or grains. Gamma rays and their Compton-scattered electrons bathe the supernova core. By disrupting the CO molecule, they cause its abundance to be very much smaller than expected from states of chemical thermal equilibrium at the ambient temperature. The small CO molecular abundance enables other non-equilibrium paths to the condensation of carbon-bearing solids. This disequilibrium can be considered to be another aspect of radiogenic luminosity causing a higher degree of excitation in the gas than would be expected in the absence of radioactivity.

Because the supernova core is hydrogen-free, chemical condensation routes utilizing H are not relevant. Polycyclic aromatic hydrocarbons, for example, do not come into play save in the envelope; but 90–95% of the mass of ejected Mg, Al and Si emerges in the H-free core. And yet abundant dust condensation is observed to have occurred in SN1987A (Wooden et al. 1993; Colgan et al. 1994), in Cas A (Arendt et al. 1999; Dunne et al. 2003) and in the Kepler remnant (Dunne et al. 2003). Apparently a few solar masses condensed in Cas A and about one solar mass in Kepler, requiring high condensation efficiency for Mg, Al and Si and even for carbon and thereby establishing supernovae as a major contributor to the budget of thermally condensed interstellar grain cores.

The traditional guideline to condensation of solids had been to assume that chemical equilibrium applies during the expansion and temperature decline of the supernova interior. Formulae yield the equilibrium condensed masses of differing minerals (Lattimer et al. 1978; Ebel and Grossman 2000). This approach can not yield grain size but only the total condensed masses. The assumption of thermal equilibrium in the solid phase exaggerates the ability of chemical reactions to maintain chemical equilibrium within the solids as the gas cools. Because of the rapid fall of density and temperature during the expansion of the supernova interior, thermal condensation must be accomplished within about 2 years, too

[3]Readers wishing familiarity with these topics can best consult later chapters of this book, or Clayton and Nittler (2004), and *Astrophysical Implications of the Laboratory Study of Pre-solar Materials* (Bernatowicz and Zinner 1997); the *Handbook of Isotopes in the Cosmos* by Clayton (2003) also presents many astrophysical consequences.

fast to maintain equilibrium. A non-equilibrium theory of condensation based on nucleation theory followed by subsequent growth has been developed by Kozasa and collaborators (Kozasa and Hasegawa 1987; Kozasa et al. 1989, 1991; Todini and Ferrara 2001). Their method identifies a *key molecular species* whose abundance controls the condensation. However, their questionable assumptions concerning gaseous mixing at the molecular level and on an outdated role for the CO molecule render their results of questionable value.

The nucleosynthesis problems posed by isotopic ratios within individual supernova grains can not be decoupled from physical questions about the chemistry of their condensation. Condensation chemistry is an essential aspect of astronomy with radioactivity. Because no single supernova mass zone is able to satisfy the isotopic ratios found in supernova grains, it has long been clear (Amari and Zinner 1997) that some type of mixing before condensation is needed to produce their chemical and isotopic compositions; but it is not clear whether that mixing represents atom-scale gaseous mixing in the very young remnant or transport of a growing grain from one composition zone into another. But the biggest question is whether the requirement that the C abundance be greater than O abundance in order to condense SiC and graphite within supernovae is a valid requirement. For their discussion of supernova stardust, for example, Travaglio et al. (1999) took the view that mixing occurs at the atomic level, prior to condensation, and that only C$>$O portions of those mixtures support condensation of carbon. They arbitrarily and instantaneously mixed regions having C$<$O with the He shell having C$>$O to find mixing fractions that retained C$>$O overall. Kozasa et al. (1989) performed similar thought experiments. Such treatment is deeply flawed, however, because it seems certain that gases can not mix at the molecular level within a few years time (see Deneault et al. 2006; Fryxell et al. 1991). The so-called mixing calculated in hydrodynamic studies of supernovae, on the other hand, represents fluids of one composition exchanging places in the homologous expansion with overlying fluids of a different composition. This is a different use of the word *mixing*, and has introduced considerable confusion into published works. Mixing at the molecular level requires very much more time than the year that is available before condensation must occur. Later the expanding remnant is too dilute, its density too small, for condensation.

To circumvent the equilibrium CO trap for carbon, Clayton et al. (1999, 2001) and Deneault et al. (2006) called upon its dissociation by the fast Compton electrons energized by supernova gamma rays. They advanced a kinetic theory of graphite growth and calculated its consequences in detail after advocating a specific nucleation model. The key idea is that small graphite test particles, if placed in a hot gas of C and O atoms, will associate with free C atoms faster than they can be oxidized by free O atoms. This is a property of reaction cross sections rather than of chemical equilibrium. Even though oxidation of carbon would surely be the ultimate end given adequate time, the expansion will terminate the chemistry after about 2 years with large graphite grains remaining. The graphite essentially is a metastable state of carbon. This theory is supported by observing that supernova 1987A ejected only $10^{-3}\,M_\odot$ of CO molecules instead of the $0.1\,M_\odot$ of CO molecules that is first formed by association reactions in the hot gas before radioactive disruption reduces its abundance (Liu and Dalgarno 1994, 1995; Gearhart et al. 1999).

Similar issues probably surround the condensation of supernova SiC. It seems plausible that radioactive liberation of free C atoms from CO molecules will also facilitate the condensation of SiC in O-rich gas; but this is hotly debated. Although a kinetic route to SiC condensation has not been layed out, Deneault et al. (2003) have formulated a physical description of the ejecta enabling them to make several relevant conclusions derived from assuming that the radioactive CO-disruption mechanism is the correct key to SiC growth. They present animations of a hydrodynamic calculation showing that a reverse shock wave launched toward inner zones by the radially increasing value of the product ρr^3 in the H envelope compresses a dense shell near radial mass coordinate $m = 3\,M_\odot$, where Si and O are the most abundant elements; but some C abundance remains there for possible carbon chemistry. They propose that SiC condenses in that zone, and they detailed several other physical processes that may produce the observed grain compositions. Mixing of a new type during condensation also occurs if the reverse shock generated by the radial expansion colliding with the pre-supernova wind arrives at the condensation zone at $m = 3\,M_\odot$ between 6 months to a year after the explosion, because that reverse shock slows the gas and forces the partially condensed SiC grains to propel forward through the decelerating gas into more ^{29,30}Si-rich regions, giving perhaps a new interpretation for mixing during condensation. Lazzati and Heger (2016) have presented a new way to think about nucleations for grain growth, but it is not clear whether it is preferable to the molecular reaction network approach. But it also replies on ^{56}Co radioactivity for the yields of dust.

Despite many uncertainties, it now appears certain that supernova grains studied by isotopic analysis will provide, through details of condensation chemistry, a new sampling spectrum of young supernova interiors, just as have gamma-ray lines and hard X rays. If so, a very rich but complex discipline of astronomy with radioactivity will follow.

2.2.5.4 Astronomy with Radioactivity Today

This introductory chapter has focussed on the history and the key physical ideas of astronomy with radioactivity. Effective research depends on a clear grasp of an interdisciplinary set of its central ideas. History itself often provides the best example by which the physical idea can be first grasped. But the remainder of this book looks forward, not backward. Every aspect of astronomy with radioactivity today involves grappling with a host of technical details. This book attempts to bring the reader to that point. Each aspect of astronomy with radioactivity today also involves grappling with the entire world of astronomy. Astronomy itself is many disciplines, and not even astronomers are always comfortable outside their own astronomical technique. How much harder it is to place the complex manifestations of radioactivity into a continuously changing astronomical fabric. But this is the direction of all scientific progress.

References

Abbott BP et al, The LIGO scientific collaboration and Virgo collaboration (2017) Phys Rev Lett 119:161101
Alexander CMO'D, Swan P, Walker RM (1990) Nature 348:715–717
Amari S, Zinner E (1997) In: Bernatowicz T, Zinner E (eds) Astrophysical implications of the laboratory study of presolar materials. American Institute of Physics, Woodbury, p 287
Amari S, Hoppe P, Zinner E, Lewis RS (1992) Astrophys J 394:L43–L46
Amari S, Lewis RS, Anders E (1994) Geochim Cosmochim Acta 58:459–470
Anders E, Zinner E (1993) Meteoritics 28:490–514
Arcavi I et al (2017) Nature 551:64
Arendt RG, Dwek E, Moseley SH (1999) Astrophys J 521:234
Arnett WD (1977) In: Papagiannis MD (ed) Eighth Texas symposium on relativistic astrophysics, vol 302, p 90
Arnett WD (1996) Supernovae and nucleosynthesis: an investigation of the history of matter from the big bang to the present. Princeton University Press, Princeton
Arnould M, Beelen W (1974) Astron Astrophys 33:215
Arnould M, Norgaard H (1978) Astron Astrophys 64:195–213
Arnould M, Norgaard H, Thielemann FK, Hillebrandt W (1980) Astrophys J 237:931–950
Beers TC, Christlieb N (2005) Annu Rev Astron Astrophys 43:531
Bernatowicz TJ, Zinner E (1997) Astrophysical implications of the laboratory study of presolar materials. American Institute of Physics, Woodbury
Bernatowicz TJ et al (1987) Nature 330:728–730
Black DC (1972) Geochim Cosmochim Acta 36:377
Bodansky D, Clayton DD, Fowler WA (1968a) Phys Rev Lett 20:161
Bodansky D, Clayton DD, Fowler WA (1968b) Astrophys J Suppl 16:299
Boggs SE et al (2015) Science 348:670
Burbidge EM, Burbidge GR, Fowler WA, Hoyle F (1957) Rev Mod Phys 29:547–650
Burris DL, Pilachowski CA, Armandroff TE, Sneden C, Cowan JJ, Roe H (2000) Astrophys J 544:302
Butcher HR (1987) Nature 328:127–131
Cameron AGW (1955) Astrophys J 121:144
Cameron AGW (1957) CRL 41, Chalk River
Cameron AGW (1959) Astrophys J 130:429
Cameron AGW, Truran JW (1977) Icarus 30:447
Clayton DD (1964) Astrophys J 139:637
Clayton DD (1968) Principles of stellar evolution and nucleosynthesis. McGraw-Hill, New York
Clayton DD (1971) Nature 234:291
Clayton DD (1974) Astrophys J 188:155
Clayton DD (1975a) Astrophys J 198:151
Clayton DD (1975b) Astrophys J 199:765–769
Clayton DD (1975c) Nature 257:36–37
Clayton DD (1977a) Icarus 32:255–269
Clayton DD (1977b) Icarus 35:398–410
Clayton DD (1978) Moon Planets 19:109–137
Clayton DD (1979) Astrophys Spa Sci 65:179–189
Clayton DD (1981) Astrophys J Lett 244:L97
Clayton DD (1982) Q J R Astron Soc 23:174–212
Clayton DD (1983) Astrophys J 268:381–384
Clayton DD (1984) Astrophys J 280:144
Clayton DD (1985) Galactic chemical evolution and nucleocosmochronology: a standard model. In: Arnett WD, Truran W (eds) Challenges and new developments in nucleosynthesis. University of Chicago Press, Chicago, p 65

Clayton DD (1987) Nature 329:397
Clayton DD (1988) Mon Not R Astron Soc 234:1
Clayton DD (1999) Meteorit Planet Sci 34:A145–A160
Clayton DD (2003) Handbook of isotopes in the cosmos. Cambridge University Press, Cambridge
Clayton DD (2007) Science 318:1876–1877
Clayton DD, Craddock W (1965) Astrophys J 142:189
Clayton DD, Fowler WA (1961) Ann Phys 16:51
Clayton DD, Hoyle F (1974) Astrophys J Lett 187:L101
Clayton DD, Hoyle F (1976) Astrophys J 203:490
Clayton DD, Nittler LR (2004) Ann Rev Astron Astrophys 42:39
Clayton DD, Silk J (1969) Astrophys J Lett 158:L43
Clayton DD, Ward RA (1978) Astrophys J 224:1000–1006
Clayton DD, Wickramasinghe NC (1976) Astrophys Spa Sci 42:463–475
Clayton DD, Fowler WA, Hull TE, Zimmerman BA (1961) Ann Phys 12:331
Clayton DD, Colgate SA, Fishman GJ (1969) Astrophys J 155:75
Clayton DD, Leising MD, The LS, Johnson WN, Kurfess JD (1992) Astrophys J 399:L141–L144
Clayton DD, Hartmann DH, Leising MD (1993) Astrophys J 415:L25
Clayton DD, Liu W, Dalgarno A (1999) Science 283:1290–1292
Clayton DD, Deneault E, Meyer BS (2001) Astrophys J 56:480–493
Colgan SWJ et al (1994) Astrophys J 427:874
Colgate SSA, Fishman GJ, Clayton DD (1969) Astrophys J 155:75
Deneault EN, Clayton DD, Heger A (2003) Astrophys J 594:312
Deneault EN, Meyer BS, Clayton DD (2006) Astrophys J 638:234–240
Diehl R et al (1995) Astron Astrophys 298:445
Diehl R et al (2014) Science 345:1162
Dunne L, Eales S, Ivison R, Morgan H, Edmunds M (2003) Nature 424:285–287
Dwek E, Moseley SH, Glaccum W, Graham JR, Loewenstein RF, Silverberg RF, Smith RK (1992) Astrophys J 389:L21–L24
Ebel D, Grossman L (2000) Geochim Cosmochim Acta 64:339–366
Fowler WA (1971) New observations and old nucleocosmochronologies. In: Reines F (ed) Cosmology, fusion and other matters. University of Colorado, Boulder, pp 67–123
Fowler WA, Hoyle F (1960) Ann Phys 10:280
Fowler WA, Wasserburg GJ, Hoyle F (1960) Phys Rev Lett 4:112
Fryxell B, Müller E, Arnett D (1991) Astrophys J 367:619–634
Gao Y, Solomon PM (2004) Astrophys J 606:271
Gearhart RA, Wheeler JC, Swartz DA (1999) Astrophys J 510:944–966
Gilroy KK, Sneden C, Pilachowski CA, Cowan JJ (1988) Astrophys J 327:298
Gray CM, Compston W (1974) Nature 251:495
Grefenstette BW et al (2014) Nature 506:339
Haxel O, Jensen JH, Suess HE (1949) Phys Rev 75:1766
Hoppe P et al (1996) Science 272:1314–1316
Hoyle F (1946) Mon Not R Astron Soc 106:343–383
Hoyle F (1954) Astrophys J Suppl 1:121–146
Hoyle F, Fowler WA (1960) Astrophys J 132:565
Hoyle F, Wickramasinghe WA (1970) Nature 226:62–63
Hudson GB, Kennedy BM, Podosek FA, Hohenberg CM (1989) Proceedings of the 19th lunar and planetary science conference, vol 19, pp 547–557
Huss GR, Meyer BM, Srinivasan G, Goswami JN, Sahijpal S (2009) Geochim Cosmochim Acta 73:4922
Iyudin AF et al (1994) Astron Astrophys 284:L1
Johnson WN, Harnden FR, Haymes RC (1972) Astrophys J 172:L1
Kaeppeler F et al (1982) Astrophys J 257:821
Kasen D et al (2017) Nature 551:80
Kozasa T, Hasegawa H (1987) Prog Theor Phys 77:1402

Kozasa T, Hasegawa H, Nomoto K (1989) Astrophys J 344:325
Kozasa T, Hasegawa H, Nomoto K (1991) Astron Astrophys 249:474
Kurfess JD et al (1992) Astrophys J Lett 399:L137
Lattimer JM, Schramm DN, Grossman L (1978) Astrophys J 219:230
Lazzati D, Heger A (2016) Astrophys J 817:134
Lee T, Papanastassious DA, Wasserburg GJ (1977) Astrophys J 211:L107
Leising MD, Share GH (1990) Astrophys J 357:638
Liu W, Dalgarno A (1994) Astrophys J 428:769
Liu W, Dalgarno A (1995) Astrophys J 454:472
MacCallum CJ, Huters JF, Stang PD, Leventhal M (1987) Astrophys J 317:877
Mahoney WA, Ling JC, Jacobson AS, Lingenfelter RE (1982) Astrophys J 262:742
Mahoney WA, Ling JC, Wheaton AS, Jacobson AS (1984) Astrophys J 286:578
McKeegan KD (2007) Meteorit Planet Sci 42:1045–1054
Nittler LR et al (1996) Astrophys J 462:L31–L34
Nittler LR et al (1997) Astrophys J 483:475–495
Pian E et al (2017) Nature 551:67
Ramaty R, Lingenfelter RE (1977) Astrophys J 213:L5
Reynolds JH (1960) Phys Rev Lett 4:8
Rutherford E (1929) Nature 123:313–314
Seeger PA, Fowler WA, Clayton DD (1965) Astrophys J 11:121
Share GH et al (1985) Astrophys J 292:L61
Smartt BJ et al (2017) Nature 551:75
Srinivasan B, Anders E (1978) Science 201:51
Srinivasan B, Sahijpal S, Ulyanov AA, Goswami JN (1996) Geochim Cosmochim Acta 60:1823–1835
Starrfield S, Truran JW, Sparks WM, Kutter GS (1972) Astrophys J 176:169
Suess HE, Urey HC (1956) Rev Mod Phys 28:53
Suntzeff NB, Phillips MM, Elias JH, Walker AR, Depoy DL (1992) Astrophys J 384:L33–L36
Tajitsu A et al (2015) Nature 518:381
The LS, Burrows A, Bussard R (1990) Astrophys J 352:731
The LS, Clayton DD, Diehl R et al (2006) Astron Astrophys 450:1037
Timmes FX, Woosley SE, Weaver TA (1995) Astrophys J Suppl 98:617
Todini P, Ferrara A (2001) Mon Not R Astron Soc 325:726
Travaglio C et al (1999) Astrophys J 510:325
Truran JW (1981) Astron Astrophys 97:391–393
Truran JW (1982) Nuclear theory of novae. In: Barnes CA, Clayton DD, Schramm DN (eds) Essays in nuclear astrophysics. Cambridge University Press, Cambridge, p 467
Truran JW, Arnett WD, Cameron AGW (1967) Can J Phys 45:2315
von Ballmoos P, Diehl R, Schonfelder V (1987) Astrophys J 318:654
Wasserburg GJ, Papanastassiou DA (1982) Some shortlived nucleides in the early solar system. In: Barnes CA, Clayton DD, Schramm DN (eds) Essays in nuclear astrophysics. Cambridge University Press, Cambridge, p 77
Weaver TA, Woosley SE (1993) Phys Rep 227:65
Wooden DH et al (1993) Astrophys J Suppl 88:477–508
Woosley SE (1999) Astrophys J 525:924
Woosley SE, Weaver TA (1980) Astrophys J 238:1017
Woosley SE, Weaver TA (1995) Astrophys J Suppl 101:181
Woosley SE, Arnett WD, Clayton DD (1973) Astrophys J Suppl 26:231
Yokoi K, Takahashi K, Arnould M (1983) Astron Astrophys 117:65
Zinner E, Tang M, Anders E (1989) Geochim Cosmochim Acta 53:3273–3290

Part II
Specific Sources of Cosmic Isotopes

The following three chapters discuss various stellar sources of new isotopes. Here we focus on the processes which take place within an astrophysical object over the course of its evolution. For stars, this can be for millions to billions of years. For explosions, evolution may take place on characteristic time scales of hours or minutes, or even as short as seconds. In these three chapters we present the material for later discussions of the evolution of larger objects ranging from stellar groups to galaxies. We grouped *stars* according to main themes related to *single stars* and *binary systems*. The section on single stars was further divided into chapters each for the *two physically-different ranges of stellar masses* defined by separating *core-collapsing* from *white-dwarf-generating* stars. Binary systems can experience *mass transfer*, which adds complexity that leads to a rich variety of phenomena. We discuss sub-categories associated with different nuclear-reaction environments, spanning a broad range from *thermonuclear supernovae* through *novae* to *X-ray bursts*.

Chapter 3
Low- and Intermediate-Mass Stars

Maria Lugaro and Alessandro Chieffi

Energy in stars is provided by nuclear reactions, which, in many cases, produce radioactive nuclei. When stable nuclei are irradiated by a flux of protons or neutrons, capture reactions push stable matter out of stability into the regime of unstable species. The ongoing production of radioactive nuclei in the deep interior of the Sun via proton-capture reactions is recorded by neutrinos emitted during radioactive decay. These neutrinos escape the inner region of the Sun and can be detected on Earth. Radioactive nuclei that have relatively long half lives may also be detected in stars via spectroscopic observations and in stardust recovered from primitive meteorites via laboratory analysis. The vast majority of these stardust grains originated from Asymptotic Giant Branch (AGB) stars. This is the final phase in the evolution of stars initially less massive than $\simeq 10\,M_{\odot}$, during which nuclear energy is produced by alternate hydrogen and helium burning in shells above the core. The long-lived radioactive nucleus ^{26}Al is produced in AGB stars by proton captures at relatively high temperatures, above 60 MK. Efficient production of ^{26}Al occurs in massive AGB stars ($> 4:5\,M_{\odot}$), where the base of the convective envelope reaches such temperatures. Several other long-lived radioactive nuclei, including ^{60}Fe, ^{87}Rb, and ^{99}Tc, are produced in AGB stars when matter is exposed to a significant neutron flux leading to the synthesis of elements heavier than iron. Here, neutron captures occur on a timescale that is typically slower than β-decay timescales, resulting in a process known as *slow* neutron captures (the *s*-process). However, when radioactive

M. Lugaro (✉)
Monash Centre for Astrophysics, Monash University, Clayton, VIC, Australia

Konkoly Observatory, Research Centre for Astronomy and Earth Sciences, Hungarian Academy of Sciences, Budapest, Hungary
e-mail: maria.lugaro@csfk.mta.hu

A. Chieffi
Istituto Nazionale Astronomia Fisica INAF, Roma, Italy
e-mail: achieffi@iasf-roma.inaf.it

R. Diehl et al. (eds.), *Astrophysics with Radioactive Isotopes*, Astrophysics and Space Science Library 453, https://doi.org/10.1007/978-3-319-91929-4_3

nuclei with half lives greater than a few days are produced, depending on the temperature and the neutron density, they may either decay or capture a neutron, thus branching up the path of neutron captures and defining the final s-process abundance distribution. The effect of these *branching points* is observable in the composition of AGB stars and stardust. This nucleosynthesis in AGB stars could produce some long-living radioactive nuclei in relative abundances that resemble those observed in the early solar system.

3.1 The Missing Element

The element with 43 protons in its nucleus, lying between molybdenum and ruthenium, was known for a long time as the *missing* element. Since the nineteenth century there had been many unsuccessful attempts at its discovery. Finally, in 1937 Italian physicist Emilio Segré and chemist Carlo Perrier found two isotopes of the missing element through measurements of radioactivity from discarded cyclotron parts: they observed several decay periods and proved they were occurring at Z=43. Hence the missing element did not exist in nature because of its instability against nuclear decay. The discoverers named the missing element technetium (Tc), from τεχνητός, which means artificial in Greek, since it was the first element produced artificially. Fifteen years later, it was shown that Tc is not only made by men but also by stars. In 1952, astronomer Paul Merrill observed the absorption lines corresponding to the atomic structure of Tc in the spectra of several giant stars. Merrill was at first cautious about this result. To start with, the element he identified did not even exist on Earth. Second, up to then it was assumed, and not proved wrong, that all stars had the same chemical composition. This was in agreement with the accepted theory of the time that the elements were all produced during the Big Bang and their abundances in the Universe were not modified by any further process. Merrill's discovery in that respect was truly revolutionary: given the relatively short half lives of the Tc isotopes (a few Myrs at most), the Tc lines were the first indisputable demonstration that this radioactive element was made *in situ* in the stars where it was observed. This finding brought a radical change in the way we understand the origin of the elements, and the theory of stellar nucleosynthesis introduced in Chap. 2 began to take shape and garnered authority. In this chapter we discuss the life of those stars that, like our Sun, evolve twice through Red Giant stages. We describe how they produce long lived radioactive nuclei, like Tc, in their interiors, how the signature of such radioactivity is carried outside the star, and how it can be observed.

3.2 The Production of Radioactive Nuclei in Stellar Interiors

In Sect. 3.2.1 we first derive the four basic equations that control the quasi-equilibrium configuration of a self-gravitating gas sphere, namely, the hydrostatic equilibrium equation (that describes the balance between the pressure gradient and gravity) and the energy transport equation (due to photons and/or convection), plus the two associated continuity equations for mass and energy flux. Then, we show that energy losses, which occur mainly from the stellar surface in stars of mass less than $\sim 10\,M_\odot$, force the gas to contract and to heat, in accordance with the virial theorem. The progressive increase of the central temperature allows the activation of nuclear processes and we describe two sequences that convert protons into ^{4}He nuclei (α particles): the PP chain and the CNO cycle. Since proton capture inevitably pushes matter out of the stability, both these sequences produce radioactive nuclei that decay by emitting neutrinos.

In Sect. 3.2.2 we briefly describe the quest for solar neutrinos and the various experiments that eventually allowed the demonstration that the lower than predicted neutrino flux from the Sun (the so-called Solar Neutrino Problem) is the consequence of neutrino oscillations among their three different flavors.

3.2.1 The Stellar Energy Source and Radioactive Isotopes

A star is, in first approximation, a spherically symmetric, gaseous cloud contracting under its own gravity and progressively heating up while losing energy from its surface in the form of photons. A strong temperature gradient, with the temperature decreasing from the centre to the surface, pushes the photon flux outward until the *mean free path*[1] of the photons eventually becomes larger than their distance from the surface, allowing their escape. The Virial theorem links the energy gained by the gravitational field $\Delta\Omega$ to that absorbed by the gas ΔU: $\Delta U = -\Delta\Omega/3(\gamma - 1)$, where γ is the ratio between two specific heats, that at constant pressure and that at constant volume, of the contracting gas. A stable quasi-equilibrium configuration exists for such a structure provided that $\gamma > 4/3$. In this case, a fraction of the energy gained by the gravitational field must be liberated from the structure before the gas cloud can contract further. In the case of a perfect gas ($\gamma = 5/3$) we obtain the classical result $\Delta U = -(1/2)\Delta\Omega$, stating that half of the gravitational energy liberated by each infinitesimal contraction is absorbed by the gas and half must be lost before an additional contraction can occur. The timescale over which the energy is lost from the system drives the timescale of contraction and keeps the structure in a quasi-equilibrium configuration. If, instead, γ drops to 4/3, all the gravitational energy is absorbed by the gas and no time delay is required before a new contraction can occur. This is an unstable situation leading to collapse. In the evolutionary phases we discuss in this chapter γ remains well above 4/3 and hence a stable quasi-equilibrium configuration is always assured.

The balancing forces required to maintain a stable stellar quasi-equilibrium configuration are due to pressure gradients and gravity. So, the first main equation of stellar structure describes the equilibrium between these two forces, at any given

[1] The average distance a particle travels between collisions.

distance from the center of the star r:

$$dP/dr = -GM\rho/r^2 \tag{3.1}$$

where P is the pressure, G the gravitational constant, M the cumulative mass inside r, and ρ the density. Associated to this equation is a continuity equation for mass:

$$dM/dr = 4\pi r^2 \rho. \tag{3.2}$$

By assuming, to zero order, that ρ is constant within the star, the integration of Eq. (3.2) implies that:

$$\rho \propto M/R^3. \tag{3.3}$$

Since the pressure at the surface of the star is much lower than where the radius approaches zero, the center of the star, the equation of hydrostatic equilibrium, (3.1), basically says that

$$P_c \propto M\rho/R_s \tag{3.4}$$

where P_c is the central pressure and R_s the stellar radius. By inserting the relation (3.3) into (3.4) one obtains:

$$P_c \propto M^2/{R_s}^4. \tag{3.5}$$

If the equation of state is that of a perfect gas, i.e. $P \propto \rho T/\mu$, relation (3.5) becomes:

$$T_c \propto \mu M/R \tag{3.6}$$

where T is the temperature and μ is the mean molecular weight. This equation provides an important basic relationship among central temperature (T_c), mass, and radius of a star, which only relies on the assumption of hydrostatic equilibrium.

The second major equation describing the structure of the quasi-equilibrium configuration of a star determines the energy flux through the structure. In stationary situations the energy is transported by photons or electrons and application of the first Fick's law leads to the well known equation:

$$dT/dr = -3\kappa\rho L/(16ac\pi r^2 T^3) \tag{3.7}$$

where κ is the opacity coefficient representing the impenetrability of a gas to light, L the luminosity representing the amount of energy radiated per unit time, a the radiation constant, and c the speed of light. Furthermore, in this case a continuity equation

$$dL/dr = 4\pi r^2 \rho\epsilon \tag{3.8}$$

controls the conservation of energy, where ϵ represents the net local energy budget, i.e., the sum of the nuclear energy production rate ϵ_{nuc}, the neutrino energy loss rate ϵ_{ν}, and the gravitational energy rate ϵ_{g}. Since the central temperature T_c is much higher than the surface temperature, it is possible to obtain a basic relationship between central temperature, mass, luminosity and radius of a star, i.e.:

$$T_c{}^4 \propto ML/R^4. \tag{3.9}$$

By combining this relation with the previous relation (3.6), derived from hydrostatic equilibrium (3.1), one eventually obtains the fundamental relation between mass and luminosity:

$$\mathrm{L} \propto \mu^4 \mathrm{M}^3. \tag{3.10}$$

Frequently, the energy produced locally cannot be transported quickly enough by radiation or conduction, and interior shells formally in an equilibrium condition can become unstable in the sense that a displacement from their equilibrium position is not fully counteracted by a restoring force. Instead, matter is accelerated even further from its original position and large scale motions of matter (*convection*) is established. Under these conditions energy is predominantly transported by buoyancy-driven motions of bulk material due to their much larger mean free path with respect to that of photons. A temperature gradient quite different from that described by Eq. (3.7) must then be used in these regions. A direct consequence of these large scale motions is that matter is mixed throughout an unstable region.

The system formed by the two basic equations related to hydrostatic equilibrium and energy transport plus the two associated continuity equations and the equation of state (supplemented by an opacity coefficient $\kappa = f(\rho, T, \text{chem.comp.})$ and a total energy generation coefficient $\epsilon = f(\rho, T, \text{chem.comp.})$) constitutes the basic set of equations that describes the internal structure of a quasi-equilibrium (non rotating) stellar configuration at a given time. The temporal evolution of such a structure is determined by the rate at which energy is lost to the surroundings: the faster the energy is lost, the faster the structure evolves. Typical stellar luminosities range between $\sim 4 \times 10^{33}$ erg/s for a star of 1 $M_{\odot}$ and $\sim 4 \times 10^{36}$ erg/s for a star of 6 $M_{\odot}$ and the associated lifetimes can be estimated by dividing the total amount of available energy by the loss rate L, i.e. the stellar luminosity. If the only energy source was the gravitational field, the lifetime of a contracting gas cloud would be of the order of a few tens of millions of years (the Kelvin-Helmholtz timescale). Instead, as was known since the 1920s from radioactive dating of terrestrial rocks, that the age of the Earth is several Gyrs, much longer than the Kelvin-Helmholtz timescale. Thus, the Sun must be powered by different means. The lifetime of most stars is much larger than permitted by their gravitational energy reservoir alone. Instead of simple contraction, energy losses are replaced by the activation of nuclear fusion reactions among charged nuclear particles near the stellar core.

The efficiency of nuclear reactions, i.e., their *rate*, depends on the abundances of the reactant nuclei and the cross section of each reaction averaged over a Maxwellian distribution of relative velocities between the target and the projectile nuclei. For charged particle reactions the rate is mainly controlled by the Coulomb barrier generated by the number of protons in a nucleus. The nuclear reactions that activate at the lowest temperatures are those involving capture of protons, i.e., the nucleus of the lightest and most abundant element: hydrogen (H). Nature, however, does not allow the build up of stable nuclei made only of protons because in order to glue nucleons (i.e., proton and neutrons) together in a nucleus via the strong nuclear force, the repulsive electromagnetic force acting between protons needs to be diluted with a certain number of neutrons. The distribution of stable nuclei in the [N=number of neutrons, Z=number of protons] plane, the *chart of nuclides*, clearly shows that the region where nuclei are stable lies close to the N=Z line (the *valley of β stability*) up to the element Ca, and then bends slightly towards the neutron-rich side as the repulsion between higher number of protons needs to be diluted with more and more neutrons. Nuclei outside this valley are radioactive, i.e., unstable, and decay towards their closest stable daughter through β-decay weak interaction reactions (as described in Chaps. 1, 2, 4, and 9). It follows that the build up of progressively heavier nuclei through the addition of protons naturally pushes the matter out of the stability valley, producing radioactive nuclei that decay back towards stability via β^+ decay.

A detailed analysis of the nuclear reactions involving the fusion of protons (H burning) foresees the existence of two processes. The PP chain activates at temperatures $\simeq$10 MK and operates through a sequence of proton captures and β decays starting with a weak interaction p+p fusion. The processes involved in the PP chain are listed in Table 3.1 together with the mass defect (Q values) in MeV and the energy carried away by neutrinos. If the neutrino emission is described by an energy continuum, the maximum energy of this spectrum is reported. A direct build up of progressively heavier nuclei through successive proton captures stops very early, at ^{3}He, because of the low cross section of the ^{3}He+p reaction. Also, proton captures on the second most abundant isotope, ^{4}He (or α particle, $N = Z = 2$), cannot even begin, because nuclei with atomic mass number A=N+Z=5 are unstable and hence rarely available as reaction targets for proton capture. In order to proceed with proton captures beyond ^{3}He it is necessary to build up enough ^{3}He nuclei

Table 3.1 The reactions of the PP chain of burning H to He

Reaction	Q_{tot} (MeV)	Q_ν (MeV)
$p + p \rightarrow d + e^+ + \nu$	1.442	0.42(spectrum)
$d + p \rightarrow^3 He$	5.494	
$^3He +^3 He \rightarrow^4 He + 2p$	12.860	
$^3He +^4 He \rightarrow^7 Be$	1.587	
$^7Be + e^- \rightarrow^7 Li + e^+ + \nu$	0.862	0.861(90%)–0.383(10%) (lines)
$^7Be + p \rightarrow^8 B \rightarrow^8 Be + e^+ + \nu \rightarrow 2\alpha$	18.209	14.060(spectrum)
$^7Li + p \rightarrow^8 Be \rightarrow \alpha + \alpha$	17.347	

to activate the capture of this nucleus by either another ^{3}He nucleus, or ^{4}He. The activation of ^{3}He captures allows to overcome the non-existence of nuclei with A=5, though it still does not allow the build up of an appreciable amount of nuclei heavier than He. In fact, the product of the ^{3}He+^{3}He reaction is an α particle plus two protons, while the product of the ^{3}He+^{4}He reaction is ^{7}Be, whose fate, either proton or electron capture leads to the formation of ^{8}Be, which very quickly decays in two α particles. In synthesis, the fusion of H mainly produces He, together with a number of radioactive nuclei that decay into their respective stable daughter nuclei emitting a neutrino. The energies and the number of neutrinos produced in these decays reflect the relative importance of the various PP-chain branches and the efficiency of the nuclear reactions in stars.

The second process converting protons to α particles is the CNO cycle. Given the high Coulomb barrier of the CNO nuclei, this cycle becomes efficient at temperatures (T $>$ 20 MK), significantly higher than those relevant to the PP chain. The main section of this sequence is characterized by the continuous conversion of C to N and viceversa. Let us start, e.g., with the capture of a proton by a ^{12}C (Table 3.2). The outcome of this fusion is the radioactive nuclide ^{13}N that quickly decays β^+ into ^{13}C. Efficient proton captures by ^{13}C lead to the synthesis of ^{14}N. Proton captures by ^{14}N produce ^{15}O, a radioactive nuclide that quickly decays in ^{15}N. The fusion of a proton and a ^{15}N particle has, as the main outcome, a ^{12}C nucleus plus an α particle. The sequence sketched above is called the CN cycle. If the temperature exceeds T $\sim$ 25–30 MK, also oxygen enters the game and the full CNO cycle activates: ^{16}O begins to capture protons forming radioactive ^{17}F that decays to ^{17}O. The capture of a proton by this particle leads to a compound nucleus that preferentially splits into ^{14}N and an α particle, and that partly turns into ^{18}F, which quickly decays to ^{18}O. Proton captures on ^{18}O produce preferentially ^{15}N

Table 3.2 The individual reactions of the CNO cycle of H burning, together with their respective reaction Q values

Reaction	Q_{tot} (MeV)
$^{12}C + p \rightarrow^{13} N$	1.944
$^{13}N \rightarrow^{13} C + e^+ + \nu$	2.220
$^{13}C + p \rightarrow^{14} N$	7.551
$^{14}N + p \rightarrow^{15} O$	7.297
$^{15}O \rightarrow^{15} N + e^+ + \nu$	2.754
$^{15}N + p \rightarrow^{12} C + \alpha$	4.966
$^{15}N + p \rightarrow^{16} O$	12.127
$^{16}O + p \rightarrow^{17} F$	0.600
$^{17}F \rightarrow^{17} O + e^+ + \nu$	2.761
$^{17}O + p \rightarrow^{14} N + \alpha$	1.192
$^{17}O + p \rightarrow^{18} F$	5.607
$^{18}F \rightarrow^{18} O + e^+ + \nu$	1.656
$^{18}O + p \rightarrow^{15} N + \alpha$	3.981
$^{18}O + p \rightarrow^{19} F$	7.994

plus an α particle. The activation of the channel $^{15}N(p,\gamma)^{16}O$ closes the NO cycle, processing material back into oxygen. The total abundance by number of the CNO isotopes remains constant with time because the proton capture on any of them (and the subsequent β^+ decays) just produce another isotope in the same set.

For T > 25–30 MK, the full CNO cycle becomes efficient and quickly reaches a quasi-equilibrium in which the abundance of each nucleus settles on a steady state value determined by the balance between its production and destruction. For example, the equilibrium abundance of ^{13}C (assuming that ^{13}N decays instantaneously) is given by:

$$\frac{dY_{^{13}C}}{dt} = Y_{^{12}C}\, Y_p\, \rho\, N_A \, <\sigma\, v>_{^{12}C+p} \; - \; Y_{^{13}C}\, Y_p\, \rho\, N_A \, <\sigma\, v>_{^{13}C+p} = 0$$

where Y_i refers to the abundance by number of a given species i, t is time, ρ is the density, N_A is Avogadro's number and $<\sigma\ v>_j$ is the Maxwellian averaged product of the velocity v times the nuclear cross section σ for a given capture j. The equilibrium condition immediately gives:

$$\frac{Y_{^{12}C}}{Y_{^{13}C}} = \frac{<\sigma\, v>_{^{13}C\, p}}{<\sigma\, v>_{^{12}C\, p}}$$

which means that the relative abundances between isotopes of the CNO cycle depend only on the ratio between the respective cross sections for proton capture. Typical isotopic and elemental ratios obtained in the temperature range $30 \leq T \leq 100$ MK are given in Table 3.3.

The neutrinos emitted by the decay of radioactive nuclei synthesized by the CNO cycle have characteristic energies different from those emitted by the PP chain. Their detection would provide precious information about the relative efficiency of the various reactions involved in the CNO cycle.

In addition to the PP chain and the CNO cycle there is another sequence of proton captures that can become efficient in stars, although it does not play a role in the energy budget. In the temperature range 40–50 MK the proton captures listed in the upper part of Table 3.4 quickly bring to their equilibrium values the

Table 3.3 Typical isotopic ratios of the fully-developed CNO cycle of hydrogen burning

Isotopic ratio	Value	Value (solar)
$Y_{^{12}C}/Y_{^{13}C}$	$\simeq$4	89
$Y_{^{14}N}/Y_{^{15}N}$	$\simeq 4 \times 10^4$–10^5	272
$Y_{^{17}O}/Y_{^{16}O}$	$\simeq 10^{-2}$–10^{-3}	3.8×10^{-4}
Y_C/Y_N	$\simeq 7 \times 10^{-3}$–2.5×10^{-2}	3.2
Y_N/Y_O	$\simeq$60–350	0.13
$Y_{^{18}O}/Y_{^{16}O}$	$\simeq 2 \times 10^{-6}$ for $T < 50 : 60\, MK$	2×10^{-3}
$Y_{^{18}O}/Y_{^{16}O}$	Declines to $\simeq 5 \times 10^{-8}$ at T$\simeq$100 MK	2×10^{-3}

Table 3.4 The reactions of the Ne-Na-Mg-Al cycle, and their reaction Q values

Reaction	Q_{tot} (MeV)
$^{20}Ne + p \rightarrow ^{21}Na$	5.979
$^{21}Na \rightarrow ^{21}Ne + e^{+} + \nu$	3.548
$^{21}Ne + p \rightarrow ^{22}Na$	6.739
$^{22}Na \rightarrow ^{22}Ne + e^{+} + \nu$	2.842
$^{22}Ne + p \rightarrow ^{23}Na$	8.794
$^{23}Na + p \rightarrow ^{20}Ne + \alpha$	2.377
$^{23}Na + p \rightarrow ^{24}Mg$	11.693
$^{24}Mg + p \rightarrow ^{25}Al$	6.548
$^{25}Al \rightarrow ^{25}Mg + e^{+} + \nu$	4.277
$^{25}Mg + p \rightarrow ^{26}Al$	6.307
$^{26}Al \rightarrow ^{26}Mg + e^{+} + \nu$	4.004
$^{26}Al + p \rightarrow ^{27}Si$	12.275
$^{26}Mg + p \rightarrow ^{27}Al$	8.271
$^{27}Si \rightarrow ^{27}Al + e^{+} + \nu$	4.812
$^{27}Al + p \rightarrow ^{28}Si$	11.585
$^{27}Al + p \rightarrow ^{24}Mg + \alpha$	1.601

abundances of ^{20}Ne, ^{21}Ne, ^{22}Ne, and ^{23}Na, forming also in this case a *NeNa* cycle. For temperature in excess of 50 MK the ^{23}Na(p,γ)^{24}Mg channel competes with the ^{23}Na(p,α)^{20}Ne so that matter from the NeNa cycle leaks towards more massive nuclei. At temperatures of order of 60 MK also the proton captures listed in the lower part of Table 3.4 fully activate so that all the nuclei between ^{20}Ne and ^{27}Al reach their equilibrium abundances. Is it worth noting that ^{26}Al, a long-lived radioactive nucleus with half life 7.17×10^5 years, is included within this sequence. ^{26}Al can be ejected into the interstellar medium by stellar outflows (winds) and its decay into ^{26}Mg can be detected as diffuse γ-ray emission (Sect. 7.4) when the metastable ^{26}Mg relaxes towards its ground state. Moreover, it can be included in dust grains that form around stars and decay within the already formed minerals. This nucleus is thoroughly discussed in Sect. 3.6.1, and in Chaps. 4, 7 and 9. Typical $Y_{^{26}Al}/Y_{^{27}Al}$ equilibrium ratios produced by H burning range between 3×10^{-2} at 60 MK and 0.8 at 100 MK. We refer the reader to the book by Cox and Giuli (1968) for a derivation of the basic stellar structure equations and detailed discussions of the physics involved in the study of stellar evolution.

Although the sequences of nuclear reactions that power stellar luminosity are now considered to be well understood, a wise Galilean approach suggests to verify experimentally (whenever possible) their occurrence in stars. Our Sun provides a unique opportunity to accomplish such verification via the detection of neutrinos produced by radioactive decay in its deep interior.

3.2.2 The Sun: Neutrinos and the Nuclear Origin of Starlight

The long-term stability of the Solar properties, in particular its luminosity and surface temperature, can be explained only if the solar energy source is of nuclear origin and specifically involves proton captures, which are associated with abundant fuel and a long time scale. As discussed above, such an energy source inevitably results in the production of radioactive nuclei, which decay into their stable daughter nuclei through weak processes, hence emitting neutrinos. The modeling of the internal structure of the Sun predicts a central temperature at present of the order of 15 MK, and hence that the PP chain dominates (99.6%) over the CNO cycle (0.4%) converting H into ^{4}He. The relative importance of the nuclear reactions in the PP chain in the Sun leads to the result that the majority (93%) of the neutrinos produced should come from $p(p, e^+\nu_e)d$ reactions (where d=deuterium, N=Z=1) and be of relatively low energy ($E \leq 0.42$ MeV, see Table 3.1), while only a minor fraction of the total neutrinos is expected to be emitted by the decay of ^{7}Be ($\simeq$7%, $E \simeq 0.86$ MeV) and ^{8}B(0.0075%, $E < 15$ MeV).

Figure 3.1 shows the rates of the nuclear reactions intervening in the PP chain and CNO cycle as a function of the mass coordinate for a 1 $M_\odot$ stellar model of

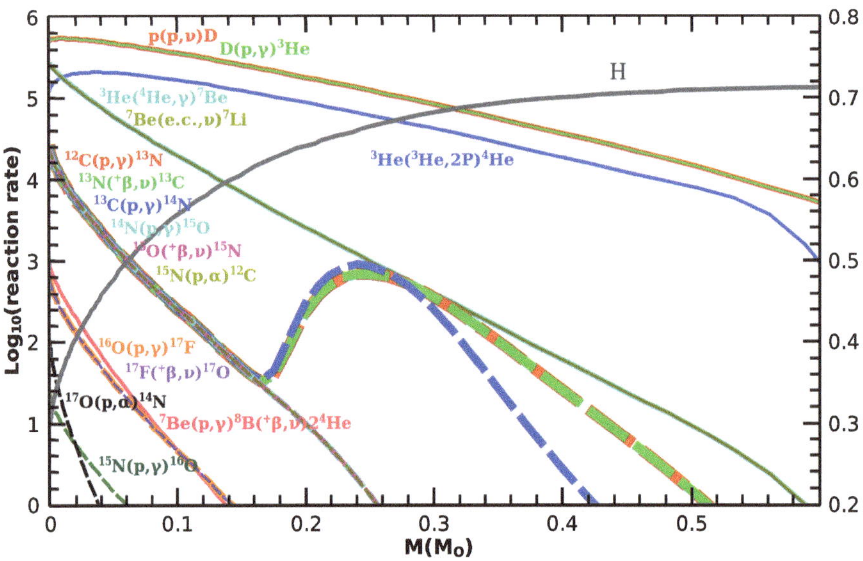

Fig. 3.1 Rates of the reactions involved in the PP chain and the CNO cycle as a function of the mass coordinate in a solar-like stellar model having approximately the age of the Sun of 4.6 Gyr. The H abundance is also plotted and its range shown on the right-side y-axis

solar "metallicity"[2] and an age close to the present age of the Sun, i.e., 4.6 Gyr. All the β^+ decays are concentrated very close to the center where the synthesis of unstable nuclei is concentrated. The relative importance of the various PP reactions, and hence of the associated neutrino fluxes, depend on the rates of the nuclear reactions involved, which, in turn, are a function of the density, temperature, and chemical composition. Since the internal structure of a model of the Sun depends on the adopted and somewhat uncertain input physics—e.g., the nuclear cross sections, the equation of state, the opacity, and the chemical composition of the gas from which the Sun formed—the detection of neutrinos from the Sun is fundamental not only to experimentally verify the nuclear origin of the Solar luminosity, but also to confirm the overall reliability of the modeling of the internal structure of the Sun and the adopted input physics.

It is therefore comprehensible that the quest for the solar neutrinos started early, more than 40 years ago, with the Davis experiment (1967–1985) (Bahcall et al. 1985; Cleveland et al. 1998). This experiment, based on the interaction between an electron neutrino and ^{37}Cl, has a threshold energy of the order of 0.8 MeV and hence could detect predominantly the ^{8}B neutrinos, which constitute a very minor fraction of the neutrino flux coming from the Sun. The Davis experiment provided two basic results, one very encouraging and another one very stimulating. First, it detected solar neutrinos, demonstrating beyond any doubt that proton captures are occurring in the interior of the Sun. Second, the detected neutrino flux was roughly one third of the predicted value. Such a result stimulated much further work and a huge amount of papers on this puzzle piled up over the decades. The discrepancy was considered by many physicists as a strong indication that the basic modeling of the Sun was wrong though, however, it must be reminded that it was confined to a very minor branching of the PP chain. The discrepancy became more serious with the advent of the GALLEX experiment (Hampel et al. 1998)—a collaboration among France, Germany, Italy, Israel, Poland, and USA, headed by MPIK Heidelberg, 1991–1997—and the SAGE experiment (Abdurashitov et al. 1999)—a Russian-American collaboration, 1990–2000. These modern sophisticated experiments were designed to detect the bulk of the neutrinos produced in the Sun, i.e., the low energy neutrinos produced by the p+p reaction. They confirmed both the detection of a firm neutrino signature from the Sun and the existence of a discrepancy between theoretical and observed fluxes. This result plunged the basics of solar modeling into a deep crisis because these experiments were sensible to the total number of electron neutrinos emitted by the Sun. This number is theoretically extremely robust, depending only on the basic assumption that the solar luminosity comes from the conversion of protons into α particles, and not on the details of the modeling of the internal structure of the Sun or of the cross sections of the nuclear reactions involved.

[2]The term metallicity indicates the abundance of metals in a star, where metals corresponds to all elements heavier than He. The metallicity of the Sun is $\simeq$0.02 by mass fraction, where abundances are normalised to a total of 1, which means that 2% of the solar matter is made up of elements heavier than He. The most abundant of these is oxygen, followed by carbon.

If the solar luminosity is powered by the conversion of protons into α particles, the total number of electron neutrinos emitted per second by the Sun must be 2.38×10^{39} ($L_{\odot}$ in MeV/s)/25 (energy provided per α nucleus in MeV) $\times$ 2 (number of ν_e produced per α nucleus) [ν_e s^{-1}], which corresponds, at one astronomical unit, to a flux equal to 6.78×10^{10} ν_e s^{-1}. Instead, the number of neutrinos detected by GALLEX and SAGE was half this predicted value.

One of the possible solutions of this puzzling result was identified since the 1960s (Gribov and Pontecorvo 1969) in the possible oscillation of neutrinos of different flavors, ν_e and ν_μ (ν_τ was discovered in 1975), a phenomenon that would have implied a non zero mass for the neutrinos. Such a solution was not considered very seriously up to the middle of the 1980s because the fit to the observed rate of the solar neutrinos would have required a very efficient oscillation among the three neutrinos flavours, a possibility thought to be quite improbable. However, Wolfenstein (1978) first and Mikheyev and Smirnov (1985) proposed what is presently called the Mikheyev-Smirnov-Wolfenstein (MSW) effect, i.e., that under proper conditions, in particular at high densities, the oscillation of the neutrinos could be strongly enhanced as a result of a coherent forward scattering when passing through matter. Such a revolutionary insight was among the reasons that led to the birth of the Sudbury Solar Neutrino Observatory (SNO) experiment (a Canadian, USA, and UK collaboration that started in 1985, McDonald et al. 2002), designed to detect all the three flavors of neutrinos. The adopted strategy was to catch neutrinos via three reactions:

$$\nu_{e^-} +^2 H \rightarrow p + p + e^-$$

$$\nu_x +^2 H \rightarrow p + n + \nu_x$$

$$\nu_x + e^- \rightarrow \nu_x + e^-$$

where x stands for e^-, μ or τ. The activation of these reactions required the building of an extremely clean detector filled with 1000 tonnes of heavy water. Data were collected in various phases between 1999 and 2006. Though the energy threshold for this experiment was around 6 MeV, so that only neutrinos emitted from the minor branch ^{7}Be(p,γ)^{8}B(e^+ ν_e)^{8}Be could be detected, this experiment showed unequivocally that neutrinos oscillate and put strong constraints on their physics. A proper inclusion of this result on the interpretation of the neutrinos emitted by the Sun led to a much better agreement between the expected and the observed neutrinos fluxes.

In spite of the great results obtained by the SNO experiment, it remained the fact that its very high energy threshold prevented the detection of the bulk of the neutrinos, that are emitted with much lower energies, of the order 200–300 keV. The following big step forward in the direction of both a better understanding of our Sun and also of the physics of the neutrinos was done with the advent of Borexino, an experiment designed to detect neutrinos over a large energy spectrum via the $\nu_x + e^- \rightarrow \nu_x + e^-$ reaction (where again x stands for e^-, μ or τ). It consists of an unsegmented liquid detector featuring 300 tonnes of well shielded ultra pure scintillator viewed by 2200 photomultipliers. It was designed to drastically reduce the energy threshold to extremely low energies: roughly 250 keV. This experiment is located in Hall C of the Laboratori Nazionali del Gran Sasso in Italy and data started to be collected since 2007 and continued up to 2013. At present all three neutrinos fluxes produced by the PP chain where measured with unprecedented accuracy: $\Phi(PP) = 6.6 \pm 0.7 \times 10^{10}\,cm^{-2}\,s^{-1}$ (BOREXINO Collaboration et al. 2014), $\Phi(^7Be) = 5.18 \pm 0.51 \times 10^9\,cm^{-2}\,s^{-1}$ (Arpesella et al. 2008) and $\Phi(^8B) = 2.4 \pm 0.4 \times 10^6\,cm^{-2}\,s^{-1}$ (Bellini et al. 2010). These fluxes are in good agreement with the latest models of the Sun. For example Serenelli et al. (2011) published a very refined standard solar model (SSM) computed for two different initial chemical compositions of the Sun, i.e., Asplund et al. (2009) and Grevesse and Sauval (1998). For the Φ(PP) they obtain 6.03 ± 0.006 and 5.98 ± 0.006 ($10^{10}\,cm^{-2}\,s^{-1}$) in the two cases, respectively. These two fluxes are very similar to each other and in excellent agreement with the one measured by the Borexino collaboration. As for the ^{7}Be branch, the two predicted fluxes are $\Phi(^7Be)= 4.56 \pm 0.07$ and 5.00 ± 0.07 ($10^9\,cm^{-2}\,s^{-1}$). In this case the flux predicted by the SSM computed by assuming the metallicity of Grevesse and Sauval (1998) is closer to the detected one. As for the $\Phi(^8B)$ the SSM predicts, for the two initial metallicities, 4.59 ± 0.14 and 5.58 ± 0.14 ($10^6\,cm^{-2}\,s^{-1}$) and in this case the flux predicted by adopting the metallicity derived by Asplund et al. (2009) is the closest to that detected (even if somewhat higher).

The Borexino experiment was also able to detect neutrinos emitted by the very inefficient reaction $p + p + e^- \;=^2 H + \nu_e^-$ (pep), and also in this case the predicted and detected fluxes converge towards a similar value: $\Phi(pep)=1.46 \pm 0.01(10^8\,cm^{-2}\,s^{-1})$. Though it was not yet possible to detect neutrinos emitted by the processes involved in the CNO cycle, this experiment was capable to put stringent limits to their fluxes.

3.2.3 Solar Flares and Radioactivities

The Sun and the phenomenon of flaring high-energy particle and photon emission shows that stellar activity also includes episodes with transient acceleration of particles. During solar flares, positron annihilation γ-rays had been discovered, with the characteristic 511 keV line (Chupp 1971). This is observational proof of radioactive isotopes being created by energetic particle interactions in the upper stellar atmosphere. This results from spallation reactions that produce intermediate-mass isotopes from Fe, as well as light-isotope nuclear reactions from protons and He isotopes, where the Coulomb barrier for nuclear reactions can be overcome, and from (secondary) neutron reactions.

The resulting radioactive species that are expected are listed in Table 3.5. Many nuclear lines occur during the *prompt* phase of flares as a result of nuclear excitation, as flare-accelerated nucleons collide with nuclei in the upper atmosphere.

Table 3.5 Radioactivities in solar flares, sorted by ascending radioactive mean lifetime in hours (from Table 1 in Tatischeff et al. 2006)

Isotope	Lifetime [h]	β^+	Isotope	Lifetime [h]	β^+
^{13}N	0.24	*	^{43}Sc	5.61	*
^{11}C	0.49	*	^{44}Sc	5.73	*
^{52}Mn	0.51	*	^{52}Fe	11.94	*
^{60}Cu	0.57	*	^{58}Co	13.04	–
^{34}Cl	0.77	*	^{24}Na	21.58	–
^{47}V	0.78	*	^{55}Co	25.29	*
^{63}Zn	0.93	*	^{55}Co	25.29	*
^{49}Cr	1.02	*	^{57}Ni	51.36	*
^{51}Mn	1.11	*	^{52}Mn	193.59	*
^{18}F	2.64	*	^{48}V	553.08	*
^{56}Mn	3.72	–	^{7}Be	1842.73	–
^{45}Ti	4.44	*	^{58}Co	2453.50	*
^{61}Cu	4.81	*	^{56}Co	2674.17	*

The last column identifies the isotopes contributing to the gamma-ray emission from positron annihilation with the characteristic line at 511 keV

Characteristic nuclear and positron annihilation emission from radioactive species would be delayed, and therefore recognised to be present after the prompt flaring phase, although many of these characteristic lines occur both prompt and from radioactivities (Dennis et al. 2007). Measuring the radioactive production, a study of the mixing of different isotopes in the outer stellar envelope thus could be inferred from characteristic γ-ray lines (Ramaty et al. 1995; Murphy and Share 2005; Murphy et al. 2005; Tatischeff et al. 2006). The particle acceleration arises from reconnection events in the magnetic field higher up in the solar corona (see review by Aschwanden 2008). The radioactive emission thus also directly relates to the energy content of the solar flare itself (Kozlovsky et al. 2002).

3.3 Evolution After the Main Sequence: The First Giant Branch

In Sect. 3.3.1 we discuss the main evolutionary properties of stars once they leave the long lasting phase of central H burning described above, commonly referred to as the Main Sequence, and enter the phase known as the First, or Red, Giant Branch (RGB). At the end of central H burning a star is composed of a H-exhausted core made primarily of He and a H-rich envelope. Hydrogen burning shifts from the center to the base of the H-rich mantle while the envelope expands causing the surface of the star to reach radii 10–1000 times the solar radius and to cool down to a few thousand K. This expansion triggers the formation of large scale convective motion extending from the surface down to deep regions in the star where partial H burning occurred during the Main Sequence. Some products of this H burning are thus brought to the surface in a process known as the 1st dredge-up. In this phase the He core grows in mass because the H-burning shell continuously converts H-rich matter into He-rich matter and deposits the ashes onto the He core. The temporal evolution of the He core depends on its initial size, i.e., the size it had just after the central H exhaustion, which is in turn mostly determined by the initial stellar mass. If the mass of the He core is less than $\simeq$0.35 $M_\odot$, which occurs for initial stellar masses less than $\sim$2 $M_\odot$, an *electron degenerate core* forms where matter reaches such extraordinarily high density, up to $\simeq 10^6$ g/cm^3, that the dominant contribution to its pressure arises from the Pauli exclusion principle, which prevents the electrons from occupying identical quantum states. This leads to an increase of the lifetime of this phase up to about 100 Myr, and forces the subsequent He ignition to occur quite far from the center. If the He core is instead more massive than 0.35 $M_\odot$, the electrons remain far from the degeneracy regime.

In Sect. 3.3.2 we discuss the conditions under which ^{7}Li, the stable daughter of radioactive ^{7}Be, may be produced, preserved, and brought to the stellar surface. This nucleus is typically destroyed in the PP chain (Table 3.1), because its destruction rate is efficient at temperatures lower than its production rate. A way to produce ^{7}Li was proposed in 1971 by Cameron and Fowler (1971): if freshly synthesized ^{7}Li is quickly brought to very low temperatures by mixing, then it can be preserved. If H burning occurs in a convective environment it is in principle possible to find a high Li abundance on the surface of a star, as observed in some stars belonging to the First Giant Branch. However, these observations are in fact difficult to explain because H burning occurs in a formally stable region well below the base of their convective envelopes. Additional mechanisms of mixing must be invoked to bring ^{7}Li-rich material into the convective envelope.

3.3.1 The First Giant Branch

During the Main Sequence phase of stellar evolution described in the previous section conversion of H into He via H burning in the centre of the star leads to a progressive increase of the mean molecular weight combined with a decrease of the amount of available fuel. The net result is a slight increase of the luminosity (because L scales with the fourth power of the molecular weight, Eq. (3.10) in Sect. 3.2.1), and a mild expansion of the surface of the star because of the formation of a molecular weight gradient (Stancliffe et al. 2009). Once H is exhausted in the central region of the star, the H-exhausted core, or He core, begins to contract on a gravitational timescale while the region of active nuclear burning smoothly shifts above the He core, where H is still abundant. Further evolution of the He core depends on its mass, which, in turn, depends on the initial total mass of the star. If the He-core is more massive than a threshold value of $\sim$0.35 $M_\odot$, which happens for an initial total mass of $\sim$2 $M_\odot$, its contraction induces strong heating of the He core itself, which quickly reaches a temperature of $\sim$100 MK at which fusion reaction of α particles (He burning) is activated. If, instead, the core is less massive than $0.35\,M_\odot$ the high densities reached render the electron gas degenerate, hence supporting the structure against gravity without the need for additional contraction.

This difference has a large impact on further evolution of the star because, in the latter case, the He core tends towards an almost isothermal configuration due to the large mean free path of degenerate electrons relative to that of photons. If the structure was isolated, as in the case of white dwarves, it would progressively cool down losing its stored energy through the surface. Instead, in the case discussed here, the degenerate structure heats up because it is surrounded by the H-burning shell, which continuously deposits freshly synthesized He onto the He core. The rate at which the maximum temperature increases with time in the degenerate He core is controlled by the growth rate of the He-core mass, which obviously coincides with the rate at which the H-burning shell converts H into He. Strong neutrino production (Itoh et al. 1989) in the center of the electron degenerate core, due to the interaction of photons with the plasma and/or to the scattering of photons on electrons, carries away energy from the core pushing the location of the maximum temperature outward in mass. The off-center location of the maximum temperature is the result of the balance between energy loss due to the neutrino emission, which scales directly with the density and pushes the maximum temperature outward, and the energy gain due to the compressional heating, which scales inversely with the density and pushes the temperature maximum back towards the center. The key stellar parameters that control the location of the maximum temperature are the CNO abundance and the initial mass of the star. The higher the CNO abundance, the faster the conversion of protons into α particles in the H-burning shell, the stronger the heating of the degenerate He core, and the closer the maximum temperature is to the center. The higher the initial mass of the star, the lower is the degree of electron degeneracy and the density in the He core, and hence the efficiency of neutrino production.

While the H-burning shell influences the evolution of the He core, the growth of the He core influences the evolution of the H-burning shell as well. In fact, the progressive heating of the core raises the temperature at the surface of the core, where H burning occurs. This results in a continuous positive feedback: the H burning shell deposits He onto the He core, which therefore heats up. Such a heating leads to an increase of the temperature and density in the H-burning shell, accelerating the H burning rate and increasing the conversion rate of H into He, and therefore the heating of the He core. As a consequence, the progressive increase of the H burning rate determined by the growth of the He core mass, forces the H rich mantle of the star to expand and to cool. The cooling of the stellar envelope triggers a large increase of the opacity because matter enters the partial ionization regime. The temperature gradient steepens, favoring the growth of convective instabilities that very quickly extend over a major part of the H-rich mantle from the surface down to near the outer border of the H-burning shell. A consequence of the growth of these convective motions within most of the H rich mantle is a large increase of the surface luminosity caused by the continuous increase of the H burning rate coupled to the fact that the convective envelope does not absorb or release energy but just transporting it outward.

Since convective motions play a fundamental role in the physical and chemical evolution of any star, we briefly sketch the basic physical reason that leads to the growth of these large scale motions. The equilibrium condition provided by counterbalancing pressure gradients and gravity in stars does not necessarily imply stationary matter: a bubble of stellar matter may be considered stable against motion if a restoring force pushes it back towards its rest position when, for any reason, it is slightly displaced from its equilibrium location. Such a restoring force is simply given by Archimede's force, i.e., it depends on the density contrast between that of the environment and that of the bubble. If the density of an element of matter displaced towards a lower/higher density region turns out to be even lower/higher than that of its new surroundings, the element will continue to *raise/sink* and move away from its rest position, otherwise it will move back towards its equilibrium location.

Changes in the physical structure of the bubble during its motion play an important role in determining its density and thus its behavior. Mechanical equilibrium with the environment is certainly well verified so that it can be safely assumed that the internal pressure within the bubble instantaneously readjusts to that of the environment. More difficult is to determine the amount of heat that the bubble can exchange with the environment while moving. In the simplest case in which the bubble does not exchange any heat with the surroundings until it has covered a certain distance (adiabatic approximation), and assuming that the region is chemically homogeneous, the critical condition for the onset of large scale motions of the matter is that the temperature gradient of the environment must exceed the adiabatic gradient (Schwarzschild criterion). While the radiative temperature gradient remains less than the adiabatic temperature gradient, an element of matter will remain more/less dense than its surroundings if displaced towards less/more dense regions (within stars these displacements are typically connected to movements outward/inward in mass), and hence it will experience a restoring force that will keep it anchored to its rest location. On the contrary, when the radiative temperature gradient exceeds the adiabatic temperature gradient stochastic motion of the matter is not hampered by a restoring force, but it is amplified leading to the growth of large scale motions. Hence, convective regions are associated with steep temperature gradients, which typically occur either close to regions where energy production is strongly concentrated, or in regions where the mean free path of the photons, which scales with the inverse of the opacity, becomes so small that radiation energy transport becomes inefficient.

The determination of the temperature gradient in convective regions is quite complex: here it suffices to say that while in the interior of a star the temperature gradient in a convective region remains very close to the adiabatic gradient, in a convective envelope the temperature gradient becomes much steeper (intermediate between the radiative and adiabatic case) because the low density in the outer envelope makes energy transport by convective eddies inefficient, so that both photons and eddies contribute to the outwards transport of thermal energy.

Since convective eddies have a very large mean free path with respect to that of photons, convection is a very efficient energy transport mechanism. In the specific case of extended convective motions that form above the H-burning shell in Red Giant stars, energy transport is so efficient that virtually all the energy produced by the burning shell is transmitted to the surface without essentially any absorption by the convective layers. It follows that a star in the H-burning shell evolutionary phase is forced to increase in size to be able to get rid of the extra energy influx, while the drop of the surface temperature is limited by the presence of a maximum temperature gradient: the adiabatic temperature gradient, which cannot be overcome by much in the largest fraction of the envelope mass.

The mere existence of stars in the RGB phase constitutes evidence of (a) the presence of an active H-burning shell, demonstrated by the breaking of the mass-luminosity relation $L \propto M^3$ that holds during the Main Sequence phase, (b) the presence of a maximum temperature gradient, demonstrated by the only minor change of the surface temperature along the RGB, (c) the continuous increase of the energy production by the H-shell burning, demonstrated by the continuous increase of the surface luminosity, and (d) the presence of an electron degenerate core (for stars with initial mass less than $\simeq 2\,M_\odot$), demonstrated by the existence of a relatively long lasting, $\sim 10^8$ years, and thus observable RGB phase, which would be prevented if the He core was gravitationally contracting.

Soon after the formation of the H burning shell, the large scale motions that grow in the H-rich envelope and rapidly extend from the surface down to just above the top of the H-burning shell, bring to the stellar surface matter partially processed by proton-capture reactions during the Main Sequence phase. This mixing, referred to as the 1st dredge-up, modifies the stellar surface composition. The amplitude of these modifications depends on the initial stellar mass and metallicity, the general rule being that the amplitude of the changes of the surface composition scales with the initial stellar mass, directly up to $3\,M_\odot$ and then inversely for higher masses, and inversely with the metallicity. Figure 3.2 shows the abundance profiles of several nuclear species as a function of the mass location for a solar-like stellar model evolved to the RGB, just before the convective envelope deeply penetrates into the star. The solid vertical line shows the maximum inward penetration of the convective envelope. Since the convective motions reach layers in which the local chemical composition has previously been modified by nuclear burning, also the surface chemical composition is modified by the mixing induced by these large scale motions. In particular the surface He abundance is slightly increased by 0.02 dex, ^{3}He increases by one order of magnitude, ^{7}Li is destroyed, the ^{12}C/ ^{13}C ratio drops from the solar value of 89 to roughly 30, the ^{14}N/ ^{15}N ratio increases from the solar value of 272 to 500 while the oxygen isotopic ratios and those of heavier nuclei remain at their solar values. A detailed quantitative determination of these changes depends on the specific stellar model considered.

The evolution of the star after the 1st dredge-up is characterized by the H-burning shell progressively converting H from the convective envelope into He, which is deposited onto the inert He core. The continuous mass transfer from the envelope to the core progressively reduces the mass of the envelope while its chemical composition does not change any more because the temperature within the convective envelope is too low to activate nuclear reactions.

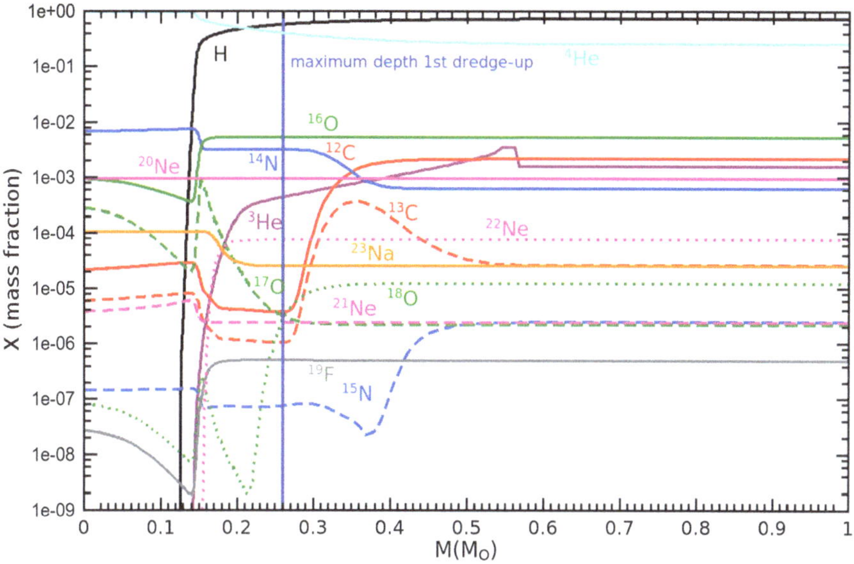

Fig. 3.2 Snapshot of the abundances of several nuclear species in a solar-like stellar model just before the onset of the 1st dredge-up. The maximum inward penetration of the convective envelope during the 1st dredge-up is marked by the vertical solid blue line

The evolution along the RGB ends when the maximum temperature in the core is high enough, $\simeq 100$ MK, to activate the burning of He via 3α reactions, during which three α particles join into a ^{12}C nucleus. If the pressure is dominated by degenerate electrons the energy released by these reactions cannot be immediately balanced by an expansion of the core. Hence, He ignition occurs through a series of *flashes*, which progressively remove the degeneracy, shifting the burning towards the center. Once the electron degeneracy is fully removed, a quiescent central He-burning phase settles in.

All along the complex, and partly still mysterious, RGB evolutionary phase that links central H to central He burning, radioactive nuclei are produced by H-shell burning mainly via the CNO cycle. Most of them, however, have negligible lifetimes, so they could only be detected through the neutrinos they emit. Unfortunately, no Red Giant star is close enough to the Earth to allow the detection of neutrinos of nuclear origin produced in its interior. However, there are two unstable nuclei, ^{7}Be and ^{13}N, whose half life may be comparable or even larger than some stellar timescales: for ^{7}Be the half life is comparable to the envelope mixing turnover time, for ^{13}N the half life is comparable to proton-capture timescale in extremely metal-poor stars, because stars of lower metallicity are more compact and hotter, due to their lower opacity.

In the next section we discuss specifically the abundance of ^{7}Li, the stable daughter of ^{7}Be, in giant stars, which could provide important clues about the presence of additional motions extending below the base of the convective envelope.

This is important because, as we described above, the modeling of large scale motions within stars is still crude and their growth, timescale, and efficiency not yet well understood.

3.3.2 The Production of Li

Lithium (Li) isotopes[3] in stars are fragile as they are easily destroyed by proton-capture reactions once the temperature exceeds 3 MK. The destruction timescale drops from 100 Myr at 3 MK to only 0.3 Myr at 5 MK while their production through fusion reactions only occurs at much higher temperatures, between 10 MK and 25 MK. The lower limit is due to the fact that the synthesis of ^{7}Li is initiated by the ^{3}He(α,γ)^{7}Be reaction, which becomes efficient only at temperature of the order of 10 MK, while the upper value is due to activation of the ^{7}Be(p,γ)^{8}B reaction, which overcomes the electron capture reaction ^{7}Be(e$^-$, ν)^{7}Li above a temperature of the order of 25 MK. Hence, Li is efficiently produced in a temperature range where it is also efficiently destroyed and therefore there seems to be no room for Li production in a star. However, there are a number of Red Giant stars observed to be Li rich (Castilho et al. 2000; Balachandran 2005; Uttenthaler et al. 2007).

A possible way out of such a puzzling situation was recognized by Cameron and Fowler (1971) and is based on the idea that instabilities, such as convection, rotation-induced instabilities, thermohaline mixing, etc., may bring freshly made ^{7}Be from its production site to more external regions, where the temperature is low enough to inhibit proton captures on ^{7}Li, on a timescale shorter than that of electron capture of ^{7}Be. Note that the electron capture rate of ^{7}Be shows a mild increase as the temperature decreases.

A typical environment in which the Cameron-Fowler mechanism operates is during the Asymptotic Giant Phase (AGB) phase (Sect. 3.4.1), if the star is more massive than 4:5 M$_\odot$. These stars develop large scale motions in the H-rich mantle that extend from the surface down to regions where the temperature is high enough (>40 MK) for some nuclear burning to occur (Hot Bottom Burning), in particular via the ^{3}He(α,γ)^{7}Be reaction. Figure 3.3 shows the cumulative turnover time from the base of the convective envelope to the region of temperature T given in the abscissa for a 6 M$_\odot$ star of solar metallicity sometimes after the beginning of the AGB phase. The horizontal dashed grey line marks the typical timescale of the ^{7}Be(e$^-$, ν)^{7}Li reaction while the vertical dotted grey line shows the threshold temperature below which the timescale of the proton capture on ^{7}Li becomes larger than 300 Myr. In this environment ^{7}Be produced above 10 MK is successfully transferred before decaying to a region where its daughter ^{7}Li can survive.

[3]Lithium has two stable isotopes ^{6}Li and ^{7}Li, of which ^{7}Li is the more abundant representing 92% of solar Li.

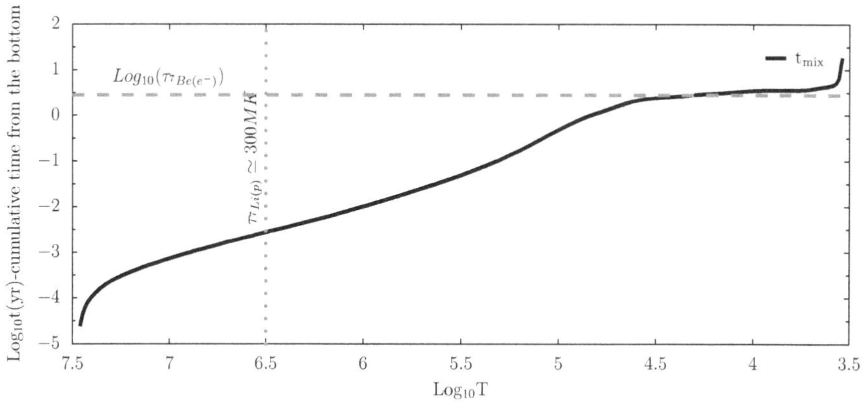

Fig. 3.3 The cumulative turnover time of the convective eddies computed from the base of the convective envelope up to the surface as a function of temperature, T. The star is a 6 $M_{\odot}$ star of solar metallicity some time after the beginning of the AGB phase. The horizontal dashed grey line marks the typical timescale of the $^{7}Be(e^{-}, \nu)^{7}Li$ reaction while the vertical dotted grey line shows the threshold temperature below which the timescale of the proton capture on ^{7}Li becomes larger than 300 Myr

An increase of the Li abundance at the surface of RGB stars is more difficult to achieve. Though the turnover time scale within the convective envelope is also in this case rather short ($\simeq$1 year at $Log_{10}(L/L_{\odot}) \simeq 2$), the temperature at the base of the convective envelope always remains well below 5 MK, too low to lead to an appreciable production of ^{7}Be. Nonetheless, observations show the existence of a small number of Li-rich RGB stars (Castilho et al. 2000; Balachandran 2005; Uttenthaler et al. 2007). Figure 3.4 shows the internal structure of the region around the H-burning shell from a solar-like stellar model on the RGB. Here, ^{7}Be is synthesized well below the region where large scale motions of the matter and hence mixing of the chemical composition occur. In this environment the Cameron-Fowler mechanism could operate only by assuming the presence of presently unidentified instabilities able to drive some mixing between the region rich in ^{7}Be and the base of the convective envelope. The main constraint on this *extra* mixing is that it must get close enough to the active H-burning shell to reach the layers where the ^{7}Be production occur, but it must not enter the region of the main nuclear burning. The reason is that the speed at which a star climbs along the RGB is regulated by the speed at which H is converted into He (see above) which, in turn, also depends on the amount of fuel that continuously enters the burning region. If this extra mixing reached the active burning region, it would inevitably bring fresh H into the burning region, therefore altering the rate at which H is consumed by the H-burning shell and hence the timescale of evolution along the RGB. This evolutionary timescale is observationally well established from counting the number of stars on the RGB in many Galactic globular clusters, and already very well reproduced by current models of these stars without extra mixing.

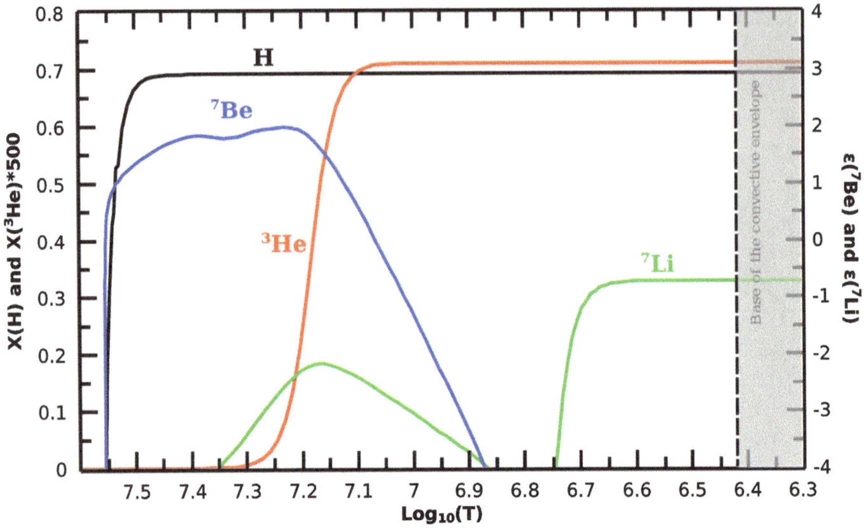

Fig. 3.4 Abundance profiles for H, ^{3}He, ^{7}Be, and ^{7}Li as function of the internal temperature in the region between the H-burning shell and the base of the convective envelope (grey area) for a solar-like stellar model on the RGB. The abundances of both ^{7}Be and ^{7}Li are given in the widely adopted logarithmic scale in which $\epsilon(X) = \mathrm{Log}_{10}(N_X/N_H) + 12$, where N_X and N_H represent the abundances of element X and of hydrogen by number. In this scale the hydrogen abundance is equal to 12

There are other hints that point towards the presence of extra-mixing phenomena in RGB stars (and perhaps in AGB stars too, as discussed in Sects. 3.6.1 and 3.6.2). The observed surface ^{12}C/^{13}C ratio and N abundance are, respectively, too low and too high with respect to the values predicted by the 1st dredge-up. Extra mixing would naturally lower the first ratio and raise the N abundance, this being the signature of H burning. A deeper mixing than predicted by current models would also reduce the abundance of ^{3}He in the stellar envelope, which is increased by the 1st dredge-up, by bringing this nucleus down to regions where it is destroyed. This reduction is needed to avoid an increase of the ^{3}He abundance in the interstellar medium, which is not observed, due to the material expelled by low-mass stars over the lifetime of the Galaxy.

Figures 3.5 and 3.6 show the isotopic abundances of several nuclei up to Al within a solar-like star while climbing the RGB (at $\mathrm{Log}(L/L_\odot) \simeq 3$). Note that we chose to use the temperature as the abscissa instead of mass to better clarify the temperature at which each nuclear species varies. The figures clearly shows that, for each given depth reached by an extra mixing process, a few nuclei are expected to be modified. For example, a drop of the oxygen abundance at the surface of an RGB star due to an extra mixing process (the depth reached by the extra mixing should extend down to at least 40 MK in this case), would also imply an increase of the surface abundances of both N and Na. Isotopes like ^{18}O and ^{22}Ne are expected to be fully destroyed, while the ^{12}C/^{13}C ratio should drop and the ^{14}N/^{15}N increase.

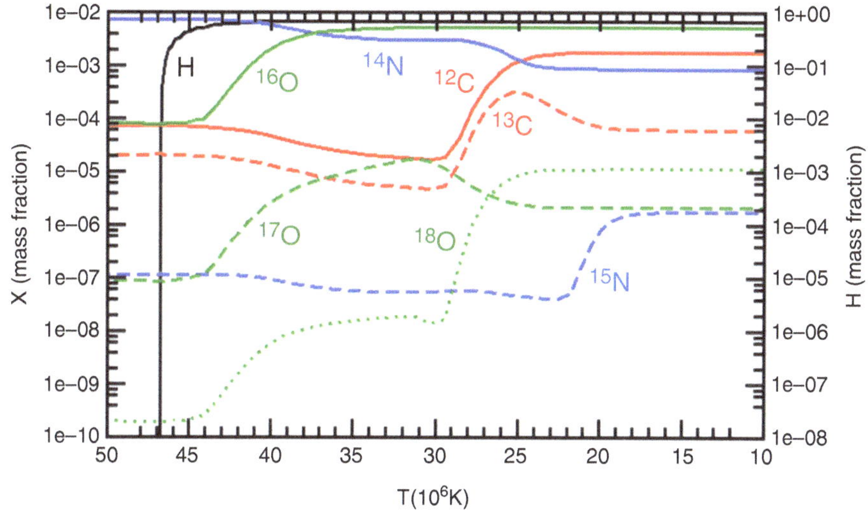

Fig. 3.5 Abundance profiles of the CNO isotopes as function of the temperature on the RGB in a solar-like stellar model

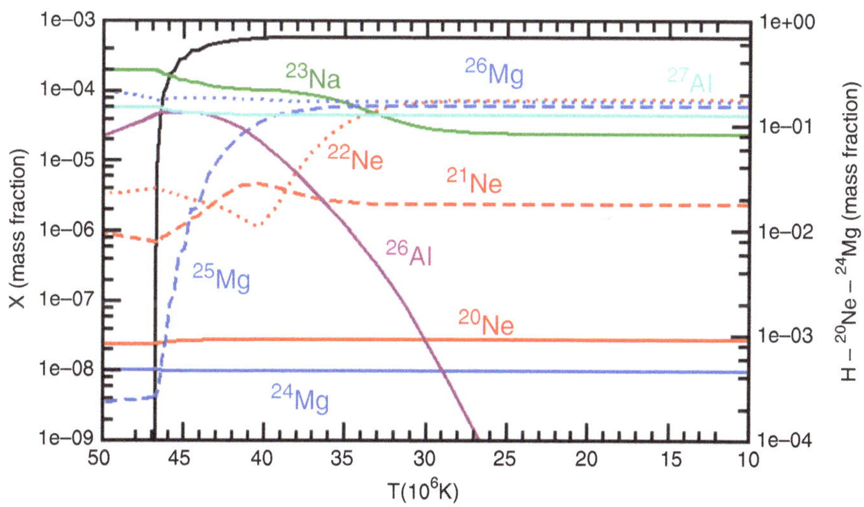

Fig. 3.6 Abundance profiles of the Ne, Na, Mg and Al isotopes as function of the temperature on the RGB in a solar-like stellar model

Note that in any case it would be very difficult to obtain a surface change of the Ne and the Mg abundances because their most abundant isotopes, ^{20}Ne and ^{24}Mg, are not modified unless the mixing reaches down to the location of main H burning.

In summary, our modeling of mixing in stars is still oversimplified and unrealistic as it is based on a simple buoyancy model. Observational evidence of stellar abundances also involving radioactive nuclei and their daughters points out that mixing of matter outside the standard convective boundaries should occur in stars. These observations can be used to improve our description of mixing phenomena in stars.

3.4 Evolution in the Double Shell Burning Phase

We start Sect. 3.4.1 by describing the central He-burning phase and the direct scaling of the mass of the convective core resulting from He burning with the mass of the He core. The mass of the convective core determines the size of the initial He-exhausted core, an important parameter for subsequent evolutionary phases. As it happens previously when H burning shifts from the centre to a shell, also when He is exhausted in the core and He burning shifts from the centre to a shell, the envelope is forced to expand and convective motions extend from the external layers deeply inward into the star. In stars more massive than 4:5 $M_\odot$ the convective envelope even penetrates within the He core reducing its mass size and carrying to the stellar surface material processed by nuclear reactions (2nd dredge-up). If the He-exhausted core does not grow above $\sim$1.1 $M_\odot$, an electron degenerate core forms again, this time made of C and O, on top of which are located two burning shells: the He-burning and the H-burning shells. This marks the beginning of the double burning shell phase: the Thermally Pulsing Asymptotic Giant Branch (TP-AGB) phase.

The two key features of this phase are that (1) the two burning shells can not be simultaneously active, but they alternate within a cycle in producing the required energy and (2) He ignition within each cycle occurs through thermal runaway (or thermal pulse, TP) that ends when enough energy is injected in the He-burning zone to convert its temperature and density profiles into a configuration that allows stable burning. The frequency of these thermal instabilities scales directly with the He-core mass. Such an abrupt injection of a quite large amount of energy ($\sim$$10^{48}$ erg) induces first the growth of a convective shell within the zone between the two shells (intershell) and second, soon after this convective region is extinguished, an expansion of the base of the H-rich envelope forces the convective envelope to penetrate well within the intershell zone (3rd dredge-up). The combination of these two successive convective episodes allows nuclei freshly synthetized by He burning to be carried up to the stellar surface. Moreover, the temperature at the base of the convective envelope scales directly with the He-core mass, and, in stars more massive that 4–5 $M_\odot$, reaches high enough values that H burning activates (Hot Bottom Burning, HBB).

In Sect. 3.4.2 we discuss Super-AGB stars, i.e., stars with initial mass in the interval between stars that develop an electron degenerate core after He is exhausted in the center and enter the AGB regime, and more massive stars that do not develop an electron degenerate core. Super-AGB stars ignite carbon out of center in semidegenerate conditions and go through a central C-burning phase. However, the

C-exhausted core is not massive enough to heat up to the Ne burning, so an electron degenerate ONeMg core forms. These stars then go through a thermally pulsing phase. The final fate of these stars depends on the capability of their ONeMg core to reach the critical mass of $\sim$1.35 $M_\odot$ required to activate electron captures on ^{24}Mg and ^{20}Ne. Stars with a core that does not reach this critical mass lose all their H-rich envelope and end their life as ONeMg white dwarves, while stars with a core that reaches this critical mass explode as *electron capture supernovae*.

We continue by briefly discussing mass loss during the AGB phase in Sect. 3.4.3. The strong increase in surface luminosity, coupled to luminosity variations and formation of dust grains in the atmospheres of AGB stars, strongly enhances the mass-loss rate in this phase with the consequence that all AGB stars lose their H-rich envelope, leaving behind the naked electron degenerate core as a cooling CO white dwarf. Finally, in Sect. 3.4.4, we discuss the different species of dust grains that form in the atmosphere of an AGB star. The key role here is played by the C/O number ratio in the atmosphere because the strong bond of the CO molecule results in trapping all of the atoms of the least abundant of the two elements. In an oxygen-rich gas (O>C) the species of dust are, for example, Al_3O_2 (corundum) and many different types of silicates (SiO, SiO_2, etc). In a carbon-rich gas (C>O), the species of dust are, for example, SiC (silicon carbide) and C itself (graphite). Some of this stellar AGB dust is now recovered from primitive meteorites, representing a real speck of an ancient AGB star under our control in the laboratory.

3.4.1 Asymptotic Giant Branch (AGB) Stars

As anticipated at the end of Sect. 3.3.1, once the central temperature in a RGB star exceeds 100 MK, He in the core starts being converted into ^{12}C via 3α reactions, and subsequently into ^{16}O via ^{12}C(α, γ) ^{16}O reactions. The cross section of the 3α reaction has a tremendous dependence on the temperature: it scales roughly as T^{23} in the range 100–300 MK, so that the energy produced by these reactions is very strongly concentrated towards the centre of the star where the temperature is at its maximum. The very large photon flux that forms in these conditions triggers the formation of large scale motions of the matter, which turn the material in the central part of the star (the convective core) in order to efficiently carry the energy outward. The mass of the convective core depends on the luminosity produced by the 3α reactions. This luminosity scales with the mass of the He core because the larger its mass, the larger is the amount of energy required to maintain the hydrostatic equilibrium (see Sect. 3.2). Hence, the size of the convective core scales directly with the mass of the He core. The mass of the He core, in turn, scales directly with the initial mass of the star, thus, in conclusion, the mass of the convective core scales directly with the initial mass of the star. Analogously to the Main Sequence central H-burning phase, the energy production in the core is dictated by the mass of the star (see Sect. 3.2.1). However, the role played by the total stellar mass in central H burning is in the central He-burning phase replaced by the He-core mass because the density contrast between the He core and the H-rich mantle is so large that the core does not feel the presence of the H rich mantle and evolves as if it was a naked He core.

In the meantime, the temperature at the He/H interface is high enough that also an efficient H-burning shell is active leading to continuous deposition of fresh He onto the He core. Moreover, large convective motions develop (in most cases) in the outer H-rich envelope. The actual extension and temporal variation of these convective regions depends on the initial mass and chemical composition of the star.

At variance with H burning, no radioactive nuclei are produced by the 3α and the $^{12}C(\alpha, \gamma)$ ^{16}O reactions because they convert matter along the valley of β stability. Radioactivity during He burning is produced instead via the sequence of reactions that convert ^{14}N into ^{22}Ne via a double α capture and the radioactive decay of ^{18}F: $^{14}N(\alpha, \gamma)$ $^{18}Fe^{+} + \nu)$ $^{18}O(\alpha, \gamma)$ ^{22}Ne. In H-exhausted regions, ^{14}N is by far the most abundant nuclear species after He because a main effect of the CNO cycle, which operated in the previous H-burning phase (see Sect. 3.2.1) is to convert most of the initial C and O, the two most abundant elements beyond H and He, into N. Hence, during He burning ^{22}Ne becomes the most abundant isotope, after C and O, once ^{14}N is fully consumed by α captures.

When He is exhausted in the centre, He burning moves smoothly outward in mass leaving behind a CO core that begins to contract on a dynamic timescale. Similar to the H-burning shell, also the He-burning shell produces more energy than required to balance gravity because energy production is controlled by the size of the underlying core, and not by the mass of the star. The CO core increases progressively in mass because of the continuous deposition of CO-rich material made in the He-burning shell, the He-burning shell increases its energy production accordingly. As a consequence, the overlying He+H-rich mantle is forced to expand substantially and to cool down so much that the H-burning shell switches off. As during the RGB phase, this expansion progressively inhibits energy transport by radiation and large scale motions of the matter progressively extend inward from the outer envelope. In stars initially more massive than 4–5 $M_\odot$ the convective envelope penetrates even inside the He core (2nd dredge-up). The main consequences of this are a change of the surface abundances and a reduction of the He-core mass. Similar to what happens during the RGB, the formation of an extended convective envelope forces the stars to expand at roughly constant surface temperature because the onset of convective motions fixes a maximum value for the temperature gradient (see Sect. 3.2.1) and increasing luminosity. This phase is called Asymptotic Giant Branch (AGB). The specific phase when the He-burning shell advances in mass eroding the He core from inside is called Early Asymptotic Giant Branch (E-AGB).

The competition between the advancing He-burning shell and the sinking of the convective envelope during the 2nd dredge-up fixes the maximum mass that the CO core (M_{CO}) reaches in this phase. If M_{CO} is larger than roughly 1.1 $M_\odot$, the core heats up to the C ignition temperature ($\sim 8 \times 10^8$ K), otherwise it turns into an electron degenerate CO core able to self-sustain against gravity without the need of additional contraction. The maximum initial stellar mass for which an electron degenerate CO core forms is of the order of 7–8 $M_\odot$, for solar metallicity stars. While the evolution of stars without an electron degenerate core is dictated by the self gravity of the core, the evolution of stars with an electron degenerate core is controlled by the burning shells. In the following we concentrate on the further evolution of the latter case, i.e., the AGB, while Chap. 4 describes the further evolution of the first case.

On the AGB three main regions may be identified: the electron degenerate CO core, a He-rich layer (also referred to as *intershell* since it is located between the He- and the H-burning shells), and a H-rich mantle, most of which forms an extended convective envelope. As the He-burning shell approaches the border of the He core, it quenches because of the steep temperature drop associated with the drastic reduction of the mean molecular weight caused by the change from a He-dominated to a H-dominated chemical composition. Being less and less supported by the extinguishing He burning shell, the mantle is forced to shrink, heat up, and progressively re-activate the H-burning shell at its base. The H-burning shell starts to deposit fresh He onto the He shell forcing the intershell to heat up again. At this point a fascinating evolutionary phase begins in which nuclear burning and instabilities coexist, realizing a unique *habitat* in which a large number of nuclear species may be synthesized: the Thermally Pulsing AGB (TP-AGB) phase.

Quite schematically, the TP-AGB phase consists of a sequence of cycles each of which may be divided in two main phases: a quiescent H-burning phase during which the He-burning shell is inactive, and a He-burning phase during which the H-burning shell is inactive. Though the two shells do not operate simultaneously, they process roughly the same amount of mass per cycle so that the intershell mass changes slowly in time. The transition from the active He-burning phase to the active H-burning phase occurs quiescently in the sense that the energy provided by the H-burning shell progressively replaces that provided by the dimming He-burning shell. Instead, the transition from the active H-burning phase to the active He-burning phase occurs in a traumatic way, which is responsible for the peculiar sequence of events that characterizes the TP-AGB phase.

The reason for such a traumatic He ignition is that the pileup of fresh He on top of an inert intershell leads to a T, ρ profile in the intershell that is controlled by the compressional heating caused by the accretion of fresh He. This T, ρ profile is quite different from the typical one determined by the presence of an active burning shell. The large amount of energy required to turn the T, ρ profile from that determined by the accretion and that required by the steady He burning, coupled to the very steep dependence of the cross section of the 3α nuclear reaction on the temperature, determines the growth of a thermal runaway (or *thermal pulse*, TP) in which a huge amount of energy is released over a very short timescale. This runaway comes to an end when enough energy has been deposited in the intershell to turn the T, ρ profile into a profile suited for quiescent He burning. As an example, Fig. 3.7 shows as a gray line the typical Log(T), Log(ρ) profile produced by the advancing H-burning shell just prior to 3α ignition, while the black line shows the typical profile at the end of the thermal runaway during steady He burning. In this specific example roughly $\sim 10^{48}$ erg must be deposited in the intershell to perform the transition between the two configurations. This amount of energy is determined by the fact that the needed

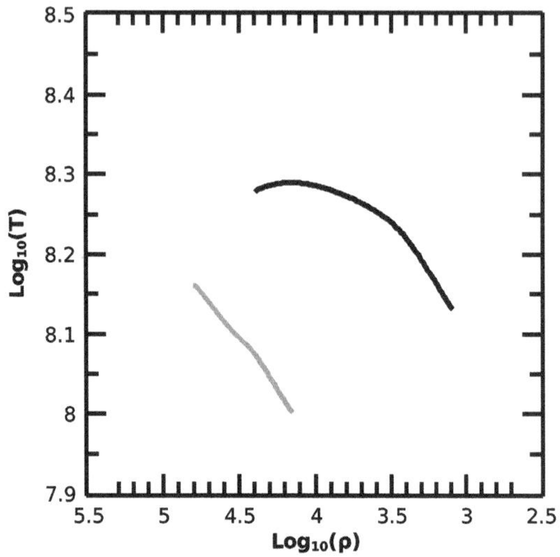

Fig. 3.7 The gray thick solid line shows the typical Log(T)-Log(ρ) profile in the intershell (in the range $10^{-3} < X_{He} < 0.9$) just prior the onset of a thermal pulse in a 3 $M_\odot$ of solar metallicity while the black thick solid line shows the typical profile in the same region at the end of the thermal runaway when the steady He burning occurs

change of the T, ρ structure in the intershell requires a reduction of the binding energy of the intershell.

The main effect of the rapid injection of energy into the intershell during the TP is the production of a very strong energy flux, which forces the growth of convective instabilities to efficiently carry the energy outward. This convective shell extends over most of the intershell region and plays a fundamental role in reshuffling the chemical composition within this region and hence influencing the detailed nucleosynthesis that occurs at this stage (see next sections). Once the TP comes to an end, the convective shell disappears and the quiescent He-burning shell phase begins. Another important side effect of the rapid energy injection caused by the TP is the expansion of the region above the He-burning shell, which forces a cooling of this region. The consequence is that the H-burning shell switches off, and the temperature gradient steepens. This favors the penetration of the convective envelope down into the intershell so that nuclei freshly synthetized in the deep interior of the star are efficiently brought up to the stellar surface (3rd dredge-up). Similar to what happens towards the end of the E-AGB phase, the He-burning shell progressively runs out of power as it approaches the border of the He-rich layer, where the temperature drops below the value necessary for the He burning. The overlying layers are forced to contract and heat so that a H-burning shell activates again and a new cycle starts.

To visually illustrate the sequence of events making up a full TP cycle and to make clear the peculiarity of the TP-AGB phase, Fig. 3.8 shows the temporal evolution of the internal structure of a typical AGB star through three consecutive TPs.

The quantitative characteristics of the TPs depend on the core and envelope masses, the general rule being that larger CO core masses correspond to higher frequencies of thermal pulses, higher temperatures, and shorter lifetimes of the He

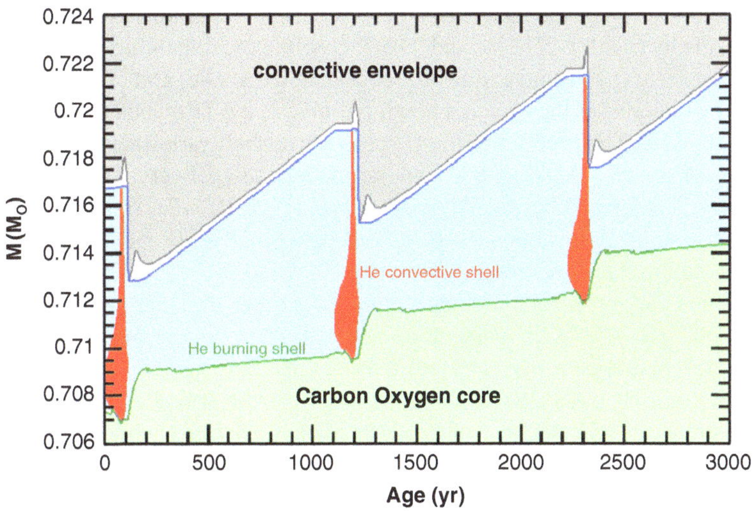

Fig. 3.8 Temporal evolution of the internal structure of a typical AGB star through three consecutive TPs. The He convective shell is shown in red while the convective envelope is grey. The H and He burning shells are shown in dark blue and dark green, respectively, while the intershell is light cyan. The timescale between the onset of the He convective shell and the maximum inward penetration of the convective envelope during the 3rd dredge-up has been increased by a factor of 100 to improve the readability of the figure

convective shell. Typical TP frequencies (determined after the first 20 TPs or so) range between 2 and 3 TPs per 10^5 years for a $3\,M_{\odot}$ star having $M_{CO} \sim 0.7\,M_{\odot}$ and more than 50 TPs per 10^5 years for a $6\,M_{\odot}$ star having $M_{CO} \sim 0.96\,M_{\odot}$, while the peak luminosity ranges between 1and10 $10^8 L_{\odot}$.

Since the two burning shells process about the same amount of matter per cycle, the average growth rate of the CO core per cycle roughly equates that of the He core. This, coupled to the fact that He burning produces (per unit mass) roughly 10% of the energy produced by H burning, and that the luminosity of these stars does not change appreciably between the two burning phases, allows us to estimate the relative burning lifetimes (t_{He}/t_H). The amount of energy produced by He burning per cycle must balance the surface losses, i.e., $\epsilon_{He} \times \Delta M_{He} = L_{surface} \times t_{He}$, where ϵ_{He} represents the amount of energy liberated by the He burning per unit mass, ΔM_{He} the amount of mass processed by the He burning, $L_{surface}$ the luminosity of the star and t_{He} the lifetime of the He burning phase. Similar for the H burning one may write that $\epsilon_H \times \Delta M_H = L_{surface} \times t_H$. If the amount of mass processed is similar in the two cases (i.e., $\Delta M_{He} \simeq \Delta M_H$), the relative lifetimes scale roughly as the two nuclear burning rates, i.e., $t_{He}/t_H \sim 1/10$. The amount of He burnt during each He-burning episode is only partial, corresponding to about 25–30% of the He present in the intershell. Of this, roughly 25% burns during the TP and the remainder during the quiescent He burning phase. The final nucleosynthetic result is that carbon is produced via the 3α reaction, but it is only marginally converted into oxygen. The typical composition of the intershell after this partial He burning is represented by matter made by roughly 75% He and 23% ^{12}C, while the remaining few percent are made up of ^{22}Ne (from conversion of ^{14}N as detailed above) and of ^{16}O. The ^{22}Ne nuclei are of interest as they act as a neutron source in the TPs when the temperature reaches 300 MK. The lifetime of the He convective shell varies between 100 years and 10 years for the 3 $M_\odot$ and 6 $M_\odot$ stellar models, respectively. Typically, the He burning shell is located between 7×10^{-3} and $1.5 \times 10^{-2} R_\odot$ from the center of the star, while the H-burning shell is located between 1×10^{-2} and $2 \times 10^{-2} R_\odot$. The intershell mass ranges roughly between 10^{-3} and 10^{-2} $M_\odot$. The surface radii of AGB stars vary between hundreds to thousands times the solar radius.

The final fate of AGB stars is to lose all their H-rich mantle before the electron degenerate core may grow to its most massive stable configuration (i.e., the Chandrasekhar mass). Such a destiny is due to the strong dependence of the mass-loss rate on the luminosity of the star and on its surface chemical composition (see Sect. 3.4.3). The maximum mass size reached by the CO core, which equates the mass of the newborn white dwarf, is determined by the competition between the speed at which the burning shells advance in mass and the efficiency of the mass loss that erodes the H-rich mantle from the surface.

The occurrence of the 3rd dredge-up significantly affects the evolutionary properties of an AGB star. First, it reduces the size of the He core anticipating the quenching of the quiescent He burning phase and hence its lifetime. Second, it slows down the overall growth rate of the CO core and the He-rich shell. Third, it carries to the stellar surface a fraction of the material freshly synthesized by partial He burning, i.e., C, ^{22}Ne, and $slow$-neutron capture (s-process) elements heavier than iron (see Sect. 3.5), drastically modifying the chemical composition of the star. In some cases, the star even changes from the usual oxygen-rich (O>C) composition to carbon-rich (C>O), with important consequences on the types of molecules and dust that can form and the ensuing mass loss (see Sect. 3.4.4). Unfortunately, the question of the maximum depth reached by the convective envelope during the 3rd dredge-up has always been highly debated and different results have been obtained over the years by different authors for AGB stars over the whole mass interval from 1.5 $M_\odot$ up to the more massive thermally pulsing stars. The reason is that, once the convective envelope enters the He core, a discontinuity in the opacity (H is much more opaque than He) determines the formation of a positive difference between the effective and adiabatic temperature gradients just at the border of the convective envelope. This is an unstable situation because the possible mixing of matter located just below the base of the convective envelope with the H-rich convective mantle is an irreversible process in the sense that these *additional* mixed layers become intrinsically convective because of the drastic increase of the opacity due to the mixing. It is therefore clear that even small different numerical techniques adopted by different authors may lead to quite different results.

Furthermore, the occurrence of the 3rd dredge-up is important because it creates a sharp discontinuity between the convective envelope and the radiative intershell. Since a sharp discontinuity is not a realistic configuration in these conditions, the occurrence of the 3rd dredge-up allows the possibility that some kind of *diffusion* of protons occurs below the formal border of the convective envelope when it reaches its maximum inward extension at the end of the 3rd dredge-up smoothing out the discontinuity (though this is not obtained by applying the standard stability criteria for mixing). However, the shape, extent, and timescale over which the diffusion of protons in the He/C intershell may occur is unknown, its modeling is still artificial and not based on self-consistent computations.

This diffusion allows the formation of regions where a small amount of protons come in contact with matter that is predominantly composed of He and C, so that the ratio $Y(H)/Y(C) \ll 1$, but does not contain any ^{14}N, since this nucleus has been fully converted into ^{22}Ne in the previous TP. When these proton-enriched layers begin to contract and to heat because of the quenching of the He burning, the CN cycle activates, but it can not go beyond the synthesis of ^{13}C due to the low proton concentration. As the temperature increases to roughly 90 MK, the $^{13}C(\alpha, \mathbf{n})$ ^{16}O reaction becomes efficient and a significant neutron flux is produced. Hence, this diffusion plays a pivotal role in the nucleosynthesis of species beyond the Fe peak via neutron captures. A detailed description of the properties of this neutron source and of its nucleosynthetic signature will be presented in Sect. 3.5.1. The lack of ^{14}N is crucial here, since this nucleus is a strong neutron poison and its presence would inhibit neutron captures by Fe and the elements heavier than Fe.

As already discussed in Sect. 3.3.2, typical temperatures at the base of the convective envelope do not exceed a few MK at most in the evolutionary phases prior to the AGB. Instead, another peculiarity of AGB stars is that during the H-burning phase the temperature at the base of the convective envelope may reach values in excess of several tens MK, and even exceed 100 MK, so that H-burning reactions activate within the convective envelope. In these conditions the coupling between burning and mixing may lead to conspicuous synthesis—and spread through the whole convective envelope up to the stellar surface—of nuclei like ^{14}N, ^{7}Li, and the long-lived radioactive nucleus ^{26}Al (discussed in detail in Sect. 3.6.1). The efficiency of this phenomenon, known as Hot Bottom Burning (HBB), scales directly with the temperature at the base of the envelope and hence with the CO-core mass, which in turn scales with the initial stellar mass. Hence, HBB is efficient in stars more massive than 4:5 $M_{\odot}$, depending on the metallicity. As the energy produced in the convective envelope sums to that produced by the H-burning shell, the core mass—luminosity relation changes (even strongly) in the presence of HBB. From a nucleosynthetic point of view the occurrence of an active H burning in a convective environment implies a redistribution of the processed material over all the convective zone, so the surface abundances turn towards the relative abundances typical of the H burning at high temperature. For example, an increase of the surface abundances of elements like ^{14}N and ^{26}Al, a temporaneous increase of ^{7}Li, a reduction of ^{12}C and of the $^{12}C/^{13}C$ ratio and the signature of the NeNa and the MgAl sequences.

We refer the reader to the review papers by Herwig (2005) and Karakas and Lattanzio (2014) and to the book on the evolution of AGB stars by Lattanzio and Wood (2004) for a thorough presentation of the evolution of these cool giant stars.

3.4.2 Super-AGB Stars

In the previous section we identified stars that go through the double shell burning of the TP-AGB phase as those that develop an electron degenerate CO core where carbon burning fails to occur. There is, however, another class of stars that experience the double shell burning phase: those with initial total mass between the maximum mass that forms an electron degenerate CO core where C does not ignite (M_{up}) and the minimum mass that does not form an electron degenerate CO core (M_{mas}). Stars more massive than M_{mas} evolve up to the final core collapse as described in Chap. 4. Depending on the initial chemical composition and the adopted physics, M_{up} ranges between 6–8 $M_{\odot}$, and M_{mas} ranges between 9–12 $M_{\odot}$. It is important at this point to recall that these limiting masses are somewhat uncertain because they depend on the size of the convective core, the carbon to oxygen ratio left by the He burning, the efficiency of the second dredge-up and the cross section of the ^{12}C plus ^{12}C nuclear reaction. Unfortunately, all these quantities are still subject to severe uncertainties.

Stars falling between these two limits form a partially electron degenerate core, but are massive enough to ignite C in the core, lift the degeneracy, and go through the C burning in the core. They are not massive enough, however, to avoid the electron degeneracy of the ONeMg core left by C burning. The evolution of stars in this relatively small mass interval, called Super-AGB stars, has not been studied extensively up to now because of the difficulty in computing the C-burning phase due to the complex removal of the degeneracy that occurs through a series of successive flashes, and the lack of massive computer power, which is needed to study this complex situation. This situation is rapidly changing and progress has been made on the computational modeling of Super-AGB stars (Doherty et al. 2017, see review by).

Since these stars form an electron degenerate core after core C burning, they also go through a double shell burning phase similar to the AGB phase experienced by their less massive counterparts. Because their degenerate cores are more massive, following the trend shown by AGB stars, the frequency of the thermal pulses is higher (up to 500 TP per 10^5 years), the He peak luminosity is lower than in the normal AGB stars (up to $\sim 4 \times 10^6 L_{\odot}$), while the base of the convective envelope may reach temperatures as high as 110 MK, hence, H burning occurs within the convective envelope (Hot Bottom Burning). Similar to what happens in the more massive AGB stars, but quantitatively more pronounced, the luminosity produced in the convective envelope adds to that produced by the radiative H burning significantly altering the core mass—luminosity relation and the surface composition is modified by the signature of H burning. The possible occurrence of the 3rd dredge-up would also shuffle the surface chemical composition with the typical products of the partial He burning, i.e. C, and *s*-process elements. The efficiency of the 3rd dredge-up is very uncertain also for these stars. In principle, one could expect a lower efficiency of the 3rd dredge-up because the amount of energy released by a TP is lower and the overall temperature is much higher than in a standard AGB star, so that it could be more difficult to expand the base of the mantle and to steepen the temperature gradient up to a value that would allow the convective envelope to penetrate the He core. Quantitative estimates of the yields of the nuclei specifically produced by the TP Super-AGB stars are now available (Doherty et al. 2017).

The final fate of a Super-AGB star depends on the competition between the advancing burning shells, which increase the size of the ONeMg core, and the mass loss, which limits its growth. Also an efficient 3rd dredge-up would contribute to limiting the growth of the core. Stars more massive than a critical value reach the threshold electron degenerate core mass for the onset of electron captures on ^{24}Mg and ^{20}Ne after a certain number of TPs and eventually explode as *electron capture supernovae*. Stars less massive than the critical value, instead, end their life as ONeMg white dwarfs. An estimate of the electron degenerate core mass above which electron captures become efficient in an ONeMg environment can be determined by considering that the threshold energy for electron capture is 6 MeV for ^{24}Mg and 8 MeV for ^{20}Ne and that the mass of a fully electron degenerate core having a Fermi energy of the order of 6 MeV is $\simeq$ 1.35 $M_{\odot}$. Thus, if the electron degenerate core grows to the threshold value of $\simeq$1.35 $M_{\odot}$ or so, electron captures are activated on ^{20}Ne and ^{24}Mg.

This process removes electrons and hence pressure from the center of the star, starting a runaway process that leads to the core collapse and final explosion known as electron capture supernovae. The explosion of these electron capture supernovae is similar to that of core collapse supernovae (see Chap. 4), with a few distinct features. During the initial collapse of the degenerate core, electron captures increase significantly the degree of neutronization of the matter, i.e., raise the global neutron over proton ratio because of the capture of the electrons by the protons. The nuclear species produced by explosive burning depend significantly on the neutron over proton ratio so that the higher the degree of neutronization of the matter the higher the production of neutron-rich nuclei: in particular ^{58}Ni becomes favored with respect to ^{56}Ni. Since the luminosity peak of a supernova correlates with the amount of ^{56}Ni produced during the explosion, a natural feature of these electron captures supernovae is a lower luminosity with respect to typical core collapse supernovae. Also, the final kinetic energy of the ejecta is expected to be of the order of 0.1×10^{51} erg, roughly one order of magnitude lower than in typical core collapse supernovae (see, e.g., Hoffman et al. 2008; Wanajo et al. 2009).

3.4.3 Winds from AGB Stars

An observed peculiarity of AGB stars is that they show strong stellar winds, which carry material away from the surface of the star into its surroundings. Nuclei newly synthetized during the AGB phase and carried to the stellar surface by the 3rd dredge-up are shed into the interstellar medium so that AGB stars contribute to the chemical make-up of their environments and of new generations of stars. The mass loss rate due to winds in AGB star increases as the star evolves along the AGB and can reach values as high as 10^{-4} $M_{\odot}$/year (to be compared, for example, to the solar mass loss rate of 10^{-11} $M_{\odot}$/year) at the end of the AGB, which is known as the *superwind* (Iben and Renzini 1983). This is a strong and dense but slow wind, with material leaving the star at relatively low speeds of 5–30 km/s.

The winds are caused by two main factors (see review by Höfner and Olofsson 2018). First, large quantity of dust form around AGB stars and radiation pressure acting on this dust contributes to driving the winds. The extended envelopes of red giant and AGB stars, where the temperature drops down to ~1000 K, are an ideal location for the formation of a large variety of molecules like CO, TiO, VO, as well as ZrO, when the gas has been enriched in heavy elements such as Zr by the *s* process and the 3rd dredge-up, and C_2, CN, and CH, when the gas has been enriched in carbon by the 3rd dredge-up. In the case of refractory elements, which have the property of condensing at high temperatures directly from gas into the solid state, the gas condenses into tiny particles, which then can grow into dust grains. Because of the large quantity of dust around them, AGB stars become obscured toward the end of their life and can only be seen as mid-infrared sources, since the dust absorbs the energy of the visual light coming from the star and reemits it as infrared light. Second, AGB stars are variable stars, meaning that their luminosity varies with time with changes occurring over relatively long periods >100 days. These luminosity variations are due to stellar pulsations, in the sense that the whole star expands and contracts. Pulsations produce changes in the stellar radius and temperature, which cause the variations in the stellar luminosity. When the pulsations attain a large amplitude they lead to strong stellar winds and a large mass-loss rate. Pulsation levitates matter above the photosphere and increases the wind density by about two orders of magnitude (Wood 1979; Sedlmayr and Dominik 1995; Dorfi et al. 2001).

The strong stellar winds driven by the combined effects of radiation pressure acting on dust and pulsation eventually erode the whole stellar envelope (Dupree 1986; Willson 2000). Hence, the winds govern the lifetime of AGB stars because when the envelope is almost completely lost the star moves away from the AGB phase into the hotter post-AGB phase. Toward the end of the post-AGB phase, the shell of material ejected by the AGB star may become illuminated by the radiation coming from the central star, and produce a planetary nebula. The former AGB stars is now referred to as a *planetary nebula nucleus* and finally turns into a cooling CO white dwarf.

3.4.4 Dust from Giant Stars and the Origin of Stardust

The specific dust species that form in the atmosphere of AGB stars depends mainly on the C/O ratio. The difference in the type of dust that can form in a carbon-rich or oxygen-rich gas is due to the strong bond of the CO molecules: if O>C, all carbon atoms are locked into CO and only oxygen-rich dust can form, viceversa, if

Table 3.6 Types and populations of meteoritic stardust grains, and their inferred origins. See also Sect. 10.2

Type	Population	Origin
Oxide and silicate grains	I	AGB stars
	II	AGB stars
	III	?
	IV	Supernovae
Silicon carbide (SiC)	Mainstream	AGB stars
	Y	AGB stars
	Z	AGB stars
	X	Supernovae
	A+B	?
	Nova grains	Novae
Silicon nitride		Supernovae
Graphite	Low-density	Supernovae
	High-density	?
Diamond		?

C>O, all oxygen atoms are locked into CO and only carbon-rich dust can form.[4] In an oxygen-rich gas (O>C) dust species are, for example, Al_3O_2 (corundum), $CaAl_{12}O_{19}$ (hibonite), $MgAl_2O_4$ (spinel), as well as many different types of silicates (SiO, SiO_2, etc). In a carbon-rich gas (C>O), dust species are, for example, SiC (silicon carbide), TiC (titanium carbide), and C itself (graphite).

Formation of dust around AGB stars is well documented by spectroscopic observations in the infrared (e.g. Treffers and Cohen 1974; Speck et al. 2000, 2009) and predicted to occur by theoretical models (e.g. Fleischer et al. 1992; Lodders and Fegley 1993; Gail and Sedlmayr 1999; Ferrarotti and Gail 2002, 2006; Nanni et al. 2013; Dell'Agli et al. 2017). It is now widely accepted that AGB stars are the most prolific source of dust in the Galaxy. When summing up the contribution of the different families of late red giant and AGB stars: i.e., spectroscopically, the M stars, the OH/IR stars,[5] and the carbon stars, it results that ~90% of all dust of stellar origin in the interstellar medium came from these sources (Whittet 2002).

Thus, it is not surprising that the vast majority of stardust grains extracted from meteorites (Chap. 2, Sects. 2.2.4 and 10.2) show the signature of an origin in AGB stars (Table 3.6). The main signatures of AGB nucleosynthesis imprinted in meteoritic stardust grains are: (1) the O isotopic composition of the majority of oxide and silicate grains showing excess in ^{17}O and deficits in ^{18}O, and known as Population I and II of stardust oxide grains (Nittler et al. 1997), which match the O isotopic ratios observed around AGB stars via spectroscopic observations of CO molecular lines (e.g. Harris et al. 1987; Hinkle et al. 2016; Abia et al. 2017), and (2) the distribution of the $^{12}C/^{13}C$ ratios of >90% of SiC grains showing a peak between

[4]This general rule is debated in the case of dust formation in supernova ejecta, see Chap. 2, Sect. 2.2.

[5]OH/IR stars are cool red giants with strong hydroxyl (OH) masers and infrared (IR) emissions.

50 and 60 (solar value is 89) and known as the *mainstream* SiC population, which match the distribution derived from spectroscopic observation of CO molecular lines in C-rich AGB stars (see Fig. 3 of Hoppe and Ott 1997). The Ne composition measured in stardust SiC grains—corresponding to the Ne-E(H) component rich in ^{22}Ne (see discussion in Chap. 2, Sect. 2.2.4) is also a clear signature of material from the intershell of AGB stars, where ^{22}Ne is abundant. Moreover, the elemental and isotopic abundances of the heavy elements Kr, Sr, Zr, Ru, Xe (the Xe-S component), Ba, Nd, Sm, W, and Pb present in trace amount and measured in SiC grains clearly show the imprint of the *s*-process, which make inevitable their connection to AGB stars. Smaller SiC Populations Y and Z ($\simeq$1% each of the total recovered stardust SiC grains) are also attributed to AGB stars, but of metallicity down to 1/3–1/5 of the solar value (Hoppe et al. 1997; Amari et al. 2001b; Zinner et al. 2006).

With regards to the remaining types and populations of stardust grains, supernovae of Type II have been invoked as the origin site of Population X of SiC grains ($\sim$1%) and the few recovered silicon nitride grains (Nittler et al. 1995), showing excesses in ^{28}Si and evidence of the early presence of ^{44}Ti, as well as low-density graphite grains and Population IV of oxide and silicate grains (with excess in ^{18}O and ^{18}Si Vollmer et al. 2008; Travaglio et al. 1999; Pignatari et al. 2013, see Chap. 4). Novae are invoked for a few SiC grains of unusual composition (excesses in ^{13}C and ^{15}N Amari et al. 2001a, see Chap. 5, Sect. 5.2), while the origin of SiC grains of Populations A+B ($\simeq$5% of all SiC grains, showing ^{13}C/^{12}C $<$10) is still unclear (Amari et al. 2001c). Oxide and silicate grains with deficits in both ^{17}O and ^{18}O, known as Population III, have been attributed to stars of metallicity lower than solar, however, the Si isotopic composition of the silicate grains belonging to this population is very close to solar, which does not support this interpretation. The origin of this population remains to be ascertained, together with the origin of high-density graphite grains and of the very abundant and extremely tiny (10^{-9} m) meteoritic diamond grains, the majority of which probably formed in the solar system. For more details in meteoritic stardust see, e.g., Clayton and Nittler (2004) and Lugaro (2005).

Given compelling evidence that most stardust came from AGB stars, the composition of these grains can be used as a stringent constraint for theoretical models of AGB stars and, viceversa, the models can be used to identify the mass and metallicity range of the parent stars of the grains. Data from the laboratory analysis of stardust are usually provided with high precision, down to a few percent errors, and for isotopic ratios. In comparison, data from spectroscopic observations of stellar atmospheres usually are provided with lower precision, errors typically $>$50%, and mostly for elemental abundances. Thus, the information from stardust grains represents a breakthrough in the study of AGB nucleosynthesis. Also, given that the abundances and isotopic compositions of elements heavier than Al and lighter than Fe, such as Si and Ti, are mostly unaltered by AGB nucleosynthesis, laboratory analysis of these elements in AGB stardust can be used to constrain in great detail the initial composition of the parent star of the grains, and in turn the chemical evolution of the Galaxy (e.g. Zinner et al. 2006).

Meteoritic stardust provides us with abundant and precise information on radioactive nuclei in stars because the initial abundance of radioactive nuclei at the time of the formation of the grains is recorded by the signature of their radioactive decay inside the grains, which is easily derived from measurements of the excesses in the abundances of their daughter nuclei. An important example is that of ^{26}Al, where the initial ^{26}Al abundance in a stardust grain is revealed by excesses in ^{26}Mg. This will be discussed in detail in Sect. 3.6.1. In general, radioactive signatures in stardust have the potential be used as clocks for the timescale of dust formation around stars and supernovae. Finally, stardust isotopic data provide a unique way to investigate the operation of the s-process in AGB stars, as will be discussed in Sect. 3.5.5.

3.5 Neutron Capture Nucleosynthesis in AGB Stars

In this section we show that:

- Free neutrons are produced in the TP-AGB phase by the ^{22}Ne$(\alpha, n)^{25}$Mg reaction, which activates at ~300 MK and operates during He burning in the intershell convective region during thermal pulses, and the ^{13}C$(\alpha, n)^{16}$O reaction, which activates at ~90 MK and operates in a radiative (and hence stable) region of the intershell during the H-burning phases. The free neutrons trigger the s-process, which produces half of the cosmic abundances of the elements heavier than iron via neutron captures mostly occurring on stable and long-lived radioactive nuclei.
- Unstable isotopes with half lives higher than a few days can also suffer neutron captures during the s-process, producing a wide variety of *branching points* on the s-process path, which define the details of the abundance distribution produced by the s-process as a function of neutron density and temperature.
- The overall s-process abundance distribution is defined by stable nuclei with a magic number of neutrons at the three s-process peaks at Sr, Ba, and Pb, and by the total amount of free neutrons available.
- Several long-lived unstable isotopes are produced by the s-process (details in Sect. 3.6.5). Among them are ^{93}Zr and ^{99}Tc. Observations of monoisotopic stable Nb (the daughter nucleus of ^{93}Zr) and of Tc itself can be used as discriminant between intrinsic (on the AGB) and extrinsic (with a former AGB binary companion) s-process-enhanced stars.

3.5.1 Neutron Sources in AGB Stars

In the double burning shell phase a nuclear reaction that may produce a copious neutron flux is ^{22}Ne$(\alpha$,n$)^{25}$Mg. ^{22}Ne is abundantly present in the intershell because it directly derives from the initial abundance of O (the most abundant nucleus after H and He) as a consequence of the operation of the CNO cycle first and of a double α capture on ^{14}N later. This means that this neutron production channel is of *secondary* origin, i.e., its efficiency scales with the initial metallicity of the star. The relatively high Coulomb barrier of Ne (Z=10) pushes the threshold temperature for α capture above 300 MK so that this process can activate only within a hot He-burning region. Since the temperature at the base of the He convective shell during thermal pulses scales directly with the mass of the H-exhausted core, only stars initially more massive than $\simeq 3\,M_{\odot}$ (Iben 1975; Iben and Truran 1978) can efficiently activate this nuclear reaction. Panel b) in Fig. 3.9, shows a typical profile of the neutron density versus time associated with this neutron source. Its shape reflects the sharp rise of the temperature caused by the growth of the thermal instability and the following quite rapid decline due to the quenching of the instability. The high activation temperature and its very short duration (a few years) lead to a very high initial neutron density

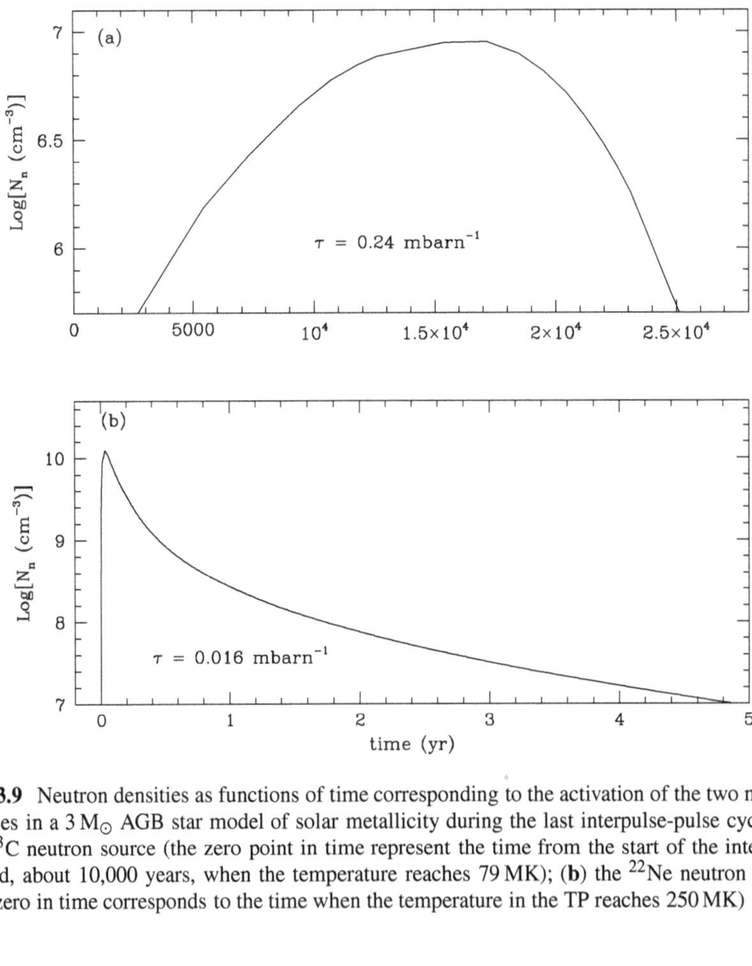

Fig. 3.9 Neutron densities as functions of time corresponding to the activation of the two neutron sources in a $3\,M_{\odot}$ AGB star model of solar metallicity during the last interpulse-pulse cycle: (**a**) the ^{13}C neutron source (the zero point in time represent the time from the start of the interpulse period, about 10,000 years, when the temperature reaches 79 MK); (**b**) the ^{22}Ne neutron source (the zero in time corresponds to the time when the temperature in the TP reaches 250 MK)

(reaching up to $N_n \simeq 10^{14}$ n/cm^3 in AGB stars of initial mass ~6 M$_\odot$) but to a small total amount of ^{22}Ne burnt per cycle, so that the total number of neutrons released, i.e., the time-integrated neutron flux, or *neutron exposure* $\tau = \int_0^t N_n v_{th} dt$, remains quite small, of the order of a few hundredth of 1/mbarn[6] (see Sect. 3.5.2). The signature of such an impulsive neutron flux on neutron capture nucleosynthesis will be discussed in the next section. We only remark here an important difference between the neutron-capture nucleosynthesis occurring during AGB thermal pulses and that occurring in the He-convective shell of a massive star (other than the fact that in the AGB case the exposure to neutrons occurs recurrently): the mass of the He-convective shell in AGB stars is orders of magnitude smaller than that of a massive star so that the smaller dilution induced by the mixing allows, in the former case, many unstable nuclei to reach a higher equilibrium concentration. This occurrence favors the synthesis of stable nuclei on the neutron-rich side of the valley of β stability.

The problem with the ^{22}Ne neutron source is that AGB stars observed to be enriched in s-process elements have been identified as AGB stars of masses lower than ~3 M$_\odot$ because (a) their relatively low luminosities (Frogel et al. 1990) match those of low-mass AGB models; (b) their surface is generally C enriched, an occurrence that rules out a significant HBB and hence an initial mass greater than 3 M$_\odot$; (c) excesses of ^{25}Mg, predicted to be produced by ^{22}Ne(α, n)^{25}Mg, and of ^{26}Mg, predicted to be produced by the twin channel ^{22}Ne(α, γ)^{26}Mg, with respect to ^{24}Mg are not observed (Smith and Lambert 1986; McWilliam and Lambert 1988); (d) the high neutron density produced by the ^{22}Ne channel, see Panel b) in Fig. 3.9, would favor the synthesis of neutron-rich nuclei like ^{96}Zr and elements as Rb, at odds with spectroscopic observations (Lambert et al. 1995; Abia et al. 2001) and the solar abundance distribution (Despain 1980). Thus, for the vast majority of s-enhanced AGB stars, another nuclear fuel for the production of neutrons has to be invoked.

Nuclei of ^{13}C are the best candidate for this role, given that the ^{13}C(α, n)^{16}O reaction activates at temperatures from approximately 90 MK, which are easily reached in low-mass AGB stars. The achievement of the threshold temperature is, however, a necessary but not sufficient condition for a nuclear reaction to be effective: an additional requirement is the presence of a sufficient amount of reactants, in this case ^{13}C. We already pointed out in the previous section that models in which no mixing is allowed in the layers in radiative equilibrium do not naturally produce a significant concentration of ^{13}C in the intershell region. In fact, the ^{13}C available in the H-exhausted zone is that corresponding to the equilibrium value provided by the CNO cycle. As a neutron source for the s-process, this ^{13}C suffers two major problems: its abundance is too low to power a significant neutron flux and its ratio with respect to ^{14}N is too low (^{13}C/^{14}N $<<$ 1). The ^{14}N(n, p)^{14}C

[6] 1 mbarn = 10^{-27} cm^2.

reaction[7] has a relatively high neutron capture cross section of $\simeq$2 mbarn (Wallner et al. 2016), with respect to typical cross section of the order of 0.1–0.01 mbarn for the light nuclei. Hence, it is a formidable poison that can even completely inhibit the *s*-process. Hence, the ^{13}C neutron source represents a valid alternative to the ^{22}Ne neutron source only if additional ^{13}C is produced in an environment depleted in ^{14}N. A way out of this problem is to assume that at the end of each 3rd dredge-up episode a small amount of protons penetrates the intershell region. The amount of protons engulfed in the He/C rich intershell must be small ($Y_p/Y_{^{12}C} << 1$) because they must allow the conversion of ^{12}C into ^{13}C, but not the conversion of ^{13}C in ^{14}N. (Note that the intershell is essentially free of ^{14}N at the end of a thermal pulse because ^{14}N nuclei have all been destroyed by α captures.) Once a small amount of protons has penetrated the intershell, the progressive heating caused by the deposition of fresh He synthetized by the H-burning shell induces the conversion of ^{12}C in ^{13}C. We can estimate the concentration of protons that allows the build up of ^{13}C, but not of ^{14}N, by considering that the production rate of ^{14}N equates that of ^{13}C when the concentration of ^{13}C rises to a value of the order of 1/4 of that of ^{12}C (see Sect. 3.2.1). Since the mass fraction of ^{12}C in the intershell is about 0.2, the two rates equate each other for a ^{13}C concentration of $X_{^{13}C} \simeq 0.20\,(13/12)/4 = 5\;10^{-2}$. If one requires the ^{13}C production rate to dominate that of ^{14}N, the ^{13}C concentration must be reduced by at least a factor of 10, so that $X_{^{13}C} \simeq 5\;10^{-3}$. This abundance of ^{13}C corresponds to a proton concentration of the order of $X_p = X_{^{13}C}/13 = 4\;10^{-4}$.

A self-consistent scenario able to produce this small amount of protons penetrating below the base of the convective envelope has not been found yet: several mechanisms have been proposed (e.g. Iben and Renzini 1982; Herwig et al. 1997; Langer et al. 1999; Denissenkov and Tout 2003) but none of them can presently be considered as widely accepted. A discussion of these alternative scenarios goes well beyond the purposes of the present discussion. What matters, and what modelers often pragmatically assume, is that a small amount of protons definitely penetrates in the intershell at the end of 3rd dredge-up. The detailed features of the ^{13}C *pocket* obtained with such a procedure are subject to large uncertainties.

Nonetheless the basic properties of the neutron flux that is obtained in this way are considered relatively well understood (Gallino et al. 1998; Goriely and Mowlavi 2000; Lugaro et al. 2003b). The activation of the $^{13}C(\alpha, n)^{16}O$ occurs well before the onset of the next thermal pulse and the *s*-process nucleosynthesis triggered by this neutron source occurs at low temperature in a radiative environment (see Sect. 3.5.2). Panel a) in Fig. 3.9 shows the temporal evolution of this neutron flux. The rather long timescale over which this neutron flux remains active is determined by the speed at which the H-burning shell accretes matter on the He core, which means a typical timescale of the order of 10^4 years. Given such a long timescale, ^{13}C is totally consumed so that the total number of neutrons released is very large,

[7]This reaction produces ^{14}C, a radioactive nucleus with a half life of 5730 years. This nucleus is not carried to the stellar surface by the 3rd dredge-up because it is destroyed by $^{14}C(\alpha, n)^{18}O$ reactions during He-burning in the thermal pulse.

with neutron exposures of the order of a tenth to a few mbarn^{-1}. The neutron density, instead, keeps to low values, up to $N_n \simeq 10^8$ n/cm^3. Let us finally remark that the neutron flux produced by the ^{13}C neutron source is of *primary* origin, i.e., independent on the initial stellar metallicity, since the ^{13}C is made from ^{12}C synthetized starting from the initial H and He.

3.5.2 The s-Process in AGB Stars

A fraction of the free neutrons produced in AGB stars by the ^{13}C and ^{22}Ne neutron sources described above is captured by Fe seed nuclei, leading to production of elements with large atomic mass numbers up to Pb (A = 208) and Bi (A= 209) via the s-process. In general, a neutron flux that irradiates the surrounding matter reproduces a situation analogous to that occurring during H burning, where matter is irradiated by a flux of protons. While during a proton flux matter is pushed out of the valley of β stability toward the proton-rich side, during a neutron flux matter is pushed out of the valley of β stability valley toward the neutron-rich side. Thus, the presence of a neutron flux is inevitably associated to the synthesis of radioactive nuclei that, sooner or later, decay back towards the valley of β stability.

During the s-process, by definition, the timescale against β decay of an unstable isotope is shorter that its timescale against neutron captures. Thus, neutron captures occur only along the valley of β stability (Fig. 3.10). For this condition to hold neutron densities must be of the order of $N_n \sim 10^7$ n/cm^3. By comparison, during the *rapid* neutron-capture process (r process), instead, neutron densities reach values as high as 10^{20}–10^{25} n/cm^3 so that neutron captures occur on a time scale less than a second (typically much shorter than that of radioactive decays) pushing matter towards very neutron-rich material. When the neutron flux is extinguished, the neutron-rich radioactive nuclei quickly decays back towards their stable isobars

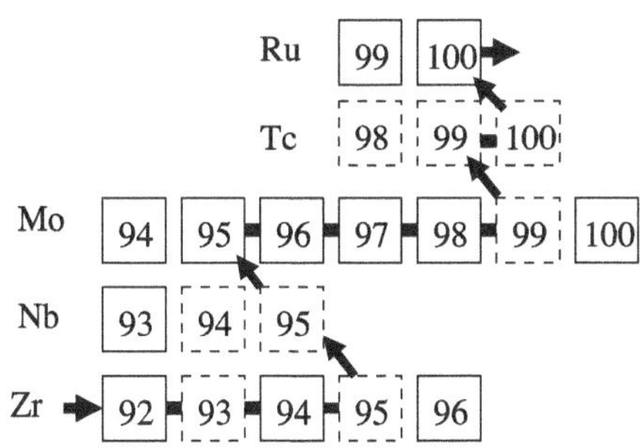

Fig. 3.10 The main s-process path along the valley of β stability from Zr to Ru is indicated by the thick solid line and arrows. Solid and dashed boxes represent stable and unstable nuclei, respectively. The radioactive nuclei ^{93}Zr and ^{99}Tc behave as stable during the s process as their half lifes (of the order of 10^5–10^6 years) are longer than the timescale of the s-process

on the valley of β stability. As presented in Chap. 4, the r-process is believed to occur in explosive conditions in supernovae.

In stellar conditions, though, neutron densities during the s-process can reach values orders of magnitude higher than 10^7 n/cm^3. Depending on the peak neutron density, as well as on the temperature and density, which can affect β-decay rates, conditions may occur for the neutron-capture reaction rate of an unstable isotope to compete with its decay rate. These unstable isotopes are known as *branching points* on the s-process path. To calculate the fraction of the s-process flux branching off the main s-process path at a given branching point a *branching factor* is defined as:

$$f_{branch} = \frac{p_{branch}}{p_{branch} + p_{main}},$$

where p_{branch} and p_{main} are the probabilities per unit time associated to the nuclear reactions suffered by the branching point nucleus and leading onto the branch or onto the main path of the s process, respectively.

There are several types of branching points: in the *classical* case p_{main} corresponds to λ, i.e., the probability per unit time of the unstable isotope to decay, and p_{branch} corresponds to p_n, i.e., the probability per unit time of the unstable isotope to capture a neutron $< \sigma v > N_n$, where N_n is the neutron density and $< \sigma v >$ is the Maxwellian averaged product of the velocity v and the neutron capture cross section σ.[8] A typical example of this case is the isotope ^{95}Zr in Fig. 3.10, which has a half life of 64 days, and can capture neutrons and produce the neutron-rich isotope ^{96}Zr, classically a product of the r-process, even during the s-process. When the branching point is a long-lived, or even stable isotope, but its β-decay rate increases with temperature, the opposite applies: $p_{branch} = \lambda$ and $p_{main} = p_n$. In even more complex situations, a radioactive isotope may suffer both β^+ and β^- decays, as well as neutron captures. In this case, three terms must be considered at denominators in the definition of the branching factor above: p_n, and λ for both β^+ and β^- decays.

Branching points have been fundamental in our understanding of the s-process conditions in AGB stars and will be discussed in more detail in Sect. 3.5.3. The low neutron density associated with the ^{13}C neutron source does not typically allow the opening of branching points. On the other hand, the high neutron density associated with the ^{22}Ne neutron source activate the operation of branching points on the s-process path, defining the details of the final abundance distribution.

It is possible to identify nuclei that can be produced only by the s process (*s-only* nuclei), which are shielded from r-process production by a stable isobar, or only by

[8]Note that σ is usually given in mbarn, corresponding to 10^{-27} cm^2, and that $< \sigma v >$ can be approximated to $\sigma \times v_{thermal}$, where $v_{thermal}$ is the thermal velocity. Neutron capture cross sections for (n, γ) reactions throughout this chapter are given at a temperature of 350 million degrees, corresponding to an energy of 30 keV, at which these rates are traditionally given. Values reported are from the Kadonis database (Karlsruhe Astrophysical Database of Nucleosynthesis in Stars, http://www.kadonis.org/) and the JINA reaclib database (http://groups.nscl.msu.edu/jina/reaclib/db/index.php), unless stated otherwise.

the r-process (r-*only* nuclei), which are not reached by neutron captures during the s-process as isotopes of the same element and same atomic mass number A-1 are unstable. Examples of s-only nuclei are ^{96}Mo and ^{100}Ru shown in Fig. 3.10, which are shielded by the r-only nuclei ^{96}Zr and ^{100}Mo, respectively. These, in turn, are not typically produced by the s-process as ^{95}Zr and ^{99}Mo are unstable. Proton-rich nuclei which cannot be reached by either the s- or the r-process must be produced via the p-*process*, i.e., proton captures or photodisintegration of heavier nuclei, and are labelled as p-*only* nuclei (e.g., ^{94}Mo in Fig. 3.10).

Models for the s-process have historically been tested against the solar system abundances of the s-only isotopes, as these were the first precise available constraints. Once a satisfactory fit is found to these abundances, the selected theoretical distribution can be used to determine the contribution from the s-process to each element and isotope. By subtracting this contribution to the total solar system abundance, an r-process contribution to each element can be obtained[9] (e.g. Kaeppeler et al. 1982; Arlandini et al. 1999), which has been widely used to test r-process models, and to compare to spectroscopic observations of stars showing the signature of the r-process (Sneden et al. 2008). For example, $\simeq$80% of the solar abundance of Ba is due to the s process, which is then classified as a typical s-process element, while $\simeq$5% of the solar abundance of Eu is due to the s process, which is then classified as a typical r-process element.

Already B^{2}FH had attributed to the operation of the s-process the three peaks in the solar abundance distribution at magic numbers of neutrons N=50, the Sr, Y, and Zr peak, N=82, the Ba and La peak, and N=126, the Pb peak. This is because nuclei with a magic number of neutrons behave with respect to neutron capture reactions in a similar way as atoms of noble gases do with respect to chemical reactions. Their energy levels, or shells, are fully populated by neutrons, in the case of magic nuclei, or by electrons, in the case of noble gases, and hence they are very stable and have a very low probability of capturing another neutron, in the case of magic nuclei, or of sharing electrons with another atom, in the case of noble gases. Nuclei with magic numbers of neutrons have small neutron-capture cross sections (of the order of a few to a few tens mbarn) with respect to other heavy nuclei, and they act as bottlenecks along the s-process path, leading to the observed abundance peaks. Nuclei located between the peaks, instead, have much higher neutron capture cross sections (of the order of a few hundred to a few thousand mbarn). The neutron-capture chain in these local regions in-between magic nuclei quickly reaches equilibrium during the s-process. During a neutron-capture process the abundance N_A of a stable isotope with atomic mass A varies with time as:

$$\frac{dN_A}{dt} = production\ term\ -\ destruction\ term$$

$$= N_{A-1} N_n \sigma_{A-1} \times v_{thermal} - N_A N_n \sigma_A \times v_{thermal}.$$

[9]The p-process contribution to elemental abundances is comparatively very small, $\simeq$1%, except in the case of Mo and Ru, which have magic and close-to-magic p-only isotopes, where it is up to $\simeq$25% and $\simeq$7%, respectively.

When replacing time with the neutron exposure τ one has:

$$\frac{dN_A}{d\tau} = N_{A-1}\sigma_{A-1} - N_A\sigma_A,$$

which, in steady-state conditions $\frac{dN_A}{d\tau} \rightarrow 0$ reached in between neutron magic nuclei, yields the simple rule to derive relative s-process abundances away from neutron magic numbers[10]:

$$N_A\sigma_A \simeq constant.$$

It follows that the relative abundances of nuclei in-between the peaks are only constrained by their neutron-capture cross sections and do not provide information on the s-process neutron exposure. On the other hand, the relative abundances of the elements belonging to the three different peaks almost uniquely constrain the s-process neutron exposure. This is the reason behind the introduction and wide usage, both theoretically and observationally, of the s-process labels *light s* (*ls*) and *heavy s* (*hs*), corresponding to the average abundances of the s-process elements belonging to the first and second peak, respectively, as well as behind the importance of the determination of the abundance of Pb, representing the third s-process peak. In AGB stars, the high neutron exposure associated with the ^{13}C neutron source drive the production of s-process elements even reaching up to the third s-process peak at Pb in low-metallicity AGB stars. On the other hand, the lower neutron exposure associated with the ^{22}Ne neutron source typically produces s-elements only up to the first s-process peak at Sr.

It is now ascertained that the s-process is responsible for the production of about half the abundances of the elements between Sr and Bi in the Universe (see, e.g., Kaeppeler et al. 1982) and that it occurs in AGB stars.[11] The first direct evidence that the s-process occurs in AGB stars—and, more generally, that nucleosynthesis is happening inside stars—was the identification in the 1950s of the absorption lines of atoms of the radioactive element Tc in the atmospheres of some cool giant stars. The longest-living isotopes of Tc are ^{97}Tc and ^{98}Tc, with a half life of 4.0 and 4.2 million years, respectively. Since these stars would have taken billions of years to evolve to the giant phase, the observed Tc could have not been present in the star initially. It follows that the Tc must have been produced by the s-process inside the stars (see also Neyskens et al. 2015). Actually, neutron captures do not produce ^{97,98}Tc, but the third longest-living isotope of this element: ^{99}Tc, with a terrestrial half life of 0.21 million years (Fig. 3.10). The presence of ^{99}Tc in AGB stars has

[10]For a detailed analytical description of the s process refer to Chapter 7 of Clayton (1968).

[11]Cosmic abundances of nuclei between Fe and Sr are also contributed by the s-process, but in this case by neutron captures occurring in massive stars during core He burning and shell C burning (Chapter 4 and, e.g., Raiteri et al. 1992; Pignatari et al. 2010).

been confirmed by measurements of the Ru isotopic composition in stardust SiC grains, as will be discussed in Sect. 3.6.5.

The presence of Tc in giant stars has also been used to classify different types of s-process enhanced stars. If a given observed s-process enriched giant star shows the lines of Tc, then it must be on the AGB and have enriched itself of s-process elements. In this case it is classified as *intrinsic* s-process enhanced star and typically belongs to one of the reddest and coolest subclasses of the spectroscopic class M: MS, S, SC, and C(N), where the different labels indicate specific spectral properties—S stars show zirconium oxide lines on top of the titanium oxide lines present in some M stars and C(N) stars have more carbon than oxygen in their atmospheres—or the transition cases between those properties—MS is the transition case between M stars and S stars and SC is the transition case between S and C(N) stars. On the other hand, if an s-process enriched giant star does not show the lines of Tc, it is classified as *extrinsic* s-process enhanced star. In this case its s-process enhancements have resulted from mass transfer from a binary companion, which was more massive and hence evolved first on the AGB phase. Stars belonging to the class of *extrinsic* s-process enhanced stars range from Ba stars in the Galactic disk, to the older halo populations of carbon-rich CH and Carbon-Enhanced Metal-Poor (CEMP) stars (e.g. Jorissen et al. 1998; Bond et al. 2000; Lucatello et al. 2005). Observations of Nb can also be used to discriminate intrinsic from extrinsic s-process enhanced stars as Nb is destroyed during the s-process, but receives a radiogenic contribution over time due to the β^- decay of ^{93}Zr, with half life 1.5 million years, which is on the s-process path (see Fig. 3.10).

The field of modelling the s process in AGB stars has boomed since 2010, with several groups presenting large sets of yields to be employed for a variety of applications, from the study of the chemical evolution of the Galaxy to the comparison to observational constraints from both stars and stardust grains. One of the most complete set of models can be found in the FRUITY database (http://fruity.oa-teramo.inaf.it/), which also includes models with stellar rotation and is based on the series of papers by Cristallo et al. (2016). Another set of models has been provided by Karakas and Lugaro (2016, and references therein). Finally, the NuGrid collaboration (http://www.nugridstars.org/) has also started to provide s-process AGB models (see, e.g., Pignatari et al. 2016). These sets of models are based on different computational tools and different choices for the implementation of the physical mechanisms behind the various features of the AGB phase, from the mass loss to the formation of the ^{13}C pocket. While comparison and analysis of these models, also in relation to the observational constraints, is still ongoing, first basic comparisons between the different model sets can be found, for example, in Karakas and Lugaro (2016). The main observation is that models of AGB stars of relatively low mass ($<4\,M_\odot$) produce results that are quite similar to each other, while for higher masses predictions diverge. A detailed analysis of the production in the different sets of new models of the long-lived radionuclei heavier than Fe discussed in Sect. 3.6.5 is the topic of future research.

3.5.3 Branchings and the s-Process in AGB Stars

Branching points at radioactive nuclei have provided for the past 50 years important tools to learn about conditions during the s-process in AGB stars. This is because branching factors depend on the neutron density and can also depend on the temperature and density of the stellar material. This happens in those cases when the decay rate of the branching nucleus is temperature and/or density dependent. These branching points are referred to as *thermometers* for the s-process. Traditionally, the solar abundances of isotopes affected by branching points were used to predict the neutron density and temperature at the s-process site using parametric models where parameters representing, e.g., the temperature and the neutron density were varied freely in order to match the observed abundances (e.g. Kaeppeler et al. 1982). Later, detailed information on branching points became available from spectroscopic observations of stellar atmospheres and from laboratory analyses of meteoritic stellar grains. At the same time, models for the s-process in AGB stars have evolved from parametric into stellar models, where the temperature and neutron density parameters governing the s-process are taken from detailed computation of the evolution of stellar structure (Gallino et al. 1998; Goriely and Mowlavi 2000; Cristallo et al. 2009; Karakas and Lugaro 2016). For these models branching points are particularly useful to constrain neutron-capture nucleosynthesis and conditions inside the thermal pulse because, typically, they open at high neutron densities during the high-temperature conditions that allow the activation of the ^{22}Ne neutron source in the convective intershell region.

As the temperature, density, and neutron density vary with time in the convective intershell region, branching factors also change over time. For example, a classical branching point, where the branching path corresponds to neutron capture, progressively opens while the neutron density reaches its maximum, and then closes again while the neutron density decreases and the main s-process path is restored. Of special interest is that toward the end of the thermal pulse the neutron density always decreases monotonically with the temperature and thus with time (Fig. 3.9) so that a *freeze out* time can be determined for a given nucleus, which represents the time after which the probability that the nucleus captures a neutron is smaller than unity and thus the abundances are *frozen* (Cosner et al. 1980). This can be calculated as the time when the neutron exposure τ left before the end of the neutron flux is $1/\sigma$, where σ is the neutron capture cross section of the nucleus.

As a general rule of thumb, branching points that have the chance of being activated at the neutron densities reached in AGB stars are those corresponding to radioactive nuclei with half lives longer than at least a couple days. These correspond to similar half lives against capturing a neutron for neutron densities $\simeq 10^9 - 10^{11}$ n/cm^3, at AGB s-process temperatures. Isotopes with half lives longer than approximately 10,000 years can be considered stable in this context as the s-process flux in AGB stars typically lasts less than this time. We refer to these isotopes as long-lived isotopes and we discuss their production in AGB stars in detail in Sect. 3.6.5. Very long-lived isotopes—half lives longer than ~10 Myr—include for example ^{87}Rb, and are considered stable in our context.

A list of unstable isotopes at which branching points that become relevant in the s-process reaction chain in AGB stars is presented in Appendix B of this book as a complete reference to be compared against observational information and as a tool for the building of s-process networks. Worth special mention are the branching points at ^{79}Se, ^{85}Kr, and ^{176}Lu for the involvement of isomeric states of these nuclei, at ^{151}Sm, one among a limited number of branching points for which an experimental estimate of the neutron-capture cross section is available, at ^{86}Rb, responsible for the production of the very long-living ^{87}Rb, and at ^{163}Dy and ^{179}Hf, which are stable nuclei in terrestrial conditions that become unstable in AGB stellar interiors.

Taken as a whole, the list of branching points that may be operating during the s-process in AGB stars sets a powerful group of constraints on our theoretical s-process scenarios. They are particularly effective when each of them is matched to the most detailed available observations of its effects. For example, some elemental abundance ratios and isotopic ratios that are affected by branching points can be measured from a stellar spectrum via identification and analysis of different emission or absorption lines. In these cases, model predictions can be compared directly to stellar observations of s-process-enhanced stars (Sect. 3.5.4). Isotopic ratios affected by branching points involving isotopes of refractory elements, but also of noble gases, have been or have the potential to be measured in meteoritic stardust SiC grains from AGB stars and provide unique constraints due to the large and expanding high-precision dataset available on the composition of stardust (Sect. 3.5.5). The values of the solar abundance ratios of s-only isotopes affected by branching points (e.g., ^{134}Ba/^{136}Ba, ^{128}Xe/^{130}Xe, and ^{176}Hf/^{176}Lu) must be matched by any s-process model. When these involve nuclei with peculiar structure, such as ^{176}Lu, combined investigation of nuclear properties and s-process models drives progress in our understanding of both.

One advantage of the computation of branching points in AGB stars is that the activation of one branching point is almost completely independent from the activation of all the other branching points because the overall neutron flux is only very marginally affected by the details of the s-process path. Thus, it is possible to include in a s-process nuclear network only the branching points of interest for a specific problem, or a specific element, hence keeping it simple and saving computational time.

One overall drawback of using branching points to understand the s-process is that for the vast majority of the radioactive nuclei involved there exist only theoretical or phenomenological determinations of their neutron-capture cross sections and of the temperature and density dependence of their decay rates. This is due to the difficulty of producing experimental data for radioactive targets (see Chap. 9) and means that there are always some uncertainties associated to model predictions of the effect of branching points. These errors and their effect need to be carefully evaluated in every single case.

3.5.4 Signatures of s-Process Branching Points: Rb, Zr, Eu

The abundance of ^{87}Rb, which can be produced in AGB stars via activation of the branching point at ^{86}Rb, is a famous example of how detailed comparison of theoretical s-process abundances to the abundances observed in s-process-enhanced stars provide a stringent test to our understanding of the s process and AGB stars. The abundance of ^{87}Rb is particularly interesting because the element Rb can be spectroscopically identified and its abundance determined in AGB stars. Overall, Rb is an r-process element—only 22% of its solar abundance can be ascribed to the s-process (Arlandini et al. 1999)—made up of two isotopes: ^{85}Rb and the very long-lived ^{87}Rb, which is treated as a stable isotope in this context. Specifically, 92% of solar ^{85}Rb is made by the r-process because this nucleus has a relatively large neutron capture cross section of 234 mbarn and thus it does not accumulate to high abundances during the s-process. On the other hand, ^{87}Rb, as described in Appendix B, has a magic number of neutrons N=50, and thus a relatively small neutron capture cross section of 15.7 mbarn. Hence, if it is reached by the s-process reaction chain via the activation of the branching points at ^{85}Kr and ^{86}Rb, it accumulates and is significantly produced. It follows that when these branching points are activated during the s-process, the abundance ^{87}Rb represents a fraction of the total abundance of s-process Rb larger than the initial solar fraction. This is illustrated in the top panel of Fig. 3.11. In the case of the massive AGB model, where the ^{22}Ne neutron source is activated, the s-process occurs at high neutron density, and branching points are open, almost half of the final total abundance of Rb is made by ^{87}Rb. In the case of the low-mass AGB model, instead, where the ^{13}C neutron source is activated, the s-process occurs at low neutron density, and branching points are closed, only a quarter of the final total abundance of Rb is made by ^{87}Rb.

The ratio of the abundance of Rb to that of a neighbouring s-process element, such as Sr, or Zr, whose overall abundance is instead not affected by the activation of branching points, can be determined in AGB stars and has been widely used as an indicator of the neutron density at which the s-process occurs. Observations of Rb/Zr ratios lower than solar in MS, S, and C stars have strongly supported the theoretical scenario where the main neutron source in these low-mass AGB stars is the $^{13}\mathrm{C}(\alpha, n)^{16}\mathrm{O}$ reaction. This is because this neutron source produces neutron densities too low to increase the Rb/Zr ratio above the solar value (see lower panel of Fig. 3.11 and Lambert et al. 1995; Abia et al. 2001).

Massive AGB stars ($> 4{:}5\,\mathrm{M}_\odot$) have only recently been identified in our Galaxy (García-Hernández et al. 2006, 2007; Pérez-Mesa et al. 2017). They belong to the group of OH/IR stars and they have been singled out as massive AGB stars on the basis of their location closer to the galactic plane, which indicates that they belong to a more massive stellar population, and their longer pulsation periods ($\simeq$400 days). Rb/Zr ratios in these stars are observed to be well above the solar value, which has given ground to the theoretical scenario where the main neutron source in these massive AGB stars must be the $^{22}\mathrm{Ne}(\alpha, n)^{25}\mathrm{Mg}$ reaction, which produces neutron

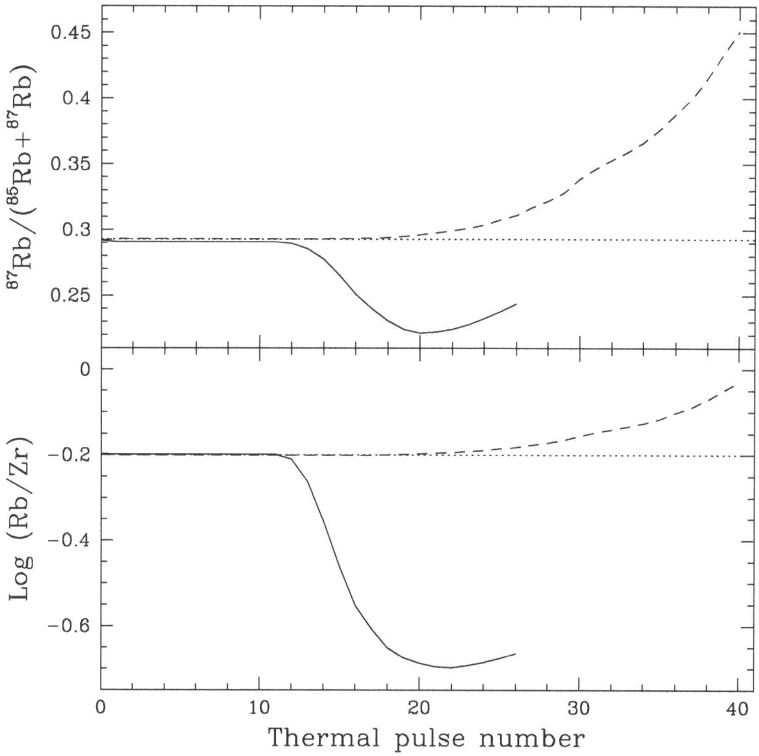

Fig. 3.11 Ratio of ^{87}Rb to the total abundance of Rb (top panel) and the Rb/Zr ratio (lower panel) computed in two solar metallicity AGB models (from van Raai et al. 2012). The dotted lines represent the initial solar ratios. The solid lines represent a massive (6.5 $M_\odot$) AGB model experiencing the activation of the ^{22}Ne neutron source only, and a low-mass (3 $M_\odot$) model experiencing the activation of the ^{13}C neutron source only (except for a marginal activation of the ^{22}Ne neutron source in the latest thermal pulses leading to the small final increase in the ^{87}Rb and Rb abundances)

densities high enough to increase the Rb/Zr ratio above the solar value (see lower panel of Fig. 3.11).

Another indicator of the neutron density measured in AGB stars is the isotopic abundance of ^{96}Zr, which is produced if the branching point at ^{95}Zr is activated. Zr isotopic ratios were determined via observations of molecular lines of ZrO in a sample of S stars (Lambert et al. 1995). No evidence was found for the presence of ^{96}Zr in these stars. This result provides further evidence that the neutron density in low-mass AGB stars must be low. A low ^{96}Zr abundance has been confirmed by high-precision data of the Zr isotopic ratios in stardust SiC grains, which are discussed in the following Sect. 3.5.5.

Isotopic information from stellar spectra has also been derived for the typical r-process element Eu in old Main Sequence stars belonging to the halo of our Galaxy and enhanced in heavy neutron-capture elements (Sneden et al. 2008). This has been possible because the atomic lines of Eu differ significantly if the Eu atoms are made of ^{151}Eu instead of ^{153}Eu, the two stable isotopes of Eu. In old stars showing overall enhancements of r-process elements, the total Eu abundance is roughly equally divided between ^{151}Eu and ^{153}Eu. This fraction is consistent with the solar fraction, and it is expected by Eu production due to the r-process (Sneden et al. 2002; Aoki et al. 2003a). On the other hand, in two old stars showing overall enhancements in s- and r-process elements, roughly 60% of Eu is in the form of ^{151}Eu (Aoki et al. 2003b). How these stars gained enhancements in the abundances of both r- and s-process elements has been an unsolved puzzle of the study of the origin of the elements heavier than iron in the Galaxy (Jonsell et al. 2006). This is because according to our current knowledge, the r- and the s-process are completely independent of each other, and occur in very different types of stars, neutron star mergers and possibly peculiar environments within core-collapse supernovae (Thielemann et al. 2011; Côté et al. 2017) and AGB stars, respectively. As will be discussed in Sect. 3.5.6, recently an $intermediate$-process origin for these peculiar abundance patterns has been proposed (Hampel et al. 2016). During the s- and the i-processes, the Eu isotopic fraction is determined mostly by the activation of the branching point at ^{151}Sm. Hence, the determination of the Eu isotopic ratios is a fundamental clue, which may help in solving the origin of these old stars. As more observations become available, the role of branching points during the s-process in AGB stars becomes more and more crucial to answering the questions on the origin of the heavy elements.

3.5.5 SiC Grains from AGB Stars and Branching Points

Stardust SiC grains from AGB stars represent a unique opportunity to study s-process conditions in the parent stars of the grains through the effect of the operation of branching points because SiC grains contain trace amounts of atoms of elements heavier than iron, which allow high-precision measurements of their isotopic ratios. Refractory heavy elements, such as Sr, Ba, Nd, and Sm, condensed from the stellar gas directly into the SiC grains while the grains were forming. Their isotopic composition have been determined from samples of meteoritic residual materials containing a large number of SiC grains using TIMS and SIMS (see Chap. 10, Sect 10.2 and Ott and Begemann 1990; Zinner et al. 1991; Prombo et al. 1993; Podosek et al. 2004). High-resolution SIMS has also been applied to derive data in single stardust SiC grains for Ba with the NanoSIMS (see Sect. 10.2 and Marhas et al. 2007) and Ba, Eu, and W with the SHRIMP (Sensitive High Resolution Ion Microprobe, e.g. Ávila et al. 2012). Isotopic ratios in a sample containing a large number of SiC grains for many elements in the mass range from Ba to Hf were also measured by ICPMS (Sect 10.2 and Yin et al. 2006).

A general drawback of these experimental methods is that they do not allow to separate ions of same mass but different elements. Hence, interferences by isotopes of the same mass (isobars) are present, which is especially problematic for the elements heavier than iron where a large number of stable isobars can be found. Branching points, in particular, by definition affect the relative abundances of isobars, thus, with the methods above it is difficult to derive precise constraints on the effect of branching points on isotopic ratios. For example, the isobars ^{96}Mo and ^{96}Zr cannot be distinguished in these measurements, and thus it is not possible to derive information on the operation of the branching point at ^{95}Zr.

Exceptions to this problem are the stable Eu isotopes, ^{151}Eu and ^{153}Eu, which do not have stable isobars and thus their ratio can be measured and used to constrain the neutron density and the temperature during the s process in the parent stars of the grains via the branching points at ^{151}Sm and ^{152}Eu (Ávila et al. 2013), and the Ba isotopes, which are not affected by isobaric interferences because their isobars, the isotopes of the noble gas Xe isotopes, are present in very low amounts in the grains and are difficult to ionize and extract from the stardust (see specific discussion below in this section.) The Ba isotopic ratios, in particular the ^{134}Ba/^{136}Ba and the ^{137}Ba/^{136}Ba ratios, can be affected by branching points at the Cs isotopes (see below and Prombo et al. 1993; Marhas et al. 2007; Liu et al. 2014b).

The application of RIMS (Sect. 10.2) to the analysis of heavy elements in SiC grains has allowed to overcome the problem of isobaric interferences, at the same time providing an experimental method of very high sensitivity, which allows the measurements of trace elements in single stardust grains (Savina et al. 2003b). Since RIMS can select which element is ionized and extracted form the grains, mass interferences are automatically avoided. The Chicago-Argonne RIMS for Mass Analysis CHARISMA has been applied to date to the measurement of Zr (Nicolussi et al. 1997; Liu et al. 2014a), Mo (Nicolussi et al. 1998a), Sr (Nicolussi et al. 1998b; Liu et al. 2015), Ba (Savina et al. 2003a; Liu et al. 2014b, 2015) and Ru (Savina et al. 2004) in large single SiC grains (average size 3 µm), providing high-precision constraints on the operation of the s-process branching points that may affect the isotopic composition of these elements. Recently, a new improved instrument has come on-line: the Chicago Instrument for Laser Ionization (CHILI) (Stephan et al. 2016), which is expected to drive an enormous improvement in the amount and quality of the data in coming years. Analysis of SiC from AGB stars with CHILI has been performed so far for Fe and Ni (Trappitsch et al. 2018). A detailed comparison between data and models (Lugaro et al. 2003a, 2014b, 2018; Liu et al. 2015; Palmerini et al. 2018) shows that AGB stellar models of low mass and roughly-solar metallicity, where the ^{13}C$(\alpha, n)^{16}$O reaction is the main neutron source and the ^{22}Ne$(\alpha, n)^{25}$Mg is only marginally activated, provide the best match to all measured isotopic ratios affected by branching points.

For example, the ^{96}Zr/^{94}Zr ratio is observed in all measured single SiC to be lower than solar by at least 50%. Low-mass AGB models can reproduce this constraint due to the low neutron density associated with the main ^{13}C neutron source, in which conditions ^{96}Zr behaves like a typical r-only nucleus and is destroyed during the neutron flux. Massive AGB stars (>4:5 $M_\odot$), on the other hand, experience high neutron densities and produce ^{96}Zr/^{94}Zr ratios higher than solar. In more detail, the ^{96}Zr/^{94}Zr ratio at the stellar surface of low-mass AGB stellar models reaches a minimum of $\simeq$90% lower than solar after roughly ten 3rd dredge-up episodes, and then may increase again, due to the marginal activation of the ^{22}Ne neutron source in the latest thermal pulses. This predicted range allows to cover most of the ^{96}Zr/^{94}Zr of single SiC grains (see Fig. 5 of Lugaro et al. 2003a).

Another interesting example is the ^{134}Ba/^{136}Ba ratio, where both isotopes are *s*-only nuclei. During the low-neutron density flux provided by the ^{13}C neutron source the branching point at ^{134}Cs is closed and the ^{134}Ba/^{136}Ba ratio at the stellar surface reaches up to $\simeq$ 20% higher than the solar ratio after roughly ten 3rd dredge-up episodes. This value is too high to match the composition of single SiC grains. However, during the marginal activation of the ^{22}Ne in the later thermal pulses, the branching point at ^{134}Cs is activated, ^{134}Ba is skipped during the *s*-process flux and the ^{134}Ba/^{136}Ba ratio at the stellar surface is lowered down to the observed values roughly 10% higher than the solar value (see Fig. 14 of Lugaro et al. 2003a).

The ^{137}Ba/^{136}Ba ratio is another indicator of the neutron density because the activation of the chain of branching points along the Cs isotopes can produce ^{137}Cs, which decays into ^{137}Ba with a half life of 30 years. Grain data do not show any contribution of ^{137}Cs to ^{137}Ba, indicating that the Cs branching points beyond ^{134}Cs are not activated in the parent stars of the grains (see Fig. 14 of Lugaro et al. 2003a). This, again, excludes massive AGB stars, with an important neutron contribution from the ^{22}Ne neutron source, as the parent stars of the grains.

Another example of the signature of the *s*-process in meteorites is represented by very small variations, of the order of parts per ten thousand, observed in various elements in different types of meteoritic samples (Dauphas and Schauble 2016, see review by). For example, osmium isotopic ratios of primitive chondritic meteorites present a fascinating anomaly that looks like a *mirror* *s*-process signature, meaning that they show exactly the opposite behaviour expected if the meteorite had a component carrying an *s*-process signature. They are thus interpreted as a sign of incomplete assimilation of stardust SiC grains within the meteorite (Brandon et al. 2005). The branching points at ^{185}W and ^{186}Re make the ^{186}Os/^{188}Os ratio a indicator of the neutron density for the *s* process and the value for this ratio observed in chondrites suggest a low neutron density of $N_n = 3 \times 10^8$ n/cm^3 (Humayun and Brandon 2007), in agreement with other evidence discussed above.

Differently from refractory elements, the noble gases He, Ne, Ar, Kr, and Xe are not chemically reactive. Still, they are found in SiC, even if in extremely low quantities. Their atoms could have been *implanted* into dust grains that already had formed earlier (Heck et al. 2007, 2009; Verchovsky et al. 2004).

It has been possible to extract noble gases from meteoritic samples by RIMS (see Chap. 11), laser gas extraction (Nichols et al. 1991) and stepped-heating combustion of the sample to high temperatures, up to 2000° (Lewis et al. 1994). In particular, for the heavy nobler gases Kr and Xe, since their abundances are very low in stardust and the stepped-heating experimental method does not provide high extraction efficiency, it has been possible to extract their ions only from a large amount of meteoritic residual material. The derived Kr and Xe isotopic data is thus the average over a large number—millions—of grains. Differential information as function of the grain sizes can still be obtained by preparing the meteoritical residual in a way that selects the size of the grains to be found in it.

The composition of Xe in SiC corresponds to the famous Xe-S component, one of the first signature of the presence of pure stellar material in primitives meteorites (see Chap. 2, section), thus named because of its obvious s-process signature: excesses in the s-only isotopes ^{128,130}Xe and deficits in the r-only and p-only isotopes ^{124,126,136}Xe (all with respect to the solar composition). The ^{134}Xe/^{130}Xe ratio may be affected by the operation of the branching point at ^{133}Xe during the s-process. This isotopic ratio in stardust SiC is very close to zero, indicating that the Xe trapped in SiC grains did not experience s-process with high neutron density (Pignatari et al. 2004). This again allows the mass and metallicity of the parent stars of the grains to be constrained to low-mass AGB stars of roughly solar metallicity, in agreement with the conclusions drawn from the composition of the refractory elements.

The situation regarding the Kr isotopic ratios measured in SiC grains is much more complex. There are two branching points affecting the Kr isotopic composition: ^{79}Se and ^{85}Kr, changing the abundances of ^{80}Kr and ^{86}Kr, respectively, and both of them are tricky to model (see description in Appendix B). Moreover, the Kr atoms in stardust SiC appear to be consistent with implantation models of this gas into the grains only if these models consider two different components of implanted Kr (Verchovsky et al. 2004). One component was ionized and implanted in SiC at low energy, corresponding to a velocity of 5–30 km s^{-1}, typical of AGB stellar winds, the other component was ionized and implanted at high energy, corresponding to a velocity of a few thousands km/s, typical of the winds driven from the central star during the planetary nebular phase. In the second situation, which is the case also for all the He, Ar, and Ne atoms found in SiC, the isotopic composition of the noble gases indicate that they must have come directly from the deep He-rich and s-process-rich layers of the star, with very small dilution with the envelope material of initial solar composition. This is consistent with the fact that at this point in time the initial envelope material would have almost completely been peeled away by the stellar winds.

While the Kr AGB component is observed to be prominent in the small grains (of average size 0.4 μm) and shows low ^{86}Kr/^{82}Kr and high ^{80}Kr/^{82}Kr ratios, in agreement again with low neutron density s-process AGB models, the Kr planetary-nebula component is observed to be prominent in the largest grains (average size 3 μm) and shows high ^{86}Kr/^{82}Kr and low ^{80}Kr/^{82}Kr ratios, as expected instead in pure He-rich intershell material due to the higher neutron density s-process occurring in the final AGB thermal pulses (Pignatari et al. 2006; Raut et al. 2013). Actually, it is difficult to reproduce the ^{86}Kr/^{82}Kr up to twice the solar value observed in the largest grains even using the final pure s-process intershell composition of low-mass and solar metallicity AGB stellar models. This high ^{86}Kr/^{82}Kr ratios may be the signature of high-neutron density s-process nucleosynthesis occurring in *late* and *very late* thermal pulses during the post-AGB phase (see, e.g., Herwig et al. 1999), rather than during the AGB phase. Detailed s-process models are currently missing for this phase of stellar evolution.

In summary, the detailed information provided by stardust data on the isotopic ratios affected by branching points at radioactive nuclei on the s-process path has allowed us to pinpoint the characteristics of the neutron flux that the parent stars of stardust SiC grains must have experienced. The vast amount of information on the composition of light and heavy elements in SiC grains has allowed us to infer with a high degree of confidence that the vast majority of these grains came from C-rich AGB stars, i.e., C(N) stars, which have C>O in their envelope, the condition for SiC grain formation, of low mass and metallicity close to solar. In turn, the stardust data has been used to refine our theoretical ideas of the s-process in these stars confirming that ^{13}C nuclei must be the main neutron source, while the ^{22}Ne neutron source is only marginally active.

3.5.6 The Intermediate Neutron-Capture Process

The traditional view of a clear separation between the main two cosmic neutron-capture processes originally proposed by B^{2}FH, the *slow* (s) and the *rapid* (r) neutron-capture processes, has been modified from the beginning of the 2010 decade by the need to introduce a new *intermediate* (i) neutron-capture process. As the name reflects, the i process is expected to occur for neutron densities intermediate between the typical s-process ($\sim 10^{7-12}$ n/cm^3) and the typical r-process neutron densities ($> 10^{20}$ n/cm^3), in a regime of $N_n \sim 10^{13} - 10^{15}$ n/cm^3. In this situation the neutron-capture flux proceeds somewhat further away to the right of the valley of β stability than in the case of the s process: all the branching points listed in Appendix B are expected to work very efficiently, as well as further branching points not listed there but with significant probability to capture a neutron at the typical i-process neutron densities. One important example is the case of the branching point at ^{135}Xe. This isotope has a half life of approximately 9 h, hence, it is not activated in s-process conditions. However, during the i-process it captures neutrons producing the stable ^{136}Xe, which has a magic number of neutrons, a very low neutron capture cross section (less than 1 mbarn), and accumulates during the production flux. The result is that Xe is a main product of the i-process, contrarily to the s process. A similar case can be described for Kr, due to accumulation of ^{86}Kr, and Rb, due to accumulation of ^{87}Rb.

While the possibility of an i process was originally proposed by Cowan and Rose (1977), the idea of it has been revived only recently when its existence in nature has become evident from several observational constraints. The first was the observed pattern of the abundances of the elements heavier than iron in Sakurai's object (Herwig et al. 2011), a post-AGB star that appears to have experienced a H-ingestion event in 1994. An H-ingestion means that some protons are mixed into a He-burning TP. This results in a "flash", i.e., a sudden release of energy due to H-burning at high temperature, and the production of the neutron source ^{13}C releasing a neutron burst at relatively short timescale, hence, high density. Another clear indication of the existence of the i-process has come from the CEMP stars originally named CEMP-s/r because of being enriched both in Ba (an s-process element) and Eu (an r-process element). The current best model to explain such pattern can be found using the i process (Dardelet et al. 2014; Hampel et al. 2016). In fact, Hampel et al. (2016) suggested to rename these stars CEMP-i.

There are several main problems currently related to our understanding of the i process. First, modelling H ingestion episodes is far from trivial, and most of the current simulations of the i process rely on some level of parametrisation. For example, the amount of ingested protons is treated as a free parameter, as well as the timescale of the process. While the former mainly determines the neutron density, to be adjusted such as it reaches the values required by the i process, the latter influences the total time-integrated number of neutrons (i.e., the neutron exposure τ). Similarly to the s process, low neutron exposures result in an i process that favours the production of the elements at the first magic neutron number beyond Fe of 50 (Kr and Rb, in this case). This is required by Sakurai's object and possibly other peculiar low-metallicity stars (Roederer et al. 2016). High neutron exposures, instead, result in an i process that favours the production of the elements at the second neutron number beyond Fe of 82 (Xe and Ba), as required by CEMP-i.

The other main problem currently related to the i process is the identification of its stellar site. For example, for CEMP-i, H-ingestion episodes similar to those reported for Sakurai's object have been found to be present in low-metallicity low-mass AGB stellar models, which presumably are the binary companions of CEMP-i stars (see, e.g., Campbell and Lattanzio 2008, and references therein). These ingestions, however, can also potentially occur in a variety of other environments including Super-AGB stars (Jones et al. 2016) and massive star, both during the pre-supernova and supernova phases. They are also seen to occur in models of rapidly accreting white dwarves (Denissenkov et al. 2017). While research is ongoing to establish more accurate scenarios for the operation, the occurrence, and the astrophysical site of the i process, its potential impact on galactic cosmic abundances will also need to be considered.

The impact of the i process on the production of long-lived isotopes affected by branching points discussed in the next section (e.g., ^{60}Fe and ^{182}Hf) remains to be investigated, although, it can be predicted that the high neutron density characteristic of the i process will results in enhanced production of such isotopes. If this effect will be important in relation to observational constraints remains to be seen.

3.6 Nucleosynthesis of Long-Lived Isotopes in AGB Stars

3.6.1 ^{26}Al

The famous long-lived radioactive nucleus ^{26}Al (with half life of 0.7 Myr), of interest from the point of view of γ-ray observations, meteoritic stellar grains, and the composition of the early solar system, can be produced in AGB stars via proton captures on ^{25}Mg, i.e., the ^{25}Mg$(p, \gamma)^{26}$Al reaction, when the temperature is above $\simeq$60 MK (Mowlavi and Meynet 2000; van Raai et al. 2008; Straniero et al. 2013). As detailed in Sect. 3.4.1, proton captures occur in AGB stars between thermal pulses in two different locations: (1) in the H-burning shell on top of the He-rich intershell, and (2) at the base of the convective envelope in massive AGB stars, above $\simeq$4 M$_\odot$ (in the process known as Hot Bottom Burning, HBB, Sect. 3.4.1).

In setting (1), the intershell material is progressively enriched in ^{26}Al as proton captures in the H-burning shell convert 80% of ^{25}Mg into ^{26}Al. The efficiency of this conversion is determined by the fraction of 20% of ^{25}Mg+p reactions producing the isomeric, rather than the ground, state of ^{26}Al, which quickly decays into ^{26}Mg with a half life of $\simeq$6 s. Most intershell ^{26}Al abundance, however, is destroyed by neutron captures before having the chance of being dredged-up to the stellar surface via the 3rd dredge-up. This is because the neutron capture cross sections of ^{26}Al, in particular the (n, p) and (n, α) channels, are very efficient: $\sigma \simeq 250$ and 180 mbarn, respectively.

Already during the interpulse period some ^{26}Al is destroyed by neutron captures. This is because in the bottom layers of the ashes of H burning the temperature reaches 90 MK, high enough for the ^{13}C(α,n)^{16}O reaction to occur using as fuel the ^{13}C nuclei in ashes of H burning produced by CNO cycling. The neutrons released in this region are of no interest to the s process because they are captured by the abundant light elements ^{14}N, and ^{26}Al itself, with relatively high neutron capture cross sections (see also Sect. 3.5.1). Then, neutron captures in the following thermal pulse destroy most of the ^{26}Al that was left over in the H-burning ashes. First, the ^{13}C nuclei that had survived in the top layers of the H-burning ashes are engulfed in the convective pulse, where the temperature quickly reaches 200 MK and the ^{13}C(α,n)^{16}O reaction is very efficiently activated. Again, these neutrons do not contribute to any s-process nucleosynthesis as they get mostly captured by ^{14}N and ^{26}Al. Second, the neutrons that may be released by the ^{22}Ne(α,n)^{25}Mg reaction later on in the convective pulse, when the temperature is higher than roughly 250 MK, contribute to further destruction of ^{26}Al. In this phase ^{26}Al can be completely destroyed, depending on the temperature reached at the base of the convective pulse, which controls the efficiency of the ^{22}Ne(α,n)^{25}Mg reaction. If the temperature reaches up to 300 MK, the ^{26}Al abundance is decreased by two orders of magnitude in the He-rich intershell at the end of the thermal pulse.

When the 3rd dredge-up occurs after the thermal pulse is extinguished, only a small mass of ^{26}Al is carried from the intershell to the stellar surface, of the order of 10^{-8} M$_\odot$, mostly coming from a tiny region (roughly 10^{-4} M$_\odot$) at the top of the intershell, which was not ingested in the convective pulse and thus did not experience the availability of free neutrons. This small abundance of ^{26}Al carried into the envelope translates into a small total contribution of the AGB winds to the abundance of ^{26}Al in the interstellar medium (also defined as *yield*) of 10^{-7} M$_\odot$, for AGB stars of masses between 1 M$_\odot$ and 4 M$_\odot$, depending on the metallicity (upper panel of Fig. 3.12), though allowing a noticeable increase in the ^{26}Al/^{27}Al ratio at the stellar surface, up to a typical value of 2×10^{-3} (lower panel of Fig. 3.12).

The situation is very different for AGB stars of masses higher than approximately 4 M$_\odot$. Proton-captures occurring in setting (2), i.e., HBB at the base of the convective envelope, combined with the 3rd dredged-up of ^{25}Mg produced from

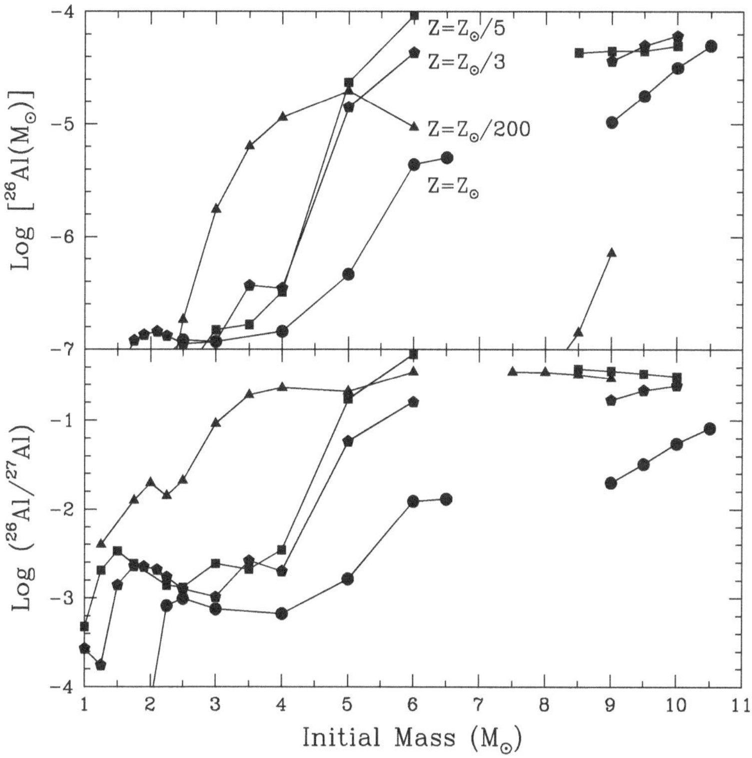

Fig. 3.12 The yields of ^{26}Al (top panel) for stellar models of different masses and metallicities (Z) from Karakas and Lattanzio (2007) for AGB stars of masses up to 6 $M_\odot$, and from Siess and Arnould (2008) for the more massive Super-AGB stars. Yields are defined as the total mass of ^{26}Al (in $M_\odot$) lost in the wind during the whole evolution of the star (calculated as the average of the time-dependent envelope composition weighed on the mass lost at each time). The ratios of the yield of ^{26}Al to the yield of ^{27}Al are also shown in the bottom panel

efficient activation of the ^{22}Ne(α,n)^{25}Mg reaction in the thermal pulse, produces large amounts of ^{26}Al. These are directly mixed to the stellar surface via the envelope convection resulting in yields up to $10^{-4}\,M_\odot$, and ^{26}Al/^{27}Al ratios up to 0.5 (Fig. 3.12). During HBB the main channel for ^{26}Al destruction is proton captures on ^{26}Al itself, i.e., ^{26}Al(p, γ)^{27}Si reactions. Also in Super-AGB stars HBB produces large quantities of ^{26}Al (Siess and Arnould 2008; Lugaro et al. 2012).

Figure 3.12 shows the yields of ^{26}Al and their ratio with the yield of ^{27}Al for a variety of AGB stars and Super-AGB of different masses and metallicities. The plot shows how the efficiency of ^{26}Al production increases with stellar mass and with decreasing metallicity of the stars. This is because the efficiency of the HBB depends on the temperature at the base of the convective envelope, which is higher for higher masses and lower metallicities. For example, a 3.5 $M_\odot$ star of metallicity 200 times lower than solar ejects the same amount of ^{26}Al than a 6.5 $M_\odot$ star at solar metallicity. The reason is that the overall temperature is controlled by the mass of the CO core, which scales directly with the initial mass and inversely with the initial metallicity (see Sect. 3.4.1). In addition, the lower opacity in lower metallicity stars keeps the structure more compact and hence hotter.

In the models of Super-AGB presented in Fig. 3.12 the 3rd dredge-up is found to be negligible and HBB produces ^{26}Al via proton captures on the ^{25}Mg initially present in the envelope, without the contribution of ^{25}Mg from the intershell. Still, these stars produce a large amount of ^{26}Al since there is a large initial amount of ^{25}Mg in the envelope due to the large envelope mass. The lower Super-AGB ^{26}Al yield at metallicity solar/200 is due to very high HBB temperatures, at which the rate of the ^{26}Al destruction reaction $^{26}\mathrm{Al}(p, \gamma)^{27}\mathrm{Si}$ is significantly enhanced.

The yields predicted for the ^{26}Al from AGB stars presented in Fig. 3.12 are quite uncertain since there are several stellar and nuclear uncertainties. First, there are uncertainties related to the modelling of HBB. In fact, the temperature reached at the base of the convective envelope, which governs the efficiency of the $^{25}\mathrm{Mg}(p, \gamma)^{26}\mathrm{Al}$ reaction, depends on the modeling of the temperature gradient within the convectively unstable region. Hence, different treatments of the convective layers may lead to significantly different efficiencies of the HBB. Second, the uncertainty in the efficiency of the 3rd dredge-up already discussed in Sect. 3.4.1 also affects the ^{26}Al yields: in the low-mass models it affects the dredge-up of ^{26}Al itself, in the massive models it affects the dredge-up of ^{25}Mg, which is then converted into ^{26}Al via HBB. Third, the mass-loss rate is another major uncertain parameter in the modelling of AGB stars. The mass-loss rate determines the stellar lifetime and thus the time available to produce ^{26}Al and the final ^{26}Al yield.

Another model uncertainty is related to the possible occurrence of extra mixing at the base of the convective envelope in the low-mass AGB models that do not experience HBB. Such extra mixing in AGB stars would be qualitatively similar to the extra mixing in red giant stars described in Sect. 3.3.2. In the hypothesis of extra mixing, material travels from the base of the convective envelope inside the radiative region close to the H-burning shell, suffers proton captures, and is taken back up into the convective envelope. If the mixed material dips into the H-burning shell, down to temperatures higher than $\simeq$50 MK, then this mechanism could produce ^{26}Al and contribute to some amount of this nucleus in the low-mass models (Nollett et al. 2003; Palmerini et al. 2011). Unfortunately, from a theoretical point of view, there is no agreement on which mechanism drives the extra mixing and on the features of the mixing. Some constraints on it, however, can be derived from the composition of MS, S, SC, and C stars as well as meteoritic stellar grains, as will be discussed in detail in Sect. 3.6.2.

As for nuclear uncertainties, the rate of the $^{26}\mathrm{Al}(p, \gamma)^{27}\mathrm{Si}$ reaction may be uncertain by three orders of magnitude in the temperature range of interest for AGB stars (Iliadis et al. 2001), with the consequence that ^{26}Al yields from AGB stars suffer from uncertainties of up to two orders of magnitude (Izzard et al. 2007; van Raai et al. 2008). New experiments and approaches to estimate this rate are needed to get a more precise determination of the production of ^{26}Al in AGB stars (see, e.g. Iliadis et al. 2010). The rate of the production reaction $^{26}\mathrm{Mg(p,}\gamma)^{26}\mathrm{Al}$ has been recently measured underground by LUNA (Laboratory for Underground Nuclear Astrophysics) at the LNGS laboratories in Italy, where the km-thick rock of the Gran Sasso mountain allows for orders-of-magnitude reduction of the background signals with respect to laboratories on the Earth surface. The rate was determined with high accuracy (Straniero et al. 2013). However, the feeding factor to the long-lived ground state of ^{26}Al is still poorly constrained, with different experiments providing a range of values and with relatively large uncertainties.

In spite of all these important uncertainties, current models do indicate that at least some AGB models produce a significant amount of ^{26}Al. These models cover a small range of stellar masses, only those suffering HBB on the AGB phase. When the yields presented in Fig. 3.12 are averaged over a Salpeter initial stellar mass function, the result is that AGB stars globally do not provide an important contribution to the present abundance of ^{26}Al in the Galaxy. This contribution sums up to only 0.24% of the contribution from massive star winds and core-collapse supernovae (Limongi and Chieffi 2006; Lugaro and Karakas 2008). Adding up the contribution of Super-AGB stars only marginally increases the contribution of AGB stars to Galactic ^{26}Al to 0.85% of the contribution coming from the more massive stars (see also Siess and Arnould 2008).

3.6.2 *Evidence of ^{26}Al in AGB Stars*

It may be possible to determine the abundance of ^{26}Al in AGB stars using molecular lines of Al-bearing molecules. This was carried out by Guelin et al. (1995) for the nearest carbon star, CW Leo, using rotational lines of AlF and AlCl molecules with different Al isotopic composition. One observed line was tentatively attributed to ^{26}AlF, and from its observed strength an upper limit of 0.04 for the ^{26}Al/^{27}Al ratio was inferred. No ^{26}AlCl lines were detected, which led to an upper limit of 0.1. These values cannot be reached by solar metallicity AGB models (Fig. 3.12), however, this detection has not been confirmed so it is doubtful if it can represent a valid model constraint.

The main observational evidence of ^{26}Al in AGB stars comes, instead, from stardust (see Fig. 3.13). Aluminium is one of the main component of most oxide stardust grains recovered to date and the initial amount of ^{26}Al present in each grain can be derived from excesses in its daughter nucleus ^{26}Mg. Magnesium is not a main component in corundum (Al_3O_2) and hibonite ($CaAl_{12}O_{19}$) grains, hence, in this cases, ^{26}Mg excesses are all attributed to ^{26}Al decay. In the case of spinel ($MgAl_2O_4$) grains, instead, Mg is a main component of the mineral and thus the contribution of ^{26}Al to ^{26}Mg needs to be more carefully evaluated by weighing the contribution of the two components. Specifically, there are two atoms of Al per each atom of Mg in spinel, which corresponds to a roughly 25 times higher ratio than in the average solar system material.

The ^{26}Al/^{27}Al ratios are observed to be different in the different populations of oxide and silicate grains (see, e.g., Fig. 8 of Nittler et al. 1997). Population I

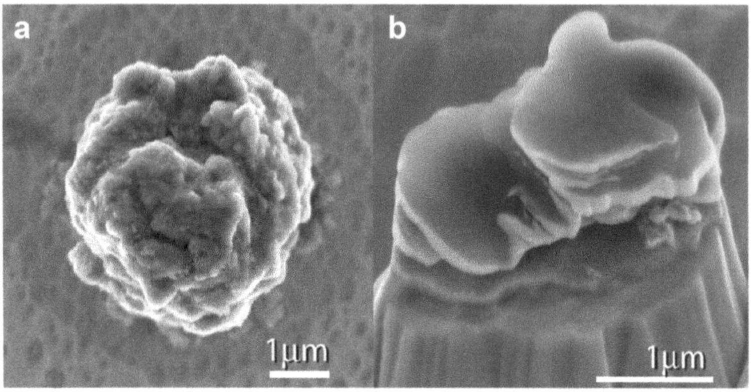

Fig. 3.13 Scanning electron microscope images of dust grains from AGB stars. (**a**) 4-μm-sized silicon carbide (SiC) grain; ubiquitous excesses in ^{26}Mg in such grains indicate prior presence of ^{26}Al. (**b**) 2-μm-sized hibonite ($CaAl_{12}O_{19}$) grain KH15 (Nittler et al. 2008). The grain is sitting on a gold pedestal created by ion-probe sputtering during isotopic analysis. Excesses of ^{26}Mg and ^{41}K indicate that the grain originally condensed with live ^{26}Al and ^{41}Ca

grains cover a wide range of ^{26}Al initial abundance, from no detection to ^{26}Al/^{27}Al $\simeq$0.02. The presence of ^{26}Al might be used to discriminate Population I oxide grains coming from red giant or from AGB stars, since ^{26}Al is expected to be present only in the winds of AGB stars. However, models of the evolution of AGB dust in the Galaxy show that most of these grains originated from AGB stars (Gail et al. 2009). The ^{26}Al/^{27}Al ratios of Population II grains lie at the upper end of the range covered by Population I grains, and reach up to $\simeq$0.1 (see also Fig. 6 of Zinner et al. 2007). This is qualitatively consistent with the strong ^{18}O deficits observed in the Population II grains, since both signatures are produced by H burning. The mysterious Population III show low or no ^{26}Al, which may indicate that these grains did not come from AGB stars. Finally, Population IV grains from supernovae show ^{26}Al/^{27}Al ratios between 0.001 and 0.01 (see Chap. 4).

The ^{26}Al/^{27}Al ratios together with the ^{18}O/^{16}O ratios in Population I and II oxide and silicate grains have provided an interesting puzzle to AGB modellers. Low-mass AGB models do not produce ^{26}Al/^{27}Al ratios high enough and ^{18}O/^{16}O ratios low enough to match the observations. Massive AGB models can produce ^{26}Al/^{27}Al ratios high enough via HBB, however, in this case the ^{18}O/^{16}O ratio is too low ($\sim 10^{-6}$) to match the observations (see Sect. 3.2.1). Grains with ^{18}O/^{16}O $< 10^{-4}$ may have been polluted by solar material during the laboratory analysis, which would have shifted the ^{18}O/^{16}O ratio to higher values with respect to the true ratio of the grain. This argument was invoked to attribute a massive AGB stars origin to a peculiar Population II spinel grain, named OC2 (Lugaro et al. 2007). However, also the ^{17}O/^{16}O ratio presents a problem for this and similar grains because at the temperature of HBB this ratio is always much higher than observed (Boothroyd et al. 1995; Lugaro et al. 2007; Iliadis et al. 2008).

The extra-mixing phenomena mentioned in Sects. 3.3.2 and 3.6.1 have been hypothesized to operate in low-mass AGB stars below the base of the formal convective envelope to explain the composition of Population I and II grains with ^{26}Al/^{27}Al ratios greater than $\simeq 10^{-3}$. This idea has been investigated in detail by Nollett et al. (2003) using a parametric model where the temperature (T_p), determined by the depth at which material is carried, and the mass circulation rate (M_{circ}) in the radiative region between the base of the convective envelope and the H-burning shell are taken as two free and independent parameters. This model was originally proposed to explain observations of AGB stars and grains showing deficits in ^{18}O (Wasserburg et al. 1995). It should be noted, however, that while AGB stars show deficits in their ^{18}O/^{16}O ratios down to roughly 10 times lower than solar, Population II oxide grains show deficits in their ^{18}O/^{16}O ratios down to roughly 100 times lower than solar (see Figs. 5 and 9 of Nittler et al. 1997).

One related problem is the fact that massive AGB stars of mass in the range suffering HBB are predicted to have produced a large fraction of the AGB stardust originally present at the formation of the Sun (Gail et al. 2009). However, if Population II oxide grains originated from low-mass AGB stars instead of massive AGB stars, no stardust grains have ever been recovered with the signature of an origin in massive AGB stars. This conundrum was solved by the recent underground measurement of the ^{17}O$p,\alpha)^{14}$N reaction rate, which controls the ^{17}O/^{16}O ratio produced by H burning. The experiment was performed by LUNA (the Laboratory for Underground Nuclear Astrophysics) taking advantage of a 15-fold reduction of the experimental background with respect to laboratories overground. The rate was found to be 2 to 2.5 higher (Bruno et al. 2016) than previously reported, which allowed Lugaro et al. (2017) to finally invoke a massive AGB stars origin for at least some of the Population II grains.

In the case of SiC grains, Al is present in the grains as a trace element in relatively large abundance, while Mg is almost absent. Again, this means that ^{26}Mg excesses represent the abundance of ^{26}Al at the time when the grains formed. Mainstream SiC grains from AGB stars were reported to show ^{26}Al/^{27}Al ratios between 10^{-4} and $\simeq 2 \times 10^{-3}$. Models of C-rich AGB stars, i.e., the low-mass models in the lower panel of Fig. 3.12, which do not suffer HBB and hence reach C/O>1 in their envelopes, match the observed upper value but did not cover the observed range down to the lower values (Zinner et al. 2007; van Raai et al. 2008). More recent analysis by Groopman et al. (2015) has produced however a more restricted range of ^{26}Al/^{27}Al ratios in mainstream SiC grains roughly between 10^{-2} and 10^{-3}, in better agreement with the models and potentially useful to constraint the reaction rates for the production and destruction of ^{26}Al.

In conclusion, observational constraints of ^{26}Al in AGB stars provide the potential to investigate some of the most uncertain input physics in the modelling of AGB nucleosynthesis: mixing phenomena and reaction rates.

3.6.3 ^{60}Fe

The other famous long-lived radioactive nucleus ^{60}Fe (with a recently revised half life of 2.6 My, Rugel et al. 2009), of interest from the point of view of γ-ray observations, meteoritic stellar grains, and the composition of the early solar system, can be produced in AGB stars (Wasserburg et al. 2006; Lugaro and Karakas 2008) via the neutron-capture chain ^{58}Fe$(n, \gamma)^{59}$Fe$(n, \gamma)^{60}$Fe, where ^{59}Fe is a branching point, and destroyed via the ^{60}Fe$(n, \gamma)^{61}$Fe reaction, whose rate has been measured experimentally by Uberseder et al. (2009). This is the same chain of reactions responsible for the production of this nucleus in massive stars (see Chap. 4). Given that ^{59}Fe is an unstable nucleus with a relatively short half life of 44 days and with a neutron capture cross section $\sigma \simeq 23$ mbarn (Rauscher and Thielemann 2000), neutron densities of at least 10^{10} n/cm^3 are needed for this branching point to open at a level of 20%, allowing production of the long-living ^{60}Fe. If the neutron density is higher than 10^{12} n/cm^3, then 100% of the neutron-capture flux goes through ^{60}Fe.

From the description of the neutron sources in AGB stars (Sect. 3.5.2), it is clear that ^{60}Fe can only be produced in the convective thermal pulses, where the neutron burst released by the ^{22}Ne neutron source can reach the high neutron density required to open the branching point at ^{59}Fe. Hence, the production of ^{60}Fe in AGB stars is almost completely determined by the activation of the ^{22}Ne neutron source. The ^{13}C neutron source may instead destroy some ^{60}Fe in the intershell (Wasserburg et al. 2006).

The AGB yields of ^{60}Fe, and their ratios with the yields of ^{56}Fe, are shown in Fig. 3.14. As the temperature at the base of the convective thermal pulses increases with increasing the stellar mass and decreasing the metallicity, the amount of ^{60}Fe delivered to the interstellar medium increases, reaching up to $10^{-5}\,M_\odot$, a value comparable to that delivered by a supernova of $\simeq 20\,M_\odot$ (Limongi and Chieffi 2006). Ratios of the ^{60}Fe and ^{56}Fe abundances at the end of the AGB phase from the AGB neutron-capture models of Wasserburg et al. (2006) also plotted in Fig. 3.14.[12]

As for ^{26}Al, also in the case of ^{60}Fe stellar and nuclear uncertainties affect the results presented in Fig. 3.14 (and different choice in the model inputs are responsible for variations in the results obtained by different authors). First, the overall mass carried to the envelope via the 3rd dredged-up is essential to the determination of the envelope ^{60}Fe abundance in AGB stars. This is because ^{60}Fe is made only via neutron captures in the He-rich intershell and needs to be mixed into the envelope in order to show up at the stellar surface and to be carried to the interstellar medium by the winds. Hence, the ^{60}Fe yield is directly related to the efficiency of the 3rd dredge-up. For example, models experiencing little or no 3rd dredge-up produce a null ^{60}Fe yield. This important point applies to all long-living radioactive nuclei produced in AGB stars, except for the case of ^{26}Al, which is made

[12]Final *abundance* ratios are equivalent to *yield* ratios because the yields reflect the composition at the end of the evolution, since more than half of the mass lost during the entire life of the star leaves the star at the end of the AGB phase in the superwind.

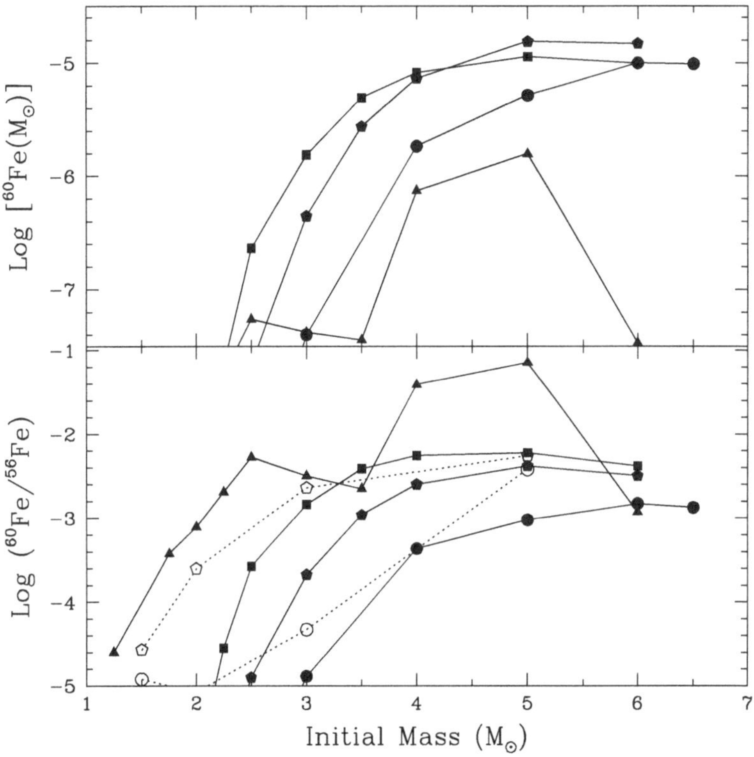

Fig. 3.14 The yields of ^{60}Fe (top panel) from Karakas and Lattanzio (2007) (see caption of Fig. 3.12 for definition of a yield) and the ratio of the yield of ^{60}Fe to the yield of ^{56}Fe (bottom panel) for stellar models of different masses and metallicities (full symbols from Karakas and Lattanzio 2007). The symbols representing the different metallicities are the same as in Fig. 3.12. For comparison, the ratios of the abundances of ^{60}Fe and ^{56}Fe at the end of the AGB evolution computed by Wasserburg et al. (2006) are also shown as open symbols

via HBB directly within the envelope. Second, the mass-loss rate affects the result as it determines the stellar lifetime and thus the number of thermal pulses and 3rd dredge-up episodes.

Nuclear physics inputs that contribute important uncertainties to the production of ^{60}Fe are the rate of the neutron source reaction $^{22}Ne(\alpha,n)^{25}Mg$, which determines how many neutrons are produced in the thermal pulses, and the neutron capture cross section of ^{60}Fe itself, which has been experimentally determined directly (Uberseder et al. 2009), and of the branching point isotope ^{59}Fe, which is estimated theoretically (Rauscher and Thielemann 2000) and via indirect measurements (Uberseder et al. 2014), as the short half life of this nucleus hampers direct experimental determinations.

3.6.4 ^{36}Cl and ^{41}Ca

Two more long-lived radioactive nuclei lighter than iron are of special interest because they are observed to be present in the early solar system and can be made by neutron captures in the intershell of AGB stars: ^{36}Cl (with half life of 0.3 Myr) and ^{41}Ca (with half life of 0.1 Myr). Differently from ^{60}Fe, production of these nuclei does not require the activation of branching points, since ^{36}Cl and ^{41}Ca are made by neutron captures on ^{35}Cl and ^{40}Ca, respectively, which are stable nuclei with relatively high solar abundances. Neutron captures also destroy ^{36}Cl and ^{41}Ca via different channels, the predominants being ^{41}Ca$(n, \alpha)^{38}$Ar, with $\sigma \simeq 360$ mbarn and ^{36}Cl$(n, p)^{36}$S, with $\sigma \simeq 118$ mbarn.

Neutrons coming from the ^{22}Ne neutron source are responsible for the production of ^{36}Cl and ^{41}Ca. As there are no branching points involved, this is not due to the high neutron density of this neutron flux, as it is for the production of ^{60}Fe, but to the fact that neutrons released by the ^{22}Ne in the thermal pulse affect the composition of the whole He-rich intershell material, where large initial quantities of the seed nuclei ^{35}Cl and ^{40}Ca are available. On the contrary, neutrons released by the ^{13}C neutron source affect a small fraction of the intershell material, being the ^{13}C-^{14}N pocket roughly 1/10th to 1/20th of the intershell (by mass) in the current models.

In general, to produce neutron-rich isotopes of elements lighter than iron by the s-process a small number of neutrons captured by seed nucleus are needed: only one in the cases of ^{36}Cl and ^{41}Ca. Hence, final abundances are determined to a higher level by the availability of seed nuclei, rather than that of free neutrons. For the light nuclei a production flux from the lighter to the heaviest elements does not occur (strictly speaking it is not correct to apply the s-process terminology in this case), instead, the nucleosynthetic process is very localized: neutron captures on the sulphur isotopes, for example, do not affect the abundances of the chlorine isotopes and so on. This is because neutron-capture cross section of nuclei lighter than iron are much smaller (by as much as 3 orders of magnitude) than those of typical nuclei heavier than iron. Hence, to produce nuclei heavier than iron by the s-process, instead, including the relatively large number of long-living radioactive nuclei lying on the s-process path discussed in Sect. 3.6.5, a production flux from the lighter to the heavier elements occurs, where many neutrons are captured by the iron seeds and it is possible to reach up to the heaviest elements. Hence, the number of free neutrons plays a dominant role in this case.

The ^{36}Cl/^{35}Cl and ^{41}Ca/^{40}Ca abundance ratios at the end of the AGB evolution computed by Wasserburg et al. (2006) and by van Raai et al. (2012) (which are based on the same codes and stellar models of Karakas and Lattanzio 2007) are plotted in Fig. 3.15. As in the case of ^{60}Fe, the main model uncertainties affecting these results is the efficiency of the 3rd dredge-up, the mass-loss rate, and the rate of the ^{22}Ne(α,n)^{25}Mg reaction.

Moreover, while experimental estimates for the neutron-capture cross section of ^{36}Cl and ^{41}Ca are available (e.g. de Smet et al. 2006), a difficult problem is to provide a reliable set of values for the electron-capture rate of ^{41}Ca, in particular as

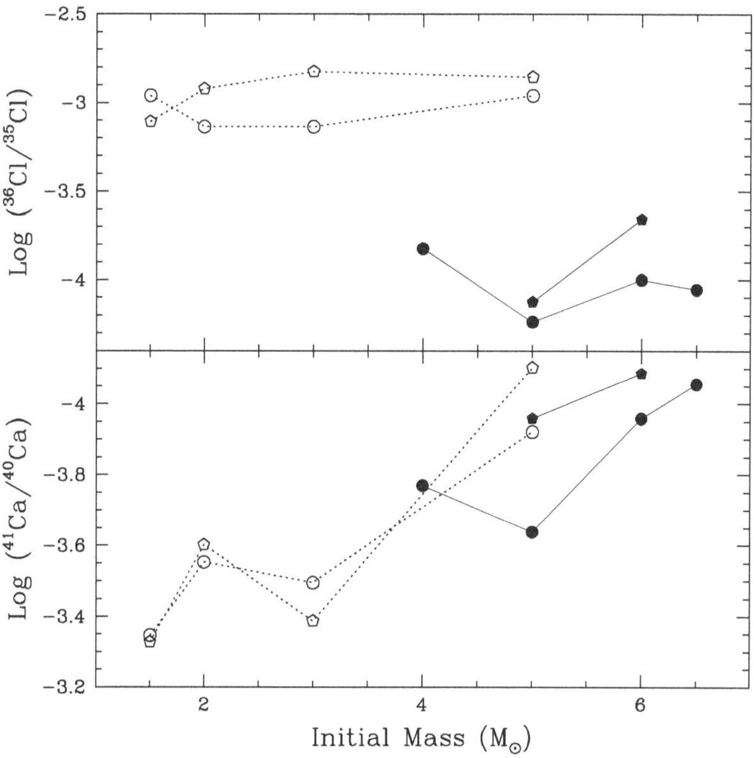

Fig. 3.15 $^{36}Cl/^{35}Cl$ and $^{41}Ca/^{40}Ca$ abundance ratios at the end of the AGB evolution computed by Wasserburg et al. (2006) (open symbols) and by van Raai et al. (2012) (full symbols). The symbols representing the different metallicities are the same as in Fig. 3.12

it is expected to vary significantly for different temperatures and densities relevant to stellar conditions (Chap. 9). As most electron captures in the ^{41}Ca atom occurs on electrons belonging to the electron shell closest to the nucleus (the K shell), when the temperature increases to 100 MK and all electrons have escaped the atom leaving the nucleus bare, the half life of ^{41}Ca increases by almost three orders of magnitude. However, if, still at a temperature of 100 MK, the density increases to 10^4 g/cm^3, electrons are forced nearby ^{41}Ca nuclei and the half life decreases back to its terrestrial value. The only set of theoretical data for this reaction are those provided by Fuller et al. (1982). Moreover, the temperature and density dependence of the electron-capture rate of ^{41}Ca has never been properly implemented in AGB stellar models, in particular it has not yet been solved coupled to convective motions, both in the thermal pulses and in the stellar envelope, where material is constantly carried from hotter denser regions to cooler less dense regions and viceversa. Given these considerations, we are far from an accurate determination of the abundance of ^{41}Ca made by AGB stars.

In summary, and in relevance to the early solar system composition of long-lived radioactive nuclei discussed in Chap. 6, AGB stars can produce some of the radioactive nuclei found to be present in the early solar system: ^{26}Al via hot bottom burning and ^{41}Ca and ^{60}Fe via neutron captures in the thermal pulse and the 3rd dredge-up. In certain mixing conditions the abundances of these nuclei can be produced by AGB stars in the same proportions observed in the early solar system (Wasserburg et al. 2006; Trigo-Rodriguez et al. 2009). On the other hand, ^{36}Cl cannot be produced in the observed amount. Uncertainties in the neutron capture cross sections of ^{35}Cl and ^{36}Cl may play a role in this context.

Finally, a characteristic signature of the AGB stars inventory of long-living radioactive nuclei, is that, unlike supernovae (Chap. 4 and Sect. 5.3), AGB stars cannot possibly produce ^{55}Mn, another nucleus of relevance to early solar system composition. This is because ^{55}Mn is a proton-rich nucleus, lying on the proton-rich side of the valley of β-stability, and thus it cannot be made by neutron captures.

3.6.5 Long-Lived Radioactive Isotopes Heavier than Fe

The *s* process in AGB stars produces significant abundances of six long-lived radioactive nuclei heavier than iron: ^{81}Kr, ^{93}Zr, ^{99}Tc, ^{107}Pd, ^{135}Cs, and ^{205}Pb. The survival of ^{135}Cs and ^{205}Pb in stellar environments is however very uncertain and can even be prevented because of the strong and uncertain temperature and density dependence of their half lives, decreasing by orders of magnitudes in stellar conditions and determined only theoretically (as in the case of ^{41}Ca. See detailed discussion by Mowlavi et al. 1998; Wasserburg et al. 2006, and also Appendix B). While ^{93}Zr, ^{99}Tc, ^{107}Pd, and ^{205}Pb are on the main *s*-process path and are produced by neutron captures on the stable isotopes ^{92}Zr, ^{98}Mo,[13] ^{106}Pd, and ^{204}Pb, respectively, ^{81}Kr, ^{181}Hf, and ^{135}Cs are not on the main *s*-process path, but can be reached via the activation of branching points at ^{79}Se and ^{80}Br, ^{134}Cs, and ^{182}Hf, respectively (as described in Appendix B).

Figure 3.16 presents the abundance ratios of long-living radioactive isotopes heavier than iron produced during the *s*-process in AGB stars to one of their nearest stable isotopes calculated by Wasserburg et al. (2006) and by van Raai et al. (2012). For all ratios, except ^{81}Kr/^{82}Kr, the inclusion of the ^{13}C neutron source for models of masses lower than $\simeq$3–4 $M_\odot$, completely changes the results, since the ^{22}Ne source is not significantly activated in these low-mass stellar models. It also makes an important difference in the absolute abundance of the all isotopes involved, with very low production factors with respect to the initial value if the ^{13}C neutron source is not included (see Table 4 of Wasserburg et al. 2006). The case of ^{81}Kr/^{82}Kr is different in that it does not feel the inclusion of the ^{13}C neutron source as much as the other ratios because, even for the low-mass stars, the marginal activation of

[13]Followed by fast decay of ^{99}Mo, with a half life of 66 h.

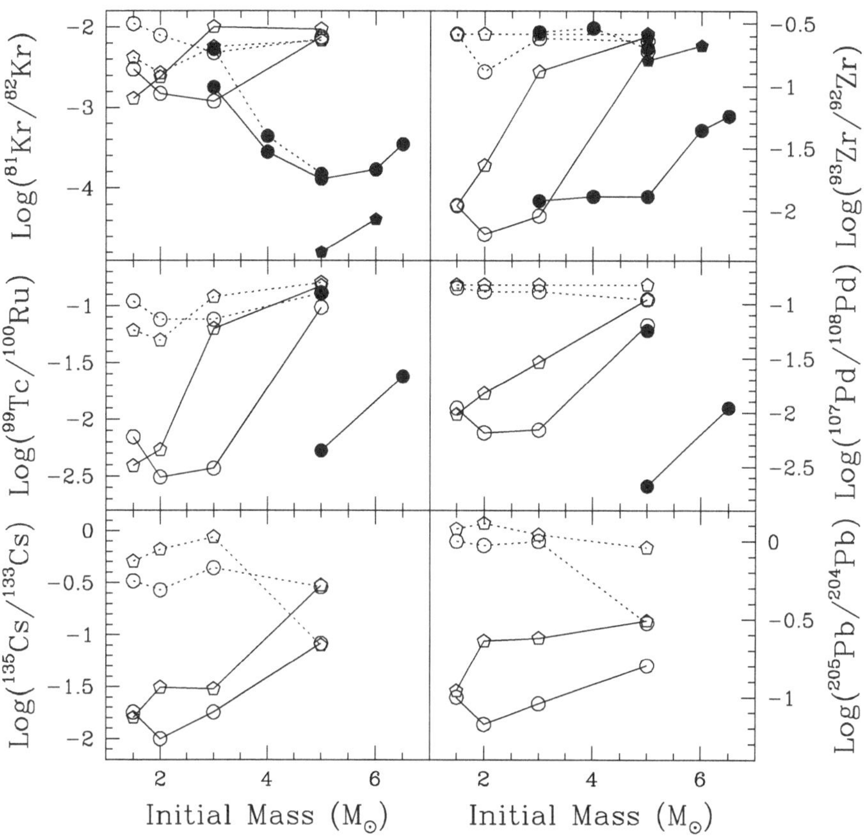

Fig. 3.16 Abundance ratios of long-lived radioactive nuclei heavier than iron, with respect to one of their nearest stable isotope, at the end of the AGB evolution computed by Wasserburg et al. (2006) (open symbols) and by van Raai et al. (2012) (full symbols). The symbols representing the different metallicities are the same as in Fig. 3.12. Symbols connected by the solid line represent models computed without the inclusion of the ^{13}C neutron source, symbols connected by the dotted lines represent models computed with the inclusion of the ^{13}C neutron source

the ^{22}Ne reaction in the latest thermal pulses affects the production of the *s*-process elements up to the first *s*-process peak, including Kr, and of ^{81}Kr in particular via the branching point at ^{79}Se.

For stellar models with initial masses higher than $\simeq$3-4 M$_\odot$, depending on the metallicity, the ^{22}Ne neutron source is mainly responsible for the activation of the *s*-process and thus the production of the heavy long-lived isotopes. Hence, in these models, the inclusion of a ^{13}C neutron source typically does not make a significant difference in the final ratios, except in the case of ^{205}Pb/^{204}Pb. This ratio is different in that it always feels the effect of the inclusion of the ^{13}C neutron source because production of the element Pb, corresponding to the third and last *s*-process peak, is possible only if very large neutron exposures are available ($\sim$ mbarn^{-1}), which can only be produced by the ^{13}C neutron source.

It is interesting to discuss in detail the results for the 3 $M_\odot$ stellar model of 1/3 solar metallicity, because this model represents an example of the transition between the two regimes of the *s* process in AGB stars: when neutrons are provided by the ^{13}C or by the ^{22}Ne source. In this model the number of free neutrons produced by the ^{22}Ne source is higher than in the solar metallicity model of the same mass partly because the temperature in the thermal pulses is slightly higher, but mostly because there is a smaller number of nuclei present to capture neutrons. Hence, the neutron flux coming from the ^{22}Ne neutron source affects the production of the long-living isotopes up to ^{99}Tc, but not that of the long-living isotopes of higher masses: for ^{107}Pd, ^{133}Cs, and ^{205}Pb, ^{13}C is still the main neutron source.

In addition to the main effect due to the shift from the ^{13}C to the ^{22}N regime with changing the initial mass and metallicity of the star, smaller variations due to the marginal effect of the ^{22}Ne neutron source in the models of low-mass are always visible in the details of the production of the heavy long-living nuclei affected by the operation of the branching points activated in thermal pulses: ^{81}Kr, ^{99}Tc, and ^{135}Cs. For example, restricting our view to the solar metallicity models of mass lower than 4 $M_\odot$ and computed with the inclusion of the ^{13}C neutron source, the ^{81}Kr/^{82}Kr ratio decreases with the stellar mass as ^{81}Kr is progressively skipped by the branching point at ^{79}Se at the higher neutron densities experienced by the higher mass models. The opposite happens for the ^{135}Cs/^{133}Cs ratio, which increases with the stellar mass as the branching point at ^{134}Cs becomes progressively more active.

When considering the effect of branching points on the production of heavy long-living radioactive nuclei by the *s*-process in AGB stars it is worth noting that ^{129}I and ^{182}Hf—two long-lived radioactive isotopes of special interest for the composition of the early solar system—were believed to not be significantly produced in AGB stars. Production of ^{129}I is not possible because the half life of ^{128}I is only 25 min (see Appendix B). Until 2014, only a marginal production of ^{182}Hf, up to ^{182}Hf/^{180}Hf$\simeq 10^{-6}$ (Wasserburg et al. 1994), was believed to be possible via activation of the branching point at ^{181}Hf (see Appendix B). This stemmed from the fact that the half life of ^{181}Hf, the branching point leading to the production of ^{182}Hf, was believed to strongly decrease from the terrestrial value of 42 days to roughly a couple of days in stellar conditions, mostly via population of an excited state at 68 keV (Takahashi and Yokoi 1987). However, the more recent, detailed experiments of Bondarenko et al. (2002) on the nuclear structure of ^{181}Hf demonstrated that this energy level does not exist. This allowed (Lugaro et al. 2014a) to attribute a large production of ^{182}Hf to the *s*-process in AGB stars, with a ^{182}Hf/^{180}Hf of the order of 0.15. This resolved the discrepancy between the abundances of ^{129}I and ^{182}Hf in the early solar system and allowed to time the latest *r*- and *s*-process events that contributed to the build-up of solar system matter to roughly 100 Myr and 10–30 Myr, respectively, before the formation of the Sun.

The main uncertainties affecting both sets of predictions shown in Fig. 3.16 are the detailed features of the proton diffusion leading to the production of the ^{13}C neutron source. Ratios that depend on the activation of the ^{22}Ne neutron source are also sensitive to the choice of the mass loss rate and of the ^{22}Ne(α,n)^{25}Mg reaction rate. The treatment of branching points is also of importance in the determination of ^{81}Kr and ^{135}Cs. For example, in the case of ^{81}Kr/^{82}Kr, the treatment of the temperature dependence of the decay rate of the branching point nucleus ^{79}Se is fundamental to the final result, as demonstrated by the fact that including the temperature dependence of these decay rate (as carried out by Wasserburg et al. 2006) produce a ^{81}Kr/^{82}Kr ratio two orders of magnitude larger than using the terrestrial value as constant (van Raai et al. 2012).

In summary, due to the *s*-process, AGB stars are a rich source of radioactive elements heavier than Fe.

In fact, the historical observation of Tc in late type giants (Sect. 3.5.2) was confirmed by the presence of ^{99}Tc in single stardust SiC grains at the time of their formation discovered via laboratory analysis of the Ru isotopic composition of these grains (Savina et al. 2004). Since both Tc and Ru are refractory elements, they were included in SiC grains as trace elements during grain formation. To match the observational stardust data both the contribution of ^{99}Ru and ^{99}Tc predicted by AGB stellar models to the total nuclear abundance at mass 99 must be considered. Radiogenic decay of ^{99}Tc occurs in the intershell in the absence of neutron fluxes, in the stellar envelope, and inside the grains. On the other hand, there is no evidence for a contribution of ^{135}Cs to ^{135}Ba when comparing AGB model predictions to laboratory data of the ^{135}Ba/^{136}Ba ratio in single SiC grains (see Fig. 16 of Lugaro et al. 2003a). This is probably because Cs is not as refractory as Ba and thus was not included in the grains at the time of their formation.

3.7 Conclusions

In summary, radioactive nuclei are both the crucial ingredient necessary to understand the physics of low- to intermediate-mass stars, and the product of the nuclear reactions that occur in their interiors. Neutrinos produced by nuclear energy generation in the Sun and observed on the Earth have allowed us both to peek in the internal layers of the Sun and validate the process of H burning in its core, and to realise the complex nature of the neutrinos themselves. During the late phases of the evolution of low- to intermediate-mass stars, off the Main Sequence and into the red giant and asymptotic giant branches, He burning together with H burning dominate the energy generation. The alternating of these two different types of burning, at different times and in different location within the star, results in a complex evolution and the production of a large variety of chemical elements, together with their radioactive isotopes. Mixing between the deep layers of the star where the nuclear reactions take place, and the surface of the star, where we can see these chemical products, is still one of the fundamental modelling uncertainty. The timescale and strength of the mass loss that carries the chemical elements away from the star, thus enriching the interstellar medium, is another long-standing uncertainty in our modelling of these stars.

In terms of observable radioactivities, H burning at the base of the envelope in relatively massive AGB stars and Super-AGB stars is responsible for the production of the long-lived ^{26}Al. Although the contribution from these stars to the Galactic enrichment of ^{26}Al is relatively small compared to that of massive stars, observational signatures of this nucleosynthesis are recorded in meteoritic stardust grains. The other crucial nucleosynthetic process proven to occur in AGB stars are neutron captures, in the form of the *s* process, producing roughly half of the cosmic abundances of the elements beyond iron. During the *s* process, radioactive nuclei act as branching points on the path of neutron captures, generating a huge diversity of possibilities in the production of the isotopes up to Bi. Implementation of the nuclear properties of these radionuclei (decay rates and neutron-capture cross sections) in nuclear reaction networks, together with stellar modelling, allow us to produce theoretical predictions. These can be compared to observational constraints coming from both spectroscopic observations (for example of Rb) and from laboratory analysis of meteoritic stardust grains (for example of the ^{96}Zr/^{94}Zr ratio) to pinpoint the features of the *s*-process in AGB stars. For example, the Rb/Sr ratios allowed us to realise that the ^{13}C(α,n)^{16}O is the main neutron source, and the ^{96}Zr/^{94}Zr ratio has shed light on the activation of the secondary neutron source ^{22}Ne(α,n)^{25}Mg. Since much of the information on the nuclear properties of radioactive nuclei is available only theoretically, this procedure allows us also to better constrain our understanding of the nuclei themselves. Finally, neutron captures in AGB stars produce many radioactive isotopes with half lives between 0.1 and 20 Myr. These are interesting as the tracer of fresh nucleosynthesis (as in the case of Tc), as well as in relation of their abundances in the early Solar System (see Chap. 6). Most likely AGB stars contributed the ^{107}Pd and ^{182}Hf that we observe to have been present in the early solar system from analysis meteoritic rocks and inclusions and that are used for dating events in the early Solar System related to the formation and evolution of planetesimals and planets.

References

Abdurashitov JN, Bowles TJ, Cherry ML, Cleveland BT, Davis R, Elliott SR, Gavrin VN, Girin SV, Gorbachev VV, Ibragimova TV, Kalikhov AV, Khairnasov NG, Knodel TV, Lande K, Mirmov IN, Nico JS, Shikhin AA, Teasdale WA, Veretenkin EP, Vermul VM, Wark DL, Wildenhain PS, Wilkerson JF, Yants VE, Zatsepin GT (1999) Measurement of the solar neutrino capture rate by SAGE and implications for neutrino oscillations in vacuum. Phys Rev Lett 83:4686–4689. https://doi.org/10.1103/PhysRevLett.83.4686, arXiv:astro-ph/9907131

Abia C, Busso M, Gallino R, Domínguez I, Straniero O, Isern J (2001) The ^{85}Kr s-Process branching and the mass of carbon stars. Astrophys J 559:1117–1134. https://doi.org/10.1086/322383, arXiv:astro-ph/0105486

Abia C, Hedrosa RP, Domínguez I, Straniero O (2017) The puzzle of the CNO isotope ratios in asymptotic giant branch carbon stars. Astron Astrophys 599:A39. https://doi.org/10.1051/0004-6361/201629969, 1611.06400

Amari S, Gao X, Nittler LR, Zinner E, José J, Hernanz M, Lewis RS (2001a) Presolar grains from novae. Astrophys J 551:1065–1072. https://doi.org/10.1086/320235, arXiv:astro-ph/0012465

Amari S, Nittler LR, Zinner E, Gallino R, Lugaro M, Lewis RS (2001b) Presolar SIC grains of type Y: origin from low-metallicity asymptotic giant branch stars. Astrophys J 546:248–266. https://doi.org/10.1086/318230

Amari S, Nittler LR, Zinner E, Lodders K, Lewis RS (2001c) Presolar SiC grains of type A and B: their isotopic compositions and stellar origins. Astrophys J 559:463–483. https://doi.org/10.1086/322397

Aoki W, Honda S, Beers TC, Sneden C (2003a) Measurement of the europium isotope ratio for the extremely metal poor, r-Process-enhanced star CS 31082-001. Astrophys J 586:506–511. https://doi.org/10.1086/367540, arXiv:astro-ph/0211617

Aoki W, Ryan SG, Iwamoto N, Beers TC, Norris JE, Ando H, Kajino T, Mathews GJ, Fujimoto MY (2003b) Europium Isotope Ratios in s-Process Element-enhanced Metal-poor Stars: A New Probe of the ^{151}Sm Branching. Astrophys J 592:L67–L70. https://doi.org/10.1086/377681, arXiv:astro-ph/0306544

Arlandini C, Käppeler F, Wisshak K, Gallino R, Lugaro M, Busso M, Straniero O (1999) Neutron capture in low-mass asymptotic giant branch stars: cross sections and abundance signatures. Astrophys J 525:886–900. https://doi.org/10.1086/307938, arXiv:astro-ph/9906266

Arpesella C, Back HO, Balata M, Bellini G, Benziger J, Bonetti S, Brigatti A, Caccianiga B, Cadonati L, Calaprice F, Carraro C, Cecchet G, Chavarria A, Chen M, Dalnoki-Veress F, D'Angelo D, de Bari A, de Bellefon A, de Kerret H, Derbin A, Deutsch M, di Credico A, di Pietro G, Eisenstein R, Elisei F, Etenko A, Fernholz R, Fomenko K, Ford R, Franco D, Freudiger B, Galbiati C, Gatti F, Gazzana S, Giammarchi M, Giugni D, Goeger-Neff M, Goldbrunner T, Goretti A, Grieb C, Hagner C, Hampel W, Harding E, Hardy S, Hartman FX, Hertrich T, Heusser G, Ianni A, Ianni A, Joyce M, Kiko J, Kirsten T, Kobychev V, Korga G, Korschinek G, Kryn D, Lagomarsino V, Lamarche P, Laubenstein M, Lendvai C, Leung M, Lewke T, Litvinovich E, Loer B, Lombardi P, Ludhova L, Machulin I, Malvezzi S, Manecki S, Maneira J, Maneschg W, Manno I, Manuzio D, Manuzio G, Martemianov A, Masetti F, Mazzucato U, McCarty K, McKinsey D, Meindl Q, Meroni E, Miramonti L, Misiaszek M, Montanari D, Monzani ME, Muratova V, Musico P, Neder H, Nelson A, Niedermeier L, Oberauer L, Obolensky M, Orsini M, Ortica F, Pallavicini M, Papp L, Parmeggiano S, Perasso L, Pocar A, Raghavan RS, Ranucci G, Rau W, Razeto A, Resconi E, Risso P, Romani A, Rountree D, Sabelnikov A, Saldanha R, Salvo C, Schimizzi D, Schönert S, Shutt T, Simgen H, Skorokhvatov M, Smirnov O, Sonnenschein A, Sotnikov A, Sukhotin S, Suvorov Y, Tartaglia R, Testera G, Vignaud D, Vitale S, Vogelaar RB, von Feilitzsch F, von Hentig R, von Hentig T, Wojcik M, Wurm M, Zaimidoroga O, Zavatarelli S, Zuzel G (2008) Direct measurement of the Be7 solar neutrino flux with 192 days of borexino data. Phys Rev Lett 101(9):091302. https://doi.org/10.1103/PhysRevLett.101.091302, 0805.3843

Aschwanden MJ (2008) Solar flare physics enlivened by TRACE and RHESSI. J Astrophys Astron 29:115–124. https://doi.org/10.1007/s12036-008-0015-0

Asplund M, Grevesse N, Sauval AJ, Scott P (2009) The chemical composition of the sun. Annu Rev Astron Astrophys 47:481–522. https://doi.org/10.1146/annurev.astro.46.060407.145222, 0909.0948

Ávila JN, Lugaro M, Ireland TR, Gyngard F, Zinner E, Cristallo S, Holden P, Buntain J, Amari S, Karakas A (2012) Tungsten isotopic compositions in stardust SiC grains from the murchison meteorite: constraints on the s-process in the Hf-Ta-W-Re-Os Region. Astrophys J 744:49. https://doi.org/10.1088/0004-637X/744/1/49, 1110.4763

Ávila JN, Ireland TR, Lugaro M, Gyngard F, Zinner E, Cristallo S, Holden P, Rauscher T (2013) Europium s-process signature at close-to-solar Metallicity in stardust SiC grains from asymptotic giant branch stars. Astrophys J 768:L18. https://doi.org/10.1088/2041-8205/768/1/L18, 1303.5932

Bahcall JN, Cleveland BT, Davis R Jr, Rowley JK (1985) Chlorine and gallium solar neutrino experiments. Astrophys J 292:L79–L82. https://doi.org/10.1086/184477

Balachandran SC (2005) Anomalous abundances in red giants: the Li-Rich stars. In: Barnes TG III, Bash FN (eds) Cosmic abundances as records of stellar evolution and nucleosynthesis, Astronomical society of the pacific conference series, vol 336, p 113

Bellini G, Benziger J, Bonetti S, Buizza Avanzini M, Caccianiga B, Cadonati L, Calaprice F, Carraro C, Chavarria A, Chepurnov A, Dalnoki-Veress F, D'Angelo D, Davini S, de Kerret H, Derbin A, Etenko A, Fomenko K, Franco D, Galbiati C, Gazzana S, Ghiano C, Giammarchi M, Goeger-Neff M, Goretti A, Guardincerri E, Hardy S, Ianni A, Ianni A, Joyce M, Korga G, Kryn D, Laubenstein M, Leung M, Lewke T, Litvinovich E, Loer B, Lombardi P, Ludhova L, Machulin I, Manecki S, Maneschg W, Manuzio G, Meindl Q, Meroni E, Miramonti L, Misiaszek M, Montanari D, Muratova V, Oberauer L, Obolensky M, Ortica F, Pallavicini M, Papp L, Perasso L, Perasso S, Pocar A, Raghavan RS, Ranucci G, Razeto A, Re A, Risso P, Romani A, Rountree D, Sabelnikov A, Saldanha R, Salvo C, Schönert S, Simgen H, Skorokhvatov M, Smirnov O, Sotnikov A, Sukhotin S, Suvorov Y, Tartaglia R, Testera G, Vignaud D, Vogelaar RB, von Feilitzsch F, Winter J, Wojcik M, Wright A, Wurm M, Xu J, Zaimidoroga O, Zavatarelli S, Zuzel G, Borexino Collaboration (2010) Measurement of the solar B8 neutrino rate with a liquid scintillator target and 3 MeV energy threshold in the Borexino detector. Phys Rev D 82(3):033006. https://doi.org/10.1103/PhysRevD.82.033006, 0808.2868

Bond H, Sion E, Murdin P (2000) CH stars and barium stars. https://doi.org/10.1888/0333750888/5413

Bondarenko V, Berzins J, Prokofjevs P, Simonova L, von Egidy T, Honzátko J, Tomandl I, Alexa P, Wirth HF, Köster U, Eisermann Y, Metz A, Graw G, Hertenberger R, Rubacek L (2002) Interplay of quasiparticle and phonon excitations in ^{181}Hf. Nucl Phys A 709:3–59. https://doi.org/10.1016/S0375-9474(02)00646-2

Boothroyd AI, Sackmann IJ, Wasserburg GJ (1995) Hot bottom burning in asymptotic giant branch stars and its effect on oxygen isotopic abundances. Astrophys J 442:L21–L24. https://doi.org/10.1086/187806

BOREXINO Collaboration, Bellini G, Benziger J, Bick D, Bonfini G, Bravo D, Caccianiga B, Cadonati L, Calaprice F, Caminata A, Cavalcante P, Chavarria A, Chepurnov A, D'Angelo D, Davini S, Derbin A, Empl A, Etenko A, Fomenko K, Franco D, Gabriele F, Galbiati C, Gazzana S, Ghiano C, Giammarchi M, Göger-Neff M, Goretti A, Gromov M, Hagner C, Hungerford E, Ianni A, Ianni A, Kobychev V, Korablev D, Korga G, Kryn D, Laubenstein M, Lehnert B, Lewke T, Litvinovich E, Lombardi F, Lombardi P, Ludhova L, Lukyanchenko G, Machulin I, Manecki S, Maneschg W, Marcocci S, Meindl Q, Meroni E, Meyer M, Miramonti L, Misiaszek M, Montuschi M, Mosteiro P, Muratova V, Oberauer L, Obolensky M, Ortica F, Otis K, Pallavicini M, Papp L, Perasso L, Pocar A, Ranucci G, Razeto A, Re A, Romani A, Rossi N, Saldanha R, Salvo C, Schönert S, Simgen H, Skorokhvatov M, Smirnov O, Sotnikov A, Sukhotin S, Suvorov Y, Tartaglia R, Testera G, Vignaud D, Vogelaar RB, von Feilitzsch F, Wang H, Winter J, Wojcik M, Wright A, Wurm M, Zaimidoroga O, Zavatarelli S, Zuber K, Zuzel G (2014) Neutrinos from the primary proton-proton fusion process in the Sun. Nature 512:383–386. https://doi.org/10.1038/nature13702

Brandon AD, Humayun M, Puchtel IS, Leya I, Zolensky M (2005) Osmium isotope evidence for an s-Process carrier in primitive chondrites. Science 309:1233–1236. https://doi.org/10.1126/science.1115053

Bruno CG, Scott DA, Aliotta M, Formicola A, Best A, Boeltzig A, Bemmerer D, Broggini C, Caciolli A, Cavanna F, Ciani GF, Corvisiero P, Davinson T, Depalo R, Di Leva A, Elekes Z, Ferraro F, Fülöp Z, Gervino G, Guglielmetti A, Gustavino C, Gyürky G, Imbriani G, Junker M, Menegazzo R, Mossa V, Pantaleo FR, Piatti D, Prati P, Somorjai E, Straniero O, Strieder F, Szücs T, Takács MP, Trezzi D, LUNA Collaboration (2016) Improved direct measurement of the 64.5 keV resonance strength in the ^{17}O (p,α)^{14}N reaction at LUNA. Phys Rev Lett 117(14):142502. https://doi.org/10.1103/PhysRevLett.117.142502, 1610.00483

Cameron AGW, Fowler WA (1971) Lithium and the s-PROCESS in red-giant stars. Astrophys J 164:111. https://doi.org/10.1086/150821

Campbell SW, Lattanzio JC (2008) Evolution and nucleosynthesis of extremely metal-poor and metal-free low- and intermediate-mass stars. I. Stellar yield tables and the CEMPs. Astron Astrophys 490:769–776. https://doi.org/10.1051/0004-6361:200809597, 0901.0799

Castilho BV, Gregorio-Hetem J, Spite F, Barbuy B, Spite M (2000) Detailed analysis of a sample of Li-rich giants. Astron Astrophys 364:674–682

Chupp EL (1971) Gamma ray and neutron emissions from the sun. Space Sci Rev 12:486–525. https://doi.org/10.1007/BF00171976

Clayton DD (1968) Principles of stellar evolution and nucleosynthesis

Clayton DD, Nittler LR (2004) Astrophysics with presolar stardust. Annu Rev Astron Astrophys 42:39–78. https://doi.org/10.1146/annurev.astro.42.053102.134022

Cleveland BT, Daily T, Davis RJ, Distel JR, Lande K, Lee CK, Wildenhain PS, Ullman J (1998) Measurement of the solar electron neutrino flux with the homestake chlorine detector. Astrophys J 496:505. https://doi.org/10.1086/305343

Cosner K, Iben I Jr, Truran JW (1980) The effects of unthermalized isomeric states and of a time-varying neutron flux on s-process branching ratios. Astrophys J 238:L91–L96. https://doi.org/10.1086/183265

Côté B, Fryer CL, Belczynski K, Korobkin O, Chruślińska M, Vassh N, Mumpower MR, Lippuner J, Sprouse TM, Surman R, Wollaeger R (2017) The origin of r-Process elements in the Milky Way. ArXiv e-prints 1710.05875

Cowan JJ, Rose WK (1977) Production of C-14 and neutrons in red giants. Astrophys J 212:149–158. https://doi.org/10.1086/155030

Cox JP, Giuli RT (1968) Principles of stellar structure. Gordon and Breach Science Publishers, New York

Cristallo S, Straniero O, Gallino R, Piersanti L, Domínguez I, Lederer MT (2009) Evolution, nucleosynthesis, and yields of low-mass asymptotic giant branch stars at different metallicities. Astrophys J 696:797–820. https://doi.org/10.1088/0004-637X/696/1/797

Cristallo S, Piersanti L, Straniero O (2016) The FRUITY database on AGB stars: past, present and future. J Phys Conf Ser 665:012019. https://doi.org/10.1088/1742-6596/665/1/012019, 1405.3392

Dardelet L, Ritter C, Prado P, Heringer E, Higgs C, Jones S, Denissenkov P, Venn K, Bertolli M, Pignatari M, Woodward P, Herwig F (2014) i process and CEMP-s/r stars. In: Nuclei in the Cosmos (NIC XIII), p PoS(NIC XIII)145, 1505.05500

Dauphas N, Schauble EA (2016) Mass fractionation laws, mass-independent effects, and isotopic anomalies. Annu Rev Earth Planet Sci 44:709–783. https://doi.org/10.1146/annurev-earth-060115-012157

de Smet L, Wagemans C, Heyse J, Vermote S, Van Gils J (2006) Experimental determination of the ^{41}Ca$(n,\alpha)^{38}$Ar reaction cross section. In: Ninth international symposium on nuclei in the Cosmos, Proceedings of science, PoS(NIC-IX)085

Dell'Agli F, García-Hernández DA, Schneider R, Ventura P, La Franca F, Valiante R, Marini E, Di Criscienzo M (2017) Asymptotic giant branch and super-asymptotic giant branch stars: modelling dust production at solar metallicity. Mon Not R Astron Soc 467:4431–4440. https://doi.org/10.1093/mnras/stx387, 1702.03904

Denissenkov PA, Tout CA (2003) Partial mixing and formation of the ^{13}C pocket by internal gravity waves in asymptotic giant branch stars. Mon Not R Astron Soc 340:722–732. https://doi.org/10.1046/j.1365-8711.2003.06284.x

Denissenkov PA, Herwig F, Battino U, Ritter C, Pignatari M, Jones S, Paxton B (2017) I-process nucleosynthesis and mass retention efficiency in he-shell flash evolution of rapidly accreting white dwarfs. Astrophys J 834:L10. https://doi.org/10.3847/2041-8213/834/2/L10, 1610.08541

Dennis BR, Hudson HS, Krucker S (2007) Review of selected RHESSI solar results. In: Klein KL, MacKinnon AL (eds) Lecture Notes in Physics, vol 725. Springer, Berlin, p 33

Despain KH (1980) A difficulty with Ne-22 as the neutron source for the solar system s-process. Astrophys J 236:L165–L168. https://doi.org/10.1086/183219

Doherty CL, Gil-Pons P, Siess L, Lattanzio JC (2017) Super-AGB stars and their role as electron capture supernova progenitors. Publ Astron Soc Aust 34:e056. https://doi.org/10.1017/pasa.2017.52, 1703.06895

Dorfi EA, Höfner S, Feuchtinger MU (2001) Pulsation and mass loss. Kluwer Academic Publisher, Dordrecht, pp 137–154

Dupree AK (1986) Mass loss from cool stars. Annu Rev Astron Astrophys 24:377–420. https://doi.org/10.1146/annurev.aa.24.090186.002113

Ferrarotti AS, Gail HP (2002) Mineral formation in stellar winds. III. Dust formation in S stars. Astron Astrophys 382:256–281. https://doi.org/10.1051/0004-6361:20011580

Ferrarotti AS, Gail HP (2006) Composition and quantities of dust produced by AGB-stars and returned to the interstellar medium. Astron Astrophys 447:553–576. https://doi.org/10.1051/0004-6361:20041198

Fleischer AJ, Gauger A, Sedlmayr E (1992) Circumstellar dust shells around long-period variables. I - Dynamical models of C-stars including dust formation, growth and evaporation. Astron Astrophys 266:321–339

Frogel JA, Mould J, Blanco VM (1990) The asymptotic giant branch of magellanic cloud clusters. Astrophys J 352:96–122. https://doi.org/10.1086/168518

Fuller GM, Fowler WA, Newman MJ (1982) Stellar weak interaction rates for intermediate mass nuclei. III - Rate tables for the free nucleons and nuclei with A= 21 to A = 60. Astrophys J Suppl 48:279–319. https://doi.org/10.1086/190779

Gail HP, Sedlmayr E (1999) Mineral formation in stellar winds. I. Condensation sequence of silicate and iron grains in stationary oxygen rich outflows. Astron Astrophys 347:594–616

Gail HP, Zhukovska SV, Hoppe P, Trieloff M (2009) Stardust from asymptotic giant branch stars. Astrophys J 698:1136–1154. https://doi.org/10.1088/0004-637X/698/2/1136

Gallino R, Arlandini C, Busso M, Lugaro M, Travaglio C, Straniero O, Chieffi A, Limongi M (1998) Evolution and nucleosynthesis in low-mass asymptotic giant branch stars. II. Neutron capture and the s-Process. Astrophys J 497:388. https://doi.org/10.1086/305437

García-Hernández DA, García-Lario P, Plez B, D'Antona F, Manchado A, Trigo-Rodríguez JM (2006) Rubidium-rich asymptotic giant branch stars. Science 314:1751. https://doi.org/10.1126/science.1133706, arXiv:astro-ph/0611319

García-Hernández DA, García-Lario P, Plez B, Manchado A, D'Antona F, Lub J, Habing H (2007) Lithium and zirconium abundances in massive Galactic O-rich AGB stars. Astron Astrophys 462:711–730. https://doi.org/10.1051/0004-6361:20065785, arXiv:astro-ph/0609106

Goriely S, Mowlavi N (2000) Neutron-capture nucleosynthesis in AGB stars. Astron Astrophys 362:599–614

Grevesse N, Sauval AJ (1998) Standard solar composition. Space Sci Rev 85:161–174. https://doi.org/10.1023/A:1005161325181

Gribov V, Pontecorvo B (1969) Neutrino astronomy and lepton charge. Physics Letters B 28:493–496. https://doi.org/10.1016/0370-2693(69)90525-5

Groopman E, Zinner E, Amari S, Gyngard F, Hoppe P, Jadhav M, Lin Y, Xu Y, Marhas K, Nittler LR (2015) Inferred initial ^{26}Al/^{27}Al ratios in presolar stardust grains from supernovae are higher than previously estimated. Astrophys J 809:31. https://doi.org/10.1088/0004-637X/809/1/31

Guelin M, Forestini M, Valiron P, Ziurys LM, Anderson MA, Cernicharo J, Kahane C (1995) Nucleosynthesis in AGB stars: observation of Mg-25 and Mg-26 in IRC+10216 and possible detection of Al-26. Astron Astrophys 297:183–196

Hampel W, Heusser G, Kiko J, Kirsten T, Laubenstein M, Pernicka E, Rau W, Roenn U, Schlosser C, Wojcik M, von Ammon R, Ebert KH, Fritsch T, Heidt D, Henrich E, Stieglitz L, Weirich F, Balata M, Hartmann FX, Sann M, Bellotti E, Cattadori C, Cremonesi O, Ferrari N, Fiorini E, Zanotti L, Altmann M, von Feilitzsch F, Moessbauer R, Berthomieu G, Schatzman E, Carmi I, Dostrovsky I, Bacci C, Belli P, Bernabei R, D'Angelo S, Paoluzi L, Bevilacqua A, Cribier M, Gosset L, Rich J, Spiro M, Tao C, Vignaud D, Boger J, Hahn RL, Rowley JK, Stoenner RW, Weneser J (1998) Final results of the ^{51}Cr neutrino source experiments in GALLEX. Phys Lett B 420:114–126

Hampel M, Stancliffe RJ, Lugaro M, Meyer BS (2016) The intermediate neutron-capture process and carbon-enhanced metal-poor stars. Astrophys J 831:171. https://doi.org/10.3847/0004-637X/831/2/171, 1608.08634

Harris MJ, Lambert DL, Hinkle KH, Gustafsson B, Eriksson K (1987) Oxygen isotopic abundances in evolved stars. III - 26 carbon stars. Astrophys J 316:294–304. https://doi.org/10.1086/165201

Heck PR, Marhas KK, Hoppe P, Gallino R, Baur H, Wieler R (2007) Presolar He and Ne isotopes in single circumstellar SiC grains. Astrophys J 656:1208–1222. https://doi.org/10.1086/510478

Heck PR, Amari S, Hoppe P, Baur H, Lewis RS, Wieler R (2009) Ne isotopes in individual presolar graphite grains from the murchison meteorite together with He, C, O, Mg-Al isotopic analyses as tracers of their origins. Astrophys J 701:1415–1425. https://doi.org/10.1088/0004-637X/701/2/1415

Herwig F (2005) Evolution of asymptotic giant branch stars. Annu Rev Astron Astrophys 43:435–479. https://doi.org/10.1146/annurev.astro.43.072103.150600

Herwig F, Bloecker T, Schoenberner D, El Eid M (1997) Stellar evolution of low and intermediate-mass stars. IV. Hydrodynamically-based overshoot and nucleosynthesis in AGB stars. Astron Astrophys 324:L81–L84. arXiv:astro-ph/9706122

Herwig F, Blöcker T, Langer N, Driebe T (1999) On the formation of hydrogen-deficient post-AGB stars. Astron Astrophys 349:L5–L8. arXiv:astro-ph/9908108

Herwig F, Pignatari M, Woodward PR, Porter DH, Rockefeller G, Fryer CL, Bennett M, Hirschi R (2011) Convective-reactive proton-^{12}C combustion in Sakurai's object (V4334 Sagittarii) and implications for the evolution and yields from the first generations of stars. Astrophys J 727:89. https://doi.org/10.1088/0004-637X/727/2/89, 1002.2241

Hinkle KH, Lebzelter T, Straniero O (2016) Carbon and oxygen isotopic ratios for nearby miras. Astrophys J 825:38. https://doi.org/10.3847/0004-637X/825/1/38, 1606.08478

Hoffman RD, Müller B, Janka H (2008) Nucleosynthesis in O-Ne-Mg supernovae. Astrophys J 676:L127–L130. https://doi.org/10.1086/587621, 0712.4257

Höfner S, Olofsson H (2018) Mass loss of stars on the asymptotic giant branch. Mechanisms, models and measurements. Astron Astrophys Rev 26:1. https://doi.org/10.1007/s00159-017-0106-5

Hoppe P, Ott U (1997) Mainstream silicon carbide grains from meteorites. In: Bernatowicz TJ, Zinner E (eds) American Institute of Physics conference series, vol 402, pp 27–58. https://doi.org/10.1063/1.53314

Hoppe P, Annen P, Strebel R, Eberhardt P, Gallino R, Lugaro M, Amari S, Lewis RS (1997) Meteoritic silicon carbide grains with unusual Si isotopic compositions: evidence for an origin in low-mass, Low-metallicity asymptotic giant branch stars. Astrophys J 487:L101. https://doi.org/10.1086/310869

Humayun M, Brandon AD (2007) s-Process implications from osmium isotope anomalies in chondrites. Astrophys J 664:L59–L62. https://doi.org/10.1086/520636

Iben I Jr (1975) Thermal pulses; p-capture, alpha-capture, s-process nucleosynthesis; and convective mixing in a star of intermediate mass. Astrophys J 196:525–547. https://doi.org/10.1086/153433

Iben I Jr, Renzini A (1982) On the formation of carbon star characteristics and the production of neutron-rich isotopes in asymptotic giant branch stars of small core mass. Astrophys J 263:L23–L27. https://doi.org/10.1086/183916

Iben I Jr, Renzini A (1983) Asymptotic giant branch evolution and beyond. Annu Rev Astron Astrophys 21:271–342. https://doi.org/10.1146/annurev.aa.21.090183.001415

Iben I Jr, Truran JW (1978) On the surface composition of thermally pulsing stars of high luminosity and on the contribution of such stars to the element enrichment of the interstellar medium. Astrophys J 220:980–995. https://doi.org/10.1086/155986

Iliadis C, D'Auria JM, Starrfield S, Thompson WJ, Wiescher M (2001) Proton-induced thermonuclear reaction rates for A=20-40 nuclei. Astrophys J Suppl 134:151–171. https://doi.org/10.1086/320364

Iliadis C, Angulo C, Descouvemont P, Lugaro M, Mohr P (2008) New reaction rate for O16(p,γ)F17 and its influence on the oxygen isotopic ratios in massive AGB stars. Phys Rev C 77(4):045802. https://doi.org/10.1103/PhysRevC.77.045802, 0803.2757

Iliadis C, Longland R, Champagne AE, Coc A, Fitzgerald R (2010) Charged-particle thermonuclear reaction rates: II. Tables and graphs of reaction rates and probability density functions. Nucl Phys A 841:31–250. https://doi.org/10.1016/j.nuclphysa.2010.04.009, 1004.4517

Itoh N, Adachi T, Nakagawa M, Kohyama Y, Munakata H (1989) Neutrino energy loss in stellar interiors. III - Pair, photo-, plasma, and bremsstrahlung processes. Astrophys J 339:354–364. https://doi.org/10.1086/167301

Izzard RG, Lugaro M, Karakas AI, Iliadis C, van Raai M (2007) Reaction rate uncertainties and the operation of the NeNa and MgAl chains during HBB in intermediate-mass AGB stars. Astron Astrophys 466:641–648. https://doi.org/10.1051/0004-6361:20066903, arXiv:astro-ph/0703078

Jones S, Ritter C, Herwig F, Fryer C, Pignatari M, Bertolli MG, Paxton B (2016) H ingestion into He-burning convection zones in super-AGB stellar models as a potential site for intermediate neutron-density nucleosynthesis. Mon Not R Astron Soc 455:3848–3863. https://doi.org/10.1093/mnras/stv2488, 1510.07417

Jonsell K, Barklem PS, Gustafsson B, Christlieb N, Hill V, Beers TC, Holmberg J (2006) The Hamburg/ESO R-process enhanced star survey (HERES). III. HE 0338-3945 and the formation of the r + s stars. Astron Astrophys 451:651–670. https://doi.org/10.1051/0004-6361:20054470. arXiv:astro-ph/0601476

Jorissen A, Van Eck S, Mayor M, Udry S (1998) Insights into the formation of barium and Tc-poor S stars from an extended sample of orbital elements. Astron Astrophys 332:877–903. arXiv:astro-ph/9801272

Kaeppeler F, Beer H, Wisshak K, Clayton DD, Macklin RL, Ward RA (1982) S-process studies in the light of new experimental cross sections - distribution of neutron fluences and r-process residuals. Astrophys J 257:821–846. https://doi.org/10.1086/160033

Karakas A, Lattanzio JC (2007) Stellar models and yields of asymptotic giant branch stars. Publ Astron Soc Aust 24:103–117. https://doi.org/10.1071/AS07021, 0708.4385

Karakas AI, Lattanzio JC (2014) The dawes review 2: nucleosynthesis and stellar yields of low- and intermediate-mass single stars. Publ Astron Soc Aust 31:e030. https://doi.org/10.1017/pasa.2014.21, 1405.0062

Karakas AI, Lugaro M (2016) Stellar yields from metal-rich asymptotic giant branch models. Astrophys J 825:26. https://doi.org/10.3847/0004-637X/825/1/26, 1604.02178

Kozlovsky B, Murphy RJ, Ramaty R (2002) Nuclear deexcitation gamma-ray lines from accelerated particle interactions. Astrophys J Suppl 141:523–541. https://doi.org/10.1086/340545

Lambert DL, Smith VV, Busso M, Gallino R, Straniero O (1995) The chemical composition of red giants. IV. The neutron density at the s-Process site. Astrophys J 450:302. https://doi.org/10.1086/176141

Langer N, Heger A, Wellstein S, Herwig F (1999) Mixing and nucleosynthesis in rotating TP-AGB stars. Astron Astrophys 346:L37–L40. arXiv:astro-ph/9904257

Lattanzio JC, Wood PR (2004) Evolution, Nucleosynthesis and pulsation of AGB stars. In: Habing HJ, Olofsson H (eds) Asymptotic giant branch stars, astronomy and astrophysics library, vol 145. Springer, Berlin, p 23

Lewis RS, Amari S, Anders E (1994) Interstellar grains in meteorites: II. SiC and its noble gases. Geochim Cosmochim Acta 58:471–494. https://doi.org/10.1016/0016-7037(94)90478-2

Limongi M, Chieffi A (2006) The nucleosynthesis of ^{26}Al and ^{60}Fe in solar metallicity stars extending in mass from 11 to 120 solar masses: the hydrostatic and explosive contributions. Astrophys J 647:483–500. https://doi.org/10.1086/505164

Liu N, Gallino R, Bisterzo S, Davis AM, Savina MR, Pellin MJ (2014a) The ^{13}C-pocket structure in AGB models: constraints from zirconium isotope abundances in single mainstream SiC grains. Astrophys J 788:163. https://doi.org/10.1088/0004-637X/788/2/163, 1405.1441

Liu N, Savina MR, Davis AM, Gallino R, Straniero O, Gyngard F, Pellin MJ, Willingham DG, Dauphas N, Pignatari M, Bisterzo S, Cristallo S, Herwig F (2014b) Barium isotopic composition of mainstream silicon carbides from murchison: constraints for s-process nucleosynthesis in asymptotic giant branch stars. Astrophys J 786:66. https://doi.org/10.1088/0004-637X/786/1/66, 1403.4336

Liu N, Savina MR, Gallino R, Davis AM, Bisterzo S, Gyngard F, Käppeler F, Cristallo S, Dauphas N, Pellin MJ, Dillmann I (2015) Correlated strontium and barium isotopic compositions of acid-cleaned single mainstream silicon carbides from murchison. Astrophys J 803:12. https://doi.org/10.1088/0004-637X/803/1/12, 1501.05883

Lodders K, Fegley B Jr (1993) Chemistry in circumstellar envelopes of carbon stars: the influence of P, T, and elemental abundances. Meteoritics 28:387

Lucatello S, Tsangarides S, Beers TC, Carretta E, Gratton RG, Ryan SG (2005) The binary frequency among carbon-enhanced, s-Process-rich, metal-poor stars. Astrophys J 625:825–832. https://doi.org/10.1086/428104, arXiv:astro-ph/0412422

Lugaro M (2005) Stardust from meteorites. An introduction to presolar grains. World Scientific Publishing, Singapore

Lugaro M, Karakas AI (2008) ^{26}Al and ^{60}Fe yields from AGB stars. New Astron Rev 52:416–418. https://doi.org/10.1016/j.newar.2008.05.005

Lugaro M, Davis AM, Gallino R, Pellin MJ, Straniero O, Käppeler F (2003a) Isotopic compositions of strontium, zirconium, molybdenum, and barium in single presolar SiC grains and asymptotic giant branch stars. Astrophys J 593:486–508. https://doi.org/10.1086/376442

Lugaro M, Herwig F, Lattanzio JC, Gallino R, Straniero O (2003b) s-Process nucleosynthesis in asymptotic giant branch stars: a test for stellar evolution. Astrophys J 586:1305–1319. https://doi.org/10.1086/367887, arXiv:astro-ph/0212364

Lugaro M, Karakas AI, Nittler LR, Alexander CMO, Hoppe P, Iliadis C, Lattanzio JC (2007) On the asymptotic giant branch star origin of peculiar spinel grain OC2. Astron Astrophys 461:657–664. https://doi.org/10.1051/0004-6361:20065768, arXiv:astro-ph/0610464

Lugaro M, Doherty CL, Karakas AI, Maddison ST, Liffman K, García-Hernández DA, Siess L, Lattanzio JC (2012) Short-lived radioactivity in the early solar system: the Super-AGB star hypothesis. Meteorit Planet Sci 47:1998–2012. https://doi.org/10.1111/j.1945-5100.2012.01411.x, 1208.5816

Lugaro M, Heger A, Osrin D, Goriely S, Zuber K, Karakas AI, Gibson BK, Doherty CL, Lattanzio JC, Ott U (2014a) Stellar origin of the ^{182}Hf cosmochronometer and the presolar history of solar system matter. Science 345:650–653. https://doi.org/10.1126/science.1253338, 1408.2050

Lugaro M, Tagliente G, Karakas AI, Milazzo PM, Käppeler F, Davis AM, Savina MR (2014b) The impact of updated Zr neutron-capture cross sections and new asymptotic giant branch models on our understanding of the S process and the origin of stardust. Astrophys J 780:95. https://doi.org/10.1088/0004-637X/780/1/95, 1311.2660

Lugaro M, Karakas AI, Bruno CG, Aliotta M, Nittler LR, Bemmerer D, Best A, Boeltzig A, Broggini C, Caciolli A, Cavanna F, Ciani GF, Corvisiero P, Davinson T, Depalo R, di Leva A, Elekes Z, Ferraro F, Formicola A, Fülöp Z, Gervino G, Guglielmetti A, Gustavino C, Gyürky G, Imbriani G, Junker M, Menegazzo R, Mossa V, Pantaleo FR, Piatti D, Prati P, Scott DA, Straniero O, Strieder F, Szücs T, Takács MP, Trezzi D (2017) Origin of meteoritic stardust unveiled by a revised proton-capture rate of ^{17}O. Nat Astron 1:0027. https://doi.org/10.1038/s41550-016-0027, 1703.00276

Lugaro M, Karakas AI, Pető M, Plachy E (2018) Do meteoritic silicon carbide grains originate from asymptotic giant branch stars of super-solar metallicity? Geochim Cosmochim Acta 221:6–20. https://doi.org/10.1016/j.gca.2017.06.006

Marhas KK, Hoppe P, Ott U (2007) NanoSIMS studies of Ba isotopic compositions in single presolar silicon carbide grains from AGB stars and supernovae. Meteorit Planet Sci 42:1077–1101

McDonald AB, Ahmad QR, Allen RC, Andersen TC, Anglin JD, Barton JC, Beier EW, Bercovitch M, Bigu J, Biller SD, Black RA, Blevis I, Boardman RJ, Boger J, Bonvin E, Boulay MG, Bowler MG, Bowles TJ, Brice SJ, Browne MC, Bullard TV, Bühler G, Cameron J, Chan YD,

Chen HH, Chen M, Chen X, Cleveland BT, Clifford ETH, Cowan JHM, Cowen DF, Cox GA, Dai X, Dalnoki-Veress F, Davidson WF, Doe PJ, Doucas G, Dragowsky MR, Duba CA, Duncan FA, Dunford M, Dunmore JA, Earle ED, Elliott SR, Evans HC, Ewan GT, Farine J, Fergani H, Ferraris AP, Ford RJ, Formaggio JA, Fowler MM, Frame K, Frank ED, Frati W, Gagnon N, Germani JV, Gil S, Graham K, Grant DR, Hahn RL, Hallin AL, Hallman ED, Hamer AS, Hamian AA, Handler WB, Haq RU, Hargrove CK, Harvey PJ, Hazama R, Heeger KM, Heintzelman WJ, Heise J, Helmer RL, Hepburn JD, Heron H, Hewett J, Hime A, Howe M, Hykawy JG, Isaac MCP, Jagam P, Jelley NA, Jillings C, Jonkmans G, Kazkaz K, Keener PT, Klein JR, Knox AB, Komar RJ, Kouzes R, Kutter T, Kyba CCM, Law J, Lawson IT, Lay M, Lee HW, Lesko KT, Leslie JR, Levine I, Locke W, Luoma S, Lyon J, Majerus S, Mak HB, Maneira J, Manor J, Marino AD, McCauley N, McDonald DS, McFarlane K, McGregor G, Drees RM, Mifflin C, Miller GG, Milton G, Moffat BA, Moorhead M, Nally CW, Neubauer MS, Newcomer FM, Ng HS, Noble AJ, Norman EB, Novikov VM, O'Neill M, Okada CE, Ollerhead RW, Omori M, Orrell JL, Oser SM, Poon AWP, Radcliffe TJ, Roberge A, Robertson BC, Robertson RGH, Rosendahl SSE, Rowley JK, Rusu VL, Saettler E, Schaffer KK, Schwendener MH, Schülke A, Seifert H, Shatkay M, Simpson JJ, Sims CJ, Sinclair D, Skensved P, Smith AR, Smith MWE, Spreitzer T, Starinsky N, Steiger TD, Stokstad RG, Stonehill LC, Storey RS, Sur B, Tafirout R, Tagg N, Tanner NW, Taplin RK, Thorman M, Thornewell PM, Trent PT, Tserkovnyak YI, van Berg R, van de Water RG, Virtue CJ, Waltham CE, Wang J, Wark DL, West N, Wilhelmy JB, Wilkerson JF, Wilson JR, Wittich P, Wouters JM, Yeh M (2002) Direct evidence for neutrino flavor transformation from neutral-current interactions in SNO. In: Elias V, Epp R, Myers RC (eds) Theoretical physics: MRST 2002, American Institute of Physics conference series, vol 646. pp 43–58. https://doi.org/10.1063/1.1524553

McWilliam A, Lambert DL (1988) Isotopic magnesium abundances in stars. Mon Not R Astron Soc 230:573–585

Mikheyev SP, Smirnov AY (1985) Resonance enhancement of oscillations in matter and solar neutrino spectroscopy. Yadernaya Fizika 42:1441–1448

Mowlavi N, Meynet G (2000) Aluminum 26 production in asymptotic giant branch stars. Astron Astrophys 361:959–976

Mowlavi N, Goriely S, Arnould M (1998) The survival of ^{205}Pb in intermediate-mass AGB stars. Astron Astrophys 330:206–214. arXiv:astro-ph/9711025

Murphy RJ, Share GH (2005) What gamma-ray deexcitation lines reveal about solar-flares. Adv Space Res 35:1825–1832. https://doi.org/10.1016/j.asr.2005.03.004

Murphy RJ, Share GH, Skibo JG, Kozlovsky B (2005) The physics of positron annihilation in the solar atmosphere. Astrophys J Suppl 161:495–519. https://doi.org/10.1086/452634

Nanni A, Bressan A, Marigo P, Girardi L (2013) Evolution of thermally pulsing asymptotic giant branch stars - II. Dust production at varying metallicity. Mon Not R Astron Soc 434:2390–2417. https://doi.org/10.1093/mnras/stt1175, 1306.6183

Neyskens P, van Eck S, Jorissen A, Goriely S, Siess L, Plez B (2015) The temperature and chronology of heavy-element synthesis in low-mass stars. Nature 517:174–176. https://doi.org/10.1038/nature14050, 1601.05640

Nichols RH Jr, Hohenberg CM, Amari S, Lewis RS (1991) ^{22}Ne-E(H) and ^{4}He Measured in individual SiC grains using laser gas extraction. Meteoritics 26:377

Nicolussi GK, Davis AM, Pellin MJ, Lewis RS, Clayton RN, Amari S (1997) s-process zirconium in presolar silicon carbide grains. Science 277:1281–1283. https://doi.org/10.1126/science.277.5330.1281

Nicolussi GK, Pellin MJ, Lewis RS, Davis AM, Amari S, Clayton RN (1998a) Molybdenum isotopic composition of individual presolar silicon carbide grains from the murchison meteorite. Geochim Cosmochim Acta 62:1093–1104. https://doi.org/10.1016/S0016-7037(98)00038-6

Nicolussi GK, Pellin MJ, Lewis RS, Davis AM, Clayton RN, Amari S (1998b) Strontium isotopic composition in individual circumstellar silicon carbide grains: a record of s-Process nucleosynthesis. Phys Rev Lett 81:3583–3586. https://doi.org/10.1103/PhysRevLett.81.3583

Nittler LR, Hoppe P, Alexander CMO, Amari S, Eberhardt P, Gao X, Lewis RS, Strebel R, Walker RM, Zinner E (1995) Silicon Nitride from Supernovae. Astrophys J 453:L25. https://doi.org/10.1086/309743

Nittler LR, Alexander CMO, Gao X, Walker RM, Zinner E (1997) Stellar sapphires: the properties and origins of presolar AL 2O 3 in meteorites. Astrophys J 483:475. https://doi.org/10.1086/304234

Nittler LR, Alexander CMO, Gallino R, Hoppe P, Nguyen AN, Stadermann FJ, Zinner EK (2008) Aluminum-, calcium- and titanium-rich oxide stardust in ordinary chondrite meteorites. Astrophys J 682:1450–1478. https://doi.org/10.1086/589430, 0804.2866

Nollett KM, Busso M, Wasserburg GJ (2003) Cool bottom processes on the thermally pulsing asymptotic giant branch and the isotopic composition of circumstellar dust grains. Astrophys J 582:1036–1058. https://doi.org/10.1086/344817, arXiv:astro-ph/0211271

Ott U, Begemann F (1990) Discovery of s-process barium in the Murchison meteorite. Astrophys J 353:L57–L60. https://doi.org/10.1086/185707

Palmerini S, La Cognata M, Cristallo S, Busso M (2011) Deep mixing in evolved stars. I. The effect of reaction rate revisions from C to Al. Astrophys J 729:3. https://doi.org/10.1088/0004-637X/729/1/3, 1011.3948

Palmerini S, Trippella O, Busso M, Vescovi D, Petrelli M, Zucchini A, Frondini F (2018) s-Processing from MHD-induced mixing and isotopic abundances in presolar SiC grains. Geochim Cosmochim Acta 221:21–36. https://doi.org/10.1016/j.gca.2017.05.030, 1711.03039

Pérez-Mesa V, Zamora O, García-Hernández DA, Plez B, Manchado A, Karakas AI, Lugaro M (2017) Rubidium and zirconium abundances in massive Galactic asymptotic giant branch stars revisited. Astron Astrophys 606:A20. https://doi.org/10.1051/0004-6361/201731245, 1706.02268

Pignatari M, Gallino R, Straniero O, Reifarth R, Käppeler F, Davis AM (2004) Stellar origin of the meteoritic Xe-S anomalous component. Memorie della Societa Astronomica Italiana 75:182

Pignatari M, Gallino R, Amari S, Davis AM (2006) Krypton in presolar mainstream SiC grains from AGB stars. Memorie della Societa Astronomica Italiana 77:897

Pignatari M, Gallino R, Heil M, Wiescher M, Käppeler F, Herwig F, Bisterzo S (2010) The weak s-Process in massive stars and its dependence on the neutron capture cross sections. Astrophys J 710:1557–1577. https://doi.org/10.1088/0004-637X/710/2/1557

Pignatari M, Wiescher M, Timmes FX, de Boer RJ, Thielemann FK, Fryer C, Heger A, Herwig F, Hirschi R (2013) Production of carbon-rich presolar grains from massive stars. Astrophys J 767:L22 https://doi.org/10.1088/2041-8205/767/2/L22, 1303.3374

Pignatari M, Herwig F, Hirschi R, Bennett M, Rockefeller G, Fryer C, Timmes FX, Ritter C, Heger A, Jones S, Battino U, Dotter A, Trappitsch R, Diehl S, Frischknecht U, Hungerford A, Magkotsios G, Travaglio C, Young P (2016) NuGrid stellar data set. I. Stellar yields from H to Bi for stars with Metallicities Z = 0.02 and Z = 0.01. Astrophys J Suppl 225:24. https://doi.org/10.3847/0067-0049/225/2/24, 1307.6961

Podosek FA, Prombo CA, Amari S, Lewis RS (2004) s-Process Sr isotopic compositions in Presolar SiC from the murchison meteorite. Astrophys J 605:960–965. https://doi.org/10.1086/382650

Prombo CA, Podosek FA, Amari S, Lewis RS (1993) S-process BA isotopic compositions in presolar SiC from the Murchison meteorite. Astrophys J 410:393–399. https://doi.org/10.1086/172756

Raiteri CM, Gallino R, Busso M (1992) S-processing in massive stars as a function of metallicity and interpretation of observational trends. Astrophys J 387:263–275. https://doi.org/10.1086/171078

Ramaty R, Mandzhavidze N, Kozlovsky B, Murphy RJ (1995) Solar atmopheric abundances and energy content in flare accelerated ions from gamma-ray spectroscopy. Astrophys J 455:L193. https://doi.org/10.1086/309841

Rauscher T, Thielemann FK (2000) Astrophysical reaction rates from statistical model calculations. At Data Nucl Data Tables 75:1–351. https://doi.org/10.1006/adnd.2000.0834, arXiv:astro-ph/0004059

Raut R, Tonchev AP, Rusev G, Tornow W, Iliadis C, Lugaro M, Buntain J, Goriely S, Kelley JH, Schwengner R, Banu A, Tsoneva N (2013) Cross-section measurements of the Kr86(γ,n) reaction to probe the s-Process branching at Kr85. Phys Rev Lett 111(11):112501. https://doi.org/10.1103/PhysRevLett.111.112501, 1309.4159

Roederer IU, Karakas AI, Pignatari M, Herwig F (2016) The diverse origins of neutron-capture elements in the metal-poor star HD 94028: possible detection of products of I-Process nucleosynthesis. Astrophys J 821:37. https://doi.org/10.3847/0004-637X/821/1/37, 1603.00036

Rugel G, Faestermann T, Knie K, Korschinek G, Poutivtsev M, Schumann D, Kivel N, Günther-Leopold I, Weinreich R, Wohlmuther M (2009) New measurement of the Fe60 half-life. Phys Rev Lett 103(7):072502. https://doi.org/10.1103/PhysRevLett.103.072502

Savina MR, Davis AM, Tripa CE, Pellin MJ, Clayton RN, Lewis RS, Amari S, Gallino R, Lugaro M (2003a) Barium isotopes in individual presolar silicon carbide grains from the Murchison meteorite. Geochim Cosmochim Acta 67:3201–3214. https://doi.org/10.1016/S0016-7037(03)00083-8

Savina MR, Pellin MJ, Tripa CE, Veryovkin IV, Calaway WF, Davis AM (2003b) Analyzing individual presolar grains with CHARISMA. Geochim Cosmochim Acta 67:3215–3225. https://doi.org/10.1016/S0016-7037(03)00082-6

Savina MR, Davis AM, Tripa CE, Pellin MJ, Gallino R, Lewis RS, Amari S (2004) Extinct technetium in silicon carbide stardust grains: implications for stellar nucleosynthesis. Science 303:649–652. https://doi.org/10.1126/science.3030649

Sedlmayr E, Dominik C (1995) Dust driven winds. Space Sci Rev 73:211–272. https://doi.org/10.1007/BF00751238

Serenelli AM, Haxton WC, Peña-Garay C (2011) Solar models with accretion. I. Application to the solar abundance problem. Astrophys J 743:24. https://doi.org/10.1088/0004-637X/743/1/24, 1104.1639

Siess L, Arnould M (2008) Production of ^{26}Al by super-AGB stars. Astron Astrophys 489:395–402. https://doi.org/10.1051/0004-6361:200810147

Smith VV, Lambert DL (1986) The chemical composition of red giants. II - helium burning and the s-process in the MS and S stars. Astrophys J 311:843–863. https://doi.org/10.1086/164823

Sneden C, Cowan JJ, Lawler JE, Burles S, Beers TC, Fuller GM (2002) Europium isotopic abundances in very metal poor stars. Astrophys J 566:L25–L28. https://doi.org/10.1086/339471, arXiv:astro-ph/0201456

Sneden C, Cowan JJ, Gallino R (2008) Neutron-capture elements in the early galaxy. Annu Rev Astron Astrophys 46:241–288. https://doi.org/10.1146/annurev.astro.46.060407.145207

Speck AK, Barlow MJ, Sylvester RJ, Hofmeister AM (2000) Dust features in the 10-mu m infrared spectra of oxygen-rich evolved stars. Astron Astrophys Suppl 146:437–464. https://doi.org/10.1051/aas:2000274

Speck AK, Corman AB, Wakeman K, Wheeler CH, Thompson G (2009) Silicon carbide absorption features: dust formation in the outflows of extreme carbon stars. Astrophys J 691:1202–1221. https://doi.org/10.1088/0004-637X/691/2/1202, 0810.2599

Stancliffe RJ, Chieffi A, Lattanzio JC, Church RP (2009) Why do low-mass stars become red giants? Publications of the Astronomical Society of Australia 26:203–208. https://doi.org/10.1071/AS08060, 0902.0406

Stephan T, Trappitsch R, Davis AM, Pellin MJ, Rost D, Savina MR, Yokochi R, Liu N (2016) CHILI - the chicago instrument for laser ionization - a new tool for isotope measurements in cosmochemistry. Int J Mass Spectrom 407:1–15. https://doi.org/10.1016/j.ijms.2016.06.001

Straniero O, Imbriani G, Strieder F, Bemmerer D, Broggini C, Caciolli A, Corvisiero P, Costantini H, Cristallo S, DiLeva A, Formicola A, Elekes Z, Fülöp Z, Gervino G, Guglielmetti A, Gustavino C, Gyürky G, Junker M, Lemut A, Limata B, Marta M, Mazzocchi C, Menegazzo R, Piersanti L, Prati P, Roca V, Rolfs C, Rossi Alvarez C, Somorjai E, Terrasi F, Trautvetter HP (2013) Impact of a revised ^{25}Mg(p, γ)^{26}Al reaction rate on the operation of the Mg-Al cycle. Astrophys J 763:100. https://doi.org/10.1088/0004-637X/763/2/100, 1211.6661

Takahashi K, Yokoi K (1987) Beta-decay rates of highly ionized heavy atoms in stellar interiors. At Data Nucl Data Tables 36:375. https://doi.org/10.1016/0092-640X(87)90010-6

Tatischeff V, Kozlovsky B, Kiener J, Murphy RJ (2006) Delayed X- and gamma-ray line emission from solar flare radioactivity. Astrophys J Suppl 165:606–617. https://doi.org/10.1086/505112, arXiv:astro-ph/0604325

Thielemann FK, Arcones A, Käppeli R, Liebendörfer M, Rauscher T, Winteler C, Fröhlich C, Dillmann I, Fischer T, Martinez-Pinedo G, Langanke K, Farouqi K, Kratz KL, Panov I, Korneev IK (2011) What are the astrophysical sites for the r-process and the production of heavy elements? Prog Part Nucl Phys 66:346–353. https://doi.org/10.1016/j.ppnp.2011.01.032

Trappitsch R, Stephan T, Savina MR, Davis AM, Pellin MJ, Rost D, Gyngard F, Gallino R, Bisterzo S, Cristallo S, Dauphas N (2018) Simultaneous iron and nickel isotopic analyses of presolar silicon carbide grains. Geochim Cosmochim Acta 221:87–108. https://doi.org/10.1016/j.gca.2017.05.031

Travaglio C, Gallino R, Amari S, Zinner E, Woosley S, Lewis RS (1999) Low-density graphite grains and mixing in type II supernovae. Astrophys J 510:325–354. https://doi.org/10.1086/306551

Treffers R, Cohen M (1974) High-resolution spectra of cool stars in the 10- and 20-micron regions. Astrophys J 188:545–552. https://doi.org/10.1086/152746

Trigo-Rodriguez JM, Anibal Garcia-Hernandez D, Lugaro M, Karakas AI, van Raai M, Garcia Lario P, Manchado A (2009) The role of massive Agb stars in the early solar system composition. Meteorit Planet Sci 44:627–641

Uberseder E, Reifarth R, Schumann D, Dillmann I, Pardo CD, Görres J, Heil M, Käppeler F, Marganiec J, Neuhausen J, Pignatari M, Voss F, Walter S, Wiescher M (2009) Measurement of the Fe60(n,γ)^{61}Fe cross section at stellar temperatures. Phys Rev Lett 102(15):151101. https://doi.org/10.1103/PhysRevLett.102.151101

Uberseder E, Adachi T, Aumann T, Beceiro-Novo S, Boretzky K, Caesar C, Dillmann I, Ershova O, Estrade A, Farinon F, Hagdahl J, Heftrich T, Heil M, Heine M, Holl M, Ignatov A, Johansson HT, Kalantar N, Langer C, Le Bleis T, Litvinov YA, Marganiec J, Movsesyan A, Najafi MA, Nilsson T, Nociforo C, Panin V, Pietri S, Plag R, Prochazka A, Rastrepina G, Reifarth R, Ricciardi V, Rigollet C, Rossi DM, Savran D, Simon H, Sonnabend K, Streicher B, Terashima S, Thies R, Togano Y, Volkov V, Wamers F, Weick H, Weigand M, Wiescher M, Wimmer C, Winckler N, Woods PJ (2014) First experimental constraint on the Fe59(n,γ)Fe60 reaction cross section at astrophysical energies via the coulomb dissociation of Fe60. Phys Rev Lett 112(21):211101. https://doi.org/10.1103/PhysRevLett.112.211101

Uttenthaler S, Lebzelter T, Palmerini S, Busso M, Aringer B, Lederer MT (2007) Low-mass lithium-rich AGB stars in the Galactic bulge: evidence for cool bottom processing? Astron Astrophys 471:L41–L45, https://doi.org/10.1051/0004-6361:20077879, 0707.1380

van Raai MA, Lugaro M, Karakas AI, Iliadis C (2008) Reaction rate uncertainties and ^{26}Al in AGB silicon carbide stardust. Astron Astrophys 478:521–526. https://doi.org/10.1051/0004-6361:20078307, 0712.3702

van Raai MA, Lugaro M, Karakas AI, García-Hernández DA, Yong D (2012) Rubidium, zirconium, and lithium production in intermediate-mass asymptotic giant branch stars. Astron Astrophys 540:A44. https://doi.org/10.1051/0004-6361/201117896, 1202.2620

Verchovsky AB, Wright IP, Pillinger CT (2004) Astrophysical significance of asymptotic giant branch stellar wind energies recorded in meteoritic SiC grains. Astrophys J 607:611–619. https://doi.org/10.1086/383230

Vollmer C, Hoppe P, Brenker FE (2008) Si isotopic compositions of presolar silicate grains from red giant stars and supernovae. Astrophys J 684:611–617. https://doi.org/10.1086/589913

Wallner A, Bichler M, Buczak K, Dillmann I, Käppeler F, Karakas A, Lederer C, Lugaro M, Mair K, Mengoni A, Schätzel G, Steier P, Trautvetter HP (2016) Accelerator mass spectrometry measurements of the ^{13}C (n,γ)^{14}C and ^{14}N(n,p)^{14}C cross sections. Phys Rev C 93(4):045803. https://doi.org/10.1103/PhysRevC.93.045803

Wanajo S, Nomoto K, Janka H, Kitaura FS, Müller B (2009) Nucleosynthesis in electron capture supernovae of asymptotic giant branch stars. Astrophys J 695:208–220. https://doi.org/10.1088/0004-637X/695/1/208, 0810.3999

Wasserburg GJ, Busso M, Gallino R, Raiteri CM (1994) Asymptotic giant branch stars as a source of short-lived radioactive nuclei in the solar nebula. Astrophys J 424:412–428. https://doi.org/10.1086/173899

Wasserburg GJ, Boothroyd AI, Sackmann IJ (1995) Deep circulation in red giant stars: a solution to the carbon and oxygen isotope puzzles? Astrophys J 447:L37. https://doi.org/10.1086/309555

Wasserburg GJ, Busso M, Gallino R, Nollett KM (2006) Short-lived nuclei in the early solar system: possible AGB sources. Nucl Phys A 777:5–69. https://doi.org/10.1016/j.nuclphysa.2005.07.015, arXiv:astro-ph/0602551

Whittet DCB (2002) Dust in the galactic environment. Series in astronomy and astrophysics, 2nd edn. Institute of Physics (IOP) Publishing, Bristol

Willson LA (2000) Mass loss from cool stars: impact on the evolution of stars and stellar populations. Annu Rev Astron Astrophys 38:573–611. https://doi.org/10.1146/annurev.astro.38.1.573

Wolfenstein L (1978) Neutrino oscillations in matter. Phys Rev D 17:2369–2374. https://doi.org/10.1103/PhysRevD.17.2369

Wood PR (1979) Pulsation and mass loss in Mira variables. Astrophys J 227:220–231. https://doi.org/10.1086/156721

Yin QZ, Lee CTA, Ott U (2006) Signatures of the s-Process in presolar silicon carbide grains: barium through hafnium. Astrophys J 647:676–684. https://doi.org/10.1086/505188

Zinner E, Amari S, Lewis RS (1991) S-process Ba, Nd, and SM in presolar SiC from the Murchison meteorite. Astrophys J 382:L47–L50. https://doi.org/10.1086/186210

Zinner E, Nittler LR, Gallino R, Karakas AI, Lugaro M, Straniero O, Lattanzio JC (2006) Silicon and carbon isotopic ratios in AGB stars: SiC grain data, models, and the galactic evolution of the Si isotopes. Astrophys J 650:350–373. https://doi.org/10.1086/506957

Zinner E, Amari S, Guinness R, Jennings C, Mertz AF, Nguyen AN, Gallino R, Hoppe P, Lugaro M, Nittler LR, Lewis RS (2007) NanoSIMS isotopic analysis of small presolar grains: Search for $Si_3 N_4$ grains from AGB stars and Al and Ti isotopic compositions of rare presolar SiC grains. Geochim Cosmochim Acta 71:4786–4813. https://doi.org/10.1016/j.gca.2007.07.012

Chapter 4
Massive Stars and Their Supernovae

Friedrich-Karl Thielemann, Roland Diehl, Alexander Heger, Raphael Hirschi, and Matthias Liebendörfer

Stars more massive than about 8–10 solar masses evolve differently from their lower-mass counterparts: nuclear energy liberation is possible at higher temperatures and densities, due to gravitational contraction caused by such high masses, until forming an iron core that ends this stellar evolution. The star collapses thereafter, as insufficient pressure support exists when energy release stops due to Fe/Ni possessing the highest nuclear binding per nucleon, and this implosion turns into either a supernova explosion or a compact black hole remnant object. Neutron stars are the likely compact-star remnants after supernova explosions for a certain stellar mass range. In this chapter, we discuss this late-phase evolution of massive stars and their core collapse, including the nuclear reactions and nucleosynthesis products. We also include in this discussion more exotic outcomes, such as magnetic

F.-K. Thielemann (✉)
Department of Physics, University of Basel, Basel, Switzerland
GSI Helmholtz Center for Heavy Ion Research, Darmstadt, Germany
e-mail: f-k.thielemann@unibas.ch

R. Diehl
Max Planck Institut für extraterrestrische Physik, Excellence Cluster 'Universe', Garching, Germany
e-mail: rod@mpe.mpg.de

A. Heger
Monash Centre for Astrophysics, School of Physics and Astronomy, Monash University, Clayton, VIC, Australia
Tsung-Dao Lee Institute, Shanghai, China

R. Hirschi
University of Keele, Keele, UK
e-mail: r.hirschi@keele.ac.uk

M. Liebendörfer
Department of Physics, University of Basel, Basel, Switzerland
e-mail: matthias.liebendoerfer@unibas.ch

R. Diehl et al. (eds.), *Astrophysics with Radioactive Isotopes*, Astrophysics and Space Science Library 453, https://doi.org/10.1007/978-3-319-91929-4_4

jet supernovae, hypernovae, gamma-ray bursts and neutron star mergers. In all cases we emphasize the viewpoint with respect to the role of radioactivities.

4.1 The Cosmic Significance of Massive Stars

Our understanding of stellar evolution and the final explosive endpoints such as supernovae or hypernovae or gamma-ray bursts relies on the combination of

(a) (magneto-) hydrodynamics
(b) nuclear reactions releasing energy and leading to changes of composition
(c) radiation transport
(d) thermodynamic properties (such as the equation of state of stellar matter).

Hydrodynamics is essentially embedded within the numerical schemes which implement the physics of processes (b)–(d). In early phases of stellar evolution, hydrodynamical processes can be approximated by a hydro*static* treatment. Nuclear energy production (b) includes all nuclear reactions triggered during stellar evolution and explosive end stages, also among unstable isotopes produced on the way. Radiation transport (c) covers atomic physics (e.g. opacities) for photon transport, but also nuclear physics and neutrino nucleon/nucleus interactions in late phases and core collapse. The thermodynamical treatment (d) addresses the mixture of *ideal gases* of (massless) photons, electrons/positrons and nuclei/ions. These are fermions and bosons, and in dilute media or at high temperatures the energies of non-massless particles can often be approximated by Maxwell-Boltzmann distributions. At very high densities, the *nuclear* equation of state is required for the relation between pressure, density, temperature, entropy, energy density, composition, etc... It exhibits a complex behavior, with transitions from individual nuclei to clusters of nucleons with a background neutron bath, homogeneous phases of nucleons, the emergence of hyperons and pions up to a possible hadron-quark phase transition.

The detailed treatment of all these ingredients and their combined application is discussed in more depth in textbooks (Kippenhahn and Weigert 1994; Maeder 2009; Arnett 1996; Iliadis 2007; Jose 2016; Branch and Wheeler 2017), and/or the preceding Chap. 3, where the evolution of low and intermediate mass stars is addressed. That chapter also includes the stellar structure equations in spherical symmetry and a discussion of opacities for photon transport. Chapters 8 and 9 (tools for modeling objects and their processes) go into more detail with regard to modeling hydro*dynamics*, (convective) instabilities and energy transport as well as the energy generation due to nuclear reactions and the determination of the latter. Here we want to focus on the astrophysical aspects, i.e. a description of the evolution of massive stars and their endpoints with a special emphasis on the composition of their ejecta (in their forms of stellar winds during the evolution, or of the explosive ejecta). This includes also aspects of metallicity, rotation, and binary evolution (see e.g. Eldridge et al. 2008; Langer 2012; Maeder and Meynet 2012; Yoon 2015; De Marco and Izzard 2017; Moe and Di Stefano 2017; Meynet and Maeder 2017).

Low and intermediate mass stars end their evolution as AGB stars, finally blowing off a planetary nebula via wind losses and leaving a white dwarf with an unburned C and O composition (see e.g. Karakas and Lattanzio 2014). In the range of 8–10 $M_\odot$, stars also undergo C-burning, but collapse due to electron capture on C-burning products in the O-Ne-Mg core, resulting in fast contraction and the formation of an Fe-core during collapse which causes finally supernova explosions, dubbed electron capture (EC) supernovae (Kitaura et al. 2006; Wanajo et al. 2009, 2011; Hüdepohl et al. 2010; Jones et al. 2013, 2016; Moriya et al. 2014; Woosley and Heger 2015b; Müller et al. 2017).

More massive stars in the mass range 10–90 $M_\odot$ evolve beyond further and experience all stellar burning stages from H over He, C, Ne, O and Si-burning, up to core collapse and explosive end stages. A major question is how the transition occurs from "regular" core-collapse supernovae (CCSNe), producing neutron stars, to the formation of a central black hole. This depends strongly on the properties of the stellar progenitor, among other properties the compactness, i.e. the central mass concentration is important (Heger and Woosley 2010; Chieffi and Limongi 2013; Karakas and Lattanzio 2014; Sukhbold et al. 2016; Ebinger et al. 2017, 2018; Nakamura et al. 2015; Pan et al. 2017). In case of fast rotation and strong magnetic fields so-called hypernovae/collapsars/long duration gamma-ray bursts (GRBs) possibly occur after black hole formation, otherwise this leads to a failed or faint supernova and "quiet" black hole formation. For the nucleosynthesis in GRBs/hypernovae, aspherical explosions are important. This is also the case if fast rotation and strong magnetic fields are present in core collapse, leading, however, finally to a central neutron star with magnetic fields as high as 10^{15} G, known as magnetar (Winteler et al. 2012; Mösta et al. 2014, 2015, 2017; Nishimura et al. 2015, 2017b; Halevi and Mösta 2018).

90–140 $M_\odot$ stars undergo pulsational nuclear instabilities at various nuclear burning stages, including O and Si-burning. 140–300 $M_\odot$ stars become pair-instability supernovae, if the mass loss is small enough to permit this final endstage. Very massive stars greater than 300 $M_\odot$ undergo core-collapse to form intermediate mass black holes. Detailed reviews and recent findings on the present understanding of end stages of such massive stars are given e.g. in Heger and Woosley (2010), Kasen et al. (2011), Nomoto et al. (2010, 2013), Woosley and Heger (2015a), and Georgy et al. (2017).

In this chapter we want to discuss the nucleosynthesis processes involved in massive stars and their explosions and the related production of radioactive nuclei[1] in more detail. This includes all hydrostatic nuclear-burning stages experienced by massive stars, the explosive burning stages when a shock wave moves outward after a successful explosion was initiated, and also final wind ejecta from the hot proto-neutron star which emerged in the collapse and explosion phase. All these ejecta

[1]We focus especially on long-lived radioactivities which can be observed with gamma-ray satellites, and refractory isotopes which can be observed in dust condensations included in meteorites.

will enter the interstellar medium in galaxies, initially appearing as gas and dust in wind bubbles and supernova remnants, later determining the evolution of the larger-scale gas composition. The interstellar gas composition will evolve with time, and the composition of newly formed stars will witness this composition at the time of their formation.

Massive stars play an important role as contributors to the gas composition of the interstellar medium via wind losses and/or explosions. In astronomical/observational terms they are the progenitors of blue supergiants (BSG), red supergiants (RSG), Wolf-Rayet (WR) and luminous blue variable (LBV) stars (Maeder and Meynet 2012) At the end of their life, they explode as core collapse supernovae (SNe), observed as SNe of type II or Ib,c (Woosley and Bloom 2006; Branch and Wheeler 2017) and also as long soft gamma-ray bursts (GRBs Piran 2004; Nakar and Piran 2017). After collapse, their cores become neutron stars or black holes. They are one of the main sites for nucleosynthesis, which takes place during both pre-SN (hydrostatic) burning stages and during explosive burning.

Neutron capture processes are mainly responsible for the heavy nuclei beyond the Fe-group, existing in the variety of a slow (s) and a rapid (r) process. A weak s-process occurs during core He-(and C-)burning (The et al. 2007; El Eid et al. 2009; Nishimura et al. 2017a), which can in case of fast rotation also be more powerful (Frischknecht et al. 2016). In past years it was also expected that an r-process occurs during the explosion (Woosley et al. 1994; Qian and Woosley 1996), but recent studies (essentially due to the neutrino interaction with the innermost ejecta, see e.g. Arcones and Thielemann 2013) indicate that only a weak r-process takes place in regular supernovae while a strong one might emerge in magnetar-producing supernovae (Winteler et al. 2012; Nishimura et al. 2017b; Halevi and Mösta 2018). A further option is the aftermath of two supernovae in binary systems, leading to neutron star mergers (see e.g. Thielemann et al. 2017b; Metzger 2017a), especially after the recent detection of GW170817 (Metzger 2017b).

Radioactive isotopes such as ^{26}Al and ^{60}Fe detected by the INTEGRAL (Diehl et al. 1997, 2006a,b; Wang et al. 2009) satellite are produced by massive stars (see e.g. Limongi and Chieffi 2006b), plus many more radioactivities from the final explosive ejecta (such as ^{44}Ti, ^{56}Ni, ^{56}Co etc., see Sect. 4.5.2 for more details). Chaps. 2 and 3 discussed also many long-lived heavy nuclei beyond Fe with half-lives larger than 10^7 and up to 10^{11}y. As massive stars are probably not the major origin of heavy s-process nuclei (see Chap. 3), we will address here those of these nuclei which are clearly identified with the r-process (^{232}Th, 1.4×10^{10} years, ^{235}U, 7×10^8 years, ^{236}U, 2.3×10^7 years, ^{238}U, 4.5×10^9 years, ^{244}Pu, 8×10^7 years, ^{247}Cm, 1.6×10^7 years) and where especially ^{232}Th and ^{238}U, with half-lives comparable to the age of the Galaxy/Universe, can also serve as chronometers (Cowan et al. 1999; Cayrel et al. 2001; Thielemann et al. 2002; Mashonkina et al. 2014)

Massive stars contribute significantly (about two thirds) to the integrated luminosity of galaxies, even though they are much less numerous than low mass stars. At high redshifts z, or low metallicities Z, they are even more important as drivers of characteristic phenomena and evolution. The first stars formed are thought to be all massive or even very massive (Karlsson et al. 2013), and to be the cause of the re-ionisation of the universe. As discussed above, if the final core collapse leads to a black hole, the endpoint of this evolution can be the origin of the subset of (long, soft) gamma ray bursts (GRBs). GRBs can be used as new *standard candles* for cosmology at high redshifts, as they are visible from higher redshifts than usual SNe (of type I or II), and thus broaden the base to constrain cosmological models. Massive stars with their large energy output can be seen out to significant (cosmological) distances—either directly through their thermal photospheric emission, or indirectly through the impact on their surroundings (ionization, or heated dust). In their *collapsar* and GRB extremes, emission is beamed into a jet (Nakar and Piran 2017), which makes them visible even at greater distances. This can also give us information on the star formation history at a very early age of the universe ($z > 10$) beyond the reach of galaxy observations. Closer to home, recent surveys of metal poor halo stars provide a rich variety of constraints for the early chemical evolution of our Galaxy and thus the nucleosynthesis ejecta (see e.g. Nomoto et al. 2013) (*astro-archeology*).

4.2 Hydrostatic and Explosive Burning

Following this motivation for studying the evolution, final fate, and remnant aspects of massive stars (previous section), we now discuss the ingredients for modeling each of these aspects. Thermonuclear energy generation is one of the key processes: It shapes the interior structure of the star, thus its evolutionary time scales, and the generation of new chemical elements with their isotopes. Without understanding this, the *feedback* from massive stars as it determines the evolution of galaxies cannot be understood in astrophysical terms.[2] Thermonuclear burning, nuclear energy generation, and the resulting nuclear abundances are determined by thermonuclear and by weak interactions. The treatment of the required nuclear and plasma physics, and a detailed technical description of reaction rates, their determination, and the essential features of composition changes and reaction networks are presented in Chap. 9. Here we discuss the types of reactions that are specific to the evolution of massive stars, their collapse stage, and the compact remnants.

Nuclear burning can in general be placed into two categories:

1. hydrostatic burning stages on timescales dictated by stellar energy loss
2. explosive burning in a specific highly-dynamic event (collapse, explosion, compact-star collision).

Massive stars (as opposed to low and intermediate mass stars) experience explosive burning (2) as a natural outcome at the end of their evolution. They also undergo

[2]Empirical descriptions derived from observations of a multitude of galaxies are often used in cosmological simulations, as a substitute to such astrophysical models.

more extended hydrostatic burning stages (1) than their low- and intermediate-mass cousins. Therefore, we first address some aspects of these nuclear burnings in a general way, before describing evolution and explosions in more detail.

Nuclear burning and the resulting composition changes (i.e. nucleosynthesis) are fundamentally characterised by (1) strong interactions (hadron reactions) and photo-disintegrations, (2) weak interactions characterised by decay half-lives of electron or positron emissions and captures, and (3) neutrino-induced reactions. These will be discussed in the following sections.[3]

4.2.1 Nuclear Burning During Hydrostatic Evolution

Hydrostatic burning stages are characterised by temperature thresholds, above which thermal Maxwell-Boltzmann distributions of (charged) particles (nuclei) to penetrate increasingly larger Coulomb barriers of electrostatic repulsion. These are (two body) reactions as discussed in Eqs. (9.6) and (9.10) of Chap. 9, representing terms of the type ${}_ir_j$ in the network equation (9.1). H-burning converts ^{1}H into ^{4}He via pp-chains or the CNO-cycles. The simplest pp-chain is initiated by ^{1}H(p,$e^+\nu$)^{2}H(p,γ)^{3}He and completed by ^{3}He(^{3}He,2p)^{4}He. The dominant CNO-cycle chain ^{12}C(p,γ)^{13}N($e^+\nu$)^{13}C(p,γ)^{14}N(p,γ)^{15}O($e^+\nu$)^{15}N(p,α)^{12}C is controlled by the slowest reaction ^{14}N(p,γ)^{15}O. Thus, the important ashes of H-burning are ^{4}He and (metallicity-dependent, as acting on prior existing ^{12}C) ^{14}N.

The major reactions in He-burning are the triple-alpha reaction ^{4}He($2\alpha, \gamma$)^{12}C and ^{12}C(α, γ)^{16}O. The triple-alpha reaction, is essentially a sequence of two-body reactions with an extremely short-lived intermediate nucleus ^{8}Be. It is thus an example for the term ${}_i\hat{r}_j$ in Eq. (9.1) in Chap. 9, which includes the product of three abundances. The other H-burning product ^{14}N is processed to ^{22}Ne via ^{14}N(α, γ)^{18}F(β^+)^{18}O(α, γ) and can act as a neutron source for the s-capture process via ^{22}Ne(α, n)^{25}Mg.

The H- and He-burning stages are encountered in massive as well as in low and intermediate mass stars, the latter leaving white dwarfs as central objects. Mixing instabilities between the H- and He-burning zone can mix ^{12}C into proton-rich environments, causing the production of ^{13}C via ^{12}C(p, γ)^{13}N(β^+)^{13}C and in further CNO-type processing also ^{14}N. When being mixed back into He-rich environments they can act as the known neutron source ^{22}Ne (discussed above) and as well as via the reaction ^{13}C(α, n)^{16}O. The first one is important for massive stars, causing in case of rotation-induced mixing instabilities also the production of ^{14}N as a primary (not metallicity-dependent) neutron source (see e.g. Frischknecht et al. 2016). ^{13}C acts via He-shell flashes as the neutron source for the main s-

[3]A review of the sources for this microphysics input is given for (1) in Chap. 9 and for (3) in Chap. 8. We will review some of the required weak interaction rates (2) in the subsections on late phases of stellar evolution/core collapse and the description of the explosion.

Table 4.1 Burning stages of a 20 $M_\odot$ star

Fuel	ρ_c ($g\,cm^{-3}$)	T_c (10^9 K)	τ (Years)	L_{phot} ($erg\,s^{-1}$)
Hydrogen	5.6(0)	0.04	1.0(7)	2.7(38)
Helium	9.4(2)	0.19	9.5(5)	5.3(38)
Carbon	2.7(5)	0.81	3.0(2)	4.3(38)
Neon	4.0(6)	1.70	3.8(−1)	4.4(38)
Oxygen	6.0(6)	2.10	5.0(−1)	4.4(38)
Silicon	4.9(7)	3.70	2 days (2)	4.4(38)

Table 4.2 Major reactions in carbon burning

(a) Basic energy generation
$^{12}C(^{12}C,\alpha)^{20}Ne$ $^{12}C(^{12}C,p)^{23}Na$
$^{23}Na(p,\alpha)^{20}Ne$ $^{23}Na(p,\gamma)^{24}Mg$ $^{12}C(\alpha,\gamma)^{16}O$
(b) Fluxes $> 10^{-2}\times$*(a)*
$^{20}Ne(\alpha,\gamma)^{24}Mg$ $^{23}Na(\alpha,p)^{26}Mg(p,\gamma)^{27}Al$
$^{20}Ne(n,\gamma)^{21}Ne(p,\gamma)^{22}Na\,(e^+\nu)^{22}Ne(\alpha,n)^{25}Mg(n,\gamma)^{26}Mg$
$^{21}Ne(\alpha,n)^{24}Mg$ $^{22}Ne(p,\gamma)^{23}Na$ $^{25}Mg(p,\gamma)^{26}Al(e^+\nu)^{26}Mg$
(c) Low temperature, high density burning
$^{12}C(p,\gamma)^{13}N(e^+\nu)^{13}C(\alpha,n)^{16}O(\alpha,\gamma)^{20}Ne$
$^{24}Mg(p,\gamma)^{25}Al(e^+\nu)^{25}Mg$
$^{21}Ne(n,\gamma)^{22}Ne(n,\gamma)^{23}Ne(e^-\bar{\nu})^{23}Na(n,\gamma)^{24}Na(e^-\nu)^{24}Mg$ + s-processing

process in low and intermediate mass stars. They are discussed in much more detail with all minor reaction pathways in Chap. 3. Important features as well as the status of nuclear cross sections involved are discussed in reviews on hydrostatic burning stages (Haxton et al. 2006; Buchmann and Barnes 2006; Costantini et al. 2009; Wiescher et al. 2010; Adelberger et al. 2011; Xu et al. 2013; deBoer et al. 2017; Bao et al. 2000; Dillmann et al. 2006, 2014).

Massive stars, the subject of the present Chapter, undergo further burning stages up to those involving the production of Fe-group nuclei. Table 4.1 lists these burning stages and their typical central densities and temperatures, their duration and luminosity in photons (from Woosley and Weaver 1995; Woosley et al. 2002) (see Sect. 4.3 for more detail).

- *Heavy-ion fusion reactions:* In C-burning the reaction $^{12}C(^{12}C,\alpha)^{20}Ne$ dominates, in O-burning it is $^{16}O(^{16}O,\alpha)^{28}Si$. The corresponding reaction rates $_i r_j$ (after integrating over a Maxwell-Boltzmann distribution of targets and projectiles) have the form given in Eq. (9.10) of Chap. 9 and contribute to the second term in Eq. (9.1). Reactions going beyond these key reactions are provided in Tables 4.2 and 4.5. Important uncertainties of nuclear cross sections are discussed in Wiescher et al. (2012) as well as the publicly available reaction libraries Kadonis, JINA Reaclib, Starlib (Dillmann et al. 2014; Cyburt et al. 2010; Sallaska et al. 2013). Extended overviews on available (also theoretical) cross section predictions can be found in the data bases Bruslib, JINA Reaclib, Kadonis,

Table 4.3 Major reactions in neon burning

(a) Basic energy generation
^{20}Ne(γ, α)^{16}O ^{20}Ne(α, γ)^{24}Mg(α, γ)^{28}Si
(b) Fluxes > $10^{-2}\times$(a)
^{23}Na(p,α)^{20}Ne ^{23}Na(α,p)^{26}Mg(α,n)^{29}Si
^{20}Ne(n,γ)^{21}Ne(α,n)^{24}Mg(n,γ)^{25}Mg(α,n)^{28}Si
^{28}Si(n,γ)^{29}Si(n,γ)^{30}Si
^{24}Mg(α,p)^{27}Al(α,p)^{30}Si
^{26}Mg(p,γ)^{27}Al(n,γ)^{28}Al($e^-\bar{\nu}$)^{28}Si
(c) Low temperature, high density burning
^{22}Ne(α,n)^{25}Mg(n,γ)^{26}Mg(n,γ)^{27}Mg($e^-\bar{\nu}$)^{27}Al
^{22}Ne left from prior neutron-rich carbon burning

and nucastro.org with their websites http://www-astro.ulb.ac.be/bruslib/, https://groups.nscl.msu.edu/jina/reaclib/db/, http://www.kadonis.org/, https://nucastro.org/reaclib.html. Further information is given in Chap. 9.

- *Photo-disintegrations:* The alternative to fusion reactions are photo-disintegrations which start to play a role at sufficiently high temperatures T when 30 kT$\approx Q$ (the Q-value or energy release of the inverse capture reaction). This ensures the existence of photons with energies $> Q$ in the Planck distribution and leads to Ne-Burning [^{20}Ne(γ, α)^{16}O, ^{20}Ne(α, γ)^{24}Mg] at $T > 1.5 \times 10^9$ K (preceding O-burning), due to a small Q-value of $\approx$4 MeV, and Si-burning at temperatures in excess of 3×10^9 K [initiated like Ne-burning by photo-disintegrations]. Such photo-disintegrations (after integrating over a thermal (Planck) distribution of photons at temperature T) have the form given in Eq. (9.4) of Chap. 9 and act similar to decays with a temperature-dependent decay constant, contributing (like decays) to the first term $_i\lambda_j$ in Eq. (9.1). In Table 4.3 we provide some of the main reactions of Ne-burning, which is initiated by the photo-disintegration of Ne.
- *Electron capture reactions:* Massive stellar cores eventually evolve to degeneracy of their electron-gas, i.e. the Pauli exclusion principle for fermions determines the population of energy states, rather than the Boltzmann statistics which applies for lower densities/higher temperatures. The Fermi energy gives a useful estimate, and would be the highest energy occupied at zero temperature from the Pauli exclusion principle of identical electron states; it is (Chandrasekhar 1957)

$$E_F = \hbar^2/2m_e(3\pi^2)^{2/3}n_e^{2/3} \tag{4.1}$$

Here n_e is the density of the electron gas $n_e = \rho N_A Y_e$, ρ denotes the matter density and N_A Avogadro's number. In late stages of O-burning, in Si-burning (and during the later collapse stage) this Fermi energy of (degenerate) electrons increases to the level of nuclear energies (MeVs).In a neutral, completely ionized plasma, the electron abundance Y_e is equal to the total proton abundance $Y_e = \sum_i Z_i Y_i$ (summing over all abundances of nuclei, including protons/hydrogen)

Table 4.4 Electron capture

$p + e^- \rightarrow \nu_e + n$ or $p(e^-, \nu_e)n$
$(A, Z) + e^- \rightarrow \nu_e + (A, Z-1)$ or $^{A}Z(e^-, \nu_e)^{A}Z\text{-}1$
$E_F(\rho Y_e = 10^7 \text{gcm}^{-3})$=0.75 MeV
$E_F(\rho Y_e = 10^9 \text{gcm}^{-3})$=4.70 MeV

Table 4.5 Major reactions in oxygen burning

(a) Basic energy generation
$^{16}O(^{16}O,\alpha)^{28}Si$ $^{16}O(^{12}O,p)^{31}P$ $^{16}O(^{16}O,n)^{31}S(e^+\nu)^{31}P$
$^{31}P(p,\alpha)^{28}Si(\alpha,\gamma)^{32}S$
$^{28}Si(\gamma,\alpha)^{24}Mg(\alpha,p)^{27}Al(\alpha,p)^{30}Si$
$^{32}S(n,\gamma)^{33}S(n,\alpha)^{30}Si(\alpha,\gamma)^{34}S$
$^{28}Si(n,\gamma)^{29}Si(\alpha,n)^{32}S(\alpha,p)^{35}Cl$
$^{29}Si(p,\gamma)^{30}P(e^+\nu)^{30}Si$
Electron captures
$^{33}S(e^-,\nu)^{33}P(p,n)^{33}S$
$^{35}Cl(e^-,\nu)^{35}S(p,n)^{35}Cl$
(b) High temperature burning
$^{32}S(\alpha,\gamma)^{36}Ar(\alpha,p)^{39}K$
$^{36}Ar(n,\gamma)^{37}Ar(e^-,\nu)^{37}Cl$
$^{35}Cl(\gamma,p)^{34}S(\alpha,\gamma)^{38}Ar(p,\gamma)^{39}K(p,\gamma)^{40}Ca$
$^{35}Cl(e^-,\nu)^{35}S(\gamma,p)^{34}S$
$^{38}Ar(\alpha,\gamma)^{42}Ca(\alpha,\gamma)^{46}Ti$
$^{42}Ca(\alpha,p)^{45}Sc(p,\gamma)^{46}Ti$
(c) Low temperature, high density burning
$^{31}P(e^-,\nu)^{31}S$ $^{31}P(n,\gamma)^{32}P$
$^{32}S(e^-,\nu)^{32}P(p,n)^{32}S$
$^{33}P(p,\alpha)^{30}Si$

and limited by the extreme values 0 (only neutrons) and 1 (only protons) with typical values during stellar evolution close to 0.5 or slightly below. Such conditions permit electron captures on protons and nuclei, if the negative Q-value of the reaction can be overcome by the electron (Fermi) energy. The general features for typical conditions are presented in Table 4.4, example reactions were already given in Table 4.5. Thus, at sufficiently high densities, electron captures—which are energetically prohibited— become possible and lead to enhanced *neutronization* of the astrophysical plasma, in addition to the role of beta-decays and electron captures with positive Q-values (Nomoto and Hashimoto 1988; Woosley et al. 2002). In degenerate Ne-O-Mg cores (after core C-burning of stars with $8 < M/M_\odot < 10$), electron captures on ^{20}Ne and ^{24}Mg cause a loss of pressure support. This introduces a collapse of the core, rather than only a contraction, which compresses all further burning stages on the short time scale of the collapse (Nomoto 1987; Kitaura et al. 2006; Wanajo et al. 2009; Jones et al. 2013). In Si-burning of more massive stars, electron capture on intermediate mass and Fe-group nuclei becomes highly important and determines

Table 4.6 Neutrino reactions

$\nu_e + n \leftrightarrow p + e^-$ or $n(\nu_e, e^-)p$	
$\bar{\nu}_e + p \leftrightarrow n + e^+$ or $p((\bar{\nu}_e, e^+)n$	
$\nu_e + (Z, A) \leftrightarrow (Z + 1, A) + e^-$ or $^A Z(\nu_e, e^-)^A Z{+}1$	
$\bar{\nu}_e + (Z, A) \leftrightarrow (Z - 1, A) + e^+$ or $^A Z(\bar{\nu}_e, e^+)^A Z{-}1$	
$(Z, A) + \nu \leftrightarrow \nu + (Z, A)^*$	

the neutronization (Y_e) of the central core. As discussed in Chap. 9, these electron captures contribute to the one-body reaction terms ${}_i\lambda_j$ in Eq. (9.1) with the effective decay constants in Eq. (9.5) being a function of T and $n_e = \rho N_A Y_e$, the electron number density, see e.g. Fuller et al. (1980, 1982), Oda et al. (1994), Langanke and Martínez-Pinedo (2001, 2003), Sampaio et al. (2003), Juodagalvis et al. (2010), and Sullivan et al. (2016).

- *Neutrino reactions:* Neutrino reaction cross section on nucleons, nuclei and electrons are minute, by comparison to above reactions, see e.g. Bruenn and Haxton (1991), Langanke and Martínez-Pinedo (2003), Kolbe et al. (2003), and Balasi et al. (2015), and for recent updates Burrows et al. (2018) and references therein. High densities of the order $\rho > 10^{12}\,\mathrm{g\,cm^{-3}}$ are therefore required for the inverse process to electron/positron capture (neutrino capture) to occur at significant rates on the relevant timescales (Table 4.6). The same holds for other processes, such as e.g. inelastic scattering, which leave a nucleus in an excited state that leads to emission of nucleons and alpha particles in exit channels. Such neutrino-induced reactions can be expressed in a similar way as photon and electron captures, integrating now over the corresponding neutrino distribution. The latter is, however, not necessarily in thermal equilibrium, therefore not just a function of temperature and neutrino densities. Neutrino energy distributions are rather the result of a variety of reactions (including scattering) with reaction rates much below those of particle collisions, and determined by (neutrino) radiation transport calculations (see Chap. 8, where also other neutrino scattering processes are discussed).

All the reaction types presented above occur at different times in the sequence of burning stages. They contribute to the three types of terms in the reaction network equation 9.1 of Chap. 9. As an illustration to show how nuclear abundances Y_i enter in this set of equations, it can also be written in the form[4]

[4]The formal difference to Eq. (9.1) is that one does not sum here over the reactions but rather over all reaction partners (see also the equation following Table 3.2 in Chap. 3). However, in total, all the terms which appear are identical. Due to the different summation indices, the P's have a slightly different notation, λ's stand for decay rates L of Chap. 9 and $< j, k >$ for $< \sigma^* v >$ of reactions between nuclei j and k, while $< j, k, l >$ includes a similar expression for three-body reactions (Nomoto et al. 1985). A survey of computational methods to solve nuclear networks is given in (Hix and Thielemann 1999a; Timmes 1999; Hix and Meyer 2006; Lippuner and Roberts 2017). (The abundances Y_i occurring in Eq. (4.2) are related—like electron abundances Y_e—to number densities $n_i = \rho N_A Y_i$ and mass fractions of the corresponding nuclei via $X_i = A_i Y_i$, where A_i is the mass number of nucleus i and $\sum X_i = 1$.)

$$\frac{dY_i}{dt} = \sum_j P_j^i \lambda_j Y_j + \sum_{j,k} P_{j,k}^i \rho N_A < j,k > Y_j Y_k$$
$$+ \sum_{j,k,l} P_{j,k,l}^i \rho^2 N_A^2 < j,k,l > Y_j Y_k Y_l. \quad (4.2)$$

Core Si-burning, the final burning stage during stellar evolution, which is initiated by the photo-disintegration ^{28}Si(γ,α)^{24}Mg close to 3×10^9 K—and followed by a large number of fusion and photo-disintegration reactions—ends with nuclear reactions in a complete *chemical equilibrium*[5] (nuclear statistical equilibrium, NSE) and an abundance distribution centered around Fe (as discussed in Chap. 9 and Eq. (9.13)). These temperatures permit photo-disintegrations with typical Q-values of 8–10 MeV as well as the penetration of Coulomb barriers in capture reaction. In such an NSE the abundance of each nucleus Y_i is only dependent on temperature T, density ρ, its nuclear binding energy B_i, and via charge conservation on $\sum_i Z_i Y_i = Y_e$. Y_e is altered by weak interactions on longer timescales. *Quasi-equilibrium* can occur, if localised nuclear mass regions are in equilibrium with the background of free neutrons, protons and alphas, but offset from other regions of nuclei and thus their NSE values (Hix and Thielemann 1996, 1999b; Hix et al. 2007). Different quasi-equilibrium regions are usually separated from each other by slow reactions with typically small Q-values. Such boundaries between QSE groups, due to slow reactions, can be related to neutron or proton shell closures, like e.g. $Z = N = 20$, separating the Si- and Fe-groups in early phases of Si-burning.

The reactions discussed above occur during all stellar burning stages, and are essentially related to processing of nuclei from H to the Fe-group, but not much beyond. The major reaction sequences shown in the previous tables and the transition to quasi-equilibria and complete NSE have been discussed in detail in Thielemann and Arnett (1985); Woosley et al. (2002); Hix and Thielemann (1996, 1999b); Hix et al. (2007).

- *Neutron captures:* Neutron capture reactions open a chance to produce heavier nuclei also during regular stellar evolution. During core and shell He-burning specific alpha-induced reactions can liberate neutrons, which then are responsible for the slow neutron capture process (s-process). One such major neutron source is the reaction ^{22}Ne(α,n)^{25}Mg, from ^{22}Ne produced via successive α-captures on the products of H-burning in the CNO cycle ^{14}N(α,γ)^{18}F(β^+)^{18}O(α,γ)^{22}Ne. If ^{12}C may be mixed into H-burning shells, this can produce an even stronger neutron source from ^{13}C(α,n)^{16}O via ^{12}C(p,γ)^{13}N(β^+)^{13}C. In massive, rotating, low metallicity stars, mixing can lead to the production of "primary" ^{14}N and ^{22}Ne, i.e. a neutron source which does not depend on the initial metallicity of ^{14}N in the CNO-cycle, and can thus be much stronger, in particular at low

[5]All strong (thermonuclear) and photo-disintegration reactions are equilibrated, while weak interaction reactions, changing Y_e, may occur on longer timescales.

metallicity. The previous Chap. 3 discussed in detail the strong s-process via a combination of ^{13}C and ^{22}Ne in He-shell flashes of low and intermediate-mass stars. In a similar way, mixing processes can also occur in massive stars due to rotation or convective instabilities. Without such mixing processes only secondary (metallicity-dependent) ^{22}Ne is available for ^{22}Ne(α, n)^{25}Mg and core He-burning as well as shell C-burning lead to a weak s-process (The et al. 2007; Käppeler et al. 2011). The s-process can in principle proceed towards elements up to Pb and Bi through a series of neutron captures and β^--decays, starting on existing heavy nuclei around Fe. Weak s-processing, based on *secondary* ^{22}Ne, does not reach beyond mass numbers of $A = 80$–90. The production of heavier nuclei is possible in massive stars if *primary* ^{14}N and ^{22}Ne are available (Pignatari et al. 2008; Frischknecht et al. 2016)

4.2.2 Nuclear Burning During the Explosion

Many of the hydrostatic nuclear-burning processes occur also under explosive conditions, at higher temperatures and on shorter timescales (see Fig. 4.1). Here, often the β-decay half-lives are longer than the explosive timescales, producing significant abundances of unstable isotopes, as the burning proceeds. This implies that additional knowledge of nuclear reactions for and among unstable nuclei is required. The fuel for explosive nucleosynthesis consists mainly of $N = Z$ nuclei such as ^{12}C, ^{16}O, ^{20}Ne, ^{24}Mg, or ^{28}Si (the ashes of previous hydrostatic burning). The results are heavier nuclei, again with $N \approx Z$. At high densities also electron captures on nuclei $e^- +^A Z \rightarrow^A Z\text{-}1 + \nu$ can occur at substantial rates, due to high-energy electrons when Fermi energies are high in the degenerate gas (Fuller et al. 1980, 1982; Oda et al. 1994; Langanke and Martínez-Pinedo 2003; Sampaio et al. 2003; Juodagalvis et al. 2010); this was already discussed above for late hydrostatic burning stages.

Explosive Si-burning is very different than its hydrostatic counterpart. It can be divided into three different regimes: (1) incomplete Si-burning, and complete Si-burning with either (2) a normal (high density, low entropy) or (3) an α-rich (low density, high entropy) freeze-out of charged-particle reactions during cooling from NSE. At high temperatures, or during a *normal* freeze-out, all abundances remain in complete NSE. Then as temperatures and/or densities fall, the NSE can break up into smaller equilibrium clusters (quasi-equilibrium, QSE); for a detailed discussion see Hix and Thielemann (1996, 1999b) and Hix et al. (2007). An example for such QSE-behavior is an alpha-rich freeze-out, caused by the inability of the triple-alpha reaction ^{4}He($2\alpha, \gamma$)^{12}C, and the ^{4}He(αn,γ)^{9}Be reaction to keep light nuclei such as n, p, and ^{4}He, and nuclei beyond $A = 12$ in NSE during declining temperatures, when densities are low. This leads to a large α-particle abundance after freeze-out of nuclear reactions. This effect, most pronounced for core collapse supernovae, depends on the entropy of the reaction environment, entropy being proportional to T^3/ρ in a radiation dominated plasma (see Fig. 4.2).

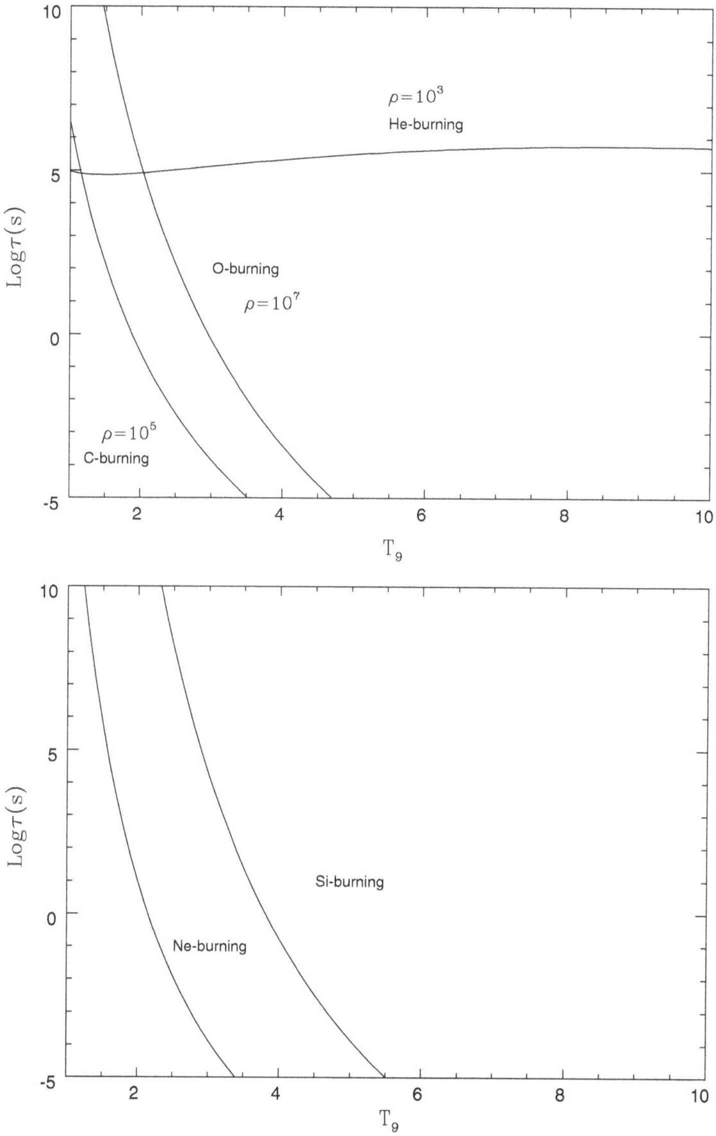

Fig. 4.1 Burning timescales τ in (log_{10}) seconds for fuel exhaustion of He-, C-, and O-burning (top) and Ne- and Si-burning (bottom), as a function of temperature, based on the expressions for dY_i/dt in Eq. (4.2) when equating this to $-1/\tau Y_i$, utilizing $Y_i = X_i/A_i = 1/A_i$ for the burning fuel. Thus, τ has no density dependence for decays and photo-disintegrations, a $1/\rho$ dependence for (two body) fusion reactions, and a $1/\rho^2$ dependence for (three body) fusion reactions. Density-dependent timescales are labeled with a chosen typical density (in g cm^{-3}). They scale with $1/\rho$ for C- and O-burning and $1/\rho^2$ for He-burning. Ne- and Si-burning, initiated by photo-disintegrations, are not density-dependent. The almost constant He-burning timescale beyond $T_9 = T/10^9\text{K} = 1$ permits efficient destruction on explosive timescales only for high densities

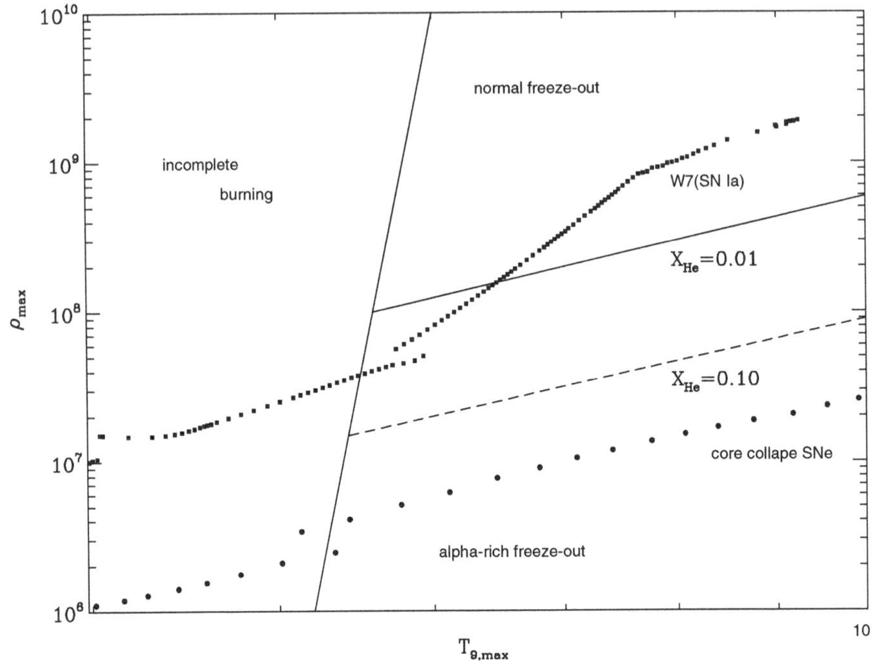

Fig. 4.2 Final results of explosive Si-burning as a function of maximum temperatures and densities attained in explosions before adiabatic expansion. For temperatures in excess of 5×10^9 K any fuel previously existing is photo-disintegrated into nucleons and alpha particles before re-assembling in the expansion. For high densities this is described by a full NSE with an Fe-group composition favouring nuclei with maximum binding energies and proton/nucleon ratios equal to Y_e. For lower densities the NSE breaks into local equilibrium groups (quasi-equilibrium, QSE) with group boundaries determined by reactions with an insufficiently fast reaction stream. Alpha-rich freeze-out (insufficient conversion of alpha-particles into nuclei beyond carbon) is such a QSE-behavior. Lines with 1% and 10% remaining alpha mass fraction are indicated as well as typical conditions in type Ia and core collapse supernovae. This division of the final outcome of explosive Si-burning goes back to Woosley et al. (1973), the type Ia and core-collapse supernova results shown here are taken from Thielemann et al. (1986, 1990) and represent still the current status (in case of type Ia supernovae for single degenerate systems)

Characteristics of the r-process (*rapid* neutron capture) relate to environments of explosive Si-burning as well. The requirement for r-process of a neutron to seed-nuclei ratio of 10 to 150 after freeze-out of charged particle reactions[6] translates into $Y_e = 0.12 - 0.3$ for a normal freeze-out. For a moderate Y_e >0.40, an extremely α-rich freeze-out is needed (see the discussion in Sect. 4.4). Under these conditions the large mass fraction in ^{4}He (with $N = Z$) permits sufficiently high ratios of remaining free neutrons to (small) abundances of heavier seed nuclei, so that r-

[6]Such neutron/seed ratio is required in order to produce all, including the heaviest, r-process nuclei via neutron capture from seed nuclei at their abundances before freeze-out.

process conditions are obtained. During the operation of the r-process, in many cases QSE-groups of neutron captures and photo-disintegrations are formed within the isotopic chains of heavy elements (see e.g. Freiburghaus et al. 1999a; Farouqi et al. 2010).

4.3 Evolution up to Core Collapse

Nuclear burning processes as relevant for massive stars have been discussed above in Sect. 4.2, including also specific individual reactions related to the main focus of this book, the production of (radioactive) nuclei in astrophysical environments. In the present section we will discuss the physics of stellar evolution and major related observational features; but we leave to review articles or textbooks a technical description of how mass, energy, and momentum conservations equations as well as energy transport (via radiation or convective motions) are treated (Maeder 2009; Maeder and Meynet 2012; Woosley et al. 2002; Heger et al. 2003; Limongi et al. 2000; Limongi and Chieffi 2003, 2006b, 2012; Umeda and Nomoto 2008; Ohkubo et al. 2008; El Eid et al. 2009; Paxton et al. 2011) (but see also the hydrostatic stellar structure/evolution equations in spherical symmetry, as presented in Chap. 3). Results of recent calculations are presented in the following subsection.

Massive star evolution is treated by a number of stellar evolution codes, a comparison to observational features has been undertaken by Martins and Palacios (2013) for the codes (MESA,STAREVOLV, GENEC, STERN, Padova, FRANEC). The main codes to follow the evolution through all burning stages until collapse are KEPLER (see e.g. Woosley et al. 2002), GENEC (Eggenberger et al. 2008; Ekström et al. 2012), FRANEC (Limongi and Chieffi 2012), and MESA (Paxton et al. 2011). A detailed comparison between predictions of MESA, KEPLER, and GENEC has recently been proved by Jones et al. (2015). There are differences in treating hydrodynamics/statics, surface layers, rotation and magnetic fields, opacities, mass loss, convection and overshooting, the equation of state, and nuclear reaction input, it is gratifying that there exists a quite good qualitative agreement, but there remain some (minor?) quantitative differences.

In the following we want to focus on results from the KEPLER and GENEC code, addressing topics like (1) adaptive reaction networks for the advanced burning stages, which are capable to follow the detailed evolution of Y_e and a large set of nuclei, (2) a discretization of the stellar-structure equations (and possible damping of instabilities occurring during the advanced stages of evolution, (3) the treatment of dynamical shear in addition to the other mixing processes (such as, e.g., horizontal turbulence, secular shear and meridional circulation), and (4) the treatment of convection/semiconvection, overshoot mixing. Both codes allow to follow the evolution of massive stars from their birth until the stage of Si-burning, including all nuclear burning stages discussed in Sect. 4.2, for a wide range of initial masses, metallicities and stellar rotation.

While there exist also first approaches to treat especially convective stages in multi dimensions (Meakin and Arnett 2007; Arnett et al. 2009; Cristini et al. 2015; Müller et al. 2016b; Edelmann et al. 2017), this can presently only be done for test cases rather than full stellar evolution calculations. Here the treatment of rotation and mixing effects is still based on spherical symmetry.

Finalizing this introduction to the further discussion on the evolution of massive stars, we want to emphasize that there have been many investigations with the FRANEC code (Limongi et al. 2000; Limongi and Chieffi 2003, 2006b, 2009, 2012; Chieffi and Limongi 2013, 2017, 2004), also with emphasis on the production of ^{26}Al, ^{44}Ti, ^{60}Fe, and ^{56}Ni, but in the discussion of the following subsection we summarize results from a large number of publications involving calculations with the KEPLER and GENEC codes (Woosley and Weaver 1995; Woosley et al. 2002; Heger and Woosley 2010, 2002; Heger et al. 2000, 2003, 2005; Woosley and Heger 2007; Jones et al. 2013, 2015, 2016, 2015; Pignatari et al. 2016, 2008; Hirschi et al. 2004, 2005, 2008; Meynet et al. 2008, 2016; Yusof et al. 2013; Georgy et al. 2013, 2017; Maeder et al. 2014; Frischknecht et al. 2016). The main emphasis lies on understanding (a) the composition of wind losses, and (b) the internal structure and composition of stars in their final stage before collapse, setting the stage for subsequent explosions.

4.3.1 Complexities of Post-Main-Sequence Evolution

The evolution of all stars is commonly visualised in the Hertzsprung-Russell (HR) diagram, which relates the stellar luminosity to the stellar surface temperature (color). Stellar evolution is initiated by core H-burning, during which the star is found on the so-called main sequence (MS) in the HR diagram. The release of nuclear binding energy mostly occurs in photons and kinetic energy, heating the star and thus producing thermal pressure against gravitational pressure. At stellar densities, photons undergo a multitude of scattering processes until they finally escape at the photosphere.[7] After the completion of core H-burning, the H-burning region continues to move outward as a burning shell. The He-core contracts and ignites core He-burning in the center, which produces C and O. In the HR diagram, this evolution is reflected by the star's position leaving the main sequence into the 'supergiant' region; the stellar radius increases due to the increased radiation pressure. Depending on the resulting surface temperature it becomes a blue or red supergiant (BSG or RSG). Radiation pressure can rise to such extreme values that stars blow off their outer parts through strong *stellar winds* of velocities up to 2000 km s^{-1}, exposing the more-interior parts of the star, the helium (or in some cases, the carbon) shell. In general, this occurs for stars more massive than

[7]It takes a photon about 10^5 years to reach the surface, after it has been launched in the hot core of, e.g., our Sun.

20–30 $M_\odot$. Such a Wolf-Rayet (WR) star loses between 10^{-6} and a few times 10^{-5} $M_\odot$ per year; for comparison, our Sun loses 10^{-14} of its $M_\odot$ per year through its solar wind. For non-rotating stars, the transition to the WR phase appears through the so-called 'Luminous Blue Variable' stars (LBVs). LBVs are massive, intrinsically bright stars which display different scales of light and color variability, ranging from rapid micro-variations to rare outbreaks of catastrophic mass loss. They represent a very short-lived (perhaps as little as 40,000 years) strongly mass-losing phase in the evolution of massive stars, during which they undergo deep erosion of the outer layers before they enter the Wolf-Rayet phase. Late burning phases progress much more rapidly than the H burning of the main sequence and He burning of the giant phase. The reason is that the energy loss, which drives the nuclear burning rate and hence the evolution, increases dramatically as neutrino production becomes a major part of the nuclear reactions; these escape immediately at the densities discussed here, and hence their share in the nuclear binding energy release is lost. The characteristics of late-burning stages are essentially identified by the size of a star's C+O-core after core He-burning.

In the following we discuss how the evolutionary phases depend on the initial properties of a star. In Fig. 4.3 we give a typical example of such an evolution for a 15 $M_\odot$ star.

The evolution of stars is governed mainly by three initial parameters: (1) its mass M (see Figs. 4.4 and 4.5), (2) its metallicity (Z, i.e. the mass fraction of pre-existing elements heavier than He from earlier stellar generations, see Fig. 4.6), and (3) the rotation rate or surface rotation velocity v_{rot} (see Figs. 4.7 and 4.8). Solar metallicity corresponds to[8] $Z = 0.02$. The evolution can also be influenced by interior magnetic fields, and by a close binary companion. Rotation significantly affects the pre-supernova state, through the impact it has on the H and He-burning evolution. Two mass groups are distinguishable: Either rotationally induced mixing dominates (for M <30 $M_\odot$), or rotationally increased mass loss dominates (for M >30 $M_\odot$). For massive stars around solar metallicity, mass loss plays a crucial role, in some cases removing more than half of the initial mass. Internal mixing, induced mainly by convection and rotation, also has a significant effect on the evolution of stars. An important result is the production of primary ^{14}N (via the CNO-cycle) and ^{22}Ne (via α-captures in He-burning), due to mixing of burning products (such as ^{12}C) with hydrogen or α's, respectively (see the discussion in Sect. 4.2).

The general impact of metallicity can be summarised in the following way: Lower metallicity implies a (slightly) lower luminosity due to the lack of CNO-cycling in hydrogen burning, which leads to slightly smaller convective cores. A lower metallicity also implies a lower opacity due to the lack of heavier elements with their many spectral lines, reducing therefore also radiation pressure and hence mass loss (as long as the chemical composition has not been changed by burning or

[8]The current value of solar metallicity is believed to be $Z = 0.014$, see Chap. 1; the value of $Z = 0.02$, which had been established before and was in common use till~2005, remains a reference for comparisons, though.

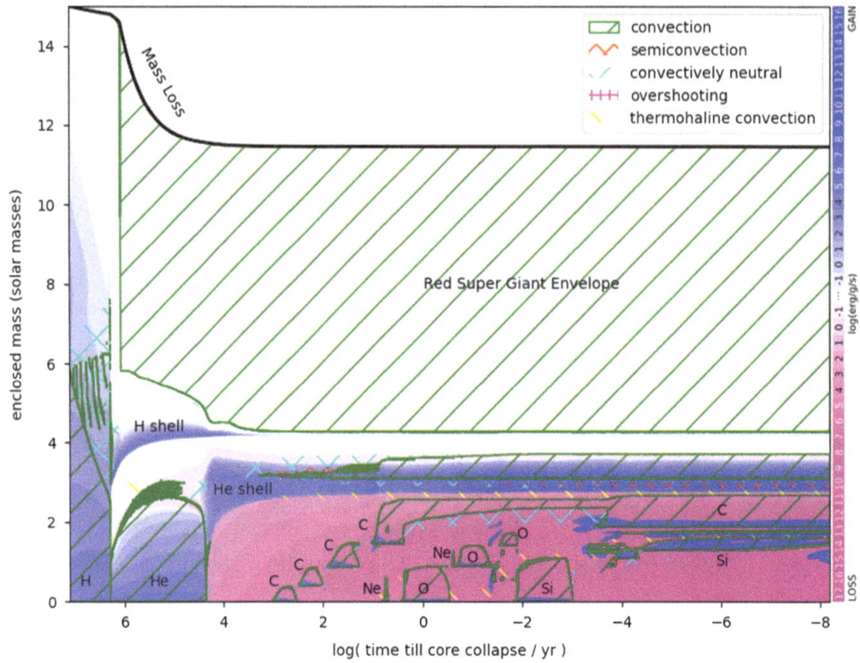

Fig. 4.3 Kippenhahn diagram of a $15\,M_\odot$ star of solar initial composition. The x-axis shoes the logarithm of the time till core collapse in years. The plots begins at about 10 million years before the star dies and ends a quarter of a second before core bounce is reached. The y-axis indicates enclosed mass relative to the center of the star, assuming a spherical symmetry. Blue and purple shading indicate net nuclear energy generation from burning minus neutrino losses. See scale on the right hand side. Green hatching indicates convection, which is the dominant mixing process. Convective regions are also outlined by a green line. Reed cross hatching indicates semi-convection, yellow hatching indicates thermohaline convection, and cyan indicates convectively neutral regions. Black labels indicate the different nuclear burning phases as a function in the core and in shells

mixing in the part of the star under consideration). This results in lower metallicity stars being more compact and experiencing less mass loss. Prescriptions for mass loss as used in the Geneva stellar evolution code are described in detail in Meynet and Maeder (2005). Mass loss rates depend on metallicity as $dM/dt \propto (Z/Z_\odot)^{0.5}$, where Z is the mass fraction of heavy elements at the surface of the star. The effects can be seen in Fig. 4.6 which shows the interior structure of stars through so-called *Kippenhahn diagrams* of $30\,M_\odot$ models for different metallicities of the stars. These diagrams indicate regions (in radial mass coordinates) where matter is unstable against convection; here the energy transport is dominated by transporting hot matter rather than through the propagation of photons. The implications of such a behavior

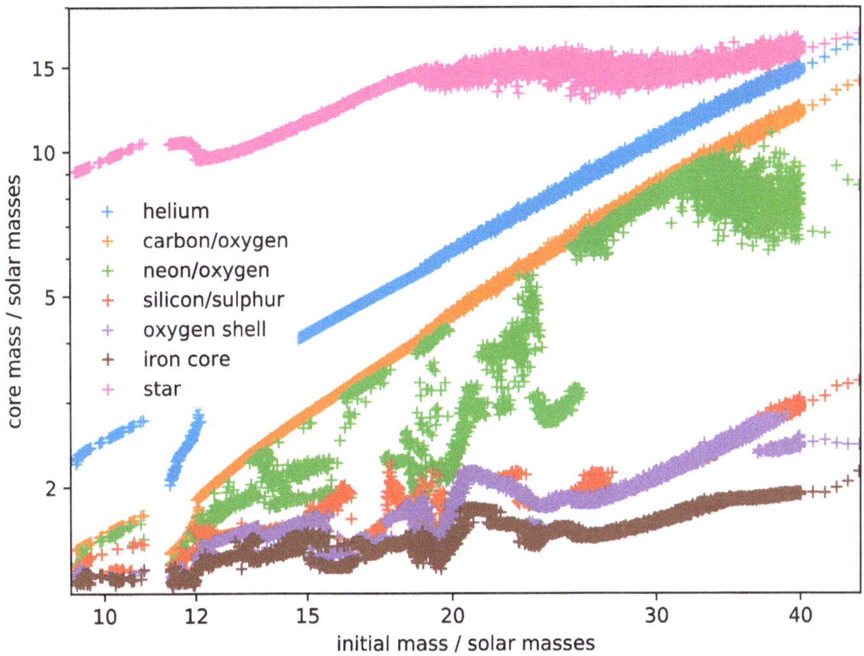

Fig. 4.4 Variation of core sizes as a function of initial mass (y-axis) for stars of solar metallicity. This is based on a compilation of 2910 pre-supernova models of solar initial composition from 9.45 $M_\odot$ to 45 $M_\odot$. Colored pluses indicate the outer edge of each core with a given composition (see figure legend). Below 9.45 $M_\odot$ stars did not undergo iron core collapse but ended up as O-Ne-Mg white dwarf stars. The mass gird is basically 0.01 $M_\odot$ solar mass resolution up to 40 $M_\odot$ and 1 $M_\odot$ above, but there are some gaps of varying size below about 11.5 $M_\odot$ due to non-convergence of the models; the stars in these gaps would still undergo iron core collapse and explode as supernova. We notice some scatter and many discontinuous changes, as well as the apparent existence of parallel branches in some cases. The discontinuous changes can be well understood due the onset of different convective shell burning stages, resulting in a non-linear behavior. Some of the noise indeed is due to numerical noise, largely seeded by semi-convection in hydrogen and helium burning, however, some of the wild variations in particular around 17–18 $M_\odot$ is due to the transition from convective to radiative carbon core burning, with many tiny shells causing an almost chaotic behavior with vastly different outcomes for small changes in initial mass (or other stellar parameters). This figure is based on models presented in Müller et al. (2016a). This is similar to results found by Sukhbold et al. (2017)

have already been described in the evolution of low and intermediate mass stars (Chap. 3), and the physical origin and treatment of these effects are addressed in Chap. 8.

With the exception of the outer convection zone, convective regions in most cases indicate *burning zones*, such as *core H-burning, core He-burning, core C-burning* etc. They testify also the ignition of outward moving burning shells of the same nuclear burning stages. When comparing models for decreasing metallicities (without rotation of Fig. 4.6) one notices only minute reductions of the core sizes,

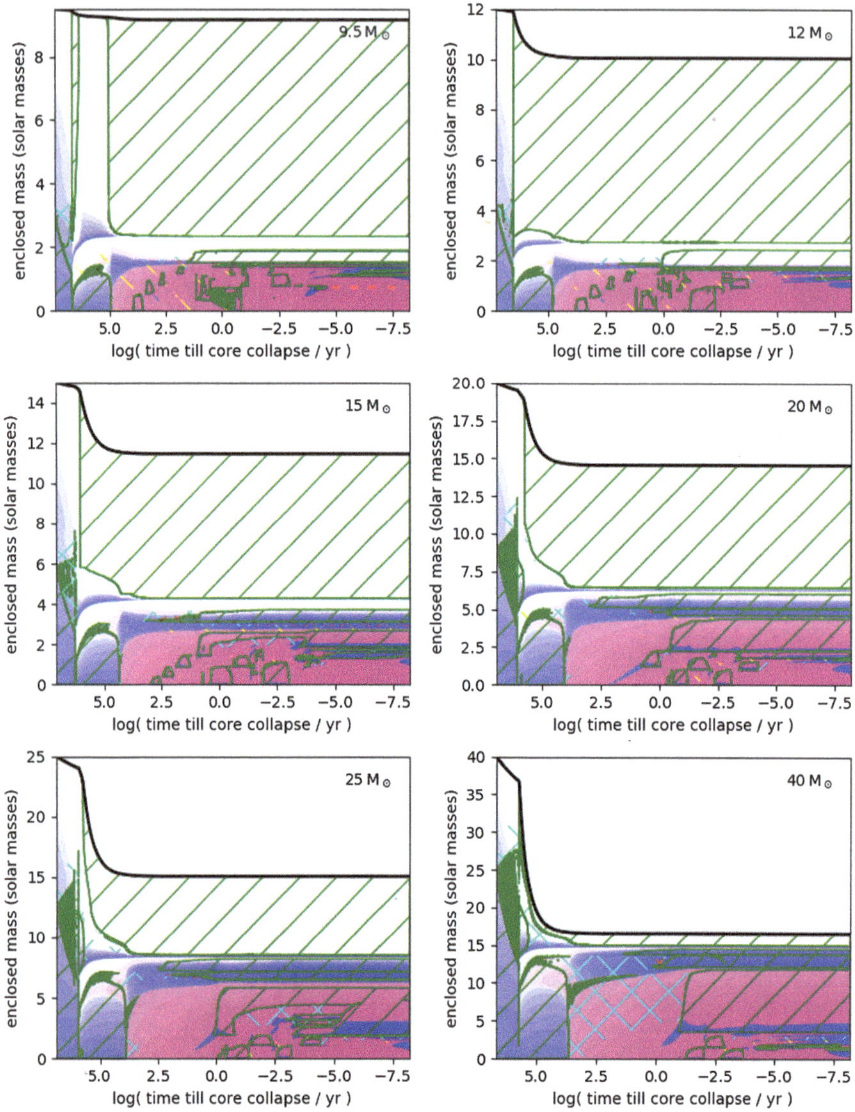

Fig. 4.5 Stellar structure (Kippenhahn) diagrams, which show the for 9.5, 12, 15, 20, 25, and 40 $M_\odot$ stars of solar composition. Coloring and axis are the same as in Fig. 4.3. For the more massive stars the mass loss form the surface significantly increases until almost all of the hydrogen envelope is lost for the 40 $M_\odot$ star. Between 15 and 20 $M_\odot$ central carbon burning transitions from convective to radiative

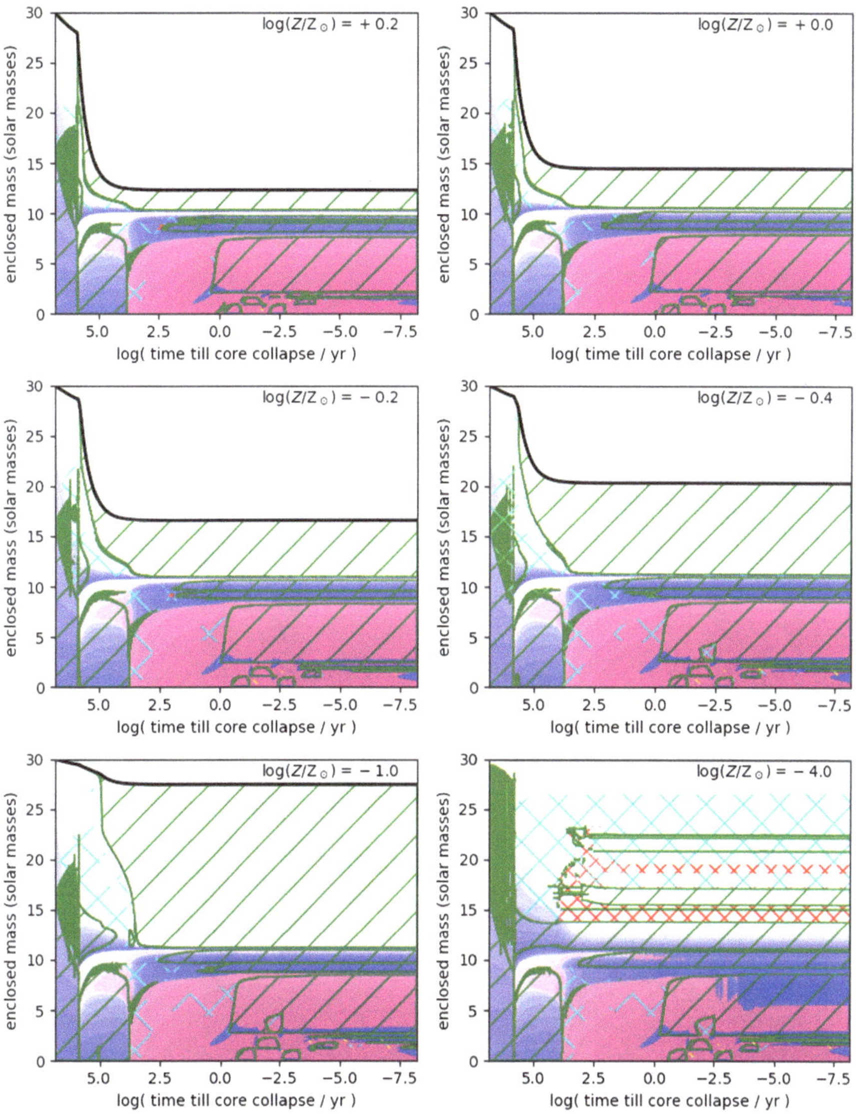

Fig. 4.6 Stellar structure (Kippenhahn) diagrams, which show the for different initial metallicities for a non-rotating $30\,M_{\odot}$ star. Each Panel is labeled with the initial metallicity relative to solar, $Z/Z_{\odot}$. At 1/10,000 solar metallicity mass loss does no longer play a role for this non-rotating model. These models, to be published in West and Heger (2018), used Lodders and Palme 2009 solar abundances as their reference

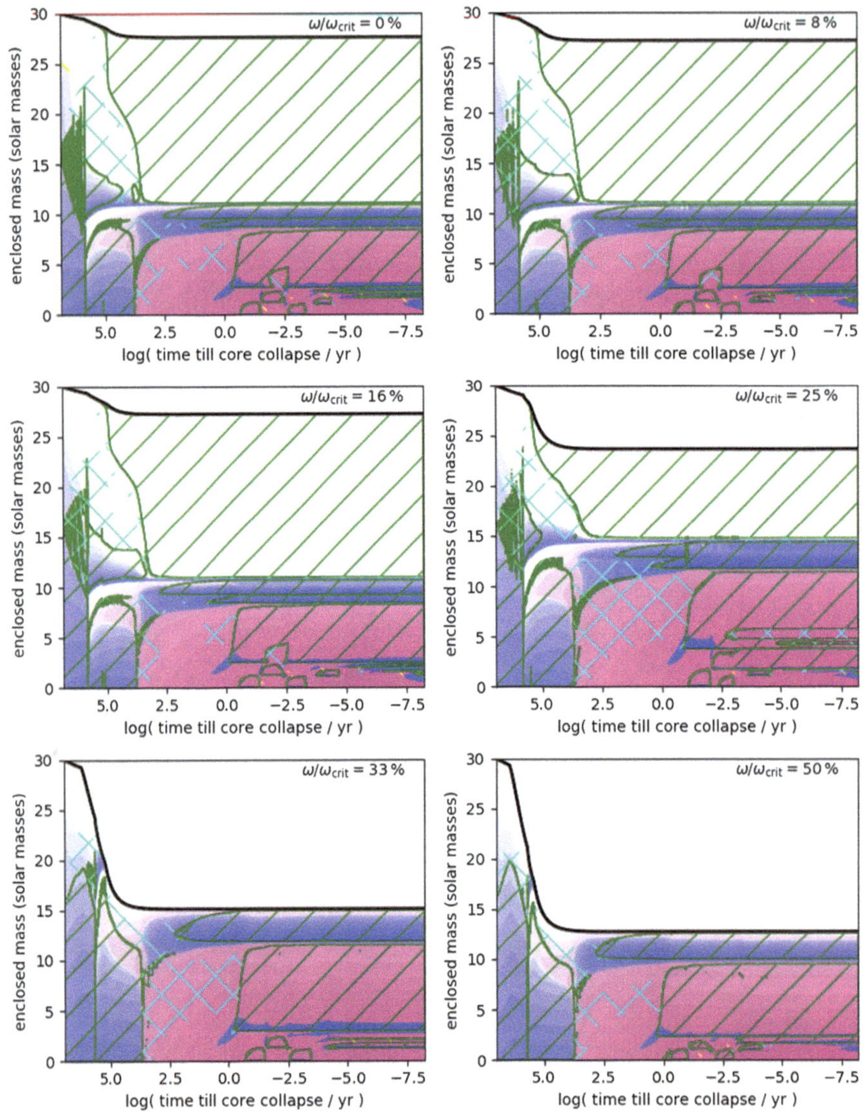

Fig. 4.7 Stellar structure (Kippenhahn) diagrams, which show the for different rotation rates for a 30 $M_\odot$ star of 1/10 solar initial metallicity. Each Panel is labeled with the equatorial surface velocity relative to critical velocity, $\omega/\omega_{\rm crit}$, at the zero-age main sequence of the star. Rotation significantly enhances mass loss from the surface of the star, and for $\gtrsim$ 30 % of critical rotation the core remains rather well-mixed during core-hydrogen burning, which can be seen as an increase of the convective core rather than a drop. These stars transition to the WNL phase during core hydrogen burning, and later become hydrogen-free WNE/WC/WO stars. This figure is based on unpublished work by Heger (2018)

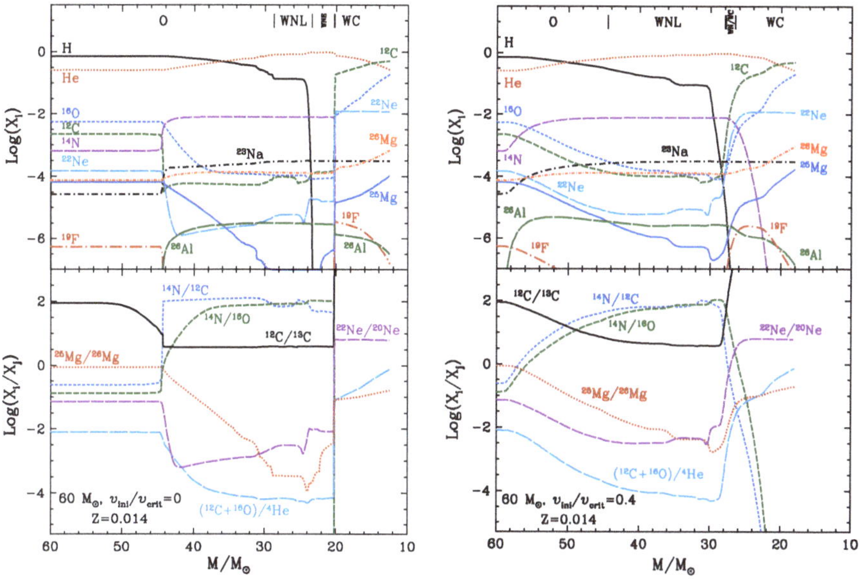

Fig. 4.8 Non-rotation (left) versus rotation-including (right) model comparison of abundance profiles versus mass fractions. Top left: evolution of the surface abundances for a non-rotating 60 $M_\odot$ model. Different evolutionary phases are indicated at the upper axis (the WNC phases occurs in a very narrow region, just before the WC phase, not labelled here). Bottom left: evolution of abundance ratios (in number) as a function of the actual mass of the star. Right panels: Same as left panels, but for the model including stellar rotation (adapted from Georgy et al. 2012)

but it is clearly seen that the outer (H-)burning shell moves further in towards smaller radial mass zones. In addition, the separation of the H- and He-burning shells becomes smaller, which can lead at even lower metallicities to a merging of these shells and a largely increased energy generation and extension of these combined burning zones.

How does rotation change this picture, and how do rotation-induced processes vary with metallicity? At all metallicities, rotation usually increases the core sizes via rotational mixing (compare all panels of Fig. 4.7 with the bottom left panel in Fig. 4.6). The supply of more H-fuel leads to more energy generation and therefore a higher luminosity. The higher luminosity increases the radiation pressure and stellar mass loss. The effect of increased core sizes (and smaller density gradients) can be viewed in all models in Fig. 4.7 with the exception for the highest rotation rate which led to extreme mass loss. Clearly the convective core sizes are increased and the shell burning zones have moved outward. For the metallicity $Z/Z_\odot = 0.1$ shown in this figure, the increased luminosity causes a sufficient increase in radiation pressure so that the mass loss is substantially enhanced (see the decrease of the stellar mass indicated by the top line). Mass loss becomes gradually unimportant for decreasing metallicities. For rotating 20 $M_\odot$ models (not shown here) the stellar fraction lost is more than 50% for solar metallicities, 13% at $Z = 0.001$, less than 3% for $Z = 10^{-5}$, and less than 0.3% for $Z = 10^{-8}$. Figure 4.8 shows the effect of rotation for a $Z = 0.014$ stellar model with an initial mass of $60\,M_\odot$ on its surface abundances. As a function of time mass loss decreases the total mass which is indicated in the abscissa. The surface abundances are given as a function of this time-dependent mass, the top panels display the actual values and the bottom panels the relative abundance ratios.

This can be different for more massive stars (Meynet et al. 2006). In Fig. 4.9, we show results for a low metallicity 60 $M_\odot$ star with $Z = 0.014$, comparing non-rotating and a fast rotating model. The surface layers of massive stars usually accelerate due to internal transport of angular momentum from the core to the envelope. Since at low Z, stellar winds are weak, this angular momentum dredged up by meridional circulation remains inside the star, and the star reaches critical rotation more easily. At the critical limit, matter can be launched into a Keplerian disk which probably dissipates under the action of the strong radiation pressure of the star. Such an effect can be seen for the $85\,M_\odot$ star, which loses in total more than 75% of its initial mass, and initially about 10% due to critical rotation. The remaining mass loss occurs during the red supergiant phase after rotation and convection have enriched the surface in primary CNO elements. We can also see that this effect becomes vanishingly small for stars with masses $M < 30\,M_\odot$. The two $20\,M_\odot$ models with varying metallicities and degrees of rotation again indicate the influence of metallicity and rotation on the compactness and mass loss of stars. In both cases the mass loss is negligible.

We have not shown here the evolution of extremely low metallicity stars. Below a metallicity of about $Z = 10^{-10}$, the CNO cycle cannot operate when H-burning stars after the star has been formed. The star therefore contracts until He-burning ignites, because the energy generation rate of H burning through the pp-chains cannot balance the effect of the gravitational force. Once enough C and O is produced, the CNO cycle can operate, and the star behaves like stars with $Z > 10^{-10}$ for the rest of the main sequence. Metal-free stellar evolution models are presented in Chieffi and Limongi (2004), Heger and Woosley (2002), Umeda and Nomoto (2005), and Ekström et al. (2008).

Including the effects of both mass loss and rotation, massive star models improve to reproduce many observables of stars with metallicities around solar Z. For example, models with rotation allow chemical surface enrichments already on the main sequence of core hydrogen burning (MS), whereas without the inclusion of rotation, self-enrichment is only possible during advanced burning evolution such as the red supergiant RSG stage (Heger and Langer 2000; Meynet and Maeder 2000). Rotating star models also better reproduce the ratio of star types, for the ones which retain their hydrogen surface layer (O stars), which lose the hydrogen layer completely (WR stars), and which even lose their helium layer. The latter affects also the appearance of later core collapse supernova explosions of massive stars. Indeed, rotation changes the supernova type due to the mass loss of the hydrogen envelope (turning such an event in optical observations from a type II supernova with a strong plateau phase to a type IIb event with a smaller plateau, or even a type Ib event for the case of complete loss of the hydrogen envelope, and a type Ic event with the additional loss of the He-envelope). This is discussed in more detail

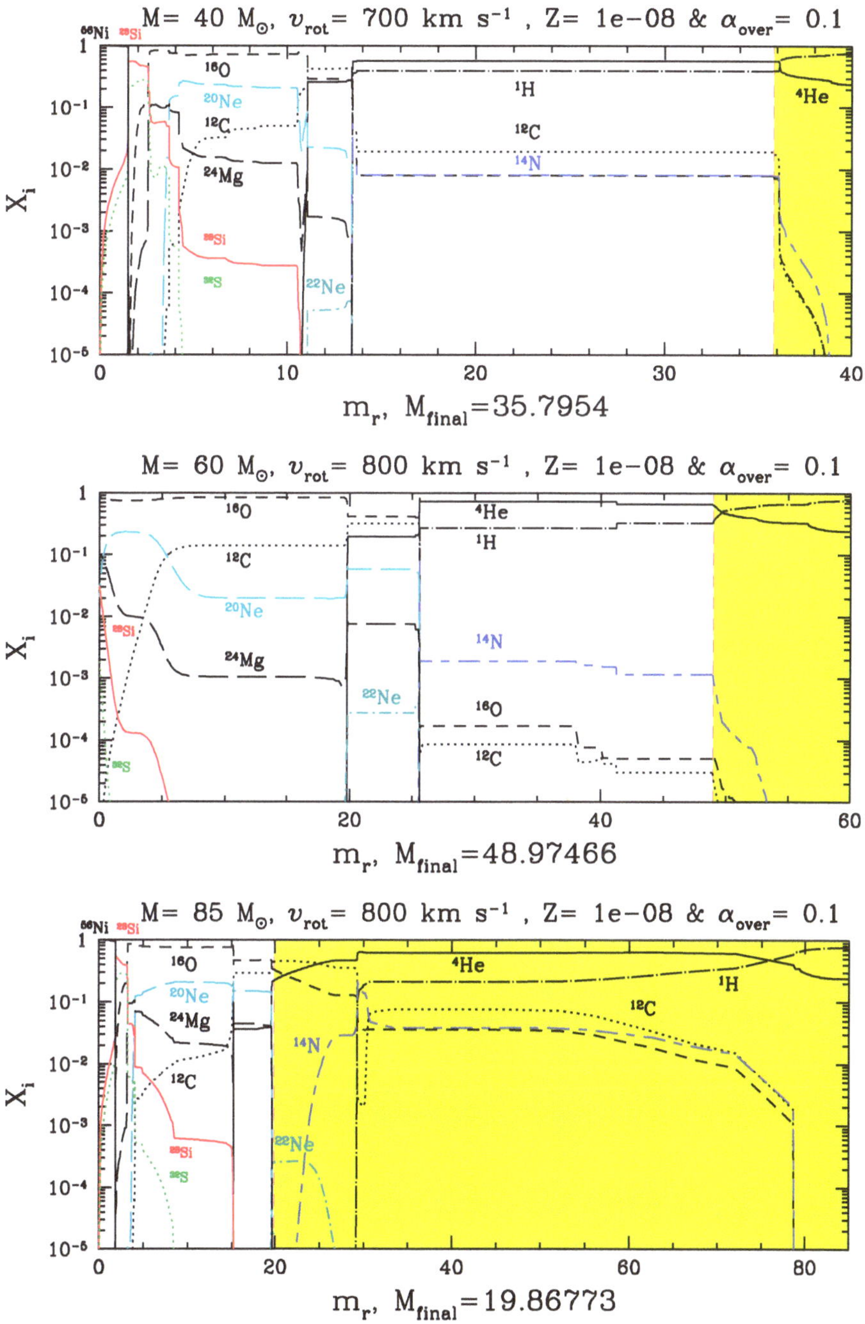

Fig. 4.9 Abundance profiles for 40 (*top*), 60 (*middle*) and 85 (*bottom*) $M_\odot$ models (Hirschi 2007). The pre-SN and wind (yellow shaded area) chemical compositions are separated by a red dashed line located at the pre-SN total mass (M_{final}), given below each graph

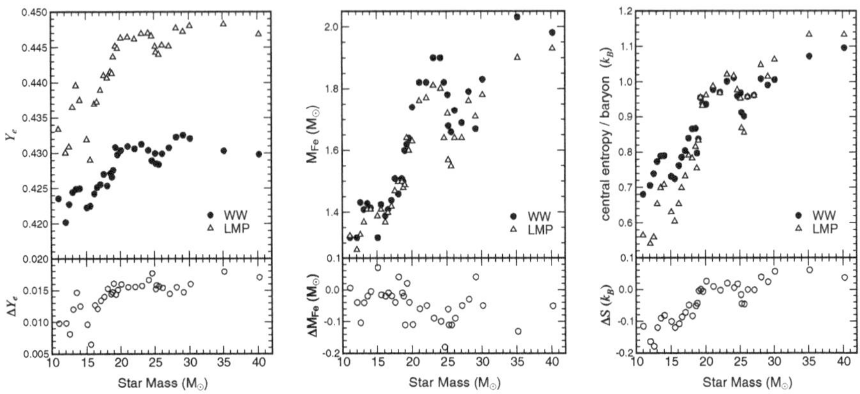

Fig. 4.10 Comparison of the center values of Y_e (left), the iron core sizes (middle) and the central entropy (right) for 11–40 $M_\odot$ stars between the WW models and the ones using the shell model weak interaction rates (LMP) (Heger et al. 2001a). The lower parts define the changes in the three quantities between the LMP and WW models

in Sect. 4.4. Both aspects, the chemical surface enrichment in MS stars as well as the ratio of type Ib+Ic to type II supernovae, as a function of metallicity, are drastically changed compared to models without rotation, which underestimate these ratios (Georgy et al. 2009; Meynet and Maeder 2005). The value of 300 km s^{-1}, used as the initial rotation velocity at solar metallicity, corresponds to an average velocity of about 220 km s^{-1} on the main sequence (MS), which is close to the average value from observations (Fukuda 1982; Meynet et al. 2008). Comparing ratios of stars of different types, as observed in the Magellanic clouds and in our Galaxy, (Maeder et al. 1999; Martayan et al. 2007), points to stars rotating more rapidly at lower metallicities. Fast initial rotation velocities in the range of 600–800 km s^{-1} (Hirschi et al. 2005) are supported by observations of very low-Z stars (Chiappini et al. 2006).

Rotation affects all burning stages and the resulting Fe-core (we will discuss this issue further in the next subsection, see also Fig. 4.10). The size of the Fe-core in turn determines the final fate, whether a supernova explosion with neutron star formation or the collapse to a black hole occurs. The effects of rotation on pre-supernova models are most spectacular for stars between 15 and 25 $M_\odot$. It changes the total size/radius of progenitors (leading to blue instead of red supergiants) and the helium and CO core (bigger by a factor of $\sim$ 1.5 in rotating models). The history of convective zones (in particular the convective zones associated with shell H-burning and core He-burning) is strongly affected by rotation induced mixing (Hirschi et al. 2005). The most important rotation induced mixing takes place while He is burning inside a convective core. Primary C and O are mixed from the convective core into the H-burning shell. Once the enrichment is strong enough, the H-burning shell is boosted (the CNO cycle depends strongly on the C and O mixing at such low initial metallicities). The shell becomes convective and leads to an important primary ^{14}N production while the convective core mass decreases, leading to a less massive CO-core after He-burning than in non-rotating models. Convective and rotational mixing brings the primary CNO to the surface with interesting consequences for the stellar yields. The yield of ^{16}O, being closely correlated with the mass of the CO-core, is reduced. At the same time the C yield is slightly increased (Hirschi et al. 2005), both due to the slightly lower temperatures in core He-burning. This is one possible explanation for the high [C/O] ratio observed in the most metal-poor halo stars (see Fig. 14 in Spite et al. (2005) and Fabbian et al. (2009)) and in damped Lyman alpha systems DLAs (Pettini et al. 2008).

The fate of rotating stars at very low Z is therefore probably the following: $M < 30$–$40\,M_{\odot}$: Mass loss is insignificant and matter is only ejected into the ISM during the SN explosion. 30–$40\,M_{\odot} < M < 60\,M_{\odot}$: Mass loss (at critical rotation and in the RSG stage) removes 10–20% of the initial mass of the star. The star probably dies as a black hole without a SN explosion and therefore the feedback into the ISM is only due to stellar winds. $M > 60\,M_{\odot}$: A strong mass loss removes a significant amount of mass and the stars enter the WR phase. These stars therefore end as type Ib/c SNe and possibly as GRBs. This behavior is displayed in Fig. 4.9. At a metallicity $Z = 10^{-8}$, corresponding to an Fe/H ratio $log_{10}[(Fe/H)/(Fe/H)_{\odot}$ =[Fe/H]~ -6.6, C and O are shown in models to be mixed into the H-burning shell during He-burning. This raises the importance of the shell, and leads to a reduction of the CO-core size. Later in the evolution, the H-shell deepens and produces large amounts of primary nitrogen. For the most massive stars ($M > 60\,M_{\odot}$), significant mass loss occurs during the red supergiant stage, caused by the surface enrichment in CNO elements from rotational and convective mixing.

The properties of non-rotating low-Z stars are presented in Heger et al. (2003), Hirschi et al. (2008), and several groups have calculated their stellar yields (Heger and Woosley 2002; Chieffi and Limongi 2004; Tominaga et al. 2007). All results for the non-rotating stars (whether at solar metallicity or for low-Z models) are consistent among these calculations, differences are understood from the treatments of convection and the rates used for $^{12}C(\alpha,\gamma)^{16}O$. The combined contributions to stellar yields by the wind and the later supernova explosion (see Sect. 4.4) are assembled separately (see, e.g., Pignatari et al. 2016; Georgy et al. 2012; Ekström et al. 2012). The results for stellar models with metallicities Z close to solar can be described as follows: Rotating stars have larger yields in their stellar winds than the non-rotating ones, because of the extra mass loss and mixing due to rotation, for masses below $\sim 30\,M_{\odot}$. The ^{12}C and ^{16}O yields are increased by a factor 1.5–2.5 by rotation. At high mass loss rates (above $\sim 30\,M_{\odot}$), the rotating and non-rotating models show similar yield values. When the wind and explosive contributions are added, the total metal production of rotating stars is larger by a factor 1.5–2.5 (see Sect. 4.4). For very massive stars, the situation varies due to the extreme mass loss, as shown in Fig. 4.9. In order to give a quantitative impression of the influence of initial mass, metallicity and rotation on the evolution of stars, we present in Tables 4.7 and 4.8 results for (a) non-rotating solar metallicity stars (Limongi and Chieffi 2006b) and (b) rotating stars for varying metallicities (Hirschi et al. 2005). Given are the initial and final mass (in order to give an impression of the

Table 4.7 Stellar properties (Limongi and Chieffi 2006a,b)

$M_{ini}/M_{\odot}$	$M_{fin}/M_{\odot}$	$M_{He}/M_{\odot}$	$M_{CO}/M_{\odot}$
11	10.56	3.47	1.75
15	13.49	5.29	2.72
20	16.31	7.64	4.35
30	12.91	12.68	8.01
40	12.52	16.49	8.98
60	17.08	25.17	12.62
80	22.62	34.71	17.41

Table 4.8 Stellar properties (Hirschi 2007)

$M_{ini}/M_{\odot}$	Z	v_{rot}	$M_{fin}/M_{\odot}$	$M_{He}/M_{\odot}$	$M_{CO}/M_{\odot}$
9	1×10^{-8}	500	9.00	1.90	1.34
20	2×10^{-2}	300	8.76	8.66	6.59
20	1×10^{-3}	0	19.56	6.58	4.39
20	1×10^{-3}	300	17.19	8.32	6.24
20	1×10^{-5}	300	19.93	7.90	5.68
20	1×10^{-5}	500	19.57	7.85	5.91
20	1×10^{-8}	300	20.00	6.17	5.18
20	1×10^{-8}	600	19.59	4.83	4.36
40	1×10^{-8}	700	35.80	13.50	12.80
60	1×10^{-8}	800	48.97	25.60	24.00
85	1×10^{-8}	800	19.87	19.90	18.80

mass loss), as well as the core size after central H-burning (the He-core) and after central He-burning (the CO-core), and in Table 4.8 also the metallicity Z and initial rotational surface velocity in km s^{-1}. As all burning stages after He-burning occur on significantly shorter timescales than the earlier burning phases, the CO-core size is the important quantity in order to determine the final outcome/fate of the star.

4.3.2 Late Burning Stages and the Onset of Core Collapse

Stars more massive than about 8 $M_{\odot}$ will, after finishing core and shell H- and He-burning, lead to CO-cores which exceed the maximum stable mass of white dwarfs (the Chandrasekhar mass). For later burning stages, when the partial or full degeneracy of the electron gas is important, this critical limit $M_{Ch}(\rho Y_e, T)$ decides upon further contraction and the central ignition of subsequent burning stages, that is C-, Ne-, O- and Si-burning. Dependent on the Fermi energy of the degenerate electron gas, electron capture on the C-burning products ^{20}Ne and ^{24}Mg can initiate a collapse, leading directly via nuclear statistical equilibrium to a central Fe-core. This evolution path occurs for stars in the range 8–10 $M_{\odot}$ (Nomoto 1987; Kitaura et al. 2006; Wanajo et al. 2009; Jones et al. 2013). More massive stars will proceed through all burning stages until Si-burning will finally produce an Fe-core. All burning stages after core H- and He-burning proceed on timescales which are shorter my orders of magnitude. The reason is that the energy carried away by freely escaping neutrinos dominates over radiation losses by photons which undergo a cascade of scattering processes before their final escape. While neutrinos are emitted during beta-decay and electron captures, discussed in Sect. 4.1, most of these neutrinos are created as central densities and temperatures increase and new degrees of freedom open up for the state of matter. The following neutrino production reactions are relevant: (1) $e^- + e^+$-pair annihilation (*pair neutrinos*), (2) electron-photon scattering with neutrino-antineutrino pair creation (photo neutrinos), and (3) neutrino-antineutrino pair creation from plasma oscillations (*plasmon neutrinos*), as presented in detail in Clayton (1968). Neutrinos dominate the energy loss in stellar evolution from this point on, and lead to increasingly shorter burning timescales, although the photon radiation luminosity of the star remains roughly constant. The timescales for the individual burning stages are given in Table 4.1 in Sect. 4.2; these values refer to a 20 $M_\odot$ star with solar metallicity and no mass loss. Effects of mass loss, rotation and metallicity can change these timescales somewhat (up to 20%). Due to the large difference in evolution timescales, the dominant mass loss by stellar winds occurs during H- and He-burning, and the final outcome of stellar evolution is determined by the CO-core size after He-burning. Therefore, given all dependencies of stellar evolution via initial metallicities and rotation, the initial main sequence mass of a star is less indicative for the final outcome than the size of its CO-core.

In the late phases of O- and Si-burning (discussed in Sect. 4.2), electrons are moderately to strongly degenerate, dependent on the initial stellar mass, and will be characterized by increasing Fermi energies. This will allow for electron captures on burning products, and will make matter more neutron-rich, i.e decrease Y_e, the electron or proton to nucleon (neutrons plus protons) ratio. In high density O-burning ($\rho > 2 \times 10^7$ g cm^{-3}) two electron capture reactions become important and lead to a decrease in Y_e, ^{33}S$(e^-, \nu)^{33}$P and ^{35}Cl$(e^-, \nu)^{35}$S. Such effects become more extensive at even higher densities in Si-burning and a large range of nuclei has been identified to be of major importance $^{55-68}$Co, $^{56-69}$Ni, $^{53-62}$Fe, $^{53-63}$Mn, $^{64-74}$Cu, $^{49-54}$Sc, $^{50-58}$V, $^{52-59}$Cr, $^{49-54}$Ti, $^{74-80}$Ga, $^{77-80}$Ge, ^{83}Se, $^{80-83}$As, $^{50-58}$V, and ^{75}Zn (Aufderheide et al. 1994). The amount of electron capture and the resulting Y_e has consequences for core sizes. (The core sizes of the late burning stages are shown in Figs. 4.3 and 4.8.) The final size of the inner Fe-core represents the maximum mass which can be supported by the pressure of the degenerate electron gas. It is a function of Y_e, but also reflects temperature effects if the electron gas is not completely degenerate (Bethe 1990), with S_e being the entropy in electrons per baryon

$$M_{Ch}(Y_e, S_e) = 1.44(2Y_e)^2[1 + (\frac{S_e}{\pi Y_e})^2]\mathrm{M}_\odot. \tag{4.3}$$

Stars with masses exceeding roughly 10 $M_\odot$ reach a point in their evolution where their Si-burning core (which will turn eventually into their Fe-core) provides no further source of nuclear energy. At this point they collapse and bounce, if not too massive, to explode in spectacular core collapse events known as type II or Ib/c supernovae. These explosions create a neutron star or black hole at the end of the life of a star. They play a preeminent role in the nucleosynthesis and chemical evolution of a galaxy.

The collapse is initiated by the capture of degenerate electrons on nuclei, which reduces the dominant contribution of the pressure (i.e. the one from the degenerate electron gas). Alternatively, for lower densities and higher temperatures (in more massive stars), the pressure supporting the core is reduced by endoergic photo-disintegrations of nuclei, reducing the thermal energy. The evolution in the core is determined by the competition of gravity (that causes the collapse of the core) and weak interaction (that determines the rate at which electrons are captured and the rate at which neutrinos are trapped during the collapse).

The early phases of this final stage of stellar evolution are known as *presupernova evolution*. They follow the late-stage stellar evolution, and proceed until core densities of about $10^{10}\,\mathrm{g\,cm^{-3}}$ and temperatures between 5 and $10\times10^{9}\,\mathrm{K}$ are reached. Until this point, modeling stellar evolution requires the consideration of extensive nuclear reaction networks, but is simplified by the fact that neutrinos need only be treated as a sink of energy and lepton number (due to their immediate escape). At later time and towards the collapse, this is no longer valid: As the weak interaction rates increase with the increasing density, the neutrino mean free paths shorten, so that the neutrinos eventually proceed from phases of free streaming, towards diffusion, and trapping. An adequate handling of the transitions between these transport regimes necessitates a detailed time- and space-dependent bookkeeping of the neutrino distributions in the core (see Chap. 8 for neutrino radiation transport and also a recent detailed discussion by Burrows et al. 2018). During collapse, electron capture, accompanied by ν_e neutrino emission, dominates over electron antineutrino emission because the positron abundance is very low under electron-degenerate conditions. Later in the evolution the electron degeneracy is partially lifted, and in addition to the electron flavor neutrinos, also heavy neutrinos, ν_μ and ν_τ and their antiparticles, are usually included in numerical simulations of core collapse and post-bounce evolution.

Advantageously, the temperature during the collapse and explosion are high enough that the matter composition is given by nuclear statistical equilibrium (NSE), i.e. without the need of reaction networks for the strong and electromagnetic interactions. The transition from a rather complex global nuclear reaction network, involving many neutron, proton and alpha fusion reactions and their inverses, to a quasi-statistical equilibrium, in which reactions are fast enough to bring constrained regions of the nuclear chart into equilibrium, to final and global nuclear statistical equilibrium is extensively discussed by Hix and Thielemann (1996, 1999b) and Hix et al. (2007). In the late stages of Si-burning and the early collapse phase, weak interactions are dominated by electron captures on protons and nuclei. These are important equally in controlling the neutronization of matter Y_e and, in a large portion, also the stellar energy loss. Due to their strong energy dependence $\propto E_e^5$, the electron capture rates increase rapidly during the collapse while the density and the temperature increase (the electron Fermi energy E_F scales with $\rho^{2/3}$, see 4.2).

The main weak interaction processes during the final evolution of a massive star are electron capture and β-decays. Their determination requires the calculation of Fermi and Gamow-Teller (GT) transitions. While the treatment of Fermi transitions (important only for β-decays) is straightforward, a correct description of the GT transitions is a difficult problem in nuclear structure physics. In astrophysical environments, nuclei are fully ionized. Therefore, electron capture occurs from the continuum of the degenerate electron plasma, and energies of the electrons are high enough to induce transitions to the Gamow-Teller resonance. Shortly after the discovery of this collective excitation, Bethe et al. (1979) recognized its importance for stellar electron capture. β^--decay converts a neutron inside the nucleus into a proton and emits an electron. In a degenerate electron gas, with fully populated levels up to the Fermi energy E_F, all decays which would produce electrons with smaller energies than E_F are not possible (*blocked*). Then, the decay rate of a given nuclear state is greatly reduced or even completely blocked at high densities. However, there is another pathway, as high temperatures populate a distribution of nuclear states: If an excited and thermally populated state of the decaying nucleus is connected by large GT transition probabilities to low-lying states in the daughter nucleus, producing electrons above the Fermi energy, such transition path can contribute significantly to the stellar β-decay rates. The importance of these states in the parent nucleus for β-decay in astrophysical environments was first recognized by Fuller et al. (1980, 1982, 1985).

Recent experimental data on GT distributions in iron group nuclei, measured in charge exchange reactions, show that the GT strength is strongly *quenched* (reduced), compared to the *independent-particle-model* value, and fragmented over many states in the daughter nucleus. An accurate understanding of these effects is essential for a reliable evaluation of the stellar weak-interaction rates, particularly for the stellar electron-capture rates (Fuller et al. 1980; Langanke and Martínez-Pinedo 2000). The nuclear *shell-model* is the only known tool to reliably describe GT distributions in nuclei. When comparing the shell-model based rates (by Langanke and Martinez-Pinedo) with the those from Fuller et al., one finds that the shell-model based rates are almost always smaller at the relevant temperatures and densities, caused by the above mentioned quenching of the Gamow-Teller strength, and by a systematic misplacement of the energy of the Gamow-Teller resonance. For an extended overview of rates utilized in astrophysical applications see Fuller et al. (1980, 1982), Oda et al. (1994), Langanke and Martínez-Pinedo (2001, 2003), Sampaio et al. (2003), and Juodagalvis et al. (2010).

The influence of these shell-model rates on the late-stage evolution of massive stars has been investigated by Heger et al. (2001a,b), and compared to earlier calculations (Woosley and Weaver 1995). Figure 4.10 illustrates the consequences of the shell model weak interaction rates for presupernova models in terms of the three decisive quantities: the central electron or proton to nucleon ratio Y_e, the entropy, and the iron core mass. The central values of Y_e at the onset of core collapse increased by 0.01–0.015 for the new rates. This is a significant effect. For example, a change from Y_e = 0.43 in the Woosley and Weaver model for a 20 $M_\odot$ star to Y_e = 0.445 in the new models increases the respective Chandrasekhar mass by about

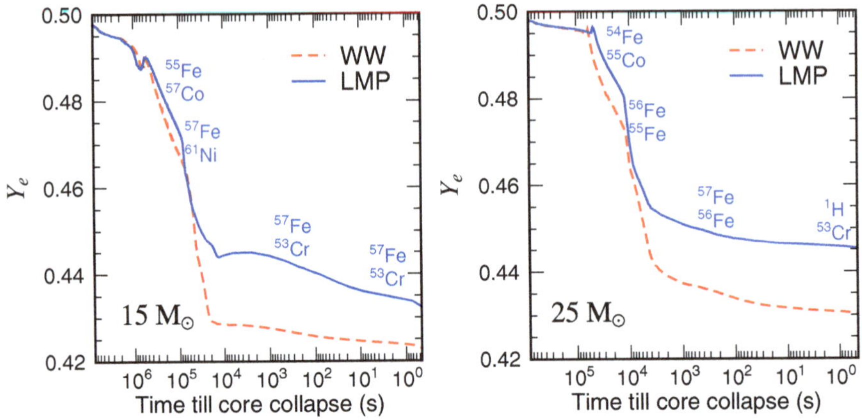

Fig. 4.11 Evolution of the Y_e value in the center of a 15 $M_\odot$ star (left panel) and a 25 $M_\odot$ star (right panel) as a function of time until bounce. The dashed line shows the evolution in the Woosley and Weaver models (WW) (Woosley and Weaver 1995), while the solid line shows the results using the shell-model based weak-interaction rates of Langanke and Martínez-Pinedo (LMP). The two most important nuclei in the determination of the total electron-capture rate, for the calculations adopting the shell model rates, are displayed as a function of stellar evolution time

$0.075\,M_\odot$ (see Eq.(4.3)). The new models also result in lower core entropies for stars with $M < 20\,M_\odot$, while for $M > 20\,M_\odot$, the new models actually have a slightly larger entropy. The Fe-core masses are generally smaller, where the effect is larger for more massive stars ($M > 20\,M_\odot$), while for the most common supernovae ($M < 20\,M_\odot$) the reduction is by about $0.05\,M_\odot$ (the Fe-core is here defined as the mass interior to the point where the composition is dominated by more than 50% of Fe-group elements with $A \geq 48$). This reduction seems opposite to the expected effect due to slower electron capture rates in the new models. It is, however, related to changes in the entropy profile during shell Si-burning which reduces the growth of the iron core just prior to collapse.

The evolution of Y_e during the presupernova phase is plotted in Fig. 4.11. Weak processes become particularly important in reducing Y_e below 0.5 after oxygen depletion ($\approx 10^7$ s and 10^6 s before core collapse for the 15 M and 25 M stars, respectively) and Y_e begins a decline which becomes precipitous during Si-burning. Initially electron captures occur much more rapidly than beta-decays. As the shell model rates are generally smaller the initial reduction of Y_e is smaller in the new models. The temperature in these models is correspondingly larger as less energy is radiated away by neutrino emission. An important feature of the new models is shown in the left panel of Fig. 4.11. For times between 10^4 and 10^3 s before core collapse, Y_e increases due to the fact that β-decay becomes competitive with electron capture after Si-depletion in the core and during shell Si-burning. The presence of an important β-decay contribution has two effects (Aufderheide et al. 1994). Obviously it counteracts the reduction of Y_e in the core, but also acts as an additional neutrino source, causing a stronger cooling of the core and a reduction in entropy. This cooling can be quite efficient, as often the average neutrino energy from the β-decays involved is larger than for the competing electron captures. As a consequence the new models have significantly lower core temperatures. At later stages of the collapse β-decay becomes unimportant again as an increased electron Fermi energy blocks/reduces its role. The shell model weak interaction rates predict the presupernova evolution to proceed along a temperature-density-Y_e trajectory where the weak processes involve nuclei rather close to stability which will permit to test these effects in the next-generation radioactive ion-beam facilities.

Figure 4.11 identifies the two most important nuclei (the ones with the largest value for the product of abundance times rate) for the electron capture during various stages of the final evolution of 15 $M_\odot$ and 25 $M_\odot$ stars. An exhaustive list of the most important nuclei for both electron capture and beta-decay during the final stages of stellar evolution for stars of different masses is given in Heger et al. (2001b). In total, the weak interaction processes shift the matter composition to smaller Y_e values and hence more neutron-rich nuclei, subsequently affecting the nucleosynthesis. Its importance for the elemental abundance distribution, however, strongly depends on the location of the mass cut in the supernova explosion. It is currently assumed that the remnant will have a larger baryonic mass than the Fe-core, but smaller than the mass enclosed by the O-shell (Woosley et al. 2002). As the reduction of Y_e occurs mainly during Si-burning, it is essential to determine how much of this material will be ejected.

4.4 Core Collapse and Supernova Explosions

4.4.1 Physics of Core Collapse, and Numerical Simulations

Supernova explosions are an application of numerical astrophysical modelling that has a long tradition. Continued improvements of the models are motivated by the following points: (1) open questions regarding the explosion mechanism; (2) availability of observations for individual supernova explosions; (3) interesting input physics that tests matter under conditions that are not accessible on earth; (4) visibility in light and other photon wavelengths, cosmic rays, neutrino emission, decay gamma-rays of radioactive products, perhaps gravitational wave emission; (5) visibility on cosmological distances with improving statistical information on the events and (6) their impact on the interstellar matter (e.g. abundances of metal-poor stars) and Galactic evolution.

As discussed in the previous sections, the death of massive stars $\approx 8 - -40\,M_\odot$ proceeds through several evolutionary and dynamical phases. At first, the modeling of a star must include the evolution through all nuclear burning stages until the resulting inner iron core grows beyond the maximum mass which can be supported by the dominant pressure of the degenerate electron gas. At this point, the inner stellar core enters a dynamical phase of gravitational collapse, during which it

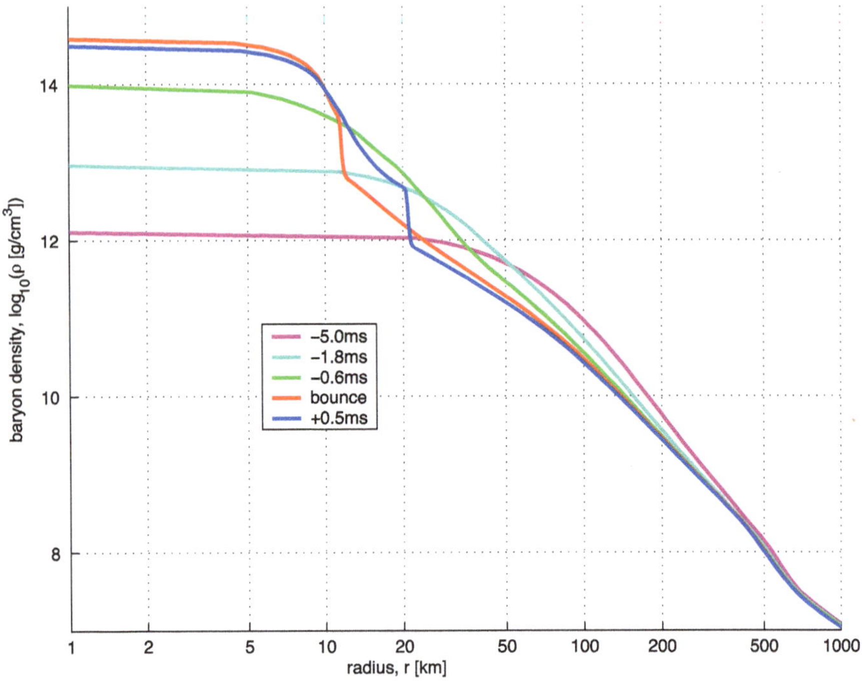

Fig. 4.12 A sequence of density profiles of a 13 $M_\odot$ star before and after core bounce in spherical symmetry. For such a relatively low mass supernova with a small Fe-core the bounce occurs at a maximum density of less than twice nuclear matter density. At the bounce one recognizes the size of the homologous core (with roughly constant density). Thereafter the emergence of an outward moving density (shock) wave can be witnessed, which, however, in spherical symmetry is not leading to an explosion

compactifies by $\sim$ 5 orders of magnitude. The nuclear saturation density (i.e. the density of stable nuclei $\approx 2 \times 10^{14}\,\mathrm{g\,cm^{-3}}$) is exceeded at the center of the collapse and a proto-neutron star (PNS) is formed. The dynamical time scale reduces from a few hundreds of milliseconds at the onset of collapse to a few milliseconds after the core has bounced back at nuclear densities (see Fig. 4.12 from a spherically symmetric simulation by Liebendörfer et al. (2003).

The ensuing accretion phase onto the proto-neutron star with fluid instabilities and radiative transfer phenomena, like the transport of neutrinos, is not, yet, fully understood, but the degree of comprehension is constantly improving. There exists a growing set of 2D and 3D CCSN explosions, see e.g. Hix et al. (2003, 2016), Liebendörfer et al. (2005), Marek et al. (2005), Burrows et al. (2006a, 2018), Sumiyoshi et al. (2007), Langanke et al. (2008), Marek and Janka (2009), Janka (2012), Burrows (2013), Takiwaki et al. (2014), Lentz et al. (2015), Melson et al. (2015), Nakamura et al. (2015), Janka et al. (2016), Bruenn et al. (2016) and the progress of active groups in Garching/Belfast/Monash, Princeton/Caltech/MSU, Oak Ridge/U. of Tennessee/Berkeley, Tokyo/Kyushu, Paris, and

Basel. The accretion may last 0.5–10 seconds and can therefore be interpreted as a second evolutionary stage (much longer than the dynamical or transport time scale). Eventually it will lead to the observed vigorous supernova explosion, a dynamic phase where heavy elements are produced by explosive nucleosynthesis in an outward propagating shock wave. The processed matter is mixed by fluid instabilities and ejected into the interstellar medium, where it contributes to Galactic evolution. The remaining PNS at the center enters another evolutionary phase during which it cools by neutrino emission and contracts or even collapses to a black hole in a last dynamical phase.

While initially such calculations were performed in spherical symmetry and therefore lacked the consistent treatment of turbulent motion, presently performed research is done with multidimensional supernova models (as discussed above). Multi-dimensional effects, including convection, turbulence and shock hydrodynamical instabilities (e.g., SASI), increase the time that a fluid particle spends inside the so-called gain region, in which electron neutrinos (ν_e's) and antineutrinos ($\bar{\nu}_e$) are absorbed and their energy is deposited, heating up matter in this region. This causes a pressure increase and shock revival. Recent multi-dimensional simulations have confirmed that these effects provide more favorable conditions for successful explosions than in spherically symmetric simulations. They rely on radiation (neutrino) transport, (relativistic) hydrodynamics, and the nuclear equation of state at such high densities (Oertel et al. 2017). In addition to non-rotating models without magnetic fields, other efforts explore their role of in multi-dimensional simulations (Winteler et al. 2012; Mösta et al. 2014, 2015, 2017; Nishimura et al. 2015, 2017b; Halevi and Mösta 2018)

Nevertheless, it is still of interest to perform spherically symmetric approaches, although the assumption of spherical symmetry is for many supernovae not compatible with observational constraints, and we now by know that—with the exception of 8–10 $M_\odot$ stars—such simulations do not lead to successful explosions. An important aspect is then to include the knowledge and effects from multi-D simulations into an effective spherically symmetric treatment, which permits to run simulations over a large range of stellar progenitor models. This will be discussed in detail in the next subsection. An important advantage of spherically symmetric models is that sophisticated treatments of the neutrino-matter interactions can be included and that the neutrino spectra and transport are correctly treated in general relativistic space-time. Models of this kind try to address the question of how many neutrinos are emerging from the compactification of an inner stellar core, how is their emission distributed as a function of time and how do these neutrino fluxes generically affect the cooling, heating, or nucleosynthesis in the outer layers of the star without the complication of 3D dynamical fluid instabilities (Liebendörfer et al. 2003, 2004; Fischer et al. 2009, 2010; Martínez-Pinedo et al. 2012). The attempt to combine all these aspects with forefront methods is ongoing in order to achieve the final goal of understanding the multi-D explosion mechanism with up to date microphysics from the equation of state to all neutrino and nuclear interactions.

This includes possible approximate treatments of neutrino transport tested in 1D and then applicable in multi-D approaches (Liebendörfer et al. 2009, 2010) and the test of equation of state effects including a quark-hadron phase transition (Sagert et al. 2009; Fischer et al. 2011; Hempel et al. 2016). A review of the tools to do so is given in Chap. 8.

The phase of stellar core collapse has intensively been studied in spherically symmetric simulations with neutrino transport. The crucial weak processes during the collapse and post-bounce evolution are $\nu + (A, Z) \leftrightarrow \nu + (A, Z)$, $\nu + e^{\pm} \leftrightarrow \nu + e^{\pm}$, $p + e^{-} \leftrightarrow n + \nu_e$, $(A, Z) + e^{-} \leftrightarrow (A, Z - 1) + \nu_e$, $\nu + N \leftrightarrow \nu + N$, $n + e^{+} \leftrightarrow p + \bar{\nu_e}$, $(A, Z) + e^{+} \leftrightarrow (A, Z + 1) + \bar{\nu_e}$, $\nu + (A, Z) \leftrightarrow \nu + (A, Z)^{*}$, $(A, Z)^{*} \leftrightarrow (A, Z) + \nu + \bar{\nu}$, $N + N \leftrightarrow N + N + \nu + \bar{\nu}$, $\nu_e + \bar{\nu_e} \leftrightarrow \nu_{\mu,\tau} + \bar{\nu_{\mu,\tau}}$, $e^{+} + e^{-} \leftrightarrow \nu + \bar{\nu}$. Here, a nucleus is symbolized by its mass number A and charge Z, N denotes either a neutron or a proton and ν represents any neutrino or antineutrino. We note that, according to the generally accepted collapse picture (Bethe 1990; Bethe et al. 1979), elastic scattering of neutrinos on nuclei is mainly responsible for the trapping, as it determines the diffusion time scale of the outwards streaming neutrinos. Shortly after trapping, the neutrinos are thermalized by energy down-scattering, experienced mainly in inelastic scattering off electrons. The relevant cross sections for these processes are discussed in Martínez-Pinedo et al. (2006). The basic neutrino opacity in core collapse is provided by neutrino scattering off nucleons. Depending on the distribution of the nucleons in space and the wavelength of the neutrinos, various important coherence effects can occur: Most important during collapse is the binding of nucleons into nuclei with a density contrast of several orders of magnitude to the surrounding nucleon gas. Coherent scattering off nuclei dominates the scattering opacity of neutrinos (and scales with A^2). Moreover, these neutrino opacities should be corrected by an ion-ion correlation function, this occurs if the neutrino wavelength is comparable to the distances of scattering nuclei and quantum mechanical interference effects appear (Sawyer 2005; Burrows et al. 2006b). Even if current core collapse models include a full ensemble of nuclei in place of the traditional approach with one representative heavy nucleus, it remains non-trivial to adequately determine correlation effects in the ion mixture. Depending on the Q-value of an electron-capturing nucleus, neutrinos are emitted with a high energy of the order of the electron chemical potential/Fermi energy. As the neutrino opacities scale with the squared neutrino energy, the initially trapped neutrinos will down-scatter to lower energies until the diffusion time scale becomes comparable to the thermalization time scale. The thermalization in current collapse models occurs through neutrino-electron scattering because the energy transfer per collision with the light electron is more efficient than with the heavier nucleons. The contribution of inelastic scattering of neutrinos off heavy nuclei depends on the individual nuclei and affects only the high-energy tail of the neutrino spectrum. For latest updates on neutrino opacities and their effect on core-collapse simulations we refer to Burrows et al. (2018).

4.4.2 Early Spherically-Symmetric Modeling

While a number of references in this subsection date back by a decade or more, they give an idea of the evolution of the field still in spherically symmetric approaches but with increasingly sophisticated microphysics included. This will be followed by a short review of present-day multi-D results, before discussing in more detail advanced 1D approximations which take into account the knowledge of the increasing number of multi-D results. Goldreich and Weber (1980) have shown that only the inner $M_{Ch}(Y_e)$ (see the definition in Eq. 4.3) undergo a homologous collapse ($v_{collapse}(r) \propto r$), while at the edge of this core the velocity becomes supersonic and a fraction of the free-fall velocity. The inner core, falling at subsonic velocities where matter can communicate with sound speed, cannot communicate with the free-falling envelope. After the neutrinos are trapped, electron captures and neutrino captures are in equilibrium ($e^- + p \leftrightarrow n + \nu_e$) and the total lepton fraction $Y_L = Y_e + Y_\nu$ stays constant. Y_e stops to decrease and M_{Ch} stops shrinking. Typical values (with the most recent electron capture rates (Langanke et al. 2003) of $Y_L \approx 0.3$ are found in numerical collapse calculations (Hix et al. 2003; Marek et al. 2005) which correspond to $M_{Ch} \approx 0.5\,M_\odot$. As soon as nuclear densities are reached at the center of the collapsing core, repulsive nuclear forces dominate the pressure in the equation of state. The collapse comes to a halt and matter bounces back to launch an outgoing pressure wave through the core. It travels through the subsonic inner core and steepens to a shock wave as soon as it faces supersonic infall velocities. Hence the matter in the PNS remains at low entropy $\sim 1.4\,k_B$ per baryon while the supersonically accreting layers become shock-heated and dissociated at entropies larger than $\sim 6\,k_B$ per baryon. Numerical simulations based on standard input physics and accurate neutrino transport exclude the possibility that the kinetic energy of the hydrodynamical bounce at nuclear densities drives a prompt supernova explosion because of dissociation and neutrino losses.

This can be seen in Fig. 4.12 presenting spherically symmetric calculations of a 13 $M_\odot$ star. The inner core contains about 0.6 $M_\odot$ of the initial Fe-core. The transition to free nucleons occurred only in this inner, homologous core and the outward moving shock runs through material consisting of Fe-group nuclei. The dissociation takes 8.7 MeV/nucleon or $8 \times 10^{18}\,\mathrm{erg\,g^{-1}}$. Based on initial shock energies of (4–8) $\times\ 10^{51}$ erg, this is sufficient for passing through 0.25–0.5 $M_\odot$ and leads in essentially all cases to a stalling of the prompt shock.

While core collapse determines the state of the *cold* nuclear matter inside the PNS, the mass of the hot mantle around the PNS grows by continued accretion. The infalling matter is heated and dissociated by the impact at the accretion front and continues to drift inward. At first, it can still increase its entropy by the absorption of a small fraction of outstreaming neutrinos (heating region). Further in, where the matter settles on the surface of the PNS, neutrino emission dominates absorption and the electron fraction and entropy decrease significantly (cooling region). The tight non-local feedback between the accretion rate and the luminosity is well captured in computer simulations in spherical symmetry that accurately solve the

Boltzmann neutrino transport equation for the three neutrino flavors. All progenitor stars between main sequence masses of 13 and 40 $M_{\odot}$ showed no explosions in simulations of the post-bounce evolution phase (Liebendörfer et al. 2003). This indicates that the neutrino flux from the PNS does not have the fundamental strength to blow off the surrounding layers for a vigorous explosion. Only recently successful (delayed) explosions could be attained in spherically symmetric models (Sagert et al. 2009; Fischer et al. 2011; Hempel et al. 2016). If a hadron-quark phase transition occurs in the collapsed core at the appropriate time, releasing additional gravitational binding energy in the form of neutrinos from this second collapse, the initially stalled prompt shock can be revived by this QCD effect.

Improved electron capture rates on heavy nuclei overcame the idealized blocking of Gamow-Teller transitions in the traditionally applied single-particle model. In the single-particle picture of nuclei the so-called pf-shell is filled for $Z = 40$ or $N = 40$ for protons or neutrons respectively. Neutron numbers beyond $N = 40$ require a filling of the gd-orbits. If during core collapse nuclei (Y_e) become so neutron-rich that nuclei with $Z < 40$ and $N > 40$ dominate the NSE composition, electron capture would require the conversion of an fp proton to a gd neutron as all pf neutron orbits are filled. This Pauli-blocked transition would lead to the dominance of electron capture on free protons rather than nuclei under such conditions. The recent finding, that configuration mixing and finite temperature effects result in unfilled pf neutron orbits, removes this Pauli-blocking and results in the fact that under these condition electron capture rates on nuclei dominate those on free protons (Langanke et al. 2003). Thus, there are two effects due to the new set of electron capture rates: (1) at low densities for less neutron-rich nuclei the total amount of electron capture is reduced with an improved description of Gamow-Teller transitions (see the discussion of the early collapse phase in Sect. 4.3), (2) at high densities in the late collapse phase the total amount of electron capture is enhanced, leading to smaller Y_e and Y_L values than before. Such changes caused a reduction of homologous core sizes down to $M_{Ch} = 0.5$ $M_{\odot}$ (see discussion above and Hix et al. (2003)). This faster deleptonization in the collapse phase in comparison to captures on free protons alone thus resulted in a 20% smaller inner core at bounce.

When applying all this improved physics in present simulations, a large range of conditions in densities ρ, electron abundance Y_e and entropy s per baryon is encountered where the equation of state or other or other microscopic physics is needed. Figure 4.13 provides this information for a simulation of a 20 $M_{\odot}$ star (Liebendörfer et al. 2009), i.e. the conditions encountered during the entire simulation in all mass zones involved.

Moreover, a comparison of the effects for a variety of equations of state (see e.g. Oertel et al. 2017) is required. In simulations of massive progenitors that do not explode and exceed the maximum stable mass of the accreting neutron star in the postbounce phase, it was demonstrated that the neutrino signal changes dramatically when the PNS collapses to a black hole (Fischer et al. 2009). Depending on the stiffness of the equation of state or the accretion rate from the external layers of the progenitor star, this can happen at very different time after bounce. Hence, the

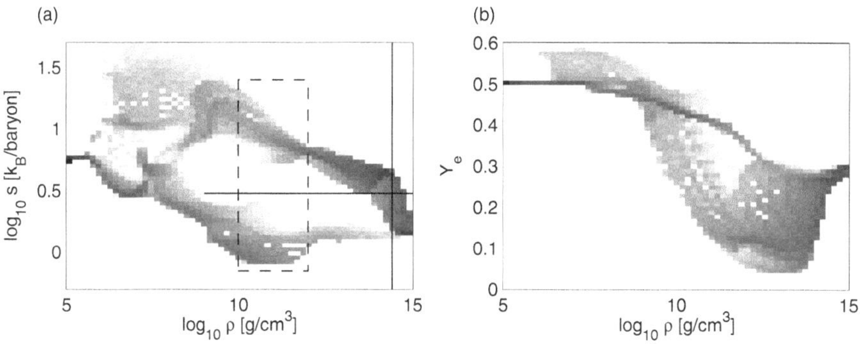

Fig. 4.13 Overview of the conditions attained in a simulation of the collapse, bounce, and explosion (artificially induced) of a 20 $M_\odot$ star. Shown are two histograms of the occurrence of conditions as a function of density ρ, specific entropy s and electron fraction Y_e. The shading of a given bin corresponds to $log_{10}(\int dmdt)$ in arbitrary units, where the integral over mass is performed over the mass dm of matter whose thermodynamic state at a given time falls into the bin. The integral over time extends over the duration of a simulation. Hence, regions of dark shading correspond to states that are experienced by considerable mass for an extended time, while light or absent shading corresponds to conditions that are rarely assumed in the supernova simulation. The vertical black line indicates the nuclear density. The horizontal black line indicates an entropy of 3 k_B/baryon beyond which ions are dissociated. It clearly separates the conditions of cold infalling matter on the lower branch from the conditions of hot shocked matter on the upper branch

neutrino signal carries a clear imprint of the stiffness of the equation of state and the accretion rate to the observer of neutrinos.

The detailed treatment of the neutrino transport and interactions is of great importance for the nucleosynthesis. High electron densities, caused by high Fermi energies of (degenerate) electrons, lead to neutron-rich conditions due to electron captures on free protons. High Y_e material is produced in subsequent ejections, following behind the earlier ejecta, which are strongly affected by neutrino and anti-neutrino captures on protons and neutrons, according to the following reaction sequences

$$\nu_e + n \rightarrow p + e^- \tag{4.4}$$

$$\bar{\nu}_e + p \rightarrow n + e^+. \tag{4.5}$$

These reactions turn matter only neutron-rich if the average anti-neutrino energy $\langle\epsilon_{\bar{\nu}_e}\rangle$ is higher than the average neutrino energy $\langle\epsilon_{\nu_e}\rangle$ by four times the neutron-proton mass difference Δ for similar (electron) neutrino L_{ν_e} and anti-neutrino $L_{\bar{\nu}_e}$ luminosities. This was pointed out initially in Qian and Woosley (1996), leading—when approaching equilibrium conditions for neutrino and antineutrino captures—to

$$Y_e = \left[1 + \frac{L_{\bar{\nu}_e}(\langle\epsilon_{\bar{\nu}_e}\rangle - 2\Delta + 1.2\Delta^2/\langle\epsilon_{\bar{\nu}_e}\rangle)}{L_{\nu_e}(\langle\epsilon_{\nu_e}\rangle + 2\Delta + 1.2\Delta^2/\langle\epsilon_{\nu_e}\rangle)}\right]^{-1}. \tag{4.6}$$

Further details and in-medium corrections for neutrons and protons in comparison to their treatment as free particles are given in Martínez-Pinedo et al. (2012) and Roberts et al. (2012). Thus, in most cases the energetically favorable first reaction (1) wins, changing Y_e from the initial (neutron-rich) conditions towards values beyond $Y_e = 0.5$. A strong νp-process is caused if $Y_e > 0.5$ conditions are attained. In such a case ^{64}Ge is produced in slightly proton-rich conditions of explosive Si-burning. The long beta-decay half-life of ^{64}Ge (which would prevent further processing via proton captures on ^{64}Ga to higher nuclear mass numbers) can be circumvented by an (n, p)-reaction, made possible by neutrons created in antineutrino captures on existing protons in this proton-rich matter. Further proton-captures and fast beta-decays can lead to the production of nuclear masses up to $A = 100$. This has been shown in number of studies (Fröhlich et al. 2006a,b; Pruet et al. 2005, 2006; Wanajo 2006). This also opens an opportunity to investigate neutrino flavor oscillations among electron, muon and tau neutrinos. On the one hand side the long term explosion runs achieve (low) density structures that allow for MSW (Mikheyev-Smirnov-Wolfenstein effect) neutrino flavor oscillations in the outer layers (Wolfenstein 1978; Mikheyev and Smirnov 1985). These may give additional hints on the expansion velocity and density distribution in case that the neutrinos can be observed from a near-by supernova. On the other hand, collective flavor transitions have recently been postulated in regions where the neutrino density exceeds the electron density (Duan et al. 2006, 2007; Fogli et al. 2007). This condition will be achieved in the evacuated zone that surrounds the PNS after the onset of an explosion. The impact of these collective neutrino flavor oscillations on the neutrino heating during the shock expansion, the neutrino wind, and the nucleosynthesis are important points that have recently led to a number of detailed investigation under consideration of accurate neutrino transport and spectra (see e.g. Wu et al. 2014, 2015, 2016b).

4.4.3 Multi-D Simulations: A Short Survey

Initially, spherically symmetric supernova models were the most realistic among all feasible computer representations of the event. With increasing observational evidence for the complexity of the explosions (Hamuy 2003) their primary purpose shifted from a realistic representation to the identification and understanding of the basic principles of the explosion mechanism. This led to the emergence of axisymmetric simulations with sophisticated and computationally intensive spectral neutrino transport (Buras et al. 2003; Walder et al. 2005).

The difficulty to reproduce explosions in spherically symmetric models of core-collapse and post-bounce evolution stimulated the consideration of numerous modifications and alternatives to this basic scenario, mostly relying on multi-dimensional effects that could not be treated in spherical symmetry. It was discussed whether convection inside the PNS could accelerate the deleptonization and increase the neutrino luminosity (Wilson and Mayle 1993). The convective overturn between the PNS and shock front was shown to increase the efficiency of neutrino energy deposition (Herant et al. 1994). Asymmetric instabilities of the standing accretion shock (Blondin et al. 2003; Foglizzo 2009) may help to push the shock to larger radii and g-mode oscillations of the PNS may contribute to neutrino heating by the dissipation of sound waves between the PNS and the shock (Burrows et al. 2006a). Moreover, it has been suggested that magnetic fields have an impact on the explosion mechanism (Kotake et al. 2006). Most of the above-mentioned modifications of the explosion mechanism are essentially of a three-dimensional nature. In order to illustrate the complexity of the crucial accretion phase we show in Fig. 4.14 a slice through a three-dimensional simulation of core-collapse and post-bounce evolution of a recent run (Liebendörfer et al. 2008). Its input physics uses the Lattimer-Swesty equation of state (Lattimer and Douglas Swesty 1991) and a parameterization of the neutrino physics for the collapse phase (Liebendörfer et al. 2005). The treatment of neutrino cooling and heating in the post-bounce phase is based on multi-group diffusion (the isotropic diffusion source approximation, IDSA, of Liebendörfer et al. 2009).

One goal of core-collapse supernova theory is to explain the mechanism of the explosion in terms of physics. We have learned that in such studies it is necessary to establish numerical simulations which include complicated multi-dimensional radiation hydrodynamics, that is, codes which incorporate the inter-related neutrino, nuclear, and gravitational physics. Then we demand that such codes reproduce explosions in a robust way. So, simulations should predict asymptotic explosion energies which are consistent with observations, the resulting neutron star masses, and last but not least the accompanying nucleosynthesis.

Many references reflect progress in this field (see Hix et al. 2003, 2016; Liebendörfer et al. 2005; Marek et al. 2005; Burrows et al. 2006a; Sumiyoshi et al. 2007; Langanke et al. 2008; Marek and Janka 2009; Janka 2012; Burrows 2013; Takiwaki et al. 2014; Lentz et al. 2015; Melson et al. 2015; Nakamura et al. 2015; Janka et al. 2016; Bruenn et al. 2016). An excellent and extended discussion has been given by Burrows et al. (2018) on successes, convergence, on similar and contradictory results. Rather than repeating such here, we draw and state as the main results that (1) 2D simulations seem to lead more easily to explosions than those in 3D (but not necessarily giving larger explosion energies), and (2) a key issue for successful explosions is the compactness of the central part of the progenitor star, or, related, the binding energy of the envelope. Explosions can be obtained for a large range of progenitor masses, even up to and beyond $40\,M_{\odot}$, while black hole formation can occur also for lighter object (e.g. in the range 20–30 $M_{\odot}$) and for the more massive ones.

An example for the complexity of such multi-D simulations is given in Fig. 4.14, indicating entropy, magnetic field strength, and matter velocities from a 3D simulation with the Basel code ELEPHANT. A detailed comparison of a set of codes (SPH, FLASH, ELEPHANT, and fully relativistic M1) actually leads to very similar results for the same input physics, which is a positive indication for convergence on numerical aspects.

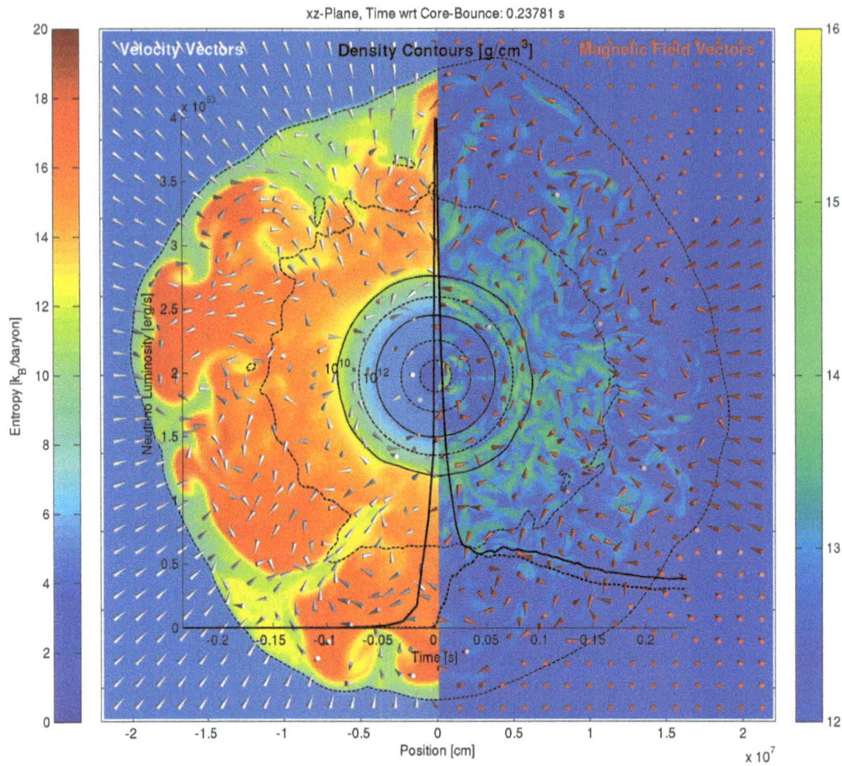

Fig. 4.14 Illustration of the early accretion phase in a three-dimensional simulation with a resolution of 600^3 zones and the isototropic diffusion source approximation for 3D neutrino transport (Liebendörfer et al. 2009). Shown are density contours as black lines for a 15 $M_{\odot}$ star from Woosley and Weaver (1995). *Left:* The color indicates the specific entropy and the cones the direction of the velocity. *Right:* The color refers to the magnetic field strength and the cones to its direction. The cool high-density interior of the PNS and the hot low-density accreted matter behind the standing accretion front are clearly distinguishable. Also shown is the luminosity of electron neutrinos (solid line) and electron antineutrinos (dashed line) as a function of time

Few and preliminary nucleosynthesis results exist from multi-dimensional simulations, e.g. Wanajo et al. (2011), Harris et al. (2017), Eichler et al. (2018), Wongwathanarat et al. (2017), and Yoshida et al. (2017). We show here the results of a detailed nucleosynthesis study by Eichler et al. (2018), based on long-term, two-dimensional core-collapse supernova simulations of an $11.2\,M_{\odot}$ star (Nakamura et al. 2015). This CCSN model shows an axisymmetric neutrino-driven explosions of a non-rotating, solar metallicity model. The numerical treatment was described in Nakamura et al. (2015), including spectral transport of electron and anti-electron neutrinos, using the isotropic diffusion source approximation (IDSA Liebendörfer et al. 2009) and a Lattimer and Swesty equation of state (Lattimer and Douglas Swesty 1991) with compressibility $K = 220\,\text{MeV}$. After a successful shock revival,

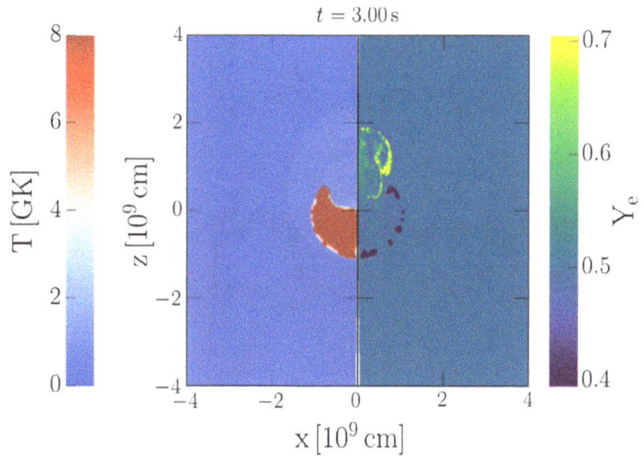

Fig. 4.15 Distribution of temperature and Y_e during the SN explosion of a 12.7 $M_\odot$ star at t = 3s, based on a 2D simulation (Nakamura et al. 2015). It can be seen that Y_e values >0.5 and <0.5 exist, dependent on the initial core-collapse phase with high electron capture effects and electron neutrino and antineutrino absorption on neutrons and protons during the explosion, see Eichler et al. (2018)

the axis-symmetric nature of the simulation causes a preference of the outflow towards polar directions (see Fig. 4.15).

This behavior is usually not observed in 3D simulations of regular CCSNe, since they do not have an imposed symmetry. Explosion times and energies are also different between 2D and 3D simulations, with 2D simulations being usually "more optimistic" to obtain successful explosions, as is shown e.g., in Takiwaki et al. (2014), Lentz et al. (2015), Melson et al. (2015), Janka et al. (2016), and Hix et al. (2016). In these simulations it is found that nuclei well beyond the iron group (up to $Z \approx 44$) can be produced, including the p-nuclei ^{92,94}Mo and ^{96,98}Ru (see Fig. 4.16). While ^{92}Mo and ^{94}Mo can be produced in slightly neutron-rich conditions, ^{96,98}Ru can only be produced efficiently via the νp-process, which depends heavily on the presence of very proton-rich material in the ejecta, obtained in polar outflows. Figure 4.15 shows the Y_e distribution of ejecta 3s after bounce. One realizes low Y_e material (<0.5) in early outflows (typically at the outer edges of ejecta) which still have values close to the ones inherited from core collapse. The effect of neutrinos on Y_e and the working of the νp-process was discussed in the previous subsection.

4.4.4 Spherically-Symmetric Simulations, New Generation

3D simulations show many effects that result from implementation details of the physics involved, and will at the end be used as a method to also describe the nucleosynthesis of the ejecta. But for a large set of supernova progenitor models they

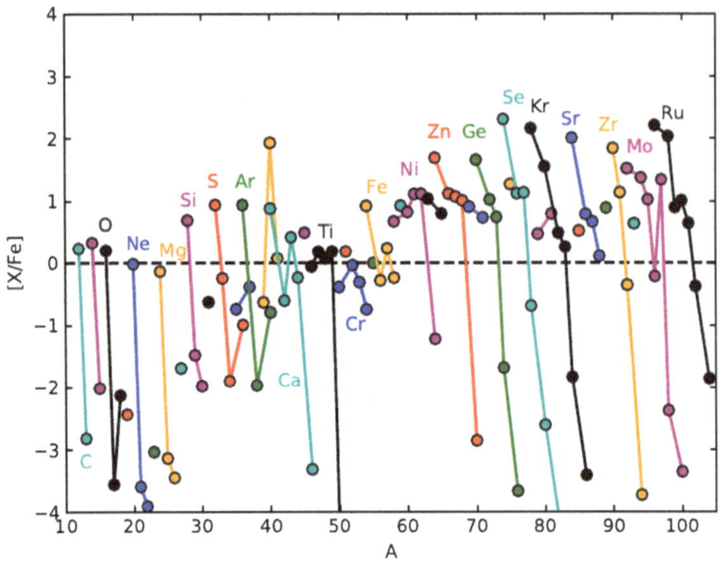

Fig. 4.16 Nucleosynthesis features of a 12.7 $M_\odot$ SN explosion from a 2D simulation (Nakamura et al. 2015), featuring a strong νp-process with the production of nuclei up to A = 100 (see also Eichler et al. 2018)

are still computationally extremely expensive. Therefore, spherically symmetric models still have several assets in supernova modeling. Taking the lessons from 3D on the explosion mechanism, explosions can be artificially launched by introducing a specific additional energy input. Early investigations with the main aim of finding the nucleosynthesis effects were just phenomenological: Artificially, explosions were induced via a "piston", or energy deposition in terms of a "thermal bomb". More realistic approaches add the energy which leads to explosions based on insight from multi-D simulation studies and their inferred behavior. This is an important step towards "self-consistent" treatment including neutrino transport, which is necessary to analyse the effect of neutrinos on the nucleosynthesis of the innermost ejecta.

The past simplified approaches artificially induced explosions with estimates for typical explosion energies. First attempts to predict supernova nucleosynthesis in this way induced the explosion energy through pistons (e.g. Woosley and Weaver 1995), or through thermal energy bomb models (e.g. Thielemann et al. 1996; Umeda and Nomoto 2008), or kinetic energy bomb models (e.g. Limongi and Chieffi 2006b, 2012; Chieffi and Limongi 2013, 2017). In the first case, the motion of a mass shell is imposed along a ballistic trajectory with a typical explosion energy of E = 1.2B = 1.2 × 10^{51} erg at a radial position related to a value of the entropy that was expected to result in the most realistic mass cut (i.e., the bifurcation between the PNS and the ejecta). In the latter cases, explosions are triggered by adding typical kinetic or thermal energies to a specific mass zone (usually in deeper zones). The mass cut was then determined by integrating the nucleosynthesis yields from the outside inwards, up to the point where the observationally indicated amount of ^{56}Ni was reached. In all cases the explosion energy is not determined self-consistently, and the physics of the explosion is not included. The mass cut and the explosion energy are free parameters of the model and have to be constrained from the yields of the innermost ejecta, separately for each progenitor. These approaches are suitable for the outer layers, where the nucleosynthesis mostly depends on the strength of the shock wave.

Interim approaches beyond piston or thermal bomb models of Woosley and Heger (2007), Heger and Woosley (2010), Limongi and Chieffi (2006b, 2012), Nomoto et al. (2006, 2013), Nomoto (2017), and Chieffi and Limongi (2017) exist by now, trying to mimic multi-D neutrino heating in a spherical approach, in order to obtain more appropriate predictions of the explosion energy, mass cut between neutron star and ejecta, as well as nucleosynthesis (including the effects of neutrinos on Y_e, the proton/nucleon ratio). This includes the "neutrino light-bulb" method, where the PNS is excised and replaced with an inner boundary condition which contains an analytical prescription for the neutrino luminosities. Suitable choices of the neutrino luminosities and energies can trigger neutrino-driven explosions (e.g. Yamasaki and Yamada 2005; Iwakami et al. 2008; Yamamoto et al. 2013). In "absorption methods" (Fröhlich et al. 2006a,b; Fischer et al. 2010) the increase in the neutrino energy deposition is obtained by modifying the neutrino opacities in spherically symmetric models with detailed Boltzmann neutrino transport, i.e. this means that neutrino-capture rates are multiplied by a factor, causing additional ν-heating, in order to trigger explosions. This led to the prediction of Y_e-values affected by the central neutrino flux (see Fig. 4.17), improving strongly the Fe-group composition due to the discovery that proton-rich conditions are attained in the innermost ejecta (see the discussion on the effect of neutrino and antineutrino capture in Sect. 4.4.2).

One of the lessons from these investigations was the discovery of the νp-process, also discussed in Sect. 4.4.2. Both aspects led to an improvement of nucleosynthesis predictions for Fe-group nuclei and opened an explanation for understanding light p-nuclei, which cannot be produced in the typical p/γ-process in explosive burning of outer stellar shells, i.e. the Ne-shell. The major drawback of these methods is, however, either the explicit modification of the electron neutrino and antineutrino luminosities or the modification of neutrino opacities, which (while apparently improving nucleosynthesis results) could directly impact the composition of the innermost ejecta in a non-consistent way.

Recent (spherically symmetric) approaches try to mimic the effect of multi-D neutrino transport in a way adapted more consistently to core-collapse and proto-neutron star accretion. They need, however, calibrations which can be provided by comparison with a variety of observations of explosion energies, deduced ejected ^{56}Ni-masses, and progenitor properties. Ugliano et al. (2012) presented a more sophisticated light-bulb method to explode spherically symmetric models, using neutrino energy deposition in post-shock layers. They used an approximate, grey neutrino transport and replaced the innermost 1.1 $M_{\odot}$ of the PNS by an inner

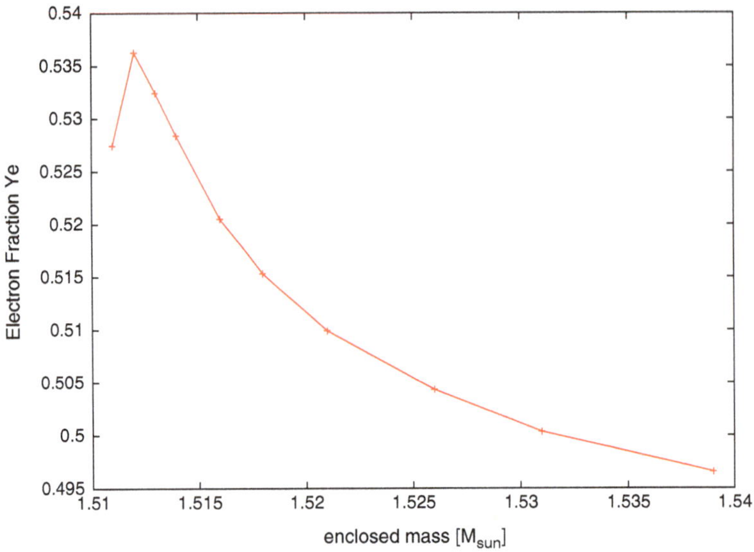

Fig. 4.17 Y_e of the innermost ejecta due to neutrino interactions with matter. At high temperatures electrons are not degenerate, thus the reduction of Y_e due to electron captures is ineffective. For similar neutrino and antineutrino spectra the neutron-proton mass difference favors $\nu_e + n \leftrightarrow p + e^-$ over $\bar{\nu}_e + p \leftrightarrow n + e^+$ (from Fröhlich et al. 2006a)

boundary. The evolution of the neutrino boundary luminosity was based on an analytic cooling model of the PNS, which depends on a set of free parameters. These parameters (within the so-called PHOTB approach) are set by fitting observational properties of SN 1987A for progenitor masses around 20 $M_\odot$ (see also Ertl et al. 2016; Sukhbold et al. 2016). Perego et al. (2015) utilize the energy in muon and tau neutrinos as an additional energy source that approximately captures the essential effects of (3D) neutrino transport (the PUSH method which is discussed in more detail below). Both approaches make it possible to predict the variation of explosion energies (and other parameters like neutron star mass cuts) as a function of stellar mass and thus can provide improved nucleosynthesis yields for chemical evolution modeling. There exist detailed results by now from PHOTB, but only PUSH includes the Y_e-effects due to neutrino interactions with nuclei, which is highly important for the Fe-group composition.

A major open question is whether core collapse leads finally to a supernova explosion with a neutron star remnant or whether the final outcome is a central black hole. Expectations from observations and their interpretation (e.g. Nomoto et al. 2006, 2013) would argue for a gradual transition between these two regimes as a function of initial stellar mass. PHOTB (Sukhbold et al. 2016) shows that both possible outcomes can occur within the same mass interval, mainly dependent on the pre-collapse stellar model and its compactness parameter. This needs further investigations and might pose questions about the stellar models. Could such scatter of the pre-core-collapse models in a narrow mass range, which is due to the properties of the last Si-shell burning stages, be avoided in general by the inclusion of rotation, leading possibly to a smearing out of radial gradients? The transition from central neutron star to black hole remnants can involve the effect of fallback of material from the innermost regions as a result of the outgoing shock wave being reflected by density gradients, causing a reverse shock. A pragmatic solution, utilizing a combination of mixing and fallback, can on the one hand minimize the effect of inconsistent Y_e-values in the innermost regions for approximated spherical models which do not include neutrino interactions (i.e. of the piston or bomb type). This mixes some Fe-group nuclei to larger radii, while the fallback reduces in total the ejected amount of Fe-group elements and can this way create overall C-rich ejecta, as observed in some extremely low metallicity stars. Such behavior is suggested to stem from some very massive stars, possible being more frequent in the early Galaxy (Nomoto et al. 2006, 2013). However, in the following we want to concentrate on the recent advances within the PUSH method.

The PUSH method provides a computationally efficient and physically motivated framework to explode massive stars in spherical symmetry. It allows to study multiple aspects related to core-collapse supernovae that require modeling of the explosion for a duration of several seconds after its onset, and also for extended sets of progenitor models. The PUSH method was primarily designed to study explosive nucleosynthesis, but it is also well suited to explore other relevant aspects, including the effects of the shock passage through the star, the neutron star mass distribution, and the distribution of the explosion energies. PUSH relies on the so-called delayed neutrino-driven mechanism as a central engine of core-collapse supernovae. In particular, it provides an artificially enhanced neutrino energy deposition inside the gain region, inspired by the increase of the net neutrino heating that a fluid element experiences due to the presence of multi-dimensional effects. Unlike other methods (that employ external energy sources or that use modified electron flavor neutrino luminosities), a fraction of the energy carried away by heavy flavor neutrinos ($\nu_x = \nu_\mu, \bar{\nu}_\mu, \nu_\tau, \bar{\nu}_\tau$) is deposited behind the shock, in order to ultimately provide successful explosion conditions. In self-consistent core-collapse models, the ν_x's present a marginal dependence on the temporal evolution of the accretion rate (e.g. Liebendörfer et al. 2004), and their contribution to the energy deposition inside the gain region is negligible. Including them in PUSH nevertheless presents a number of advantages towards a more realistic nucleosynthesis: As one of these, the properties of the ν_x emission, which includes dynamical feedback from accretion history, as well as cooling properties of the forming compact object, correlate significantly with the main features of the ν_e and $\bar{\nu}_e$ emission (O'Connor and Ott 2013). As another, the accretion luminosity depends not only on the accretion rate but also on the evolution of the mass and radius of the PNS, which is treated accurately and self-consistently in this method. This achieves an explosion trigger in 1D simulations without modifying ν_e and $\bar{\nu}_e$ luminosities nor changing charged current reactions. It increases the accuracy of electron fraction treatment for the innermost ejecta, which is a crucial ingredient for nucleosynthesis. In addition, unlike the electron (anti-) neutrino luminosities, which decrease suddenly once the shock has been revived in spherically symmetric models, ν_x luminosities are only marginally affected by the development of an explosion. This allows PUSH to continue injecting energy inside the expanding shock for a few hundreds of milliseconds after the explosion has set in. A first implementation of the PUSH method was presented in Perego et al. (2015). The hydrodynamical evolution in spherical symmetry uses the general relativistic hydrodynamics code AGILE (Liebendörfer et al. 2001). For the stellar collapse, the deleptonization scheme of Liebendörfer (2005) was applied. The Isotropic Diffusion Source Approximation (IDSA) was employed for the electron neutrino and anti-neutrino transport (Liebendörfer et al. 2009), while the heavy-lepton flavor neutrinos were modelled by an Advanced Spectral Leakage scheme (ASL, Perego et al. 2016). An extended Equation of State (EOS), comprising both nuclear statistical equilibrium (NSE) and non-NSE conditions, was included in the model. For the former, the tabulated microphysical EOS HS(DD2) was employed (Hempel and Schaffner-Bielich 2010), while for the latter an ideal gas of electrons, positrons, photons and 25 representative nuclei from neutrons and protons to iron-group nuclei was used. For conditions not permitting to use NSE, the changes in nuclear composition were followed by an approximate α-network.

PUSH is not a fully self-consistent method, it requires a calibration of the free parameters $k_{\rm push}$ and $t_{\rm rise}$. This has been done comparing explosion and nucleosynthetic properties with those observed for SN 1987A (Woosley et al. 2002). Such calibration ensures that the artificially increased heating efficiency has an empirical foundation and the model has a predictive power in the sense of an effective model.

The analysis of explosion models with a broad parameter exploration revealed a clear dependence on the compactness of the progenitors. This is defined as ξ_M (O'Connor and Ott 2011), with $\xi_M \equiv (M/M_\odot)\,/\,(R(M)/1000\mathrm{km})$, where $R(M)$ is the radius of the enclosed mass M, computed for $M = 1.75\,\mathrm{M}_\odot$ at the onset of the collapse. The requirement of an explosion energy around 1 Bethe was achieved only by progenitors with a moderately high compactness parameter. Explosions tend to set in earlier, with lower explosion energies and lower remnant masses in the case of progenitors with low compactness. For high-compactness progenitors, explosions are more difficult to achieve and require more time to develop. However, when they occur, they are more energetic and produce more massive proto-neutron stars. These differences plausibly relate to different accretion histories: For high compactness, neutrino luminosities are larger and neutrino spectra are harder, due to the larger accretion rates. In order to overcome the stronger ram pressure at the shock front, a more intense neutrino energy deposition is required.

All the simulations start from a progenitor star including all mass up to the helium shell, corresponding to a radius of $R \approx (1.3\text{–}1.5) \times 10^{10}\,\mathrm{cm}$. Simulations were run for a total time of 5 s, corresponding to $\gtrsim 4.6\,\mathrm{s}$ after core bounce. During this time, the shock was always contained inside the computational domain. Tracer mass elements (shells) were extracted from the simulation and post-processed with the `WINNET` (Winteler et al. 2012) nuclear network.

Initial calibration attempts found a systematic overproduction of ^{56}Ni for runs with an explosion energy around and above 1 Bethe. This discrepancy was cured

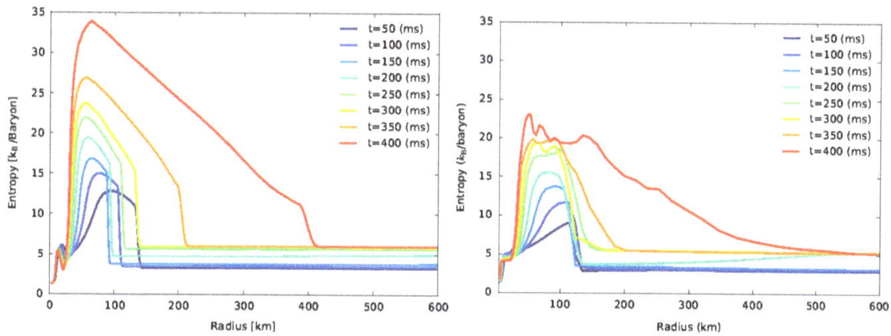

Fig. 4.18 Radial profiles of the spherically averaged entropy per baryon obtained from a spherically symmetric simulation with PUSH (left panel) and from a axisymmetric Flash simulation (right panel) of a 20 $M_{\odot}$ progenitor

by invoking a relatively large amount of fallback (0.1 $M_{\odot}$). The more recent implementations achieve a satisfactory agreement with SN 1987A observables without invoking any fallback (Ebinger et al. 2017, 2018), when using dedicated blue supergiant progenitors and slightly different parameters than those in Perego et al. (2015). As a result, the SN 1987A remnant should host a neutron star with a baryonic mass of $\sim$1.66 $M_{\odot}$, corresponding to a gravitational mass of $\sim$1.50 $M_{\odot}$ for a cold neutron star with the HS(DD2) EOS. The formation of a black hole seems unlikely, as a very large fallback ($\gtrsim$ 0.5 $M_{\odot}$) would be required to reach the observed neutron star maximum mass limit, and an even larger fallback ($\sim$1.3 $M_{\odot}$) to reach the maximum neutron star baryonic mass as predicted by the HS(DD2) EOS.

In Fig. 4.18 the spherically averaged entropy per baryon as a function of radius is shown, obtained (a) from a 2D Flash simulation (Pan et al. 2016), see right panel, and (b) from a 1D simulation performed with PUSH for the same progenitor (see left panel). The PUSH method presents a behavior more consistent with multi-dimensional models than older methods (e.g. pistons and thermal bombs) (Ebinger et al. 2017, 2018).

This method also allows to explore other important stellar collapse features, such as their explodability versus the formation of a stellar black hole (BH) from a failed supernova, over a large set of progenitors. Figure 4.19 shows the explosion energy obtained with PUSH for a set of progenitor models that encompasses a large portion of the mass range for core collapse progenitors (IMF). We see a rise in explosion energies up to about 18–20 $M_{\odot}$ (accompanied by an increasing compactness of the stellar models). Similar to Sukhbold et al. (2016) one can see that there is a transition to BH formation beyond 20 $M_{\odot}$, but that still individual models above that mass range can experience explosions, while BHs can also form for some lower mass progenitors. The window above 20 $M_{\odot}$, where BHs are formed, is followed by a range where explosions are possible again, continuing to a general tendency to black hole formation above 35 $M_{\odot}$. This procedure has been carried through for a number of stellar model samples, also from Heger and Woosley (and for different metallicities). The results differ in detail; dependent on the compactness of stellar

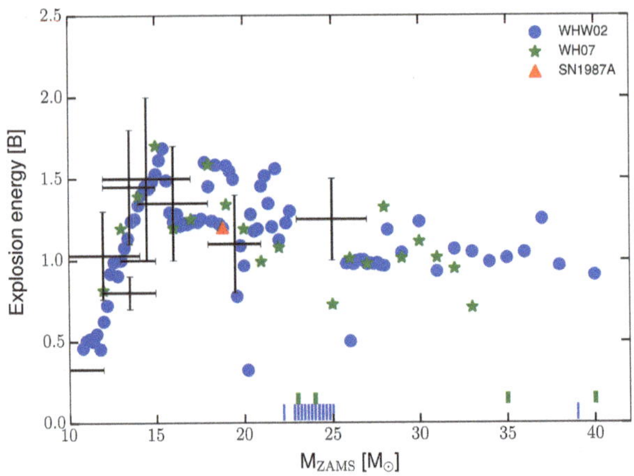

Fig. 4.19 Explosion energies as a function of progenitor mass deduced from observations (Nomoto et al. 2013) (black crosses) and predicted from PUSH simulations (Ebinger et al. 2017, 2018) for two sets of progenitor models (blue dots Woosley et al. 2002) and (green stars Woosley and Heger 2007). Dashed vertical lines close to the abscissa indicate no explosions (in blue and green for the two sets, mentioned above), i.e. black hole formation in PUSH simulations in the range 22–25 $M_{\odot}$ and beyond 33 $M_{\odot}$

models, black hole formation can also occur in some cases for lower initial masses, but the general tendency remains (Ebinger et al. 2017, 2018). Clearly, there is strong dependence of the core-collapse outcome on the detailed structure of progenitor models, hence on accurate modeling of the entire stellar evolution.

We conclude to state that improved spherically-symmetric modeling approaches such as PHOTB and PUSH provide for the first time results with self-consistent values for supernova explosion energies and resulting neutron star masses, as well as the transition to black hole formation. Thus, this modeling also permits to predict the transition from regular CCSNe to faint or failed supernovae. But, as the present models do not include stellar rotation nor magnetic fields, predictions of the hypernova branch, leading to long-duration gamma-ray bursts, cannot be made within this approach. On the other hand, the correct inclusion of electron neutrino and antineutrino interactions with matter permit also to predict reliable values of the electron fraction Y_e and therefore also a reliable prediction for nucleosynthesis with the detailed isotopic composition of the Fe-group.

4.5 Exotic Explosions: Hypernovae, Gamma-Ray Bursts, Neutron Star Mergers

Massive stars in the range of 8–$\sim$ 130 $M_{\odot}$ undergo core-collapse at the end of their evolution and become Type II and Ib/c supernovae, unless the entire star collapses into a black hole with no mass ejection (Heger et al. 2003). Such Type

II and Ib/c supernovae (as well as Type Ia supernovae, see Chap. 5) release large explosion energies and eject matter which experienced explosive nucleosynthesis. So, they have a strong dynamical, thermal, and chemical influence on the evolution of interstellar matter and of galaxies as a whole.

The explosion energies of core-collapse supernovae are fundamentally important quantities, and an estimate of $E \sim 1 \times 10^{51}$ erg has often been used in calculating nucleosynthesis and the impact on the interstellar medium. (Here we use the explosion energy E for the final kinetic energy of the explosion.) A close-by example is SN1987A in the Large Magellanic Cloud, whose energy is estimated to be $E = (1.0 - 1.5) \times 10^{51}$ ergs from its early light curve.

But without rotation and magnetic fields, simulations show that more massive objects will end up as black holes, even in multi-D simulations. In that case a major question is, at which progenitor mass the turnover takes place from successful explosions with neutron star formation to black holes as the final outcome (if there exits such a clear limit!). The previous section addressed this transition to faint or failed supernovae by advanced recent spherically symmetric approaches (i.e., emulating 3D effects). Here we show a case of a 2D simulation which leads to black hole formation after core-collapse of a non-rotating 40 $M_\odot$ progenitor star (Pan et al. 2017). This study also makes use of the isotropic diffusion source approximation (IDSA) for the transport of electron flavor neutrinos and a modified gravitational potential for general relativistic effects. It was performed for four different neutron star equations of state (EoS), including LS220, SFHo, BHB$\Lambda\Phi$ and DD2, examining the impact of the equation of state on black hole formation dynamics and gravitational wave emissions. The simulations utile the FLASH code (Fryxell et al. 2000).

It is found that the black hole formation time is sensitive to the equation of state and is delayed in multiple dimensions due to finite entropy effects that enlarge the maximum proto-neutron star (PNS) mass and via proto-neutron star convection. Depending on the equation of state, these simulations also show the possibility that the shock is revived together with the formation of a black hole. Figure 4.20 displays the behavior of the average shock radius as a function of time. In Fig. 4.21 the multi-D behavior is indicated, utilizing the EoS SFHo (Oertel et al. 2017). Convective regions, where neutrino heating takes place, are characterized by negative Brunt-Väisälä frequencies. Here not only the average but also the maximum shock radius is given as a function of time. We see that the radius of the PNS recedes, finally forming a black hole.

There exist similar results with the aid of fully relativistic 3D simulations (Kuroda et al. 2018), which indicate stellar mass black hole formation in failed supernovae for models without rotation, being responsible for the faint branch of the so-called Nomoto plot (Nomoto et al. 2013).

The key questions that remain are:

- do 8–10 $M_\odot$ stars which produce an Fe-core in a collapse initiated via electron capture after core He-burning (electron capture supernovae) have a different explosion mechanism than more massive stars? Is here only a small amount of material involved outside the collapsing C-core and little Ni-ejection occurring?

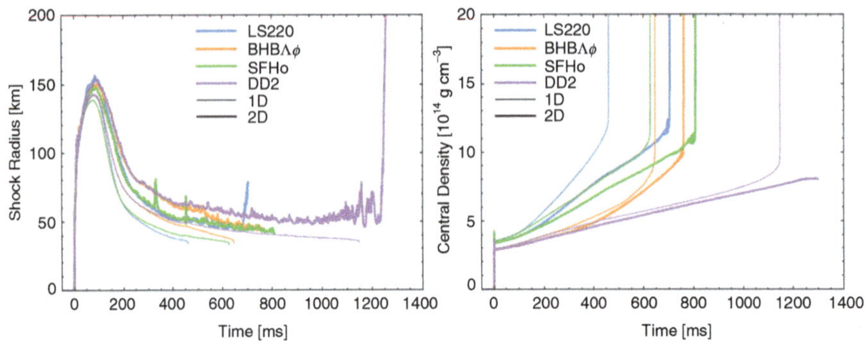

Fig. 4.20 Time evolution of averaged shock radius (left) and central density (right). Different colors represent simulations with different nuclear EoS. Thick and thin lines indicate simulations in 2D and 1D respectively. Except for EoS DD2, all utilized EoSs lead to black hole formation (Pan et al. 2017)

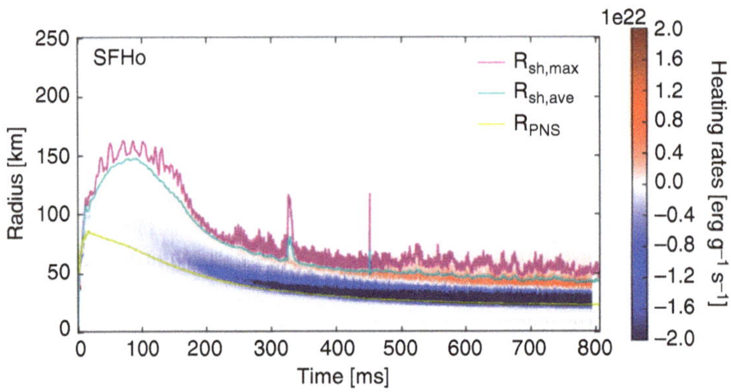

Fig. 4.21 Color maps of the Brunt-Väisälä frequency as a function of time and radius. Negative values, indicated by the red color represent unstable regions based on the Ledoux criterion for convection. These are the regions where convective overturn takes place and neutrino heating is active. The magenta and light blue lines show the maximum and averaged shock radius. The yellow line indicates the radius of the PNS (Pan et al. 2017)

- for which stellar progenitor masses do we have a transition from the formation of neutron stars to the formation of black holes after collapse?
- to which extent is this transition region influenced by the nuclear equation of state?
- for which transition region are initially neutron stars formed, causing a regular supernova explosion, but followed by fall back from the reverse shock that swallows inner matter, leading to a small final Ni-ejection and faint light curves?
- for which progenitor masses are black holes formed directly during collapse and how can this be observed?

- what is the role of rotation and magnetic fields with respect gamma-ray bursts/hypernovae?
- can we give reliable nucleosynthesis yields for such events?

So far we have covered most of these items, except for the last two. Thus, how can the hypernova branch of existing observations with high to very explosion energies, even for massive stars, be explained?

4.5.1 Hypernovae/Long Duration Gamma-Ray Bursts/Collapsars

One of the most interesting developments in the study of supernovae (SNe) is the discovery of some very energetic supernovae (see e.g. Nomoto et al. 2006), whose kinetic energy (KE) (in spherically symmetric analysis, see also Piran 2004) exceeds 10^{52} erg, about 10 times the KE of normal core-collapse SNe (hereafter $E_{51} = E/10^{51}$ erg). The most luminous and powerful of these objects, the Type Ic supernova (SN Ic) 1998bw, was probably linked to the gamma-ray burst GRB 980425, thus establishing for the first time a connection between (long-duration) gamma-ray bursts (GRBs) and the well-studied phenomenon of core-collapse SNe. However, SN 1998bw was exceptional for a SN Ic: it was as luminous at peak as a SN Ia, indicating that it synthesized $\sim 0.5\,M_{\odot}$ of ^{56}Ni, and its KE was estimated at $E \sim 3 \times 10^{52}$ erg.

Before going into a too involved discussion of the causes of these events, let us first consider the possible effect which higher energy explosions have on the ejecta, i.e. nucleosynthesis products. Here we use the term 'hypernova' to describe an extremely energetic supernova with $E \geq 10^{52}$ erg without specifying the explosion mechanism (Nomoto et al. 2001). Following SN 1998bw, other *hypernovae* of Type Ic have been discovered or recognised. Nucleosynthesis features in such hyper-energetic supernovae must show some important differences in comparison to normal supernova explosions. The higher explosion energies could lead to larger ejected ^{56}Ni masses, as observed in such explosions. They also cause higher entropies in the innermost ejecta, which result in a more extreme alpha-rich freeze-out from explosive Si-burning. Such conditions permit the sizable production of Fe-group nuclei beyond ^{56}Ni, up to ^{64}Ge which decays to ^{64}Zn (Nakamura et al. 2001). This feature could have an influence on abundance patterns observed in extremely metal-poor halo stars. In fact, the observational finding that Zn behaves like an Fe-group element in galactic evolution - and was underproduced in existing supernova models (which were not including the νp-process)—was used as a strong argument that a large fraction of massive stars explode as hypernovae (Nomoto et al. 2006, 2013; Kobayashi et al. 2006; Nomoto 2017).

We know that for non-rotating cases only the supernova branch (with neutron stars as final outcome) can be attained, and the faint or failed supernova branch (leading eventually to black holes, but not to gamma-ray bursts and high ejecta masses). Thus, massive stars, which fail to explode as CCSNe via neutrino-powered explosions, will eventually experience the formation of central black hole (BH) remnants. However, rotating BHs and the formation of accretion disks with accretion rates of about $\approx 0.1\,M_\odot$/s can lead—for certain conditions (strong magnetic fields)—to long duration gamma-ray bursts (lGRBs) or hypernovae (Nomoto's hypernova branch), also dubbed collapsars as they result from a core-collapse to the formation of a black hole. Many authors have contributed to the discovery and shaping of first ideas for theoretical explanations (e.g. Klebesadel, Bloom, Paczynski, Mezaros, Rees, Piran, see the review by Piran (2004)). The collapsar model was proposed by Woosley, MacFadyen and others (see also MacFadyen and Woosley 1999; MacFadyen et al. 2001; Nagataki et al. 2007; Sekiguchi and Shibata 2011; Nagataki 2011), based on neutrino heating from the accretion disk and/or the winding of strong magnetic fields and MHD jets (McKinney et al. 2013; Ono et al. 2012). Hydrodynamic simulations (injecting explosion energies artificially) were performed by Umeda, Nomoto, Maeda, Iwamoto, Nagataki, either by introducing high explosion energies (up to 10^{52} erg) in a spherically symmetric way or aspherically in order to understand jet-like explosions (Nakamura et al. 2001; Nomoto et al. 2006, 2013; Nomoto 2017). The role of weak interactions and resulting nucleosynthesis was the focus of contributions by Pruet, Surman, McLaughlin, Hoffman, Hix, Janka, Ruffert, Fuller, Lemoine, Inoue, Fujimoto, Beloborodov, Janiuk (for specific nucleosynthesis results see e.g. Surman et al. 2006; Janiuk 2014). The basic (consensus) picture is the following: explosion energies can be found up to 5×10^{52} erg, ^{56}Ni ejecta up to $0.5\,M_\odot$, and the ejecta are beamed with relativistic jets. Many attempts have been undertaken to model such events. There exists uncertainty in predicting Y_e, due to weak interactions and especially neutrino transport in disks and jets, but there exists also the constraint of high ^{56}Ni ejecta. Therefore, the dominant Y_e in matter has to be of the order of 0.5. High explosion energies lead to high entropies and a strong alpha-rich freeze-out, including interesting amounts of ^{45}Sc, ^{64}Zn (from ^{64}Ge-decay) and other Fe-group elements. For general and more detailed considerations see Nomoto (2017), concluding that larger abundance ratios for (Zn,Co,V,Ti)/Fe and smaller (Mn,Cr)/Fe ratios are expected than for normal SNe, which seems to be consistent with observations in extremely metal-poor (EMP) stars, as will be discussed later.

4.5.2 MHD-Driven Supernovae/Magnetars

Recent observations (and their interpretation) (Greiner et al. 2015; Bernardini 2015) underline that there exist core-collapse supernova explosions whose light curves are not determined by (large) amounts of ^{56}Ni ejecta, but rather by the energy release of a fast rotating neutron star (pulsar) with extremely strong magnetic fields of the order 10^{15} G (magnetars). The question is how can neutron stars of such extremely high magnetic fields (in comparison to the typical 10^{12} G) emerge from a supernova explosion? A reasonable assumption is that such objects originate from massive stars which are fast rotators with initially also rather strong magnetic fields. Such objects, with assumptions made for the initial rotation rate and magnetic fields, have been modeled extensively (Fujimoto et al. 2007, 2008; Winteler et al. 2012; Mösta et al. 2014, 2015, 2017; Nishimura et al. 2015, 2017b; Halevi and Mösta 2018) and will be called here magneto-rotational or MHD-jet supernovae. The result is typically (when starting with very large initial fields of the order 10^{12} G) that the winding up of magnetic fields results in strong magnetic pressure along the polar rotation axis and jet-like ejection of matter. This matter has experienced high densities (and thus degenerate electrons with high Fermi energies), leading via electron captures on protons and nuclei to strongly neutron-rich matter with a Y_e of the order 0.1–0.15. The fast ejection along the poles avoids that the interaction with neutrinos and anti-neutrinos causes a major rise of Y_e (see Eqs. (4.5)–(4.6)). Such conditions permit a strong r-process, dependent on the initial magnetic field strength and rotation rate (Winteler et al. 2012; Nishimura et al. 2015, 2017b; Mösta et al. 2017; Halevi and Mösta 2018). In this process, due to a high fraction of neutrons per nucleus after the freeze-out of charged-particle reactions during the expansion at $\approx 3 \times 10^9$ K, rapid neutron capture can occur, leading to nuclei far from stability with extremely short beta-decay half-lives and the production of heavy nuclei up to the actinides. In Fig. 4.22 we show the nucleosynthesis results of the 3D collapse of a fast rotator with a strong initial magnetic field of 5×10^{12} G in z-direction before core collapse. A 15 $M_\odot$ progenitor with an initial shellular rotation with period of 2 s at a 1000 km radius results in a rare class of supernovae with a central magnetar and negligible amounts of Fe-group ejecta. Initial results (Winteler

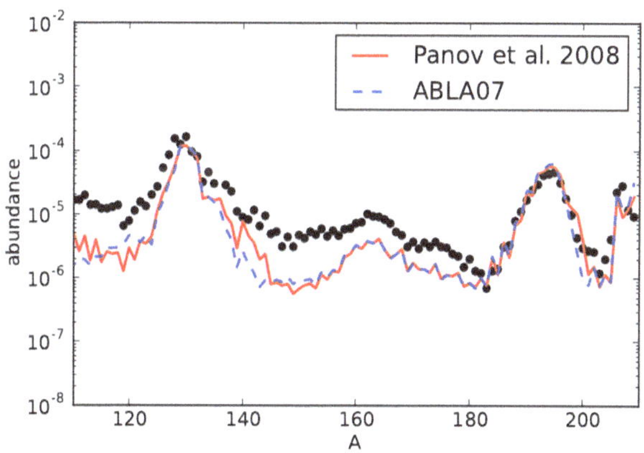

Fig. 4.22 Nucleosynthesis features of an MHD-jet supernova, originating from a 15 $M_\odot$ progenitor with a 5×10^{12} G magnetic field of the collapsing Fe-core. The original simulation (Winteler et al. 2012) showed that the second and third t-process peak could be reproduced very well with a $Y_e = 0.1 - 0.15$ of the ejecta, but big troughs occurred below and above the second r-process peak at A = 130. Utilizing more modern fission fragment distributions (here ABLA07 and Panov08, Kelic et al. 2008; Panov et al. 2008) improved these features strongly for not extremely neutron-rich environments with only weak fission-cycling

et al. 2012) made use of older fission fragment distributions. Figure 4.22 shows the effect of more modern fission yields which avoid the abundance troughs below and above the A = 130 peak. This result should be compared to other investigations (Shibagaki et al. 2016), which rely on quite different abundance features for MHD-jet supernovae and neutron star mergers, based on fission barriers which introduce fission in the r-process only above A = 300.

A fully self-consistent treatment would require high resolution simulations which can resolve magneto-rotational instabilities (MRI) in order to predict reliably the possible amplification of magnetic fields during the explosion. While the latter is actually possible by now (Nishimura et al. 2017b), present calculations depend on the uncertain and therefore assumed initial conditions, that either cause strong jet ejection or can develop kink instabilities of the jets (Mösta et al. 2015, 2017). Based on initial conditions, either neutrino heating or magnetic pressure is causing the supernova explosion, for which the production of heavy neutron capture elements varies strongly. This is shown in Fig. 4.23, which underlines that results can range from a weak r-process, barely proceeding up to the second peak at A = 130, to a full r-process with a strong contribution to the actinides.

In terms of applications to galactic chemical evolution it should be noticed that the MHD-jet supernovae discussed here are expected to occur as a fraction of 0.1–1% of all CCSNe, probably being somewhat metallicity dependent. Higher metallicities lead to stronger stellar wind loss which will be accompanied by a loss of angular momentum, thus reducing the fast rotation necessary for this type of SN explosions. Another feature is that these events can lead to small amounts of Fe-group ejecta for the cases of strong r-processing (Nishimura et al. 2015, 2017b). Plotting the results of Fig. 4.24 with respect to the relative influence of neutrino heating vs. magnetic field effects, one can see that the Ni/Eu-ratio (and similarly the Fe/Eu-ratio) varies strongly (see Fig. 4.24).Thus, if these types of supernovae would contribute already at low metallicities, they alone would be able to provide a large spread in Eu/Fe and might even explain the variations in actinides vs. Eu, seen in a number of cases at low metallicities (see e.g. Wehmeyer et al. 2015; Thielemann et al. 2017a).

4.5.3 Pair-Instability Supernovae (PISNe)

We limit our discussion of massive-star fates in this Chapter to stars below 130 $M_\odot$ which still undergo core collapse and do not explode via explosive O-burning like the so-called pair-creation supernovae (Heger et al. 2003). Such explosions seem theoretically possible, provided that these massive cores can result from stellar evolution. The apparent absence of predicted abundance patterns in low metallicity stars, as well as our current understanding of massive stars with the effects of stellar rotation (Maeder and Meynet 2012) cast some doubts, and the existence of PISNe remains an exciting possibility to be proven.

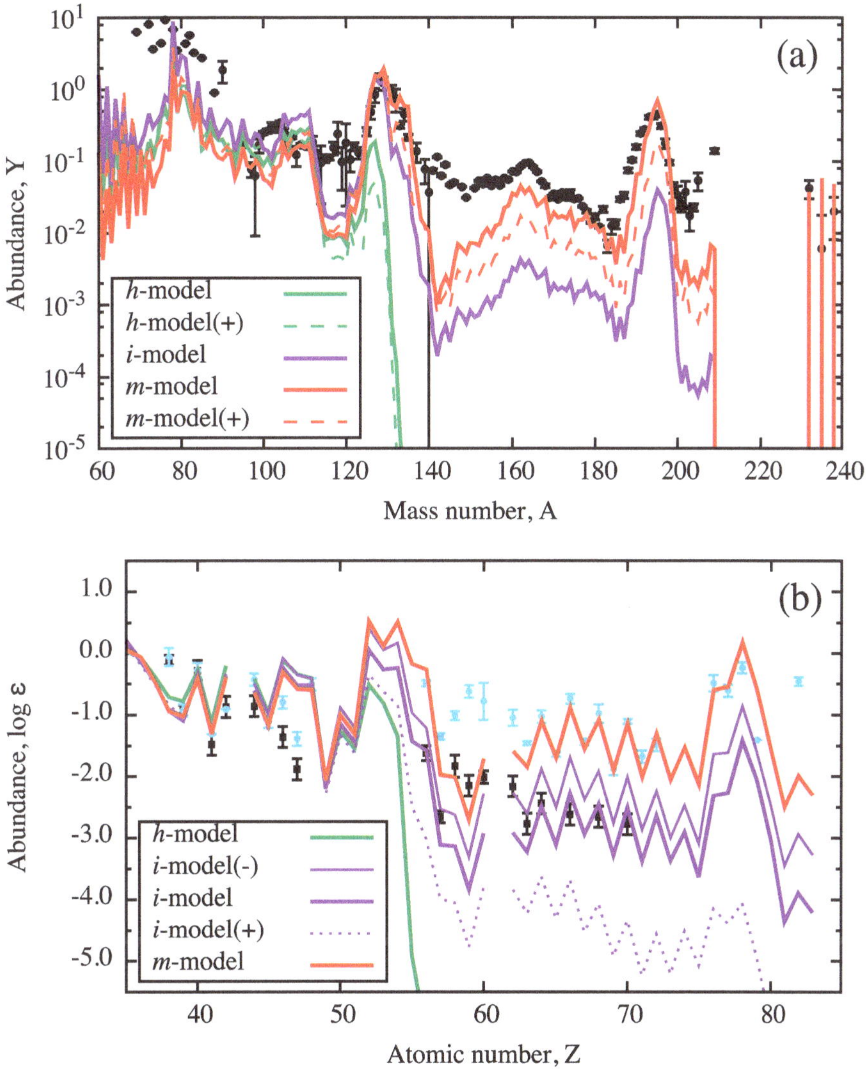

Fig. 4.23 Abundances from nucleosynthesis calculations with varying ratios of magnetic field strength with respect to the neutrino heating mechanism of regular CCSNe, increasing for the models h, i−, i, i+, and m (for details see Nishimura et al. 2017b). For comparison also (**a**) solar r-process abundances are shown (black dots Arlandini et al. 1999), as well as (**b**) abundances from metal-poor stars with a weak r-process, i.e. HD122563 (black dots Honda et al. 2006), and solar-type r-process observations from CS22892-052 (blue dots Sneden et al. 1996). Abundances are normalized for Z = 40 of HD122563. Observations of low metallicity stars with strong r-process contributions vary for abundances below Z = 50 (Sneden et al. 2008)

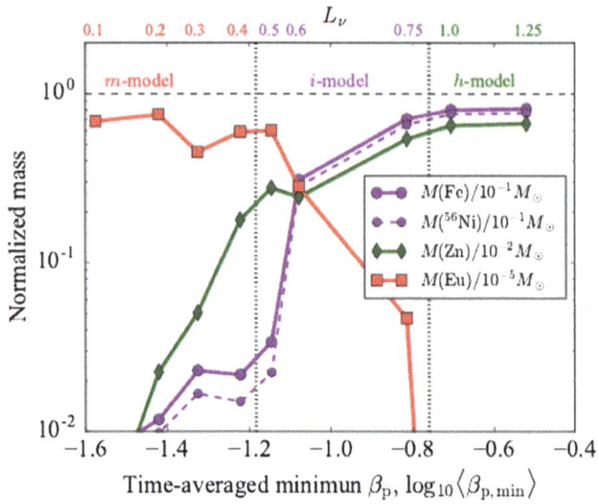

Fig. 4.24 Nucleosynthesis features of rotating CCSN models (h, i−, i, i+, m) with varying ratios of neutrino luminosity and magnetic field strengths as in Fig. 4.23. Model m represents a strong MHD-jet supernova. One can see the transition from a regular CCSN pattern, dominated by ^{56}Ni, total Fe (after decay), and Zn to a strong r-process pattern with a high Eu abundance (for details see Nishimura et al. 2017b)

4.5.4 Neutron Star Mergers

Short-duration GRBs (sGRBs, with a light curve decline of less than about 2 s) are due to relativistic jets created by the merger of two compact stellar objects (specifically two neutron stars or a neutron star and a black hole). Mergers of this kind are also expected to produce significant quantities of neutron-rich radioactive species, whose decay should result in a faint transient, known as kilonova,[9] in the days following the burst. Recent calculations suggest that much of the kilonova energy should appear in the near-infrared, because of the high optical opacity created by these heavy r-process elements. Optical and near-infrared observations of such an event, accompanying the short-duration GRB130603B have been reported in recent years by Tanvir et al. (2013), (see also Barnes et al. 2016). The first gravitational wave detection of such an event (GW170817), combined with a short GRB (GRB170817A) and the optical and infrared afterglow (see e.g. Abbott et al. 2017) has clearly underlined this r-process production site (Metzger 2017a). After the first detailed nucleosynthesis predictions (following ideas of Lattimer and Schramm (1974) and Eichler et al. (1989) of such an event) by Freiburghaus et al. (1999b), many more and more sophisticated investigations have been undertaken (for a review see Thielemann et al. 2017b). Because this book chapter is dedicated to nucleosynthesis in supernovae, we do not want to treat such neutron star merger events in great detail, just to mention them here as a valid (and possibly dominant) site of the astrophysical r-process.

[9]The term 'kilonova' appears to imply luminosities of 10^3 times those of novae; therefore, many scientists prefer the term 'macronova' for these transients, with their luminosities in between novae ad supernovae.

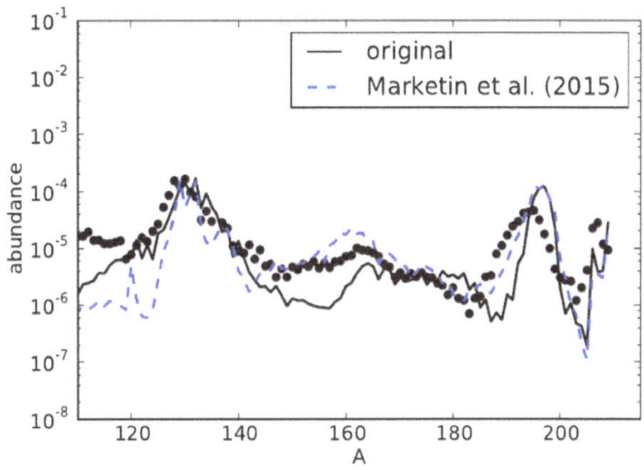

Fig. 4.25 r-process abundance distribution in dynamical ejecta of neutron star mergers, utilizing different mass models, fission fragment distributions, and half-lives (Eichler et al. 2015). Here the effect of improved half-life calculations (Marketin et al. 2016) in comparison to the original FRDM half-lives (Möller et al. 2003) is shown (from Eichler et al. 2015)

Figure 4.25 displays the early dynamical ejecta and their dependence on nuclear properties. Non-relativistic simulations (e.g. Korobkin et al. 2012; Rosswog et al. 2014) lead to large amounts of ejecta of the order $10^{-2}\,M_\odot$ with very small Y_e-values of 0.04 and less. This causes a very strong r-process with fission cycling. The utilized mass model, beta-decay half-lives, fission barriers, and fission yield prescriptions have a strong effect on the final abundance distribution. While during the r-process (when reactions are still in $n, \gamma - \gamma, n$ equilibrium) the second and third r-process peak are exactly at the right position, the neutron capture of large amounts of fission neutrons (after freeze-out from this equilibrium) has some effect on abundances below A = 165, and the third peak seems always shifted to heavier nuclei. Deviations (troughs) in the mass range A = 130 − 165 can be improved with modern fission fragment distribution (Kelic et al. 2008, ABLA07,), but not the shift of the third peak. One option to remedy this effect are variations in beta-decay rates. Shorter half-lives (Marketin et al. 2016) of heavies release neutrons from fission earlier, when $n, \gamma - \gamma, n$ equilibrium is still in place and can avoid or strongly reduce the late shift of the 3rd peak (Eichler et al. 2015). This effect is also seen with the HFB mass model (see also Goriely 2015; Goriely et al. 2015), while Mendoza-Temis et al. (2015) analyzed additional nuclear uncertainties and showed especially an improvement with the Duflo-Zuker mass model (Duflo and Zuker 1995). Alternative ways to cure the problems of the third r-process peak discussed above, come from full general relativistic modeling of the merger event. This leads to deeper gravitational potentials, higher temperatures (including neutrino energies), electron-positron pairs, which—via positron captures and neutrino interaction with nuclei/nucleons—increase Y_e to values of the order 0.15 in equatorial dynamic ejecta, comparable to those mentioned above in MHD-supernova (magnetar) jets (Wanajo et al. 2014; Goriely et al. 2015; Martin et al. 2017), where the third peak shift did not occur (Fig. 4.26).

After ballistic/hydrodynamic ejection of matter (i.e. the dynamic ejecta), dependent on the equation of state, a hot and massive combined neutron star forms and, before collapsing to a black hole, evaporates a neutrino wind (Rosswog et al. 2014;

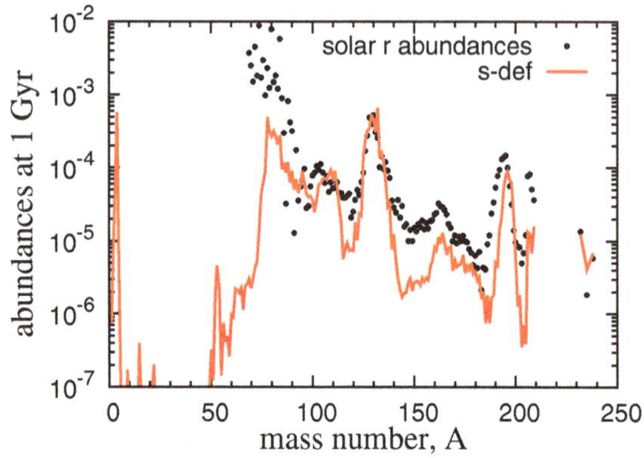

Fig. 4.26 Resulting r-process abundances (in comparison to solar values—black dots) from black hole accretion disk simulations (Wu et al. 2016a), making use of a black hole mass of 3 $M_\odot$, a disk mass of 0.03 $M_\odot$, an initial Y_e of 0.1, entropy per baryon of $8k_b$, an alpha parameter of the viscous disk of 0.03, and a vanishing black hole spin

Perego et al. 2014; Martin et al. 2015) which contributes matter in more polar directions with Y_e's up to 0.4. These winds contain also the lighter r-process nuclei. Outflow from the black hole accretion disk, also powered by neutrinos (and viscous disk heating), can provide the abundance component of light r-process nuclei, in addition to heavy r-process nuclei (Wanajo et al. 2014; Just et al. 2015; Goriely et al. 2015; Wu et al. 2016a).

4.6 Nucleosynthesis in Explosions from Massive Stars

4.6.1 Nuclear Burning During Explosions

Despite considerable improvements of stellar models and numerical simulations in recent years, some fundamental problems remain in nucleosynthesis predictions. It has become evident that certain evolution aspects can only be followed in models going beyond one-dimensional simulations, such as convection, rotation, and the explosion mechanism. There exist also a few multi-D nucleosynthesis studies by now (e.g. Harris et al. 2017; Eichler et al. 2018), and we gave also a few examples of these in Sect. 4.5. However, it is still not feasible to directly couple full reaction networks, containing several thousand nuclei, to multi-dimensional hydrodynamic calculations due to the lack of required computing power, even in modern computers. Thus, post-processing after explosion models with parameterized networks still remains an important approach. One-dimensional models can directly accommodate increasingly larger networks but they cannot capture all of the necessary physics. On the other hand, It has been shown that the delayed neutrino mechanism works combined with a multi-D convection treatment for unstable layers (possibly with the aid of rotation, magnetic fields and/or still existent uncertainties in neutrino opacities). As outlined in the previous sections, a truly self-consistent treatment of core collapse supernovae in 1D does not lead to successful explosions, when using presently known input physics while 2D models show some promise. Therefore, hybrid approaches using certain parameterizations or approximations have been and are still necessary when predicting the nucleosynthetic yields required for the application described above. Intelligently added energy via neutrino absorption, guided by existing multi-D results of e.g. PHOTB and PUSH, can recover many shortcomings of the early spherically symmetric results. We will discuss explosive nucleosynthesis still in this framework, but present the improvements since the early "piston" and "thermal bomb" approaches.

Supernova nucleosynthesis predictions have a long tradition. All of these predictions relied on an artificially introduced explosion, either via a piston or a thermal bomb introduced into the progenitor star model. The mass cut between the ejecta and the remnant does not emerge from this kind of simulations but has to be determined from additional conditions. While the usage of artificially introduced explosions is justifiable for the outer stellar layers, provided we know the correct explosion energy to be dumped into the shock front (on the order of 10^{51} erg seen in observations), it clearly is incorrect for the innermost ejected layers which should be directly related to the physical processes causing the explosion. This affects the Fe-group composition, which has been recognized as a clear problem by many groups (Woosley and Weaver 1995; Thielemann et al. 1990, 1996; Nakamura et al. 1999, 2001; Nomoto et al. 2006, 2013; Fröhlich et al. 2006a,b; Pruet et al. 2005, 2006). The problem is also linked to the so-called neutrino wind, emitted seconds after the supernova explosion, which was considered as a possible source of the r-process to produce the heaviest elements via neutron captures (Woosley et al. 1994; Woosley and Weaver 1994; Takahashi et al. 1994; Qian and Woosley 1996; Hoffman et al. 1997; Arcones and Thielemann 2013; Farouqi et al. 2010).

Given the above detailed discussion of the physics, problems and options regarding core collapse supernovae, we will adopt the following approach in order to predict the most reliable nucleosynthesis predictions for the ejecta in a 1D spherically symmetric treatment, based on the PUSH approach discussed in Sect. 4.5, which can mimic the enhanced energy deposition which multi-D models show. The free parameters are tuned to give correct explosion energies and ^{56}Ni yields for a number of well known supernovae. This approach provides clear predictions for the mass cut between the remaining neutron star and the ejecta. It also includes the effect neutrinos can have on the correct Y_e in the ejecta and the related nucleosynthesis. In the outer explosively burning layers, essentially only the energy in the shock front matters, but opposite to earlier piston or thermal bomb approaches "self-consistent" explosion energies as a function of progenitor mass can be determined. The behavior of these zones can be easily understood from the maximum temperatures attained in the radiation bubble and for a first discussion we will just focus on these features, which can also be obtained with an artificially induced thermal bomb treatment.

For a given/known Y_e and density ρ, the most significant parameter in explosive nucleosynthesis is the temperature, and a good prediction for the composition can already be made by only knowing T_{max}, without having to perform complex nucleosynthesis calculations. Weaver and Woosley (1980) already recognized, that matter behind the shock front is strongly radiation dominated. Assuming an almost homogeneous density and temperature distribution behind the shock (which is approximately correct), one can equate the supernova energy with the radiation energy inside the radius r of the shock front

$$E_{SN} = \frac{4\pi}{3} r^3 a T^4(r). \tag{4.7}$$

This equation can be solved for r. With $T = 5 \times 10^9$ K, the lower bound for explosive Si-burning with complete Si-exhaustion, and an induced thermal bomb energy of $E_{SN} = 10^{51}$ erg, the result is $r \approx 3700$ km. For the evolutionary model by Nomoto and Hashimoto (1988) of a 20 $M_\odot$ star this radius corresponds to 1.7 $M_\odot$, in excellent agreement with the exact hydrodynamic calculation. Temperatures which characterize the edge of the other explosive burning zones correspond to the following radii: incomplete Si-burning (T_9=4, r=4980 km), explosive O-burning (3.3, 6430), and explosive Ne/C-burning (2.1, 11750). This relates to masses of 1.75, 1.81, and 2.05 $M_\odot$ in case of the 20 $M_\odot$ star. The radii mentioned are model independent and vary only with the supernova energy. In the following we present a number of plots which show the different mass fractions $X_i = A_i Y_i$ as a function of radial mass $M(r)/M_\odot$, passing outwards through a 20 $M_\odot$ star through all explosive burning regions.

Matter between the mass cut $M(r)$=M_{cut} and the mass enclosed in the radius corresponding to explosive Si-burning with complete Si-exhaustion is indicated with M(ex Si-c). Then follows the zone of incomplete Si-burning until M(ex Si-i), explosive O-burning until M(ex O), explosive Ne/C-burning until M(ex Ne), and unprocessed matter from the C/Ne-core is ejected until M(C-core). In these early calculations the mass cut was artificially determined by integrating the ^{56}Ni-yields inwards down to a radius when the observed 0.07 $M_\odot$ were obtained. We will discuss improvements, resulting from the PUSH approach below. The zones beyond explosive Ne/C-burning ($T_{max} < 2.1 \times 10^9$ K) are essentially unaltered and the composition is almost identical to the pre-explosive one. When performing such calculations for a variety of progenitors over a range of initial stellar masses, one can analyze the dependence of the mass involved in these different burning regimes as a function initial stellar mass (see Sect. 4.5).

For pedagogical reasons, here early results for a 20 $M_\odot$ star (Nomoto and Hashimoto 1988) are still given as examples for the abundance behavior in a series of Figs. 4.27, 4.28, 4.29 and 4.30. It should be mentioned here that these still resulted a simplified thermal bomb treatment for the pre-collapse model rather than from a 1D spherically symmetric simulation with modified neutrino energy absorption, i.e. the PUSH approach which ensures more realistic explosion conditions. The explosion energy used corresponds to a supernova energy of 10^{51} erg. As mentioned before, this treatment could not predict a self-consistent explosion and the position of the mass cut between neutron star and ejecta. Only the observation of 0.07 ± 0.01 $M_\odot$ of ^{56}Ni in SN1987A (a 20 $M_\odot$ star) gives an important constraint, because ^{56}Ni is produced in the innermost ejected zones. The explosive nucleosynthesis due to burning in the shock front is shown in Fig. 4.27 for a few major nuclei. Inside

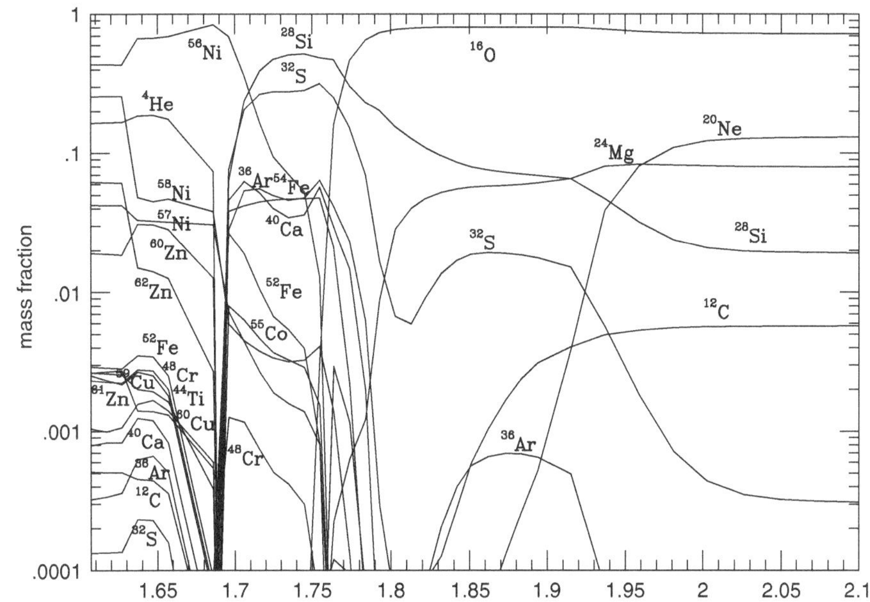

Fig. 4.27 Mass fractions of a few major nuclei after passage of the supernova shockfront through a star with an intimal mass of 20 $\mathrm{M}_\odot$ obtained with a thermal bomb approach. Matter outside 2 $\mathrm{M}_\odot$ is essentially unaltered. Mass zones further in experience explosive Si, O, Ne, and C-burning. The strange variation of yields in the innermost zones result from utilizing the initial Y_e existing in the progenitor star before the explosion, which would be affected by the and during the explosion. For ejecting 0.07 $\mathrm{M}_\odot$ of ^{56}Ni the mass cut between neutron star and ejecta is required to be located at 1.6 $\mathrm{M}_\odot$

1.7 $\mathrm{M}_\odot$ all Fe-group nuclei are produced in *explosive* Si-burning during the SN II event. At 1.63 $\mathrm{M}_\odot$ Y_e changes from 0.494 to 0.499 and leads to a smaller ^{56}Ni abundance further inside, where more neutron-rich Ni-isotopes share the abundance with ^{56}Ni. This is an artefact of the Y_e gradient in the pre-collapse model which can be changed in a consistent explosion treatment via neutrino interactions with this matter (Fig. 4.31).

For comparison we show also recent PUSH results (Sinha et al. 2017; Curtis et al. 2018) for a 16 $\mathrm{M}_\odot$ model which leads to an explosion energy of 1.5×10^{51} erg (see Fig. 4.19) avoids the Y_e-problem of the innermost ejecta in the early piston or thermal bomb approaches.

In explosive Si-burning only alpha-rich freeze-out and incomplete Si-burning are encountered. Contrary to SNe Ia, densities in excess of $10^8\,\mathrm{g\,cm^{-3}}$, which would result in a normal freeze-out, are not attained in the ejecta (see also Fig. 4.2). The most abundant nucleus in the alpha-rich freeze-out is ^{56}Ni. For the less abundant nuclei the final alpha-capture plays a dominant role transforming nuclei like ^{56}Ni, ^{57}Ni, and ^{58}Ni into ^{60}Zn, ^{61}Zn, and ^{62}Zn (see Fig. 4.28).

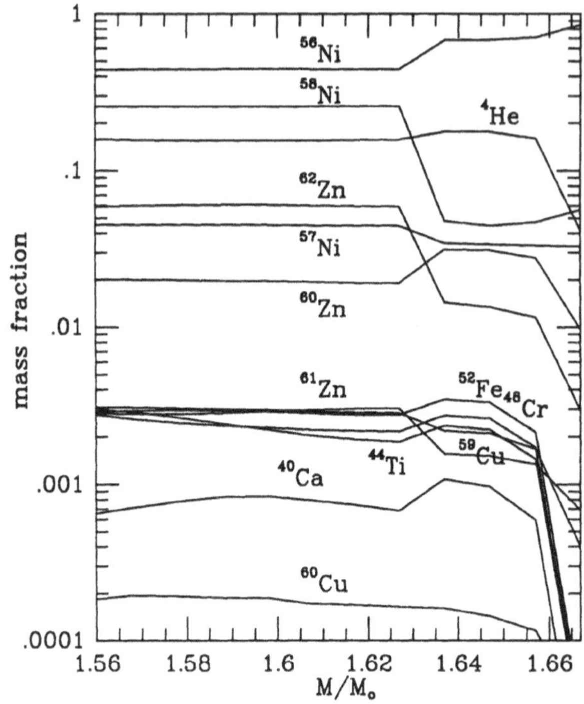

Fig. 4.28 Mass fractions of the dominant nuclei in zones which experience alpha-rich freeze-out. Notice the relatively large amounts of Zn and Cu nuclei, which originate from alpha-captures on Ni and Co. One can recognize their strong decrease beyond 1.66 $M_\odot$, which goes parallel with the decrease of the ^{4}He-abundance and other alpha-nuclei like ^{40}Ca, ^{44}Ti, ^{48}Cr, and ^{52}Fe. Nuclei which would dominate in a nuclear statistical equilibrium like ^{56,57,58}Ni stay constant or increase even slightly. The increase of all nuclei with $N = Z$ at 1.63 $M_\odot$ and the decrease of nuclei with N>Z is due to the change in Y_e in the original stellar model before collapse, neglecting changes which would have taken place during a realistic explosion simulation (see also Fig. 4.27)

The region of the previously discussed 20 $M_\odot$ star which experiences incomplete Si-burning starts at 1.69 $M_\odot$ and extends out to 1.74 $M_\odot$. In the innermost zones with temperatures close to 4×10^9 K there exists still a contamination by the Fe-group nuclei ^{54}Fe, ^{56}Ni, ^{52}Fe, ^{58}Ni, ^{55}Co, and ^{57}Ni. Explosive O-burning occurs in the mass zones up to 1.8 $M_\odot$ (see Fig. 4.29). The main burning products are ^{28}Si, ^{32}S, ^{36}Ar, ^{40}Ca, ^{38}Ar, and ^{34}S. With mass fractions less than 10^{-2} also ^{33}S, ^{39}K, ^{35}Cl, ^{42}Ca, and ^{37}Ar are produced. Explosive Ne-burning leads to an ^{16}O-enhancement over its hydrostatic value in the mass zones up to 2 $M_\odot$ (see Fig. 4.30).

Complete explosive nucleosynthesis predictions for a range of progenitor stars with induced explosions have been given by a number of authors in recent years (Rauscher et al. 2002; Woosley et al. 2002; Chieffi and Limongi 2004, 2017; Nomoto et al. 2006, 2013; Limongi and Chieffi 2006b, 2012; Woosley and Heger 2007; Umeda and Nomoto 2008; Nomoto 2017), updating some of the discussions

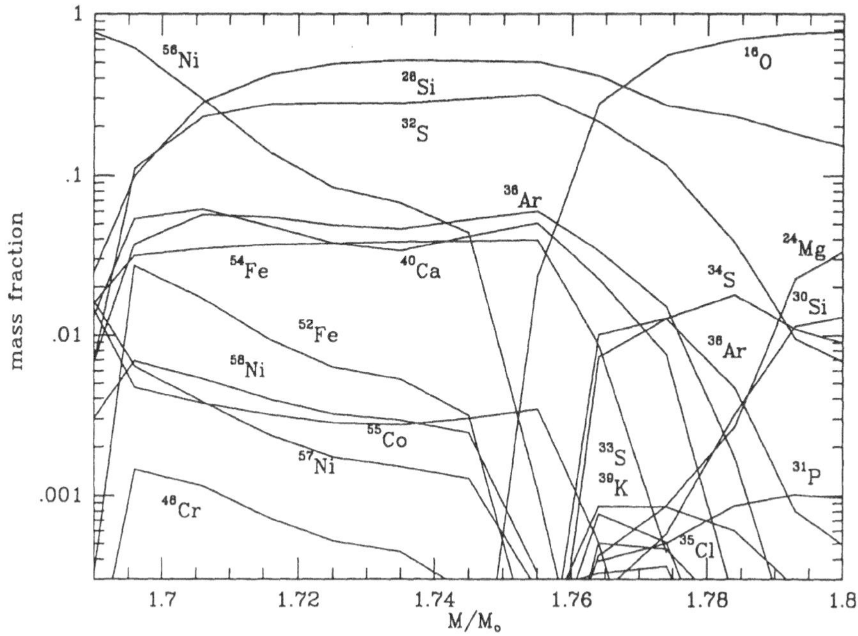

Fig. 4.29 Mass fractions of nuclei in the zones of incomplete Si-burning $M<1.74\,M_{\odot}$ and explosive O-burning $M<1.8\,M_{\odot}$. The Si-burning zones are characterized by important quantities of Fe-group nuclei besides ^{28}Si, ^{32}S, ^{36}Ar, and ^{40}Ca. Explosive O-burning produces mostly the latter, together with more neutron-rich nuclei like ^{30}Si, ^{34}S, ^{38}Ar etc

made above, based on earlier models (Woosley and Weaver 1995; Thielemann et al. 1996). Also specific investigations were undertaken for Pop III low metallicity stars (Umeda and Nomoto 2005; Chieffi and Limongi 2004; Tominaga et al. 2007; Ohkubo et al. 2008).

New spherically symmetric nucleosynthesis predictions have recently become available (Ertl et al. 2016; Sukhbold et al. 2016; Sinha et al. 2017; Curtis et al. 2018), which also provide complete isotopic yields. They are based on PHOTB and PUSH approaches discussed above. Especially with PUSH the Y_e of the innermost zones is treated more consistently, and is expected to result in a more realistic Fe-group composition. One of the possible tests is to compare the composition of explosive ejecta with abundances found in low metallicity stars. Their abundances are determined by the explosive yields of massive stars before type Ia supernovae can set in during the evolution of the Galaxy (being delayed due their origin in lower/intermediate mass stars producing white dwarfs and the effects of binary evolution). Figure 4.32 presents the observed abundance ratios of Fe-group elements (Sneden et al. 2016) for the very metal-poor main sequence turnoff star HD 84937. PUSH yields (Sinha et al. 2017; Curtis et al. 2018) are shown along with piston yields (Woosley and Weaver 1995) and thermal bomb yields (Thielemann et al. 1996) for a $20\,M_{\odot}$ model from Woosley and Heger (2007) progenitor set.

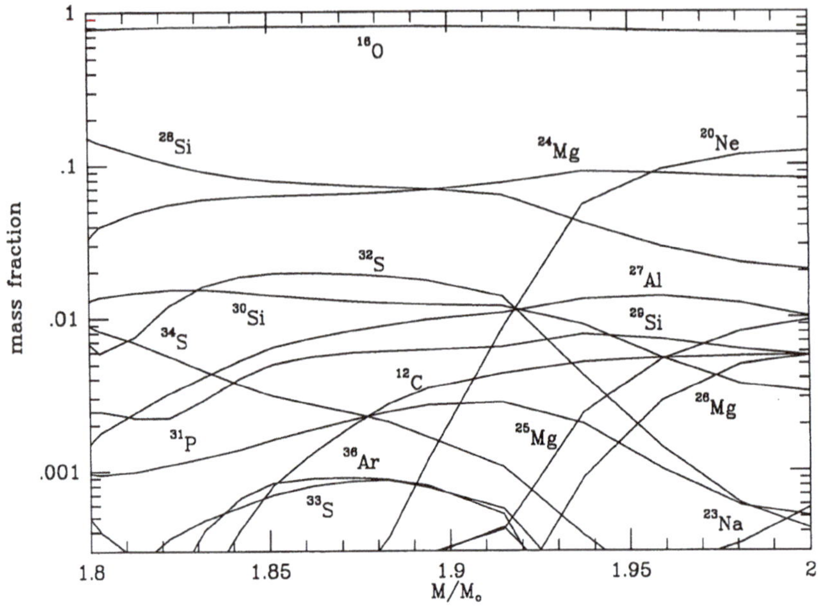

Fig. 4.30 Composition in mass zones of explosive Ne and C-burning. The dominant products are ^{16}O, ^{24}Mg, and ^{28}Si. Besides the major abundances, mentioned above, explosive Ne-burning supplies also substantial amounts of ^{27}Al, ^{29}Si, ^{32}S, ^{30}Si, and ^{31}P. Explosive C-burning contributes in addition the nuclei ^{20}Ne, ^{23}Na, ^{24}Mg, ^{25}Mg, and ^{26}Mg

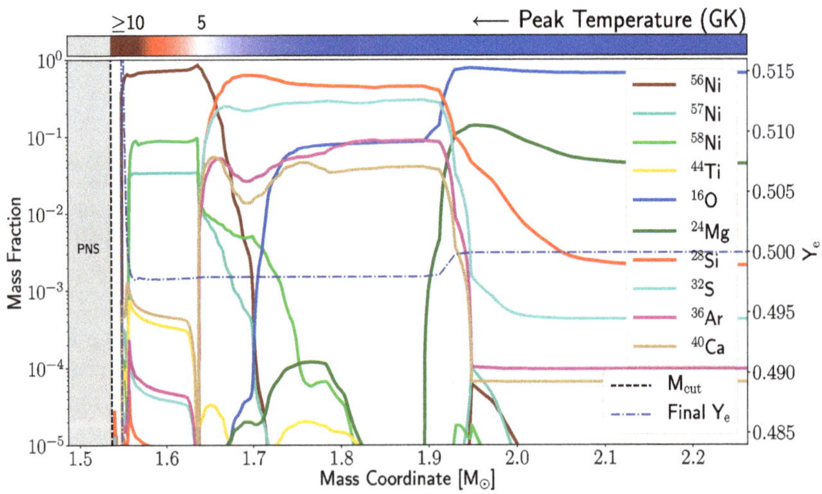

Fig. 4.31 Composition in ejected mass zones after passage of the supernova explosion front for a 16 $M_{\odot}$ star, utilizing the PUSH approach (Sinha et al. 2017; Curtis et al. 2018). In contrast to what is shown in Fig. 4.27, the unrealistic Y_e-caused features disappear here in the innermost ejected zones

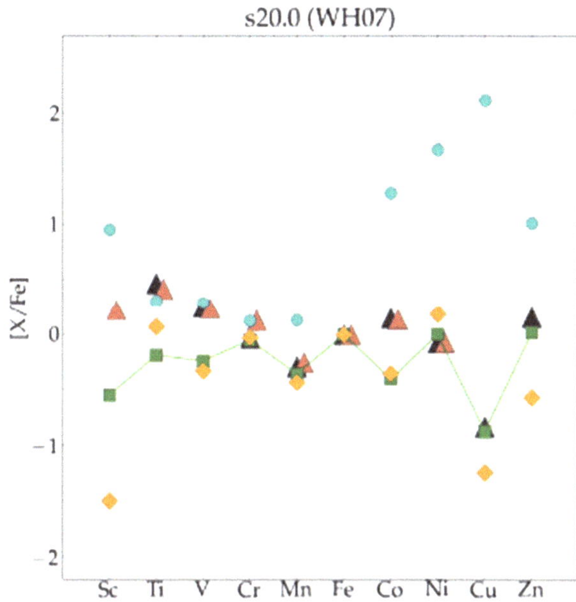

Fig. 4.32 Observed abundances of HD 84937 (Sneden et al. 2016) (black and magenta) compared with the yields obtained with progenitor models of a 20 $M_\odot$ star from the Woosley and Heger (2007) progenitor series. Shown are piston results (Woosley and Weaver 1995) (blue), thermal bomb results (Thielemann et al. 1996) (yellow), and the recent PUSH results (Sinha et al. 2017; Curtis et al. 2018) (green). Due to consistent explosion energies and Y_e-values based on neutrino interactions during the explosion, it can be recognized that especially Sc and Zn show highly improved values in comparison to these older approaches (a first improvement,when including neutrino interactions during the explosion was already obtained in Fröhlich et al. (2006a). Cu is mostly due to s-processing during stellar evolution, which is not part of the explosion calculation

After this discussion of the explosive conditions and their outcome in detail, plus referring the reader to presently available isotopic yields, we now summarize a few major aspects, related to dominant isotopes and those of specific interest. This includes a reflection about the radial-zone (or burning zone) origin of these species. This information can be found in Table 4.9. Especially the innermost alpha-rich freeze-out ejecta are influenced by neutrino interactions with matter during the explosion which determine Y_e (and the entropy).

The basic pattern given in Table 4.9 always applies. The abundances from incomplete Si-burning and explosive O-burning can explain galactic chemical evolution (see e.g. Nomoto et al. 2013; Mishenina et al. 2017), and it seems that the Fe-group elements also fit with the improved inner Y_e treatment of PUSH (see Fig. 4.32). A question is whether the Zn (from ^{64}Ge-decay) produced in regular supernovae is sufficient to explain Zn abundances in the very early Galaxy, or whether hypernovae have to be invoked as the main contributors in this early phase of the Galaxy. As mentioned in Sect. 4.4, the classical p-process takes place in

Table 4.9 Main products of explosive and hydrostatic burning

Burning type	(Main) products, major radioactivities
α-rich compl. Si-burning	He, ^{56}Ni, ^{60}Zn, ^{40}Ca, ^{44}Ti. ^{48}Cr, ^{52}Fe dependent on Y_e also ^{57,58}Ni, ^{61,62}Zn, ^{59}Cu and/or ^{64}Ge and νp-process or weak r-process?
Expl. incompl. Si-burning	^{56}Ni, ^{28}Si, ^{32}S, ^{36}Ar, ^{40}Ca, ^{54}Fe, ^{52}Fe, ^{55}Co, ^{57,58}Ni
Expl. O-burning	O, Si, S, Ar, Ca
Expl. Ne-burning	O, Mg, Si, Ne; ^{26}Al, p-process
Hydr. He-burning products in C/O-core	O, Ne, Mg, Si, s-process

explosive Ne-burning via photo-disintegrations of pre-existing heavy nuclei, but even with the best nuclear input the underproduction of light p-nuclei cannot be solved. The solution can be obtained by adding a *light (heavy) element primary process* (LEPP, Travaglio et al. 2004) where the best candidate is the νp-process, discussed briefly in Sect. 4.5 and further below. Thus, the classical p-process isotopes have possibly to be explained by a superposition of the innermost proton-rich complete Si-burning ejecta with those of explosive Ne-burning in outer zones.

The scheme indicated in Table 4.9 is too simplified when considering the ejecta of outer layers, whose composition was produced during stellar evolution and ejected essentially unaltered during the explosion. However, the *CO-core* scheme is not sufficient to describe massive-star yields. While it includes all matter which underwent He-burning, it does not differentiate between core He-burning and shell He-burning. The latter occurs at higher temperatures and has specific features different from core He-burning. In a similar way, the NeO-core contains all matter which underwent C-burning during stellar evolution, but also here, no difference is made between core C-burning and higher temperature shell C-burning. The same is true for Ne-burning.

4.6.2 Production of Long-Lived Radioactivities ^{44}Ti, ^{26}Al and ^{60}Fe

The radioactive isotopes ^{44}Ti, ^{26}Al and ^{60}Fe are sufficiently long-lived so that their decay occurs outside the star and its explosion. Hence they provide a messenger through characteristic decay gamma-rays of their nucleosynthesis environments, discussed in Sect. 4.7 below. ^{44}Ti probably originates from explosive burning in inner core-collapse regions, whereas the ^{26}Al and ^{60}Fe that is ejected by the explosion mostly is produced in pre-supernova burning stages, with some modifications by the explosion. Because of their special role for this book, their production is discussed in this subsection in more detail.

4.6.2.1 ^{44}Ti

The material which collapsed towards the proto-neutron star is decomposed into nucleons and α particles while being heated during the explosion, then reassembled during the expansion and cooling via the nuclear reactions in the complete Si-burning regime with alpha-rich freeze-out. This produce α-multiples and thus also ^{44}Ti, dependent on the entropy attained. Its total amount depends on the matter experiencing this burning and being ejected. In one-dimensional models only regions outside the mass cut contribute, and this mass cut was in principle unknown without successful explosion calculations. Its position was either derived based on total ^{56}Ni ejecta in thermal bombs (Thielemann et al. 1996; Nomoto et al. 2006) or on entropy jumps in the pre-collapse models in piston-induced explosions (Woosley and Weaver 1995). Depending on how the explosion is modelled, by (1) a thermal bomb or (2) a piston, ^{44}Ti productions are either larger or smaller than 5×10^{-5} $M_{\odot}$. This apparently leads to higher entropies in the first case and a more intense alpha-rich freeze-out. The values of Thielemann et al. (1996) range between 2×10^{-5} and 1.5×10^{-4} $M_{\odot}$. An interesting point here is that variations in the Y_e-structure can lead to changes up to a factor of 2. Rauscher et al. (2002) and Tur et al. (2010) find smaller values (either $1.5 - 5 \times 10^{-5}$ and $3.5 - 6 \times 10^{-5}$ $M_{\odot}$). The latter is based on a readjustment of the triple-alpha and $^{12}C(\alpha, \gamma)^{16}O$-rates to most recent experimental values, which does not have a drastic influence, however. Indirectly, core sizes and other stellar properties, including explosion energies, can enter. Chieffi and Limongi (2017) discuss the effect of rotating (but still spherically symmetric) progenitor models and come to the conclusion that actually rotating models reduce the amount of alpha-rich to incomplete Si-burning ejecta, therefore also the ^{44}Ti/^{56}Ni ratio. Also the improved, but still spherically symmetric, PHOTB and PUSH approaches give results of this order (Sukhbold et al. 2016; Sinha et al. 2017; Curtis et al. 2018). Summing up the results of spherically symmetric investigations, whether from early piston or thermal bomb approaches or from improved approaches with adapted effective neutrino heating to mimic multi-D effects, all these result in too low ^{44}Ti yields. Also investigations on the reactions producing ($^{40}Ca(\alpha, \gamma)^{44}Ti$) and destroying ($^{44}Ti(\alpha, p)^{47}V$) ^{44}Ti could not change this result. The radioactive half-life of ^{44}Ti has also been re-measured with much better precision, after Cas A had been detected in ^{44}Ti γ-rays, to (59 ± 0.3) years (Ahmad et al. 1998). But as complete explosive Si-burning always results from a freeze-out from equilibrium, Hoffman et al. (1999), even a rate change by a factor of 6 changes the ^{44}Ti production only by a factor of 1.3.

Early, still artificially induced multi-D explosion models (Nagataki et al. 1998; Nagataki 2000), were able to obtain higher ^{44}Ti yields. In recent self-consistent 3D models with CHIMERA (Harris et al. 2017), an important finding is that due to filamentary inflow and ejection occurring simultaneously, the radioactive ejecta containing ^{44}Ti (the same argument holds for ^{56}Ni) may actually have rather high characteristic velocities, and the conceptual view of ^{44}Ti being slow because originating from near the mass cut may be considered obsolete (Harris et al. 2017);

in fact, the concept of a 'mass cut' is intrinsically a one-dimensional view, and probably misleading here. This intrinsic 3D effect, being highly important not only for the heating and explosion mechanism, but also for the total integrated ejecta from deeper layers, seems to be able to solve the ^{44}Ti puzzle.

4.6.2.2 ^{26}Al

Long-lived ^{26}Al ($\tau \sim 1.04 \times 10^6$ years) is produced in core and shell H-burning via the NaMgAl-cycle (see Chap. 3) through the ^{25}Mg$(p,\gamma)^{26}$Al reaction. Ejection into the interstellar medium may occur through the stellar wind during the Wolf-Rayet phase, and in the supernova. The amount of ^{26}Al ejected into the interstellar medium is very sensitive to metallicity, initial stellar mass, rotation and mass loss rate, related to one or more of the physical effects discussed above. Results of detailed calculations can be found in Langer et al. (1995), Meynet et al. (1997), Palacios et al. (2005), Limongi and Chieffi (2006b), Woosley and Heger (2007), Tur et al. (2010), Ekström et al. (2012), and Chieffi and Limongi (2013); see also Sect. 4.7 below. The dominant source of ^{26}Al production during stellar evolution is the ^{25}Mg$(p,\gamma)^{26}$Al reaction. Therefore the resulting ^{26}Al abundance depends (1) on this reaction rate converting ^{25}Mg into ^{26}Al, (2) on the amount of ^{25}Mg available, i.e. the total amount of matter in the NeNaMgAl-cycle (either in terms of the abundance/metallicity or in terms of the H-core size), and finally (3) on the amount of ^{26}Al destruction. In the part of the He-core which undergoes He-burning, (α,n)-reactions efficiently produce neutrons. These destroy the ^{26}Al produced earlier (in H burning) via ^{26}Al$(n,p)^{26}$Mg and ^{26}Al$(n,\alpha)^{23}$Na. A further reduction of the final ^{26}Al yield arises from β^+-decay with a half-life of 7.17×10^5 years between the time of its synthesis and the time when matter is ejected in winds (i.e. mass loss) during later stellar evolution. He-burning, with its neutrons released, is destructive for ^{26}Al, but shell C-burning is again a source of ^{26}Al, also via ^{25}Mg$(p,\gamma)^{26}$Al, which is effective due to protons released in ^{12}C(^{12}C,p)^{23}Na (see Table 4.2 in Sect. 4.2). Convection in the C-burning shell brings in fresh ^{12}C fuel and ^{25}Mg which has been also produced in prior He-burning in the ^{22}Ne$(\alpha,n)^{25}$Mg reaction. ^{26}Al production may be effective also in Ne-burning, based on ^{25}Mg left over from C-burning and protons released via ^{23}Na$(\alpha,p)^{26}$Mg (see Table 4.3). This ^{26}Al only survives if rapidly convected outwards to lower temperature environments (^{26}Al may decay rapidly in hot regions due to thermal population of its short-lived isomeric state; cmp. Fig. 1.3 in Chap. 1).

Some of the ^{26}Al produced during stellar evolution will again be destroyed, when a shock front is released in the supernova explosion and propagates through the stellar envelope. Such explosive post-processing will affect mainly the products from hydrostatic C and Ne-burning, being close to the Fe-core. But there are also source processes for explosive ^{26}Al production. The total yields, hydrostatic-evolution yields combined with the destruction and contribution from explosive burning are given in Fig. 4.36.

The explosive production of ^{26}Al occurs in the regions of explosive Ne/C-burning. Under these conditions ^{25}Mg is produced via ^{24}Mg$(n,\gamma)^{25}$Mg and the protons arise from ^{23}Na$(\alpha,p)^{26}$Mg, similar to the reaction pattern shown in Table 4.3 for hydrostatic Ne-burning (and partially also C-burning). Under explosive conditions at temperatures of the order 2.3×10^9 K, these reactions act in a combined way, and the temperatures are also sufficiently high to utilize the released protons for the ^{25}Mg$(p,\gamma)^{26}$Al reaction. However neutrons are also produced abundantly (see Table 4.3), and act as the main destructive species via (n,p) and (n,α) reactions. The mass involved in explosive Ne/C burning strongly depends on the progenitor mass. Thus, we expect a dramatic increase of ^{26}Al yields with increasing initial stellar mass. Limongi and Chieffi (2006b) have analyzed in detail the contributions from (1) wind ejecta during stellar evolution, (2) hydrostatic burning products ejected during the explosion, and (3) explosive Ne/C-burning. The latter dominates up to about 60 M$_\odot$ and increases from initially about 2×10^{-5} M$_\odot$ per event to $2{-}3\times10^{-4}$ M$_\odot$. Then wind ejecta start to take over and flatten out close to 10^{-3} M$_\odot$ at initial stellar masses of 120–140 M$_\odot$. The latter are subject to effects stimulated by stellar rotation (Langer et al. 1995; Meynet et al. 1997; Palacios et al. 2005), and increase with higher rotation rates (see Sect. 4.3). Tur et al. (2010) have confirmed this trend in the lower mass range from 15 to 25 M$_\odot$. They also found that the result does not depend much on the He-burning reaction rates of triple-alpha and ^{12}C$(\alpha,\gamma)^{16}$O. They show nicely how ^{26}Al production starts early, in H-burning, but the final explosion produces close to a factor of 10 more than the initial H-burning yields. Yields from different studies have been assembled in Fig. 4.36.

4.6.2.3 ^{60}Fe

^{60}Fe is produced by neutron captures on ^{59}Fe, and destroyed again via ^{60}Fe$(n,\gamma)^{61}$Fe, i.e. in an s-process. Generally, slow capture of neutrons released from the ^{22}Ne$(\alpha,n)^{25}$Mg reaction in core He-burning leads to the so-called *weak* s-process, producing nuclei up to nuclear mass numbers of around A = 90. ^{59}Fe is beta-unstable, thus requires a typical neutron density of about 3×10^{10} cm^{-3} in order for neutron capture rates to equate the beta-decay rates. These are relatively high neutron densities for an s-process, which then also make the destruction of ^{60}Fe via neutron captures dominate over its decay; the decay half-life is 2.6×10^6 years (Rugel et al. 2009, see also Chap. 7). Core He-burning will not provide sufficiently high-temperatures for the ^{22}Ne$(\alpha,n)^{25}$Mg reaction to produce such high neutron densities, whereas the conditions in shell He-burning could do so. Apparently, conditions are most favorable at late times during shell He-burning when central O-burning and a C-burning shell are both already active (see Woosley and Weaver 1995; Rauscher et al. 2002; Limongi and Chieffi 2006a,b; Tur et al. 2010). ^{60}Fe yields are very sensitive to uncertainties in He-destruction reactions (such as the 3α-rate and ^{12}C$(\alpha,\gamma)^{16}$O) which compete with the neutron source reaction ^{22}Ne$(\alpha,n)^{25}$Mg and neutron(-capture) *poisons*, which compete with the production and destruction rates of ^{60}Fe via neutron captures (Rauscher et al. 2002; Tur et al. 2010; Giron et al. 2010; Uberseder et al. 2009). Such uncertainties add up to factors of up to 5, from individual rate uncertainties. Another possible uncertainty which has not been studied yet is the amount of ^{22}Ne available in He-burning. Here, an important effect in low metallicity stars is the production of primary ^{14}N (not inherited from CNO of previous stellar generations, but produced inside the star due to mixing of He-burning products with H); this causes the production of ^{22}Ne in He-burning and can permit sizable s-processing at low metallicities (with small seed abundances of Fe), thus affecting the abundance of ^{60}Fe.

The overall production ranges from 2×10^{-6} to 8×10^{-5} $M_\odot$ for initial stellar masses between 10 and 40 $M_\odot$. This result depends significantly on the He-burning reactions (triple-alpha and $^{12}C(\alpha, \gamma)^{16}O$), as they compete with the neutron producing reaction $^{22}Ne(\alpha, n)^{25}Mg$. Additional uncertainties in $^{59}Fe(n, \gamma)^{60}Fe$ and $^{60}Fe(n, \gamma)^{61}Fe$ cause yield uncertainties by a factor of up to 5. If the star experiences strong mass loss, the He-burning shell does not encounter the higher density conditions required for the high neutron density of 3×10^{10} cm^{-3}. Thus for initial stellar masses in excess of 40 $M_\odot$, the mass loss treatment can also lead to variations in predicted yields of more than a factor of 10.

4.6.3 *Explosive Burning Off the Regime of Nuclear Stability*

In Sect. 4.2 we introduced in a short way all hydrostatic and explosive burning processes (with the exception of the νp-process). In Sect. 4.4 we gave a specific example of the νp-process in combination with multi-D modeling of core-collapse supernovae, and in Sect. 4.5 we addressed already the r-process in simulations of MHD supernovae/magnetars as well as neutron star mergers. All those were given as short presentations of results with the focus on multi-D (magneto-)hydrodynamic modeling. In the present subsection we give a more in depth general presentation, which applies to the spherically symmetric simulations of the present section as well as the multi-D simulations of previous sections.

4.6.3.1 The p-Process

Up to now we discussed the production of heavy nuclei beyond the Fe-group only via slow neutron captures (the s-process) in hydrostatic stellar evolution. A number of proton-rich (p-)isotopes of naturally occurring stable heavy nuclei cannot be produced by neutron captures along the line of stability. The currently most favored production mechanism for those 35 p-isotopes between Se and Hg is photodisintegration (γ-process) of intermediate and heavy elements at high temperatures in late (explosive) evolution stages of massive stars (Woosley and Howard 1978; Rayet et al. 1990). However, not all p-nuclides can be produced satisfactorily, yet. A well-known deficiency in the model is the underproduction of the Mo-Ru region, but the region 151<A<167 is also underproduced, even in recent calculations (Rauscher et al. 2002, 2013; Arnould and Goriely 2003; Rapp et al. 2006; Dillmann et al. 2008). There exist deficiencies in astrophysical modeling and the employed nuclear physics. Recent investigations have shown that there are still considerable uncertainties in the description of nuclear properties governing the relevant photo-disintegration rates. This has triggered a number of experimental efforts to directly or indirectly determine reaction rates and nuclear properties for the p/γ-process (Rauscher 2006, 2013). Here it is important to investigate the sensitivity of the location of the γ-process path with respect to reaction rate uncertainties.

Concerning the astrophysical modeling, only a range of temperatures has to be considered which are related to the explosive Ne/O-burning zones of a supernova explosion (see Figs. 4.29 and 4.30), where partial (but not complete) photo-disintegration of pre-existing nuclei occurs (from prior hydrostatic evolution or inherited metallicity), i.e. at $\approx 2 - 3\times10^9$ K. The γ-process starts with the photo-disintegration of stable seed nuclei that are present in the stellar plasma. During the photo-disintegration period, neutron, proton, and alpha-emission channels compete with each other and with beta-decays further away from stability. In general, the process, acting like "spallation" of pre-existing nuclei, commences with a sequence of (γ, n)-reactions, moves the abundances to the proton-rich side. At some point in a chain of isotopes, (γ, p) and/or (γ, α)-reactions become faster than neutron emissions, and the flow branches and feeds other isotopic chains. At late times photo-disintegrations become less effective, when decreasing temperatures shift the branching points and make beta-decays more important. Finally the remaining unstable nuclei decay back to stability. The branchings established by the dominance of proton and/or alpha-emission over neutron emission are crucial in determining the radioactive progenitors of the stable p-nuclei and depend on the ratios of the involved reaction rates. Numerous experimental and theoretical efforts have been undertaken to improve the reaction input, especially with respect to open questions in optical potentials for alpha particles and protons (Gyürky et al. 2006; Kiss et al. 2007, 2008; Yalçın et al. 2009). One of the major open problems was the apparent alpha-potential mystery, leading to reduced (α, γ) cross sections in comparison to theoretical predictions. Rauscher (2013) could show that the consistent inclusion of Coulomb excitations solves this discrepancy.

Applications of p-process network calculations to the temperature profiles of initiated explosions have been performed by Rayet et al. (1995), Rapp et al. (2006), and Dillmann et al. (2008). Here, in Fig. 4.33 we present the results of a 25 $M_\odot$ mass model (Dillmann et al. 2008) with two reaction rate libraries without and with inclusion of all experimental improvements, existing at that point. It is noticed that the nuclear uncertainties cannot change the underproduction of especially the light p-nuclei. Another process seems to be required to supply these missing abundances.

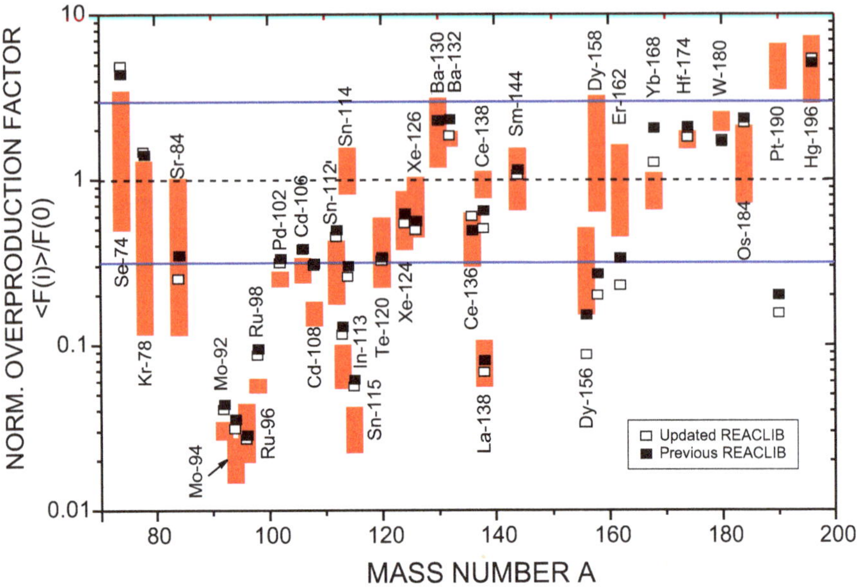

Fig. 4.33 Normalized overproduction factors of p-process nuclei derived with the Rapp et al. (2006) (open squares) and Dillmann et al. (2008) (full squares) reaction library. In addition, the results from a range of stellar models (10–25 $M_\odot$) from Rayet et al. (1995) are given for comparison. A value equal to unity corresponds to relative solar abundances

4.6.3.2 The ν p-Process

Neutron-deficient nuclei can also be produced by two other astrophysical nucleosynthesis processes: the rp-process in X-ray bursts (which, however, does not eject matter into the interstellar medium (Wallace and Woosley 1981; Schatz et al. 1998; Fisker et al. 2008; Cyburt et al. 2010) and the νp-process in core-collapse supernovae, discovered by Fröhlich et al. (2006a,b), Pruet et al. (2006), and Wanajo (2006). The νp-process occurs in explosive environments when proton-rich matter is ejected under the influence of strong neutrino fluxes. This includes the innermost ejecta of core-collapse supernova (see Sect. 4.4) and possibly also ejecta from black hole accretion disks (see Sect. 4.5). The discussion of these innermost ejected mass zones has been skipped above, presenting the results for explosive nucleosynthesis in a 20 $M_\odot$ star, but we mentioned already (together with the PUSH approach) the neutrino effects on Y_e, the effect on the composition of the Fe-group, and entered in Table 4.9 an entry νp-process for proton-rich conditions in the alpha-rich complete Si-burning. The PUSH treatment guaranteed a Y_e that is consistently determined by all weak interactions processes and led to a Y_e enhanced beyond 0.5 (see Fig. 4.17).

The matter in these ejecta is heated to temperatures well above 10^{10} and becomes fully dissociated into protons and neutrons. The ratio of protons to neutrons is mainly determined by neutrino and antineutrino absorptions on neutrons and protons, respectively. Similar neutrino and antineutrino energy spectra and fluxes produce proton-dominated matter in the reactions $\nu_e + n \leftrightarrow p + e^-$ and $\bar{\nu}_e + p \leftrightarrow n + e^+$, due to the n-p mass difference. When the matter expands and cools, the free neutrons and protons combine into α-particles. Later, at temperatures around 5×10^9 K, alpha-particles assemble into heavier nuclei via unstable intermediate nuclei, e.g. the triple-α reaction via unstable ^{8}Be, but—depending on the entropy and the expansion of matter—only a fraction of those form iron-group nuclei (alpha-rich freeze-out). In case of a proton-rich environment, there are also still free protons available at the time of the alpha freeze-out. Once the temperature drops to about 2×10^9 K, the composition of the ejecta consists mostly of ^{4}He, protons, and iron group nuclei with N≈Z (mainly ^{56}Ni) in order of decreasing abundance. Without neutrinos, synthesis of nuclei beyond the iron peak becomes very inefficient due to bottleneck (mainly even-even $N = Z$) nuclei with long beta-decay half-lives and small proton-capture cross sections. Such a nucleus is ^{64}Ge. Thus, with the Y_e determined by neutrino interactions with free neutrons and protons in the early very hot phase of dissociated nuclei, the nucleosynthesis leads to an alpha- and proton-rich freeze-out which does not stop at ^{56}Ni but continues up to ^{64}Ge (which later decays to ^{64}Zn. This part of the story enables core collapse yields which produce Fe-group nuclei up to essentially ^{64}Zn. The effect is seen in the upper portion Fig. 4.34 from the original publications (Fröhlich et al. 2006a,b).

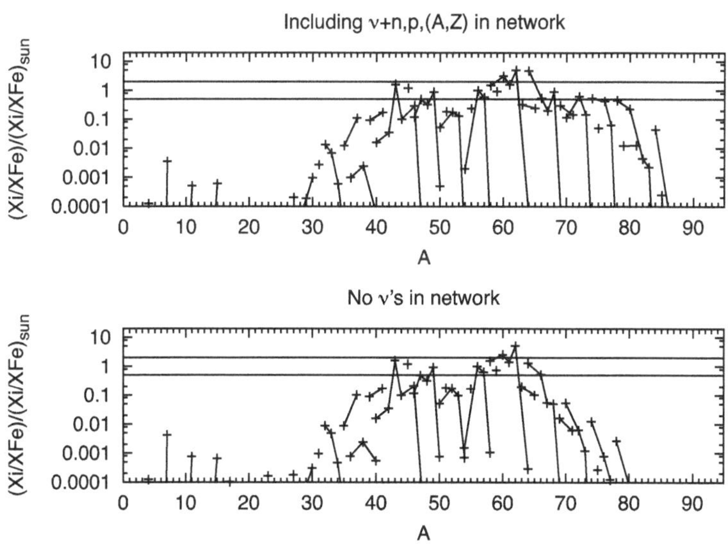

Fig. 4.34 Final abundances in mass zones in the innermost ejecta which experienced neutrino irradiation, leading to proton-rich conditions ($Y_e > 0.5$). The upper part of the figure shows the nucleosynthesis results in the innermost ejecta of explosive, after alpha-rich and proton-rich freeze-out from Si-burning, normalized to solar after decay. The bottom part of the figure also includes the interaction of anti-electron neutrinos with protons ($\bar{\nu}_e + p \rightarrow n + e^+$) which produces neutrons, permitting the late change of ^{64}Ge via ^{64}Ge$(n, p)^{64}$Ga. This feature permits further proton captures to produce heavier nuclei (the so-called νp-process. Here matter up to $A = 85$ is produced

However, the matter is subject to a large neutrino/antineutrino flux from the proto-neutron star. While neutrons are bound in neutron-deficient $N = Z$ nuclei and neutrino captures on these nuclei are negligible due to energetics, antineutrinos are readily captured both on free protons and on heavy nuclei on a timescale of a few seconds. As protons are more abundant than heavy nuclei, antineutrino captures occur predominantly on protons, leading to residual neutron densities of 10^{14}–$10^{15}\,\mathrm{cm}^{-3}$ for several seconds. These neutrons are easily captured by heavy neutron-deficient nuclei, for example ^{64}Ge, inducing (n, p) reactions with time scales much shorter than the beta-decay half-life. This permits further proton captures and allows the nucleosynthesis flow to continue to heavier nuclei (see lower part of Fig. 4.34). The νp-process (Fröhlich et al. 2006b) is this sequence of (p, γ)-reactions, followed by (n, p)-reactions or beta-decays, where the neutrons are supplied by antineutrino captures on free protons.

Similar effects were found by Weber et al. (2008) with improved experimental nuclear input on the proton-rich side of stability for the isotopes ^{87,88}Sr, ^{88}Tc, ^{89}Y, and ^{90,91}Zr. The recent PUSH results (Sinha et al. 2017; Curtis et al. 2018) find sizable abundances up to A = 120.

4.6.3.3 The r-Process

A rapid neutron-capture process (r-process) in an explosive environment is traditionally believed to be responsible for the nucleosynthesis of about half of the heavy elements above Fe. While in recent history the high entropy (neutrino) wind (HEW) of core-collapse supernovae had been considered to be one of the most promising sites, hydrodynamical simulations with the appropriate input physics (see Sect. 4.4.2) encounter difficulties to reproduce the astrophysical conditions under which this process occurs and more appropriate sites are probably MHD supernovae/magnetars or neutron star mergers (see Sect. 4.5 and Thielemann et al. 2017b).

For a pedagogical understanding the classical *waiting-point* approximation, with the basic assumptions of an Fe-group seed, an $(n, \gamma) - (\gamma, n)$-equilibrium for constant neutron densities n_n at a chosen temperature T, over a process duration τ, and an instantaneous freeze-out, has helped to gain improved insight into the systematics of an r-process in terms of its dependence on nuclear-physics input and astrophysical conditions (Cowan et al. 1991; Kratz et al. 1993, 2007). This corresponds to a set of quasi-equilibria with each QSE group being represented by an isotopic chain. Taking a specific seed nucleus, the solar r-process pattern peaks can be reproduced by a variation/superposition of neutron number densities n_n and durations τ. Whether the solar r-process abundances are fully reproduced in each astrophysical event, i.e., whether each such event encounters the full superposition of conditions required, is a matter of debate (Freiburghaus et al. 1999a; Farouqi et al. 2010). In realistic astrophysical environments with time variations in n_n and T, it has to be investigated whether at all and for which time duration τ the supposed $(n, \gamma) - (\gamma, n)$-equilibrium of the classical approach will hold and how freeze-out effects change this behavior. In general, late neutron captures may alter the final abundance distribution. In this case neutron capture reactions will be important. Also β-delayed neutrons can play a role in forming and displacing the peaks after freeze-out.

There have been a number of suggestions for sites in which the strong r-process originates, being related (1) to the innermost ejecta of regular neutrino-driven core-collapse supernovae (Woosley et al. 1994; Takahashi et al. 1994; Hoffman et al. 1997; Qian and Wasserburg 2007; Farouqi et al. 2010; Roberts et al. 2010, 2012; Martínez-Pinedo et al. 2012; Arcones and Thielemann 2013; Mirizzi et al. 2015), (2) ejecta from binary neutron star mergers (Lattimer and Schramm 1976; Eichler et al. 1989, 2015; Freiburghaus et al. 1999b; Goriely et al. 2011, 2015; Bauswein et al. 2013; Rosswog et al. 2014; Wanajo et al. 2014; Just et al. 2015; Ramirez-Ruiz et al. 2015; Thielemann et al. 2017b), and (3) a special class of core collapse supernovae (MHD-jet supernovae) with fast rotation, high magnetic fields and neutron-rich jet ejecta along the poles (Fujimoto et al. 2008; Ono et al. 2012; Winteler et al. 2012; Mösta et al. 2014, 2015, 2017; Nishimura et al. 2015, 2017b).

Supernovae have been thought to be the origin of a strong r-process for many years, with the intrinsic expectation that the innermost ejecta, coming from regions close to the neutron star, should be neutron-rich (see e.g. Cowan et al. 1991; Sumiyoshi et al. 2001). Even when prompt explosion were realized to fail, early detailed and full-fledged r-process calculations in the neutrino wind, emerging from the hot proto neutron star, still underlined this expectation (Woosley and Hoffman 1992; Meyer et al. 1992; Woosley et al. 1994; Takahashi et al. 1994; Hoffman et al. 1997; Qian and Wasserburg 2007; Roberts et al. 2012; Arcones and Thielemann 2013) and parameterized simulations led to quite impressive results (Freiburghaus et al. 1999a; Farouqi et al. 2010; Kratz et al. 2014). Figure 4.35 is taken from the latter reference and shows a close to excellent fit to solar r-process abundances, especially when utilizing modern input from nuclear mass models.

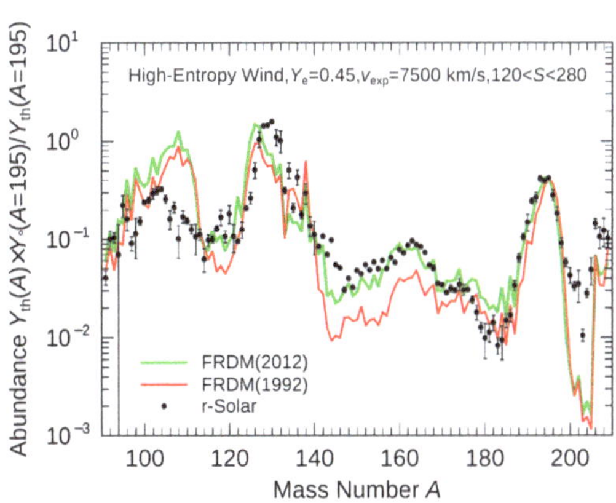

Fig. 4.35 Results from an r-process calculation, assuming an initial Y_e of 0.45, the adiabatic expansion of matter in a so-called neutrino wind with a given expansion speed v_{exp} of ejected mass shells, and that a superposition of entropies S between 120 and 280 k_B/baryon can be attained. The abundance plot assumes that similar amounts of matter are ejected per entropy interval and indicates the changes which occur due to utilizing an improved nuclear mass model (Möller et al. 2012, 2016)

However, in order to obtain this result, a superposition of entropies of up to 280 k_B per baryon is needed, which present simulations of neutrino-driven core collapse supernovae do not support, (see e.g. Fischer et al. 2010; Hüdepohl et al. 2010; Arcones and Thielemann 2013; Roberts et al. 2012; Mirizzi et al. 2016). Thus, while a real high-entropy wind would be able to lead to a strong r-process, presently there is no indication that the required entropies can be attained in realistic core-collapse supernova simulations. An exception might be so-called electron-capture supernovae in the stellar mass range 8–10 $M_{\odot}$, which could lead to a weak r-process (Kitaura et al. 2006; Janka et al. 2008; Wanajo et al. 2009, 2011), possibly producing nuclei up to Eu, but not up to and beyond the third r-process peak (for more details see Mirizzi et al. 2016).

4.7 The Aftermath of Explosions

In the preceding sections we have given an overview of hydrostatic and explosive burning processes in massive stars, including the individual phases of stellar evolution, and their final stages such as core collapse, and supernovae explosion. Variations in outcome will occur if core collapse ends in neutron star and black hole formation, related possibly to hypernovae or gamma-ray bursts. In the sections that follow we aim to (1) get a complete picture of how hydrostatic/wind and explosive contributions add up to the complete yields that could be observed, then (2) verify such models and simulations through specific types of observations, such as light-curves and spectroscopic signatures, and last (3) integrate all such events/stellar yields over a mass distribution and metallicity range as encountered in the evolution of galaxies, in order to discuss comparisons with observations which average over several/many stellar generations.

4.7.1 Overall Nucleosynthesis Yields

In Sect. 4.4 we have introduced in Eq. (4.7) a simplified rule which determined at which radius certain temperatures are attained in the explosion, assuming that the explosion energy is distributed at all times in a homogenous bubble within the radius of the present supernova shock position. If one knew the radial mass distribution $M(r)$ in pre-explosion models, which the shock front will traverse, one would know the amount of matter which encountered certain burning conditions. In Table 4.10 we put together a set of guiding information for different initial stellar masses at (up to) which radial mass position explosive (complete and incomplete) Si-burning, O-burning, Ne/C-burning would occur (upper portion), and on the size of these regions involved herein (in $M_{\odot}$; lower portion); still based on models from Nomoto and Hashimoto (1988). In addition, we give the size of the CO-core from prior He-burning in the stellar evolution. To first order, matter is ejected

Table 4.10 Mass zones and products in explosive and hydrostatic burning

$M(r)$	*Burning site*	$13\,M_\odot$	$15\,M_\odot$	$20\,M_\odot$	$25\,M_\odot$
Fe-core	Hydr. Si-burning	1.18	1.28	1.40	1.61
Mass cut	(Expl. mechanism)	?	?	?	?
Ex Si-c	Expl. compl. Si-burning	1.40	1.44	1.69	1.80
Ex Si-i	Expl. incompl. Si-burning	1.47	1.51	1.75	1.89
Ex O	Expl. O-burning	1.54	1.57	1.81	2.00
Ex C/Ne	Expl. Ne-burning	1.65	1.70	2.05	2.40
CO-core	Hydr. He-burning	1.75	2.02	3.70	5.75

unchanged from between the explosive C/Ne-burning region and the stellar surface. This simplified treatment ignores every detail of the explosion mechanism which produced the energy of the explosion, and also the separation between compact remnant and ejecta (the mass cut) is not known, and therefore also not the total amount of complete Si-burning material. The listed core sizes (e.g., CO-core after He-burning) also do not distinguish whether this matter resulted from initial core burning or subsequent outward propagating shell burning (e.g. shell He-burning or shell C-burning). Each of these produce specific isotopes of interest in this book, such as ^{26}Al, ^{60}Fe, as discussed in Sect. 4.3 and Chap. 3.

Initially we want to focus here on the explosive burning phases. We also want to add that the radial masses given in Table 4.10 show consistent values with (a) the simplified Eq.(4.7) applied to the appropriate stellar model ($M(r)$) and as well as with (b) results from an actual explosion calculation (initiated via a thermal bomb), as obtained in Thielemann et al. (1996). When comparing these numbers, we see a quite close agreement, except for the most massive stars where non-negligible deviations are encountered.

Based on this information we want to discuss complete nucleosynthesis yields, including explosive processing (also the νp-process, affected by neutrinos in the innermost ejecta, as well as the p-process in explosive Ne/O-burning), hydrostatic yields from the outer layers (including s-process) which are ejected unaltered, and prior wind yields lost during stellar evolution. Then we concentrate on the long-lived radioactivities ^{26}Al, ^{60}Fe, ^{44}Ti, other Fe-group and lightcurve-determining nuclei, including their origin which is e.g. important for ^{26}Al and ^{60}Fe, which have hydrostatic burning as well as explosive origins. The r-process in the neutrino wind or possibly polar jets has been presented qualitatively with entropy, Y_e and expansion timescale as free parameters or expansion timescale of neutron star matter as a free parameter. Presently no realistic explosion models are available to discuss this in sufficient detail for a link to observations, but a short discussion of long-lived radioactive chronometers is presented.

Table 4.10 leads to the following conclusions: The amount of ejected mass from the unaltered (essentially only hydrostatically processed) CO-core varies strongly over the progenitor mass range. The variation is still large for the ashes from explosive Ne/C-burning, while the amounts from explosive O- and Si-burning are almost the same for all massive stars. Therefore, the ejecta mass from the unaltered CO-core and from explosive Ne/C-burning (C, O, Ne, Mg) varies strongly over the progenitor mass range, while the amount of ejecta from explosive O- and Si-burning (S, Ar, and Ca) is almost the same for all massive stars. Si has some contribution from hydrostatic burning and varies by a factor of 2–3. It should be mentioned that the present numbers are obtained, assuming for all progenitors the same explosion energy of 10^{51} erg. Thus, changing explosion energies, i.e. with the compactness of the progenitor will have an effect as well. Increasing compactness with progenitor mass would add (as correction) an additional progenitor mass dependence. The amount of ejecta with Fe-group nuclei depends directly on the explosion mechanism, which also affects the Y_e in these inner zones. These numbers are already available for the PHOTB simulations (Sukhbold et al. 2016) and will be soon available for the PUSH simulations (Ebinger et al. 2017, 2018; Sinha et al. 2017; Curtis et al. 2018).

Thus, we essentially have three types of nucleosynthesis products, which test different aspects of supernovae, when comparing with individual observations. The first set (C, O, Ne, Mg) tests the stellar progenitor models, the second (Si, S, Ar, Ca) the progenitor models and the explosion energy in the shock wave, and the Fe-group (beyond Ti) in addition probes the actual supernova mechanism. Thus, we require that all three aspects of the predicted abundance yields are based on secure modeling (stellar evolution, explosion energy, and explosion mechanism) in order to be secure for their application in lightcurve modeling, radioactivities in remnants as well as the in chemical evolution of galaxies (Nomoto et al. 2013; Mishenina et al. 2017).

As an example, we show in Fig. 4.36 the total yield of massive stars, versus their initial mass, and also the different components contributed from hydrostatic burning and ejected with stellar wind, and from the explosive release, which includes explosive nuclear burning contributions as well. Although over-all, different implementations of stellar evolution and explosion find yields which agree within factors of a few, there clearly are characteristic yield behaviors which depend on the implementation of the model. Unlike believed about 20 years ago, the wind contributions will likely not dominate the total ^{26}Al yields, although they remain a key contribution. Then, depending on the explodability of stars with masses above 25–40 $M_\odot$, the more-massive end of the stellar mass distribution may only contribute through their wind phases.

4.7.2 *Spectroscopic Observations of Nucleosynthesis Products*

The supernova explosion ejects freshly-produced nuclei, among those radioactive isotopes, which decay in and outside the expanding remnant. They can be observed rather directly through their nuclear and atomic lines, in X- and gamma-rays.

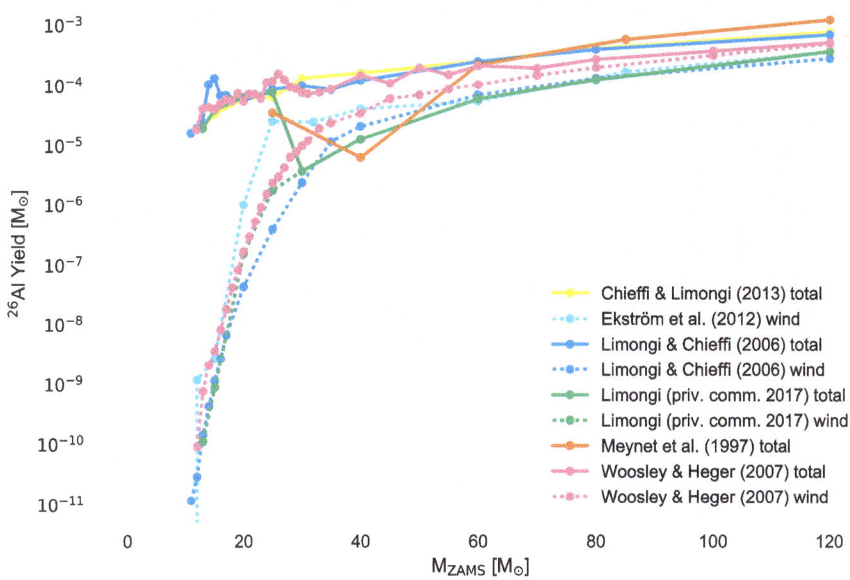

Fig. 4.36 The ^{26}Al yields from wind ejections and from the explosive release in the supernova, as a function of the initial mass of the star (from Meynet et al. 1997; Woosley and Heger 2007; Limongi and Chieffi 2006b; Ekström et al. 2012; Chieffi and Limongi 2013; Limongi, private communication)

4.7.2.1 SN1987A

The γ-ray line discovery only a few weeks after SN1987A had occurred in the Large Magellanic Cloud in February 1987 marks the first case to identify characteristic γ-rays from a single nucleosynthesis source and a specific isotope, through ^{56}Co decay lines at 847 and 1238 keV. These lines were seen in data of the Gamma-Ray Spectrometer on the Solar Maximum Mission (Matz et al. 1988; Leising and Share 1990). Their appearance was significantly earlier than expected from a spherically stratified distribution of elements, where the Fe-group nuclei are produced in the center. This is interpreted as due to deviations from spherical symmetry in the expanding remnant, bringing Ni-rich clumps to the surface earlier by convective instabilities, mixing ^{56}Ni/^{56}Co to outer layers at early times. Gamma-ray line profiles measured with high spectral resolution (a Ge detector balloon instrument launched quickly to exploit this unique opportunity, Tueller et al. 1990) indicated Doppler broadening of the lines from their ejecta motion. Several years later, after the Compton Gamma Ray Observatory (CGRO) was launched, the OSSE instrument on CGRO detected ^{57}Co (τ ~1.1 years) through its gamma-ray line at 122 keV, as another direct proof of iron group nuclei produced in the right amounts (Kurfess et al. 1992). A tentative detection of ^{44}Ti gamma-rays had been reported from INTEGRAL's two instruments (Grebenev et al. 2012); the flux inferred from this observation seemed surprisingly high, and suggested some contamination may have contributed to this signal. But with NuSTAR, finally, also the fainter gamma-ray lines from decay of ^{44}Ti (τ ~89 years) could be clearly detected (Boggs et al. 2015), with an inferred ^{44}Ti amount of $(1.5 \pm 0.3) \times 10^{-4}$ $M_\odot$. SN1987A clearly is the best-observed core-collapse supernova, also in gamma rays that directly relate to a specific isotope.

4.7.2.2 Cas A

The discovery of the 1157 keV ^{44}Ti γ-ray line emission from the youngest Galactic SNR Cas A with COMPTEL indexsupernova!Cas A (Iyudin et al. 1994) was the first direct proof that this isotope is indeed produced in a supernova explosion from a massive star. Many follow-up observations were able to confirm this: With the two lines at 68 and 78 keV, respectively, from the initial stage of ^{44}Ti decay to ^{44}Sc, also X-ray instruments such as BeppoSAX/PDS and later NuSTAR were able to measure the in-situ decay of ^{44}Ti in this 350-year-old (or 'young') supernova remnant. By combining observations, Vink et al. (2001) deduced a ^{44}Ti yield of $(1.5 \pm 1.0) \times 10^{-4}$ $M_\odot$. This could later be refined, adding the high signal-to-noise NuSTAR measurement and the measurement of all 3 decay lines from ^{44*}Sc and also ^{44*}Ca de-excitations with INTEGRAL/SPI (Siegert et al. 2015). The consolidated value of ^{44}Ti yield of $(1.4 \pm 0.2) \times 10^{-4}$ $M_\odot$ seemed higher than the predictions of most models, and provides a constraint to be met by model variants.

The ^{44}Ti detection in Cas A stimulated several activities in related fields: The decay time was re-investigated in nuclear laboratories, through both measurements tracing the decay as well as the absolute radioactive activity, both challenging; the uncertainty of 10–20 % at the time of the Cas A detection could thus be reduced to below 1 % and a consolidated decay time of (85.12 ± 0.43) years (Ahmad et al. 2006). Then, nuclear reactions involved in ^{44}Ti production were re-investigated. The α-rich freeze-out from Si burning involves many reactions, and detailed nuclear network analysis is required (Magkotsios et al. 2010). Finally, the supernova explosion from the collapse of a massive star is a complex, 3-dimensional phenomenon, as discussed above. Simulations and models have begun to explore the impacts of deviations from spherical symmetry, which, as discussed above (Sect. 4.6), are probably very significant for ^{44}Ti ejection (Nagataki et al. 1998; Wongwathanarat et al. 2015, 2017; Harris et al. 2017), unlike many other supernova observables. From the three different γ-ray lines resulting from the ^{44}Ti decay chain, constraints for kinematic Doppler broadening can be derived: The Doppler broadening being a linear function of energy, it would broaden the 1157 keV line to values in the few to tens of keV range, which can be measured with Ge spectrometers; the lower-energy lines at 68 and 78 keV would not show significant kinematic broadening. INTEGRAL/SPI spectrometer data show that there is some tension among constraints from the different ^{44}Ti lines; but all confirm that the inner ejecta have velocities faster than 1500 km s^{-1}, clearly contradicting an earlier belief

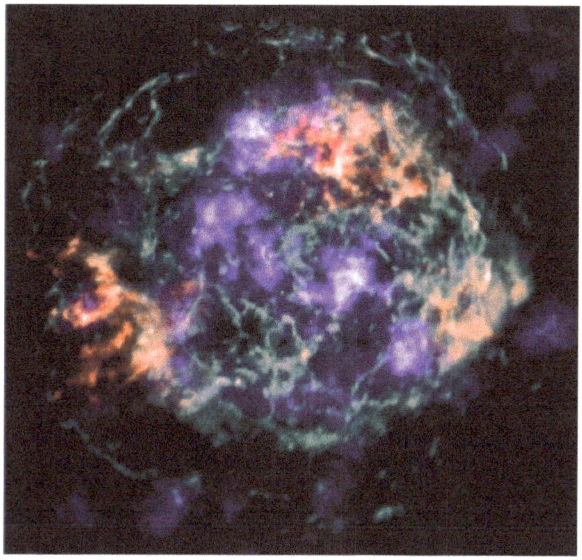

Fig. 4.37 The Cas A supernova remnant image in emission from ^{44}Ti decay (purple), superimposed on an X-ray image showing recombination lines (iron in red, silicon in green) (from Grefenstette et al. 2014)

that inner ejecta from close to the mass cut should be slow: The ‘mass cut’ concept over-simplifies the true 3D nature of the explosion.

The observations of the ^{44}Sc lines from ^{44}Ti decay with NuSTAR have obtained another important piece of information on the Cas A explosion: NuSTAR’s X-ray mirror and field of view were just right to obtain an *image* of Cas A in line emission from ^{44}Ti decay, shown in Fig. 4.37 (Grefenstette et al. 2014). As emission from ^{44}Ti decay does not depend on thermodynamic variables nor ionisation state, the purple features in the image clearly show that these ejecta appear in clumps. Further analysis of spatial and velocity information leads to a picture where much of the ejected ^{44}Ti occurs away from our line of sight (Grefenstette et al. 2017), possibly in momentum balance with the motion of the remnant neutron star. Comparing with recombination emission that had been observed and imaged earlier with the Chandra X-ray satellite (red structures in Fig. 4.37), it appears that some of the iron that would be expected to be ejected with ^{44}Ti is not seen in X-rays, likely because the reverse shock has not reached those clumps to ionise the iron and thus make them shine in X-ray line emission. This demonstrates the complementary views of different observables, and also reminds us that observational bias could be severe and thus could have distorted our view of how supernovae explode and shed their ejecta.

Many supernova remnants show mixing in their ejecta (Vink 2005). There are theoretical indications that this could arise from instabilities as the propagating shock wave traverses the envelope; however, steep density gradients such as the H/He interface also have the opposite effect, as the shock is accelerated when passing such an interface. Also, the expansion of the supernova into an inhomogeneous medium such as, for example, structured by the pre-supernova wind phase, will lead to deviations from spherical symmetry. SN1987A and its rings is a prominent such example (as reviewed in McCray and Fransson 2016). This illustrates the special role of ^{44}Ti and ^{56}Ni radioactive products, where 3D effects are much more directly related to the physics of the inner collapse and explosion.

4.7.2.3 Other Young Supernova Remnants

If ^{44}Ti ejection as seen in the Cas A event was typical for core-collapse events, the *gamma*-ray surveys made with COMPTEL (Dupraz et al. 1997; Iyudin et al. 1999) and with INTEGRAL (Renaud et al. 2006; Tsygankov et al. 2016) should have seen several objects along the plane of the Galaxy through their ^{44}Ti decay emission. But beyond Cas A, no firm detection of ^{44}Ti has been established since, though several candidates have been suggested: SNR RXJ0852.0-46.22 (Aschenbach 1998; Aschenbach et al. 1999; Slane et al. 2001), SNR G1.9+0.3 (Borkowski et al. 2010, 2013), and a source candidate in Perseus (Dupraz et al. 1997). None of these could be consolidated. From this, it had been concluded that ^{44}Ti ejection is rather a characteristic of a rare subclass of core-collapse supernovae (The et al. 2006; Dufour and Kaspi 2013).

The Vela region includes the Gum nebula and the Vela Supernova Remnant, two remnants from supernovae which are bright in X- and radio emissions. Both are relatively nearby, in the foreground of the Vela molecular ridge which is one of the nearest star-forming regions and located in about 700($\pm$200) kpc distance (Massi et al. 2007). At this distance, even for supernova remnants much older than the ^{44}Ti radioactive lifetime, a search for ^{26}Al seemed promising. With COMPTEL, diffuse and extended emission had been recognized from this direction (Diehl et al. 1995). But possibly-underlying extended ^{26}Al emission limits the interpretation of this measurement towards calibration of a single supernova in ^{26}Al; the ^{26}Al gamma-ray flux attributed to this supernova remnant is 0.5–2.7 10^{-5} ph cm^{-2} s^{-1}, well within expectations of an ^{26}Al yield of 10^{-4} $M_{\odot}$. Also for the much younger supernova remnant RXJ0852.0-46.22 or *Vela Junior*, no ^{26}Al emission was found. This is plausible, as the supernova remnant is probably older than 1000 y and more distant than 740 pc (Katsuda et al. 2009).

4.7.2.4 The ^{60}Fe to ^{26}Al Ratio

Cumulative emission from massive-star nucleosynthesis and its long-lived by-products ^{26}Al and ^{60}Fe has also been analysed. As pointed out from theorists (see, e.g. Woosley and Heger 2007), the same massive-star population likely is responsible for the bulk of these isotopes in the current Galaxy. As discussed above, ^{26}Al and ^{60}Fe originate, however, in very different zones and evolutionary phases of massive stars. ^{60}Fe only is ejected in the supernova, while ^{26}Al is both ejected in the supernova and in an earlier wind for very massive (Wolf Rayet) stars. The ratio measured from gamma rays is $\sim$ (15 $\pm$ 5) % (Wang et al. 2007).

Apparently a high mass loss rate is required to not overproduce ^{60}Fe in high mass stars $M > 40\ M_{\odot}$ (Limongi and Chieffi 2006b) with respect to these γ-ray line constraints (see discussion in Woosley and Heger 2007; Wang et al. 2007, and Chap. 7).

4.7.2.5 Fe-Group and Beyond, r-Process Ejecta

Matter close to the newly-forming neutron star is expected to be exposed to conditions ranging from nuclear statistical equilibrium, with Fe-group nuclei as major outcomes, to intense irradiation of available seed nuclei with free nucleons and α's, leading to an r-process or p-process, depending on the Y_e, and α captures in the nucleosynthesis and freeze-out phases.

In this picture, innermost matter which mainly experienced the shock only is thus ejected earlier. A typical example for the composition of such ejecta is shown in Fig. 4.28, which displays the zones of complete Si-burning with alpha-rich freeze-out. The change in abundances at mass coordinate $M = 1.63\,M_{\odot}$ is due to a change in Y_e in the underlying stellar-evolution model result, which was utilized for explosive nucleosynthesis predictions here, by introducing a simply thermal bomb of 10^{51} erg energy in the inner region. Early neutrino-wind models assumed neutron-rich material in the region behind those ejecta, shaped by the evaporising neutron star. But if one accounts correctly for the neutrino interactions during collapse and explosion, this matter even turns out to be slightly proton-rich ($Y_e > 0.5$). This leads to the νp-process discussed in Sect. 4.4 and Figs. 4.16 and 4.34 (Fröhlich et al. 2006a,b; Pruet et al. 2005, 2006; Wanajo 2006). This process then results also in a production of substantial fractions of ^{64}Ge (decaying to ^{64}Zn via ^{64}Ga. Both isotopes have a short half-life (minutes to seconds), and are therefore not of interest in terms of radioactivities which power an afterglow, or which may decay with characteristic gamma rays being observable. But in this νp-process also to the production of Sr and heavier nuclei can occur. The isotopic ratios ^{58}Ni, and ^{60,61,62}Zn are strongly modified by this process. Alpha-nuclei abundances such as ^{40}Ca, ^{44}Ti, ^{48}Cr, and ^{52}Fe are affected in characteristic ways as well. Higher entropies (such as expected in very energetic explosions and hypernovae) and higher Y_e-values, that is, close to 0.5, both increase the fraction of these alpha-nuclei, and this could proceed beyond ^{56}Ni up to radioactive ^{64}Ge that decays to the most-abundant Zn isotope ^{64}Zn, as discussed in Sects, 4.4 and 4.5. From observational constraints compared to chemical-evolution models, probably only a small fraction of black-hole producing events should lead to hypernovae, rather than $\sim$50% as assumed in some cases (Nomoto et al. 2006; Kobayashi et al. 2006; Kobayashi and Nakasato 2011)). We caution, however: There are degeneracies, and a reduced ^{64}Ge production from hypernovae can well be balanced by a larger ^{64}Ge production in regular supernovae, which may follow with correct inclusion of neutrino-interactions and their effect on increasing Y_e to values larger than 0.5.

After the supernova shock has been launched, the neutrino wind evaporises the surface layers of the proto-neutron star, and also energises the shock from below to launch the explosion. This is an area of biggest uncertainty. It was believed that a neutron-rich wind which could trigger a rather complete, full r-process (Qian and Woosley 1996; Woosley et al. 1994). But after early questioning from other investigations (Takahashi et al. 1994; Liebendörfer et al. 2003), more-detailed calculations of neutrino physics and detailed transport modeling find the opposite behavior, i.e. proton-rich conditions for more than the first 10s after the explosion (Fischer et al. 2010). This is discussed above Sect. 4.6) together with the νp-process. A major question is if and how this environment might change again to be neutron-rich in late phases of the explosion, and what physics underlies this change (such as the nuclear equation-of-state, or neutrino interactions; see Fischer et al. 2016) a following question then is how the high entropies can be attained to produce also the heaviest nuclei. Observations of low metallicity stars show huge variations in heavy r-process content relative to iron. This indicates that in most supernova explosions no r-process is taking place, leaving the *full* r-process to a rare subclass with special settings. Typical supernovae might only provide an environment for a weak r-process. Whether either high entropies are only attained in exceptional cases or other origins of low entropy, highly neutron-rich matter (neutron star mergers or neutron-rich jets from rotating core collapse supernovae Freiburghaus et al. 1999a; Cameron 2003) cause the main r-process has to be explored, including the still-remaining challenges of nuclear physics far from stability.

In the preceding sections we have shown that a fundamental understanding about the nuclear working of the r-process is available, and that it is possible to reproduce solar system r-process elemental abundance pattern though superpositions of components with different environmental conditions. Yet, it seems impossible to clearly identify *the* responsible astrophysical site.

One can use radioactivity and nucleochronocosmology to advance (as also discussed in Chap. 2): Adopting such a composite, superimposed, r-process fit as setting *zero-age* abundances, that is, with production ratios for ^{232}Th/^{238}U or other actinide (chronometer) nuclei, one can evaluate how such ratios evolve as a function of decay time to later and present-age abundances. This allows to identify formation ages of very metal-poor stars, born with a fresh ingestion of an r-process pattern at that time (see, e.g., Cowan et al. 1991, 1999; Thielemann et al. 2002; Kratz et al. 2007, and references therein). A typical result of such studies is that these chronometers suggest an age of the oldest stars in our Galaxy of 14–15 Gyr, with an uncertainty of about 3–4 Gyr.

4.7.3 Radiogenic Luminosity and Late Lightcurves

Supernova light curves are powered by radioactive decays. Very early interpretations of supernova lightcurves related them to the radioactive decay of ^{254}Cf (Burbidge et al. 1957). In fact, a strong r-process (with fission-cycling) would cause observable features based on the decay of heavy radioactive nuclei. This question re-appeared, as a potential optical transient (macronova) was predicted from a neutron star merger and its r-process radioactivities (see Metzger et al. 2010, and below). Supernova lightcurves, however, are dominated by Fe-group ejecta. In addition to abundant ^{56}Ni, there are a number of radioactive nuclei which will decay on time scales of ms up to 10^7y. Here we only want to concentrate on a few nuclei, which by a combination of their abundances and half-lives, can be of importance. These nuclei are ^{56}Co (^{56}Ni), ^{57}Co (^{57}Ni), ^{55}Fe (^{55}Co), ^{44}Ti, and ^{22}Na. For a 20 $M_\odot$ star like SN 1987A they were predicted with total masses of 0.07, 3.12×10^{-3}, 3.03×10^{-4}, 1.53×10^{-4}, and $1.33\times10^{-7}\,M_\odot$ (Thielemann et al. 1990, 1996).

Observations of light curves in radiation which reflects the thermalized energy of this radioactivity constrained these values to $M(^{56}\text{ Ni})\approx 0.071\,M_\odot$ (e.g., Suntzeff and Bouchet 1990) and $M(^{57}\text{ Ni})\approx 3.3\times10^3\,M_\odot$ (Fransson and Kozma 1993, and references therein). A very careful analysis extracted an upper limit on ^{44}Ti of the order $1.1\times10^{-4}\,M_\odot$ (Lundqvist et al. 2001). A recent re-assessment, including radioactivity inputs from gamma-rays, positrons, as well as electron captures (Auger electrons) (Seitenzahl et al. 2012) finds a ^{44}Ti amount of $(0.55\pm0.2)\times10^{-4}\,M_\odot$ to best fit the detailed late light curve constraints up to more than 4000 days. This is significantly below the value derived from the hard X-ray lines of ^{44}Ti decay directly. Uncertainties from the radioactive energy deposit in the expanding remnant exist, and certainly it is important to have calibrations from more than a few objects with measurements of direct radioactivity as well as its re-radiated energy deposit.

Let us discuss the various steps of radioactive energy deposition in more detail: Generally, after beta-decay or electron capture, a daughter nucleus is produced in an excited state (^{55}Fe is a notable exception, see below). The ground state is reached by one or several gamma transitions, observable by current gamma-ray detectors for nearby sources (see Sect. 10.1). Photons, positron-electron annihilations following β^+-decays, and the kinetic energy given to the decay products can contribute to the light curve at later times. The number of photons released for each of the transitions, occurring in the daughter nucleus after beta-decay, is equal to the number of decays N_d, multiplied with the appropriate percentage of the occurrence (*branching ratio*) for the specific transition. The total energy released corresponds to the product of the number of decays with the decay Q-value:

$$N_d(t) = -\frac{dN}{dt}(t) = \lambda N_o \exp(-\lambda t) \quad \frac{dE}{dt}(t) = QN_d(t) = Q\lambda N_o \exp(-\lambda t), \quad (4.8)$$

where $\lambda = \ln 2/t_{1/2}$ is the decay rate of the nucleus. The initial number of radioactive nuclei can be calculated from their total mass by $N_o = M/Am_u$, with A being the nucleon number of the nucleus, m_u the atomic mass unit, and M the mass given above. When using the radioactivity half-lives of relevant isotopes expected in supernova ejecta (i.e., 78.76d, 271.3d, 2.7y, 54.2y, and 2.602y, and atomic Q-values of 4.566, 0.835, 0.232, 3.919, and 2.842 MeV) we can estimate radioactive-energy generation rates in erg s^{-1} and the total number of decays per sec. The Q-value used for ^{44}Ti combines the subsequent decays of ^{44}Ti and ^{44}Sc.

Q-values include all available energies, i.e. the kinetic energy of the decay products, the energy in photons, the annihilation energy of positron-electron pairs in β^+-decays, and the neutrino energy. At densities prevailing in the expanding remnant, neutrinos will escape freely and their energy has to be subtracted, which leaves corrected values for the appropriate energy deposits of 3.695, 0.136, 0.0, 2.966, and 2.444 MeV. Because the electron capture on ^{55}Fe does only lead to an energetic neutrino, there is no local energy deposition from this isotope.[10] Gamma transitions following the decays of the other isotopes under consideration obtain candidate γ-rays at (rounded to full percent values): ^{56}Co, 847 keV (100%), 1038 keV (14%), 1238 keV (68%), 1772 keV (16%), 2599 keV (17%); ^{57}Co, 122 keV (86%), 136 keV (11%); ^{44}Ti, 78 keV (93%), 68 keV (88%), 147 keV (9%), 1157 keV (100%); ^{22}Na, 1275 keV (100%; branching ratios given as percentages per decay). If positrons from β^+-decay slow down and annihilate with electrons locally within the supernova envelope, the full neutrino-loss corrected energy corresponding to the reaction Q-value will be deposited in the envelope. Observable signatures include high energy photons such as the ones from the gamma transitions, and their Compton scattered and completely thermalized descendants.[11]

Then the sum of all individual contributions discussed above would make up the bolometric lightcurve of the supernova (see Fig. 4.38). The *light curve*, i.e. the brightness as a function of time, will be dominated first by the decay of ^{56}Co, and then ^{57}Co and ^{44}Ti, if one neglects possible radiation from a pulsar. ^{22}Na never plays a significant role. At lower densities (and later times), escaping high energy photons or positrons lead to a reduction of the brightness of *bolometric* emission. This can be seen towards the later-time observations, as shown e.g. in Leibundgut and Suntzeff (2003) (see Fig. 4.38). An important consistency check is to compare this bolometric light curve (which includes only optical, UV and IR emission, hence thermalized gas and dust components) to the high-energy photons more directly reflecting radioactive decays. At late times, those high energy photons escape freely.

Other indications came from the modeling of the optical light curve. The best agreement between calculated and observed light curves were obtained for a composition which mixed a small fraction of Ni all the way into the 10 $M_\odot$ hydrogen envelope and hydrogen into the deeper layers, containing mostly heavy elements (see, e.g., Benz and Thielemann 1990).

[10] This contribution was recently re-evaluated by Seitenzahl et al. (2009). The electron capture occurs from an electron in an atomic orbit, leaving a hole which can be filled by other electrons cascading down to fill this hole, thus emitting photons—X-rays— or depositing the energy in ejecting outer electrons—Auger electrons. Thus, in cases where only ground-state to ground-state electron capture occurs and the energy is emitted in an escaping neutrino only Auger electrons or X-rays can contribute to local energy deposition.

[11] Deposition of energy from radioactive decay involves absorption of high-energy photons, slowing down of ~MeV-type energy electrons and positrons, and proper treatment of temporary energy reservoirs such as ionization and inhibited radioactive decay from completely-ionized nuclei (see, e.g., Sim et al. 2009; Mochizuki et al. 1999; Kerzendorf and Sim 2014).

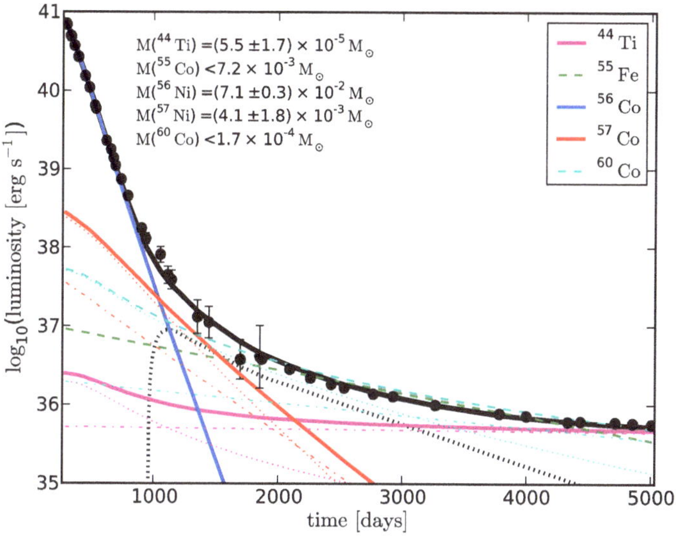

Fig. 4.38 Reconstructed bolometric lightcurve of SN1987A for 5000 days past explosion, using observational data from different bands, cross calibrated and merged (Seitenzahl et al. 2012). The model with its different radioactivity components is shown to provide a satisfactory fit

The lightcurve from SN1987A could be reproduced with theoretical modelling, including the effects of X-ray and γ-ray escape, as well as mixing of ^{56}Ni (see Fig. 4.38). SNe Ib and Ic events, believed to be core collapse events without an overlying hydrogen envelope have to be treated accordingly. The combination of small masses involved (only He-cores or C-cores without H-envelope) and the assumption of mixing can reproduce the steeper decline than found in massive SNe II. A typical case of a type Ic supernova is SN 1998bw, associated with GRB 980425. The straight-forward modeling of the observed lightcurve (Sollerman et al. 2002) led to interpretations of a largely non-solar ^{56}Ni/Fe to ^{56}Ni/Fe ratio. The inclusion of internal conversion and Auger electrons, as suggested by Seitenzahl et al. (2009) could naturally explain the observed slowdown of the lightcurve without invoking such extreme abundance ratios.

We note that in recent years photon transport calculations have reached major improvements, and are now able to consistently reproduce both light curves and spectra from SNIa, and also from core collapse supernovae (e.g. Kerzendorf and Sim 2014; Kasen et al. 2008; Dessart et al. 2015, 2017, for descriptions of the method). Presently, systematic uncertainties of the method are being investigated, and appear rather well understood (at least for SNIa (see, e.g., Woosley et al. 2007). As optical-to-IR light curves and spectra will be collected in abundance through large telescope survey programs for cosmological studies, it is likely that those (more indirect) measurements of core-collapse supernova nucleosynthesis will generate the tightest constraints to learn more about these events and their internal nuclear processes.

We have shown how supernova lightcurves reflect explosion properties and ejecta, though indirectly. More detailed information is contained in the evolution of spectra following an explosion, as the cooling of initially ionised gas and its dilution from expansion provides a characteristic spectral evolution, sometimes called 'tomography'. The receding photosphere in terms of radial mass in an expanding, radiation-filled bubble can give clear indications of the element composition (as a function of time equivalent to declining Lagrangian mass). The problem of type II supernovae is that the huge H-envelope does not really contain much information in terms of nucleosynthesis. Type Ib and Ic supernovae, which lost their H- and possibly He-envelope, reveal much more information of the compact inner part, which experienced explosive processing (e.g., Matheson et al. 2001; Sauer et al. 2006; Branch et al. 2002). This is similar to type Ia supernovae, originating from exploding white dwarfs, which have been extensively utilized for abundance diagnostics.

4.7.3.1 Macronovae from Neutron Star Mergers

Energy deposition from radioactive decays are understood to power the light seen at UV to optical to IR wavelengths from a supernova (see Fig. 4.38 for SN1987A). Such radiation transport is complex, starting from MeV gamma-rays and particles, which have a significant path length until they interact with envelope material to lose their energy, and multiple such interactions then deposit thermal energy. In the case of supernovae, the composition of the envelope is rather well known, as is the expected radioactivity, dominated by ^{56}Ni and ^{44}Ti, as shown above. In the case of neutron star mergers, the situation is much less clear (Kasen et al. 2015). First, the r-process may proceed into the A$\sim$130 region, or may proceed up to the third r-process peak at A$\sim$195 from efficient fission cycling see above). Then, the composition of the envelope where radioactive energy is deposited is unclear as well: Ejecta are expected from the dynamical interaction as the neutron stars approach each other, then from an accretion disk around the newly-forming black hole, and here a wind may be launched for even more complex envelope structure. Hence, it is not straightforward how to learn about those various unknowns; light curves in re-radiated light result from atomic-line multitudes that are largely unknown. Unless such an event would come fortunately-close to measure its radioactivity more directly in nuclear lines (which would require distances below a Mpc), modeling and cross-calibrations of radiogenic luminosities are the tedious way forward.

4.7.4 Material Deposits on Earth and Moon

Ejecta from massive stars are predominant drivers of chemical evolution, as they cool and are incorporated into next generations of stars forming from material enriched by products of nucleosynthesis (see Chapter on Chemical Evolution). Tracing the flow of ejected plasma as it cools and propagates into the star-forming dense cores of molecular clouds is difficult, as it occurs over times of scale 100 million years, and phase changes from plasma to atoms and molecules add complexity, as do accompanying physical and chemical reactions under varying environmental conditions. However, dust grains formed near the exploding nucleosynthesis site can propagate in different ways, ballistic trajectories and the solid-state form simplifying some of the processing along its journey. Analysing interstellar material which can reach us for example through cosmic rays and meteorites thus provides a valuable 'material astronomy'.

'Stardust' is a major messenger from nucleosynthesis sources (see Chap. 1, Fig. 1.7 and Table 1.1, Chap. 2 for stardust science history, and Chap. 10 on instrumentation). The formation of dust in a supernova envelope is driven by the cooling properties of expanding gas and by nucleation initiating the formation of dust grains from molecules. The composition of such a dust grain then reflects the extent of mixing in the expanding and cooling remnant when chemical reactions and dust formation set in. Clayton and Nittler (2004) review what has been learned from stardust. But, as one example, Liu et al. (2017) discuss the issues as ambiguities among nova or supernova origins remain, while abundances of C, N, and Si isotopes indicate characteristic signatures.

^{60}Fe has been discovered through accelerator-mass spectroscopy (AMS) analyses of ocean crust and lunar material (Knie et al. 2004) (Fig. 10.7 in Chap. 10). If taken from places on Earth which are remote from any antropogeneous contamination, such as in deep parts of the Pacific ocean, they provide a record of past composition of ocean water. Manganese crusts grow very slowly from sedimentation. Therefore, a rather small sample will cover tens of My of sedimentation history within a few cm of depth. ^{60}Fe production from cosmic ray irradiation in the atmosphere is unlikely, other systematic contaminations also seem low. The age of each depth layer can be determined from Be isotopes produced by cosmic rays in the atmosphere of the Earth, also ingested into ocean water with other atmospheric gas and dust. The AMS method is one of the most-sensitive techniques to detect small amounts of specific isotopes, reaching a sensitivity of 10^{-16}.

Meanwhile, ^{60}Fe has been measured in a variety of ocean crust and sediment samples from different deep-sea locations around Earth, and also in lunar material probes; see Sect. 7.3.3 for discussion of how this might relate to nearby supernovae, or to sweeping up by the interstellar shell bounding the Local Bubble cavity of cumulative nucleosynthesis ejecta (Schulreich et al. 2017).

Such terrestrial samples also have been investigated for signs from r-process material, specifically ^{244}Pu, also using the sensitive AMS analysis method. Recently, a positive detection of ^{244}Pu was achieved (Wallner et al. 2015). From the detected amount of nuclei, knowing its age sampling, and estimating transport from interstellar dust to ocean crust settling using cosmic-ray produced trace materials, Wallner et al. (2015) find that their number of nuclei should have been two orders of magnitude larger if r process nucleosynthesis occurred at the same rate as the synthesis of iron group or lighter nuclei. From this, either r-process ejecta are produced by rare events only, such as neutron star mergers (see Sect. 4.5 and Thielemann et al. 2017b, for a review), or else their transport in interstellar medium is very different from how ejecta from normal massive-star nucleosynthesis is transported to reach Earth.

References

Abbott BP, Abbott R, Abbott TD, Acernese F, Ackley K, Adams C, Adams T, Addesso P, Adhikari RX, Adya VB et al (2017) Multi-messenger observations of a binary neutron star merger. Astrophys J 848:L12. https://doi.org/10.3847/2041-8213/aa91c9. arXiv:1710.05833

Adelberger EG, García A, Robertson RGH, Snover KA, Balantekin AB, Heeger K, Ramsey-Musolf MJ, Bemmerer D, Junghans A, Bertulani CA, Chen JW, Costantini H, Prati P, Couder M, Uberseder E, Wiescher M, Cyburt R, Davids B, Freedman SJ, Gai M, Gazit D, Gialanella L, Imbriani G, Greife U, Hass M, Haxton WC, Itahashi T, Kubodera K, Langanke K, Leitner D, Leitner M, Vetter P, Winslow L, Marcucci LE, Motobayashi T, Mukhamedzhanov A, Tribble RE, Nollett KM, Nunes FM, Park TS, Parker PD, Schiavilla R, Simpson EC, Spitaleri C, Strieder F, Trautvetter HP, Suemmerer K, Typel S (2011) Solar fusion cross sections. II. The pp chain and CNO cycles. Rev Mod Phys 83:195–246. https://doi.org/10.1103/RevModPhys.83.195. arXiv:1004.2318

Ahmad I, Bonino G, Castagnoli GC, Fischer SM, Kutschera W, Paul M (1998) Three-laboratory measurement of the ^{44}Ti half-life. Phys Rev Lett 80:2550–2553. https://doi.org/10.1103/PhysRevLett.80.2550

Ahmad I, Greene JP, Moore EF, Ghelberg S, Ofan A, Paul M, Kutschera W (2006) Improved measurement of the Ti44 half-life from a 14-year long study. Phys Rev C 74(6):065803. https://doi.org/10.1103/PhysRevC.74.065803

Arcones A, Thielemann FK (2013) Neutrino-driven wind simulations and nucleosynthesis of heavy elements. J Phys G Nucl Phys 40(1):013201. https://doi.org/10.1088/0954-3899/40/1/013201. arXiv:1207.2527

Arlandini C, Käppeler F, Wisshak K, Gallino R, Lugaro M, Busso M, Straniero O (1999) Neutron capture in low-mass asymptotic giant branch stars: cross sections and abundance signatures. Astrophys J 525:886–900. https://doi.org/10.1086/307938. arXiv:astro-ph/9906266

Arnett D (1996) 2D simulations of supernovae. In: IAU Colloquium. Supernovae and supernova remnants, vol 145, pp 91–+

Arnett D, Meakin C, Young PA (2009) Turbulent convection in stellar interiors. II. The velocity field. Astrophys J 690:1715–1729. https://doi.org/10.1088/0004-637X/690/2/1715. arXiv:0809.1625

Arnould M, Goriely S (2003) The p-process of stellar nucleosynthesis: astrophysics and nuclear physics status. Phys Rep 384:1–84. https://doi.org/10.1016/S0370-1573(03)00242-4

Aschenbach B (1998) Discovery of a young nearby supernova remnant. Nature 396:141–142. https://doi.org/10.1038/24103

Aschenbach B, Iyudin AF, Schönfelder V (1999) Constraints of age, distance and progenitor of the supernova remnant RX J0852.0-4622/GRO J0852-4642. Astron Astrophys 350:997–1006. arXiv:astro-ph/9909415

Aufderheide MB, Fushiki I, Woosley SE, Hartmann DH (1994) Search for important weak interaction nuclei in presupernova evolution. Astrophys J Suppl 91:389–417. https://doi.org/10.1086/191942

Balasi KG, Langanke K, Martínez-Pinedo G (2015) Neutrino-nucleus reactions and their role for supernova dynamics and nucleosynthesis. Prog Part Nucl Phys 85:33–81. https://doi.org/10.1016/j.ppnp.2015.08.001. arXiv:1503.08095

Bao ZY, Beer H, Käppeler F, Voss F, Wisshak K, Rauscher T (2000) Neutron cross sections for nucleosynthesis studies. At Data Nucl Data Tables 76:70–154. https://doi.org/10.1006/adnd.2000.0838

Barnes J, Kasen D, Wu MR, Martínez-Pinedo G (2016) Radioactivity and thermalization in the ejecta of compact object mergers and their impact on kilonova light curves. Astrophys J 829:110. https://doi.org/10.3847/0004-637X/829/2/110. arXiv:1605.07218

Bauswein A, Goriely S, Janka HT (2013) Systematics of dynamical mass ejection, nucleosynthesis, and radioactively powered electromagnetic signals from neutron-star mergers. Astrophys J 773:78. https://doi.org/10.1088/0004-637X/773/1/78. arXiv:1302.6530

Benz W, Thielemann F (1990) Convective instabilities in SN 1987A. Astrophys J 348:L17–L20. https://doi.org/10.1086/185620

Bernardini MG (2015) Gamma-ray bursts and magnetars: observational signatures and predictions. J High Energy Astrophys 7:64–72. https://doi.org/10.1016/j.jheap.2015.05.003

Bethe HA (1990) Supernova mechanisms. Rev Mod Phys 62:801–866. https://doi.org/10.1103/RevModPhys.62.801

Bethe HA, Brown GE, Applegate J, Lattimer JM (1979) Equation of state in the gravitational collapse of stars. Nucl Phys A 324:487–533. https://doi.org/10.1016/0375-9474(79)90596-7

Blondin JM, Mezzacappa A, DeMarino C (2003) Stability of standing accretion shocks, with an eye toward core-collapse supernovae. Astrophys J 584:971–980. https://doi.org/10.1086/345812. arXiv:astro-ph/0210634

Boggs SE, Harrison FA, Miyasaka H, Grefenstette BW, Zoglauer A, Fryer CL, Reynolds SP, Alexander DM, An H, Barret D, Christensen FE, Craig WW, Forster K, Giommi P, Hailey CJ, Hornstrup A, Kitaguchi T, Koglin JE, Madsen KK, Mao PH, Mori K, Perri M, Pivovaroff MJ, Puccetti S, Rana V, Stern D, Westergaard NJ, Zhang WW (2015) ^{44}Ti gamma-ray emission lines from SN1987A reveal an asymmetric explosion. Science 348:670–671. https://doi.org/10.1126/science.aaa2259

Borkowski KJ, Reynolds SP, Green DA, Hwang U, Petre R, Krishnamurthy K, Willett R (2010) Radioactive scandium in the youngest galactic supernova remnant G1.9+0.3. Astrophys J 724:L161–L165. https://doi.org/10.1088/2041-8205/724/2/L161. arXiv:1006.3552

Borkowski KJ, Reynolds SP, Hwang U, Green DA, Petre R, Krishnamurthy K, Willett R (2013) Supernova ejecta in the youngest galactic supernova remnant G1.9+0.3. Astrophys J 771:L9. https://doi.org/10.1088/2041-8205/771/1/L9. arXiv:1305.7399

Branch D, Wheeler JC (2017) Supernova explosions. https://doi.org/10.1007/978-3-662-55054-0

Branch D, Benetti S, Kasen D, Baron E, Jeffery DJ, Hatano K, Stathakis RA, Filippenko AV, Matheson T, Pastorello A, Altavilla G, Cappellaro E, Rizzi L, Turatto M, Li W, Leonard DC, Shields JC (2002) Direct analysis of spectra of type Ib supernovae. Astrophys J 566:1005–1017. https://doi.org/10.1086/338127. arXiv:astro-ph/0106367

Bruenn SW, Haxton WC (1991) Neutrino-nucleus interactions in core-collapse supernovae. Astrophys J 376:678–700. https://doi.org/10.1086/170316

Bruenn SW, Lentz EJ, Hix WR, Mezzacappa A, Harris JA, Messer OEB, Endeve E, Blondin JM, Chertkow MA, Lingerfelt EJ, Marronetti P, Yakunin KN (2016) The development of explosions in axisymmetric ab initio core-collapse supernova simulations of 12–25 M stars. Astrophys J 818:123. https://doi.org/10.3847/0004-637X/818/2/123. arXiv:1409.5779

Buchmann LR, Barnes CA (2006) Nuclear reactions in stellar helium burning and later hydrostatic burning stages. Nucl Phys A 777:254–290. https://doi.org/10.1016/j.nuclphysa.2005.01.005

Buras R, Rampp M, Janka H, Kifonidis K (2003) Improved models of stellar core collapse and still no explosions: what is missing? Phys Rev Lett 90(24):241, 101–+. https://doi.org/10.1103/PhysRevLett.90.241101. arXiv:astro-ph/0303171

Burbidge EM, Burbidge GR, Fowler WA, Hoyle F (1957) Synthesis of the elements in stars. Rev Mod Phys 29:547–650. https://doi.org/10.1103/RevModPhys.29.547

Burrows A (2013) Colloquium: perspectives on core-collapse supernova theory. Rev Mod Phys 85:245–261. https://doi.org/10.1103/RevModPhys.85.245. arXiv:1210.4921

Burrows A, Livne E, Dessart L, Ott CD, Murphy J (2006a) A new mechanism for core-collapse supernova explosions. Astrophys J 640:878–890. https://doi.org/10.1086/500174. arXiv:astro-ph/0510687

Burrows A, Reddy S, Thompson TA (2006b) Neutrino opacities in nuclear matter. Nucl Phys A 777:356–394. https://doi.org/10.1016/j.nuclphysa.2004.06.012. arXiv:astro-ph/0404432

Burrows A, Vartanyan D, Dolence JC, Skinner MA, Radice D (2018) Crucial physical dependencies of the core-collapse supernova mechanism. Space Sci Rev 214:33. https://doi.org/10.1007/s11214-017-0450-9

Cameron AGW (2003) Some nucleosynthesis effects associated with r-process jets. Astrophys J 587:327–340. https://doi.org/10.1086/368110

Cayrel R, Hill V, Beers TC, Barbuy B, Spite M, Spite F, Plez B, Andersen J, Bonifacio P, François P, Molaro P, Nordström B, Primas F (2001) Measurement of stellar age from uranium decay. Nature 409:691–692. https://doi.org/10.1038/35055507. arXiv:astro-ph/0104357

Chandrasekhar S (1957) An introduction to the study of stellar structure

Chiappini C, Hirschi R, Meynet G, Ekström S, Maeder A, Matteucci F (2006) A strong case for fast stellar rotation at very low metallicities. Astron Astrophys 449:L27–L30. https://doi.org/10.1051/0004-6361:20064866. arXiv:astro-ph/0602459

Chieffi A, Limongi M (2004) Explosive yields of massive stars from Z = 0 to $Z = Z_{solar}$. Astrophys J 608:405–410. https://doi.org/10.1086/392523. arXiv:astro-ph/0402625

Chieffi A, Limongi M (2013) Pre-supernova evolution of rotating solar metallicity stars in the mass range 13-120 M Ł and their explosive yields. Astrophys J 764:21. https://doi.org/10.1088/0004-637X/764/1/21

Chieffi A, Limongi M (2017) The synthesis of ^{44}Ti and ^{56}Ni in massive stars. Astrophys J 836:79. https://doi.org/10.3847/1538-4357/836/1/79. arXiv:1701.02914

Clayton DD (1968) Principles of stellar evolution and nucleosynthesis. McGraw-Hill, New York, 1968

Clayton DD, Nittler LR (2004) Astrophysics with presolar stardust. Annu Rev Astron Astrophys 42:39–78. https://doi.org/10.1146/annurev.astro.42.053102.134022

Costantini H, Formicola A, Imbriani G, Junker M, Rolfs C, Strieder F (2009) LUNA: a laboratory for underground nuclear astrophysics. Rep Prog Phys 72(8):086,301–+. https://doi.org/10.1088/0034-4885/72/8/086301. arXiv:0906.1097

Cowan JJ, Thielemann F, Truran JW (1991) The R-process and nucleochronology. Phys Rep 208:267–394. https://doi.org/10.1016/0370-1573(91)90070-3

Cowan JJ, Pfeiffer B, Kratz K, Thielemann F, Sneden C, Burles S, Tytler D, Beers TC (1999) R-process abundances and chronometers in metal-poor stars. Astrophys J 521:194–205. https://doi.org/10.1086/307512. arXiv:astro-ph/9808272

Cristini A, Hirschi R, Georgy C, Meakin C, Arnett D, Viallet M (2015) Linking 1D stellar evolution to 3D hydrodynamic simulations. In: Meynet G, Georgy C, Groh J, Stee P (eds) New windows on massive stars. IAU symposium, vol 307, pp 98–99. https://doi.org/10.1017/S1743921314006371. arXiv:1410.7672

Curtis S, Ebinger K, Fröhlich C, Hempel M, Perego A, Liebendörfer M, Thielemann F-K (2018) PUSHing core-collapse supernovae to explosions in spherical symmetry III: nucleosynthesis yields. arXiv:1805.00498

Cyburt RH, Amthor AM, Ferguson R, Meisel Z, Smith K, Warren S, Heger A, Hoffman RD, Rauscher T, Sakharuk A, Schatz H, Thielemann FK, Wiescher M (2010) The JINA REACLIB database: its recent updates and impact on type-I X-ray bursts. Astrophys J Suppl 189:240–252. https://doi.org/10.1088/0067-0049/189/1/240

De Marco O, Izzard RG (2017) Dawes review 6: the impact of companions on stellar evolution. Publ Astron Soc Aust 34:e001. https://doi.org/10.1017/pasa.2016.52. arXiv:1611.03542

deBoer RJ, Görres J, Wiescher M, Azuma RE, Best A, Brune CR, Fields CE, Jones S, Pignatari M, Sayre D, Smith K, Timmes FX, Uberseder E (2017) The $^{12}C(\alpha,\gamma)^{16}O$ reaction and its implications for stellar helium burning. Rev Mod Phys 89(3):035007. https://doi.org/10.1103/RevModPhys.89.035007. arXiv:1709.03144

Dessart L, Hillier DJ, Woosley S, Livne E, Waldman R, Yoon SC, Langer N (2015) Radiative-transfer models for supernovae IIb/Ib/Ic from binary-star progenitors. Mon Not R Astron Soc 453:2189–2213. https://doi.org/10.1093/mnras/stv1747. arXiv:1507.07783

Dessart L, John Hillier D, Yoon SC, Waldman R, Livne E (2017) Radiative-transfer models for explosions from rotating and non-rotating single WC stars. Implications for SN 1998bw and LGRB/SNe. Astron Astrophys 603:A51. https://doi.org/10.1051/0004-6361/201730873. arXiv:1703.08932

Diehl R, Bennett K, Bloemen H, Dupraz C, Hermsen W, Knoedlseder J, Lichti G, Morris D, Oberlack U, Ryan J, Schoenfelder V, Steinle H, Varendorff M, Winkler C (1995) 1.809 MeV gamma-rays from the VELA region. Astron Astrophys 298:L25+

Diehl R, Wessolowski U, Oberlack U, Bloemen H, Georgii R, Iyudin A, Knodlseder J, Lichti G, Hermsen W, Morris D, Ryan J, Schonfelder V, Strong A, von Ballmoos P, Winkler C (1997) 26Al and the COMPTEL 60Fe data. In: Dermer CD, Strickman MS, Kurfess JD (eds) Proceedings of the fourth compton symposium. American institute of physics conference series, vol 410, pp 1109–+. https://doi.org/10.1063/1.54176

Diehl R, Halloin H, Kretschmer K, Lichti GG, Schönfelder V, Strong AW, von Kienlin A, Wang W, Jean P, Knödlseder J, Roques JP, Weidenspointner G, Schanne S, Hartmann DH, Winkler C, Wunderer C (2006a) Radioactive ^{26}Al from massive stars in the galaxy. Nature 439:45–47. https://doi.org/10.1038/nature04364. arXiv:astro-ph/0601015

Diehl R, Prantzos N, von Ballmoos P (2006b) Astrophysical constraints from gamma-ray spectroscopy. Nucl Phys A 777:70–97. https://doi.org/10.1016/j.nuclphysa.2005.02.155. arXiv:astro-ph/0502324

Dillmann I, Heil M, Käppeler F, Plag R, Rauscher T, Thielemann FK (2006) KADoNiS- the Karlsruhe astrophysical database of nucleosynthesis in stars. In: Woehr A, Aprahamian A (eds) Capture gamma-ray spectroscopy and related topics. American institute of physics conference series, vol 819, pp 123–127. https://doi.org/10.1063/1.2187846

Dillmann I, Rauscher T, Heil M, Käppeler F, Rapp W, Thielemann F (2008) p-Process simulations with a modified reaction library. J Phys G Nucl Phys 35(1):014,029–+. https://doi.org/10.1088/0954-3899/35/1/014029. arXiv:0805.4756

Dillmann I, Szücs T, Plag R, Fülöp Z, Käppeler F, Mengoni A, Rauscher T (2014) The Karlsruhe astrophysical database of nucleosynthesis in stars project - status and prospects. Nucl Data Sheets 120:171–174. https://doi.org/10.1016/j.nds.2014.07.038. arXiv:1408.3688

Duan H, Fuller GM, Qian Y (2006) Collective neutrino flavor transformation in supernovae. Phys Rev D 74(12):123,004–+. https://doi.org/10.1103/PhysRevD.74.123004. arXiv:astro-ph/0511275

Duan H, Fuller GM, Carlson J, Qian Y (2007) Neutrino mass hierarchy and stepwise spectral swapping of supernova neutrino flavors. Phys Rev Lett 99(24):241,802–+. https://doi.org/10.1103/PhysRevLett.99.241802. arXiv:0707.0290

Duflo J, Zuker AP (1995) Microscopic mass formulas. Phys Rev C 52:R23–R27. https://doi.org/10.1103/PhysRevC.52.R23. arXiv:nucl-th/9505011

Dufour F, Kaspi VM (2013) Limits on the number of galactic young supernova remnants emitting in the decay lines of ^{44}Ti. Astrophys J 775:52. https://doi.org/10.1088/0004-637X/775/1/52. arXiv:1308.4859

Dupraz C, Bloemen H, Bennett K, Diehl R, Hermsen W, Iyudin AF, Ryan J, Schoenfelder V (1997) COMPTEL three-year search for galactic sources of ^44^Ti gamma-ray line emission at 1.157MeV. Astron Astrophys 324:683–689

Ebinger K, Sinha S, Fröhlich C, Perego A, Hempel M, Eichler M, Casanova J, Liebendörfer M, Thielemann FK (2017) Explosion dynamics of parametrized spherically symmetric core-collapse supernova simulations. In: Kubono S, Kajino T, Nishimura S, Isobe T, Nagataki S, Shima T, Takeda Y (eds) 14th international symposium on nuclei in the cosmos (NIC2016), p 020611. https://doi.org/10.7566/JPSCP.14.020611. arXiv:1610.05629

Ebinger K, Curtis S, Fröhlich C, Hempel M, Perego A, Liebendörfer M, Thielemann F-K (2018) PUSHing core-collapse supernovae to explosions in spherical symmetry II: explodability and global properties. arXiv:1804.03182

Edelmann PVF, Röpke FK, Hirschi R, Georgy C, Jones S (2017) Testing a one-dimensional prescription of dynamical shear mixing with a two-dimensional hydrodynamic simulation. Astron Astrophys 604:A25. https://doi.org/10.1051/0004-6361/201629873. arXiv:1704.06261

Eggenberger P, Meynet G, Maeder A, Hirschi R, Charbonnel C, Talon S, Ekström S (2008) The Geneva stellar evolution code. Astrophys Space Sci 316:43–54. https://doi.org/10.1007/s10509-007-9511-y

Eichler D, Livio M, Piran T, Schramm DN (1989) Nucleosynthesis, neutrino bursts and gamma-rays from coalescing neutron stars. Nature 340:126–128. https://doi.org/10.1038/340126a0

Eichler M, Arcones A, Kelic A, Korobkin O, Langanke K, Marketin T, Martinez-Pinedo G, Panov I, Rauscher T, Rosswog S, Winteler C, Zinner NT, Thielemann FK (2015) The role of fission in neutron star mergers and its impact on the r-process peaks. Astrophys J 808:30. https://doi.org/10.1088/0004-637X/808/1/30. arXiv:1411.0974

Eichler M, Nakamura K, Takiwaki T, Kuroda T, Kotake K, Hempel M, Cabezón R, Liebendörfer M, Thielemann F (2018) Nucleosynthesis in 2D core-collapse supernovae of 11.2 and 17.0 M_⊙ progenitors: implications for Mo and Ru production. J Phys G Nucl Phys 48(1):014001. https://doi.org/10.1088/1361-6471/aa8891. arXiv:1708.08393

Ekström S, Meynet G, Chiappini C, Hirschi R, Maeder A (2008) Effects of rotation on the evolution of primordial stars. Astron Astrophys 489:685–698. https://doi.org/10.1051/0004-6361:200809633. arXiv:0807.0573

Ekström S, Georgy C, Eggenberger P, Meynet G, Mowlavi N, Wyttenbach A, Granada A, Decressin T, Hirschi R, Frischknecht U, Charbonnel C, Maeder A (2012) Grids of stellar models with rotation. I. Models from 0.8 to 120 M_⊙ at solar metallicity (Z = 0.014). Astron Astrophys 537:A146. https://doi.org/10.1051/0004-6361/201117751. arXiv:1110.5049

El Eid MF, The L, Meyer BS (2009) Massive stars: input physics and stellar models. Space Sci Rev 147:1–29. https://doi.org/10.1007/s11214-009-9517-6

Eldridge JJ, Izzard RG, Tout CA (2008) The effect of massive binaries on stellar populations and supernova progenitors. Mon Not R Astron Soc 384:1109–1118. https://doi.org/10.1111/j.1365-2966.2007.12738.x. arXiv:0711.3079

Ertl T, Janka HT, Woosley SE, Sukhbold T, Ugliano M (2016) A Two-parameter criterion for classifying the explodability of massive stars by the neutrino-driven mechanism. Astrophys J 818:124. https://doi.org/10.3847/0004-637X/818/2/124. arXiv:1503.07522

Fabbian D, Nissen PE, Asplund M, Pettini M, Akerman C (2009) The C/O ratio at low metallicity: constraints on early chemical evolution from observations of galactic halo stars. Astron Astrophys 500:1143–1155. https://doi.org/10.1051/0004-6361/200810095. arXiv:0810.0281

Farouqi K, Kratz KL, Pfeiffer B, Rauscher T, Thielemann FK, Truran JW (2010) Charged-particle and neutron-capture processes in the high-entropy wind of core-collapse supernovae. Astrophys J 712:1359–1377. https://doi.org/10.1088/0004-637X/712/2/1359. arXiv:1002.2346

Fischer T, Whitehouse SC, Mezzacappa A, Thielemann F, Liebendörfer M (2009) The neutrino signal from protoneutron star accretion and black hole formation. Astron Astrophys 499:1–15. https://doi.org/10.1051/0004-6361/200811055. arXiv:0809.5129

Fischer T, Whitehouse SC, Mezzacappa A, Thielemann FK, Liebendörfer M (2010) Protoneutron star evolution and the neutrino-driven wind in general relativistic neutrino radiation hydrodynamics simulations. Astron Astrophys 517:A80. https://doi.org/10.1051/0004-6361/200913106. arXiv:0908.1871

Fischer T, Sagert I, Pagliara G, Hempel M, Schaffner-Bielich J, Rauscher T, Thielemann FK, Käppeli R, Martínez-Pinedo G, Liebendörfer M (2011) Core-collapse supernova explosions triggered by a Quark-Hadron phase transition during the early post-bounce phase. Astrophys J Suppl 194:39. https://doi.org/10.1088/0067-0049/194/2/39. arXiv:1011.3409

Fischer T, Huther L, Lohs A, Martínez-Pinedo G (2016) Early protoneutron star deleptonization - consistent modeling of weak processes and equation of state. J Phys Conf Ser 665:012069. https://doi.org/10.1088/1742-6596/665/1/012069

Fisker JL, Schatz H, Thielemann F (2008) Explosive hydrogen burning during type I X-ray bursts. Astrophys J Suppl 174:261–276. https://doi.org/10.1086/521104

Fogli G, Lisi E, Marrone A, Mirizzi A (2007) Collective neutrino flavor transitions in supernovae and the role of trajectory averaging. J Cosmol Astropart Phys 12:10–+. https://doi.org/10.1088/1475-7516/2007/12/010. arXiv:0707.1998

Foglizzo T (2009) A simple toy model of the advective-acoustic instability. I. Perturbative approach. Astrophys J 694:820–832. https://doi.org/10.1088/0004-637X/694/2/820. arXiv:0809.2302

Fransson C, Kozma C (1993) The freeze-out phase of SN 1987A - implications for the light curve. Astrophys J 408:L25–L28. https://doi.org/10.1086/186822

Freiburghaus C, Rembges J, Rauscher T, Kolbe E, Thielemann F, Kratz K, Pfeiffer B, Cowan JJ (1999a) The astrophysical r-process: a comparison of calculations following adiabatic expansion with classical calculations based on neutron densities and temperatures. Astrophys J 516:381–398. https://doi.org/10.1086/307072

Freiburghaus C, Rosswog S, Thielemann FK (1999b) R-process in neutron star mergers. Astrophys J 525:L121–L124. https://doi.org/10.1086/312343

Frischknecht U, Hirschi R, Pignatari M, Maeder A, Meynet G, Chiappini C, Thielemann FK, Rauscher T, Georgy C, Ekström S (2016) s-Process production in rotating massive stars at solar and low metallicities. Mon Not R Astron Soc 456:1803–1825. https://doi.org/10.1093/mnras/stv2723. arXiv:1511.05730

Fröhlich C, Hauser P, Liebendörfer M, Martínez-Pinedo G, Thielemann F, Bravo E, Zinner NT, Hix WR, Langanke K, Mezzacappa A, Nomoto K (2006a) Composition of the innermost core-collapse supernova ejecta. Astrophys J 637:415–426. https://doi.org/10.1086/498224. arXiv:astro-ph/0410208

Fröhlich C, Martínez-Pinedo G, Liebendörfer M, Thielemann F, Bravo E, Hix WR, Langanke K, Zinner NT (2006b) Neutrino-induced nucleosynthesis of a greater 64 nuclei: the νp process. Phys Rev Lett 96(14):142502. https://doi.org/10.1103/PhysRevLett.96.142502. arXiv:astro-ph/0511376

Fryxell B, Olson K, Ricker P, Timmes FX, Zingale M, Lamb DQ, MacNeice P, Rosner R, Truran JW, Tufo H (2000) FLASH: an adaptive mesh hydrodynamics code for modeling astrophysical thermonuclear flashes. Astrophys J Suppl 131:273–334. https://doi.org/10.1086/317361

Fujimoto Si, Hashimoto Ma, Kotake K, Yamada S (2007) Heavy-element nucleosynthesis in a collapsar. Astrophys J 656:382–392. https://doi.org/10.1086/509908. arXiv:astro-ph/0602460

Fujimoto Si, Nishimura N, Hashimoto Ma (2008) Nucleosynthesis in magnetically driven jets from collapsars. Astrophys J 680:1350–1358. https://doi.org/10.1086/529416. arXiv:0804.0969

Fukuda I (1982) A statistical study of rotational velocities of the stars. Publ Astron Soc Pac 94:271–284. https://doi.org/10.1086/130977

Fuller GM, Fowler WA, Newman MJ (1980) Stellar weak-interaction rates for sd-shell nuclei. I - nuclear matrix element systematics with application to Al-26 and selected nuclei of importance to the supernova problem. Astrophys J Suppl 42:447–473. https://doi.org/10.1086/190657

Fuller GM, Fowler WA, Newman MJ (1982) Stellar weak interaction rates for intermediate-mass nuclei. II - A = 21 to A = 60. Astrophys J 252:715–740. https://doi.org/10.1086/159597

Fuller GM, Fowler WA, Newman MJ (1985) Stellar weak interaction rates for intermediate-mass nuclei. IV - interpolation procedures for rapidly varying lepton capture rates using effective log (ft)-values. Astrophys J 293:1–16. https://doi.org/10.1086/163208

Georgy C, Meynet G, Walder R, Folini D, Maeder A (2009) The different progenitors of type Ib, Ic SNe, and of GRB. Astron Astrophys 502:611–622. https://doi.org/10.1051/0004-6361/200811339. arXiv:0906.2284

Georgy C, Ekström S, Meynet G, Massey P, Levesque EM, Hirschi R, Eggenberger P, Maeder A (2012) Grids of stellar models with rotation. II. WR populations and supernovae/GRB progenitors at Z = 0.014. Astron Astrophys 542:A29. https://doi.org/10.1051/0004-6361/201118340. arXiv:1203.5243

Georgy C, Ekström S, Eggenberger P, Meynet G, Haemmerlé L, Maeder A, Granada A, Groh JH, Hirschi R, Mowlavi N, Yusof N, Charbonnel C, Decressin T, Barblan F (2013) Grids of stellar models with rotation. III. Models from 0.8 to 120 $M_\odot$ at a metallicity Z = 0.002. Astron Astrophys 558:A103. https://doi.org/10.1051/0004-6361/201322178. arXiv:1308.2914

Georgy C, Meynet G, Ekström S, Wade GA, Petit V, Keszthelyi Z, Hirschi R (2017) Possible pair-instability supernovae at solar metallicity from magnetic stellar progenitors. Astron Astrophys 599:L5. https://doi.org/10.1051/0004-6361/201730401. arXiv:1702.02340

Giron S, Hammache F, de Séréville N, Beaumel D, Burgunder J, Caceres L, Clement E, Duchene G, Flavigny F, de France G, Franchoo S, Fernandez B, Galaviz-Redondo D, Gasques L, Gibelin J, Gillibert A, Grevy S, Guillot J, Heil M, Kiener J, Lapoux V, Maréchal F, Matta A, Matea Y, Moukaddam M, Nalpas L, Obertelli A, Perrot L, Raabe R, Scarpaci JA, Sorlin O, Stefan I, Stoedel C, Takechi M, Thomas JC, Togano Y (2010) Study of ^{60}Fe(n,γ)^{61}Fe reaction of astrophysical interest via d(^{60}Fe,pγ) indirect reaction. In: Spitaleri C, Rolfs C, Pizzone RG (ed) American institute of physics conference series, vol 1213, pp 201–204. https://doi.org/10.1063/1.3362577

Goldreich P, Weber SV (1980) Homologously collapsing stellar cores. Astrophys J 238:991–997. https://doi.org/10.1086/158065

Goriely S (2015) Towards more accurate and reliable predictions for nuclear applications. Eur Phys J A 51:172. https://doi.org/10.1140/epja/i2015-15172-2

Goriely S, Bauswein A, Janka HT (2011) r-process nucleosynthesis in dynamically ejected matter of neutron star mergers. Astrophys J 738:L32. https://doi.org/10.1088/2041-8205/738/2/L32. arXiv:1107.0899

Goriely S, Bauswein A, Just O, Pllumbi E, Janka HT (2015) Impact of weak interactions of free nucleons on the r-process in dynamical ejecta from neutron star mergers. Mon Not R Astron Soc 452:3894–3904. https://doi.org/10.1093/mnras/stv1526. arXiv:1504.04377

Grebenev SA, Lutovinov AA, Tsygankov S, Winkler C (2012) Hard-X-ray emission lines from the decay of ^{44}ti in the remnant of supernova 1987A. Nature 490:373–375. https://doi.org/10.1038/nature11473. arXiv:1211.2656

Grefenstette BW, Harrison FA, Boggs SE, Reynolds SP, Fryer CL, Madsen KK, Wik DR, Zoglauer A, Ellinger CI, Alexander DM, An H, Barret D, Christensen FE, Craig WW, Forster K, Giommi P, Hailey CJ, Hornstrup A, Kaspi VM, Kitaguchi T, Koglin JE, Mao PH, Miyasaka H, Mori K, Perri M, Pivovaroff MJ, Puccetti S, Rana V, Stern D, Westergaard NJ, Zhang WW (2014) Asymmetries in core-collapse supernovae from maps of radioactive ^{44}Ti in CassiopeiaA. Nature 506:339–342. https://doi.org/10.1038/nature12997. arXiv:1403.4978

Grefenstette BW, Fryer CL, Harrison FA, Boggs SE, DeLaney T, Laming JM, Reynolds SP, Alexander DM, Barret D, Christensen FE, Craig WW, Forster K, Giommi P, Hailey CJ, Hornstrup A, Kitaguchi T, Koglin JE, Lopez L, Mao PH, Madsen KK, Miyasaka H, Mori K, Perri M, Pivovaroff MJ, Puccetti S, Rana V, Stern D, Westergaard NJ, Wik DR, Zhang WW, Zoglauer A (2017) The distribution of radioactive ^{44}Ti in Cassiopeia A. Astrophys J 834:19. https://doi.org/10.3847/1538-4357/834/1/19. arXiv:1612.02774

Greiner J, Mazzali PA, Kann DA, Krühler T, Pian E, Prentice S, Olivares E F, Rossi A, Klose S, Taubenberger S, Knust F, Afonso PMJ, Ashall C, Bolmer J, Delvaux C, Diehl R, Elliott J, Filgas R, Fynbo JPU, Graham JF, Guelbenzu AN, Kobayashi S, Leloudas G, Savaglio S, Schady P, Schmidl S, Schweyer T, Sudilovsky V, Tanga M, Updike AC, van Eerten H, Varela K (2015) A very luminous magnetar-powered supernova associated with an ultra-long γ-ray burst. Nature 523:189–192. https://doi.org/10.1038/nature14579. arXiv:1509.03279

Gyürky G, Kiss GG, Elekes Z, Fülöp Z, Somorjai E, Palumbo A, Görres J, Lee HY, Rapp W, Wiescher M, Özkan N, Güray RT, Efe G, Rauscher T (2006) α-induced cross sections of Cd106 for the astrophysical p process. Phys Rev C 74(2):025,805–+. https://doi.org/10.1103/PhysRevC.74.025805. arXiv:nucl-ex/0605034

Halevi G, Mösta P (2018) r-Process nucleosynthesis from three-dimensional jet-driven core-collapse supernovae with magnetic misalignments. Mon Not R Astron Soc 477:2366–2375. https://doi.org/10.1093/mnras/sty797. arXiv:1801.08943

Hamuy M (2003) Observed and physical properties of core-collapse supernovae. Astrophys J 582:905–914. https://doi.org/10.1086/344689. arXiv:astro-ph/0209174

Harris JA, Hix WR, Chertkow MA, Lee CT, Lentz EJ, Messer OEB (2017) Implications for post-processing nucleosynthesis of core-collapse supernova models with lagrangian particles. Astrophys J 843:2. https://doi.org/10.3847/1538-4357/aa76de. arXiv:1701.08876

Haxton WC, Parker PD, Rolfs CE (2006) Solar hydrogen burning and neutrinos. Nucl Phys A 777:226–253. https://doi.org/10.1016/j.nuclphysa.2005.02.088. arXiv:nucl-th/0501020
Heger A (2018) Private communication
Heger A, Langer N (2000) Presupernova evolution of rotating massive stars. II. Evolution of the surface properties. Astrophys J 544:1016–1035. https://doi.org/10.1086/317239. arXiv:astro-ph/0005110
Heger A, Woosley SE (2002) The nucleosynthetic signature of population III. Astrophys J 567:532–543. https://doi.org/10.1086/338487. arXiv:astro-ph/0107037
Heger A, Woosley SE (2010) Nucleosynthesis and evolution of massive metal-free stars. Astrophys J 724:341–373. https://doi.org/10.1088/0004-637X/724/1/341. arXiv:0803.3161
Heger A, Langer N, Woosley SE (2000) Presupernova evolution of rotating massive stars. I. Numerical method and evolution of the internal stellar structure. Astrophys J 528:368–396. https://doi.org/10.1086/308158. arXiv:astro-ph/9904132
Heger A, Langanke K, Martínez-Pinedo G, Woosley SE (2001a) Presupernova collapse models with improved weak-interaction rates. Phys Rev Lett 86:1678–1681. https://doi.org/10.1103/PhysRevLett.86.1678. arXiv:astro-ph/0007412
Heger A, Woosley SE, Martínez-Pinedo G, Langanke K (2001b) Presupernova evolution with improved rates for weak interactions. Astrophys J 560:307–325. https://doi.org/10.1086/324092. arXiv:astro-ph/0011507
Heger A, Fryer CL, Woosley SE, Langer N, Hartmann DH (2003) How massive single stars end their life. Astrophys J 591:288–300. https://doi.org/10.1086/375341. arXiv:astro-ph/0212469
Heger A, Woosley SE, Spruit HC (2005) Presupernova evolution of differentially rotating massive stars including magnetic fields. Astrophys J 626:350–363. https://doi.org/10.1086/429868. arXiv:astro-ph/0409422
Hempel M, Schaffner-Bielich J (2010) A statistical model for a complete supernova equation of state. Nucl Phys A 837:210–254. https://doi.org/10.1016/j.nuclphysa.2010.02.010. arXiv:0911.4073
Hempel M, Heinimann O, Yudin A, Iosilevskiy I, Liebendörfer M, Thielemann FK (2016) Hot third family of compact stars and the possibility of core-collapse supernova explosions. Phys Rev D 94(10):103001. https://doi.org/10.1103/PhysRevD.94.103001. arXiv:1511.06551
Herant M, Benz W, Hix WR, Fryer CL, Colgate SA (1994) Inside the supernova: a powerful convective engine. Astrophys J 435:339–361. https://doi.org/10.1086/174817. arXiv:astro-ph/9404024
Hirschi R (2007) Very low-metallicity massive stars: pre-SN evolution models and primary nitrogen production. Astron Astrophys 461:571–583. https://doi.org/10.1051/0004-6361:20065356. arXiv:astro-ph/0608170
Hirschi R, Meynet G, Maeder A (2004) Stellar evolution with rotation. XII. Pre-supernova models. Astron Astrophys 425:649–670. https://doi.org/10.1051/0004-6361:20041095. arXiv:astro-ph/0406552
Hirschi R, Meynet G, Maeder A (2005) Stellar evolution with rotation. XIII. Predicted GRB rates at various Z. Astron Astrophys 443:581–591. https://doi.org/10.1051/0004-6361:20053329. arXiv:astro-ph/0507343
Hirschi R, Frischknecht U, Thielemann F, Pignatari M, Chiappini C, Ekström S, Meynet G, Maeder A (2008) Stellar evolution in the early universe. In: Hunt LK, Madden S, Schneider R (ed) IAU symposium, vol 255, pp 297–304. https://doi.org/10.1017/S1743921308024976
Hix WR, Meyer BS (2006) Thermonuclear kinetics in astrophysics. Nucl Phys A 777:188–207. https://doi.org/10.1016/j.nuclphysa.2004.10.009. arXiv:astro-ph/0509698
Hix WR, Thielemann F (1996) Silicon burning. I. Neutronization and the physics of quasi-equilibrium. Astrophys J 460:869–+. https://doi.org/10.1086/177016. arXiv:astro-ph/9511088
Hix WR, Thielemann F (1999a) Computational methods for nucleosynthesis and nuclear energy generation. J Comput Appl Math 109:321–351. arXiv:astro-ph/9906478
Hix WR, Thielemann F (1999b) Silicon burning. II. Quasi-equilibrium and explosive burning. Astrophys J 511:862–875. https://doi.org/10.1086/306692. arXiv:astro-ph/9808203

Hix WR, Messer OE, Mezzacappa A, Liebendörfer M, Sampaio J, Langanke K, Dean DJ, Martínez-Pinedo G (2003) Consequences of nuclear electron capture in core collapse supernovae. Phys Rev Lett 91(20):201,102–+. https://doi.org/10.1103/PhysRevLett.91.201102. arXiv:astro-ph/0310883

Hix WR, Parete-Koon ST, Freiburghaus C, Thielemann F (2007) The QSE-reduced nuclear reaction network for silicon burning. Astrophys J 667:476–488. https://doi.org/10.1086/520672

Hix WR, Lentz EJ, Bruenn SW, Mezzacappa A, Messer OEB, Endeve E, Blondin JM, Harris JA, Marronetti P, Yakunin KN (2016) The multi-dimensional character of core-collapse supernovae. Acta Phys Pol B 47:645. https://doi.org/10.5506/APhysPolB.47.645. arXiv:1602.05553

Hoffman RD, Woosley SE, Qian Y (1997) Nucleosynthesis in neutrino-driven winds. II. Implications for heavy element synthesis. Astrophys J 482:951–+. https://doi.org/10.1086/304181. arXiv:astro-ph/9611097

Hoffman RD, Woosley SE, Weaver TA, Rauscher T, Thielemann FK (1999) The reaction rate sensitivity of nucleosynthesis in type II supernovae. Astrophys J 521:735–752. https://doi.org/10.1086/307568. arXiv:astro-ph/9809240

Honda S, Aoki W, Ishimaru Y, Wanajo S, Ryan SG (2006) Neutron-capture elements in the very metal poor star HD 122563. Astrophys J 643:1180–1189. https://doi.org/10.1086/503195. arXiv:astro-ph/0602107

Hüdepohl L, Müller B, Janka H, Marek A, Raffelt GG (2010) Neutrino signal of electron-capture supernovae from core collapse to cooling. Phys Rev Lett 104(25):251,101–+. https://doi.org/10.1103/PhysRevLett.104.251101. arXiv:0912.0260

Iliadis C (2007) Nuclear physics of stars. Wiley-VCH Verlag, Weinheim. ISBN 978-3-527-40602-9

Iwakami W, Kotake K, Ohnishi N, Yamada S, Sawada K (2008) Three-dimensional simulations of standing accretion shock instability in core-collapse supernovae. Astrophys J 678:1207–1222. https://doi.org/10.1086/533582. arXiv:0710.2191

Iyudin AF, Diehl R, Bloemen H, Hermsen W, Lichti GG, Morris D, Ryan J, Schoenfelder V, Steinle H, Varendorff M, de Vries C, Winkler C (1994) COMPTEL observations of Ti-44 gamma-ray line emission from CAS A. Astron Astrophys 284:L1–L4

Iyudin AF, Schönfelder V, Bennett K, Bloemen H, Diehl R, Hermsen W, Knödlseder J, Lichti GG, Oberlack U, Ryan J, Strong AW, Winkler C (1999) COMPTEL all-sky survey in ^{44}Ti line emission. Astrophys Lett Commun 38:383–+

Janiuk A (2014) Nucleosynthesis of elements in gamma-ray burst engines. Astron Astrophys 568:A105. https://doi.org/10.1051/0004-6361/201423822. arXiv:1406.4440

Janka HT (2012) Explosion mechanisms of core-collapse supernovae. Annu Rev Nucl Part Sci 62:407–451. https://doi.org/10.1146/annurev-nucl-102711-094901. arXiv:1206.2503

Janka HT, Müller B, Kitaura FS, Buras R (2008) Dynamics of shock propagation and nucleosynthesis conditions in O-Ne-Mg core supernovae. Astron Astrophys 485:199–208. https://doi.org/10.1051/0004-6361:20079334. arXiv:0712.4237

Janka HT, Melson T, Summa A (2016) Physics of core-collapse supernovae in three dimensions: a sneak preview. Annu Rev Nucl Part Sci 66:341–375. https://doi.org/10.1146/annurev-nucl-102115-044747. arXiv:1602.05576

Jones S, Hirschi R, Nomoto K, Fischer T, Timmes FX, Herwig F, Paxton B, Toki H, Suzuki T, Martínez-Pinedo G, Lam YH, Bertolli MG (2013) Advanced burning stages and fate of 8-10 M $_{\odot}$ stars. Astrophys J 772:150. https://doi.org/10.1088/0004-637X/772/2/150. arXiv:1306.2030

Jones S, Hirschi R, Pignatari M, Heger A, Georgy C, Nishimura N, Fryer C, Herwig F (2015) Code dependencies of pre-supernova evolution and nucleosynthesis in massive stars: evolution to the end of core helium burning. Mon Not R Astron Soc 447:3115–3129. https://doi.org/10.1093/mnras/stu2657. arXiv:1412.6518

Jones S, Ritter C, Herwig F, Fryer C, Pignatari M, Bertolli MG, Paxton B (2016) H ingestion into He-burning convection zones in super-AGB stellar models as a potential site for intermediate neutron-density nucleosynthesis. Mon Not R Astron Soc 455:3848–3863. https://doi.org/10.1093/mnras/stv2488. arXiv:1510.07417

Jose J (2016) Stellar explosions: hydrodynamics and nucleosynthesis. CRC Press, Boca Raton. https://doi.org/10.1201/b19165

Juodagalvis A, Langanke K, Hix WR, Martínez-Pinedo G, Sampaio JM (2010) Improved estimate of electron capture rates on nuclei during stellar core collapse. Nucl Phys A 848:454–478. https://doi.org/10.1016/j.nuclphysa.2010.09.012. arXiv:0909.0179

Just O, Bauswein A, Pulpillo RA, Goriely S, Janka HT (2015) Comprehensive nucleosynthesis analysis for ejecta of compact binary mergers. Mon Not R Astron Soc 448:541–567. https://doi.org/10.1093/mnras/stv009. arXiv:1406.2687

Käppeler F, Gallino R, Bisterzo S, Aoki W (2011) The s process: Nuclear physics, stellar models, and observations. Rev Mod Phys 83:157–194. https://doi.org/10.1103/RevModPhys.83.157. arXiv:1012.5218

Karakas AI, Lattanzio JC (2014) The Dawes review 2: nucleosynthesis and stellar yields of low- and intermediate-mass single stars. Publ Astron Soc Aust 31:e030. https://doi.org/10.1017/pasa.2014.21. arXiv:1405.0062

Karlsson T, Bromm V, Bland-Hawthorn J (2013) Pregalactic metal enrichment: the chemical signatures of the first stars. Rev Mod Phys 85:809–848. https://doi.org/10.1103/RevModPhys.85.809. arXiv:1101.4024

Kasen D, Thomas RC, Röpke F, Woosley SE (2008) Multidimensional radiative transfer calculations of the light curves and spectra of type Ia supernovae. J Phys Conf Ser 125:012007. https://doi.org/10.1088/1742-6596/125/1/012007

Kasen D, Woosley SE, Heger A (2011) Pair instability supernovae: light curves, spectra, and shock breakout. Astrophys J 734:102. https://doi.org/10.1088/0004-637X/734/2/102. arXiv:1101.3336

Kasen D, Fernández R, Metzger BD (2015) Kilonova light curves from the disc wind outflows of compact object mergers. Mon Not R Astron Soc 450:1777–1786. https://doi.org/10.1093/mnras/stv721. arXiv:1411.3726

Katsuda S, Tsunemi H, Mori K (2009) Is Vela Jr. a young supernova remnant? Adv Space Res 43:895–899. https://doi.org/10.1016/j.asr.2009.01.004

Kelic A, Ricciardi MV, Schmidt KH (2008) New insight into the fission process from experiments with relativistic heavy-ion beams. In: Kliman J, Itkis MG, Gmuca Š (eds) Dynamical aspects of nuclear fission, pp 203–215. https://doi.org/10.1142/9789812837530_0016

Kerzendorf WE, Sim SA (2014) A spectral synthesis code for rapid modelling of supernovae. Mon Not R Astron Soc 440:387–404. https://doi.org/10.1093/mnras/stu055. arXiv:1401.5469

Kippenhahn R, Weigert A (1994) Stellar structure and evolution. Springer, Berlin

Kiss GG, Gyürky G, Elekes Z, Fülöp Z, Somorjai E, Rauscher T, Wiescher M (2007) Ge70(p,γ)As71 and Ge76(p,n)As76 cross sections for the astrophysical p process: sensitivity of the optical proton potential at low energies. Phys Rev C 76(5):055,807–+. https://doi.org/10.1103/PhysRevC.76.055807. arXiv:0711.1079

Kiss GG, Rauscher T, Gyürky G, Simon A, Fülöp Z, Somorjai E (2008) Coulomb suppression of the stellar enhancement factor. Phys Rev Lett 101(19):191,101–+. https://doi.org/10.1103/PhysRevLett.101.191101. arXiv:0809.2676

Kitaura FS, Janka H, Hillebrandt W (2006) Explosions of O-Ne-Mg cores, the crab supernova, and subluminous type II-P supernovae. Astron Astrophys 450:345–350. https://doi.org/10.1051/0004-6361:20054703. arXiv:astro-ph/0512065

Knie K, Korschinek G, Faestermann T, Dorfi EA, Rugel G, Wallner A (2004) ^{60}Fe anomaly in a deep-sea manganese crust and implications for a nearby supernova source. Phys Rev Lett 93(17):171,103–+. https://doi.org/10.1103/PhysRevLett.93.171103

Kobayashi C, Nakasato N (2011) Chemodynamical simulations of the milky way galaxy. Astrophys J 729:16. https://doi.org/10.1088/0004-637X/729/1/16. arXiv:1012.5144

Kobayashi C, Umeda H, Nomoto K, Tominaga N, Ohkubo T (2006) Galactic chemical evolution: carbon through zinc. Astrophys J 653:1145–1171. https://doi.org/10.1086/508914. arXiv:astro-ph/0608688

Kolbe E, Langanke K, Martínez-Pinedo G, Vogel P (2003) Neutrino nucleus reactions and nuclear structure. J Phys G Nucl Phys 29:2569–2596. https://doi.org/10.1088/0954-3899/29/11/010. arXiv:nucl-th/0311022

Korobkin O, Rosswog S, Arcones A, Winteler C (2012) On the astrophysical robustness of the neutron star merger r-process. Mon Not R Astron Soc 426:1940–1949. https://doi.org/10.1111/j.1365-2966.2012.21859.x. arXiv:1206.2379

Kotake K, Sato K, Takahashi K (2006) Explosion mechanism, neutrino burst and gravitational wave in core-collapse supernovae. Rep Prog Phys 69:971–1143. https://doi.org/10.1088/0034-4885/69/4/R03. arXiv:astro-ph/0509456

Kratz K, Bitouzet J, Thielemann F, Moeller P, Pfeiffer B (1993) Isotopic r-process abundances and nuclear structure far from stability - implications for the r-process mechanism. Astrophys J 403:216–238. https://doi.org/10.1086/172196

Kratz K, Farouqi K, Pfeiffer B, Truran JW, Sneden C, Cowan JJ (2007) Explorations of the r-processes: comparisons between calculations and observations of low-metallicity stars. Astrophys J 662:39–52. https://doi.org/10.1086/517495. arXiv:astro-ph/0703091

Kratz KL, Farouqi K, Möller P (2014) A high-entropy-wind r-process study based on nuclear-structure quantities from the new finite-range droplet model FRDM(2012). Astrophys J 792:6. https://doi.org/10.1088/0004-637X/792/1/6. arXiv:1406.2529

Kurfess JD, Johnson WN, Kinzer RL, Kroeger RA, Strickman MS, Grove JE, Leising MD, Clayton DD, Grabelsky DA, Purcell WR, Ulmer MP, Cameron RA, Jung GV (1992) Oriented scintillation spectrometer experiment observations of Co-57 in SN 1987A. Astrophys J 399:L137–L140. https://doi.org/10.1086/186626

Kuroda T, Kotake K, Takiwaki T, Thielemann FK (2018) A full general relativistic neutrino radiation-hydrodynamics simulation of a collapsing very massive star and the formation of a black hole. ArXiv e-prints 1801.01293

Langanke K, Martínez-Pinedo G (2000) Shell-model calculations of stellar weak interaction rates: II. Weak rates for nuclei in the mass range/A=45-65 in supernovae environments. Nucl Phys A 673:481–508. https://doi.org/10.1016/S0375-9474(00)00131-7. arXiv:nucl-th/0001018

Langanke K, Martínez-Pinedo G (2001) Rate tables for the weak processes of pf-SHELL nuclei in stellar environments. At Data Nucl Data Tables 79:1–46. https://doi.org/10.1006/adnd.2001.0865

Langanke K, Martínez-Pinedo G (2003) Nuclear weak-interaction processes in stars. Rev Mod Phys 75:819–862. https://doi.org/10.1103/RevModPhys.75.819. arXiv:nucl-th/0203071

Langanke K, Martínez-Pinedo G, Sampaio JM, Dean DJ, Hix WR, Messer OE, Mezzacappa A, Liebendörfer M, Janka H, Rampp M (2003) Electron capture rates on nuclei and implications for stellar core collapse. Phys Rev Lett 90(24):241,102–+. https://doi.org/10.1103/PhysRevLett.90.241102. arXiv:astro-ph/0302459

Langanke K, Martínez-Pinedo G, Müller B, Janka H, Marek A, Hix WR, Juodagalvis A, Sampaio JM (2008) Effects of inelastic neutrino-nucleus scattering on supernova dynamics and radiated neutrino spectra. Phys Rev Lett 100(1):011,101–+. https://doi.org/10.1103/PhysRevLett.100.011101. arXiv:0706.1687

Langer N (2012) Presupernova evolution of massive single and binary stars. Annu Rev Astron Astrophys 50:107–164. https://doi.org/10.1146/annurev-astro-081811-125534. arXiv:1206.5443

Langer N, Braun H, Fliegner J (1995) The production of circumstellar ^{26}Al by massive stars. Astrophys Space Sci 224:275–278. https://doi.org/10.1007/BF00667858

Lattimer JM, Douglas Swesty F (1991) A generalized equation of state for hot, dense matter. Nucl Phys A 535:331–376. https://doi.org/10.1016/0375-9474(91)90452-C

Lattimer JM, Schramm DN (1974) Black-hole-neutron-star collisions. Astrophys J 192:L145–L147. https://doi.org/10.1086/181612

Lattimer JM, Schramm DN (1976) The tidal disruption of neutron stars by black holes in close binaries. Astrophys J 210:549–567. https://doi.org/10.1086/154860

Leibundgut B, Suntzeff NB (2003) Optical light curves of supernovae. In: Weiler K (ed) Supernovae and gamma-ray bursters. Lecture notes in physics, vol 598. Springer, Berlin, pp 77–90

Leising MD, Share GH (1990) The gamma-ray light curves of SN 1987A. Astrophys J 357:638–648. https://doi.org/10.1086/168952

Lentz EJ, Bruenn SW, Hix WR, Mezzacappa A, Messer OEB, Endeve E, Blondin JM, Harris JA, Marronetti P, Yakunin KN (2015) Three-dimensional core-collapse supernova simulated using a 15 $M_\odot$ Progenitor. Astrophys J 807:L31. https://doi.org/10.1088/2041-8205/807/2/L31. arXiv:1505.05110

Liebendörfer M (2005) A simple parameterization of the consequences of deleptonization for simulations of stellar core collapse. Astrophys J 633:1042–1051. https://doi.org/10.1086/466517. arXiv:astro-ph/0504072

Liebendörfer M, Mezzacappa A, Thielemann FK (2001) Conservative general relativistic radiation hydrodynamics in spherical symmetry and comoving coordinates. Phys Rev D 63(10):104,003–+. https://doi.org/10.1103/PhysRevD.63.104003. arXiv:astro-ph/0012201

Liebendörfer M, Mezzacappa A, Messer OEB, Martinez-Pinedo G, Hix WR, Thielemann F (2003) The neutrino signal in stellar core collapse and postbounce evolution. Nucl Phys A 719:144–+. https://doi.org/10.1016/S0375-9474(03)00984-9. arXiv:astro-ph/0211329

Liebendörfer M, Messer OEB, Mezzacappa A, Bruenn SW, Cardall CY, Thielemann F (2004) A finite difference representation of neutrino radiation hydrodynamics in spherically symmetric general relativistic spacetime. Astrophys J Suppl 150:263–316. https://doi.org/10.1086/380191. arXiv:astro-ph/0207036

Liebendörfer M, Rampp M, Janka H, Mezzacappa A (2005) Supernova simulations with Boltzmann neutrino transport: a comparison of methods. Astrophys J 620:840–860. https://doi.org/10.1086/427203. arXiv:astro-ph/0310662

Liebendörfer M, Fischer T, Fröhlich C, Thielemann F, Whitehouse S (2008) Nuclear physics with spherically symmetric supernova models. J Phys G Nucl Phys 35(1):014,056–+. https://doi.org/10.1088/0954-3899/35/1/014056. arXiv:0708.4296

Liebendörfer M, Whitehouse SC, Fischer T (2009) The isotropic diffusion source approximation for supernova neutrino transport. Astrophys J 698:1174–1190. https://doi.org/10.1088/0004-637X/698/2/1174. arXiv:0711.2929

Liebendörfer M, Fischer T, Hempel M, Käppeli R, Pagliara G, Perego A, Sagert I, Schaffner-Bielich J, Scheidegger S, Thielemann F, Whitehouse SC (2010) Neutrino radiation-hydrodynamics: general relativistic versus multidimensional supernova simulations. Prog Theor Phys Suppl 186:87–92. https://doi.org/10.1143/PTPS.186.87

Limongi M, Chieffi A (2003) Evolution, explosion, and nucleosynthesis of core-collapse supernovae. Astrophys J 592:404–433. https://doi.org/10.1086/375703. arXiv:astro-ph/0304185

Limongi M, Chieffi A (2006a) Nucleosynthesis of ^{60}Fe in massive stars. New Astron Rev 50:474–476. https://doi.org/10.1016/j.newar.2006.06.005. arXiv:astro-ph/0512598

Limongi M, Chieffi A (2006b) The nucleosynthesis of ^{26}Al and ^{60}Fe in solar metallicity stars extending in mass from 11 to 120 $M_\odot$: the hydrostatic and explosive contributions. Astrophys J 647:483–500. https://doi.org/10.1086/505164. arXiv:astro-ph/0604297

Limongi M, Chieffi A (2009) Presupernova evolution and explosion of massive stars: the role of mass loss during the Wolf-Rayet stage. Mem Soc Astron Ital 80:151

Limongi M, Chieffi A (2012) Presupernova evolution and explosive nucleosynthesis of zero metal massive stars. Astrophys J Suppl 199:38. https://doi.org/10.1088/0067-0049/199/2/38. arXiv:1202.4581

Limongi M, Straniero O, Chieffi A (2000) Massive stars in the range 13-25 M_{solar}: evolution and nucleosynthesis. II. The solar metallicity models. Astrophys J Suppl 129:625–664. https://doi.org/10.1086/313424. arXiv:astro-ph/0003401

Lippuner J, Roberts LF (2017) SkyNet: A modular nuclear reaction network library. Astrophys J Suppl 233:18. https://doi.org/10.3847/1538-4365/aa94cb. arXiv:1706.06198

Liu N, Nittler LR, Pignatari M, O'D Alexander CM, Wang J (2017) Stellar origin of ^{15}N-rich presolar SiC grains of type AB: supernovae with explosive hydrogen burning. Astrophys J 842:L1. https://doi.org/10.3847/2041-8213/aa74e5. arXiv:1705.08222

Lodders K, Palme H (2009) Solar system elemental abundances in 2009. Meteor Planet Sci Suppl 72:5154

Lundqvist P, Kozma C, Sollerman J, Fransson C (2001) ISO/SWS observations of SN 1987A. II. A refined upper limit on the mass of ^{44}Ti in the ejecta of SN 1987A. Astron Astrophys 374:629–637. https://doi.org/10.1051/0004-6361:20010725. arXiv:astro-ph/0105402

MacFadyen AI, Woosley SE (1999) Collapsars: gamma-ray bursts and explosions in "Failed supernovae". Astrophys J 524:262–289. https://doi.org/10.1086/307790. arXiv:astro-ph/9810274

MacFadyen AI, Woosley SE, Heger A (2001) Supernovae, jets, and collapsars. Astrophys J 550:410–425. https://doi.org/10.1086/319698. arXiv:astro-ph/9910034

Maeder A (2009) Physics, formation and evolution of rotating stars. Springer, Berlin. https://doi.org/10.1007/978-3-540-76949-1

Maeder A, Meynet G (2012) Rotating massive stars: from first stars to gamma ray bursts. Rev Mod Phys 84:25–63. https://doi.org/10.1103/RevModPhys.84.25

Maeder A, Grebel EK, Mermilliod J (1999) Differences in the fractions of Be stars in galaxies. Astron Astrophys 346:459–464. arXiv:astro-ph/9904008

Maeder A, Przybilla N, Nieva MF, Georgy C, Meynet G, Ekström S, Eggenberger P (2014) Evolution of surface CNO abundances in massive stars. Astron Astrophys 565:A39. https://doi.org/10.1051/0004-6361/201220602. arXiv:1404.1020

Magkotsios G, Timmes FX, Hungerford AL, Fryer CL, Young PA, Wiescher M (2010) Trends in ^{44}Ti and ^{56}Ni from core-collapse supernovae. Astrophys J Suppl 191:66–95. https://doi.org/10.1088/0067-0049/191/1/66. arXiv:1009.3175

Marek A, Janka H (2009) Delayed neutrino-driven supernova explosions aided by the standing accretion-shock instability. Astrophys J 694:664–696. https://doi.org/10.1088/0004-637X/694/1/664. arXiv:0708.3372

Marek A, Janka H, Buras R, Liebendörfer M, Rampp M (2005) On ion-ion correlation effects during stellar core collapse. Astron Astrophys 443:201–210. https://doi.org/10.1051/0004-6361:20053236. arXiv:astro-ph/0504291

Marketin T, Huther L, Martínez-Pinedo G (2016) Large-scale evaluation of β -decay rates of r -process nuclei with the inclusion of first-forbidden transitions. Phys Rev C 93(2):025805. https://doi.org/10.1103/PhysRevC.93.025805. arXiv:1507.07442

Martayan C, Frémat Y, Hubert A, Floquet M, Zorec J, Neiner C (2007) Effects of metallicity, star-formation conditions, and evolution in B and Be stars. II. Small magellanic cloud, field of NGC330. Astron Astrophys 462:683–694. https://doi.org/10.1051/0004-6361:20065076. arXiv:astro-ph/0609677

Martin D, Perego A, Arcones A, Thielemann FK, Korobkin O, Rosswog S (2015) Neutrino-driven winds in the aftermath of a neutron star merger: nucleosynthesis and electromagnetic transients. Astrophys J 813:2. https://doi.org/10.1088/0004-637X/813/1/2. arXiv:1506.05048

Martin D, Perego A, Kastaun W, Arcones A (2017) The role of weak interactions in dynamic ejecta from binary neutron star mergers. ArXiv e-prints 1710.04900

Martínez-Pinedo G, Liebendörfer M, Frekers D (2006) Nuclear input for core-collapse models. Nucl Phys A 777:395–423. https://doi.org/10.1016/j.nuclphysa.2006.02.014. arXiv:astro-ph/0412091

Martínez-Pinedo G, Fischer T, Lohs A, Huther L (2012) Charged-Current weak interaction processes in hot and dense matter and its impact on the spectra of neutrinos emitted from protoneutron star cooling. Phys Rev Lett 109(25):251104. https://doi.org/10.1103/PhysRevLett.109.251104. arXiv:1205.2793

Martins F, Palacios A (2013) A comparison of evolutionary tracks for single galactic massive stars. Astron Astrophys 560:A16. https://doi.org/10.1051/0004-6361/201322480. arXiv:1310.7218

Mashonkina L, Christlieb N, Eriksson K (2014) The hamburg/ESO R-process enhanced star survey (HERES). X. HE 2252-4225, one more r-process enhanced and actinide-boost halo star. Astron Astrophys 569:A43. https://doi.org/10.1051/0004-6361/201424017. arXiv:1407.5379

Massi F, de Luca M, Elia D, Giannini T, Lorenzetti D, Nisini B (2007) Star formation in the Vela molecular ridge. Large scale mapping of cloud D in the mm continuum. Astron Astrophys 466:1013–1023. https://doi.org/10.1051/0004-6361:20066438. arXiv:astro-ph/0702687

Matheson T, Filippenko AV, Li W, Leonard DC, Shields JC (2001) Optical spectroscopy of type IB/C supernovae. Astron J 121:1648–1675. https://doi.org/10.1086/319390. arXiv:astro-ph/0101119

Matz SM, Share GH, Leising MD, Chupp EL, Vestrand WT (1988) Gamma-ray line emission from SN1987A. Nature 331:416–418. https://doi.org/10.1038/331416a0

McCray R, Fransson C (2016) The remnant of supernova 1987A. Annu Rev Astron Astrophys 54:19–52. https://doi.org/10.1146/annurev-astro-082615-105405

McKinney JC, Tchekhovskoy A, Blandford RD (2013) Alignment of magnetized accretion disks and relativistic jets with spinning black holes. Science 339:49. https://doi.org/10.1126/science.1230811. arXiv:1211.3651

Meakin CA, Arnett D (2007) Turbulent convection in stellar interiors. I. Hydrodynamic simulation. Astrophys J 667:448–475. https://doi.org/10.1086/520318. arXiv:astro-ph/0611315

Melson T, Janka HT, Bollig R, Hanke F, Marek A, Müller B (2015) Neutrino-driven explosion of a 20 solar-mass star in three dimensions enabled by strange-quark contributions to neutrino-nucleon scattering. Astrophys J 808:L42. https://doi.org/10.1088/2041-8205/808/2/L42. arXiv:1504.07631

Mendoza-Temis JdJ, Wu MR, Langanke K, Martínez-Pinedo G, Bauswein A, Janka HT (2015) Nuclear robustness of the r process in neutron-star mergers. Phys Rev C 92(5):055805. https://doi.org/10.1103/PhysRevC.92.055805

Metzger BD (2017a) Kilonovae. Living Rev Relat 20:3. https://doi.org/10.1007/s41114-017-0006-z. arXiv:1610.09381

Metzger BD (2017b) Welcome to the multi-messenger era! lessons from a neutron star merger and the landscape ahead. arXiv:1710.05931

Metzger BD, Martínez-Pinedo G, Darbha S, Quataert E, Arcones A, Kasen D, Thomas R, Nugent P, Panov IV, Zinner NT (2010) Electromagnetic counterparts of compact object mergers powered by the radioactive decay of r-process nuclei. Mon Not R Astron Soc 406:2650–2662. https://doi.org/10.1111/j.1365-2966.2010.16864.x. arXiv:1001.5029

Meyer BS, Mathews GJ, Howard WM, Woosley SE, Hoffman RD (1992) R-process nucleosynthesis in the high-entropy supernova bubble. Astrophys J 399:656–664. https://doi.org/10.1086/171957

Meynet G, Maeder A (2000) Stellar evolution with rotation. V. Changes in all the outputs of massive star models. Astron Astrophys 361:101–120. arXiv:astro-ph/0006404

Meynet G, Maeder A (2005) Stellar evolution with rotation. XI. Wolf-Rayet star populations at different metallicities. Astron Astrophys 429:581–598. https://doi.org/10.1051/0004-6361:20047106. arXiv:astro-ph/0408319

Meynet G, Maeder A (2017) Supernovae from rotating stars. ArXiv e-prints 1711.07740

Meynet G, Arnould M, Prantzos N, Paulus G (1997) Contribution of Wolf-Rayet stars to the synthesis of ^{26}Al. I. The γ-ray connection. Astron Astrophys 320:460–468

Meynet G, Ekström S, Maeder A (2006) The early star generations: the dominant effect of rotation on the CNO yields. Astron Astrophys 447:623–639. https://doi.org/10.1051/0004-6361:20053070. arXiv:astro-ph/0510560

Meynet G, Ekström S, Georgy C, Maeder A, Hirschi R (2008) Massive star evolution: from the early to the present day universe. In: Deng L, Chan KL (ed) IAU symposium, vol 252, pp 317–327. https://doi.org/10.1017/S1743921308023119

Meynet G, Maeder A, Eggenberger P, Ekstrom S, Georgy C, Chiappini C, Privitera G, Choplin A (2016) Impact of rotation on stellar models. Astron Nachr 337:827. https://doi.org/10.1002/asna.201612380. arXiv:1512.00767

Mikheyev SP, Smirnov AY (1985) Resonance enhancement of oscillations in matter and solar neutrino spectroscopy. Yad Fiz Sov J Nucl Phys 42:1441–1448, 913

Mirizzi A, Mangano G, Saviano N (2015) Self-induced flavor instabilities of a dense neutrino stream in a two-dimensional model. Phys Rev D 92(2):021702. https://doi.org/10.1103/PhysRevD.92.021702. arXiv:1503.03485

Mirizzi A, Tamborra I, Janka HT, Saviano N, Scholberg K, Bollig R, Hüdepohl L, Chakraborty S (2016) Supernova neutrinos: production, oscillations and detection. Nuovo Cimento Rivista Serie 39:1–112. https://doi.org/10.1393/ncr/i2016-10120-8. arXiv:1508.00785

Mishenina T, Pignatari M, Côté B, Thielemann FK, Soubiran C, Basak N, Gorbaneva T, Korotin SA, Kovtyukh VV, Wehmeyer B, Bisterzo S, Travaglio C, Gibson BK, Jordan C, Paul A, Ritter C, Herwig F, NuGrid Collaboration (2017) Observing the metal-poor solar neighbourhood: a comparison of galactic chemical evolution predictions. Mon Not R Astron Soc 469:4378–4399. https://doi.org/10.1093/mnras/stx1145. arXiv:1705.03642

Mochizuki Y, Takahashi K, Janka HT, Hillebrandt W, Diehl R (1999) 44Ti: its effective decay rate in young supernova remnants, and its abundance in Cassiopeia A. Astron Astrophys 346:831–842. arXiv:astro-ph/9904378

Moe M, Di Stefano R (2017) Mind your Ps and Qs: the interrelation between period (P) and mass-ratio (Q) distributions of binary stars. Astrophys J Suppl 230:15. https://doi.org/10.3847/1538-4365/aa6fb6. arXiv:1606.05347

Möller P, Pfeiffer B, Kratz KL (2003) New calculations of gross β-decay properties for astrophysical applications: speeding-up the classical r process. Phys Rev C 67(5):055802. https://doi.org/10.1103/PhysRevC.67.055802

Möller P, Myers WD, Sagawa H, Yoshida S (2012) New finite-range droplet mass model and equation-of-state parameters. Phys Rev Lett 108(5):052501. https://doi.org/10.1103/PhysRevLett.108.052501

Möller P, Sierk AJ, Ichikawa T, Sagawa H (2016) Nuclear ground-state masses and deformations: FRDM(2012). At Data Nucl Data Tables 109:1–204. https://doi.org/10.1016/j.adt.2015.10.002. arXiv:1508.06294

Moriya TJ, Tominaga N, Langer N, Nomoto K, Blinnikov SI, Sorokina EI (2014) Electron-capture supernovae exploding within their progenitor wind. Astron Astrophys 569:A57. https://doi.org/10.1051/0004-6361/201424264. arXiv:1407.4563

Mösta P, Richers S, Ott CD, Haas R, Piro AL, Boydstun K, Abdikamalov E, Reisswig C, Schnetter E (2014) Magnetorotational core-collapse supernovae in three dimensions. Astrophys J 785:L29. https://doi.org/10.1088/2041-8205/785/2/L29. arXiv:1403.1230

Mösta P, Ott CD, Radice D, Roberts LF, Schnetter E, Haas R (2015) A large-scale dynamo and magnetoturbulence in rapidly rotating core-collapse supernovae. Nature 528:376–379. https://doi.org/10.1038/nature15755. arXiv:1512.00838

Mösta P, Roberts LF, Halevi G, Ott CD, Lippuner J, Haas R, Schnetter E (2017) R-process nucleosynthesis from three-dimensional magnetorotational core-collapse supernovae. ArXiv e-prints 1712.09370

Müller B, Heger A, Liptai D, Cameron JB (2016a) A simple approach to the supernova progenitor-explosion connection. Mon Not R Astron Soc 460:742–764. https://doi.org/10.1093/mnras/stw1083. arXiv:1602.05956

Müller B, Viallet M, Heger A, Janka HT (2016b) The last minutes of oxygen shell burning in a massive star. Astrophys J 833:124. https://doi.org/10.3847/1538-4357/833/1/124. arXiv:1605.01393

Müller B, Wanajo S, Janka HT, Heger A, Gay D, Sim SA (2017) Simulations of electron capture and low-mass iron core supernovae. ArXiv e-prints 1710.02641

Nagataki S (2000) Effects of jetlike explosion in SN 1987A. Astrophys J Suppl 127:141–157. https://doi.org/10.1086/313317. arXiv:astro-ph/9907109

Nagataki S (2011) Grb-Sn connection:. central engine of long GRBs and explosive nucleosynthesis. Int J Mod Phys D 20:1975–1978. https://doi.org/10.1142/S0218271811020032

Nagataki S, Hashimoto MA, Sato K, Yamada S, Mochizuki YS (1998) The high ratio of 44Ti/56Ni in Cassiopeia A and the axisymmetric collapse-driven supernova explosion. Astrophys J 492:L45+. https://doi.org/10.1086/311089. arXiv:astro-ph/9807015

Nagataki S, Takahashi R, Mizuta A, Takiwaki T (2007) Numerical study of gamma-ray burst jet formation in collapsars. Astrophys J 659:512–529. https://doi.org/10.1086/512057. arXiv:astro-ph/0608233

Nakamura K, Takiwaki T, Kuroda T, Kotake K (2015) Systematic features of axisymmetric neutrino-driven core-collapse supernova models in multiple progenitors. Publ Astron Soc Jpn 67:107. https://doi.org/10.1093/pasj/psv073. arXiv:1406.2415

Nakamura T, Umeda H, Nomoto K, Thielemann F, Burrows A (1999) Nucleosynthesis in type II supernovae and the abundances in metal-poor stars. Astrophys J 517:193–208. https://doi.org/10.1086/307167. arXiv:astro-ph/9809307

Nakamura T, Umeda H, Iwamoto K, Nomoto K, Hashimoto M, Hix WR, Thielemann F (2001) Explosive nucleosynthesis in hypernovae. Astrophys J 555:880–899. https://doi.org/10.1086/321495. arXiv:astro-ph/0011184

Nakar E, Piran T (2017) The observable signatures of GRB cocoons. Astrophys J 834:28. https://doi.org/10.3847/1538-4357/834/1/28. arXiv:1610.05362

Nishimura N, Takiwaki T, Thielemann FK (2015) The r-process nucleosynthesis in the various jet-like explosions of magnetorotational core-collapse supernovae. Astrophys J 810:109. https://doi.org/10.1088/0004-637X/810/2/109. arXiv:1501.06567

Nishimura N, Hirschi R, Rauscher T, St J Murphy A, Cescutti G (2017a) Uncertainties in s-process nucleosynthesis in massive stars determined by Monte Carlo variations. Mon Not R Astron Soc 469:1752–1767. https://doi.org/10.1093/mnras/stx696. arXiv:1701.00489

Nishimura N, Sawai H, Takiwaki T, Yamada S, Thielemann FK (2017b) The Intermediate r-process in core-collapse supernovae driven by the magneto-rotational instability. Astrophys J 836:L21. https://doi.org/10.3847/2041-8213/aa5dee. arXiv:1611.02280

Nomoto K (1987) Evolution of 8-10 solar mass stars toward electron capture supernovae. II - collapse of an O + NE + MG core. Astrophys J 322:206–214. https://doi.org/10.1086/165716

Nomoto K (2017) Nucleosynthesis in hypernovae associated with gamma ray bursts. In: Alsabti AW, Murdin P (eds) Handbook of supernovae. Springer, Berlin. https://doi.org/10.1007/978-3-319-20794-0_86-1

Nomoto K, Hashimoto M (1988) Presupernova evolution of massive stars. Phys Rep 163:13–36. https://doi.org/10.1016/0370-1573(88)90032-4

Nomoto K, Thielemann F, Miyaji S (1985) The triple alpha reaction at low temperatures in accreting white dwarfs and neutron stars. Astron Astrophys 149:239–245

Nomoto K, Mazzali PA, Nakamura T, Iwamoto K, Danziger IJ, Patat F (2001) The properties of hypernovae: SNe Ic 1998bw, 1997ef, and SN IIn 1997cy. In: Livio M, Panagia N, Sahu (eds) Supernovae and gamma-ray bursts: the greatest explosions since the big bang, pp 144–170

Nomoto K, Tominaga N, Umeda H, Kobayashi C, Maeda K (2006) Nucleosynthesis yields of core-collapse supernovae and hypernovae, and galactic chemical evolution. Nucl Phys A 777:424–458. https://doi.org/10.1016/j.nuclphysa.2006.05.008. arXiv:astro-ph/0605725

Nomoto K, Tanaka M, Tominaga N, Maeda K (2010) Hypernovae, gamma-ray bursts, and first stars. New A Rev 54:191–200. https://doi.org/10.1016/j.newar.2010.09.022

Nomoto K, Kobayashi C, Tominaga N (2013) Nucleosynthesis in stars and the chemical enrichment of galaxies. Annu Rev Astron Astrophys 51:457–509. https://doi.org/10.1146/annurev-astro-082812-140956

O'Connor E, Ott CD (2011) Black hole formation in failing core-collapse supernovae. Astrophys J 730:70. https://doi.org/10.1088/0004-637X/730/2/70. arXiv:1010.5550

O'Connor E, Ott CD (2013) The progenitor dependence of the pre-explosion neutrino emission in core-collapse supernovae. Astrophys J 762:126. https://doi.org/10.1088/0004-637X/762/2/126. arXiv:1207.1100

Oda T, Hino M, Muto K, Takahara M, Sato K (1994) Rate tables for the weak processes of sd-shell nuclei in stellar matter. At Data Nucl Data Tables 56:231–403. https://doi.org/10.1006/adnd.1994.1007

Oertel M, Hempel M, Klähn T, Typel S (2017) Equations of state for supernovae and compact stars. Rev Mod Phys 89(1):015007. https://doi.org/10.1103/RevModPhys.89.015007. arXiv:1610.03361

Ohkubo T, Umeda H, Maeda K, Nomoto K, Suzuki T, Tsuruta S, Rees MJ (2008) Evolution of core-collapse very massive population III stars. In: O'Shea BW, Heger A (eds) First stars III. American institute of physics conference series, vol 990, pp 244–246. https://doi.org/10.1063/1.2905553

Ono M, Hashimoto M, Fujimoto S, Kotake K, Yamada S (2012) Explosive nucleosynthesis in magnetohydrodynamical jets from collapsars. II — heavy-element nucleosynthesis of s, p, r-processes. Prog Theor Phys 128:741–765. arXiv:1203.6488

Palacios A, Meynet G, Vuissoz C, Knödlseder J, Schaerer D, Cerviño M, Mowlavi N (2005) New estimates of the contribution of Wolf-Rayet stellar winds to the galactic ^{26}Al. Astron Astrophys 429:613–624. https://doi.org/10.1051/0004-6361:20041757. arXiv:astro-ph/0409580

Pan KC, Liebendörfer M, Hempel M, Thielemann FK (2016) Two-dimensional core-collapse supernova simulations with the isotropic diffusion source approximation for neutrino transport. Astrophys J 817:72. https://doi.org/10.3847/0004-637X/817/1/72. arXiv:1505.02513

Pan KC, Liebendörfer M, Couch SM, Thielemann FK (2017) Equation of state dependent dynamics and multimessenger signals from stellar-mass black hole formation. arXiv:1710.01690

Panov IV, Korneev IY, Thielemann FK (2008) The r-Process in the region of transuranium elements and the contribution of fission products to the nucleosynthesis of nuclei with A $\leq$ 130. Astron Lett 34:189–197. https://doi.org/10.1007/s11443-008-3006-1

Paxton B, Bildsten L, Dotter A, Herwig F, Lesaffre P, Timmes F (2011) Modules for experiments in stellar astrophysics (MESA). Astrophys J Suppl 192:3. https://doi.org/10.1088/0067-0049/192/1/3. arXiv:1009.1622

Perego A, Rosswog S, Cabezón RM, Korobkin O, Käppeli R, Arcones A, Liebendörfer M (2014) Neutrino-driven winds from neutron star merger remnants. Mon Not R Astron Soc 443:3134–3156. https://doi.org/10.1093/mnras/stu1352. arXiv:1405.6730

Perego A, Hempel M, Fröhlich C, Ebinger K, Eichler M, Casanova J, Liebendörfer M, Thielemann FK (2015) PUSHing core-collapse supernovae to explosions in spherical symmetry i: the model and the case of SN 1987A. Astrophys J 806:275. https://doi.org/10.1088/0004-637X/806/2/275. arXiv:1501.02845

Perego A, Cabezón RM, Käppeli R (2016) An advanced leakage scheme for neutrino treatment in astrophysical simulations. Astrophys J Suppl 223:22. https://doi.org/10.3847/0067-0049/223/2/22. arXiv:1511.08519

Pettini M, Zych BJ, Steidel CC, Chaffee FH (2008) C, N, O abundances in the most metal-poor damped Lyman alpha systems. Mon Not R Astron Soc 385:2011–2024. https://doi.org/10.1111/j.1365-2966.2008.12951.x. arXiv:0712.1829

Pignatari M, Gallino R, Meynet G, Hirschi R, Herwig F, Wiescher M (2008) The s-process in massive stars at low metallicity: the effect of primary ^{14}N from fast rotating stars. Astrophys J 687:L95–L98. https://doi.org/10.1086/593350. arXiv:0810.0182

Pignatari M, Herwig F, Hirschi R, Bennett M, Rockefeller G, Fryer C, Timmes FX, Ritter C, Heger A, Jones S, Battino U, Dotter A, Trappitsch R, Diehl S, Frischknecht U, Hungerford A, Magkotsios G, Travaglio C, Young P (2016) NuGrid stellar data set. I.Stellar yields from H to Bi for stars with metallicities Z = 0.02 and Z = 0.01. Astrophys J Suppl 225:24. https://doi.org/10.3847/0067-0049/225/2/24. arXiv:1307.6961

Piran T (2004) The physics of gamma-ray bursts. Rev Mod Phys 76:1143–1210. https://doi.org/10.1103/RevModPhys.76.1143. arXiv:astro-ph/0405503

Pruet J, Woosley SE, Buras R, Janka H, Hoffman RD (2005) Nucleosynthesis in the hot convective bubble in core-collapse supernovae. Astrophys J 623:325–336. https://doi.org/10.1086/428281. arXiv:astro-ph/0409446

Pruet J, Hoffman RD, Woosley SE, Janka H, Buras R (2006) Nucleosynthesis in early supernova winds. II. The role of neutrinos. Astrophys J 644:1028–1039. https://doi.org/10.1086/503891. arXiv:astro-ph/0511194

Qian YZ, Wasserburg GJ (2007) Where, oh where has the r-process gone? Phys Rep 442:237–268. https://doi.org/10.1016/j.physrep.2007.02.006. arXiv:0708.1767

Qian YZ, Woosley SE (1996) Nucleosynthesis in neutrino-driven winds. I. The physical conditions. Astrophys J 471:331–+. https://doi.org/10.1086/177973. arXiv:astro-ph/9611094

Ramirez-Ruiz E, Trenti M, MacLeod M, Roberts LF, Lee WH, Saladino-Rosas MI (2015) Compact stellar binary assembly in the first nuclear star clusters and r-process synthesis in the early universe. Astrophys J 802:L22. https://doi.org/10.1088/2041-8205/802/2/L22. arXiv:1410.3467

Rapp W, Görres J, Wiescher M, Schatz H, Käppeler F (2006) Sensitivity of p-process nucleosynthesis to nuclear reaction rates in a 25 M_{solar} supernova model. Astrophys J 653:474–489. https://doi.org/10.1086/508402. arXiv:astro-ph/0608341

Rauscher T (2006) Branchings in the γ process path revisited. Phys Rev C 73(1):015,804–+. https://doi.org/10.1103/PhysRevC.73.015804. arXiv:astro-ph/0510710

Rauscher T (2013) Solution of the α-potential mystery in the γ process and its impact on the Nd/Sm ratio in meteorites. Phys Rev Lett 111(6):061104. https://doi.org/10.1103/PhysRevLett.111.061104. arXiv:1307.4921

Rauscher T, Heger A, Hoffman RD, Woosley SE (2002) Nucleosynthesis in massive stars with improved nuclear and stellar physics. Astrophys J 576:323–348. https://doi.org/10.1086/341728. arXiv:astro-ph/0112478

Rauscher T, Dauphas N, Dillmann I, Fröhlich C, Fülöp Z, Gyürky G (2013) Constraining the astrophysical origin of the p-nuclei through nuclear physics and meteoritic data. Rep Prog Phys 76(6):066201. https://doi.org/10.1088/0034-4885/76/6/066201. arXiv:1303.2666

Rayet M, Arnould M, Prantzos N (1990) The p-process revisited. Astron Astrophys 227:271–281

Rayet M, Arnould M, Hashimoto M, Prantzos N, Nomoto K (1995) The p-process in type II supernovae. Astron Astrophys 298:517–+

Renaud M, Vink J, Decourchelle A, Lebrun F, Terrier R, Ballet J (2006) An INTEGRAL/IBIS view of young galactic SNRs through the ^{44}Ti gamma-ray lines. New Astron Rev 50:540–543. https://doi.org/10.1016/j.newar.2006.06.061. arXiv:astro-ph/0602304

Roberts LF, Woosley SE, Hoffman RD (2010) Integrated nucleosynthesis in neutrino-driven winds. Astrophys J 722:954–967. https://doi.org/10.1088/0004-637X/722/1/954. arXiv:1004.4916

Roberts LF, Reddy S, Shen G (2012) Medium modification of the charged-current neutrino opacity and its implications. Phys Rev C 86(6):065803. https://doi.org/10.1103/PhysRevC.86.065803. arXiv:1205.4066

Rosswog S, Korobkin O, Arcones A, Thielemann FK, Piran T (2014) The long-term evolution of neutron star merger remnants - I. The impact of r-process nucleosynthesis. Mon Not R Astron Soc 439:744–756. https://doi.org/10.1093/mnras/stt2502. arXiv:1307.2939

Rugel G, Faestermann T, Knie K, Korschinek G, Poutivtsev M, Schumann D, Kivel N, Günther-Leopold I, Weinreich R, Wohlmuther M (2009) New measurement of the Fe60 half-life. Phys Rev Lett 103(7):072,502–+. https://doi.org/10.1103/PhysRevLett.103.072502

Sagert I, Fischer T, Hempel M, Pagliara G, Schaffner-Bielich J, Mezzacappa A, Thielemann F, Liebendörfer M (2009) Signals of the QCD phase transition in core-collapse supernovae. Phys Rev Lett 102(8):081101. https://doi.org/10.1103/PhysRevLett.102.081101. arXiv:0809.4225

Sallaska AL, Iliadis C, Champange AE, Goriely S, Starrfield S, Timmes FX (2013) STARLIB: A next-generation reaction-rate library for nuclear astrophysics. Astrophys J Suppl 207:18. https://doi.org/10.1088/0067-0049/207/1/18. arXiv:1304.7811

Sampaio JM, Langanke K, Martínez-Pinedo G, Kolbe E, Dean DJ (2003) Electron capture rates for core collapse supernovae. Nucl Phys A 718:440–442. https://doi.org/10.1016/S0375-9474(03)00832-7. arXiv:nucl-th/0209057

Sauer DN, Mazzali PA, Deng J, Valenti S, Nomoto K, Filippenko AV (2006) The properties of the 'standard' type Ic supernova 1994I from spectral models. Mon Not R Astron Soc 369:1939–1948. https://doi.org/10.1111/j.1365-2966.2006.10438.x. arXiv:astro-ph/0604293

Sawyer RF (2005) Effects of ion and electron correlations on neutrino scattering in the infall phase of a supernova. Phys Lett B 630:1–6. https://doi.org/10.1016/j.physletb.2005.09.032. arXiv:astro-ph/0505520

Schatz H, Aprahamian A, Goerres J, Wiescher M, Rauscher T, Rembges JF, Thielemann F, Pfeiffer B, Moeller P, Kratz K, Herndl H, Brown BA, Rebel H (1998) rp-Process nucleosynthesis at extreme temperature and density conditions. Phys Rep 294:167–264. https://doi.org/10.1016/S0370-1573(97)00048-3

Schulreich MM, Breitschwerdt D, Feige J, Dettbarn C (2017) Numerical studies on the link between radioisotopic signatures on Earth and the formation of the local bubble. I. ^{60}Fe transport to the solar system by turbulent mixing of ejecta from nearby supernovae into a locally homogeneous interstellar medium. Astron Astrophys 604:A81. https://doi.org/10.1051/0004-6361/201629837. arXiv:1704.08221

Seitenzahl IR, Taubenberger S, Sim SA (2009) Late-time supernova light curves: the effect of internal conversion and Auger electrons. Mon Not R Astron Soc 400:531–535. https://doi.org/10.1111/j.1365-2966.2009.15478.x. arXiv:0908.0247

Seitenzahl I, Timmes F, Magkotsios G (2012) Sn87a late lightcurve analysis. Astrophys J 792(1):7

Sekiguchi Y, Shibata M (2011) Formation of black hole and accretion disk in a massive high-entropy stellar core collapse. Astrophys J 737:6. https://doi.org/10.1088/0004-637X/737/1/6. arXiv:1009.5303

Shibagaki S, Kajino T, Mathews GJ, Chiba S, Nishimura S, Lorusso G (2016) Relative contributions of the weak, main, and fission-recycling r-process. Astrophys J 816:79. https://doi.org/10.3847/0004-637X/816/2/79. arXiv:1505.02257

Siegert T, Diehl R, Krause MGH, Greiner J (2015) Revisiting INTEGRAL/SPI observations of ^{44}Ti from Cassiopeia A. Astron Astrophys 579:A124. https://doi.org/10.1051/0004-6361/201525877. arXiv:1505.05999

Sim SA, Kromer M, Roepke FK, Sorokina EI, Blinnikov SI, Kasen D, Hillebrandt W (2009) Monte Carlo radiative transfer simulations: applications to astrophysical outflows and explosions. ArXiv e-prints 0911.1549

Sinha S, Fröhlich C, Ebinger K, Perego A, Hempel M, Eichler M, Liebendörfer M, Thielemann FK (2017) PUSHing core-collapse supernovae to explosions in spherical symmetry: nucleosynthesis yields. In: Kubono S, Kajino T, Nishimura S, Isobe T, Nagataki S, Shima T, Takeda Y (eds) 14th international symposium on nuclei in the cosmos (NIC2016), p 020608. https://doi.org/10.7566/JPSCP.14.020608. arXiv:1701.05203

Slane P, Hughes JP, Edgar RJ, Plucinsky PP, Miyata E, Tsunemi H, Aschenbach B (2001) RX J0852.0-4622: another nonthermal shell-type supernova remnant (G266.2-1.2). Astrophys J 548:814–819. https://doi.org/10.1086/319033. arXiv:astro-ph/0010510

Sneden C, McWilliam A, Preston GW, Cowan JJ, Burris DL, Armosky BJ (1996) The ultra–metal-poor, neutron-capture–rich giant star CS 22892-052. Astrophys J 467:819. https://doi.org/10.1086/177656

Sneden C, Cowan JJ, Gallino R (2008) Neutron-capture elements in the early galaxy. Annu Rev Astron Astrophys 46:241–288. https://doi.org/10.1146/annurev.astro.46.060407.145207

Sneden C, Cowan JJ, Kobayashi C, Pignatari M, Lawler JE, Den Hartog EA, Wood MP (2016) Iron-group abundances in the metal-poor main-sequence turnoff star HD˜84937. Astrophys J 817:53. https://doi.org/10.3847/0004-637X/817/1/53. arXiv:1511.05985

Sollerman J, Holland ST, Challis P, Fransson C, Garnavich P, Kirshner RP, Kozma C, Leibundgut B, Lundqvist P, Patat F, Filippenko AV, Panagia N, Wheeler JC (2002) Supernova 1998bw - the final phases. Astron Astrophys 386:944–956. https://doi.org/10.1051/0004-6361:20020326. arXiv:astro-ph/0204498

Spite M, Cayrel R, Plez B, Hill V, Spite F, Depagne E, François P, Bonifacio P, Barbuy B, Beers T, Andersen J, Molaro P, Nordström B, Primas F (2005) First stars VI - abundances of C, N, O, Li, and mixing in extremely metal-poor giants. Galactic evolution of the light elements. Astron Astrophys 430:655–668. https://doi.org/10.1051/0004-6361:20041274. arXiv:astro-ph/0409536

Sukhbold T, Ertl T, Woosley SE, Brown JM, Janka HT (2016) Core-collapse supernovae from 9 to 120 solar masses based on neutrino-powered explosions. Astrophys J 821:38. https://doi.org/10.3847/0004-637X/821/1/38. arXiv:1510.04643

Sukhbold T, Woosley S, Heger A (2017) High resolution study of presupernova compactness. arXiv:1710.03243

Sullivan C, O'Connor E, Zegers RGT, Grubb T, Austin SM (2016) The sensitivity of core-collapse supernovae to nuclear electron capture. Astrophys J 816:44. https://doi.org/10.3847/0004-637X/816/1/44. arXiv:1508.07348

Sumiyoshi K, Terasawa M, Mathews GJ, Kajino T, Yamada S, Suzuki H (2001) r-Process in prompt supernova explosions revisited. Astrophys J 562:880–886. https://doi.org/10.1086/323524. arXiv:astro-ph/0106407

Sumiyoshi K, Yamada S, Suzuki H (2007) Dynamics and neutrino signal of black hole formation in nonrotating failed supernovae. I. Equation of state dependence. Astrophys J 667:382–394. https://doi.org/10.1086/520876. arXiv:0706.3762

Suntzeff NB, Bouchet P (1990) The bolometric light curve of SN 1987A. I - Results from ESO and CTIO U to Q0 photometry. Astron J 99:650–663. https://doi.org/10.1086/115358

Surman R, McLaughlin GC, Hix WR (2006) Nucleosynthesis in the outflow from gamma-ray burst accretion disks. Astrophys J 643:1057–1064. https://doi.org/10.1086/501116. arXiv:astro-ph/0509365

Takahashi K, Witti J, Janka H (1994) Nucleosynthesis in neutrino-driven winds from protoneutron stars II. The r-process. Astron Astrophys 286:857–869

Takiwaki T, Kotake K, Suwa Y (2014) A comparison of two- and three-dimensional neutrino-hydrodynamics simulations of core-collapse supernovae. Astrophys J 786:83. https://doi.org/10.1088/0004-637X/786/2/83. arXiv:1308.5755

Tanvir NR, Levan AJ, Fruchter AS, Hjorth J, Hounsell RA, Wiersema K, Tunnicliffe RL (2013) A 'kilonova' associated with the short-duration γ-ray burst GRB 130603B. Nature 500:547–549. https://doi.org/10.1038/nature12505. arXiv:1306.4971

The L, El Eid MF, Meyer BS (2007) s-Process nucleosynthesis in advanced burning phases of massive stars. Astrophys J 655:1058–1078. https://doi.org/10.1086/509753. arXiv:astro-ph/0609788

The LS, Clayton DD, Diehl R, Hartmann DH, Iyudin AF, Leising MD, Meyer BS, Motizuki Y, Schönfelder V (2006) Are ^{44}Ti-producing supernovae exceptional? Astron Astrophys 450:1037–1050. https://doi.org/10.1051/0004-6361:20054626. arXiv:astro-ph/0601039

Thielemann FK, Arnett WD (1985) Hydrostatic nucleosynthesis - part two - core neon to silicon burning and presupernova abundance yields of massive stars. Astrophys J 295:604–+. https://doi.org/10.1086/163403

Thielemann FK, Nomoto K, Yokoi K (1986) Explosive nucleosynthesis in carbon deflagration models of Type I supernovae. Astron Astrophys 158:17–33

Thielemann F, Hashimoto M, Nomoto K (1990) Explosive nucleosynthesis in SN 1987A. II - composition, radioactivities, and the neutron star mass. Astrophys J 349:222–240. https://doi.org/10.1086/168308

Thielemann FK, Nomoto K, Hashimoto MA (1996) Core-collapse supernovae and their ejecta. Astrophys J 460:408–+. https://doi.org/10.1086/176980

Thielemann F, Hauser P, Kolbe E, Martinez-Pinedo G, Panov I, Rauscher T, Kratz K, Pfeiffer B, Rosswog S, Liebendörfer M, Mezzacappa A (2002) Heavy elements and age determinations. Space Sci Rev 100:277–296

Thielemann FK, Eichler M, Panov IV, Pignatrari M, Wehmeyer B (2017a) Making the heaviest elements in a rare class of supernovae. In: Alsabti AW, Murdin P (eds) Handbook of supernovae. Springer, Berlin. https://doi.org/10.1007/978-3-319-20794-0_81-1

Thielemann FK, Eichler M, Panov IV, Wehmeyer B (2017b) Neutron star mergers and nucleosynthesis of heavy elements. Annu Rev Nucl Part Sci 67. https://doi.org/10.1146/annurev-nucl-101916-123246. arXiv:1710.02142

Timmes FX (1999) Integration of nuclear reaction networks for stellar hydrodynamics. Astrophys J Suppl 124:241–263. https://doi.org/10.1086/313257

Tominaga N, Umeda H, Nomoto K (2007) Supernova nucleosynthesis in population III 13-50 M_{solar} stars and abundance patterns of extremely metal-poor stars. Astrophys J 660:516–540. https://doi.org/10.1086/513063

Travaglio C, Gallino R, Arnone E, Cowan J, Jordan F, Sneden C (2004) Galactic evolution of Sr, Y, and Zr: a multiplicity of nucleosynthetic processes. Astrophys J 601:864–884. https://doi.org/10.1086/380507. arXiv:astro-ph/0310189

Tsygankov SS, Krivonos RA, Lutovinov AA, Revnivtsev MG, Churazov EM, Sunyaev RA, Grebenev SA (2016) Galactic survey of ^{44}Ti sources with the IBIS telescope onboard INTEGRAL. Mon Not R Astron Soc 458:3411–3419. https://doi.org/10.1093/mnras/stw549. arXiv:1603.01264

Tueller J, Barthelmy S, Gehrels N, Teegarden BJ, Leventhal M, MacCallum CJ (1990) Observations of gamma-ray line profiles from SN 1987A. Astrophys J 351:L41–L44. https://doi.org/10.1086/185675

Tur C, Heger A, Austin SM (2010) Production of ^{26}Al, ^{44}Ti, and ^{60}Fe in core-collapse supernovae: sensitivity to the rates of the triple alpha and ^{12}C(α, γ)^{16}O reactions. Astrophys J 718:357–367. https://doi.org/10.1088/0004-637X/718/1/357. arXiv:0908.4283

Uberseder E, Reifarth R, Schumann D, Dillmann I, Pardo CD, Görres J, Heil M, Käppeler F, Marganiec J, Neuhausen J, Pignatari M, Voss F, Walter S, Wiescher M (2009) Measurement of the Fe60(n,γ)^{61}Fe cross section at stellar temperatures. Phys Rev Lett 102(15):151,101–+. https://doi.org/10.1103/PhysRevLett.102.151101

Ugliano M, Janka HT, Marek A, Arcones A (2012) Progenitor-explosion connection and remnant birth masses for neutrino-driven supernovae of iron-core progenitors. Astrophys J 757:69. https://doi.org/10.1088/0004-637X/757/1/69. arXiv:1205.3657

Umeda H, Nomoto K (2005) Variations in the abundance pattern of extremely metal-poor stars and nucleosynthesis in population III supernovae. Astrophys J 619:427–445. https://doi.org/10.1086/426097. arXiv:astro-ph/0308029

Umeda H, Nomoto K (2008) How much ^{56}Ni can be produced in core-collapse supernovae? Evolution and explosions of 30-100 M_{solar} stars. Astrophys J 673:1014–1022. https://doi.org/10.1086/524767. arXiv:0707.2598

Vink J (2005) Gamma-ray observations of explosive nucleosynthesis products. Adv Space Res 35:976–986. https://doi.org/10.1016/j.asr.2005.01.097. arXiv:astro-ph/0501645

Vink J, Laming JM, Kaastra JS, Bleeker JAM, Bloemen H, Oberlack U (2001) Detection of the 67.9 and 78.4 keV lines associated with the radioactive decay of ^{44}Ti in Cassiopeia A. Astrophys J 560:L79–L82. https://doi.org/10.1086/324172. arXiv:astro-ph/0107468

Walder R, Burrows A, Ott CD, Livne E, Lichtenstadt I, Jarrah M (2005) Anisotropies in the neutrino fluxes and heating profiles in two-dimensional, time-dependent, multigroup radiation hydrodynamics simulations of rotating core-collapse supernovae. Astrophys J 626:317–332. https://doi.org/10.1086/429816. arXiv:astro-ph/0412187

Wallace RK, Woosley SE (1981) Explosive hydrogen burning. Astrophys J Suppl 45:389–420. https://doi.org/10.1086/190717

Wallner A, Faestermann T, Feige J, Feldstein C, Knie K, Korschinek G, Kutschera W, Ofan A, Paul M, Quinto F, Rugel G, Steier P (2015) Abundance of live ^{244}Pu in deep-sea reservoirs on earth points to rarity of actinide nucleosynthesis. Nat Commun 6:5956. https://doi.org/10.1038/ncomms6956. arXiv:1509.08054

Wanajo S (2006) The rp-process in neutrino-driven winds. Astrophys J 647:1323–1340. https://doi.org/10.1086/505483. arXiv:astro-ph/0602488

Wanajo S, Nomoto K, Janka H, Kitaura FS, Müller B (2009) Nucleosynthesis in electron capture supernovae of asymptotic giant branch stars. Astrophys J 695:208–220. https://doi.org/10.1088/0004-637X/695/1/208. arXiv:0810.3999

Wanajo S, Janka HT, Müller B (2011) Electron-capture supernovae as the origin of elements beyond iron. Astrophys J 726:L15. https://doi.org/10.1088/2041-8205/726/2/L15. arXiv:1009.1000

Wanajo S, Sekiguchi Y, Nishimura N, Kiuchi K, Kyutoku K, Shibata M (2014) Production of all the r-process nuclides in the dynamical ejecta of neutron star mergers. Astrophys J 789:L39. https://doi.org/10.1088/2041-8205/789/2/L39. arXiv:1402.7317

Wang W, Harris MJ, Diehl R, Halloin H, Cordier B, Strong AW, Kretschmer K, Knödlseder J, Jean P, Lichti GG, Roques JP, Schanne S, von Kienlin A, Weidenspointner G, Wunderer C (2007) SPI observations of the diffuse ^{60}Fe emission in the galaxy. Astron Astrophys 469:1005–1012. https://doi.org/10.1051/0004-6361:20066982. arXiv:0704.3895

Wang W, Lang MG, Diehl R, Halloin H, Jean P, Knödlseder J, Kretschmer K, Martin P, Roques JP, Strong AW, Winkler C, Zhang XL (2009) Spectral and intensity variations of Galactic ^{26}Al emission. Astron Astrophys 496:713–724. https://doi.org/10.1051/0004-6361/200811175. arXiv:0902.0211

Weaver TA, Woosley SE (1980) Evolution and explosion of massive stars. In: Ehlers J, Perry JJ, Walker M (ed) Ninth texas symposium on relativistic astrophysics. N Y Acad Sci Ann, 336:335–357. https://doi.org/10.1111/j.1749-6632.1980.tb15942.x

Weber C, Elomaa V, Ferrer R, Fröhlich C, Ackermann D, Äystö J, Audi G, Batist L, Blaum K, Block M, Chaudhuri A, Dworschak M, Eliseev S, Eronen T, Hager U, Hakala J, Herfurth F, Heßberger FP, Hofmann S, Jokinen A, Kankainen A, Kluge H, Langanke K, Martín A, Martínez-Pinedo G, Mazzocco M, Moore ID, Neumayr JB, Novikov YN, Penttilä H, Plaß WR, Popov AV, Rahaman S, Rauscher T, Rauth C, Rissanen J, Rodríguez D, Saastamoinen A, Scheidenberger C, Schweikhard L, Seliverstov DM, Sonoda T, Thielemann F, Thirolf PG, Vorobjev GK (2008) Mass measurements in the vicinity of the rp-process and the νp-process paths with the penning trap facilities JYFLTRAP and SHIPTRAP. Phys Rev C 78(5):054,310–+. https://doi.org/10.1103/PhysRevC.78.054310. arXiv:0808.4065

Wehmeyer B, Pignatari M, Thielemann FK (2015) Galactic evolution of rapid neutron capture process abundances: the inhomogeneous approach. Mon Not R Astron Soc 452:1970–1981. https://doi.org/10.1093/mnras/stv1352. arXiv:1501.07749

West C, Heger A (2018, in preparation)

Wiescher M, Görres J, Pignatari M (2010) Experimental status of reactions in H- and He-burning. Annu Rev Nucl Part Sci 60:175–251. https://doi.org/10.1146/annurev.nucl.60.1.175

Wiescher M, Käppeler F, Langanke K (2012) Critical reactions in contemporary nuclear astrophysics. Annu Rev Astron Astrophys 50:165–210. https://doi.org/10.1146/annurev-astro-081811-125543

Wilson JR, Mayle RW (1993) Report on the progress of supernova research by the Livermore group. Phys Rep 227:97–111. https://doi.org/10.1016/0370-1573(93)90059-M

Winteler C, Käppeli R, Perego A, Arcones A, Vasset N, Nishimura N, Liebendörfer M, Thielemann FK (2012) Magnetorotationally driven supernovae as the origin of early galaxy r-process elements? Astrophys J 750:L22. https://doi.org/10.1088/2041-8205/750/1/L22. arXiv:1203.0616

Wolfenstein L (1978) Neutrino oscillations in matter. Phys Rev D 17:2369–2374. https://doi.org/10.1103/PhysRevD.17.2369

Wongwathanarat A, Müller E, Janka HT (2015) Three-dimensional simulations of core-collapse supernovae: from shock revival to shock breakout. Astron Astrophys 577:A48. https://doi.org/10.1051/0004-6361/201425025. arXiv:1409.5431

Wongwathanarat A, Janka HT, Müller E, Pllumbi E, Wanajo S (2017) Production and distribution of ^{44}Ti and ^{56}Ni in a three-dimensional supernova model resembling Cassiopeia A. Astrophys J 842:13. https://doi.org/10.3847/1538-4357/aa72de. arXiv:1610.05643

Woosley SE, Bloom JS (2006) The supernova gamma-ray burst connection. Annu Rev Astron Astrophys 44:507–556. https://doi.org/10.1146/annurev.astro.43.072103.150558. arXiv:astro-ph/0609142

Woosley SE, Heger A (2007) Nucleosynthesis and remnants in massive stars of solar metallicity. Phys Rep 442:269–283. https://doi.org/10.1016/j.physrep.2007.02.009. arXiv:astro-ph/0702176

Woosley SE, Heger A (2015a) The deaths of very massive stars. In: Vink JS (ed) Very massive stars in the local universe. Astrophysics and space science library, vol 412, p 199. https://doi.org/10.1007/978-3-319-09596-7_7. arXiv:1406.5657

Woosley SE, Heger A (2015b) The remarkable deaths of 9-11 solar mass stars. Astrophys J 810:34. https://doi.org/10.1088/0004-637X/810/1/34. arXiv:1505.06712

Woosley SE, Hoffman RD (1992) The alpha-process and the r-process. Astrophys J 395:202–239. https://doi.org/10.1086/171644

Woosley SE, Howard WM (1978) The p-process in supernovae. Astrophys J Suppl 36:285–304. https://doi.org/10.1086/190501

Woosley SE, Weaver TA (1994) Sub-Chandrasekhar mass models for type IA supernovae. Astrophys J 423:371–379. https://doi.org/10.1086/173813

Woosley SE, Weaver TA (1995) The evolution and explosion of massive stars. II. Explosive hydrodynamics and nucleosynthesis. Astrophys J Suppl 101:181–+. https://doi.org/10.1086/192237

Woosley SE, Arnett WD, Clayton DD (1973) The explosive burning of oxygen and silicon. Astrophys J Suppl 26:231–+. https://doi.org/10.1086/190282

Woosley SE, Wilson JR, Mathews GJ, Hoffman RD, Meyer BS (1994) The r-process and neutrino-heated supernova ejecta. Astrophys J 433:229–246. https://doi.org/10.1086/174638

Woosley SE, Heger A, Weaver TA (2002) The evolution and explosion of massive stars. Rev Mod Phys 74:1015–1071. https://doi.org/10.1103/RevModPhys.74.1015

Woosley SE, Kasen D, Blinnikov S, Sorokina E (2007) Type Ia supernova light curves. Astrophys J 662:487–503. https://doi.org/10.1086/513732. arXiv:astro-ph/0609562

Wu MR, Fischer T, Huther L, Martínez-Pinedo G, Qian YZ (2014) Impact of active-sterile neutrino mixing on supernova explosion and nucleosynthesis. Phys Rev D 89(6):061303. https://doi.org/10.1103/PhysRevD.89.061303. arXiv:1305.2382

Wu MR, Qian YZ, Martínez-Pinedo G, Fischer T, Huther L (2015) Effects of neutrino oscillations on nucleosynthesis and neutrino signals for an 18 M? supernova model. Phys Rev D 91(6):065016. https://doi.org/10.1103/PhysRevD.91.065016. arXiv:1412.8587

Wu MR, Fernández R, Martínez-Pinedo G, Metzger BD (2016a) Production of the entire range of r-process nuclides by black hole accretion disc outflows from neutron star mergers. Mon Not R Astron Soc 463:2323–2334. https://doi.org/10.1093/mnras/stw2156. arXiv:1607.05290

Wu MR, Martínez-Pinedo G, Qian YZ (2016b) Linking neutrino oscillations to the nucleosynthesis of elements. Eur Phys J Web Conf 109, 06005. https://doi.org/10.1051/epjconf/201610906005. arXiv:1512.03630

Xu Y, Takahashi K, Goriely S, Arnould M, Ohta M, Utsunomiya H (2013) NACRE II: an update of the NACRE compilation of charged-particle-induced thermonuclear reaction rates for nuclei with mass number A<16. Nucl Phys A 918:61–169. https://doi.org/10.1016/j.nuclphysa.2013.09.007. arXiv:1310.7099

Yalçın C, Güray RT, Özkan N, Kutlu S, Gyürky G, Farkas J, Kiss GG, Fülöp Z, Simon A, Somorjai E, Rauscher T (2009) Odd p isotope In113: measurement of α-induced reactions. Phys Rev C 79(6):065,801–+. https://doi.org/10.1103/PhysRevC.79.065801. arXiv:0906.4041

Yamamoto Y, Fujimoto Si, Nagakura H, Yamada S (2013) Post-shock-revival evolution in the neutrino-heating mechanism of core-collapse supernovae. Astrophys J 771:27. https://doi.org/10.1088/0004-637X/771/1/27. arXiv:1209.4824

Yamasaki T, Yamada S (2005) Effects of rotation on the revival of a stalled shock in supernova explosions. Astrophys J 623:1000–1010. https://doi.org/10.1086/428496. arXiv:astro-ph/0412625

Yoon SC (2015) Evolutionary models for type Ib/c supernova progenitors. Publ Astron Soc Aust 32:e015. https://doi.org/10.1017/pasa.2015.16. arXiv:1504.01205

Yoshida T, Suwa Y, Umeda H, Shibata M, Takahashi K (2017) Explosive nucleosynthesis of ultra-stripped Type Ic supernovae: application to light trans-iron elements. Mon Not R Astron Soc 471:4275–4285. https://doi.org/10.1093/mnras/stx1738. arXiv:1707.02685

Yusof N, Hirschi R, Meynet G, Crowther PA, Ekström S, Frischknecht U, Georgy C, Abu Kassim H, Schnurr O (2013) Evolution and fate of very massive stars. Mon Not R Astron Soc 433:1114–1132. https://doi.org/10.1093/mnras/stt794. arXiv:1305.2099

Chapter 5
Binary Systems and Their Nuclear Explosions

Jordi Isern, Margarita Hernanz, and Jordi José

5.1 Accretion onto Compact Objects and Thermonuclear Runaways

The nuclear energy supply of a typical star like the Sun would be $\sim 10^{52}$ erg if all the hydrogen could be incinerated into iron peak elements. Since the gravitational binding energy is $\sim 10^{49}$ erg, it is evident that the nuclear energy content is more than enough to blow up the Sun. However, stars are stable thanks to the fact that their matter obeys the equation of state of a classical ideal gas that acts as a thermostat: if some energy is released as a consequence of a thermal fluctuation, the gas expands, the temperature drops and the instability is quenched. The first researchers to discuss the scenario under which stars could explosively release their nuclear energy were Hoyle and Fowler (1960). They showed that this could occur under conditions of dynamic compression, as a consequence of collapse, or under electron degeneracy. They also pointed out in their seminal paper that hydrogen could only be responsible for mild explosions, like novae, as a consequence of the necessity to convert two protons into two neutrons, and that only the thermonuclear fusion of carbon could be energetic enough to feed a strong explosion. They did not consider helium because by this epoch the He-burning mechanism was not yet known.

J. Isern · M. Hernanz (✉)
Institute of Space Sciences (ICE, CSIC), Barcelona, Spain

Institut d'Estudis Espacials de Catalunya (IEEC), Barcelona, Spain
e-mail: isern@ieec.uab.es; hernanz@ieec.uab.es

J. José
Universitat Politècnica de Catalunya (UPC), Barcelona, Spain

Institut d'Estudis Espacials de Catalunya (IEEC), Barcelona, Spain
e-mail: jordi.jose@upc.edu

R. Diehl et al. (eds.), *Astrophysics with Radioactive Isotopes*, Astrophysics and Space Science Library 453, https://doi.org/10.1007/978-3-319-91929-4_5

Intermediate and low-mass stars ($M < 10$–$12\,M_{\odot}$) are able to get rid of their envelope and end their life as a white dwarf. On the contrary, massive stars form an iron core that grows until it reaches the Chandrasekhar mass and collapses to a neutron star or a black hole. The degenerate core of white dwarfs can have different chemical compositions, He, C-O or O-Ne, and different size depending on the mass and the single or binary nature of the progenitor. Single stars with masses in the range of 8–9 to 10–12 $M_{\odot}$ produce oxygen-neon cores, those in the mass range 0.5 to 8–9 $M_{\odot}$ produce carbon-oxygen cores, while stars with a mass in the range 0.08–0.5 $M_{\odot}$ produce helium cores, but the lifetime of such stars is so large that they cannot produce a white dwarf in a Hubble time. Members of a close binary system can be strongly perturbed by their companion and thereby produce different outcomes. For instance, stars with a mass of the order of 2.5 $M_{\odot}$ can end their life as He white dwarfs with a mass of the order of 0.4 $M_{\odot}$.

The destiny of isolated white dwarfs is to cool forever. However, if they are members of a close binary system, they can revive as a consequence of mass transfer from their companion. As the mass grows, the radius of the white dwarf shrinks and the density increases, as can be derived from simple dimensional arguments. The hydrostatic and the degenerate, non-relativistic electron pressures have the functional form $P \sim M^2 R^{-4}$, and $P \sim M^{5/3} R^{-5}$, respectively. Thus it is always possible to find an equilibrium configuration defined by a mass-radius relation $R \sim M^{-1/3}$. However, as the density grows, the Fermi energy increases and electrons become relativistic. In the extreme case, the electron pressure takes the form $P \sim M^{4/3} R^{-4}$ and, since it has the same dependence on R as hydrostatic pressure, there is no longer a definite lengthscale. Furthermore, according to the virial theorem, stars supported by relativistic particles are not gravitationally bound and the injection or removal of small amounts of energy can cause a large expansion or contraction of the star.

The behavior of the different cores depends on the net rate at which energy is injected by the burning front or removed by electron captures on the ashes that were left from the previous burning cycle. Both quantities depend on the chemical composition of the stellar cores. Helium cores always experience a thermonuclear explosion because of the large energy content and the extreme flammability of He. Carbon-oxygen cores can explode or collapse, depending on the ignition density (Canal et al. 1990). If this density is larger than some critical value, $\sim$(5–8) $\times$ 10^9 g/cm^3, the electron captures become dominant and they collapse to a neutron star (Bravo and García-Senz 1999). ONe-cores ignite at such a density that they always tend to collapse (Nomoto and Kondo 1991; Gutierrez et al. 1996) and Fe-cores always collapse because of their inability to release nuclear energy. Recently, however, 3D models of explosion have cast some doubts to this picture and the possibility that C-O and O-Ne-Mg cores could experience a mild explosion leaving a gravitationally bound remnant made of iron-peak elements has emerged.

5.1.1 Evolution of Degenerate Cores Before Ignition

The behavior of the white dwarf interior during the accretion phase depends on the competition between the physical processes that increase the temperature of the material (compression, nuclear reactions in the inner core and possible burning of the freshly accreted matter) and those that cool the star (via neutrino and photon losses). Since the energy transport is dominated by electron conduction, one of the relevant timescales is the time taken by a thermal signal to cross the star, given by Henyey and L'Ecuyer (1969):

$$\tau_{\rm TH} = \frac{3\kappa\rho^2 c_{\rm P}}{64\sigma T^3} l^2 \qquad (5.1)$$

where κ, ρ, T, σ and $c_{\rm P}$ have their usual meanings and l is the linear extent of the region considered, the radius of the white dwarf in this case.

The effects of the compression induced by the accreted mass can be separated into two terms (Nomoto 1982). The first term is due to the increase in density at a fixed mass fraction as a consequence of the increase in mass, and its effects are quite uniform throughout the whole star. The second term corresponds to compression as matter moves inward in mass fraction space. It is negligible in the inner, strongly degenerate regions, where the major part of the compression work is invested in increasing the Fermi energy of electrons, but is very large in the external semi-degenerate layers. This means that a thermal wave generates in the outer layers and propagates towards the interior. A rough estimate of the compression-induced luminosity is: $L_{\rm c}/L_\odot = 1.4 \times 10^{-3} T_7 \dot{M}_{10}$, where T_7 is the temperature in units of 10^7 K, and $\dot{M}_{10}$ is the mass accretion rate in units of 10^{-10} M$_\odot$/year.

The effects of this thermal wave on the physical state of the white dwarf interior depend on the time this wave takes to reach the center of the star, $\tau_{\rm TH}$, as compared with the time required for the star to reach the Chandrasekhar mass, $\tau_{\rm comp}$ (Hernanz et al. 1988). For low accretion rates, $\dot{M} \leq 3 \times 10^{-10}$ M$_\odot$/year, the thermal wave has time to reach the center, but normal cooling through the photosphere is dominant and the white dwarf evolves nearly isothermally with a temperature determined by the balance between compression and cooling, with the contribution of neutrinos and nuclear reactions. For high accretion rates, $5 \times 10^{-8} \leq \dot{M} \leq 10^{-6}$ M$_\odot$/year, compression heating dominates but, if the mass of the white dwarf is large enough, the thermal wave has no time to reach the center and, since there $\tau_{\rm TH} \gg \tau_{\rm comp}$, these layers evolve with an adiabatic index:

$$\Gamma_3 - 1 = \frac{0.815 + 0.251\Gamma^{1/4}}{0.945 + 0.646\Gamma^{1/4}} \qquad (5.2)$$

where Γ is the Coulomb coupling constant. For typical values of Γ ~100–200, this index is ~0.5 and degenerate matter is heated gently. For intermediate accretion rates, the thermal wave has sufficient time to arrive at the central layers and they

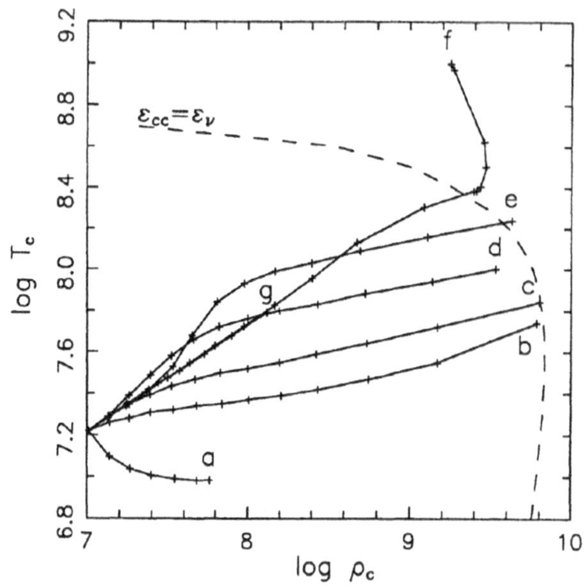

Fig. 5.1 Evolution of the center of a C-O white dwarf in the log T-log ρ plane for several accretion rates of pure C-O: (a) 10^{-10} $M_\odot$/year; (b) $5\,10^{-10}$ $M_\odot$/year; (c) 10^{-9} $M_\odot$/year; (d) $5\,10^{-9}$ $M_\odot$/year; (e) $5\,10^{-8}$ $M_\odot$/year; (f) $5\,10^{-7}$ $M_\odot$/year; (g) $5\,10^{-6}$ $M_\odot$/year (Bravo et al. 1996). The dashed line represents the ^{12}C ignition curve

experience a sudden heating, followed by an evolution in the ρ-T diagram governed by the balance between the heating and cooling agents already mentioned (see Fig. 5.1).

As the temperature increases, fusion reactions start to become important. At low temperatures, neutrino emission is able to control them, but due to the different temperature dependences the energy production by nuclear reactions overwhelms neutrino losses and matter burning becomes unstable. This critical temperature, commonly called ignition temperature, T_{ig}, is defined as $\epsilon_{CC}(T_{ig}) = \epsilon_\nu(T_{ig})$. If ignition happens under degenerate conditions, a thermonuclear runaway occurs. The nature of this instability can be understood with the following argument. Assume that $P = P_e(\rho) + P_i(\rho, T)$, where P_e and P_i are the electron and ion pressure, respectively, and that ions behave as an ideal gas. Also assume that nuclear reactions release isochorically an amount of energy δq, and that matter expands adiabatically until pressure equilibrium is reached. The corresponding density change is:

$$\frac{\delta\rho}{\rho} = -\frac{2}{3\Gamma_1}\frac{P_i}{P_e + P_i}\frac{Q}{kT}\frac{\delta N}{N} \tag{5.3}$$

where Γ_1 is the first adiabatic index, Q is the energy released per fused nucleus and δN is the number of nuclei that have fused. Since $Q \sim 1\,\mathrm{MeV}$ and $kT \sim 1\,\mathrm{keV}$, if the ideal gas is dominant a small energy release will cause a large expansion with an associated cooling. On the contrary, if $P_e \gg P_i$ adiabatic cooling is not efficient and matter will heat until $P_e \sim P_i$. At this point, if $\tau_{nuc} \ll \tau_{HD}$, nuclear reactions will continue until incineration of matter is complete. Here $\tau_{nuc}^{-1} = d\ln(\epsilon_{nuc})/dt$, and $\tau_{HD} = l/c_s$ is the hydrodynamic time, l the dimension of the burning region

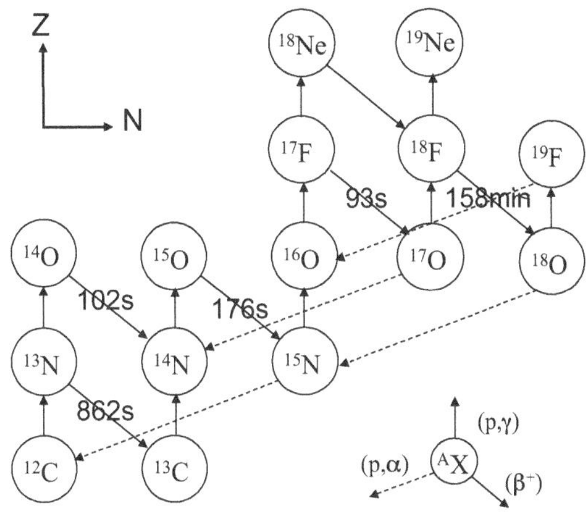

Fig. 5.2 Scheme of the carbon-nitrogen-oxygen (CNO) cycle of hydrogen burning, which operates out of equilibrium in stellar explosions. The lifetimes of the β^+-unstable nuclei, which act as *bottlenecks* of the cycle, are displayed

and c_s the sound velocity. Under hydrostatic equilibrium, $\tau_{\rm HD} \sim \tau_{\rm ff}$, where $\tau_{\rm ff} = (24\pi G\rho)^{-1/2} \sim 444\rho^{-1/2}$ is the free-fall time.

It is important to realize here that in the case of H-burning two protons have to be converted into two neutrons and that β-decays will control the total rate. At high temperatures, the longest β-decay timescale is that of ^{15}O, with a mean lifetime of $\tau_{15O} = 178$ s (see Fig. 5.2) and the maximum energy production rate is:

$$\epsilon_{CNO} \leq 1.3 \times 10^{14} \frac{X_{CNO}}{0.01} \text{ erg/g/s} \tag{5.4}$$

for an assumed energy release of 28 MeV per reaction (Mazurek and Wheeler 1980). Therefore, complete burning cannot be achieved in a short time in comparison to the hydrodynamic time and H-driven explosions cannot involve all of the star.

5.1.2 *The Thermonuclear Runaway*

When the central regions cross the ignition line, the temperature starts to rise and nuclear reactions accelerate. Conduction is rapidly overwhelmed by energy production and a convective core forms. This core grows very quickly as a consequence of the energy release enhancement and cannot prevent the continuous rise of the temperature. When the turnover timescale of the convective eddies is longer than the heating timescale, one or several bubbles enter into the dynamical regime (Nomoto et al. 1984; García-Senz and Bravo 2005; Woosley et al. 2004), a thermonuclear runaway occurs, and a flame begins to propagate (Timmes and Woosley 1992).

The igniting zone can be imagined as a highly turbulent region where the evolution of turbulent elements towards the thermonuclear runaway is governed by the balance between heating by nuclear burning and collision of turbulent eddies, and cooling by electron conduction and expansion pdV-work. In principle it is possible to assume a distribution of fluctuations characterized by their size, δ, and their temperature excess, ΔT. These fluctuations will be able to grow only if the conductive cooling is not able to evacuate the nuclear energy generated at the center of the bubble. Consequently their size has to be larger than:

$$\delta = \sqrt{\frac{2\sigma \Delta T}{\rho \epsilon_{\rm nuc}}} \tag{5.5}$$

where σ is the thermal conductivity and $\epsilon_{\rm nuc}$ is the nuclear energy generation rate. For background temperatures in the range $(6-8) \times 10^8$ K, fluctuations must have a minimum size of 4 m–30 cm, respectively, to be able to grow. When this condition is satisfied, the temperature increases, the burning propagates by conduction (see next section) and the buoyancy accelerates the bubble to a substantial fraction of the sound speed (García-Senz and Bravo 2005). During this time other bubbles can develop similar runaways, grow and float away when they reach a critical size of $\sim$1 km, such that the final outcome is an asynchronous ignition at multiple points.

The runaway of hydrogen is responsible for nova explosions. The mechanism for such explosions can be better understood after evaluating some relevant timescales (Starrfield 1989): the accretion timescale, defined as $\tau_{\rm acc} \sim {\rm M_{acc}}/\dot{\rm M}$ (which is of the order of 10^4–10^5 years, depending on the accretion rate $\dot{M}$ and accreted mass $M_{\rm acc}$), the nuclear timescale $\tau_{\rm nuc} \sim c_{\rm p}T/\epsilon_{\rm nuc}$ (which is as small as a few seconds at peak burning) and the hydrodynamic timescale ($\tau_{\rm HD} \sim H_{\rm p}/c_{\rm s} \sim (1/g)\sqrt{P/\rho}$; $H_{\rm p}$ is the pressure scale height). During the accretion phase, $\tau_{\rm acc} \leq \tau_{\rm nuc}$, accretion proceeds and the envelope mass increases. When degenerate ignition conditions are reached, degeneracy prevents envelope expansion and the thermonuclear runaway occurs. As temperature increases, the sudden release of energy would lift degeneracy in the envelope and ultimately halt the thermonuclear runaway, but this is not the case because $\tau_{\rm nuc} \ll \tau_{\rm HD}$. Therefore, the value of the nuclear timescale is crucial for the development of the thermonuclear runaway (TNR) and its final fate. In fact there are two main types of nuclear timescales: those related to β^+-decays, τ_{β^+}, and those related to proton capture reactions, $\tau_{({\rm p},\gamma)}$. In the early evolution towards the TNR, $\tau_{\beta^+} < \tau_{({\rm p},\gamma)}$ and the CNO cycle operates at equilibrium. But as temperature increases up to $\sim 10^8$ K, the reverse situation occurs, ($\tau_{\beta^+} > \tau_{({\rm p},\gamma)}$), and thus the CNO cycle is β-limited (see Fig. 5.2). In addition, since the large energetic output produced by nuclear reactions can not be evacuated by radiation only, convection sets in and transports the β^+-unstable nuclei to the outer cooler regions where they are preserved from destruction and where they will decay later on ($\tau_{\rm conv} < \tau_{\beta^+}$), leading to envelope expansion, luminosity increase and mass ejection if the attained velocities are larger than escape velocity.

5.1.3 Physics of the Burning Front

An explosion is the mechanical disruption of a system as a consequence of a rapid release of energy. In the case of an exploding white dwarf, enough mass, $\sim 0.3\,M_\odot$, has to be quickly embraced by the burning region to unbind the star. This can be accomplished either through detonation (Arnett 1969) or through deflagration (Nomoto et al. 1976). A detonation is shock-induced burning propagating supersonically into an unburned medium, while a deflagration is a burning front that propagates by thermal conduction at subsonic velocities. Both, detonation and deflagration, are driven by a physical mechanism. However, there is a third possibility: spontaneous burning. This case occurs when the ignition conditions are reached nearly simultaneously in several points in such a way that burning spreads over a large region without any transport mechanism (Blinnikov and Khokhlov 1986; Woosley and Weaver 1986). The propagation velocity, a phase velocity in fact, can be estimated as $v_{sb} = (d\tau_{nuc}/dr)^{-1}$, where τ_{nuc} plays a critical role at the onset of burning. This velocity increases when the absolute values of the temperature and density gradients decrease. Thus, regions with $v_{sb} \geq c_s$ ignite spontaneously and the burning front propagates supersonically. Because of the strong dependence on T, the most important factor is the temperature profile.

In order to describe the properties of the burning front, either supersonic or subsonic, it is usually assumed (Landau and Lifshitz 1959) that the unburned material is separated from the combustion products by a region of width δ where reactions take place. If $\delta \ll l$, where l is the typical scale length of the system, it is possible to connect both sides of the front by means of conservation laws of mass, momentum and energy. In the frame associated with the front, these equations, known as the Rankine-Hugoniot jump conditions, can be written as (Landau and Lifshitz 1959; Mazurek and Wheeler 1980):

$$\rho_1 u_1 = \rho_0 u_0 \tag{5.6}$$

$$P_1 + \rho_1 u_1^2 = P_0 + \rho_0 u_0^2 \tag{5.7}$$

$$\varepsilon_1 + \frac{P_1}{\rho_1} + \frac{u_1^2}{2} = \varepsilon_0 + \frac{P_0}{\rho_0} + \frac{u_0^2}{2} + q \tag{5.8}$$

that are similar to those describing shock waves except for the presence of the term q that represents the amount of energy released by reactions. The subscripts 0 and 1 denote fuel and ashes, respectively, u is the matter velocity, ε is the specific internal energy, and the remaining symbols have their usual meaning. The mass flux crossing the front is given by:

$$j = \rho_0 u_0 = \rho_1 u_1 \tag{5.9}$$

which can be written, using the mass and momentum conservation equations (Eqs. (5.6) and (5.7) respectively), as:

$$j^2 = -\frac{P_0 - P_1}{V_0 - V_1} \tag{5.10}$$

where $V = 1/\rho$ is the specific volume. The mass flux (and the velocity of the front with respect to the unburned material) is determined by the ratio between the difference of pressures and specific volumes at both sides of the burning front. Therefore, real solutions must satisfy: $(P_1 > P_0,\ V_1 < V_0)$ or $(P_1 < P_0,\ V_1 > V_0)$. The first solution corresponds to a detonation and the second one to a deflagration (Landau and Lifshitz 1959).

The velocity at which a detonation propagates can be obtained from the energy conservation equation. Equation (5.8) can be written as:

$$\varepsilon_0 + q - \varepsilon_1 + \frac{1}{2}(P_0 + P_1)(V_0 - V_1) = 0 \tag{5.11}$$

which is called the detonation adiabat (the case $q = 0$ is called the shock adiabat). The final state is obtained equating (5.10) and (5.11), once the properties of the front have been specified (see Fig. 5.3). The physical meaning of this solution is clear. A shock heats and compresses the material to a state $(P_{\rm sh}, V_{\rm sh})$ given by the intersection of Eq. (5.10) with the shock adiabat. Because of the temperature increment, material burns and reaches the state (P_1, V_1), defined by the intersection of Eq. (5.10) with the detonation adiabat. Since $q \geq 0$, then $P_1 < P_{\rm sh}$, $V_1 > V_{\rm sh}$ and the post-shock burning produces a rarefaction.

The family of solutions obtained from Eqs. (5.10) and (5.11) and j as a free parameter has an extremum for which j and the front velocity are minima. This solution, called the Chapman-Jouguet detonation, corresponds to the case where Eq. (5.10) is tangent to (5.11). This extremal solution has the following properties: (1) it is only determined by the thermodynamic properties of the material, including

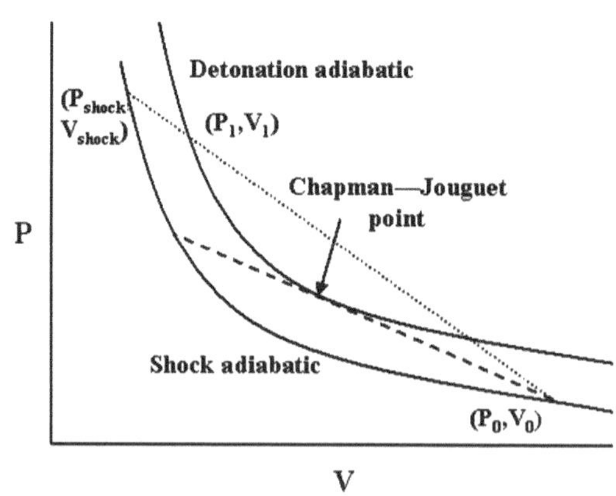

Fig. 5.3 Adiabatic of a shock and a detonation in the P–V diagram. The dotted and dashed lines represent the condition of conservation of mass and momentum ($-j^2$ = constant)

q, (2) the entropy is maximum and (3) the velocity with respect to the unburned material is minimal and equal to the sound velocity of the material behind the front. All the remaining solutions, called strong detonations, move supersonically with respect to the burned material and subsonically with respect to the unburned material. Therefore, if a detonation starts at the center of the white dwarf, all the material, from the center to very near the surface, will be incinerated to ^{56}Ni.

Strong detonations are not allowed in stars. Since material must be at rest at the centre, the velocity has to decrease from a positive value behind the front to zero at the centre. This means that a rarefaction wave has to follow the detonation. Since the velocity of a rarefaction wave is equal to the sound velocity of the material, it is necessary that the front moves at least with the sound velocity with respect to the burned material in order to not be overtaken by the rarefaction wave. Thus, due to the boundary conditions, the only acceptable detonations in stars are those of Chapman-Jouguet type.

In the case of deflagration solutions matter is subsonic on both sides of the front. Thus, any perturbation behind or ahead the front can affect it. As an example, consider a spherically symmetric burning front propagating outwards with a velocity D and the unburned matter at rest. From Eqs. (5.6) and (5.7) it is possible to write:

$$V_0(P_1 - P_0) = v_1 D \tag{5.12}$$

where v_1 is the velocity measured in the frame fixed to the center of the star. Since $P_1 - P_0 < 0$ in deflagrations, and $D > 0$, the velocity of the burned matter must be negative, $v_1 < 0$, in contradiction to the boundary condition that requires matter to be at rest at the center. Thus a deflagration can only exist if it generates a shock precursor that boosts matter outwards (Mazurek and Wheeler 1980).

The speed of the laminar flame can be estimated as follows (Landau and Lifshitz 1959): the velocity of the burning front is $D \sim \delta/\tau_{\rm burn}$, where δ is the width of the front, and $\tau_{\rm burn}$ is the lasting time of the burning ($\tau_{\rm burn} \sim \varepsilon/\varepsilon_{\rm nuc}$, where $\varepsilon_{\rm nuc}$ has to be evaluated at the critical temperature). Since a stationary flame can only exist if $\tau_{\rm burn} \sim \tau_{\rm diff}$, where the diffusion time is given by $\tau_{\rm diff} \sim \delta^2/\chi$ and χ is the thermometric conductivity, the width of the front has to be

$$\delta \sim \sqrt{\frac{\varepsilon\chi}{\varepsilon_{\rm nuc}}} \tag{5.13}$$

and the laminar velocity

$$D \sim \sqrt{\frac{\varepsilon_{\rm nuc}\chi}{\varepsilon}} \tag{5.14}$$

In the case when a white dwarf is near the Chandrasekhar limit, $\delta \sim 10^{-4}$ cm, $D \sim 10^7$ cm/s $\sim 10^{-2}c_{\rm s}$ and the density contrast between burned and unburned matter is $\Delta\rho/\rho \sim 0.2$. These values relax to 1 cm, 10^4 cm/s and 0.5 respectively when $\rho \sim 10^7$ g/cm^3.

As a consequence of turbulence induced by instabilities like the Rayleigh-Taylor buoyancy-driven or the Kelvin-Helmholtz shear-driven instability, for instance, the flame surface is wrinkled and stretched in such a way that, despite the flame continuing to propagate at the laminar velocity, the effective burning rate is strongly enhanced. Buoyancy induces the formation of burning bubbles that rise into the fuel and generate turbulent motions. The turbulent motions decay downward to the smaller Kolmogorov scale and the eddies of this cascade interact with the flame, further wrinkling and stretching its surface, thereby further enhancing the burning rate. This effect acts down to the Gibson scale, defined as the size of the eddy that can turn over in a nuclear burning time. Below this scale, the laminar velocity is larger than the turbulent velocity and fuel is burned before the eddies are able to change the shape of the flame. If the Gibson length is large as compared with the width of the flame, the internal structure of the flame is not altered (the flamelet regime). In the opposite case, the turbulent motion is able to modify the internal structure of the flame and burning enters the so called distributed regime.

5.1.4 Scenarios Leading to a Thermonuclear Runaway

Possible scenarios leading to a thermonuclear runaway can be classified according to the chemical composition of the donor (H, He,C+O, O+Ne) and the nature of the accretor, a white dwarf made of He, C+O, or O+Ne, or a neutron star. Some of the possible combinations are very rare, if not forbidden, and have not yet been associated with any observed astronomical event. Within the category of accreting white dwarfs it is possible to adopt the following scenarios (Iben and Tutukov 1985; Nomoto 1982; Webbink 1984; Whelan and Iben 1973):

Hydrogen accretion There are many astronomical objects containing a white dwarf that accretes hydrogen rich matter from a non-degenerate companion and that could suffer a thermonuclear runaway. The nature and the intensity of this instability depend on the accretion rate and the mass of the object. If the accretion rate is smaller than $\sim 10^{-8-9}\,\mathrm{M_\odot}$/year, hydrogen accumulates on the surface of the white dwarf and becomes degenerate. When the accumulated mass reaches a critical value, $\Delta M_{\rm H} \sim 10^{-4}$–$10^{-5}\,\mathrm{M_\odot}$, the exact values depending on the properties of the binary system, it experiences a strong flash that can be identified with the nova phenomenon (see Sect. 5.2). This flash expels almost all the accreted mass or even erodes the mass of the accreting object, for which reason the white dwarf is unable to reach the Chandrasekhar mass, except in the case when $M_{WD} > 1.3\,\mathrm{M_\odot}$. However, since the chemical composition of such white dwarfs is a mixture of oxygen and neon, the fate of such scenario is collapse to a neutron star.

For intermediate rates, $10^{-8-9} \leq \dot{M}_{\rm H}(\mathrm{M_\odot/year}) \leq 5 \times 10^{-7}$, hydrogen burns steadily or through mild flashes, and helium accumulates on the surface of the star. If the accretion rate is high enough, this helium is converted into carbon and oxygen through weak flashes or steady burning and the white dwarf approaches the Chandrasekhar mass. But, if the effective accretion rate of helium is in the range $10^{-9} \leq \dot{M}_{\rm H}({\rm M}_\odot/{\rm year}) \leq 5 \times 10^{-8}$, the helium layer becomes degenerate and when it reaches a critical mass, $\Delta M_{\rm He} \sim 0.3\,{\rm M}_\odot$, it ignites under degenerate conditions and experiences a thermonuclear runaway that can trigger the explosive destruction of the complete star. This scenario has been proposed for type Ia supernova progenitors (see Sect. 5.3.2).

If the accretion rate is larger than $\sim 5 \times 10^{-7}\,{\rm M}_\odot$/year a red giant-like envelope forms, a strong wind appears and the mass accumulates over the degenerate core at a rate (Hachisu et al. 1999):

$$\dot{M}_{\rm cr} \simeq 5.3 \times 10^{-7} \frac{1.7 - X}{X} (M_{\rm WD}/{\rm M}_\odot - 0.4)\ \ {\rm M}_\odot/{\rm year} \tag{5.15}$$

where X is the mass fraction of H in the accreted matter. As before, hydrogen and helium burn peacefully and the white dwarf has the possibility to reach the Chandrasekhar mass. Typical examples are cataclysmic variables, classical novae, recurrent novae, symbiotic stars and supersoft X-ray sources.

Helium accretion There are at least two scenarios in which a white dwarf can directly accrete helium from the companion. One consists of two degenerate objects, a primary made of carbon-oxygen and a secondary composed of helium, that merge as a consequence of the emission of gravitational waves. Since the mass of the secondary is small $\sim$0.3–0.4 ${\rm M}_\odot$, the process of merging is self-regulated. The second scenario consists of a C+O white dwarf plus a non-degenerate helium star and the mass transfer is powered by helium burning in the secondary. As mentioned above, if $10^{-9} \leq \dot{M}_{\rm H}({\rm M}_\odot/{\rm year}) \leq 5 \times 10^{-8}$, helium ignites at the base of the freshly accreted mantle under degenerate conditions and can trigger the thermonuclear explosion of the accreting white dwarf despite the fact that its mass is smaller than the Chandrasekhar limit (see Sect. 5.3.2). A typical example is that of the AM CVn systems.

Carbon-oxygen accretion Close enough binary systems formed by two intermediate mass stars can experience two episodes of common envelope evolution that result in the formation of two C/O white dwarfs with a separation that is smaller than the initial one.

Depending on the parameters of the system, one possibility is that the first newly formed C/O white dwarf merges with the core of the AGB companion during the second common envelope evolution. The merger has a mass of the order of the Chandrasekhar's mass and explodes after some time (Kashi and Soker 2011; Aznar-Siguán et al. 2015). Another possibility is that the two C/O white dwarfs are left close enough to allow an important loss of angular momentum via the emission of gravitational waves that induces an additional reduction of the orbital radius at a rate:

$$\dot{r} = -\frac{64 G^3 m_1 m_2 (m_1 + m_2)}{5 c^5 r^3}$$

where G is the gravitational constant, c is the speed of light and r is the separation of both stars. If the separation of the two white dwarfs is smaller than $\sim 3R_{\odot}$, nothing can prevent their merging in less than a Hubble time and the primary will start accreting a mixture of carbon and oxygen.
During the merging process, the secondary is destroyed in a few orbital periods after filling its Roche lobe (Benz et al. 1990) and forms a hot and thick accretion disk around the primary. The impact is not able to induce prompt ignition (Guerrero et al. 2004) and the final outcome depends on the subsequent evolution of the disk. If the accretion rate is spherically symmetric and larger than about $\dot{M} \geq 2.7 \times 10^{-6}\,M_{\odot}$/year, carbon ignites off-center, the flame propagates conductively inwards and the white dwarf is converted into an O-Ne white dwarf before central carbon ignition. Upon further accretion the white dwarf collapses to a neutron star (Nomoto and Kondo 1991). Recent calculations (Yoon and Langer 2005) indicate that, at least in some cases, neutrino cooling is able to quench off-center carbon ignition. An open question is the effective rate at which matter is accreted (Piersanti et al. 2003a,b) since it also contains angular momentum that prevents the contraction of the primary unless it is dissipated. Therefore, the interplay between disk and star is crucial for understanding the outcome of such a scenario.

Concerning the category of accreting neutron stars, van Horn and Hansen (van Horn and Hansen 1974; Hansen and van Horn 1975) were the first to point out that nuclear burning on their surface can also be unstable. The regimes of unstable burning have been extensively discussed elsewhere (Fujimoto et al. 1981). To summarize, for a chemical mixture with Z(CNO)$\sim$ 0.01: mixed H/He-burning is expected for $\dot{M} < 2 \times 10^{-10}\,M_{\odot}$/year, triggered by thermally unstable H-ignition; pure He-shell ignition for $2 \times 10^{-10} < \dot{M}\ (M_{\odot}/\text{year}) < (4.4 - 11.1) \times 10^{-10}$, following completion of H-burning; and mixed H/He-burning for $\dot{M} > (4.4-11.1) \times 10^{-10}\,M_{\odot}$/year, triggered by thermally unstable He ignition. A reduction of the CNO content lowers the critical accretion rates and substantially narrows the range for pure He bursts.

5.2 Classical Novae

The origin of the term *nova* comes from the Latin *nova stella*, meaning that a *new star* appeared in the sky. But it has been known for a long time that the new star is in fact not new, and that a nova is more properly defined as an existing star that suddenly increases its luminosity—by more than $\sim$10 magnitudes, i.e., by a factor larger than 10^4—and then returns to its previous faint state in a few months, or even years. In fact, already Newton in the seventeenth to eighteenth century talked about *temporary stars* shining suddenly and then vanishing. It was not until the twentieth century that novae and supernovae were distinguished from each other, once the distances to the *nebulae* where they had been discovered were better known, and thus some extragalactic objects turned out to be novae with much larger intrinsic brightness (*super-novae*). An interesting and complete historical perspective of novae can be found in Bode and Evans (2008).

The discovery of the binarity of classical novae was made by Walker (1954), who observed DQ Her (a nova that exploded in 1934) and deduced that it was an eclipsing binary with a very short period. Later, Kraft showed that this was a common property of novae and of cataclysmic variables in general (Kraft 1964). It is now well known that nova explosions occur on white dwarfs accreting hydrogen-rich matter from a main sequence star companion, in a close binary system of the cataclysmic variable type. Accumulation of matter on the white dwarf leads to hydrogen ignition in degenerate conditions, which prevents the self-adjustment of the envelope through expansion. Therefore, a thermonuclear runaway ensues (see Sect. 5.1.2) and the final consequence is mass ejection at large velocities (hundreds to thousands of km s^{-1}) and a large increase of luminosity, even reaching the Eddington luminosity of the white dwarf (10^{34}–10^{35} erg/s).

In contrast to type Ia supernovae, which also occur on white dwarfs in binary systems, novae do not experience a complete disruption of the white dwarf, because the outburst only affects the external hydrogen-rich layers, i.e. 10^{-4}–$10^{-5}\,M_{\odot}$. Therefore, the nova phenomenon is expected to recur, with periods of a few tens or hundreds thousand years, which is the typical accretion time required to build-up again a critical H-rich envelope ready to explode. Mass transfer onto the white dwarfs in cataclysmic variables is a long-lasting phase, so that many nova explosions on a given white dwarf must occur. However, for historical reasons the term *recurrent nova* is reserved for another type of eruptive phenomena, those that have more than one *recorded* nova outburst. These systems also correspond to white dwarfs experiencing a thermonuclear runaway of their H-rich envelope, but the companion star is in general a red giant, instead of a main sequence star. The binary system is not a cataclysmic variable anymore; both its period—and the related binary separation—and the mass transfer rate onto the white dwarf are larger, thus allowing for a faster build-up of the critical mass and thus a shorter recurrence period (decades rather than thousands of years).

It is worth mentioning that the term *recurrent novae* is also applied to white dwarf explosions, with similar outburst properties and recurrence periods than *genuine* recurrent novae, but with a non thermonuclear origin. A completely different case are the so-called *dwarf novae*, which have much smaller outburst amplitudes and which are produced by accretion disk instabilities in cataclysmic variables. Here we are only concerned with novae from thermonuclear explosions, i.e., *classical novae* and the sub-class of *recurrent novae* with thermonuclear origin.

The long term evolution of the white dwarfs in classical and recurrent novae is debated, since it is not clear if the mass of the white dwarf grows towards the Chandrasekhar mass or decreases after each explosion. Observations of nova ejecta often show overabundances with respect to solar of elements such as carbon, oxygen, neon, among others, indicating that some mixing between the core and the accreted envelope occurs. Then, some core mass is in principle ejected indicating that the white dwarf mass might in fact decrease. However, in *recurrent novae* no large overabundances are observed. On the other hand, *recurrent novae* should take place on very massive white dwarfs, which only need a very small amount of added mass to explode; this combined with a larger accretion rate leads to the very short recurrence period observed. All in all, recurrent novae are one of the possible scenarios of type Ia supernova progenitors, although their internal composition (likely ONe instead of CO) presents a problem for this scenario.

Interestingly enough, there is a remarkable recurrent nova in the Andromeda Galaxy, M31N 2008-12a, with an extremely short recurrence period—of about 1 year—the shortest known to date (Darnley et al. 2015; Henze et al. 2015; Darnley et al. 2016a,b). This is the best candidate for a type Ia supernova explosion, because the deduced mass of the white dwarf is extremely close to the Chandrasekhar limit (Kato et al. 2014; Hachisu et al. 2016), but its chemical composition should be CO and not ONe. From the observational point of view, HST observations of the 2015 eruption of M31N 2008-12a yielded non detection of Neon, which may be indicative of a CO white dwarf (Darnley et al. 2017a). Theoretically, the white dwarf of M31N 2008-12a could reach near-Chandrasekhar-mass through successive eruptions with an initial CO core (Hillman et al. 2016). All in all, it is predicted that the M31N 2008-12a white dwarf could reach the Chandrasekhar mass and thus explode as a SNIa in less than 20 kyr (Darnley et al. 2017b).

5.2.1 Observational Properties

Most of the galactic classical novae have been discovered optically by amateur astronomers. In addition, some robotic telescopes, mainly devoted to search for optical counterparts of GRBs or to perform surveys are also finding novae and supernovae. Around 5 novae per year are being discovered in recent years in our galaxy, and several novae have been found as well in external galaxies (see Shara in Bode and Evans (2008) and references therein). However, most of the galactic novae suffer from large optical extinction (reddening) by interstellar dust, and hence the real nova rate is expected to be much larger; it should be determined from extrapolations, either from extragalactic or from galactic data. In the first case, the dependence of the nova rate on the type of galaxy has been derived, indicating that early-type galaxies are more prolific nova producers; the derived nova rate is 15–24 year^{-1} to 27 ± 8 year^{-1} (Della Valle and Livio 1994; Shafter et al. 2000). Larger rates are obtained when galactic data are extrapolated, taking into account the amount and distribution of galactic dust: 35 ± 11 year^{-1} or 41 ± 20 year^{-1} (Hatano et al. 1997; Shafter 1997).

From optical light curves of classical novae one finds an increase in luminosity corresponding to a decrease of m_V (apparent visual magnitude) of more than 9 magnitudes occurring in just a few days, and a pre-maximum halt 2 magnitudes before maximum, in some cases (Warner (1995) and references therein). Nova light curves are classified according to their speed class, defined from either t_2 or t_3, i.e., the time needed to decay by 2 or 3 visual magnitudes after maximum. Speed classes range from very fast ($t_2 < 10$ days) and fast ($t_2 \sim 11$–25 days) to very slow ($t_2 \sim 151$–250 days) (Payne Gaposchkin 1957). Some examples are the fast nova N Cyg 1992, which had $t_2 \sim 12$ days, the even faster nova N Her 1991 ($t_2 \sim 2$ days), and the slow nova N Cas 1993, which had $t_2 \sim 100$ days. An empirical relationship between the absolute magnitude at maximum M_V and the speed class of novae shows that brighter novae have shorter decay times (t_2 or t_3). The theoretical explanation of this relationship (Livio 1992) relies on novae reaching a maximum luminosity close to the Eddington limit and ejecting roughly all their envelope in a period similar to t_3. It was established that L_{max} is an increasing function of M_{wd} and that t_3 is a decreasing function of M_{wd}. From these two relationships an expression relating M_V at maximum and t_3 is deduced. This empirical relation, valid both in the V and B photometric bands, is very often used to determine distances to novae, once visual extinction is known. Different calibrations of the maximum magnitude-rate of decline relationship (MMRD) exist, with that from Della Valle and Livio (1995) being the most commonly employed form.

It is worth mentioning that in fact the MMRD relationship has not been proven extensively, and can't be considered as universal. In fact, the extensive grid of nova numerical simulations by Yaron et al. (2005) first predicted that some classical novae might deviate significantly from the MMRD relation. On the observational side, Kasliwal et al. (2011) discovered a new photometric sub-class of faint and fast classical novae in the Andromeda Galaxy, M31, inconsistent with the canonical MMRD relationship. They suggested that the MMRD, characterized only by the white dwarf mass, was probably an oversimplification. Six years later, Shara et al. (2017) found a similar class of *faint, fast novae* in the giant elliptical galaxy M87; their conclusion was that the MMRD relationship should not be used to determine cosmic distances or distances to Galactic novae.

There have been efforts to improve the quality of nova optical light curves. In this respect, the *Solar Mass Ejection Imager* satellite, SMEI, has done an important contribution regarding the quality, because it provides good precision visible-light photometry at 102-min cadence; Hounsell et al. (2010, 2016) report on the SMEI nova observations between 2004 and 2009. Additionally, there is a recent catalog of 97 very-well-observed nova light curves (Strope et al. 2010), mainly from the *American Association of Variable Star Observers*, AAVSO, database, which has led to a new more sophisticated classification system, based not only on the speed class (time to decline by a given number of magnitudes), but also on the shape of the light curves. They use designations S for smooth light curves (38% of the novae), P for plateaus (21%), D for dust dips (18%), C for cusp-shaped secondary maxima (1%), O for quasi-sinusoidal oscillations superposed on an otherwise smooth decline (4%), F for flat-topped light curves (2%), and J for jitters or flares superposed on the decline (16%). Their classification consists of the corresponding single letter followed by the t_3 value in parentheses.

The optical light curves were extended with space-based observations to energy ranges not observable from the ground. An important step forward was the discovery of the luminosity increase in the ultraviolet when the optical started to decline, thanks to the IUE satellite (International Ultraviolet Explorer); the reason is that the spectral energy distribution shifts to higher energies when deeper and thus hotter regions of the expanding envelope are revealed (the photosphere recedes as a consequence of the decreasing opacity). On the other end of the spectrum, infrared observations (especially for novae which form dust) indicate an increase in luminosity once the ultraviolet luminosity starts to decline, which is interpreted as the resulting re-radiation (in the infrared) by dust grains of the ultraviolet energy they absorbed. Therefore, during optical decline the bolometric luminosity of classical novae remains constant, during a period of time which depends on the mass of the H-rich envelope remaining on the white dwarf after the nova explosion, which is expected to burn steadily. Evidence for residual H-burning came from observations in the supersoft X-ray range with ROSAT (Krautter et al. 1996; Balman et al. 1998; Orio et al. 2001), which revealed the related very hot photosphere.

After the ROSAT era, the Swift Gamma-Ray Burst satellite, launched in 2004, has been and continues to be an excellent facility for the study of novae in soft X-rays, mainly with its X-ray telescope instrument, XRT, thanks to its rapid response and its scheduling flexibility. Observations can start promptly, at about 9 h from discovery. It has observed 73 Galactic and Magellanic Clouds novae within 11 years of outburst (data up to June 2017); 43 of them were detected in X-rays and 12 have been observed more than 100 ks. Novae in M31 and M33 have also been observed. See Ness et al. (2007), Schwarz et al. (2011) and Osborne (2015) for reports on X-ray observations of novae with Swift.

Prior to Swift, the ESA and NASA X-ray satellites XMM-Newton and Chandra, both launched in 1999, have crucially contributed to the continued study of novae in X-rays, but they are not suited to do a systematic study of the duration of the supersoft X-ray phase of novae as Swift is. They have instead provided the highest-resolution spectra available for novae (see for instance the review Ness (2012) and references therein). In the supersoft X-ray range, these reveal a wealth of absorption lines, related to the hot white dwarf photospheric emission, much more complicated than the often assumed black body model. Also, plasma related emission lines have been revealed; a complete explanation of the whole X-ray spectra is still lacking. The Japanese satellite Suzaku has also provided important data, in the harder X-ray energy range not reachable with XMM-Newton and Chandra (see for instance Takei et al. (2009)). Finally, NuSTAR has also observed a few novae, searching for the prompt hard X-ray emission related to shocks (Orio et al. 2015; Mukai et al. 2017).

The bolometric luminosity deduced from observations is close to or even larger than the Eddington luminosity, and thus radiation pressure is probably responsible for ejection of nova envelopes (Kato and Hachisu 1994).

A very important result deduced from nova observations in all spectral wavelengths is that their ejecta are often enriched in carbon, nitrogen and oxygen—as well as neon in many objects (around 1/3 of the total); the global metallicities in nova ejecta are well above solar metallicities (see Gehrz et al. 1998 for a review). This observational fact is one of the main drivers of theoretical models, which should be able to explain it. These metallicity enhancements are not likely to be produced in the TNR, because the temperatures achieved in nova explosions are not high enough. An alternative and more widely accepted explanation is that there is some mixing between accreted matter, assumed to be of solar composition, and the underlying CO or ONe core (Starrfield et al. 1978b; Prialnik et al. 1978, 1979). In fact, such enrichment is also required to power the nova explosion itself except for very slow novae.

It is important to point out that there are two distinct nova populations: disk novae, which are in general fast and bright ($M_V(max) \simeq -8$), and bulge novae, slower and dimmer ($M_V(max) \simeq -7$). This was first suggested by Della Valle et al. (1992) and later corroborated by their early post-outburst spectra (Williams 1992) and based on the stronger group of emission lines they display (either FeII lines or He and N lines); FeII-type novae evolve more slowly and have a lower level of ionization, whereas He/N-type novae have larger expansion velocities and a higher level of ionization. It has been deduced that the faster and brighter He/N novae are concentrated closer to the galactic plane than those of the slower and dimmer FeII type, which would preferentially belong to the bulge population (Della Valle and Livio 1998; Della Valle 2002).

Novae have not been detected yet in γ-rays from radioactivities, but they have been, however, detected in high-energy γ-rays (energy larger than 100 MeV), with the Large Area Telescope (LAT) instrument onboard the *Fermi* satellite, launched by NASA in 2008. The first nova detected by Fermi/LAT was V407 Cyg (Abdo et al. 2010). This source is a binary system with a white dwarf and a Mira pulsating red giant companion; the emission lasted for about 2 weeks after the nova eruption. Other novae have been detected with Fermi/LAT, almost one per year on average (Ackermann et al. 2014; Cheung et al. 2016).

The main mechanisms responsible for the production of high-energy γ-rays are pion decay and Inverse Compton; neutral pions come from proton-proton collisions, when accelerated protons exist (hadronic process), whereas Inverse Compton relies on the existence of relativistic electrons (leptonic process). Protons and electrons are accelerated in the shock wave formed when the nova ejecta interacts with a dense ambient medium, either the wind of its the red giant companion (case of symbiotic recurrent novae, like RS Oph (Tatischeff and Hernanz 2007)) or with the nova ejecta itself (case of classical novae, not well understood yet; see for instance Chomiuk et al. (2014), Vurm and Metzger (2018) and references therein).

5.2.2 *Modeling Classical Novae*

The scenario of classical nova explosions consists of a white dwarf (either CO or ONe) accreting hydrogen-rich matter in a cataclysmic binary system, as a result of Roche lobe overflow from its main sequence companion. For accretion rates low enough, e.g. $\dot{M} \sim 10^{-9}$–$10^{-10}\,M_\odot$/year, accreted hydrogen is compressed to degenerate conditions until ignition occurs, thus leading to a thermonuclear runaway (TNR, see Sect. 5.1.2). Explosive hydrogen burning synthesizes some β^+-unstable nuclei of short lifetimes (e.g. ^{13}N, ^{14}O, ^{15}O, ^{17}F, with τ = 862, 102, 176, and 93 s respectively, see Fig. 5.2) which are transported by convection to the outer envelope, where they are saved from destruction. These decays lead to a large energy release in the outer shells which causes the nova outburst, i.e. a visual luminosity increase accompanied by mass ejection with typical velocities 10^2–10^3 km s^{-1}. Another important effect of convection is that it transports unburned material to the burning shell (see Starrfield et al. (2016), Jose (2016), for recent reviews on nova modeling).

Mixing at the core-envelope interface turned out to be an essential ingredient in the simulations, both to power the TNR and to explain observed enhancements in metals in many novae. Several mechanisms have been suggested to explain this process, operating prior or during the thermonuclear runaway, but none of them is completely satisfactory (see an extensive review in Livio (1994)). Diffusion induced convection, first discussed by Prialnik and Kovetz (1984) and Kovetz and Prialnik (1985), can explain moderate enrichments but has difficulties to account for some of the largest metallicities inferred (Kovetz and Prialnik 1997). Other possibilities are shear mixing, convection induced shear mixing, and convective overshooting induced flame propagation. Two approaches have been adopted to simulate mixing in one-dimensional simulations: parameterization (Starrfield et al. 1998; José and Hernanz 1998) or follow-up of many successive eruptions, with inclusion of diffusion (Prialnik and Kovetz 1995; Yaron et al. 2005). The latter is in principle self-consistent, but the treatment of mass-loss between successive outbursts is quite uncertain.

Despite many observational features that characterize the nova phenomenon have been successfully reproduced by hydrodynamic simulations under the assumption of spherical symmetry, certain aspects like the way in which a thermonuclear runaway sets in and propagates, the treatment of convective transport, and most likely, the mixing at the core-envelope interface, clearly require a multidimensional approach.

Early semianalytic estimates, focused on the onset of localized TNRs on the surface of white dwarfs, suggested that heat transport was not efficient enough to spread a localized TNR to the entire surface, concluding that localized, volcanic-like TNRs were likely expected (Shara 1982). The first studies that tackled this question in the context of truly multidimensional nova simulations were conducted by Glasner et al. (Glasner and Livne 1995; Glasner et al. 1997): indeed, two-dimensional simulations were performed with the code VULCAN, an arbitrary Lagrangian Eulerian (ALE) code. To this end, a box (0.1 $\pi^{\rm rad}$) in spherical-polar coordinates, with reflecting boundary conditions, was adopted. The resolution adopted near the envelope base was 5 km × 5 km. In the simulations, the evolution of an accreting 1 $M_\odot$ CO white dwarf was initially followed with a one-dimensional code. The structure was subsequently mapped into a two-dimensional domain, when the temperature at the envelope base reached 100 MK. The two-dimensional simulations, that relied on a 12-isotope network, showed good agreement with the main results obtained with one-dimensional models: specifically, the critical role played by the β-unstable nuclei ^{13}N, ^{14,15}O, and ^{17}F in the expansion and ejection stages, and therefore, the presence of large amounts of ^{13}C, ^{15}N, and ^{17}O in the ejecta. Nevertheless, some remarkable differences were also identified: on one hand, the TNR was initiated as a handful of irregular, localized eruptions that set in at the envelope base, caused by convection-driven temperature fluctuations.

This suggested that combustion likely proceeds as a chain of multiple localized flames, rather than as a thin front, each surviving only a few seconds. However, turbulent diffusion efficiently dissipates any local burning, such that the fast stages of the TNR cannot be localized. Therefore, the runaway must finally spread along the stellar surface. On the other hand, the core-envelope interface is now convectively unstable, providing a source for the metallicity enhancement through Kelvin-Helmholtz instabilities (a mechanism that bears similarities with convective overshooting (Woosley 1986)).

Results from other 2D (and 3D) simulations were published, shortly after, by Kercek et al. (1998, 1999). They found substantially less violent outbursts (i.e., longer TNRs with lower peak temperatures and ejection velocities) caused by large differences in the convective flow patterns. Indeed, whereas in Glasner et al., a few, large convective eddies dominated the flow, most of the early TNR reported by Kercek et al. was governed by small, very stable eddies, which led to very limited dredge-up and mixing episodes. In fact, Kercek et al. concluded that mixing must take place prior to the TNR, in contrast with the simulations reported by Glasner et al. In summary, two independent studies, based upon the same initial model, yielded opposite results on the strength of the runaway and its capability to power a fast nova. The differences between both studies were carefully analyzed by Glasner et al. (2005), who concluded that the early stages of the TNR, when the evolution is quasi-static, are extremely sensitive to the adopted outer boundary conditions. Indeed, they showed that Lagrangian simulations, in which the envelope was allowed to expand and mass was conserved, led to consistent explosions. In contrast, in Eulerian schemes with a free outflow outer boundary condition (the choice adopted in Kercek et al.), the outburst was artificially quenched.

The feasibility of this mechanism was further explored by Casanova et al. through a set of independent two-dimensional simulations (Casanova et al. 2010, 2011a), proving that even in an Eulerian scheme (e.g., the FLASH code) with a proper choice of outer boundary conditions, Kelvin-Helmholtz instabilities (see Fig. 5.4) can naturally lead to self-enrichment of the accreted envelope with core material, at levels that agree with observations. It is worth noting, however, that convective transport cannot be described accurately by means of two-dimensional prescriptions. In fact, the conservation of vorticity (a measure of the local spinning motion of the particles in a fluid), imposed by a 2D geometry, forces all small convective cells to merge into large eddies, with a size of the order of the pressure scale height of the envelope. In sharp contrast, eddies become unstable in 3D in fully developed turbulent convection, and consequently break up, transferring their energy to progressively smaller scales (Pope 2000; Shore 2007). The resulting structures (e.g., vortices and filaments) undergo a similar fate down to approximately the Kolmogorov scale. In this framework, a pioneering three-dimensional simulation of mixing at the core-envelope interface during nova explosions (Casanova et al. 2011b) has shed light into the nature of the highly fragmented, chemically enriched, and inhomogeneous nova shells, observed in high-resolution spectra: such features have been interpreted as relics of the hydrodynamic instabilities that develop during the initial ejection stage, as predicted by the Kolmogorov theory of turbulence.

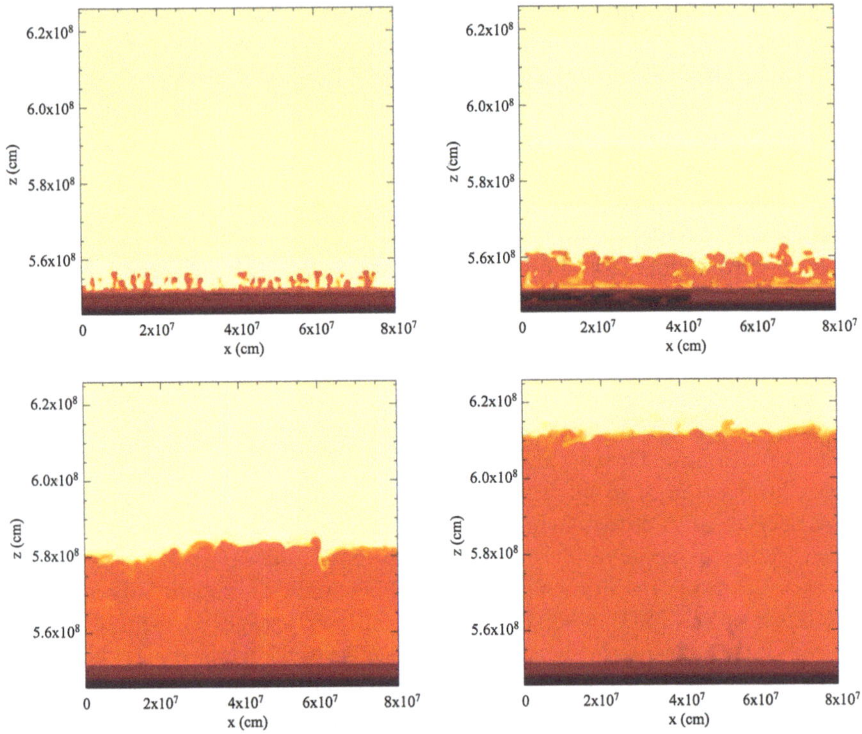

Fig. 5.4 Two-dimensional plots of the development of hydrodynamic instabilities, during a 3-D simulation of mixing at the core-envelope interface during a nova explosion, calculated with the hydrodynamic code FLASH

The inhomogeneities inferred from the ejecta have been usually attributed to uncertainties in the observational techniques, but they may represent a real signature of the turbulence generated during the thermonuclear runaway. More recently, similar results have also been reported for ONe-rich substrates (3D models; Casanova et al. 2016) and for different white dwarf masses (2D models; Casanova et al. 2018), proving that higher degrees of mixing (and therefore, more energetic outbursts) are found in ONe-rich than in CO-rich substrates, and for more massive white dwarfs.

5.2.3 Nucleosynthesis in Classical Novae

Nova outbursts eject much less mass than supernova explosions, but novae are much more frequent events than supernovae in the Galaxy; this has raised the issue of the potential contribution of such stellar cataclysms to Galactic abundances. Although the mass injected into the ISM per novae is small, detailed numerical simulations have indicated novae as major players in the synthesis of some specific nuclear species, largely overabundant in their ejecta, such as ^{13}C, ^{15}N, and ^{17}O, with a minor contribution to Galactic levels of other nuclei with A $<$ 40, such as ^{7}Li, ^{19}F, or ^{26}Al (Starrfield et al. 1998; José and Hernanz 1998).

Radioactivities present in nova ejecta, previously synthesized during the explosion, also constitute a major source of positrons. Indeed, ^{13}N and ^{18}F, and to a lesser extent, ^{22}Na, are the major contributors. The synthesis of ^{13}N and ^{18}F naturally occurs during the operation of the CNO cycle. Actually, the triggering reaction that powers the onset of the thermonuclear runaway is ^{12}C(p,γ)^{13}N, leading to ^{13}N synthesis. The exact amount transported to the outer envelope and contributing to γ-ray emission once transparency allows for the escape of photons, depends on details of the evolution, specially on convection. Therefore, detection of positrons from ^{13}N, through the associated electron-positron annihilation emission, would provide an important diagnostic of the dynamics of nova explosions.

The synthesis of ^{18}F in novae proceeds through the hot CNO cycle. Regardless of the nature of the white dwarf hosting the explosion (CO or ONe), the initial abundance of ^{16}O is large, and thus ^{16}O is the main source for ^{18}F formation, which can take place either through the reaction chain ^{16}O(p,γ)^{17}F(p,γ)^{18}Ne(β^+)^{18}F or via ^{16}O(p,γ)^{17}F(β^+)^{17}O(p,γ)^{18}F. The ^{18}F yields are severely constrained by its destruction mode, whatever the production channel is. During the runaway, ^{18}F destruction by beta decays can be neglected when compared to its destruction by proton captures (mainly through ^{18}F(p,α)^{15}O, which is faster than ^{18}F(p,γ)^{19}Ne (Hernanz et al. 1999). Other nuclear reactions affecting ^{18}F synthesis are proton captures on ^{17}O, i.e. ^{17}O(p,γ)^{18}F and ^{17}O(p,α)^{14}N (Coc et al. 2000).

Another interesting isotope likely produced during nova outbursts is ^{7}Li. Its synthesis is believed to proceed through the so-called *beryllium transport mechanism* (Cameron 1955), in which the previously synthesized ^{7}Be transforms into ^{7}Li through electron capture (τ = 77 days, see Table 5.1) releasing a γ-ray photon of 478 keV. For this mechanism to be effective, ^{7}Be has to be transported to the outer, cooler envelope layers, with a timescale shorter than its decay time, in order to preserve its fragile daughter ^{7}Li from destruction. This mechanism requires a dynamic situation like the one encountered in novae.

Table 5.1 Radioactive isotopes synthesized in explosive events

Isotope	Decay chain	Disintegration process	Lifetime	Line energy (keV)
^{7}Be	^{7}Be $\rightarrow$ ^{7}Li	e^--capture	77 days	478
^{22}Na	^{22}Na $\rightarrow$ ^{22}Ne	β^+	3.8 years	1275
^{26}Al	^{26}Al $\rightarrow$ ^{26}Mg	β^+	1.0 $\times$ 10^6 years	1809
^{44}Ti	^{44}Ti $\rightarrow$ ^{44}Sc $\rightarrow$ ^{44}Ca	e^--capture, β^+	89 years (5.4 h)	78, 68, 1157
^{56}Ni	^{56}Ni $\rightarrow$ ^{56}Co	e^--capture	8.8 days	158, 812, 750, 480
^{56}Co	^{56}Co $\rightarrow$ ^{56}Fe	β^+	111 days	847, 1238
^{57}Ni	^{57}Ni $\rightarrow$ ^{57}Co $\rightarrow$ ^{57}Fe	e^--capture	(52 h) 390 days	122, 136
^{60}Fe	^{60}Fe $\rightarrow$ ^{60}Co $\rightarrow$ ^{60}Ni	β^-	2.0 $\times$ 10^6 years (7.6 years)	1173, 1332

The production of ^{7}Li in novae was in some way debated during years, but the recent detections of both ^{7}Li and ^{7}Be in novae (first in 2015, see below) have confirmed that novae produce it. The first studies based on parameterized one-zone models (Arnould and Norgaard 1975), were followed by hydrodynamic computations (Starrfield et al. 1978b), which did not follow the accretion phase (i.e., they had an initial envelope already in place). These models indicated that the final amount of ^{7}Li synthesized depends on the initial abundance of ^{3}He and on the treatment of convection. Later works based on one or two-zone models (Boffin et al. 1993) pointed out the critical role played by the photodisintegration reaction ^{8}B(γ,p)^{7}Be. Finally, a complete hydrodynamic study, following both the accretion and the explosion phases, was performed (Hernanz et al. 1996). Formation of ^{7}Be proceeds through α-captures on ^{3}He, ^{3}He(α, γ)^{7}Be, since (p,γ) reactions can not bridge the A=5 gap; destruction occurs via ^{7}Be(p,γ)^{8}B; however, at high temperatures ($T \approx 10^8$ K) this rate achieves quasi-equilibrium with the inverse photodisintegration reaction, ^{8}B(γ,p)^{7}Be. Indeed, it was shown that a critical issue is the amount of ^{7}Be surviving the TNR, thanks to the efficient role played by ^{8}B photodisintegration (Hernanz et al. 1996). ^{7}Li formation is favored in CO novae with respect to ONe novae, because their faster evolution prior to the TNR (driven by the larger amount of ^{12}C injected in the envelope) favors ^{7}Be survival and thus final ^{7}Li production.

A tentative detection of ^{7}Li in the optical band, through the LiI doublet at 6708 Å, was reported for Nova Vel 1999 (Della Valle et al. 2002), but it was later suggested that the spectral feature could instead correspond to a doublet from neutral nitrogen (Shore et al. 2003). More than 10 years later, Tajitsu et al. (2015) provided the first observational evidence of ^{7}Li synthesis in novae (see comment in Hernanz (2015)). They detected ^{7}Be, the parent nucleus of ^{7}Li, during the nova explosion of Nova Del 2013 (V339 Del) with the High Dispersion Spectrograph (HDS) of the 8.2 m Subaru Telescope, in Mauna Kea (Hawaï). Observations were done at four epochs from 2013 September to October 7: 38, 47, 48 and 52 days after maximum. The HDS provided high spectral resolution in the near-UV (from 60,000 to 90,000), allowing them to distinguish the near-UV absorption lines of the resonance doublet of singly ionized ^{7}BeII at 313.0583 and 313.1228 nm, from those of the ^{9}BeII doublet at 313.0422 and 313.1067 nm, ruling out ^{9}Be. The ^{7}BeII lines were observed with blueshifts of 1103 and 1268 km s^{-1}, the same as for the Hη and CaII K lines.

Tajitsu et al. (2016) reported additional detections of the ^{7}BeII doublet with Subaru, in Nova Sgr. 2015 No.2 (V5668 Sgr) and in Nova Oph 2015 (V2944 Oph). The same year, Molaro et al. (2016) detected the same ^{7}BeII doublet with UVES, the high-resolution spectrograph of the ESO Very Large Telescope (VLT). They reported on the detection of highly blueshifted resonance lines of ^{7}BeII at 313.0583 and 313.1228 nm in Nova Sgr. 2015 No.2 (V5668 Sgr).

It is remarkable that ^{7}Li itself has also been detected for the first time in 2015 (Izzo et al. 2015): the detection of the ^{7}LiI doublet at 6708 Å, in Nova Cen 2013 (V1369 Cen), thanks to early observations getting high resolution spectra, was reported. Alternative identifications, however, are not discarded by the authors.

Large overabundances of ^{7}Be-^{7}Li with respect to solar are obtained in general from most of the observations, larger by factors that can reach 10 than the theoretically predicted ones (see below). However, these abundances from observations are not absolute but relative to Ca; also, in some cases—like for Nova Sgr 2015 No.2 (V5668 Sgr)—the optical light curve showed several maxima, which makes the computation of the time origin for the ^{7}Be decay into ^{7}Li ambiguous, impacting the value of the final ^{7}Li abundance.

Overproduction factors of ^{7}Li with respect to solar values around 1000 are predicted by CO nova models, meaning that novae can be important contributors to the Galactic ^{7}Li (Hernanz et al. 1996) (up to 20% of the Galactic content) and may help to reproduce the steep rise of the observed lithium abundance between the formation of the Solar System and the present (Romano et al. 1999; Alibés et al. 2002).

Classical nova explosions are also sources of two important radioactive isotopes: ^{22}Na and ^{26}Al. In the pioneering work by Clayton and Hoyle (1974), it was mentioned that novae are potential emitters of 1275 keV γ-rays resulting from ^{22}Na decay. They assumed ^{22}Na mass fractions in the ejecta of the order of 10^{-3}, from the conversion of ^{20}Ne to ^{22}Na. In the last 15 years it has been shown that this conversion is not so efficient, but interestingly, the current accepted ^{22}Na yields in the most prolific novae are not far from those historic predictions. The synthesis of ^{22}Na in novae proceeds through ^{20}Ne(p,γ)^{21}Na followed either by the decay of ^{21}Na to ^{21}Ne, i.e. ^{21}Na(β^{+})^{21}Ne(p,γ)^{22}Na, or by a proton capture on ^{21}Na, i.e. ^{21}Na(p,γ)^{22}Mg(β^{+})^{22}Na (José et al. 1999).

The amount of ^{22}Na synthesized during nova explosions has not yet been determined reliably. The first hydrodynamic models of nova outbursts did not include complete nuclear reaction networks covering the Ne-Na and Al-Mg regions. In the 80's, the crucial role played by some uncertain nuclear reaction rates for the yields of ^{22}Na (and ^{26}Al, see below) was finally pointed out, and extensive nova nucleosynthesis models were computed, with parameterized models (i.e. through simplified one-zone models) with representative temperature-density temporal profiles taken from evolutionary nova models (Starrfield et al. 1978a). In the 90s, new one-zone models for nova nucleosynthesis were developed, adopting various initial compositions which included the possibility of mixing with massive white dwarf cores. These models (Weiss and Truran 1990; Nofar et al. 1991) investigated in detail the synthesis of ^{22}Na and ^{26}Al, in view of the then recent detection of galactic ^{26}Al (and non detection of ^{22}Na). Prompted by the recent discovery of large enrichments of neon in the spectra of some novae, these calculations explored the outcome of nova outbursts on massive, ONeMg white dwarfs. Interestingly, Weiss and Truran (1990) obtained ^{22}Na yields as large as 10^{-4}, which combined with envelope masses of $2 \times 10^{-5}\,M_\odot$ gave $4 \times 10^{-9}\,M_\odot$ of ^{22}Na ejected into the interstellar medium. The most recent hydrodynamic models of ONe (or ONeMg) nova outbursts on masses larger than $1.0\,M_\odot$ provide ^{22}Na yields in the range 10^{-4} to $10^{-3}\,M_\odot$ (José et al. 1999; Politano et al. 1995; Starrfield et al. 1998). It is worth mentioning that mixing can occur at various depths inside the stratified ONe white dwarf: inner ONe core, outer CO buffer or middle transition zone (José et al. 2003). If mixing occurs with the unburned CO buffer on top of the ONe core, no ^{22}Na would be expected (José et al. 2003).

^{26}Al production is complicated by the presence of a short-lived (half-life 6.3 s) isomer state, ^{26}Al^m. In fact, when the temperature is smaller than 4×10^8 K (as is the case in novae), the ground (^{26}Al^g) and isomeric states must be treated as two separate isotopes, because they do not reach thermal equilibrium (Ward and Fowler 1980).

The first calculations of ^{26}Al synthesis during explosive hydrogen burning (Hillebrandt and Thielemann 1982; Wiescher et al. 1986) suggested that novae are likely sites for the synthesis of this radioactive isotope, but not in very large amounts; these computations relied on solar or CNO-enhanced white dwarf envelopes. Later computations, again one-zone models, demonstrated the need for initial envelope enrichment in O, Ne and Mg, dredged-up from the white dwarf cores, to obtain larger amounts of ^{26}Al (Weiss and Truran 1990; Nofar et al. 1991).

The major seed nuclei for ^{26}Al synthesis are ^{24,25}Mg. At the early phases of the thermonuclear runaway (burning shell temperatures around 5×10^7 K), the dominant reaction is ^{25}Mg(p,γ)^{26}Al^{g,m}; the subsequent reaction ^{26}Al^m(β^+)^{26}Mg(p,γ)^{27}Al produces the stable isotope ^{27}Al. At larger temperatures ($\sim 10^8$ K), the nuclear path ^{24}Mg(p,γ)^{25}Al(β^+)^{25}Mg dominates, with again ^{25}Mg(p,γ)^{26}Al^{g,m}. When temperature reaches 2×10^8K, (p,γ) reactions proceed very efficiently and reduce the amount of ^{25}Al, leading to the formation of ^{26}Si (^{25}Al(p,γ)^{26}Si) which decays into ^{26}Al^m, thus by-passing ^{26}Al^g formation. Also ^{26}Al itself (in both states) is destroyed to ^{27}Si which decays into ^{27}Al. In summary, the final amount of ^{26}Al^g and the ratio ^{26}Al^g/^{27}Al mainly depend on the competition between the two nuclear paths ^{24}Mg(p,γ)^{25}Al(β^+)^{25}Mg(p,γ)^{26}Al^{g,m} and ^{24}Mg(p,γ)^{25}Al(p,γ)^{26}Si. The first channel is the only one producing ^{26}Al^g, whereas both channels produce ^{27}Al (through ^{26}Al^{g,m}(p,γ)^{27}Si(β^+)^{27}Al or ^{26}Si(β^+)^{26}Al^m(β^+) ^{26}Mg(p,γ)^{27}Al (José et al. 1999).

The final ^{26}Al yields from novae sensitively depend on the initial mass of the white dwarf and on the degree of mixing between the accreted envelope and the core. Recent hydrodynamic models of ONe (or ONeMg) nova outbursts on masses larger than 1.0 $M_\odot$ suggest ^{26}Al yields in the range 10^{-4} to 10^{-2} (José et al. 1999; Politano et al. 1995; Starrfield et al. 1998). If mixing occurs with the CO buffer on top of the bare ONe nuclei (or in the transition zone), some ^{26}Al would be expected (but no ^{22}Na), since there is a non negligible amount of the seed nucleus ^{25}Mg both in the CO buffer and in the transition zone (José et al. 2003).

5.3 SNIa Explosions

The pyrotechnical displays of the total explosion of stars are called supernovae. They are characterized by a sudden rise of the luminosity, followed by a steep decline lasting several weeks and eventually followed by a more gradual decline lasting many years. The total electromagnetic output, obtained by integrating the light curve, is $\sim 10^{49}$ erg, while the luminosity at maximum can be as high as $\sim 10^{10}\,L_\odot$. The kinetic energy of supernovae can be estimated from the expansion velocity of the ejecta, $v_{\rm exp} \sim$ 5000–10,000 km s^{-1}, and turns out to be $\sim 10^{51}$ erg. Such an amount of energy can only be obtained from the gravitational collapse of an electron degenerate core, forming a proto-neutron star or a black hole, or from its thermonuclear incineration to iron-peak isotopes. In the former case, the gravitational binding energy of a neutron star of $\sim 1.4\,M_\odot$ and ~ 10 km radius is of the order of $\sim 10^{53}$ ergs and a weak coupling between the source of energy and matter is enough to fulfill the energetic requirements (Zwicky 1938). In the second case, the nuclear specific energy of a carbon oxygen mixture is $q \sim 10^{18}$ erg/g, sufficient to obtain the required energy from burning $\sim 1\,M_\odot$ (Hoyle and Fowler 1960).

Supernovae are classified according to their spectrum at maximum light. If hydrogen lines are absent, they are called Type I supernovae or SNI. If these lines are present, they are referred to as Type II or SNII (Minkowski 1941). Also according to their spectra, SNI are further divided into three categories. If a prominent line of SiII is present, they are labeled SNIa. If this line is absent but there is a prominent HeI line, they are denoted SNIb. If both SiII and HeI lines are absent, the classification label is SNIc (Wheeler and Harkness 1990).

The light curves of SNIa are characterized by a rapid rise in luminosity, up to an average maximum $M_V \approx M_B \approx -19.30 \pm 0.03 + 5\log(H_0/60)$ (Riess et al. 1999) in about 20 days, where H_O is the Hubble constant, followed by a comparatively gentle decline divided into two different epochs. The first epoch after maximum light lasts ~ 30 days and the luminosity typically drops by $\sim 3^{mag}$, while the second phase is characterized by a slow decline with a characteristic time of ~ 70 days. Infrared photometry shows that in the J, H and K bands there is a well defined minimum at ~ 20 days after maximum and a secondary peak ~ 30 days after maximum (Elias et al. 1985), although in some cases this secondary peak is absent.

This behavior can be understood in terms of the deposition of a large amount of energy, $\sim 10^{51}$ ergs, in the interior of a stellar envelope (Mochkovitch 1994). If it is assumed that half of this energy is invested into the expansion of the debris and half into their internal energy, the average temperature is:

$$T = 6.3 \times 10^4 \left(\frac{E_{SN,51}}{R_{15}^3} \right)^{1/4} \tag{5.16}$$

where $E_{\rm SN,51}$ is the energy released by the supernova in units of 10^{51} erg and R_{15} the radius in units of 10^{15} cm. Consequently:

$$\frac{P_R}{P_G} = \frac{{}^1\!/_3 a T^4}{\frac{\Re}{\mu} T \rho} \approx 1.6 \times 10^4 \frac{\left(R_{15} E_{SN,51}\right)^{3/4}}{M} \tag{5.17}$$

which means that a supernova is, in a certain sense, just a ball of light. Typical expansion velocities are:

$$v \approx 10^9 \sqrt{\left(\frac{E_{SN}}{10^{51}} \cdot \frac{M_\odot}{M}\right)} cm/s \tag{5.18}$$

Since the internal energy is dominated by radiation and the expansion is nearly adiabatic, $T \propto R^{-1}$, the total energy will evolve as $E_{th} \propto VT^4 \propto R^{-1}$. If the initial structure is compact, $\sim 10^8$ cm, the typical size of a relativistic degenerate stellar core, the energy decreases from $\sim 10^{51}$ to $\sim 10^{44}$ erg when the radius is $\sim 10^{15}$ cm, the typical radius at maximum light; i.e. the energy deposited by the shock is invested in the adiabatic expansion of material. On the contrary, if the initial structure is extended,

$$E_{th} \approx E_{th,0} \frac{R_0}{R_{env}} \approx \frac{E_{SN}}{2} \frac{R_0}{R_{env}} \tag{5.19}$$

the luminosity will be given by

$$L_P \approx \frac{E_{th}}{\tau_{diff}} \approx \frac{2\pi c}{9k_{Th}} \frac{E_{SN}}{M_{env}} R_0 \tag{5.20}$$

where the diffusion time has been estimated as:

$$\tau_{diff} \approx \frac{3R_{env}^2}{\lambda c} \approx \frac{9\kappa M_{env}}{4\pi c R_{env}} \tag{5.21}$$

and the thermal energy will provide a luminosity plateau or a broad peak after maximum.

Therefore, the explosion of a compact object is able to account for the light curve of Type Ia supernovae, but an additional source of energy is necessary to explain the tail. Although in the past some other possibilities were considered, there is now broad consensus that this energy source is provided by the radioactive decay of ^{56}Ni (Truran et al. 1967; Colgate and McKee 1969):

$$^{56}\text{Ni} \rightarrow ^{56}\text{Co} \rightarrow ^{56}\text{Fe}$$

with $q_{Ni} \sim 7 \times 10^{49}$ erg/M$_\odot$, $q_{Co} \sim 1.5 \times 10^{50}$ erg/M$_\odot$ and $\tau_{1/2}(Ni) = 6.1$ days, $\tau_{1/2}(Co) = 77.1$ days. This hypothesis has been recently confirmed with the observation by *INTEGRAL* of the emission of ^{56}Ni around the maximum of the light curve (Diehl et al. 2014; Churazov et al. 2015; Isern et al. 2016), see Fig. 5.5, and that of ^{56}Co 60 days after the explosion (Churazov et al. 2014) of SN2014J.

The majority of the energy released by the decay of ^{56}Ni and ^{56}Co is in the form of γ-rays of $\sim$1 MeV that are scattered and eventually thermalized via Compton scattering and photoelectric absorption. The resulting thermal photons diffuse and

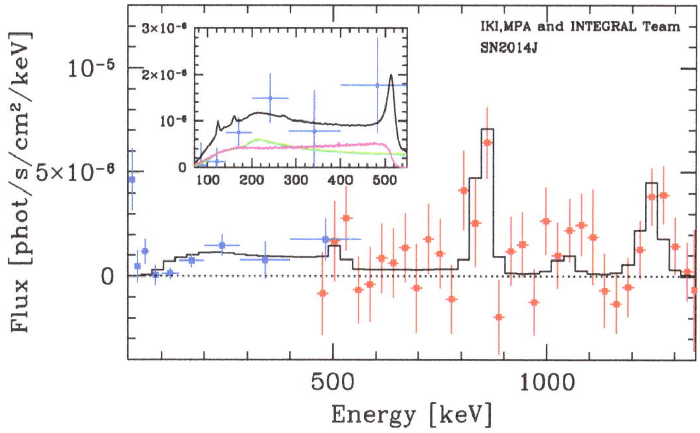

Fig. 5.5 ^{56}Co gamma lines from SN 2014J obtained with the *INTEGRAL* instruments, with data points from INTEGRAL SPI (red) and IBIS (blue) instruments; the black histogram is a fiducial model of the spectrum expected for day 75 after the explosion (Churazov et al. 2015)

eventually escape. The observed light curve thus results from a competition of two time scales describing diffusive energy transport and dynamic expansion. As before, the diffusion time scale is dominated by Thomson scattering and by absorption from bound electrons. The contribution of true absorption to the total opacity is a complicated issue because of the departures from LTE, the Doppler effect introduced by the expansion velocity and the uncertainties introduced by the chemical composition and the energy levels of different ions. Nevertheless, the opacity seems to be confined within the range 0.2–0.03 cm^2/g. If it is assumed that the envelope expands with constant velocity, $R_{env} \sim R_0 + v_{exp}t$, the hydrodynamic time scale is $\tau_h \sim R_0/v_{exp}$.

Initially the ejecta are opaque, $\tau_{diff} \gg \tau_h$, and the luminosity is small. As time goes by, $\tau_{diff} \sim \tau_h$, and photons begin to escape. Since the energy output decreases exponentially, a peak appears in the light curve that is equal to the instantaneous deposition of energy and therefore $L_{max} \propto M_{Ni}$. After the peak, the radiation trapped in the envelope diffuses outwards and the luminosity exceeds the instantaneous energy deposition rate. The width of the peak is determined by an effective diffusion time:

$$\tau_m = \sqrt{2\tau_{diff}\tau_h} \propto \kappa^{1/2} M_{env}^{3/4} M_{Ni}^{-1/4} \tag{5.22}$$

Later, when the density is small enough, an increasing fraction of the γ-photons (and later also positrons) can escape and, consequently, the luminosity is smaller than the energy output of radioactive decays. Some radioactive energy may be stored in the form of ionization, and defer luminosity originating from radioactive energy. See Arnett (1996) and references therein for a complete discussion.

The observation of the bolometric light curves together with these simple relationships allow estimates of the mass expelled and the mass of the radioactive elements synthesized by SNIa: $M_{env} = 0.4$–$1.4\,M_{\odot}$ and $M_{Ni} = 0.1$–$1\,M_{\odot}$ (Stritzinger et al. 2006). One of the most striking properties of SNIa is their photometric homogeneity i.e. the light curve of the majority shows a very small dispersion at maximum, $\sigma_M \leq 0.3^{mag}$ (Cadonau et al. 1985; Hamuy et al. 1996) when they are normalized to the peak. All these properties together immediately suggest that the most plausible scenario is the explosion of a CO white dwarf near the Chandrasekhar mass in a close binary system. This hypothesis has been confirmed in the case of SN 2011fe, which shows that the properties of the early light curve are only compatible with the explosion of a white dwarf (Bloom et al. 2012)

Spectroscopic observations at different epochs enable tomography of supernovae. During maximum light, the spectra of SNIa are characterized by various lines of neutral and singly ionized atoms of Si, Ca, Mg, S and O moving at high velocities ($v \sim 8000$–$30{,}000\,\mathrm{km\,s^{-1}}$) indicating that the outer layers are mainly composed of intermediate mass elements, i.e. that thermonuclear burning was not complete (Filippenko et al. 1992). Two weeks after maximum, permitted FeII lines are prominent, indicating that the photosphere has reached regions where the star was able to completely incinerate matter (Harkness 1991). The nebular phase starts 1 month after the maximum, roughly when the tail of the lightcurve begins. During this epoch, the spectrum is dominated by forbidden FeII, FeIII and CoIII emission lines (Axelrod 1980). The decrease of the Co lines, together with the relative intensity of CoIII with FeIII (Kuchner et al. 1994) provide support for the idea of a light curve tail powered by the decay of ^{56}Co. In general, the lines of different elements have different expansion velocities, indicating a layered structure where the central regions are occupied by completely burned material, i.e. the iron peak elements. This property rules out the hypothesis of a prompt detonation since in this case the white dwarf would be completely converted to ^{56}Ni.

Despite their remarkable photometric homogeneity, SNIa do exhibit some degree of heterogeneity. Already in 1973, it was proposed (Barbon et al. 1973) to sub-divide SNIa into a fast and slow class according to the rate of decline of their light curve just after maximum, the transition from the peak to the tail, and the decline of the tail. The slow class is characterized by a broader and more luminous peak than the fast class. The most extreme cases are SN1991T, considered until recently the most energetic event with the broadest peak, and SN1991bg and SN1992K, which are the reddest, fastest and most subluminous Type Ia supernovae known to date (Phillips et al. 1992; Ruiz-Lapuente et al. 1992). The difference in brightness between the extreme cases is $\sim 2^{mag}$. The large majority of SNIa also display a remarkable spectroscopic homogeneity (Branch et al. 1993) not only during the maximum but also during subsequent months. They are classified as *Branch-normal* and represent 85% of the total, although there are suggestions that this value should be reduced to 70%. The prototypes are SN1972E, 1981B, 1989B and 1994D. On the contrary, 91T-like events display FeIII lines before maximum while the 91bg-like supernovae lack the characteristic secondary maximum in the infrared. At present the question is to decide if they can be considered as different subtypes or just extreme cases of the normal events

This mildly inhomogeneous set of SNIa exhibits a correlation between peak magnitude, the width of the peak, and the expansion velocity, in the sense that the brightest SNIa show the largest expansion velocities (Pskovskii 1977; Branch 1981). The correlation between the brightness and the shape of the light curve was settled definitively when a clear relationship between the maximum of light and the magnitude decline during the first 15 days after maximum (Δm_{15}) was firmly established (Phillips 1993). This correlation (which can be parameterized in terms of the decline rate (Phillips 1993; Hamuy et al. 1996), the stretch parameter (Perlmutter et al. 1997), or via a multi-parameter fit of colors (Riess et al. 1996)) is used to renormalize the peak magnitudes and thereby substantially reduces the dispersion of the absolute magnitudes, making SNIa one of the most powerful tools for measuring cosmological distances.

In principle, these variations in the shape of the light curve can be understood in terms of the total amount of ^{56}Ni synthesized, if the ejected mass is kept constant. Since the maximum of the luminosity is proportional to the ^{56}Ni mass, the brightest events are those that have synthesized the largest amount of this material and consequently have larger expansion velocities and broader peaks since the opacity of iron peak elements is very large. In any case, this diversity of properties poses the question whether there are two explosion mechanisms, one for the *Branch-normal* supernovae and another for the peculiar events, or if there is a unique mechanism able to account for the broad range of behaviors.

The recent systematic searches of supernovae have revealed the existence of new subtypes besides the Branch-normal ones (Taubenberger 2017). The most relevant are the so called SNIax, the Super-Chandrasekhar, and the CSM types but several other exist although their relevance has not been elucidated yet. One example of the last ones is that of the Ca-rich transients, which have a luminosity in between novae and normal supernovae and display prominent Ca lines in the late spectra. All of them are thought to have a thermonuclear origin.

Type Iax supernovae (SNIax) were proposed by Silverman et al. (2012) as a true subclass. The prototype is SN 2002cx, their spectrum at maximum is similar to that of SN 1991T but they are as subluminous as SN 1991bg. The expansion velocity at maximum is $\sim$6000 km s^{-1}, that is half of that of normal supernovae, indicating a smaller kinetic energy per unit mass and do not display a secondary maximum in the infrared. The amount of ^{56}Ni synthesized in this case is $\sim$0.25 M$_\odot$. They also present a correlation between the luminosity at maximum and the early decline of the light curve, but is different from that of the normal ones. Their maximum luminosity and expansion velocity lie in the range $-14.2 > M_V > -18.4$ and $2000 < v < 8000$ km s^{-1} respectively, suggesting a large range in explosion energies, ejected masses and ^{56}Ni masses. It is suspected that their contribution to the total number of SNIa could be as large as 1/3.

The 'Super-Chandrasekhar' SNIa are $\sim$1 mag brighter than the normal ones and have low expansion velocities, $\sim$8000 km s^{-1}, around the maximum. According to the Arnett's rule the mass of ^{56}Ni synthesized during the event should be $\sim$1.3 M$_\odot$.

This value, together with the low velocities suggest that the mass of the object that exploded was larger than the Chandrasekhar's mass. This class contains few events, being SN 2003gz (Howell et al. 2006) the first one to be discovered. This, together with their high luminosity suggests that their frequency is very low.

SNIa-CSM are characterized by the coexistence of a normal SNIa spectrum, often 91T-like, with a blue continuum and the presence of hydrogen Balmer lines, suggesting an interaction between the supernova and the circumstellar material. The first event discovered was SN 2002ic (Hamuy et al. 2003). Since then, several different events have been added to the group, but the sample is still small, suggesting they are rare events that not represent more than $\sim 1\%$ of the total.

The frequency and the impact on the chemical content of galaxies provide additional constraints on the different supernova mechanisms. The rate of supernovae in galaxies is usually normalized to the galaxy blue luminosity (Tammann 1970) or to the mass assuming an average M/L-ratio for each galaxy type (Cappellaro et al. 2003). The most striking feature is that SNIb/c and SNII only appear in spiral—and irregular galaxies, and are associated with young populations, while SNIa can appear in all galaxy types, indicating that they are related to the old stellar populations. Nevertheless, the SNIa rate per unit mass is almost three times larger in late spirals than in ellipticals, thus implying that at least a fraction of SNIa must be related to the young stellar population (Cappellaro et al. 2003). Furthermore, there is some evidence that, on average, SNIa in red or early type galaxies are dimmer, have faster light curves and slower expansion velocities than those in blue or late type galaxies (Hamuy et al. 1995, 1996; Branch et al. 1996). On the other hand, the frequency of supernovae in the Milky Way has been estimated (van den Bergh and Tammann 1991) to be: $R_{\rm II} = 3.32 \times 10^{-2}$ year^{-1}, $R_{\rm Ib/c} = 0.65 \times 10^{-2}$ year^{-1} and $R_{\rm Ia} = 0.41 \times 10^{-2}$ year^{-1}. Taking into account that the mass of ^{56}Ni ejected per event is roughly 0.07, 0.3 and 0.6 $M_\odot$ for SNII, SNIb/c and SNIa, respectively, it turns out that nearly half of the galactic iron was synthesized by Type Ia supernovae. This means that SNIa have to produce the right amount of iron peak isotopes to account for the observed isotopic Solar abundances.

5.3.1 *Chandrasekhar-Mass Models*

As discussed in Sect. 5.1.3, the outcome of carbon ignition under degenerate conditions in a white dwarf near the Chandrasekhar limit can be a detonation or a deflagration, depending on the particular structure at the moment of ignition, represented by density, temperature, chemical composition and velocity profiles. For instance, it is easier to generate the overpressure necessary to launch a detonation at low densities, $\sim 3 \times 10^7$ g/cm^3, than at high densities due to the degeneracy dependence on density and temperature.

The Prompt Detonation Model Even though a pure detonation seems possible from a physical point of view (Blinnikov and Khokhlov 1986), this kind of explosion cannot account for the observed SNIa spectra at maximum light.

At densities above $\sim 10^7$ g/cm^3, the fuel is completely incinerated to Fe-peak elements and it leaves only a few hundredths of a solar mass of intermediate mass elements, which is not enough to produce the characteristic strong SiII line of SNIa. The rejection of a pure detonation as the SNIa mechanism is a consequence of the simplicity of this burning mode. As discussed in Sect. 5.1.3, in the absence of external perturbations like a piston, the Chapman-Jouguet detonation is the only stable solution (other than a deflagration) of the Rankine-Hugoniot equations that define the burning front. Thus, there are no free parameters left, no time for modification of the fuel pre-combustion structure, no diversity, and pure detonations always produce the wrong result. Notice however that the presence of a shallow thermal gradient close to the ignition profile might induce the formation of shocks that could burn a large mass in a short time, starting the dynamical phase of the supernova (Blinnikov and Khokhlov 1986; Bravo et al. 1996) and a mixture of deflagration and detonation regimes might be the result.

The Deflagration Model Deflagrations are less constrained from a physical point of view, but their properties are strongly conditioned by the hydrodynamics of the explosion process itself. As described in Sect. 5.1.3, at the microscopic scale the speed of the flame only depends on the local physical conditions. Thus, the laminar velocity of the flame can be determined as a function of density (Timmes and Woosley 1992): $v'_{\mathrm{lam}} = \alpha\rho_9^{\beta}$ cm/s, where ρ_9 is the density in units of 10^9 g/cm^3, and α and β are fit parameters. For $\rho_9 < 0.36$, $\alpha = 5.68 \times 10^6$ and $\beta = 1.46$, while for $2 \geq \rho_9 \geq 0.36$, $\alpha = 3.68 \times 10^6$ and $\beta = 1.03$. A further correction can be obtained taking into account the effect of Coulomb interactions: $v_{\mathrm{lam}} = K_{\mathrm{cc}}v'_{\mathrm{lam}}$, where $K_{\mathrm{cc}} = 0.894 - 0.0316\log(\rho_9)$ (Bravo and García-Senz 1999). The situation becomes extremely complicated when the flame is accelerated by the deformation induced by hydrodynamic instabilities. This acceleration is difficult to describe because it implies many length scales, from the global length scale, $\sim 10^7$–10^8 cm, to the microscopic width of the laminar flame, which strongly depends on the density as shown before.

One possibility (Woosley and Weaver 1986) is to parameterize the velocity of the deflagration as a function of flame radius, r, as: $v_{\mathrm{def,W}} = Av_{\mathrm{sound}}\left(1 - \mathrm{e}^{-Br}\right)$. The parameters A and B are constrained by the condition that the flame should propagate at a small Mach number close to the center but should reach velocities as high as 0.1–0.5 Mach in the outer layers of the white dwarf. A second possibility (Khokhlov 1995), assumes that the rate of surface creation by the turbulence is balanced by the rate of surface destruction due to flame propagation. The deflagration velocity should then be given by $v_{\mathrm{def,K}} = 0.5\sqrt{g_{\mathrm{eff}}L}$, where L is the driving scale, $g_{\mathrm{eff}} = g\mathrm{At}$, g is the gravitational acceleration and At is the Atwood number (Timmes and Woosley 1992). Such kind of self-regulating regime has been implemented in several ways in many multidimensional simulations of SNIa (Gamezo et al. 2003; García-Senz and Bravo 2005). A third possibility (Niemeyer and Woosley 1997) is that the deflagration moves at the speed of the Rayleigh-Taylor bubbles in the nonlinear scale, the so-called Sharp-Wheeler model, in which the velocity increases linearly with time t, $v_{\mathrm{def,NW}} = 0.1g_{\mathrm{eff}}t$. Finally, the concept of a subgrid-scale model that takes into account the dissipation of turbulent energy at microscopic scales has been adopted in many multidimensional simulations performed to date (Schmidt and Niemeyer 2006; Röpke et al. 2006). In spite of the differences in the treatment of the flame, most three-dimensional simulations of SNIa produce quite homogeneous results.

The success or failure of a deflagration model depends on its ability to consume the fuel with the same speed as the front engulfs it, such that it does not leave unburned pockets of carbon and oxygen behind (Niemeyer and Woosley 1997). High-resolution simulations aimed to explore the multipoint ignition scenario (Röpke et al. 2006) indicate that when the number of initial seeds increases, the ignition volume becomes saturated and the gross features of the explosion converge towards a unique solution. The optimal number of flame seeds is estimated to be in the range $\sim$100–400 per octant distributed in radius following a Gaussian up to $\sim$100–150 km. However, even in the most favorable case it is difficult to obtain more than 0.7 $M_\odot$ of ^{56}Ni and a kinetic energy above 0.7×10^{51} ergs, which is too small to account for the bulk of bright-normal SNIa. In addition, the deflagration always leaves a large mass of carbon and oxygen unburned, $M_{\rm ub} > 0.57\ M_\odot$ (Schmidt et al. 2006).

The present three-dimensional simulations of Type Ia supernovae based on a pure deflagration algorithm have to face the following problems when confronted with observations: (i) although the amount of Fe-group elements synthesized in the explosion is sufficient, the mass of ^{56}Ni is not. (ii) the final kinetic energy is always smaller than the canonical value of 10^{51} ergs. (iii) the synthesis of intermediate-mass elements is scarce. (iv) the ejecta lack chemical stratification. (v) big clumps of radioactive ^{56}Ni are present at the photosphere at the time of maximum luminosity.

Before discarding deflagrations as the main mode of Type Ia supernovae it is necessary to examine some still poorly understood aspects. For instance, it might be that the theoretical description of subsonic flames included in the hydrodynamical codes and used in the simulations is incomplete as is the case when they enter the distributed regime at densities lower than $\sim(1\text{–}3) \times 10^7$ g/cm^3. It is also important to notice that the influence of the initial conditions at the onset of the ignitions has not yet been clarified.

The Delayed Detonation Model In 1974, a burning regime was proposed (Ivanova et al. 1974) in which the initial flame was not able to unbind the star, leading to a pulsation and a delayed transformation of the deflagration into a detonation (deflagration-detonation transition or DDT). The DDT concept was later extended (Khokhlov 1991; Khokhlov et al. 1993) to include the possibility of a transition to detonation without an intervening pulsation. The essential ingredient for the formation of a detonation is the existence of a non-uniformly preheated region with a level of fluctuations of temperature, density, and chemical composition such that a sufficiently large mass would be burnt before a sonic wave could cross it. The thermal gradient needed (Khokhlov 1991) is:

$$\nabla T < \frac{\Theta T}{A v_{\rm sound} \tau_i} \tag{5.23}$$

where A is a numerical coefficient, $A \sim 0.2$–5.0, $\tau_i = T/\dot{T}$ is the induction time at the temperature T, and $\Theta \sim 0.04$–0.05 is the Frank-Kameneetskii factor:

$$\Theta = -\frac{\partial \ln \tau_i}{\partial \ln T} \tag{5.24}$$

There are several mechanisms that can produce such fluctuations: adiabatic pre-compression in front of a deflagration wave, shock heating, mixing of hot ashes with fresh fuel (Khokhlov 1991), accumulation of pressure waves due to a topologically complex geometrical structure of the flame front, or transition to the distributed burning regime (Niemeyer and Woosley 1997). Among these possibilities the turbulence pre-conditioning has received more attention. In the case of a white dwarf expanding as a consequence of a deflagration, the turbulent velocity has to exceed the laminar flame velocity by a factor 1–8 at a length scale comparable to the detonation wave thickness (Khokhlov et al. 1997). This criterion is fulfilled for flame densities in the range 5×10^6 g/cm^3 $< \rho <$ $(2 - 5) \times 10^7$ g/cm^3 for reasonable assumptions. At densities above 10^8 g/cm^3, a DDT transition is unlikely (Khokhlov et al. 1997) although the bubble fragmentation could increase the flame surface and facilitate a DDT at $\rho \sim 2 \times 10^8$ g/cm^3 (Zingale and Dursi 2007). Despite the difficulties to justify the DDT models, the one-dimensional delayed detonation simulations are particularly successful in reproducing many key observational characteristics of SNIa (Hoeflich and Khokhlov 1996), like the light curves and photospheric expansion velocities.

The Pulsational Delayed Detonation Model The 3D formulation of the pulsational delayed detonation model is the so-called Pulsational Reverse Detonation model or PRD. In this scenario the detonation is triggered by an accretion shock that forms above a quasi-hydrostatic core composed mainly of C-O with a mass of 0.8–1.15 $M_\odot$. Heating by the accretion shock ignites the fuel slightly below the core's surface. Because of the inertial confinement provided by the material falling through the accretion shock, the core cannot expand and cool efficiently. As a consequence, a detonation propagating inwards forms and burns most of the core. The resulting energetics as well as the nucleosynthesis are roughly in agreement with observations, specially concerning the observed stratification.

5.3.2 *Super-Chandrasekhar Models, Sub-Chandrasekhar, and White Dwarf-White Dwarf Collisions*

The only way known to push the mass of a degenerate structure beyond the Chandrasekhar limit is through rotation. If it is assumed that the white dwarf rotates as a rigid body, it is possible to delay the ignition up to masses of the order of 1.4–1.5 $M_\odot$ (Piersanti et al. 2003a,b). If differential rotation is allowed, the ignition can be delayed up to masses $\sim$2 $M_\odot$ (Piersanti et al. 2003a). A follow-up of rigid rotation models has been calculated with a 1D hydrodynamic code modified to take into account the centrifugal force (Domínguez et al. 2006) and a weak dependence on the rotation period has been found. The problem is that these calculations assume that the transition from deflagration to detonation occurs at a fixed density and it is not known how rotation affects this change of regime of the burning front.

Sub-Chandrasekhar models assume a white dwarf with a mass $M_{WD} \leq 1.1$–$1.2\,M_{\odot}$ that accretes helium rich matter at a rate in the range $10^{-9} \leq \dot{M} \leq 5 \times 10^{-8}\,M_{\odot}$/year. These rates allow the formation of a degenerate helium mantle around the initial CO core. When this mantle reaches a critical mass, $\sim$0.2–0.3 $M_{\odot}$, a thermonuclear runaway starts at its bottom and triggers the explosion of the star before reaching the Chandrasekhar limit. Notice that white dwarfs with an initial mass larger than 1.2 $M_{\odot}$ could reach the Chandrasekhar mass before exploding and experience central ignition.

One dimensional models indicate that before the thermonuclear runaway occurs, the base of the helium layers becomes convective and transports energy and part of the reactants away from the inner core boundary in such a way that He ignites above the interface. The high flammability of helium together with the low density of the envelope guarantees the formation of a detonation that incinerates the envelope and launches a shock wave inwards through the CO core. Because of the focusing effect of the spherical symmetry, this shock strengthens and induces the detonation of C in the central region that leads to a supernova explosion.

These explosions reproduce the gross features of SNIa explosions, specially sub-luminous ones like SN 1991bg and allow to explain with a single parameter, the initial mass of the white dwarf, their observed diversity. Despite such advantages, Sub-Chandrasekhar models were not the favorite to account for SNIa outbursts. The reason was that they predicted the existence of a very fast moving layer composed of ^{56}Ni and ^{4}He that is not observed, as well resulting in light curves that rise too fast (Hoeflich and Khokhlov 1996). The situation changed when it was shown (Bildsten et al. 2007; Shen et al. 2010; Shen and Moore 2014) that it was possible to induce a detonation with a He-envelope as small as $\simeq 10^{-2}\,M_{\odot}$, thus removing the constrain introduced by the lack of ^{56}Ni lines in the spectrum at maximum light. Multidimensional calculations by Fink et al. (2007), Sim et al. (2007, 2010, 2012) in 2D, and Moll and Woosley (2013) in 3D, have confirmed these results and, furthermore the recent work by Blondin et al. (2017) and Goldstein and Kasen (2018) has shown that this scenario can reproduce the faint end of the Phillips relationship. From the observational point of view it has been shown that the companion of the white dwarf star that exploded as SN 2012Z, a Type Iax supernova, was probably a He-star (McCully et al. 2014) and that the protuberance recently found in the early light curve of SN 2017a could be due to the presence of ^{56}Ni in the outer layers (Jiang et al. 2017). To these evidences it has to be added the excess of gamma-ray emission near the maximum of the optical light curve that seems to be produced by the presence of ^{56}Ni in the outer layers (Diehl et al. 2014; Isern et al. 2016).

Another possibility that has emerged within the context of the double degenerate scenario is the collision, not the merging, of two white dwarfs with velocities of the order of the free-fall time. Such collisions can produce events that can be assimilated to standard and non standard SNIa (Benz et al. 1989; Raskin et al. 2009; Rosswog et al. 2009a; Lorén-Aguilar et al. 2010; Aznar-Siguán et al. 2013, 2014; García-Senz et al. 2013; Kushnir et al. 2013), but it was believed they were very rare and only could occur inside dense ambients like globular clusters. However, very recently it has been suggested that triple systems containing an inner close binary white dwarf could experience significant pericenter changes with time scales of the order of the orbital period that could end with a violent collision (Antonini and Perets 2012), although the frequency of such scenario has not been elucidated yet.

The outcome of the collision strongly depends on the size of the He-layer of white dwarfs. Papish and Perets (2016) have found that if the mass of the helium layer is larger than $\sim$0.1 $M_\odot$, the detonation of He propagates on the white dwarf surface before triggering the central ignition of the core. Since the burning of helium at these densities is not efficient enough, important amounts of ^{44}Ti and ^{48}Cr are synthesized. If the He-shells are low mass, there is not a helium detonation, but helium burning precedes the detonation of the C/O core, and important amounts of material enriched with intermediate mass elements are ejected at high velocities.

5.3.3 Nucleosynthesis in Thermonuclear Supernovae

The abundances of the elements synthesized in SNIa events depend on the peak temperature reached by the material and on the excess of neutrons versus protons. Roughly speaking, the SNIa material undergoes four burning regimes: (i) nuclear statistical equilibrium (NSE), (ii) incomplete Si burning, (iii) incomplete O burning and (iv) incomplete C-Ne burning (Woosley 1986). The neutron excess depends on the initial abundance and distribution of the neutron rich isotopes like ^{22}Ne, which depend on the metallicity and thermal history of the white dwarf, and on the extent of electron captures on the burned material, which mainly depends on the ignition density and on the burning regime. Another complication comes from the additional degrees of freedom introduced by 3D flames that open a variety of possible ignition modes as well as the possibility of leaving pockets of unburned material. Finally, the adopted nuclear reaction and electron capture rates are an important source of uncertainty. Despite all these factors it is possible to obtain some insight into the problem using parameterized 1D models with different propagation velocities of the burning front, different ignition densities and different initial metallicities. Three dimensional models still need some additional work.

The chemical composition of matter in nuclear statistical equilibrium (NSE) with equal number of protons and neutrons, i.e. with electron mole number $Y_e = 0.5$, peaks around ^{56}Ni. When Y_e takes values in the range 0.470–0.485, the peak moves towards ^{58}Ni, and ^{54}Fe; values in the interval 0.46–0.47 produce predominantly ^{56}Fe; values in the range of 0.45–0.43 are responsible of the formation of ^{58}Fe, ^{54}Cr, ^{50}Ti and ^{64}Ni, while values below 0.43–0.42 are responsible for ^{48}Ca. Parameterized models indicate that the amount of mass with $Y_{\rm e} < 0.45$ depends on the ignition density, while that with $0.47 < Y_{\rm e} < 0.485$ depends on the deflagration velocity.

Therefore ^{58}Fe, ^{54}Cr, ^{50}Ti, ^{64}Ni and ^{48}Ca are a measure of ρ_{ig} while ^{58}Ni, ^{54}Fe are a measure of v_{def} (Thielemann et al. 2004). It is important to realize that the change from using the Fuller et al. (1985) rates to the Langanke and Martínez-Pinedo (2000) rates strongly alleviates the chronic problem of producing an excess of neutronized species. In any case, to correctly evaluate the implications of the nucleosynthesis resulting from the different mechanisms and explosion scenarios it is necessary to integrate them into a galactic chemical evolution model that takes into account the contributions of all the iron-peak producers (Bravo et al. 1992; Matteucci et al. 1993). The nucleosynthesis yields also depend on a more subtle parameter, the abundance profiles of carbon and oxygen, which in turn are a function of the mass and initial metallicity of the progenitor of the white dwarf. In general, low mass progenitors produce white dwarfs with oxygen abundances in the center as large as $X_O = 0.7$ (D'Antona and Mazzitelli 1989; Salaris et al. 1997). This abundance can be enhanced as a consequence of the sedimentation induced by crystallization (Canal et al. 1990) if the white dwarf has had time to solidify before the start of the accretion phase. The differences in the energetic contents of carbon and oxygen nuclei translates into different ^{56}Ni yields (Domínguez et al. 2001). A similar effect is produced by the abundance and distribution of ^{22}Ne across the star (Bravo et al. 1992). These differences in the final production of ^{56}Ni translate into a dispersion of the peak SNIa luminosities of ~0.2 magnitudes. This value is smaller than the observed differences at high redshift and thus does not invalidate the use of SNIa for measuring distances but introduces some caution in the context of their use to determine the cosmological equation of state (Domínguez et al. 2001).

5.4 X-ray Bursts and Superbursts

X-ray bursts were serendipitously discovered in 1975 by Grindlay and Gursky (1976), and independently, by Belian et al. (1976). In contrast to standard transient sources, characterized by lifetimes ranging from weeks to months, these new cosmic X-ray sources (a subset of the low-mass X-ray binary class, or LMXB) exhibit brief *bursts*, lasting from seconds to minutes (see Bildsten 1998; Lewin et al. 1993, 1995; Strohmayer and Bildsten 2006, for reviews).

The two bursting episodes reported by Grindlay et al. were based on observations performed with the Astronomical Netherlands Satellite (ANS) on a previously known X-ray source, 3U 1820-30, located in the globular cluster NGC 6624. Similar events were reported by Belian et al., from X-ray observations of sources in the Norma constellation, performed with two military *Vela-5* satellites, covering the 15-month period from May 1969 to August 1970.

One year later, three additional burst sources, one of them, the enigmatic *Rapid Burster* (XBT 1730-335) (Fig. 5.6), were identified within a few degrees of the Galactic center (Lewin et al. 1976). Within a year, 20 additional burst sources were discovered, mainly by SAS-3 and OSO-8 satellites. To date, ~110 Galactic (Type

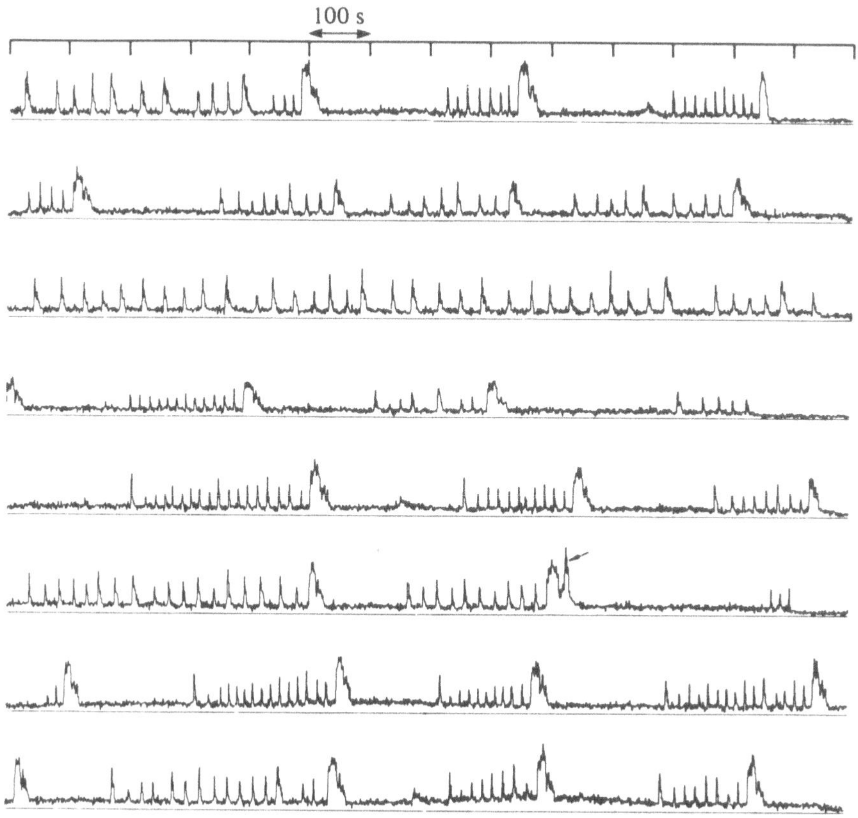

Fig. 5.6 Type II bursts from the Rapid Burster, based on SAS-3 observations performed in March 1976. The burst pinpointed with an arrow is actually a type I burst. Image from Lewin (1977)

I) X-ray burst sources (hereafter, XRBs) have been identified[1] with burst durations of ~10–100 s, and recurrence periods ranging typically from hours to days. Some bursts have been reported with extremely short recurrence times, ranging from 4 to 10 min; their ignition has been linked to rotational mixing (Keek et al. 2010). On the other hand, longer duration bursts have also been identified more recently (Galloway et al. 2008; Keek and in't Zand 2008): intermediate-duration bursts, for instance, can last for about 15–40 min and are characterized by a total energy output of $\sim 10^{40}$–10^{41} erg, and recurrence periods of tens of days (Linares et al. 2009; in't Zand et al. 2005; Falanga et al. 2008); superbursts, in turn, have typical durations of about a day, a total energy output of $\sim 10^{42}$ erg, and recurrence periods of about a year (Cornelisse et al. 2000; Strohmayer and Brown 2002). These differences have been suggested to result from different fuels and ignition depths.

[1] See http://www.sron.nl/$\sim$jeanz/bursterlist.html, for an updated list of known bursting sources.

5.4.1 The Nature of Type I X-ray Bursts

Maraschi and Cavaliere (1977), and independently, Woosley and Taam (1976), were the first to suggest the possibility that XRBs are powered by thermonuclear runaways on the surface of accreting neutron stars. However, it was soon realized that the quick succession of flashes exhibited by the Rapid Burster (with recurrent times as short as $\sim$10 s), didn't match the general pattern shown by these bursting sources. A major breakthrough in the understanding of the nature of these cataclysmic events was the discovery of two different kinds of bursts associated with the Rapid Burster (Hoffman et al. 1978): a classification of type I and type II bursts was then established, the former associated with thermonuclear flashes, the later linked to accretion instabilities. In fact, during many years, type II bursts were unequivocally linked with the Rapid Burster, the only object that showed both type I and type II bursts. More recently, a second member of the type II class, the bursting pulsar GRO J1744-28, has been identified. Hereafter, we will focus on type I X-ray bursts, the most frequent type of thermonuclear stellar explosion in the Galaxy (the third, in terms of total energy output after supernovae and classical novae).

The first evidence of the thermonuclear origin of type I XRBs came from lightcurve analysis, in particular the ratio between time-integrated persistent and burst fluxes, α. It was soon realized that the ratio between the gravitational potential energy released by matter falling onto a neutron star (G $M_{NS}/R_{NS} \sim$ 200 MeV/nucleon) during the accretion stage and the nuclear energy liberated during the burst ($\sim$5 MeV/nucleon, for a solar mixture transformed into Fe-group nuclei), match the values inferred for α, in the range $\sim$40–100.

The spatial distribution of type I XRBs matches that of LMXBs, with a clear concentration towards the Galactic center. A significant fraction of XRBs is indeed found in globular clusters. This pattern suggests that they are associated with an old stellar population (Lewin et al. 1993). The donors transferring material onto the neutron stars in XRBs are faint, low-mass stars ($M < 1\,M_\odot$), either Main Sequence or Red Giant stars. Recently, the first extragalactic XRBs were discovered in two globular cluster source candidates of the Andromeda galaxy (M31) (Pietsch and Haberl 2005).

It is believed that mass transfer episodes are driven by Roche-lobe overflow, hence leading to the build-up of an accretion disk around the neutron star. The maximum mass-accretion rate is set by the Eddington limit ($\dot{M}_{Edd} \sim 2 \times 10^{-8}\,M_\odot$/year, for H-rich accretion onto a 1.4 $M_\odot$ neutron star). Typically, XRB sources have orbital periods ranging from 1 to 15 h (White et al. 1995).

The nature of the underlying primary star was initially a matter of debate. A model involving accretion onto massive black holes ($>100\,M_\odot$) was proposed in the 70s (Grindlay and Gursky 1976). Nevertheless, XRB observations in globular clusters (from which reasonably accurate distance estimates can be obtained), performed with the OSO-8 satellite, best fitted with a blackbody spectrum with kT $\sim$ 0.87–2.3 keV (Swank et al. 1977), suggested a source with much smaller dimensions than a super-massive black hole (either a neutron star or a stellar mass black hole).

Other features, such as the harder X-ray spectra of XRB sources compared with most of the X-ray transients hosting black hole candidates, as well as the masses inferred from those systems, point towards a neutron star primary (van Paradijs and McClintock 1995).

Indeed, the masses inferred for neutron stars in XRBs are quite uncertain. However, two lines observed in the XRB spectra of EXO 0748-676 (suggested to be H- and He-like Fe lines; see Cottam et al. 2002), plus the measurement of a 45 Hz neutron star spin frequency in the same source, allowed mass estimates in the range $1.5 < M_{NS}(M_\odot) < 2.3$, with a best fit for 1.8 $M_\odot$ (Villarreal and Strohmayer 2004).

Light curves from X-ray bursts show a large variety of shapes (with single, double, or triple-peaked bursts) (Fig. 5.7). Generally speaking, they are characterized

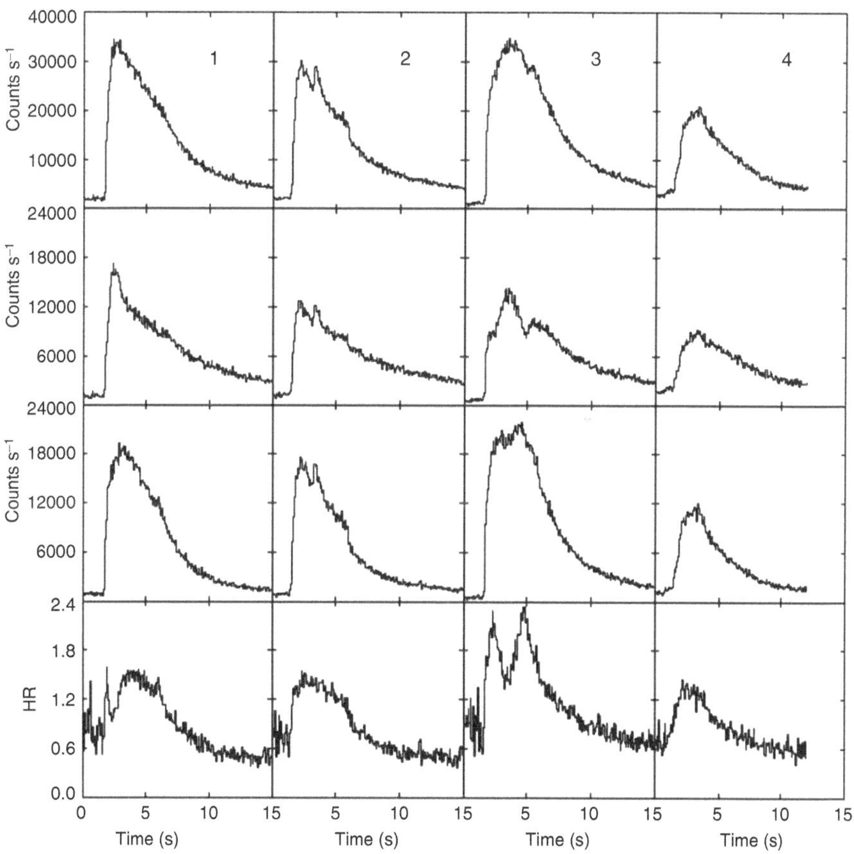

Fig. 5.7 A suite of XRB lightcurves from the LMXB source 4U1728-34 as observed with the RXTE satellite. Each sequence (top to bottom), shows the overall count rates in the energy bands 2–60, 2–6, 6–30 keV, and the ratio (6–30 keV)/(2–6 keV). Figure from Strohmayer and Bildsten (2006)

by a fast rise (~1–10 s), a peak luminosity of $\sim 3 \times 10^{38}\,\mathrm{erg\,s^{-1}}$ (Galloway et al. 2008; Lewin et al. 1993; Kuulkers et al. 2003), followed by a slower (sometimes exponential-like) decline (~10–100 s).

An interesting feature, observed in the spectra of many XRBs, is a 4.1 keV emission line (Waki et al. 1984), interpreted as Lyman α lines of helium-like Fe atoms, broadened by Doppler and gravitational effects, likely originating at the inner edge of the accretion disk. Indeed, it has been suggested that time-resolved spectroscopy can in principle allow measurements of the surface gravitational redshift (Damen et al. 1990; Smale 2001).

The fact that XRB sources do not exhibit X-ray pulsations suggest that the underlying neutron stars have weak magnetic fields ($< 10^{11}$ G). Indeed, pulsations are assumed to result from misalignment between the magnetic axis and the rotation axis of the neutron star. Moreover, it is unlikely that XRBs will originate from highly magnetized neutron stars, as a strong magnetic field would funnel the infalling charged plasma towards a small fraction of the neutron star surface, close to the magnetic caps; the effective accretion rate (per unit area) would be so high, that suppression of thermonuclear flashes be expected (Joss 1978; Taam and Picklum 1978).

The understanding of the nature of XRBs requires also multiwavelength observations beyond the X-ray domain: in 1978, the first simultaneous optical/X-ray burst was detected from the source 1735-444 (Grindlay et al. 1978). The fluence in the optical burst was $\sim 2 \times 10^{-5}$ times that of the X-ray band, too large to be explained by the low-energy tail of the blackbody X-ray burst emission (Lewin et al. 1993). More important, the optical burst was delayed by ~3 s with respect to the X-ray peak (McClintock et al. 1979). A similar delay (~1.4 s) was also reported from Ser X-1 (Hackwell et al. 1979), and later, from many other sources (Lewin et al. 1993). These results suggest that optical emission observed from XRBs corresponds to reprocessing of X-rays in material within a few light-seconds from the source. Likely sites for this reprocessing include the accretion disk that surrounds the neutron star as well as the hemisphere of the secondary star directly illuminated by the X-ray source. Hence, the delay in the optical wavelengths results from travel-time differences between the X-rays leading directly to the observer and those that first intercept the disk, lose energy (becoming optical photons), and finally reach the observer.

The situation is less clear at other wavelengths: infrared emission has been suggested to accompany type I X-ray bursts. Indeed, detection of infrared burst from the Rapid Burster has been claimed in the past (Kulkarni et al. 1979), although an unambiguous confirmation is lacking (Lewin et al. 1980). Also, although radio bursts have been reported from the Rapid Burster, no X-ray bursts were seen simultaneously (Hayakawa 1981). More detailed observations at these wavelengths are required to disentangle the controversy.

5.4.2 Modeling X-ray Bursts

Modeling of type I XRBs and their associated nucleosynthesis has been extensively addressed by different groups. This reflects the astrophysical interest in determining the nuclear processes that power such explosions as well as in providing reliable estimates for the composition of the neutron star surface (Maraschi and Cavaliere 1977; Woosley and Taam 1976; Joss 1977). Nonetheless, several thermal, radiative, electrical, and mechanical properties of the neutron star depend critically on the specific abundance pattern of its outer layers. Moreover, the diversity of shapes of XRB lightcurves is also probably due to different nuclear histories (see Heger et al. (2007), for a detailed analysis of the interplay between long bursts and the extension of the rp-process), suggesting that the final chemical composition, at the end of the burst, is not unique.

The properties of the bursts are also affected by *compositional inertia*; that is, they are sensitive to the fact that accretion proceeds onto the ashes of previous bursts (Taam 1980; Woosley et al. 2004). Indeed, this compositional inertia reduces the recurrence times between bursts, especially for scenarios involving accretion of metal-poor matter. Another critical quantity is the emerging heat flux from deeper layers of the neutron star (Ayasli and Joss 1982; Fushiki and Lamb 1987; Brown 2000), which proved critical to the burst properties of pure He bursts (Bildsten 1995).

The first studies of localized TNRs on neutron stars (Shara 1982) suggested that heat transport was too inefficient to spread a local flame to the overall stellar envelope. Therefore, localized, volcanic-like explosions were predicted during X-ray bursts. However, it is worth noting that these studies relied only on radiative and conductive transport, ignoring the crucial role played by convection on the lateral thermalization of a TNR. The scenario was revisited by Fryxell and Woosley (1982b), who suggested that the most likely outcome involves TNRs propagated by small-scale turbulences, in a deflagrative regime, leading to the horizontal spread of the front at typical velocities of $\sim 5 \times 10^6\,\mathrm{cm\,s^{-1}}$. Such speeds suggest that the time required for a flame to engulf the entire stellar surface is much longer than the characteristic spin periods of accreting neutron stars (∼milliseconds). Hence, it was predicted that fast rotation of the neutron star could modulate localized burning regions, eventually allowing for a direct observation of the neutron star spin. Indeed, the discovery of high-frequency, burst oscillations in the X-ray source 4U1728-34 (360–600 Hz; see Strohmayer et al. 1996) provided first observational evidence for millisecond rotation periods in accreting neutron stars. Since then, burst oscillations have been claimed for many additional sources. Studies to constrain neutron star properties based on modeling of such oscillations are currently underway.

5.4.3 *Nucleosynthesis in Type I X-ray Bursts*

In contrast to classical nova outbursts, where the main nuclear activity is driven by proton-capture reactions in competition with β^+-decays, X-ray bursts are triggered by a combination of nuclear reactions, including H-burning (via rp-process) and He-burning (that initiates with the triple α-reaction, and is followed both by the breakout of the CNO cycle through ^{14,15}O+α, plus a competition of proton captures and (α,p) reactions—the so-called αp-process). Moreover, with a neutron star as the underlying compact object hosting the explosion, temperatures and densities in the accreted envelope reach quite high values: $T_{peak} > 10^9$ K (an order of magnitude higher than in nova outbursts), and $\rho \sim 10^6$ g cm^{-3}. As a result, detailed nucleosynthesis studies require the use of hundreds of isotopes, up to the SnSbTe mass region (Schatz et al. 2001) (Fig. 5.8), or even beyond (the nuclear activity in Koike et al. (2004) reaches ^{126}Xe), and thousands of nuclear interactions extending to the proton drip line. In sharp contrast, the main nuclear activity for classical novae is limited to Ca, and runs close to the valley of stability.

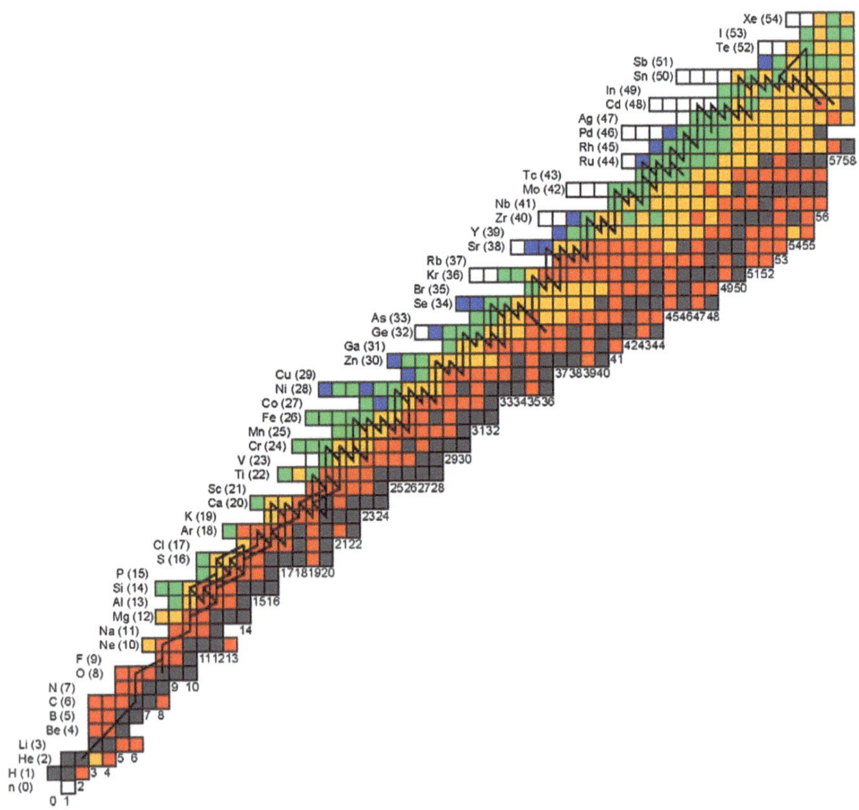

Fig. 5.8 Main nuclear path during a typical X-ray burst. Figure from Schatz et al. (1999)

Because of computational limitations, studies of XRB nucleosynthesis have usually been performed with limited nuclear reaction networks. More recently (Schatz et al. 1999, 2001), detailed nucleosynthesis calculations have been carried out with networks containing more than 600 isotopes (up to Xe, in Schatz et al. 2001), but using a one-zone approach, or also one-zone nucleosynthesis calculations with temperature and density profiles obtained with spherically symmetric evolutionary codes, linked to a 1270-isotope network extending up to ^{198}Bi (Koike et al. 2004). Other attempts (Parikh et al. 2008) include one-zone nucleosynthesis calculations, with temperature and density profiles obtained from the literature, and a large nuclear reaction network, containing 606 isotopes (up to ^{113}Xe) and more than 3500 nuclear processes. Note however that different numerical approaches and approximations have been adopted in all those works (hydrodynamic simulations with limited networks or one-zone calculations with detailed networks) and hence, the predicted nucleosynthesis in each case has to be taken with caution. Indeed, recent attempts have been made to couple hydrodynamic stellar calculations (in 1-D) and detailed networks (with $\sim$300 isotopes, up to ^{107}Te (Fisker et al. 2008; Tan et al. 2007), with 1392 nuclear processes and 325 isotopes, up to ^{107}Te (José et al. 2010), or with networks containing up to 1300 isotopes in an adaptive framework (Woosley et al. 2004).

To date, no fully multidimensional calculation for realistic XRB conditions has been performed. So far, a number of efforts have focused on the analysis of flame propagation on the envelopes accreted onto neutron stars and on *convection-in-a-box* studies aimed at characterizing convective transport prior to ignition. Some of the pioneering studies of thermonuclear flame propagation on neutron stars (Shara 1982) suggested that the onset of a localized ignition on a neutron star would likely propagate as a deflagration front, incinerating the whole envelope in a timescale of 100 s. The dichotomy between detonations and deflagrations was subsequently explored by different groups (Fryxell and Woosley 1982a,b; Simonenko et al. 2012a,b; Zingale et al. 2001), but their results depend critically on the assumed initial density at the ignition point (frequently, too extreme and therefore not representative of X-ray burst conditions). The current consensus, however, suggests that TNRs in X-ray bursts propagate subsonically (i.e., a deflagration front). The early development of convection in the stages preceding thermonuclear ignition has been recently analyzed in a multidimensional framework, in an attempt to assess the possibility of dredge up of ashes enriched in heavy elements to the neutron star photosphere (in't Zand and Weinberg 2010). Several efforts in this direction have been undertaken by different groups. In particular, simulations of pure He bursts and mixed H/He bursts in two and three dimensions (Malone et al. 2011, 2014; Zingale et al. 2015) have been performed in the last years with the MAESTRO code. The latter assumed an outer envelope composed of a mixture of H and He, slightly overabundant in CNO nuclei with respect to solar values, on top of an inert nickel substrate. The simulation assumed a plane-parallel geometry on a uniform grid, with a spatial resolution of only $\sim$6 cm. Comparison between 2D and 3D turbulent convection shows similar peak temperatures and Mach numbers, but different convective patterns, with evidence of the energy cascade into smaller scales that characterizes 3D convection. Further multidimensional studies of X-ray bursts under realistic conditions are really needed to shed light into the way ignition initiates and propagates, as well as in the way convection sets in and extends throughout the accreted envelope.

The relevant nuclear reaction path in XRBs has been extensively discussed in the literature (José et al. 2010; Iliadis 2007; Fisker et al. 2008): the most interesting nucleosynthesis is achieved for mixed H/He bursts, because of the complex nuclear reaction interplay (see details in Fisker et al. 2008; José et al. 2010). For illustrative purposes, we describe the main nuclear activity achieved for typical XRB conditions: a 1.4 $M_\odot$ neutron star, accreting solar-like material at a constant rate of 1.75×10^{-9} $M_\odot$/year. In general, such bursts are initiated by H-burning, specifically the cold mode of the CNO cycle (mainly through $^{12}C(p,\gamma)^{13}N(\beta^+)^{13}C(p,\gamma)^{14}N$). At moderate temperatures, the main nuclear flow proceeds close to the valley of stability. When $T \sim 2 \times 10^8$ K, the nuclear activity already reaches ^{40}Ca, with the most abundant species being H, ^4He, and $^{14,15}O$. When T approaches $\sim 3 \times 10^8$ K, the 3α reaction dominates the nuclear flow, together with a combination of (p,γ) and (p,α) reactions, and some β^+ decays (mainly $^{32,33}Cl$). At $T \sim 4 \times 10^8$ K, CNO-breakout ensues, initially led by $^{15}O(\alpha,\gamma)^{19}Ne$ (see Fisker et al. (2006), for a study of the impact of the $^{15}O(\alpha,\gamma)$ rate on the bursting behavior of an accreting neutron star), and followed by two consecutive proton-captures on ^{20}Na and ^{21}Mg, where the flow recedes due to the strong photodisintegration reactions on ^{22}Al. Following ^{21}Mg-decay, the flow shifts to $^{21}Na(p,\gamma)^{22}Mg$, moving away from the valley of stability, towards the proton-drip line. As the rise of the temperature continues, and enough ^{14}O is build-up through the triple-α reaction, followed by $^{12}C(p,\gamma)^{13}N(p,\gamma)^{14}O$, the alternative path through $^{14}O(\alpha,p)^{17}F$ dominates the flow (Champagne and Wiescher 1992; Woosley et al. 2004), by-passing the $^{15}O(\alpha,\gamma)^{19}Ne$ link to ^{21}Na through $^{17}F(p,\gamma)^{18}Ne(\alpha,p)^{21}Na$, with $^{18}Ne(\alpha,p)^{21}Na$ representing the main path towards heavier species.

When $T \sim 1.5 \times 10^9$ K, the most abundant species in the envelope become ^{18}Ne, $^{21,22}Mg$ (from $^{18}Ne(\alpha,p)^{21}Na(p,\gamma)^{22}Mg$), ^{25}Si, ^{28}S-^{28}P, ^{33}Ar-^{33}Cl, and ^{37}K, the first isotope that achieves an abundance of 10% by mass. At this stage, the flow has reached ^{64}Ge. Shortly after, the envelope achieves peak temperature, $T_{peak} \sim 1.7 \times 10^9$ K. The most abundant isotope (except for H) is ^{54}Ni, and later, ^{64}Ge and ^{68}Se. During the subsequent decline, the nuclear flow is dominated by a cascade of β-decays. The final composition of the envelope, which is not ejected by the TNR, is essentially composed of elements with $A = 60$–70, mainly ^{64}Zn (originally as ^{64}Ge, and ^{64}Ga), and ^{68}Zn (^{68}Se), with traces of other species. Explosions in lower metallicity envelopes are characterized by an extension of the main nuclear path by the rp-process, much beyond ^{56}Ni, up to the SnSbTe region (Schatz et al. 2001) or beyond, according to a recent reanalysis of the role of photodisintegration reactions in this mass region (Elomaa et al. 2009).

Most of the reaction rates required for these extensive nucleosynthesis calculations rely on theoretical estimates from statistical models, and may be affected by significant uncertainties. Efforts to quantify the impact of such nuclear uncertainties on the overall abundance pattern accompanying XRBs have been undertaken by different groups (Iliadis et al. 1999; Thielemann et al. 2001; Amthor et al. 2006), revealing a complex interplay between the nuclear activity and the shape of the light curve (Hanawa et al. 1983; Woosley et al. 2004). The most extensive work to date (Parikh et al. 2008), has helped to identify the most influential nuclear processes: ^{65}As(p, γ)^{66}Se, ^{61}Ga(p, γ)^{62}Ge, ^{12}C(α, γ)^{16}O, ^{96}Ag(p, γ)^{97}Cd, and in a lesser extent, ^{30}S(α, p)^{33}Cl, ^{56}Ni(α, p)^{59}Cu, ^{59}Cu(p, γ)^{60}Zn, ^{86}Mo(p, γ)^{87}Tc, ^{92}Ru(p, γ)^{93}Rh, ^{102}In(p, γ)^{103}Sn, and ^{103}In(p, γ)^{104}Sn.

A major drawback in the modeling of X-ray bursts comes from the lack of observational nucleosynthetic constraints (beyond the obvious implications for the physics of the neutron star crusts, outlined at the beginning of this section). The potential impact of XRB nucleosynthesis on Galactic abundances is still a matter of debate: although ejection from a neutron star is unlikely because of its large gravitational potential, radiation-driven winds during photospheric radius expansion may lead to ejection of a tiny fraction of the envelope, containing nuclear processed material (Weinberg et al. 2006; MacAlpine et al. 2007). Indeed, although it has been claimed that XRBs may help to explain the Galactic abundances of the problematic light *p-nuclei* (Schatz et al. 1998), new calculations have ruled out this possibility (Bazin et al. 2008; José et al. 2010). Finally, it has been proposed that a way to overcome the lack of observational constraints may come from the identification of gravitationally redshifted atomic absorption lines, which could be identified through high-resolution X-ray spectra (Bildsten et al. 2003; Chang et al. 2005, 2006; Weinberg et al. 2006). Indeed, although specific features have been reported in the spectra of 28 XRBs detected from EXO 0748-676 during a 335 ks observation with XMM-Newton (Cottam et al. 2002), interpreted as gravitationally redshifted absorption lines of Fe XXVI (during the early phase of the bursts), Fe XXV, and perhaps O VIII (during the late stages), no evidence for such spectral features was found neither after a 200 ks observation of GS 1826-24, from which 16 XRBs were detected (Kong et al. 2007), nor after a 600 ks observation of the original source EXO 0748-676 (Cottam et al. 2008; Rauch et al. 2008).

5.4.4 Superbursts

Whereas regular, type I XRBs are characterized by common features in terms of duration, energetics, and recurrence times, a few extremely energetic events have recently been detected thanks to better performances in monitoring achieved with X-ray satellites (i.e., BeppoSAX, Chandra, or XMM-Newton). These rare and rather violent events are known as *superbursts* (see Kuulkers 2004; Cumming 2005; in't Zand 2017, for reviews). The first observation of a superburst was reported by Cornelisse et al. (2000) in the framework of a "common" type I bursting source (c.f., the BeppoSAX source 4U1735-44) (Fig. 5.9).

About 26 superbursts from 15 different bursting sources have been discovered, including GX 17+2 and 4U 1636-536, for which 4 superbursts have been identified (in't Zand 2017). Although the term *superburst* was first used by Wijnands (2001)

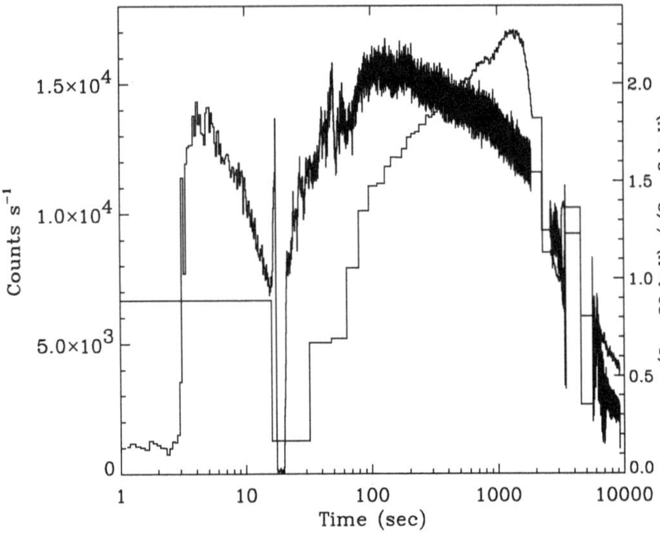

Fig. 5.9 Two superbursts observed with the RXTE satellite, in the (2–30) keV band. Figure from Strohmayer and Brown (2002)

to describe these very long X-ray bursts, historically the same name was applied to a relatively strong type I XRB reported from 4U 1728-34 by Basinska et al. (1984).

Superbursts represent some sort of extreme X-ray bursts: they have long durations, with a typical (exponential) decay time ranging from 1 to 3 h (including an extreme case, KS 1731-260, that lasted for more than 10 h; see Kuulkers et al. 2002), extremely energetic (about $\sim$1000 times more than a typical XRB, that is, $\sim 10^{42}$ erg), and with much longer recurrence periods (4.7 years for the system 4U 1636-53, for which two superbursts have been observed to date; see Wijnands 2001). Although superburst sources also exhibit regular type I XRBs, their occurrence is quenched for about a month after each superburst.

The duration and energetics of superbursts suggest that they result from thermonuclear flashes occurring in deeper fuel layers than those from typical X-ray bursts (at densities exceeding 10^9 g cm^{-3}; see Cumming and Bildsten (2001)), more likely, in the C-rich ashes resulting from type I X-ray bursts (first proposed by Woosley and Taam 1976; see also Brown and Bildsten 1998; Cumming and Bildsten 2001; Weinberg and Bildsten 2007; Keek et al. 2012).

Controversy remains over how much carbon is left after a type I burst: some studies (Schatz et al. 1999, 2001) have indeed shown than most of the C is burnt during the previous H/He burning episodes. However, other analyses (Cumming and Bildsten 2001) led to the conclusion that even small amounts of carbon are enough to power a superburst (especially in neutron star oceans enriched from the heavy ashes driven by the rp-process). Recent studies suggest that both stable and unstable burning of the accreted H/He mixture are required to power a superburst (in't Zand et al. 2003). Alternative models have also been proposed to account for the origin of such superbursts, including TNRs on strange quark matter stars (Page and Cumming 2005).

5.5 Observational Diagnostics of Binary-Systems

5.5.1 Gamma-Rays from Radioactivity

Novae and supernovae emit γ-rays because some of the nuclei they synthesize and eject into the interstellar medium are radioactive, either β^+-unstable (i.e., emitting a positron when decaying) or undergoing electron captures. Radioactive isotopes decay to excited states of their daughter nuclei, which de-excite to their ground states by emitting γ-ray photons with energies around one MeV, over a wide range of timescales. Table 5.1 shows the most relevant radioactive isotopes produced in novae and supernovae. Two additional isotopes, the β^+-unstable ^{13}N and ^{18}F (τ=862 s and 158 min, respectively), are also important in the case of novae. The emitted γ-rays can be potentially detected, either in individual objects or as diffuse emission from the cumulative γ-ray output of many objects in the galaxy, whenever the lifetime of a given isotope is longer than the average period between two successive events producing it (see Sect. 5.5.1.3). In addition, the positrons emitted when β^+-unstable nuclei decay annihilate with electrons and thus emit γ-rays, powering a 511 keV line plus a continuum below this energy.

The shape and intensity of the γ-ray output of novae and supernovae, as well as its temporal evolution, depend not only on the number of γ-ray photons produced, but also on how they propagate through the expanding envelope and ejecta. The first step to compute the spectrum is to generate γ-rays according to the decay schemes of the corresponding radioactive isotopes. The number of photons generated in a particular object depends on the isotopic abundances and decay rates of the relevant nuclei. In addition to these *direct* γ-ray photons, positrons emitted as a consequence of the radioactive decays of the β^+-unstable nuclei (see Table 5.1) should be traced. Once photons are generated, their trip across the expanding ejecta should be simulated by taking into account the various interaction processes affecting their propagation, i.e., Compton scattering, e^-–e^+ pair production, and photoelectric absorption.

The treatment of positron annihilation deserves particular attention. The role of magnetic fields is crucial, but it is not well known how to handle it. Thus, some drastic approximations are often made. When a positron is emitted, it can either escape without interacting with the expanding envelope or annihilate with an ambient electron. In nova envelopes, it is safe to assume that positrons thermalize before annihilating. This approximation is wrong in less than 1% of cases in an electronic plasma (Leising and Clayton 1987). In a neutral envelope, the excitation cross-section dominates any other interaction at energies above ~100 eV (Bussard et al. 1979), and thus positrons lose energy until they reach this value. In order to reproduce this braking effect, positrons should be propagated until they cross an equivalent column of $\sim$0.2 g cm^{-2}, measured along a straight line (Chan and Lingenfelter 1993). This is the mean range expected for a 0.6 MeV positron slowing to energies $\sim$100 eV through elastic collisions with the surrounding medium, when the effect of magnetic fields on its propagation is neglected. Once thermalized, the positron covers a negligible distance and then annihilates.

For densities and temperatures typical of novae and SNIa envelopes, positrons form positronium (positron-electron system) in $\sim$90% of annihilations (Leising and Clayton 1987), while in the remaining 10% of cases they annihilate directly. Positronium is formed in the singlet state 25% of the time, leading to the emission of two 511 keV photons, and in the triplet state 75% of the time, leading to a three-photon annihilation continuum. The spectrum of photons produced from the triplet state was obtained by Ore and Powell (1949). Therefore, once a positron is produced, its trip should be followed until it escapes or covers the average energy-loss distance. In the latter case it produces positronium 90% of the time, resulting in triplet or singlet annihilations in a 3:1 ratio, while in 10% of the cases it annihilates directly. Monte Carlo codes, based for instance on the method described in Pozdnyakov et al. (1983) and Ambwani and Sutherland (1988), are well suited to compute the γ-ray output of novae and type Ia supernovae (Gómez-Gomar et al. 1998b,a).

5.5.1.1 Gamma-Ray Emission from Individual Classical Novae

The potential of novae as γ-ray emitters was first pointed out by Clayton and Hoyle (1974), who stated that observable γ-rays from novae would come from electron-positron annihilation, with positrons from ^{13}N, ^{14}O, ^{15}O and ^{22}Na decays, as well as a result of the decay of ^{14}O and ^{22}Na to excited states of ^{14}N and ^{22}Ne nuclei, which de-excite by emitting photons at 2.312 and 1.274 MeV respectively. Some years later, Clayton (1981) noticed that another γ-ray line could be expected from novae when ^{7}Be transforms (through an electron capture) to an excited state of ^{7}Li, which de-excites by emitting a photon of 478 keV. The original idea came from Audouze and Reeves (1982), and both works were inspired by the contemporaneous papers mentioning the possibility of ^{7}Li synthesis in novae (Arnould and Norgaard 1975; Starrfield et al. 1978b). In fact, ^{7}Li production in novae was, and continuous to be, a crucial topic (Hernanz et al. 1996), since Galactic ^{7}Li is not well accounted for by other sources, either stellar (AGB stars), interstellar (spallation reactions by cosmic rays) or cosmological (Big Bang). The main ideas presented in these pioneering studies have remained unchanged; but some aspects have changed in the last years, mainly related to new detailed nucleosynthesis studies of novae.

The γ-ray signatures of classical novae depend on their yields of radioactive nuclei (see the reviews Leising (1991, 1993), Hernanz (2002)). CO and ONe novae differ in their production of ^{7}Be, ^{22}Na and ^{26}Al, while they synthesize similar amounts of ^{13}N and ^{18}F. In both nova types, there should be line emission at 511 keV related to e^-–e^+ annihilation, and a continuum produced by Comptonized 511 keV emission and positronium decay.

The yields of radioactive nuclei adopted to compute the γ-ray spectra and light curves presented here are from José (unpublished) based on Iliadis et al. nuclear reaction rates; see Hernanz (2014). The main change with respect to previous models is that ^{18}F yields are lower, thus impacting the 511 keV line and the continuum below it, as seen when compared with γ-ray spectra and light curves from the first edition of this book (also published in Hernanz (2012)).

The temporal evolution of the whole γ-ray spectrum of four representative nova models is shown in Fig. 5.10. The most prominent features of the spectra are the annihilation line at 511 keV and the continuum at energies between 20–30 keV and 511 keV (in both nova types), the ^{7}Be line at 478 keV in CO novae, and the ^{22}Na line at 1275 keV in ONe novae. Therefore, the main difference between spectra of CO and ONe novae are the long-lived lines, which directly reflect the different chemical composition of the expanding envelope (^{7}Be-rich in CO novae and ^{22}Na-rich in ONe ones).

The early γ-ray emission, or *prompt* emission, of novae is related to the disintegration of the very short-lived radioisotopes ^{13}N and ^{18}F. The radiation is emitted as a line at 511 keV (direct annihilation of positrons and singlet state positronium), plus a continuum (Gómez-Gomar et al. 1998a; Hernanz et al. 2002). The continuum is related to both the triplet state positronium continuum and the Comptonization of the photons emitted in the line. There is a sharp cut-off

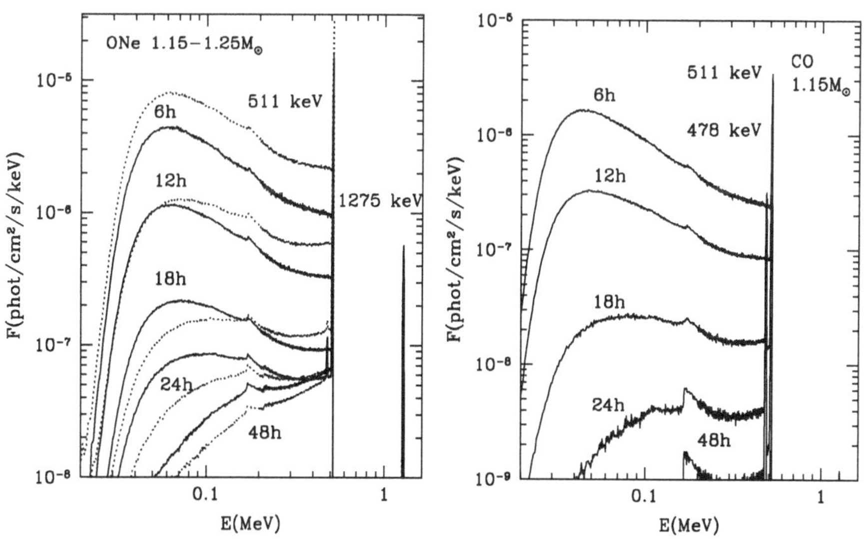

Fig. 5.10 Left: spectra of ONe novae of masses 1.15 (solid) and 1.25 $M_\odot$ (dotted) at different epochs after T_{max} (labels for dotted lines follow the same sequence as those for solid lines: from top to bottom 6, 12, 18, 24 and 48 h). Right: same for a CO nova of mass 1.15 $M_\odot$ (solid). Distance is 1 kpc

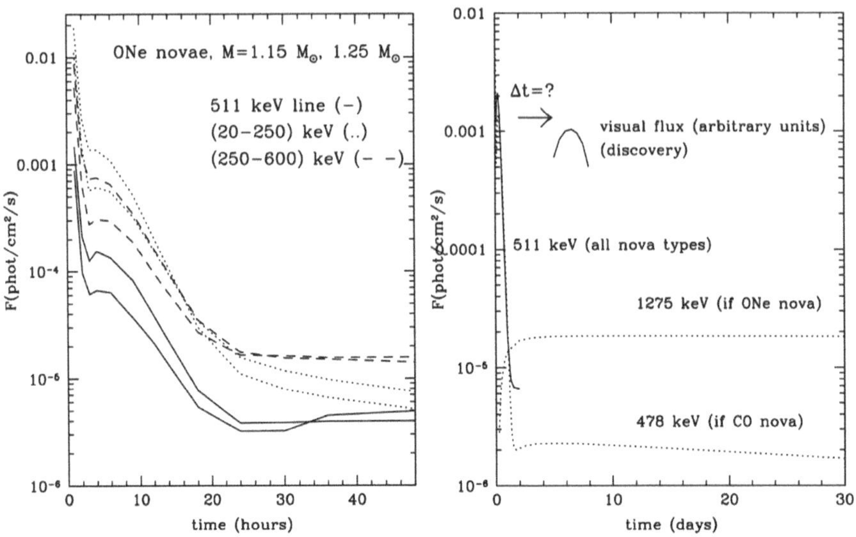

Fig. 5.11 Left: light curve of two continuum bands below 511 keV for ONe novae. The upper curves correspond to the larger mass, at early times; but at later epochs the most massive nova emits a slightly smaller flux, because of larger transparency. The light curve of the 511 keV line is also shown. Distance is 1 kpc. Right: nova γ-ray light curves, as compared with visual ones. The vertical scale for the visual light curve is arbitrary

at energies 20–30 keV (the exact value depending on the envelope composition) because of photoelectric absorption (see Fig. 5.10). The largest flux is emitted in the (20–250) keV range, since the continuum has its maximum at ~60 keV (ONe novae) and at ~45 keV (CO novae), followed by the flux in the (250–511) keV range (excluding the 511 keV line) and the flux in the 511 keV line (see Fig. 5.11). The two maxima in the light curves of the 511 keV line correspond to ^{13}N and ^{18}F decays, but the first maximum is difficult to resolve because its duration is extremely short; in addition, it is very model dependent: only ^{13}N in the outermost zones of the envelope could be seen in γ-rays because of limited transparency at very early epochs and, therefore, the intensity of the first maximum depends on the efficiency of convection. This first maximum thus provides important insight into the dynamics of the envelope after the peak temperature is attained at its base.

The annihilation emission is the most intense γ-ray feature expected from novae, but unfortunately it has a very short duration, because of the short lifetime of the main positron producers (^{13}N and ^{18}F). There are also positrons available from ^{22}Na decay in ONe novae, but these contribute much less (they are responsible for the *plateau* at a low level, between 10^{-6} and 10^{-5} phot cm^{-2} s^{-1}, for d=1 kpc; see Fig. 5.11). However, after roughly 1 week the envelope is so transparent that ^{22}Na positrons escape freely without annihilating. In summary, annihilation radiation lasts only ~1 day at a high level, and 1–2 weeks at a lower level *plateau* (the latter only in ONe novae). Another fact preventing easy detection is the early (before

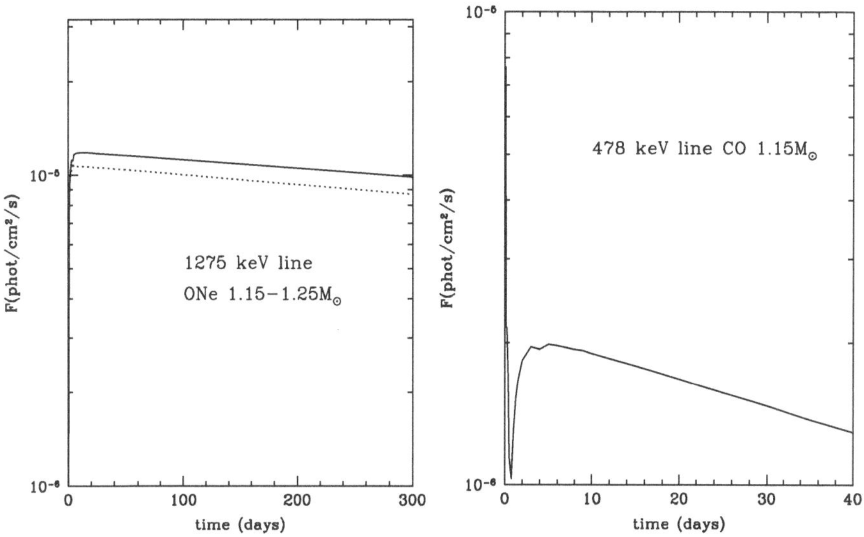

Fig. 5.12 Left: light curve of the 1275 keV line for two ONe nova models. Right: light curve of the 478 keV line for a CO nova model. Distance is 1 kpc

the nova is discovered optically) appearance of γ-rays from electron-positron annihilation (see Fig. 5.11).

The most distinctive feature in the γ-ray spectra of CO novae is line emission at 478 keV, related to de-excitation of the ^{7}Li which results from an electron capture on ^{7}Be. The light curves of the 478 keV line are shown in Fig. 5.12: the flux reaches its maximum at day 13 and 5 in the more and less opaque models, with total masses 0.8 and 1.15 $M_\odot$, respectively. The width of the line is 3 and 8 keV for the 0.8 and 1.15 $M_\odot$ CO novae, respectively. The maximum flux is around 10^{-6} phot cm^{-2} s^{-1}, for d=1 kpc.

The ^{22}Na line at 1275 keV appears only in ONe novae, because CO novae do not synthesize this isotope. The rising phase of the 1275 keV line light curves (see Fig. 5.12) lasts between 10 (1.25 $M_\odot$) and 20 days (1.15 $M_\odot$). Soon after the maximum, the line flux declines with the lifetime of ^{22}Na, 3.75 years. The line intensities directly reflect the amount of ^{22}Na ejected mass during this phase. The corresponding fluxes at maximum are typically around 10^{-5} phot cm^{-2} s^{-1}, at d=1 kpc, and the width of the line is around 20 keV, which poses severe problems for its detectability with instruments having high energy resolution, like SPI onboard INTEGRAL.

There have been many unsuccessful attempts to detect γ-rays from novae. The main efforts have focused on the 1275 keV line from ^{22}Na in individual objects, but searches of the cumulative emission have also been performed. The annihilation line has also been searched for whenever wide field of view instruments were available, scanning zones of the sky where novae had exploded.

The most recent observational search for the 1275 keV line from novae was performed with the COMPTEL instrument onboard the Compton Gamma-Ray Observatory (CGRO) (Iyudin et al. 1995). COMPTEL observed a number of recent novae during the period 1991–1993, five of which of the neon type (i.e. those expected to emit the 1275 keV line). None was detected. The average 2σ upper limit for any nova of the ONe type in the galactic disk was around 3×10^{-5} phot cm^{-2} s^{-1}, which translated into an upper limit of the ejected ^{22}Na mass around 3.7×10^{-8} M$_{\odot}$, for the adopted distances. This limit was constraining for models available at the time (Starrfield et al. 1992, 1993; Politano et al. 1995), but is not so for current models (José and Hernanz 1998). The main reason for the discrepancy between models of different groups (José and Hernanz (1998) versus Politano et al. (1995) and Starrfield et al. (1998)) is the following: old models were based on the explosion on ONeMg white dwarfs, with some mixing between the accreted H-rich matter and the underlying material, whereas recent models adopt ONe white dwarfs as underlying cores, because more recent evolutionary calculations of stellar evolution predict much lower magnesium abundances (Ritossa et al. 1996; Dominguez et al. 1993). The smaller initial content of neon and magnesium makes ^{22}Na synthesis much less favored. Different reaction networks also have an impact on the final yields obtained by different groups.

The first search for the 478 keV line from the galactic center and from some particular novae was performed with SMM/GRS (Harris et al. 1991), yielding upper limits around 10^{-3} phot cm^{-2} s^{-1}, corresponding to ^{7}Be ejected masses around 10^{-7} M$_{\odot}$. These fluxes and masses are well above the current theoretical predictions and thus do not constrain the models. More recent analyses of novae during the period 1995–1997, have been possible thanks to the Transient Gamma-Ray Spectrometer (TGRS) on board the Wind satellite. The flux limits from TGRS were a factor of 10 smaller than those from SMM observations, but the upper limits on ^{7}Be ejected masses did not improve by the same factor, mainly because novae observed with TGRS were at larger distances than those observed with SMM (Harris et al. 2001).

It is worth noticing that the detection of the 478 keV line from ^{7}Be in novae would provide unambiguously the amount of ^{7}Be, and thus of ^{7}Li, ejected by the corresponding nova, without the problems mentioned from UV-optical detections: these give relative amounts, e.g., with respect to Ca, and also depend on the delicate process of abundance determinations from equivalent widths of the different lines of ^{7}BeII.

As mentioned above, the emission resulting from e^{-}-e^{+} annihilation is the most intense γ-ray outcome of classical novae, but γ-rays are emitted well before the visual maximum of the nova, i.e. typically before the nova is discovered, and have a very short duration (see Fig. 5.11). Therefore, they can not be detected through observations pointing to a particular nova already discovered. Wide field of view instruments monitoring the sky in the appropriate energy range, like the Burst and Transient Source Experiment (BATSE) on board CGRO or TGRS on board Wind, are best suited for the search of the 511 keV line and the continuum below it.

TGRS was very convenient to search for the 511 keV line, because of its large field of view, and also because its germanium detectors had enough spectral resolution to separate the cosmic 511 keV line from the nova line, provided that the latter is a bit blueshifted (this happens only at the beginning of the emission phase, when material is not completely transparent yet) (Harris et al. 1999). TGRS's field of view contained five new novae during the period 1995–1997; upper limits were obtained by Harris et al. (1999), who deduced that their method was sensitive enough to detect novae occurring out to about 0.8 kpc, for any nova type (CO and ONe).

Another instrument that was well suited for the detection of the prompt γ-ray emission from novae was BATSE on board CGRO. Before the launch of CGRO in 1991, a prediction was made (Fishman et al. 1991) on the detectability of low-energy γ-rays from novae with the BATSE instrument, based on the models of γ-ray emission from Leising and Clayton (1987). BATSE had the advantage of continuously covering almost the whole sky, but on the other hand it was less sensitive and had poor energy resolution. More recently, a posteriori analyses of the background data at the explosion epoch of all classical novae discovered optically during the whole period of CGRO operation (1991–2000), searching for some signal, were performed (Hernanz et al. 2000). The 3-σ sensitivity using the 511 keV data only is similar to that with WIND/TGRS (Harris et al. 1999), but TGRS's sensitivity required a particular line blueshift, whereas BATSE is independent of it. The 3-σ sensitivity using the (250–511) keV data is a little more than a factor of 2 better than that from TGRS (Harris et al. 1999).

The 2002 launch of the ESA satellite International Gamma-Ray Laboratory, INTEGRAL, opened new perspectives for the detection of γ-rays from explosive events, with its two major instruments, the spectrometer SPI and the imager IBIS. SPI is made of 19 germanium detectors; its 3σ sensitivity at 1 MeV, for 10^6 s observation time and narrow lines, is around 2.4×10^{-5} phot cm^{-2} s^{-1}, with 2 keV energy resolution. However, this sensitivity degrades considerably for broad lines. Detection of γ-rays from novae with INTEGRAL is not too likely, because its detectability distance limits are small and, therefore, few novae are expected (Hernanz and José 2004). This is due to both the small fluxes expected and the reduced (with respect to pre-launch estimates) inflight measured sensitivities at the relevant energies. Very small distances are needed to obtain a secure detection: around 0.2 kpc for the 478 keV line from ^{7}Be and around 0.7 kpc for the 1275 keV line from ^{22}Na.

There is a new mission concept, named *e-ASTROGAM*, presented to ESA call M5, which if accepted would represent an important step forward for γ-ray astrophysics in the MeV and GeV range. Sensitivities better by factors of around 10 would be reached for the MeV lines, leading to detectability distances larger by factors of about 3, with respect to those with INTEGRAL/SPI. The mission proposal description, with a detailed insight into the instrumentation, can be found in De Angelis et al. (2017b) and Tatischeff et al. (2016), whereas a more detailed view of the science is reported on the Science White Book (De Angelis et al. 2017a).[2]

[2] e-ASTROGAM was not selected by ESA for feasibility study (M5 call), in May 2018.

5.5.1.2 Gamma-Ray Emission from Individual Type Ia Supernovae

In Type Ia supernova ejecta, the dominant radioactive chains are $^{56}\mathrm{Ni} \rightarrow ^{56}\mathrm{Co} \rightarrow ^{56}\mathrm{Fe}$ and $^{57}\mathrm{Ni} \rightarrow ^{57}\mathrm{Co} \rightarrow ^{57}\mathrm{Fe}$ (see Table 5.1). The amount of radioactive material, its distribution within the ejecta as well as the density, velocity and chemical composition profiles are different for each model described in Sect. 5.3 and these differences affect the total intensity and the evolution of the different lines, as well as the importance and extension of the continuum component of the spectrum (Burrows and The 1990; Kumagai and Nomoto 1997; Gómez-Gomar et al. 1998b).

In 1D geometry, the predicted γ-emission can be roughly described as follows: Twenty days after the explosion all models involving a prompt or a delayed detonation display strong lines because their high expansion rates induce a rapid decrease of the density, as shown in Fig. 5.13. Lines are particularly intense for those models containing ^{56}Ni and ^{56}Co in the outer layers (pure detonation and sub-Chandrasekhar models). The maximum intensity of these lines is model dependent since it is a function of the expansion rate and of the distribution of ^{56}Ni. Pure deflagration models only display a continuum since they efficiently Comptonize high energy γ-rays. The shape of the continuum at low energies is limited in all models by the competing photoelectric absorption, which imposes a cut-off below 40–100 keV. The energy of the cut-off is determined by the chemical composition of the external layers where most of the emergent continuum is formed at this epoch. Consequently, the continuum of those models containing low Z elements in the outer layers will extend to lower energies than that of those containing high Z elements. Therefore, it is possible to use these differences to discriminate among the different burning modes.

Two months after the explosion, Fig. 5.13, the ^{56}Ni lines have disappeared and the emission is dominated by the ^{56}Co lines, which reach their maximum of intensity roughly 2 months after the explosion in all models except for the pure deflagration ones. At maximum, the intensity of the lines in pure detonations, delayed detonations and sub-Chandrasekhar models is determined by the total mass

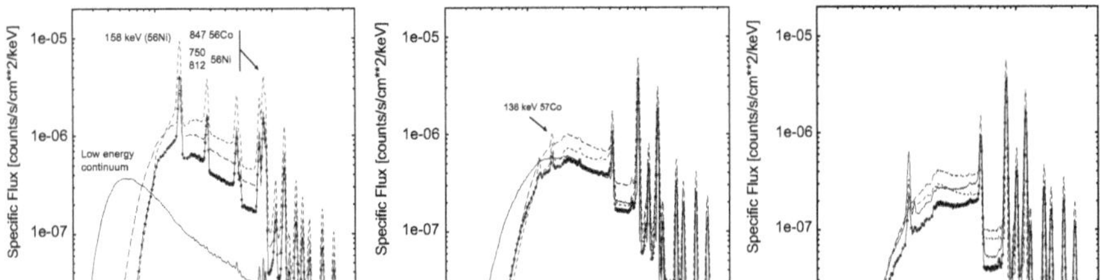

Fig. 5.13 Gamma-ray spectrum for four models of SNIa explosion at 5 Mpc 20, 60, and 120 days (from left to right) after the explosion. Pure deflagration model (solid line), delayed detonation model (long-dashed line), detonation model (dashed line) and sub-Chandrasekhar model (starred line) (Gómez-Gomar et al. 1998b)

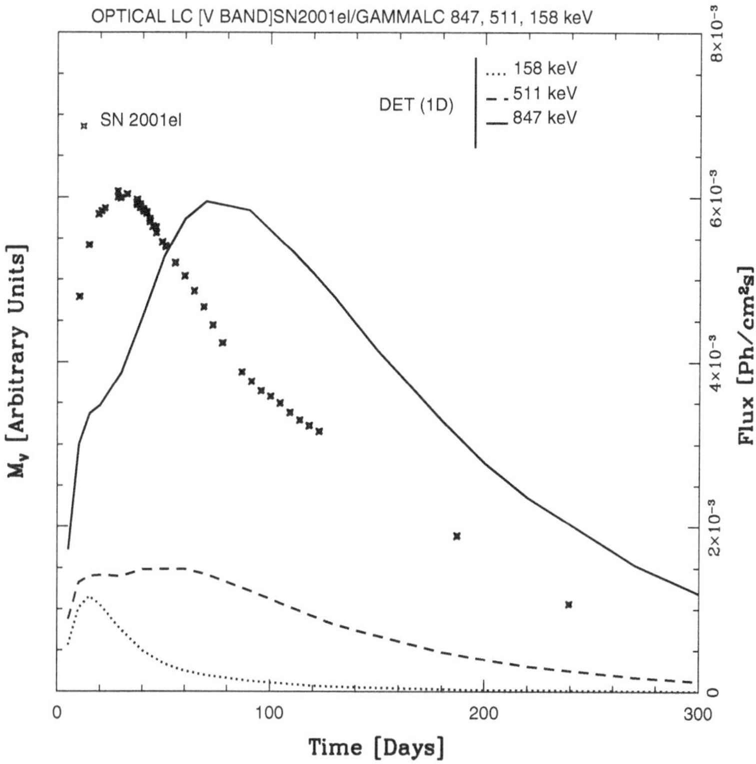

Fig. 5.14 Evolution of the different lines as a function of time for a typical delayed detonation model. The distance is assumed to be 1 Mpc. The optical light curve in the visual has also been included in order to provide a time reference. Courtesy of A. Hirschmann

of radioactive isotopes, while the differences caused by the expansion velocities are secondary. The 122 and 136 keV lines of ^{57}Co are already visible but faint.

Four months after the explosion, the ejecta are optically thin in all cases and the intensity of the lines is proportional to the parent isotopes (Fig. 5.13). The continuum is faint and dominated by the positronium annihilation component which shows a step below 170 keV, the energy of the backscattered 511 keV photons, plus a contribution of photons scattered once.

Figure 5.14 displays the temporal behavior of the ^{56}Ni and ^{56}Co lines. The 158 and 812 keV ^{56}Ni-lines peak very early, near the maximum of light and, because of absorption, they are much weaker than those of ^{56}Co Therefore, an early detection can provide information about the location of ^{56}Ni in the debris. The most prominent spectral feature is the 847 keV ^{56}Co line, which reaches maximum intensity roughly 2 months after the explosion in all models except for the pure deflagration case. Since the intensity of this line at maximum is essentially determined by the total mass of the radioactive isotopes it can be used to measure the total amount of ^{56}Ni synthesized during the explosion.

Figure 5.14 also displays the evolution of one of the most prominent lines, the 511 keV annihilation one. Positrons emitted during the decay of ^{56}Co thermalize because of ionization and excitation energy losses as well as other mechanisms, and eventually they annihilate either directly or through the formation of positronium. The degree of ionization and the structure of the magnetic field is crucial to determine the fraction of positrons that escape from SNIa. After 200 days almost all the high energy photons escape and the energy deposited by the annihilation of positrons is the only available source to power the light curve. Therefore, a careful determination of the 511 keV line is fundamental to understand the evolution of the supernova debris (Milne et al. 2001).

The decay chain ^{44}Ti→^{44}Sc→^{44}Ca ($\tau_{1/2} = 60.0$ years and $\tau_{1/2} = 3.97$ h, respectively) offers an additional opportunity to obtain information about the explosion although, as a consequence of the relatively long lifetime of ^{44}Ti, the observation has to be performed in young supernova remnants. This isotope is synthesized during the α-rich freeze out in a low density ambient, similar to those found in delayed detonation and sub-Chandrasekhar models.

The early X-ray emission in the 6–8 keV region can also provide an important diagnostic for discriminating between Chandrasekhar and sub-Chandrasekhar models. The γ-rays produced by disintegration of ^{56}Ni and ^{56}Co, together with the thermalized photons with energies above ∼7 keV, induce strong emission of the K_α-lines of Fe, Co and Ni. In the case of the Chandrasekhar models, ^{56}Ni is so deeply placed that these photons are absorbed before escaping while in the sub-Chandrasekhar models they are produced in the outermost layers from where they freely escape producing a distinctive feature in the spectra. The total expected flux in the 5–10 keV band at 15 Mpc is $\sim 2 \times 10^{-7}$ photons cm^{-2} s^{-1}, which means that it could be detected from a reasonably close supernova (Pinto et al. 2001). Another feature that could also be used to distinguish among these two families of models is the 14.4 keV emission of ^{57}Co, which is only expected in sub-Chandrasekhar models.

XMM and Chandra allow high spectroscopic and angular resolution studies of some galactic remnants of SNIa. In particular, for the X-rays from the Tycho supernova remnant the best fit is obtained with a one dimensional delayed detonation characterized by a quite high density transition placed in the range $(2.2 - 2.5) \times 10^7$ g/cm^3 (Badenes et al. 2006), while in the case of G337.2-07 the best fit is obtained for a pulsational delayed detonation with a density transition at 7.7×10^6 g/cm^3 (Rakowski et al. 2006). In both cases, the X-ray spectrum strongly suggests a high degree of chemical stratification, a property that is lacking in most current three-dimensional models of SNIa.

Interestingly enough, DDT models also provide the best fit to the X-ray spectra of 22 clusters of galaxies (de Plaa et al. 2007) and is the only model able to match the observed Ar/Ca and the Ca/Fe ratios. It is important to remember that the chemical composition of the intracluster medium is representative of the average supernova yields, since it is the result of the contributions from many supernovae during the clusters life time.

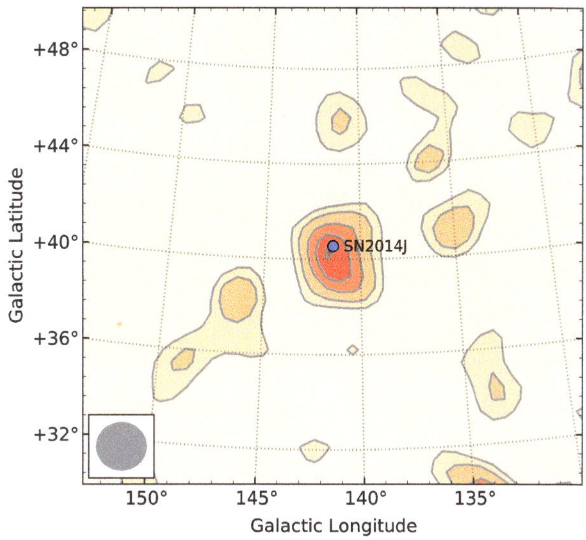

Fig. 5.15 Gamma-ray signature of SN 2014J in the SPI data during the period 16–35 days after the explosion. The excess in the supernova position is 5σ (Isern et al. 2016)

SN2014J was discovered by Fossey et al. (2014) on January 21st 2014 in M82 ($d = 3.5 \pm 0.3$ Mpc). The moment of the explosion was estimated to be on January 14.72 UT 2014 or JD 2456672.22 (Zheng et al. 2014) and was observed three times by the *INTEGRAL* instruments SPI and IBIS/ISGRI. During the first observation run, 16.5–35.2 days after the explosion, *INTEGRAL* obtained a robust detection of the gamma emission near the maximum of light, as illustrated in Fig. 5.15 (Churazov et al. 2014; Diehl et al. 2014; Churazov et al. 2015; Isern et al. 2016). Effectively, the analysis of the data obtained by INTEGRAL during this epoch, in the position of the supernova, revealed the existence of emission excesses with a significance of $\sim 5\,\sigma$ in the energy bands 70–190 (SPI and IBIS) and 650–1380 keV (SPI) that were not present during the observations performed before the explosion. The excess found at low energies is associated to the ^{56}Ni 158 keV gamma ray line and has completely unexpected properties. Diehl et al. (2014) found that the 158 keV ^{56}Ni line was very near to the laboratory value, the Doppler shift was below 2100 km s^{-1} and the broadening modest, less than 6000 km s^{-1}, suggesting the existence of a disk or ring containing ^{56}Ni placed almost perpendicularly to the line of sight. A behavior consistent with these values was also found in the 812 keV line. On the contrary, Isern et al. (2016) found a broad and redshifted feature associated to this emission excess. When the temporal evolution of the spectrum was taken into account and the secondary photons were removed, a line of intensity $(1.59 \pm 0.57) \times 10^{-4}$ photons cm^{-2} s^{-1} centered at 154 ± 0.64 keV and a width of 3.7 ± 1.5 keV appeared. This line almost disappeared during the period 22.6–28.2 days after the explosion and reappeared in the period 28.6–35.2 after the explosion, although at this epoch is too weak to obtain any conclusion. Interestingly enough, the IBIS/ISGRI displayed a similar behavior in the energy band 67.5–189 keV, during the same time intervals, but with a better signal to noise ratio. This behavior suggested a blobby ring receding from the observer.

The observations during the late period, 50–162 days after the explosion allowed the detection of the ^{56}Co 847 an1238 keV lines at 4.7 σ and 4.3 σ of confidence level, thus confirming the hypothesis that the light curves of SNIa are powered by the ^{56}Ni decay chain. The spectra and the light curve obtained in this way, were broadly consistent with the standard spherical deflagration or delayed detonation models near the Chandrasekhar's mass (Churazov et al. 2014, 2015). The mass of ^{56}Ni obtained from the intensity of this line is completely consistent with the value obtained from the Arnett's rule.

5.5.1.3 Contribution of Classical Novae to Diffuse Radioactivities

Some radioactive nuclei have lifetimes larger than the typical time elapsed between successive novae or type Ia supernova explosions in the Galaxy. For such cases, diffuse emission resulting from the cumulative effect of several sources is expected. This kind of emission should trace the galactic distribution of the corresponding sources of the given isotope. If detected, it would give a valuable information, not available from observations at other wavelengths because of interstellar extinction. The Galactic nova spatial distribution and the nova rate are in fact poorly known, since their determination relies on observations of novae in other galaxies or on extrapolations of observations in our Galaxy, taking into account the distribution of extinction related to interstellar dust (Della Valle and Livio 1994; Shafter 1997, 2002). ^{22}Na and ^{26}Al from novae are potential contributors to diffuse emission at 1275 and 1809 keV, respectively. For ^{22}Na, there is the advantage that only novae are expected to contribute to its galactic content, whereas for ^{26}Al massive stars and AGBs also clearly contribute. Therefore, the galactic 1275 keV emission from ^{22}Na, should trace directly the spatial nova distribution; but unfortunately, as analyzed below, the predicted emission is too low for the performances of the current instruments. Concerning ^{26}Al, since its emission has been detected in the Galaxy, an estimate of the nova contribution to the global line flux is needed (Diehl et al. 1995; Prantzos and Diehl 1996).

The global flux at 1275 keV depends on the amount of ^{22}Na ejected per nova explosion and on the distribution and rate of ONe novae in the Galaxy (since only ONe produce ^{22}Na). A detailed study of the diffuse galactic 1275 keV line emission from novae showed that contributions from a few young and close novae dominate, yielding a very irregular distribution versus galactic longitude (Higdon and Fowler 1987). A comparison with the upper limits from HEAO 3 observations (Mahoney et al. 1982) gave 5.6×10^{-7} $M_\odot$ as upper limit to the mean ^{22}Na yield per nova, for a disk nova population. It was clear from this work that the results were subject to many uncertainties, such as their galactic distribution, the bulge/disk ratio, their global rate and the fraction of ONe versus CO novae. A recent analysis of the cumulative emission at 1275 keV from novae shows that the ejected ^{22}Na masses needed for a detection of this emission with the SPI spectrometer, onboard the

INTEGRAL satellite, are far above what current theoretical models predict ($\sim 10^{-7}$ versus a few $10^{-9}\,M_\odot$) (Jean et al. 2000).

The production of ^{26}Al by classical novae occurs again mainly in ONe novae, with low mass white dwarfs more prolific producers of ^{26}Al than massive ones. A crude estimate of the global contribution of novae to the ^{26}Al content in the Galaxy can be made, assuming that all novae contribute with the same amount of ^{26}Al, $M_{ejec}(^{26}Al)$, and that ^{26}Al is active during a time equal to its lifetime τ. Then the Galactic mass of ^{26}Al coming from novae would be (Weiss and Truran 1990; José et al. 1997)

$$M(^{26}Al)(M_\odot) = M_{ejec}(^{26}Al)\,\tau\,R_{nova}\,f_{ONe} = 0.12\frac{M_{ejec}}{10^{-8}\,M_\odot}\frac{R_{nova}}{35\,\text{year}^{-1}}\frac{f_{ONe}}{0.33} \tag{5.25}$$

where R_{nova} is the total galactic nova rate and f_{ONe} is the fraction of ONe novae. Adopting typical ^{26}Al ejected masses (i.e., $2 \times 10^{-8}\,M_\odot$), the contribution of novae to galactic ^{26}Al would be $\sim 0.2\,M_\odot$, more than a factor of 10 below the observed mass, in agreement with the current idea (deduced from the observed 1.809 MeV line sky map) that galactic ^{26}Al comes mainly from massive stars (Knödlseder 1999) A complete analysis of the global contribution of novae to the ^{26}Al in the Galaxy was carried out by Kolb and Politano (1997), applying galactic nova population models, adopting the ^{26}Al yields from Politano et al. (1995) and taking very large ejected masses (larger than those from typical hydrodynamic models). The authors concluded that the nova contribution could range between 0.15 and $3\,M_\odot$, but this number largely depended on the unknown degree of mixing in novae, which largely influences their ^{26}Al yield, in addition to other parameters of the population synthesis code, like for instance the mass ratio (primary versus secondary star masses) distribution in zero-age main sequence binaries.

5.5.2 *Dust from Novae and Thermonuclear Supernovae*

Astrophysics has basically relied on electromagnetic radiation (collected by ground-based telescopes as well as by space-borne observatories) as the basic tool to determine stellar properties. But since the mid-80s, new methods that rely on matter rather than on radiation, have become available as well.

5.5.2.1 Stardust Mineralogy

Back in 1973, A.G.W. Cameron speculated in a seminal paper (Cameron 1973) that primitive carbonaceous chondrites may host *presolar grains*, tiny spherules of *stardust* condensed in the outflows of stars in advanced stages or in the ejecta of stellar explosions, containing a record of the nuclear history of their stellar parent bodies. Indeed, presolar grains have been isolated from meteorites, suggesting that the chemical processes that affected some meteoritic bodies were apparently mild and non-destructive to the grains.

The stellar paternity of these grains can be assessed by their anomalous isotopic composition, significantly different from that of the Solar System, and attributed to a suite of nucleosynthetic processes that took place in their parent stellar sources. In turn, the discovery of isotopically anomalous grains embedded in meteorites provided evidence of the chemical heterogeneity of the solar nebula (Cameron 1962). Moreover, although grains are difficult to date because of their low content in radioactive species, their large isotopic anomalies, including $^{14}N/^{15}N$, $^{12}C/^{13}C$, or silicon ratios far beyond the values reported from any other Solar System sample, suggest an ancient origin, with an age older than the Solar System itself (thus the label *presolar*).

Diamonds were the first presolar grains isolated from meteorites (Lewis et al. 1987). This was followed by the isolation of SiC grains (Bernatowicz et al. 1987; Tang and Anders 1988), and graphite (Amari et al. 1990). These three carbonaceous phases were identified because of their isotopically anomalous noble gas (Xe, Ne) components. So far, silicon carbide (SiC), graphite (C), diamond (C), silicon nitride (Si_3N_4), silicates (Messenger et al. 2003; Nguyen and Zinner 2004; Mostefaoui and Hoppe 2004), and oxides, such as corundum (Al_2O_3), or spinel ($MgAl_2O_4$), have been identified as presolar grains (Table 5.2). In fact, all SiC grains extracted from meteorites are of presolar origin; approximately half of the graphite grains are presolar; only $\sim$2% of the spinel grains, and scarcely 0.001–0.02% of the silicates, are presolar.

Those grains, identified and extracted from meteorites, are systematically analyzed in the laboratory with ever improving precision. Such laboratory analyses revealed a variety of isotopic signatures that point towards several stellar progenitors, such as asymptotic giant branch stars and supernovae (see Clayton and Nittler 2004; Lodders 2005; Meyer and Zinner 2006, for recent reviews).

Several meteoritic bodies have been used to study presolar grains, basically very primitive, mildly metamorphosed, carbonaceous chondrites, such as Murchison, or Allende. Indeed, the anomalous size of Murchison's grains (Lodders 2005; Zinner 2005), much larger than those isolated from other meteorites (for reasons not yet understood), as well as the large number of samples available, made Murchison one of the favorite targets for studies of presolar grains.

5.5.2.2 Silicon Carbide Grains

SiC grains have been most extensively studied. They are often classified into different populations (presumably reflecting different stellar birthplaces) on the basis of their C, N, and Si isotopic ratios (Hoppe and Ott 1997).

It is widely accepted that about $\sim$93% of all SiC grains, the so-called *mainstream population*, are formed in the winds accompanying solar-metallicity AGB stars

Table 5.2 Inventory of known presolar grain types (adapted from Zinner 2005; Lodders 2005)

Grain type	Characteristic size	Potential stellar sources[a]	Discovery papers
Nanodiamond	2 nm	AGB SN	Lewis et al. (1987)
SiC	0.1–20 μm	AGB, SN, J-stars, CN	Bernatowicz et al. (1987), Tang and Anders (1988)
Graphite	1–20 μm	SN, AGB, CN	Amari et al. (1990)
Corundum	0.2–3 μm	RGB, AGB, SN	Hutcheon et al. (1994), Nittler et al. (1994)
Spinel	0.2–3 μm	RGB, AGB, SN	Nittler et al. (1997), Choi et al. (1998)
Hibonite	0.2–3 μm	RGB, AGB, SN	Choi et al. (1999)
Si_3N_4	0.3–1 μm	AGB, SN	Nittler et al. (1995)
Silicates (olivine, pyroxene)	0.1–0.3 μm	RGB, AGB, SN	Messenger et al. (2003), Nguyen and Zinner (2004)

[a]Acronyms: *AGB* Asymptotic Giant Branch Stars, *SN* Supernovae, *CN* Classical Novae, *RGB* Red Giant Branch Stars

(Gallino et al. 1993; Lugaro et al. 2003; Ott and Begemann 1990). About ∼1% correspond to *X grains*, which are characterized by moderate excesses of ^{12}C and ^{15}N, large $^{26}Al/^{27}Al$ ratios, and excesses of ^{28}Si, features pointing towards a supernova origin (Amari et al. 1992; Hoppe et al. 2000; see also Sect. 5.3.3). In addition, a variety of carbon-rich J-type stars are expected to account for ∼4–5% of the overall SiC grains, the so-called *A* and *B grains* (with born-again AGB stars, such as the Sakurai's object V4334 Sgr, or other C-rich stellar types, like R- or CH-stars, not being totally excluded; see Amari et al. (2001c)). Other populations include *Y* (∼1%) and *Z grains* (∼1%), whose origin is probably linked to low-metallicity AGB stars (Amari et al. 2001b; Hoppe et al. 1997). A rare variety of SiC grains (<1%), together with a couple of graphite grains, that exhibit a suite of isotopic signatures characteristic of classical nova outbursts, have been reported in recent years (Amari et al. 2001a; Amari 2002) (Fig. 5.16).

5.5.2.3 Supernova Grains

SiC grains of type X, low-density graphites, and the very rare silicon nitrates are believed to originate in ejecta accompanying supernovae. Many of the isotopic signatures of these grains (namely, moderate excesses of ^{12}C, and ^{15}N, large $^{26}Al/^{27}Al$ ratios, and excesses of ^{28}Si) are qualitatively consistent with supernova models, although some of these features can also be produced by other stellar sources. Both thermonuclear and core-collapse supernovae have been proposed as potential sources for these grains, although type II supernova models seem to be favored (Nittler and Ciesla 2016). A clear fingerprint of their supernova origin is the excess of ^{44}Ca (attributed to in situ decay of ^{44}Ti), present in ∼10–20% of the X grains (Amari et al. 1992), unaccompanied by anomalies in other stable calcium

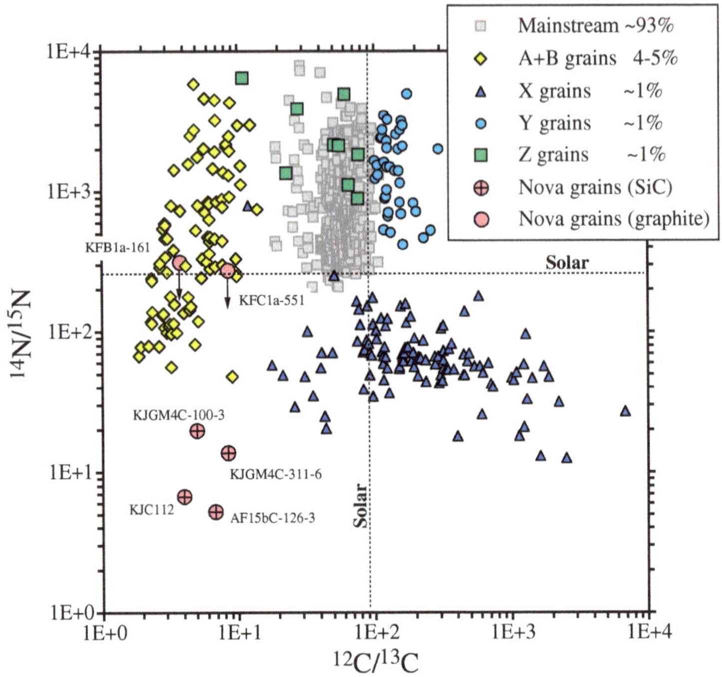

Fig. 5.16 Carbon and nitrogen isotopic ratios for the different SiC grain populations. Error bars are smaller than the symbols

isotopes. Other isotopic signatures that suggest a supernova origin include long-decayed species, such as ^{32}Si, ^{26}Al, ^{41}K, and ^{49}V (Nittler and Ciesla 2016).

A major problem to quantitatively match the grain data with supernova models (type II, in particular) is the need for selective mixing between different stellar layers (Lodders 2005; Zinner 2005), at a much larger scale than that suggested by observations and/or simulations. Efforts to quantitatively match grain data by means of supernova models have revealed a number of discrepancies. For instance, supernova SiC grains are systematically ^{15}N rich and ^{54}Fe poor with respect to model predictions. Recent simulations suggest that many isotopic features of supernova SiC and graphite grains can be reproduced by H-ingestion in the He-rich shells, that is, by the presence of residual H during explosive He-burning in supernova models (Pignatari et al. 2013; Liu et al. 2016, 2017, 2018). An alternative to selective mixing suggests the formation of supernova SiC and graphite grains in the O-rich layers of a massive star during a type II supernova explosion, where radiation may play a key role in dissociating the very stable CO molecules, hence freeing C atoms (Clayton et al. 1999, 2001; Clayton 2013). However, some isotopic features predicted in such models disagree with current presolar grain data (Nittler and Ciesla 2016). The supernova paternity of some presolar grains can also be settled by microstructural and mineralogical studies. In particular, supernova SiC and Si_3N_4 grains frequently appear as aggregates of small crystals while AGB SiC grains are typically single crystals. On the other hand, supernova graphites often present TiC subgrains that exhibit evidence of past ion irradiation (see details in Nittler and Ciesla (2016)).

A major, unsolved question is which fraction of the dust synthesized during core-collapse supernovae survives the passage of reverse shocks before being injected into the interstellar medium. Such issue will help to shed light into the specific contribution of supernovae to the dust we observe in the Universe.

5.5.2.4 Nova Grains

Infrared and ultraviolet observations have revealed dust forming episodes in the shells ejected during classical nova outbursts (Gehrz et al. 1998; Gehrz 2002). Their relatively high frequency (about ~30–35 nova explosions per year, just in our Galaxy (Shafter 2002)), has raised the issue of the potential contribution of novae to the different grain populations.

Since the pioneering studies of dust formation in novae by D.D. Clayton and F. Hoyle (Clayton and Hoyle 1976) (a concept already suggested by A.G.W. Cameron in 1973), all efforts devoted to the identification of potential nova grains relied mainly on the search for low $^{20}Ne/^{22}Ne$ ratios (since noble gases, such as Ne, do not condense into grains, ^{22}Ne is frequently attributed to in situ ^{22}Na decay, a clear imprint of a classical nova explosion). Indeed, Clayton and Hoyle pointed out several isotopic signatures (large overproduction of $^{13,14}C$, ^{18}O, ^{22}Na, ^{26}Al or ^{30}Si), that may help in the identification of such nova candidate grains. Forty years later, most of these signatures still hold, in view of our current understanding of nova explosions (see Jose 2016; Starrfield et al. 2016, for recent reviews), except ^{14}C, bypassed by the main nuclear path in novae, and ^{18}O, slightly overproduced by novae although grains nucleated in this environment are expected to be much more anomalous in ^{17}O (Kovetz and Prialnik 1997; José and Hernanz 1998; Starrfield et al. 1998, 2009).

A major step forward in the discovery of presolar nova candidate grains was achieved by Amari et al. (2001a), Amari (2002) (Fig. 5.17), who reported on several SiC and graphite grains, isolated from the Murchison and Acfer 094 meteorites, with an abundance pattern qualitatively similar to nova model predictions: low $^{12}C/^{13}C$ and $^{14}N/^{15}N$ ratios, high $^{30}Si/^{28}Si$, and close-to-solar $^{29}Si/^{28}Si$; and high $^{26}Al/^{27}Al$ and $^{22}Ne/^{20}Ne$ ratios for some of the grains (José et al. 2004). But in order to quantitatively match the grain data, one had to assume a mixing process between material newly synthesized in the nova outburst and more than ten times as much unprocessed, isotopically close-to-solar, material before grain formation.

Concerns about the likely nova paternity of these grains have been raised (Nittler and Hoppe 2005), after three additional micron-sized SiC grains were also isolated from the Murchison meteorite with similar trends (in particular, low $^{12}C/^{13}C$ and $^{14}N/^{15}N$ ratios), but with additional imprints (mainly non-solar Ti features), from which a supernova origin cannot be excluded. It is not clear, however, whether

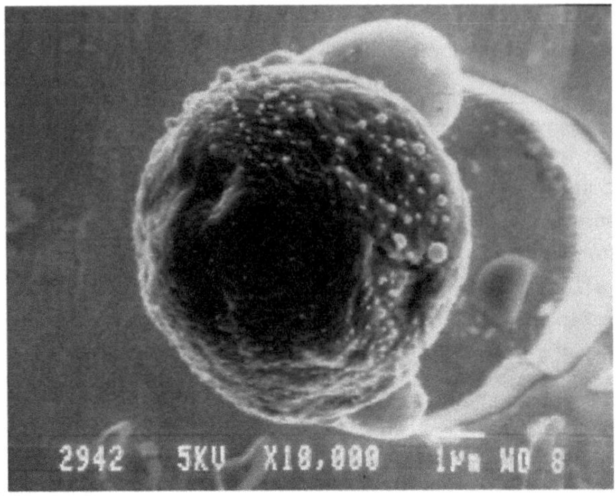

Fig. 5.17 The nova candidate graphite grain KFC1a-511 after Secondary Ion Mass Spectrometry (SIMS). Image courtesy of S. Amari

both samples (hereafter, A01 and NH05, respectively) correspond to the same progenitor. After all, their isotopic signatures are not identical: for instance, grains from the NH05 sample have much larger $^{26}Al/^{27}Al$ ratios and are more heavily depleted in ^{29}Si than grains from the A01 sample. Furthermore, grain M11-334-2 (NH05 sample) is deficient in ^{30}Si with respect to solar (whereas ^{30}Si excesses, characteristic of the A01 sample, are expected in the ejecta from ONe novae). Moreover, it must be stressed that the presence of anomalous Ti does not necessarily rule out a possible nova paternity (with the exception of ^{44}Ti, attributed to in situ decay of ^{44}Ca, an isotope clearly linked to a supernova explosion): titanium is, indeed, very close to the nucleosynthetic endpoint for novae (calcium), and hence, it could easily be reached by a slightly more violent outburst. This could be driven by explosions in cooler white dwarfs, or following lower mass-accretion rate episodes (José and Hernanz 2007a; Glasner and Truran 2009). Furthermore, explosions in metal-deficient envelopes, such as those expected for primordial nova systems, could have a similar effect (José et al. 2007; José and Hernanz 2007b).

Other nova candidate grains have been proposed in the last decades: for instance, the SiC grain 240-1 (Nittler et al. 2006), also isolated from Murchison, exhibits both $^{12}C/^{13}C$ and $^{14}N/^{15}N$ ratios lower than for any other presolar grain reported so far. These isotopic features are consistent with pure nova ejecta from a white dwarf with a mass ranging between 1.0 and 1.2 $M_\odot$. However, the ^{29}Si excesses measured in this grain do not match the usual predictions from nova models (which usually reflect ^{29}Si deficits, with respect to solar). A putative nova origin has also been attributed to the oxide grain T54 (Nittler 1997), with $^{16}O/^{17}O \sim 71$ and $^{16}O/^{18}O \sim 2000$, likely condensed in the shells ejected from a nova outburst on a 0.8 $M_\odot$ CO white dwarf. Unfortunately, no additional isotopic determinations were carried out on this grain. The inventory of nova candidate grains includes as well other oxide grains (two alumina and two spinel; Gyngard et al. 2011). Again, mixing between the nova ejecta and material of solar composition was required to match the composition of those grains. Finally, the presolar graphite grain LAP-149, isolated from the primitive meteorite LaPaz Icefield 031117, has also been suggested as a potential nova grain, as it exhibits one of the lowest $^{12}C/^{13}C$ ratios observed among presolar grains. Being extremely ^{13}C-rich and ^{15}N-poor, its origin suggests condensation in the ejecta of a low-mass CO nova (Haenecour et al. 2016).

The main difficulty faced in the unambiguous identification of presolar nova grains is the need for a simultaneous match of multiple isotopic ratios. Furthermore, the requirement of dilution of nova material with large amounts of unprocessed, isotopically close-to-solar material before grain formation, is considered another drawback for a proper identification of nova candidate grains. A very recent effort in this regard (Iliadis et al. 2018), based on a Monte Carlo technique, that involves the random sampling over the most important nova model parameters, has led to the identification of 18 presolar grains with measured isotopic signatures consistent with a CO nova origin, without assuming any dilution of the ejecta. Among these, the grains G270-2, M11-334-2, G278, M11-347-4, M11-151-4, and Ag2-6 have the highest plausibility of a (CO) nova paternity.

Future nova candidate grains will reveal more clues on the mechanisms powering nova explosions. To achieve this, cosmochemists will have to rely on a much wider range of isotopic determinations for proper identification of the stellar source (to disentangle, for instance, which grains are formed in supernova blasts and which in nova explosions). Novae hosting very massive white dwarfs (around 1.35 $M_{\odot}$) likely imprint additional signatures in the grains condensing in their ejecta (in particular, a suite of sulfur anomalies as well as severe ^{31}P overproduction). New techniques for laboratory analysis need to be developed to unambiguously identify such signatures, avoiding potential contamination of the samples by sulfuric acid, one of the standard methods used during the separation process.

5.6 Accretion in Binaries: Special Cases

It was seen in the case of the merging of two white dwarfs that if the total mass was larger than the Chandrasekhar mass, the final outcome could either be a SNIa or a collapse to a neutron star. However in the large majority of cases, the total mass is smaller than the critical value and the final result is a white dwarf "born again". The fraction of the secondary that is expelled is not yet known and, consequently, the influence of such systems on the chemical evolution of the Galaxy has not been yet elucidated. The total amount of freshly synthesized elements during the impact is small (Guerrero et al. 2004), except in the case of a secondary made of helium that shows an enhancement of Ca, Mg, Si, S and Fe, and confined to a corona around the primary.

In the collapse case, since the primary is rapidly rotating, as expected from the transfer of angular momentum from the disk to the star, a centrifugally supported disk made of heavily neutronized species, $Y_e \sim 0.1$, will form around the proto-neutron star. As a consequence of the neutrino irradiation, electron neutrino captures will increase the electron mole number to a value $Y_e \sim 0.5$ and α-particles will form inducing a wind that will blow away the disk. During this process it is expected that $\sim 10^{-2}\,M_\odot$ of ^{56}Ni will be synthesized and that the event will look as a dim SNIa-like transient (Metzger et al. 2009).

Merging of close binaries as well as close encounters in densely populated stellar systems, like globular clusters or galactic nuclei, can also provide violent scenarios able to trigger a nucleosynthetic activity other than the conventional thermonuclear explosions described up to now. The center of the Milky Way, for instance, contains a massive black hole surrounded by a swarm of stars, many of them white dwarfs. Close encounters are very common and tidal torques can produce extreme deformations of the stars or even trigger an explosion.

The tidal interaction between a white dwarf and a black hole is characterized by three length scales (Carter and Luminet 1983; Rosswog et al. 2009b): (i) the stellar radius, $R_{\rm WD}$, (ii) the gravitational radius of the black hole $R_{rmg} = 2GM_{\rm BH}/c^2 \simeq 3 \times 10^{11} M_{\rm BH,6}$, where $M_{\rm BH,6}$ is the mass of the black hole in units of $10^6\,M_\odot$, and (iii) the tidal radius, $R_\tau \simeq 1.2 \times 10^{11} M_{\rm BH,6}^{1/3} (R_{\rm WD}/10^9\,{\rm cm})(M_{\rm WD}/0.6\,M_\odot)^{-1/3}$ cm, that is the distance from the black hole at which $M_{\rm BH}/R_\tau^3$ equals the mean density of the passing star.

The strength of the tidal encounter can be estimated from the dimensionless parameter $\beta = R_\tau/R_{\rm P}$, where $R_{\rm P}$ is the pericenter distance, assuming a parabolic orbit. When $\beta \geq 1$, the star is disrupted in a single flyby. The energy to tear apart the star (the binding energy of the star) is supplied by the orbital energy. In the case of white dwarfs, the ratio between the total disruption radius and the gravitational radius is

$$\beta_{\rm g} = \frac{R_\tau}{R_{\rm g}} \simeq 0.4 M_{\rm BH,6}^{-2/3} \left(\frac{R_{\rm WD}}{10^9\,{\rm cm}}\right) \left(\frac{M_{\rm WD}}{0.6\,M_\odot}\right)^{-1/3} \qquad (5.26)$$

therefore, if the mass of the black hole is high enough, the tidal radius is inside the gravitational radius and the white dwarf is swallowed without being disrupted. This critical mass is

$$M_{\rm BH,lim} \simeq 2.5 \times 10^5 \left(\frac{R_{\rm WD}}{10^9\,{\rm cm}}\right)^{3/2} \left(\frac{M_{\rm WD}}{0.6\,M_\odot}\right)^{-1/2} \qquad (5.27)$$

The dynamics of the encounter can be described as follows (Carter and Luminet 1983). When the star is far from the black hole, the tidal interaction is negligible and the white dwarf is in hydrostatic equilibrium. As soon as it enters the Roche lobe, tidal interaction quickly grows in strength. As a consequence, matter is compressed by the flattening of the star. When the white dwarf is flat enough, the internal energy becomes dominant and the star experiences a bounce that reduces the pressure and makes the tidal interaction dominant once more. Depending on the parameters of the encounter (M_{BH}, M_{WD}, β) and on the chemical composition of the white dwarf, a vigorous thermonuclear burning can occur during the compression phase that can even produce a substantial amount of iron peak elements (Rosswog et al.

2009b). Finally, when the star is far enough, self-gravitation recovers control and, depending on the balance between internal and gravitational energies, matter is partially swallowed by the black hole, partially ejected to the interstellar medium and partially remains bound. During the expansion phase the radioactive debris, the β-decay of ^{56}Ni, mainly, can emit light and produce some kind of peculiar, sub-luminous SNIa (Rosswog et al. 2009b).

The recent success of LIGO in detecting gravitational waves from the merging of two compact objects has triggered the interest on the neutron star -neutron star (NS+NS) and neutron star—black hole (NS+BH) merging. There are several reasons for such interest: the possibility to test the equation of state of nuclear matter, a site for the synthesis of r-elements, and electromagnetic events displaying a wide range of time scales and wavelengths (Fernández and Metzger 2016).

This electromagnetic counterpart can manifest itself as a beam of electromagnetic radiation, i.e. as a short Gamma Ray Burst (Narayan et al. 1992; Mochkovitch et al. 1993) and as a *kilonova* or *macronova*, i.e. a transient powered by the radioactive decay of the r-process elements produced in the expanding ejecta that can be detected at optical and infrared wavelengths (Metzger 2017). One of the main characteristics of these transients is an increase of their duration due to the presence of Lanthanid an Actynid isotopes ($A > 140$) that extraordinarily increase the opacity as compared with the usual case of iron peak elements. Since the opacity and the velocity of the ejecta control the diffusion time, magnitude, color and duration of the kilonova will depend on the composition, geometry and kinematics of the ejecta (Fernández and Metzger 2016). The first confirmed kilonova was associated to the gamma-ray burst GRB 130603B (Tanvir et al. 2013; Berger et al. 2013; Fan et al. 2013), and the first detection of the gravitational emission of a NS+NS merger was GW170817 (Abbott et al. 2017), which was associated to the short GRB 170817A and was observed at almost all wavelengths (Antonini and Perets 2012)

In principle there are two potential sources of debris: material expelled on dynamical time scales, with typical velocities of 0.1–0.3c and material coming from the remnant disk. The relative importance of both components depends on the parameters of the system (Fernández and Metzger 2016). In the case of NS-BH mergers, the dynamical ejection is induced by the tidal forces and material is confined near the equatorial region while in the case of NS+NS, the ejection occurs in the contact interface and material is ejected over a wider solid angle. Part of this material is heated by a shock wave and irradiated with neutrinos. The chemical composition nicely fits the observed abundances of r-elements observed in the Solar System. Outflows from the remnant disk can be produced on longer time scales, depending on neutrino heating, state of the disk, presence of magnetic fields and so on. The velocity of this material is smaller, $\sim 0.05c$, than that of the dynamic component and less neutron rich as a consequence of a longer exposure to weak interaction.

References

Abbott BP, Abbott R, Abbott TD, Acernese F, Ackley K, Adams C, Adams T, Addesso P, Adhikari RX, Adya VB, et al (2017) GW170817: observation of gravitational waves from a binary neutron star inspiral. Phys Rev Lett 119(16):161101. https://doi.org/10.1103/PhysRevLett.119.161101, arXiv:1710.05832

Abdo AA, Ackermann M, Ajello M, Atwood WB, Baldini L, Ballet J, Barbiellini G, Bastieri D, Bechtol K, Bellazzini R, et al (2010) Gamma-ray emission concurrent with the nova in the symbiotic binary V407 Cygni. Science 329:817–821. https://doi.org/10.1126/science.1192537, arXiv:1008.3912

Ackermann M, Ajello M, Albert A, Baldini L, Ballet J, Barbiellini G, Bastieri D, Bellazzini R, Bissaldi E, Blandford RD, Bloom ED, Bottacini E, Brandt TJ, Bregeon J, Bruel P, Buehler R, Buson S, Caliandro GA, Cameron RA, Caragiulo M, Caraveo PA, Cavazzuti E, Charles E, Chekhtman A, Cheung CC, Chiang J, Chiaro G, Ciprini S, Claus R, Cohen-Tanugi J, Conrad J, Corbel S, D'Ammando F, de Angelis A, den Hartog PR, de Palma F, Dermer CD, Desiante R, Digel SW, Di Venere L, do Couto e Silva E, Donato D, Drell PS, Drlica-Wagner A, Favuzzi C, Ferrara EC, Focke WB, Franckowiak A, Fuhrmann L, Fukazawa Y, Fusco P, Gargano F, Gasparrini D, Germani S, Giglietto N, Giordano F, Giroletti M, Glanzman T, Godfrey G, Grenier IA, Grove JE, Guiriec S, Hadasch D, Harding AK, Hayashida M, Hays E, Hewitt JW, Hill AB, Hou X, Jean P, Jogler T, Jóhannesson G, Johnson AS, Johnson WN, Kerr M, Knödlseder J, Kuss M, Larsson S, Latronico L, Lemoine-Goumard M, Longo F, Loparco F, Lott B, Lovellette MN, Lubrano P, Manfreda A, Martin P, Massaro F, Mayer M, Mazziotta MN, McEnery JE, Michelson PF, Mitthumsiri W, Mizuno T, Monzani ME, Morselli A, Moskalenko IV, Murgia S, Nemmen R, Nuss E, Ohsugi T, Omodei N, Orienti M, Orlando E, Ormes JF, Paneque D, Panetta JH, Perkins JS, Pesce-Rollins M, Piron F, Pivato G, Porter TA, Rainò S, Rando R, Razzano M, Razzaque S, Reimer A, Reimer O, Reposeur T, Saz Parkinson PM, Schaal M, Schulz A, Sgrò C, Siskind EJ, Spandre G, Spinelli P, Stawarz Ł, Suson DJ, Takahashi H, Tanaka T, Thayer JG, Thayer JB, Thompson DJ, Tibaldo L, Tinivella M, Torres DF, Tosti G, Troja E, Uchiyama Y, Vianello G, Winer BL, Wolff MT, Wood DL, Wood KS, Wood M, Charbonnel S, Corbet RHD, De Gennaro Aquino I, Edlin JP, Mason E, Schwarz GJ, Shore SN, Starrfield S, Teyssier F, Fermi-LAT Collaboration (2014) Fermi establishes classical novae as a distinct class of gamma-ray sources. Science 345:554–558. https://doi.org/10.1126/science.1253947, arXiv:1408.0735

Alibés A, Labay J, Canal R (2002) Galactic cosmic rays from superbubbles and the abundances of lithium, beryllium, and boron. Astrophys J 571:326–333. https://doi.org/10.1086/339937, arXiv:astro-ph/0202097

Amari S (2002) Presolar grains from novae: their isotopic ratios and radioactivities. New Astron Rev 46:519–524. https://doi.org/10.1016/S1387-6473(02)00194-X

Amari S, Anders A, Virag A, Zinner E (1990) Interstellar graphite in meteorites. Nature 345:238–240. https://doi.org/10.1038/345238a0

Amari S, Hoppe P, Zinner E, Lewis RS (1992) Interstellar SiC with unusual isotopic compositions – grains from a supernova? Astrophys J 394:L43–L46. https://doi.org/10.1086/186468

Amari S, Gao X, Nittler LR, Zinner E, José J, Hernanz M, Lewis RS (2001a) Presolar grains from novae. Astrophys J 551:1065–1072. https://doi.org/10.1086/320235, arXiv:astro-ph/0012465

Amari S, Nittler LR, Zinner E, Gallino R, Lugaro M, Lewis RS (2001b) Presolar SIC grains of Type Y: origin from low-metallicity asymptotic giant branch stars. Astrophys J 546:248–266. https://doi.org/10.1086/318230

Amari S, Nittler LR, Zinner E, Lodders K, Lewis RS (2001c) Presolar SiC grains of Type A and B: their isotopic compositions and stellar origins. Astrophys J 559:463–483. https://doi.org/10.1086/322397

Ambwani K, Sutherland P (1988) Gamma-ray spectra and energy deposition for Type IA supernovae. Astrophys J 325:820–827. https://doi.org/10.1086/166052

Amthor MA, Galaviz D, Heger A, Sakharuk A, Schatz H, Smith K (2006) Sensitivity of Type I X-ray bursts to rp-process reaction rate. In: International symposium on nuclear astrophysics – nuclei in the cosmos

Antonini F, Perets HB (2012) Secular evolution of compact binaries near massive black holes: gravitational wave sources and other exotica. Astrophys J 757:27. https://doi.org/10.1088/0004-637X/757/1/27, arXiv:1203.2938

Arnett WD (1969) A possible model of supernovae: detonation of ^{12}C. Astrophys Space Sci 5:180–212. https://doi.org/10.1007/BF00650291

Arnett D (1996) Supernovae and nucleosynthesis: an investigation of the history of matter from the big bang to the present. Princeton University Press, Princeton

Arnould M, Norgaard H (1975) The explosive thermonuclear formation of 7Li and 11B. Astron Astrophys 42:55–70

Audouze J, Reeves H (1982) The origin of the light elements. In: Essays in Nuclear Astrophysics. Cambridge University Press, Cambridge, pp 355–375

Axelrod TS (1980) Late time optical spectra from the Ni-56 model for Type 1 supernovae. PhD thesis, California University, Santa Cruz

Ayasli S, Joss PC (1982) Thermonuclear processes on accreting neutron stars – a systematic study. Astrophys J 256:637–665. https://doi.org/10.1086/159940

Aznar-Siguán G, García-Berro E, Lorén-Aguilar P, José J, Isern J (2013) Detonations in white dwarf dynamical interactions. Mon Not R Astron Soc 434:2539–2555. https://doi.org/10.1093/mnras/stt1198, arXiv:1306.6559

Aznar-Siguán G, García-Berro E, Magnien M, Lorén-Aguilar P (2014) On the possible observational signatures of white dwarf dynamical interactions. Mon Not R Astron Soc 443:2372–2383. https://doi.org/10.1093/mnras/stu1309, arXiv:1407.0200

Aznar-Siguán G, García-Berro E, Lorén-Aguilar P, Soker N, Kashi A (2015) Smoothed particle hydrodynamics simulations of the core-degenerate scenario for Type Ia supernovae. Mon Not R Astron Soc 450:2948–2962. https://doi.org/10.1093/mnras/stv824, arXiv:1503.02444

Badenes C, Borkowski KJ, Hughes JP, Hwang U, Bravo E (2006) Constraints on the physics of Type Ia supernovae from the X-ray spectrum of the Tycho supernova remnant. Astrophys J 645:1373–1391. https://doi.org/10.1086/504399, arXiv:astro-ph/0511140

Balman S, Krautter J, Oegelman H (1998) The X-ray spectral evolution of classical Nova V1974 Cygni 1992: a reanalysis of the ROSAT data. Astrophys J 499:395–406

Barbon R, Ciatti F, Rosino L (1973) Light curves and characteristics of recent supernovae. Astron Astrophys 29:57–67

Basinska EM, Lewin WHG, Sztajno M, Cominsky LR, Marshall FJ (1984) X-ray observations of the burst source MXB 1728-34. Astrophys J 281:337–353. https://doi.org/10.1086/162103

Bazin D, Montes F, Becerril A, Lorusso G, Amthor A, Baumann T, Crawford H, Estrade A, Gade A, Ginter T, Guess CJ, Hausmann M, Hitt GW, Mantica P, Matos M, Meharchand R, Minamisono K, Perdikakis G, Pereira J, Pinter J, Portillo M, Schatz H, Smith K, Stoker J, Stolz A, Zegers RGT (2008) Production and β decay of rp-process nuclei Cd96, In98, and Sn100. Phys Rev Lett 101(25):252501/1–252501/4. https://doi.org/10.1103/PhysRevLett.101.252501, arXiv:0810.3597

Belian RD, Conner JP, Evans WD (1976) The discovery of X-ray bursts from a region in the constellation Norma. Astrophys J 206:L135–L138. https://doi.org/10.1086/182151

Benz W, Thielemann FK, Hills JG (1989) Three-dimensional hydrodynamical simulations of stellar collisions. II – White dwarfs. Astrophys J 342:986–998. https://doi.org/10.1086/167656

Benz W, Cameron AGW, Press WH, Bowers RL (1990) Dynamic mass exchange in doubly degenerate binaries. I – 0.9 and 1.2 solar mass stars. Astrophys J 348:647–667. https://doi.org/10.1086/168273

Berger E, Fong W, Chornock R (2013) An r-process Kilonova associated with the short-hard GRB 130603B. Astrophys J 774:L23. https://doi.org/10.1088/2041-8205/774/2/L23, arXiv:1306.3960

Bernatowicz T, Fraundorf G, Ming T, Anders E, Wopenka B, Zinner E, Fraundorf P (1987) Evidence for interstellar SiC in the Murray carbonaceous meteorite. Nature 330:728–730. https://doi.org/10.1038/330728a0

Bildsten L (1995) Propagation of nuclear burning fronts on accreting neutron stars: X-ray bursts and sub-hertz noise. Astrophys J 438:852–875. https://doi.org/10.1086/175128

Bildsten L (1998) Thermonuclear burning on rapidly accreting neutron stars. In: Buccheri R, van Paradijs J, Alpar A (eds) NATO ASIC Proc. 515: the many faces of neutron stars, p 419

Bildsten L, Chang P, Paerels F (2003) Atomic spectral features during thermonuclear flashes on neutron stars. Astrophys J 591:L29–L32. https://doi.org/10.1086/377066, arXiv:astro-ph/0303147

Bildsten L, Shen KJ, Weinberg NN, Nelemans G (2007) Faint thermonuclear supernovae from AM Canum Venaticorum binaries. Astrophys J 662:L95–L98. https://doi.org/10.1086/519489, arXiv:astro-ph/0703578

Blinnikov SI, Khokhlov AM (1986) Development of detonations in degenerate stars. Sov Astron Lett 12:131–133

Blondin S, Dessart L, Hillier DJ, Khokhlov AM (2017) Evidence for sub-Chandrasekhar-mass progenitors of Type Ia supernovae at the faint end of the width-luminosity relation. Mon Not R Astron Soc 470:157–165. https://doi.org/10.1093/mnras/stw2492, arXiv:1706.01901

Bloom JS, Kasen D, Shen KJ, Nugent PE, Butler NR, Graham ML, Howell DA, Kolb U, Holmes S, Haswell CA, Burwitz V, Rodriguez J, Sullivan M (2012) A compact degenerate primary-star progenitor of SN 2011fe. Astrophys J 744:L17. https://doi.org/10.1088/2041-8205/744/2/L17, arXiv:1111.0966

Bode MF, Evans A (2008) Classical Novae. Cambridge University Press, Cambridge

Boffin HMJ, Paulus G, Arnould M, Mowlavi N (1993) The explosive thermonuclear formation of Li-7 revisited. Astron Astrophys 279:173–178

Branch D (1981) Some statistical properties of Type I supernovae. Astrophys J 248:1076–1080. https://doi.org/10.1086/159237

Branch D, Fisher A, Nugent P (1993) On the relative frequencies of spectroscopically normal and peculiar Type IA supernovae. Astron J 106:2383–2391. https://doi.org/10.1086/116810

Branch D, Romanishin W, Baron E (1996) Statistical connections between the properties of Type IA supernovae and the B-V colors of their parent galaxies, and the value of H 0. Astrophys J 465:73–78. https://doi.org/10.1086/177402, arXiv:astro-ph/9510071

Bravo E, García-Senz D (1999) Coulomb corrections to the equation of state of nuclear statistical equilibrium matter: implications for SNIa nucleosynthesis and the accretion-induced collapse of white dwarfs. Mon Not R Astron Soc 307:984–992. https://doi.org/10.1046/j.1365-8711.1999.02694.x

Bravo E, Isern J, Canal R, Labay J (1992) On the contribution of Ne-22 to the synthesis of Fe-54 and Ni-58 in thermonuclear supernovae. Astron Astrophys 257:534–538

Bravo E, Tornambe A, Dominguez I, Isern J (1996) Clues to Type IA SN progenitors from degenerate carbon ignition models. Astron Astrophys 306:811–822

Brown EF (2000) Nuclear heating and melted layers in the inner crust of an accreting neutron star. Astrophys J 531:988–1002. https://doi.org/10.1086/308487, arXiv:astro-ph/9910215

Brown EF, Bildsten L (1998) The ocean and crust of a rapidly accreting neutron star: implications for magnetic field evolution and thermonuclear flashes. Astrophys J 496:915–933. https://doi.org/10.1086/305419, arXiv:astro-ph/9710261

Burrows A, The L (1990) X- and gamma-ray signatures of Type IA supernovae. Astrophys J 360:626–638. https://doi.org/10.1086/169150

Bussard RW, Ramaty R, Drachman RJ (1979) The annihilation of galactic positrons. Astrophys J 228:928–934. https://doi.org/10.1086/156920

Cadonau R, Tammann GA, Sandage A (1985) Type I supernovae as standard candles. In: Bartel N (ed) Supernovae as distance indicators. Lecture notes in physics, vol 224. Springer, Berlin, pp 151–165. https://doi.org/10.1007/3-540-15206-7-56

Cameron AGW (1955) Origin of anomalous abundances of the elements in giant stars. Astrophys J 121:144–160. https://doi.org/10.1086/145970

Cameron AGW (1962) The formation of the sun and planets. Icarus 1:13–69. https://doi.org/10.1016/0019-1035(62)90005-2
Cameron AGW (1973) Interstellar grains in museums? In: Greenberg JM, van de Hulst HC (eds) Interstellar dust and related topics, IAU symposium, vol 52, pp 545–547
Canal R, Isern J, Labay J (1990) The origin of neutron stars in binary systems. Annu Rev Astron Astrophys 28:183–214. https://doi.org/10.1146/annurev.aa.28.090190.001151
Cappellaro E, Barbon R, Turatto M (2003) Supernova statistics. Astrophysics. arXiv:astro-ph/0310859
Carter B, Luminet J (1983) Tidal compression of a star by a large black hole. I Mechanical evolution and nuclear energy release by proton capture. Astron Astrophys 121:97–113
Casanova J, José J, García-Berro E, Calder A, Shore SN (2010) On mixing at the core-envelope interface during classical nova outbursts. Astron Astrophys 513:L5. https://doi.org/10.1051/0004-6361/201014178, arXiv:1004.2792
Casanova J, José J, García-Berro E, Calder A, Shore SN (2011a) Mixing in classical novae: a 2-D sensitivity study. Astron Astrophys 527:A5. https://doi.org/10.1051/0004-6361/201015895, arXiv:1012.3199
Casanova J, José J, García-Berro E, Shore SN, Calder AC (2011b) Kelvin-Helmholtz instabilities as the source of inhomogeneous mixing in nova explosions. Nature 478:490–492. https://doi.org/10.1038/nature10520
Casanova J, José J, García-Berro E, Shore SN (2016) Three-dimensional simulations of turbulent convective mixing in ONe and CO classical nova explosions. Astron Astrophys 595:A28. https://doi.org/10.1051/0004-6361/201628707
Casanova J, José J, Shore SN (2018) Two-dimensional simulations of mixing in classical novae: the effect of the white dwarf composition and mass. Astron Astrophys. https://doi.org/10.1051/0004-6361/201628707
Champagne AE, Wiescher M (1992) Explosive hydrogen burning. Annu Rev Nucl Part Sci 42: 39–76. https://doi.org/10.1146/annurev.ns.42.120192.000351
Chan K, Lingenfelter RE (1993) Positrons from supernovae. Astrophys J 405:614–636. https://doi.org/10.1086/172393
Chang P, Bildsten L, Wasserman I (2005) Formation of resonant atomic lines during thermonuclear flashes on neutron stars. Astrophys J 629:998–1007. https://doi.org/10.1086/431730, arXiv:astro-ph/0505062
Chang P, Morsink S, Bildsten L, Wasserman I (2006) Rotational broadening of atomic spectral features from neutron stars. Astrophys J 636:L117–L120. https://doi.org/10.1086/499428, arXiv:astro-ph/0511246
Cheung CC, Jean P, Shore SN, Stawarz Ł, Corbet RHD, Knödlseder J, Starrfield S, Wood DL, Desiante R, Longo F, Pivato G, Wood KS (2016) Fermi-LAT gamma-ray detections of classical Novae V1369 Centauri 2013 and V5668 Sagittarii 2015. Astrophys J 826:142. https://doi.org/10.3847/0004-637X/826/2/142, arXiv:1605.04216
Choi B, Huss GR, Wasserburg GJ, Gallino R (1998) Presolar corundum and spinel in ordinary chondrites: origins from AGB stars and a supernova. Science 282:1284–1289. https://doi.org/10.1126/science.282.5392.1284
Choi B, Wasserburg GJ, Huss GR (1999) Circumstellar hibonite and corundum and nucleosynthesis in asymptotic giant branch stars. Astrophys J 522:L133–L136. https://doi.org/10.1086/312239
Chomiuk L, Linford JD, Yang J, O'Brien TJ, Paragi Z, Mioduszewski AJ, Beswick RJ, Cheung CC, Mukai K, Nelson T, Ribeiro VARM, Rupen MP, Sokoloski JL, Weston J, Zheng Y, Bode MF, Eyres S, Roy N, Taylor GB (2014) Binary orbits as the driver of γ-ray emission and mass ejection in classical novae. Nature 514:339–342. https://doi.org/10.1038/nature13773, arXiv:1410.3473
Churazov E, Sunyaev R, Isern J, Knödlseder J, Jean P, Lebrun F, Chugai N, Grebenev S, Bravo E, Sazonov S, Renaud M (2014) Cobalt-56 γ-ray emission lines from the Type Ia supernova 2014J. Nature 512:406–408. https://doi.org/10.1038/nature13672, arXiv:1405.3332

Churazov E, Sunyaev R, Isern J, Bikmaev I, Bravo E, Chugai N, Grebenev S, Jean P, Knödlseder J, Lebrun F, Kuulkers E (2015) Gamma-rays from Type Ia Supernova SN2014J. Astrophys J 812:62. https://doi.org/10.1088/0004-637X/812/1/62, arXiv:1502.00255

Clayton DD (1981) Li-7 gamma-ray lines from novae. Astrophys J 244:L97. https://doi.org/10.1086/183488

Clayton DD (2013) Analytic approximation of carbon condensation issues in Type II Supernovae. Astrophys J 762:5. https://doi.org/10.1088/0004-637X/762/1/5

Clayton DD, Hoyle F (1974) Gamma-ray lines from novae. Astrophys J 187:L101+. https://doi.org/10.1086/181406

Clayton DD, Hoyle F (1976) Grains of anomalous isotopic composition from novae. Astrophys J 203:490–496. https://doi.org/10.1086/154104

Clayton DD, Nittler LR (2004) Astrophysics with presolar stardust. Annu Rev Astron Astrophys 42:39–78. https://doi.org/10.1146/annurev.astro.42.053102.134022

Clayton DD, Liu W, Dalgarno A (1999) Condensation of carbon in radioactive supernova gas. Science 283:1290. https://doi.org/10.1126/science.283.5406.1290

Clayton DD, Deneault EAN, Meyer BS (2001) Condensation of carbon in radioactive supernova gas. Astrophys J 562:480–493. https://doi.org/10.1086/323467

Coc A, Hernanz M, José J, Thibaud J (2000) Influence of new reaction rates on ^{18}F production in novae. Astron Astrophys 357:561–571. arXiv:astro-ph/0003166

Colgate SA, McKee C (1969) Early supernova luminosity. Astrophys J 157:623–644. 10.1086/150102

Cornelisse R, Heise J, Kuulkers E, Verbunt F, in't Zand JJM (2000) The longest thermonuclear X-ray burst ever observed? A BeppoSAX Wide Field Camera observation of 4U 1735-44. Astron Astrophys 357:L21–L24. arXiv:astro-ph/0003454

Cottam J, Paerels F, Mendez M (2002) Gravitationally redshifted absorption lines in the X-ray burst spectra of a neutron star. Nature 420:51–54. https://doi.org/10.1038/nature01159, arXiv:astro-ph/0211126

Cottam J, Paerels F, Méndez M, Boirin L, Lewin WHG, Kuulkers E, Miller JM (2008) The Burst Spectra of EXO 0748-676 during a Long 2003 XMM-Newton Observation. Astrophys J 672:504–509. https://doi.org/10.1086/524186, arXiv:0709.4062

Cumming A (2005) Superbursts: a new probe of the rp-process. Nucl Phys A 758:439–446. https://doi.org/10.1016/j.nuclphysa.2005.05.081

Cumming A, Bildsten L (2001) Carbon flashes in the heavy-element ocean on accreting neutron stars. Astrophys J 559:L127–L130. https://doi.org/10.1086/323937, arXiv:astro-ph/0107213

D'Antona F, Mazzitelli I (1989) The fastest evolving white dwarfs. Astrophys J 347:934–949. https://doi.org/10.1086/168185

Damen E, Magnier E, Lewin WHG, Tan J, Penninx W, van Paradijs J (1990) X-ray bursts with photospheric radius expansion and the gravitational redshift of neutron stars. Astron Astrophys 237:103–109

Darnley MJ, Henze M, Steele IA, Bode MF, Ribeiro VARM, Rodríguez-Gil P, Shafter AW, Williams SC, Baer D, Hachisu I, Hernanz M, Hornoch K, Hounsell R, Kato M, Kiyota S, Kučáková H, Maehara H, Ness JU, Piascik AS, Sala G, Skillen I, Smith RJ, Wolf M (2015) A remarkable recurrent nova in M31: discovery and optical/UV observations of the predicted 2014 eruption. Astron Astrophys 580:A45. https://doi.org/10.1051/0004-6361/201526027, arXiv:1506.04202

Darnley MJ, Henze M, Bode MF, Hachisu I, Hernanz M, Hornoch K, Hounsell R, Kato M, Ness JU, Osborne JP, Page KL, Ribeiro VARM, Rodríguez-Gil P, Shafter AW, Shara MM, Steele IA, Williams SC, Arai A, Arcavi I, Barsukova EA, Boumis P, Chen T, Fabrika S, Figueira J, Gao X, Gehrels N, Godon P, Goranskij VP, Harman DJ, Hartmann DH, Hosseinzadeh G, Horst JC, Itagaki K, José J, Kabashima F, Kaur A, Kawai N, Kennea JA, Kiyota S, Kučáková H, Lau KM, Maehara H, Naito H, Nakajima K, Nishiyama K, O'Brien TJ, Quimby R, Sala G, Sano Y, Sion EM, Valeev AF, Watanabe F, Watanabe M, Williams BF, Xu Z (2016a) M31N 2008-12a – the remarkable recurrent nova in M31: panchromatic observations of the 2015 eruption. Astrophys J 833:149. https://doi.org/10.3847/1538-4357/833/2/149, arXiv:1607.08082

Darnley MJ, Henze M, Steele IA, Bode MF, Ribeiro VARM, Rodríguez-Gil P, Shafter AW, Williams SC, Baer D, Hachisu I, Hernanz M, Hornoch K, Hounsell R, Kato M, Kiyota S, Kučáková H, Maehara H, Ness JU, Piascik AS, Sala G, Skillen I, Smith RJ, Wolf M (2016b) A remarkable recurrent nova in M31: discovery and optical/UV observations of the predicted 2014 eruption (Corrigendum). Astron Astrophys 593:C3. https://doi.org/10.1051/0004-6361/201526027e

Darnley MJ, Hounsell R, Godon P, Perley DA, Henze M, Kuin NPM, Williams BF, Williams SC, Bode MF, Harman DJ, Hornoch K, Link M, Ness JU, Ribeiro VARM, Sion EM, Shafter AW, Shara MM (2017b) Inflows, outflows, and a giant donor in the remarkable recurrent nova M31N 2008-12a? – Hubble space telescope photometry of the 2015 eruption. Astrophys J 849:96. https://doi.org/10.3847/1538-4357/aa9062, arXiv:1709.10145

Darnley MJ, Hounsell R, Godon P, Perley DA, Henze M, Kuin NPM, Williams BF, Williams SC, Bode MF, Harman DJ, Hornoch K, Link M, Ness JU, Ribeiro VARM, Sion EM, Shafter AW, Shara MM (2017a) No neon, but jets in the remarkable recurrent nova M31N 2008-12a? – Hubble space telescope spectroscopy of the 2015 eruption. Astrophys J 847:35. https://doi.org/10.3847/1538-4357/aa8867, arXiv:1708.06795

De Angelis A, Tatischeff V, Grenier IA, McEnery J, Mallamaci M, Tavani M, Oberlack U, Hanlon L, Walter R, Argan A, et al (2017a) Science with e-ASTROGAM (A space mission for MeV-GeV gamma-ray astrophysics). arXiv:1711.01265

De Angelis A, Tatischeff V, Tavani M, Oberlack U, Grenier I, Hanlon L, Walter R, Argan A, von Ballmoos P, Bulgarelli A, Donnarumma I, Hernanz M, Kuvvetli I, Pearce M, Zdziarski A, Aboudan A, Ajello M, Ambrosi G, Bernard D, Bernardini E, Bonvicini V, Brogna A, Branchesi M, Budtz-Jorgensen C, Bykov A, Campana R, Cardillo M, Coppi P, De Martino D, Diehl R, Doro M, Fioretti V, Funk S, Ghisellini G, Grove E, Hamadache C, Hartmann DH, Hayashida M, Isern J, Kanbach G, Kiener J, Knödlseder J, Labanti C, Laurent P, Limousin O, Longo F, Mannheim K, Marisaldi M, Martinez M, Mazziotta MN, McEnery J, Mereghetti S, Minervini G, Moiseev A, Morselli A, Nakazawa K, Orleanski P, Paredes JM, Patricelli B, Peyré J, Piano G, Pohl M, Ramarijaona H, Rando R, Reichardt I, Roncadelli M, Silva R, Tavecchio F, Thompson DJ, Turolla R, Ulyanov A, Vacchi A, Wu X, Zoglauer A (2017b) The e-ASTROGAM mission. Exploring the extreme Universe with gamma rays in the MeV–GeV range. Exp Astron 44:25–82. https://doi.org/10.1007/s10686-017-9533-6, arXiv:1611.02232

de Plaa J, Werner N, Bleeker JAM, Vink J, Kaastra JS, Méndez M (2007) Constraining supernova models using the hot gas in clusters of galaxies. Astron Astrophys 465:345–355. https://doi.org/10.1051/0004-6361:20066382, arXiv:astro-ph/0701553

Della Valle M (2002) Nova populations. In: Hernanz M, José J (eds) Classical nova explosions, American Institute of Physics conference series, vol 637, pp 443–456. https://doi.org/10.1063/1.1518244

Della Valle M, Livio M (1994) On the nova rate in the Galaxy. Astron Astrophys 286:786–788

Della Valle M, Livio M (1995) The calibration of novae as distance indicators. Astrophys J 452:704–+. https://doi.org/10.1086/176342

Della Valle M, Livio M (1998) The spectroscopic differences between disk and thick-disk/bulge novae. Astrophys J 506:818–823. https://doi.org/10.1086/306275

Della Valle M, Bianchini A, Livio M, Orio M (1992) On the possible existence of two classes of progenitors for classical novae. Astron Astrophys 266:232–236

Della Valle M, Pasquini L, Daou D, Williams RE (2002) The evolution of Nova V382 Velorum 1999. Astron Astrophys 390:155–166. https://doi.org/10.1051/0004-6361:20020611, arXiv:astro-ph/0205135

Diehl R, Dupraz C, Bennett K, Bloemen H, Hermsen W, Knoedlseder J, Lichti G, Morris D, Ryan J, Schoenfelder V, Steinle H, Strong A, Swanenburg B, Varendorff M, Winkler C (1995) COMPTEL observations of Galactic ^{26}Al emission. Astron Astrophys 298:445–+

Diehl R, Siegert T, Hillebrandt W, Grebenev SA, Greiner J, Krause M, Kromer M, Maeda K, Röpke F, Taubenberger S (2014) Early ^{56}Ni decay gamma rays from SN2014J suggest an unusual explosion. Science 345:1162–1165. https://doi.org/10.1126/science.1254738, arXiv:1407.3061

Dominguez I, Tornambe A, Isern J (1993) On the formation of O-Ne white dwarfs in metal-rich close binary systems. Astrophys J 419:268–+. https://doi.org/10.1086/173480
Domínguez I, Höflich P, Straniero O (2001) Constraints on the progenitors of Type Ia supernovae and implications for the cosmological equation of state. Astrophys J 557:279–291. https://doi.org/10.1086/321661, arXiv:astro-ph/0104257
Domínguez I, Piersanti L, Bravo E, Tornambé A, Straniero O, Gagliardi S (2006) Rotating Type Ia SN progenitors: explosion and light curves. Astrophys J 644:21–29. https://doi.org/10.1086/503534
Elias JH, Matthews K, Neugebauer G, Persson SE (1985) Type I supernovae in the infrared and their use as distance indicators. Astrophys J 296:379–389. https://doi.org/10.1086/163456
Elomaa V, Vorobjev GK, Kankainen A, Batist L, Eliseev S, Eronen T, Hakala J, Jokinen A, Moore ID, Novikov YN, Penttilä H, Popov A, Rahaman S, Rissanen J, Saastamoinen A, Schatz H, Seliverstov DM, Weber C, Äystö J (2009) Quenching of the SnSbTe cycle in the rp process. Phys Rev Lett 102(25):252501/1–252501/4. https://doi.org/10.1103/PhysRevLett.102.252501
Falanga M, Chenevez J, Cumming A, Kuulkers E, Trap G, Goldwurm A (2008) Intermediate long X-ray bursts from the ultra-compact binary candidate SLX 1737-282. Astron Astrophys 484:43–50. https://doi.org/10.1051/0004-6361:20078982, arXiv:0711.0328
Fan YZ, Yu YW, Xu D, Jin ZP, Wu XF, Wei DM, Zhang B (2013) A supramassive magnetar central engine for GRB 130603B. Astrophys J 779:L25. https://doi.org/10.1088/2041-8205/779/2/L25, arXiv:1311.7185
Fernández R, Metzger BD (2016) Electromagnetic signatures of neutron star mergers in the advanced LIGO era. Ann Rev Nucl Part Sci 66:23–45. https://doi.org/10.1146/annurev-nucl-102115-044819, arXiv:1512.05435
Filippenko AV, Richmond MW, Branch D, Gaskell M, Herbst W, Ford CH, Treffers RR, Matheson T, Ho LC, Dey A, Sargent WLW, Small TA, van Breugel WJM (1992) The subluminous, spectroscopically peculiar Type IA supernova 1991bg in the elliptical galaxy NGC 4374. Astron J 104:1543–1556. https://doi.org/10.1086/116339
Fink M, Hillebrandt W, Röpke FK (2007) Double-detonation supernovae of sub-Chandrasekhar mass white dwarfs. Astron Astrophys 476:1133–1143. https://doi.org/10.1051/0004-6361:20078438, arXiv:0710.5486
Fishman GJ, Wilson RB, Meegan CA, Brock MN, Horack JM, Paciesas WS, Pendleton GN, Harmon BA, Leising M (1991) Detectability of early low-energy gamma rays from nearby novae by BATSE/GRO. In: Durouchoux P, Prantzos N (eds) Gamma-ray line astrophysics. American Institute of Physics conference series, vol 232, pp 190–192. https://doi.org/10.1063/1.40956
Fisker JL, Görres J, Wiescher M, Davids B (2006) The importance of $^{15}O(\alpha,\gamma)^{19}Ne$ to X-Ray bursts and superbursts. Astrophys J 650:332–337. https://doi.org/10.1086/507083, arXiv:astro-ph/0410561
Fisker JL, Schatz H, Thielemann F (2008) Explosive hydrogen burning during Type I X-ray bursts. Astrophys J Suppl Ser 174:261–276. https://doi.org/10.1086/521104
Fossey SJ, Cooke B, Pollack G, Wilde M, Wright T (2014) Supernova 2014J in M82 = Psn J09554214+6940260. Central Bureau Electronic Telegrams 3792
Fryxell BA, Woosley SE (1982a) A two-dimensional model for gamma-ray bursts. Astrophys J 258:733–739. https://doi.org/10.1086/160121
Fryxell BA, Woosley SE (1982b) Finite propagation time in multidimensional thermonuclear runaways. Astrophys J 261:332–336. https://doi.org/10.1086/160344
Fujimoto MY, Hanawa T, Miyaji S (1981) Shell flashes on accreting neutron stars and X-ray bursts. Astrophys J 247:267–278. https://doi.org/10.1086/159034
Fuller GM, Fowler WA, Newman MJ (1985) Stellar weak interaction rates for intermediate-mass nuclei. IV – interpolation procedures for rapidly varying lepton capture rates using effective log (ft)-values. Astrophys J 293:1–16. https://doi.org/10.1086/163208
Fushiki I, Lamb DQ (1987) New insights from a global view of X-ray bursts. Astrophys J 323:L55–L60. https://doi.org/10.1086/185056

Gallino R, Raiteri CM, Busso M (1993) Carbon stars and isotopic BA anomalies in meteoritic SiC grains. Astrophys J 410:400–411. https://doi.org/10.1086/172757

Galloway DK, Muno MP, Hartman JM, Psaltis D, Chakrabarty D (2008) Thermonuclear (Type I) X-ray bursts observed by the Rossi X-ray timing explorer. Astrophys J Suppl Ser 179:360–422. https://doi.org/10.1086/592044, arXiv:astro-ph/0608259

Gamezo VN, Khokhlov AM, Oran ES, Chtchelkanova AY, Rosenberg RO (2003) Thermonuclear supernovae: simulations of the deflagration stage and their implications. Science 299:77–81. https://doi.org/10.1126/science.1078129, arXiv:astro-ph/0212054

García-Senz D, Bravo E (2005) Type Ia Supernova models arising from different distributions of igniting points. Astron Astrophys 430:585–602. https://doi.org/10.1051/0004-6361:20041628, arXiv:astro-ph/0409480

García-Senz D, Cabezón RM, Arcones A, Relaño A, Thielemann FK (2013) High-resolution simulations of the head-on collision of white dwarfs. Mon Not R Astron Soc 436:3413–3429. https://doi.org/10.1093/mnras/stt1821, arXiv:1309.6884

Gehrz RD (2002) Infrared and radio observations of classical novae: physical parameters and abundances in the ejecta. In: Hernanz M, José J (eds) Classical nova explosions. American Institute of Physics conference series, vol 637, pp 198–207. https://doi.org/10.1063/1.1518200

Gehrz RD, Truran JW, Williams RE, Starrfield S (1998) Nucleosynthesis in classical novae and its contribution to the interstellar medium. Publ Astron Soc Pac 110:3–26. https://doi.org/10.1086/316107

Glasner SA, Livne E (1995) Convective hydrogen burning down a nova outburst. Astrophys J 445:L149–L151. https://doi.org/10.1086/187911

Glasner SA, Truran JW (2009) Carbon-nitrogen-oxygen "breakout" and nucleosynthesis in classical novae. Astrophys J 692:L58–L61. https://doi.org/10.1088/0004-637X/692/1/L58, arXiv:0812.3984

Glasner SA, Livne E, Truran JW (1997) Reactive flow in nova outbursts. Astrophys J 475:754–762. https://doi.org/10.1086/303561

Glasner SA, Livne E, Truran JW (2005) The sensitivity of multidimensional nova calculations to the outer boundary condition. Astrophys J 625:347–350. https://doi.org/10.1086/429482, arXiv:astro-ph/0504054

Goldstein DA, Kasen D (2018) Evidence for sub-Chandrasekhar mass Type Ia supernovae from an extensive survey of radiative transfer models. Astrophys J 852:L33. https://doi.org/10.3847/2041-8213/aaa409, arXiv:1801.00789

Gómez-Gomar J, Hernanz M, Jose J, Isern J (1998a) Gamma-ray emission from individual classical novae. Mon Not R Astron Soc 296:913–920. https://doi.org/10.1046/j.1365-8711.1998.01421.x, arXiv:astro-ph/9711322

Gómez-Gomar J, Isern J, Jean P (1998b) Prospects for Type IA supernova explosion mechanism identification with gamma rays. Mon Not R Astron Soc 295:1–+. https://doi.org/10.1046/j.1365-8711.1998.29511115.x, arXiv:astro-ph/9709048

Grindlay J, Gursky H (1976) Scattering model for X-ray bursts – massive black holes in globular clusters. Astrophys J 205:L131–L133. https://doi.org/10.1086/182106

Grindlay JE, McClintock JE, Canizares CR, Cominsky L, Li FK, Lewin WHG, van Paradijs J (1978) Discovery of optical bursts from an X-ray burst source, MXB1735-44. Nature 274:567–+. https://doi.org/10.1038/274567a0

Guerrero J, García-Berro E, Isern J (2004) Smoothed particle hydrodynamics simulations of merging white dwarfs. Astron Astrophys 413:257–272. https://doi.org/10.1051/0004-6361:20031504

Gutierrez J, Garcia-Berro E, Iben I Jr, Isern J, Labay J, Canal R (1996) The final evolution of ONeMg electron-degenerate cores. Astrophys J 459:701–705. https://doi.org/10.1086/176934

Gyngard F, Nittler LR, Zinner E, Jose J, Cristallo S (2011) New reaction rates and implications for nova nucleosynthesis and presolar grains. In: Lunar and planetary science conference, vol 42, p 2675

Hachisu I, Kato M, Nomoto K (1999) A wide symbiotic channel to Type IA supernovae. Astrophys J 522:487–503. https://doi.org/10.1086/307608, arXiv:astro-ph/9902304

Hachisu I, Saio H, Kato M (2016) Shortest recurrence periods of forced novae. Astrophys J 824:22. https://doi.org/10.3847/0004-637X/824/1/22, arXiv:1604.02965
Hackwell JA, Grasdalen GL, Gehrz RD, Cominsky L, Lewin WHG, van Paradijs J (1979) The detection of an optical burst coincident with an X-ray burst from MXB 1837 + 05 /Ser X-1/. Astrophys J 233:L115–L119. https://doi.org/10.1086/183088
Haenecour P, Floss C, José J, Amari S, Lodders K, Jadhav M, Wang A, Gyngard F (2016) Coordinated analysis of two graphite grains from the CO3.0 LAP 031117 meteorite: first identification of a CO nova graphite and a presolar iron sulfide subgrain. Astrophys J 825:88. https://doi.org/10.3847/0004-637X/825/2/88, arXiv:1606.08310
Hamuy M, Phillips MM, Maza J, Suntzeff NB, Schommer RA, Aviles R (1995) A Hubble diagram of distant Type IA supernovae. Astron J 109:1–13. https://doi.org/10.1086/117251
Hamuy M, Phillips MM, Suntzeff NB, Schommer RA, Maza J, Antezan AR, Wischnjewsky M, Valladares G, Muena C, Gonzales LE, Aviles R, Wells LA, Smith RC, Navarrete M, Covarrubias R, Williger GM, Walker AR, Layden AC, Elias JH, Baldwin JA, Hernandez M, Tirado H, Ugarte P, Elston R, Saavedra N, Barrientos F, Costa E, Lira P, Ruiz MT, Anguita C, Gomez X, Ortiz P, Della Valle M, Danziger J, Storm J, Kim Y, Bailyn C, Rubenstein EP, Tucker D, Cersosimo S, Mendez RA, Siciliano L, Sherry W, Chaboyer B, Koopmann RA, Geisler D, Sarajedini A, Dey A, Tyson N, Rich RM, Gal R, Lamontagne R, Caldwell N, Guhathakurta P, Phillips AC, Szkody P, Prosser C, Ho LC, McMahan R, Baggley G, Cheng K, Havlen R, Wakamatsu K, Janes K, Malkan M, Baganoff F, Seitzer P, Shara M, Sturch C, Hesser J, Hartig ANP, Hughes J, Welch D, Williams TB, Ferguson H, Francis PJ, French L, Bolte M, Roth J, Odewahn S, Howell S, Krzeminski W (1996) BVRI light curves for 29 Type IA supernovae. Astron J 112:2408–2437. https://doi.org/10.1086/118192, arXiv:astro-ph/9609064
Hamuy M, Phillips MM, Suntzeff NB, Maza J, González LE, Roth M, Krisciunas K, Morrell N, Green EM, Persson SE, McCarthy PJ (2003) An asymptotic-giant-branch star in the progenitor system of a Type Ia supernova. Nature 424:651–654. https://doi.org/10.1038/nature01854, arXiv:astro-ph/0306270
Hanawa T, Sugimoto D, Hashimoto M (1983) Nucleosynthesis in explosive hydrogen burning and its implications in ten-minute interval of X-ray bursts. Publ Astron Soc Jpn 35:491–506
Hansen CJ, van Horn HM (1975) Steady-state nuclear fusion in accreting neutron-star envelopes. Astrophys J 195:735–741. https://doi.org/10.1086/153375
Harkness R (1991) Type Ia Supernovae. In: Danziger IJ, Kjaer K (eds) European southern observatory astrophysics symposia, vol 37, pp 447–456
Harris MJ, Leising MD, Share GH (1991) A search for the 478 keV line from the decay of nucleosynthetic Be-7. Astrophys J 375:216–220. https://doi.org/10.1086/170183
Harris MJ, Naya JE, Teegarden BJ, Cline TL, Gehrels N, Palmer DM, Ramaty R, Seifert H (1999) Transient gamma ray spectrometer observations of gamma-ray lines from novae. I. Limits on the positron annihilation line in five individual novae. Astrophys J 522:424–432. https://doi.org/10.1086/307625, arXiv:astro-ph/0004164
Harris MJ, Teegarden BJ, Weidenspointner G, Palmer DM, Cline TL, Gehrels N, Ramaty R (2001) Transient gamma-ray spectrometer observations of gamma-ray lines from novae. III. The 478 keV line from ^{7}Be decay. Astrophys J 563:950–957. https://doi.org/10.1086/323951
Hatano K, Branch D, Fisher A, Starrfield S (1997) On the spatial distribution and occurrence rate of Galactic classical novae. Mon Not R Astron Soc 290:113–118
Hayakawa S (1981) Galactic X-rays observed with X-ray astronomy satellite 'Hakucho'. Space Sci Rev 29:221–290. https://doi.org/10.1007/BF00229297
Heger A, Cumming A, Galloway DK, Woosley SE (2007) Models of Type I X-ray bursts from GS 1826-24: a probe of rp-process hydrogen burning. Astrophys J 671:L141–L144. https://doi.org/10.1086/525522, arXiv:0711.1195
Henyey L, L'Ecuyer J (1969) Studies in stellar evolution. VIII. The time scale for the diffusion of energy in the stellar interior. Astrophys J 156:549–558. https://doi.org/10.1086/149988
Henze M, Ness JU, Darnley MJ, Bode MF, Williams SC, Shafter AW, Sala G, Kato M, Hachisu I, Hernanz M (2015) A remarkable recurrent nova in M 31: the predicted 2014 outburst in

X-rays with Swift. Astron Astrophys 580:A46. https://doi.org/10.1051/0004-6361/201526028, arXiv:1504.06237
Hernanz M (2002) Gamma-ray emission from classical novae. In: Hernanz M, José J (eds) Classical nova explosions. American Institute of Physics conference series, vol 637, pp 399–408. https://doi.org/10.1063/1.1518237
Hernanz M (2012) Novae in γ-rays. Bull Astron Soc India 40:377. arXiv:1301.1660
Hernanz M (2014) Gamma-ray emission from nova outbursts. In: Woudt PA, Ribeiro VARM (eds) Stellar novae: past and future decades. Astronomical Society of the Pacific conference series, vol 490, p 319. arXiv:1305.0769
Hernanz M (2015) Astrophysics: a lithium-rich stellar explosion. Nature 518:307–308. https://doi.org/10.1038/518307a
Hernanz M, José J (2004) γ-Rays from classical novae: expectations from present and future missions. New Astron Rev 48:35–39. https://doi.org/10.1016/j.newar.2003.11.005
Hernanz M, Isern J, Canal R, Labay J, Mochkovitch R (1988) The final stages of evolution of cold, mass-accreting white dwarfs. Astrophys J 324:331–344. https://doi.org/10.1086/165898
Hernanz M, Jose J, Coc A, Isern J (1996) On the synthesis of 7Li and 7Be in novae. Astrophys J 465:L27+. https://doi.org/10.1086/310122
Hernanz M, José J, Coc A, Gómez-Gomar J, Isern J (1999) Gamma-ray emission from novae related to positron annihilation: constraints on its observability posed by new experimental nuclear data. Astrophys J 526:L97–L100. https://doi.org/10.1086/312372, arXiv:astro-ph/9910111
Hernanz M, Smith DM, Fishman J, Harmon A, Gómez-Gomar J, José J, Isern J, Jean P (2000) BATSE observations of classical novae. In: McConnell ML, Ryan JM (eds) American Institute of Physics conference series, vol 510, pp 82–86. https://doi.org/10.1063/1.1303179
Hernanz M, Gómez-Gomar J, José J (2002) The prompt gamma-ray emission of novae. New Astron Rev 46:559–563. https://doi.org/10.1016/S1387-6473(02)00201-4, arXiv:astro-ph/0110203
Higdon JC, Fowler WA (1987) Gamma-ray constraints on Na-22 yields in nova explosions. Astrophys J 317:710–716. https://doi.org/10.1086/165317
Hillebrandt W, Thielemann F (1982) Nucleosynthesis in novae – a source of Ne-E and Al-26. Astrophys J 255:617–623. https://doi.org/10.1086/159864
Hillman Y, Prialnik D, Kovetz A, Shara MM (2016) Growing white dwarfs to the Chandrasekhar limit: the parameter space of the single degenerate SNIa channel. Astrophys J 819:168. https://doi.org/10.3847/0004-637X/819/2/168, arXiv:1508.03141
Hoeflich P, Khokhlov A (1996) Explosion models for Type IA supernovae: a comparison with observed light curves, distances, H 0, and Q 0. Astrophys J 457:500. https://doi.org/10.1086/176748, arXiv:astro-ph/9602025
Hoffman JA, Marshall HL, Lewin WHG (1978) Dual character of the rapid burster and a classification of X-ray bursts. Nature 271:630–633. https://doi.org/10.1038/271630a0
Hoppe P, Ott U (1997) Mainstream silicon carbide grains from meteorites. In: Bernatowicz TJ, Zinner E (eds) American Institute of Physics conference series, vol 402, pp 27–58. https://doi.org/10.1063/1.53314
Hoppe P, Annen P, Strebel R, Eberhardt P, Gallino R, Lugaro M, Amari S, Lewis RS (1997) Meteoritic silicon carbide grains with unusual Si isotopic compositions: evidence for an origin in low-mass, low-metallicity asymptotic giant branch stars. Astrophys J 487:L101–L104. https://doi.org/10.1086/310869
Hoppe P, Strebel R, Eberhardt P, Amari S, Lewis RS (2000) Isotopic properties of silicon carbide X grains from the Murchison meteorite in the size range 0.5–1.5 μm. Meteoritics and Planetary Science 35:1157–1176
Hounsell R, Bode MF, Hick PP, Buffington A, Jackson BV, Clover JM, Shafter AW, Darnley MJ, Mawson NR, Steele IA, Evans A, Eyres SPS, O'Brien TJ (2010) Exquisite nova light curves from the Solar Mass Ejection Imager (SMEI). Astrophys J 724:480–486. https://doi.org/10.1088/0004-637X/724/1/480, arXiv:1009.1737

Hounsell R, Darnley MJ, Bode MF, Harman DJ, Surina F, Starrfield S, Holdsworth DL, Bewsher D, Hick PP, Jackson BV, Buffington A, Clover JM, Shafter AW (2016) Nova light curves from the Solar Mass Ejection Imager (SMEI) – II. The extended catalog. Astrophys J 820:104. https://doi.org/10.3847/0004-637X/820/2/104, arXiv:1512.03321

Howell DA, Sullivan M, Nugent PE, Ellis RS, Conley AJ, Le Borgne D, Carlberg RG, Guy J, Balam D, Basa S, Fouchez D, Hook IM, Hsiao EY, Neill JD, Pain R, Perrett KM, Pritchet CJ (2006) The Type Ia supernova SNLS-03D3bb from a super-Chandrasekhar-mass white dwarf star. Nature 443:308–311. https://doi.org/10.1038/nature05103, arXiv:astro-ph/0609616

Hoyle F, Fowler WA (1960) Nucleosynthesis in supernovae. Astrophys J 132:565–590. https://doi.org/10.1086/146963

Hutcheon ID, Huss GR, Fahey AJ, Wasserburg GJ (1994) Extreme Mg-26 and O-17 enrichments in an Orgueil corundum: identification of a presolar oxide grain. Astrophys J 425:L97–L100. https://doi.org/10.1086/187319

Iben I Jr, Tutukov AV (1985) On the evolution of close binaries with components of initial mass between 3 solar masses and 12 solar masses. Astrophys J Suppl Ser 58:661–710. https://doi.org/10.1086/191054

Iliadis C (2007) Nuclear physics of stars. Wiley-VCH, Weinheim

Iliadis C, Endt PM, Prantzos N, Thompson WJ (1999) Explosive hydrogen burning of ^{27}Si, ^{31}S, ^{35}Ar, and ^{39}Ca in novae and X-ray bursts. Astrophys J 524:434–453. https://doi.org/10.1086/307778

Iliadis C, Downen LN, José J, Nittler LR, Starrfield S (2018) On presolar stardust grains from CO classical novae. Astrophys J, arXiv:1801.09568

in't Zand J (2017) Understanding superbursts. In: Serino M, Shidatsu M, Iwakiri W, Mihara T (eds) 7 years of MAXI: monitoring X-ray transients, p 121. arXiv:1702.04899

in't Zand JJM, Weinberg NN (2010) Evidence of heavy-element ashes in thermonuclear X-ray bursts with photospheric superexpansion. Astron Astrophys 520:A81. https://doi.org/10.1051/0004-6361/200913952, arXiv:1001.0900

in't Zand JJM, Kuulkers E, Verbunt F, Heise J, Cornelisse R (2003) A superburst from 4U 1254-69. Astron Astrophys 411:L487–L491. https://doi.org/10.1051/0004-6361:20031586, arXiv:astro-ph/0310364

in't Zand JJM, Cumming A, van der Sluys MV, Verbunt F, Pols OR (2005) On the possibility of a helium white dwarf donor in the presumed ultracompact binary 2S 0918-549. Astron Astrophys 441:675–684. https://doi.org/10.1051/0004-6361:20053002, arXiv:astro-ph/0506666

Isern J, Jean P, Bravo E, Knödlseder J, Lebrun F, Churazov E, Sunyaev R, Domingo A, Badenes C, Hartmann DH, Hoeflich P, Renaud M, Soldi S, Elias-Rosa N, Hernanz M, Domínguez I, García-Senz D, Lichti GG, Vedrenne G, Von Ballmoos P (2016) Gamma-ray emission from SN2014J near maximum optical light. Astron Astrophys 588:A67. https://doi.org/10.1051/0004-6361/201526941, arXiv:1602.02918

Ivanova LN, Imshennik VS, Chechetkin VM (1974) Pulsation regime of the thermonuclear explosion of a star's dense carbon core. Astrophys Space Sci 31:497–514. https://doi.org/10.1007/BF00644102

Iyudin AF, Bennett K, Bloemen H, Diehl R, Hermsen W, Lichti GG, Morris D, Ryan J, Schoenfelder V, Steinle H, Strong A, Varendorff M, Winkler C (1995) COMPTEL search for ^{22}Na line emission from recent novae. Astron Astrophys 300:422–428

Izzo L, Della Valle M, Mason E, Matteucci F, Romano D, Pasquini L, Vanzi L, Jordan A, Fernandez JM, Bluhm P, Brahm R, Espinoza N, Williams R (2015) Early optical spectra of nova V1369 Cen show the presence of lithium. Astrophys J 808:L14. https://doi.org/10.1088/2041-8205/808/1/L14, arXiv:1506.08048

Jean P, Hernanz M, Gómez-Gomar J, José J (2000) Galactic 1.275-MeV emission from ONe novae and its detectability by INTEGRAL/SPI. Mon Not R Astron Soc 319:350–364. https://doi.org/10.1046/j.1365-8711.2000.03587.x, arXiv:astro-ph/0004126

Jiang JA, Doi M, Maeda K, Shigeyama T, Nomoto K, Yasuda N, Jha SW, Tanaka M, Morokuma T, Tominaga N, Ivezić Ž, Ruiz-Lapuente P, Stritzinger MD, Mazzali PA, Ashall C, Mould J, Baade D, Suzuki N, Connolly AJ, Patat F, Wang L, Yoachim P, Jones D, Furusawa H, Miyazaki

S (2017) A hybrid Type Ia supernova with an early flash triggered by helium-shell detonation. Nature 550:80–83. https://doi.org/10.1038/nature23908, arXiv:1710.01824
Jose J (2016) Stellar explosions: hydrodynamics and nucleosynthesis. CRC Press/Taylor and Francis, Boca Raton. https://doi.org/10.1201/b19165
José J, Hernanz M (1998) Nucleosynthesis in classical novae: CO versus ONe white dwarfs. Astrophys J 494:680–690. https://doi.org/10.1086/305244, arXiv:astro-ph/9709153
José J, Hernanz M (2007a) The origin of presolar nova grains. Meteorit Planet Sci 42:1135–1143
José J, Hernanz M (2007b) TOPICAL REVIEW: nucleosynthesis in classical nova explosions. J Phys G Nucl Phys 34:431–458. https://doi.org/10.1088/0954-3899/34/12/R01
José J, Hernanz M, Coc A (1997) New results on 26Al production in classical novae. Astrophys J 479:L55–L58. https://doi.org/10.1086/310575, arXiv:astro-ph/9701181
José J, Coc A, Hernanz M (1999) Nuclear uncertainties in the NENA-MGAL cycles and production of ^{22}Na and ^{26}Al during nova outbursts. Astrophys J 520:347–360. https://doi.org/10.1086/307445, arXiv:astro-ph/9902357
José J, Hernanz M, García-Berro E, Gil-Pons P (2003) The impact of the chemical stratification of white dwarfs on the classification of classical novae. Astrophys J 597:L41–L44. https://doi.org/10.1086/379782, arXiv:astro-ph/0309451
José J, Hernanz M, Amari S, Lodders K, Zinner E (2004) The imprint of nova nucleosynthesis in presolar grains. Astrophys J 612:414–428. https://doi.org/10.1086/422569, arXiv:astro-ph/0405332
José J, García-Berro E, Hernanz M, Gil-Pons P (2007) The first nova explosions. Astrophys J 662:L103–L106. https://doi.org/10.1086/519521
José J, Moreno F, Parikh A, Iliadis C (2010) Hydrodynamic models of Type I X-ray bursts: metallicity effects. Astrophys J Suppl Ser 189:204–239. https://doi.org/10.1088/0067-0049/189/1/204, arXiv:1005.4767
Joss PC (1977) X-ray bursts and neutron-star thermonuclear flashes. Nature 270:310–314. https://doi.org/10.1038/270310a0
Joss PC (1978) Helium-burning flashes on an accreting neutron star – a model for X-ray burst sources. Astrophys J 225:L123–L127. https://doi.org/10.1086/182808
Kashi A, Soker N (2011) A circumbinary disc in the final stages of common envelope and the core-degenerate scenario for Type Ia supernovae. Mon Not R Astron Soc 417:1466–1479. https://doi.org/10.1111/j.1365-2966.2011.19361.x, arXiv:1105.5698
Kasliwal MM, Cenko SB, Kulkarni SR, Ofek EO, Quimby R, Rau A (2011) Discovery of a new photometric sub-class of faint and fast classical novae. Astrophys J 735:94. https://doi.org/10.1088/0004-637X/735/2/94, arXiv:1003.1720
Kato M, Hachisu I (1994) Optically thick winds in nova outbursts. Astrophys J 437:802–826. https://doi.org/10.1086/175041
Kato M, Saio H, Hachisu I, Nomoto K (2014) Shortest recurrence periods of novae. Astrophys J 793:136. https://doi.org/10.1088/0004-637X/793/2/136, arXiv:1404.0582
Keek L, in't Zand JJM (2008) On burning regimes and long duration X-ray bursts. In: Proceedings of the 7th INTEGRAL workshop, p 32. arXiv:0811.4574
Keek L, Galloway DK, in't Zand JJM, Heger A (2010) Multi-instrument X-ray observations of thermonuclear bursts with short recurrence times. Astrophys J 718:292–305. https://doi.org/10.1088/0004-637X/718/1/292, arXiv:1005.3302
Keek L, Heger A, in't Zand JJM (2012) Superburst models for neutron stars with hydrogen- and helium-rich atmospheres. Astrophys J 752:150. https://doi.org/10.1088/0004-637X/752/2/150, arXiv:1204.1343
Kercek A, Hillebrandt W, Truran JW (1998) Two-dimensional simulations of the thermonuclear runaway in an accreted atmosphere of a C+O White Dwarf. Astron Astrophys 337:379–392. arXiv:astro-ph/9801054
Kercek A, Hillebrandt W, Truran JW (1999) Three-dimensional simulations of classical novae. Astron Astrophys 345:831–840. arXiv:astro-ph/9811259
Khokhlov AM (1991) Delayed detonation model for Type IA supernovae. Astron Astrophys 245:114–128

Khokhlov AM (1995) Propagation of turbulent flames in supernovae. Astrophys J 449:695–713. https://doi.org/10.1086/176091
Khokhlov A, Mueller E, Hoeflich P (1993) Light curves of Type IA supernova models with different explosion mechanisms. Astron Astrophys 270:223–248
Khokhlov AM, Oran ES, Wheeler JC (1997) Deflagration-to-detonation transition in thermonuclear supernovae. Astrophys J 478:678–688. https://doi.org/10.1086/303815, arXiv:astro-ph/9612226
Knödlseder J (1999) Implications of 1.8 MEV gamma-ray observations for the origin of ^{26}Al. Astrophys J 510:915–929. https://doi.org/10.1086/306601
Koike O, Hashimoto M, Kuromizu R, Fujimoto S (2004) Final products of the rp-process on accreting neutron stars. Astrophys J 603:242–251. https://doi.org/10.1086/381354
Kolb U, Politano M (1997) The contribution of O-Ne-Mg novae to the ^{26}Al production in the Galaxy. Astron Astrophys 319:909–922
Kong AKH, Miller JM, Méndez M, Cottam J, Lewin WHG, Paerels F, Kuulkers E, Wijnands R, van der Klis M (2007) Nondetection of gravitationally redshifted absorption lines in the X-ray burst spectra of GS 1826-24. Astrophys J 670:L17–L20. https://doi.org/10.1086/524137, arXiv:0708.0413
Kovetz A, Prialnik D (1985) CNO abundances resulting from diffusion in accreting nova progenitors. Astrophys J 291:812–821. https://doi.org/10.1086/163117
Kovetz A, Prialnik D (1997) The composition of nova ejecta from multicycle evolution models. Astrophys J 477:356–367. https://doi.org/10.1086/303675
Kraft RP (1964) Are all novae binary stars? Leaflet Astronom Soc Pac 9:137–144
Krautter J, Oegelman H, Starrfield S, Wichmann R, Pfeffermann E (1996) ROSAT X-ray observations of nova V1974 Cygni: the rise and fall of the brightest supersoft X-ray source. Astrophys J 456:788–+. https://doi.org/10.1086/176697
Kuchner MJ, Kirshner RP, Pinto PA, Leibundgut B (1994) Evidence for Ni-56 yields Co-56 yields Fe-56 decay in Type IA supernovae. Astrophys J 426:L89–L92. https://doi.org/10.1086/187347
Kulkarni PV, Ashok NM, Apparao KMV, Chitre SM (1979) Discovery of IR bursts from Liller I/MXB1730-333. Nature 280:819–820. https://doi.org/10.1038/280819a0
Kumagai S, Nomoto K (1997) Gamma-rays and X-rays from Type Ia supernovae. In: Ruiz-Lapuente P, Canal R, Isern J (eds) NATO ASIC Proc. 486: thermonuclear supernovae, pp 515–540
Kushnir D, Katz B, Dong S, Livne E, Fernández R (2013) Head-on collisions of white dwarfs in triple systems could explain Type Ia supernovae. Astrophys J 778:L37. https://doi.org/10.1088/2041-8205/778/2/L37, arXiv:1303.1180
Kuulkers E (2004) The observers' view of (very) long X-ray bursts: they are super! Nucl Phys B Proc Suppl 132:466–475. https://doi.org/10.1016/j.nuclphysbps.2004.04.081, arXiv:astro-ph/0310402
Kuulkers E, in't Zand JJM, van Kerkwijk MH, Cornelisse R, Smith DA, Heise J, Bazzano A, Cocchi M, Natalucci L, Ubertini P (2002) A half-a-day long thermonuclear X-ray burst from KS 1731-260. Astron Astrophys 382:503–512. https://doi.org/10.1051/0004-6361:20011654, arXiv:astro-ph/0111261
Kuulkers E, den Hartog PR, in't Zand JJM, Verbunt FWM, Harris WE, Cocchi M (2003) Photospheric radius expansion X-ray bursts as standard candles. Astron Astrophys 399:663–680. https://doi.org/10.1051/0004-6361:20021781, arXiv:astro-ph/0212028
Landau LD, Lifshitz EM (1959) Fluid mechanics. Pergamon Press, Oxford
Langanke K, Martínez-Pinedo G (2000) Shell-model calculations of stellar weak interaction rates: II. Weak rates for nuclei in the mass range /A=45-65 in supernovae environments. Nucl Phys A 673:481–508. https://doi.org/10.1016/S0375-9474(00)00131-7, arXiv:nucl-th/0001018
Leising MD (1991) Gamma-ray lines from classical novae. In: Durouchoux P, Prantzos N (eds) Gamma-ray line astrophysics, American Institute of Physics conference series, vol 232, pp 173–182. https://doi.org/10.1063/1.40932
Leising MD (1993) Hard emission from classical novae. Astron Astrophys Suppl Ser 97:299–301

Leising MD, Clayton DD (1987) Positron annihilation gamma rays from novae. Astrophys J 323:159–169. https://doi.org/10.1086/165816

Lewin WHG (1977) X-ray burst sources. In: Papagiannis MD (ed) Eighth Texas symposium on relativistic astrophysics. New York Academy Sciences annals, vol 302, pp 210–227. https://doi.org/10.1111/j.1749-6632.1977.tb37050.x

Lewin WHG, Clark G, Doty J (1976) X-ray bursts. IAU Circ 2922:1

Lewin WHG, Cominsky LR, Walker AR, Robertson BSC (1980) Simultaneous IR and X-ray burst observation of SER X-1. Nature 287:27–28. https://doi.org/10.1038/287027a0

Lewin WHG, van Paradijs J, Taam RE (1993) X-ray bursts. Space Sci Rev 62:223–389. https://doi.org/10.1007/BF00196124

Lewin WHG, van Paradijs J, Taam RE (1995) X-ray bursts. In: Lewin WHG, van Paradijs J, van den Heuvel EPJ (eds) X-ray binaries. Cambridge University Press, Cambridge, pp 175–232

Lewis RS, Ming T, Wacker JF, Anders E, Steel E (1987) Interstellar diamonds in meteorites. Nature 326:160–162. https://doi.org/10.1038/326160a0

Linares M, Watts AL, Wijnands R, Soleri P, Degenaar N, Curran PA, Starling RLC, van der Klis M (2009) The Swift capture of a long X-ray burst from XTE J1701-407. Mon Not R Astron Soc 392:L11–L15. https://doi.org/10.1111/j.1745-3933.2008.00572.x, arXiv:0808.3950

Liu N, Nittler LR, O'D Alexander CM, Wang J, Pignatari M, José J, Nguyen A (2016) Stellar origins of extremely 13C- and 15N-enriched Presolar SiC grains: novae or supernovae? Astrophys J 820:140. https://doi.org/10.3847/0004-637X/820/2/140, arXiv:1602.05252

Liu N, Nittler LR, Pignatari M, O'D Alexander CM, Wang J (2017) Stellar Origin of ^{15}N-rich presolar SiC grains of Type AB: supernovae with explosive hydrogen burning. Astrophys J 842:L1. https://doi.org/10.3847/2041-8213/aa74e5, arXiv:1705.08222

Liu N, Nittler LR, Alexander COD, Wang J (2018) Late formation of silicon carbide in Type II supernovae. Astrophys J. arXiv:1801.06463

Livio M (1992) Classical novae and the extragalactic distance scale. Astrophys J 393:516–522. https://doi.org/10.1086/171524

Livio M (1994) Topics in the theory of cataclysmic variables and X-ray binaries. In: Shore SN, Livio M, van den Heuvel EPJ (eds) Saas-Fee Advanced Course 22: interacting binaries, pp 135–262

Lodders K (2005) Presolar grains from meteorites: remnants from the early times of the solar system. Chem Erde-Geochem 65:93–166. https://doi.org/10.1016/j.chemer.2005.01.001, arXiv:astro-ph/0501430

Lorén-Aguilar P, Isern J, García-Berro E (2010) Smoothed particle hydrodynamics simulations of white dwarf collisions and close encounters. Mon Not R Astron Soc 406:2749–2763. https://doi.org/10.1111/j.1365-2966.2010.16878.x, arXiv:1004.4783

Lugaro M, Davis AM, Gallino R, Pellin MJ, Straniero O, Käppeler F (2003) Isotopic compositions of strontium, zirconium, molybdenum, and barium in single presolar SiC grains and asymptotic giant branch stars. Astrophys J 593:486–508. https://doi.org/10.1086/376442

MacAlpine GM, Ecklund TC, Lester WR, Vanderveer SJ, Strolger L (2007) A spectroscopic study of nuclear processing and the production of anomalously strong lines in the crab nebula. Astron J 133:81–88. https://doi.org/10.1086/509504, arXiv:astro-ph/0609803

Mahoney WA, Ling JC, Jacobson AS, Lingenfelter RE (1982) Diffuse galactic gamma-ray line emission from nucleosynthetic Fe-60, Al-26, and Na-22 – preliminary limits from HEAO 3. Astrophys J 262:742–748. https://doi.org/10.1086/160469

Malone CM, Nonaka A, Almgren AS, Bell JB, Zingale M (2011) Multidimensional modeling of Type I X-ray bursts. I. Two-dimensional convection prior to the outburst of a pure ^{4}He accretor. Astrophys J 728:118. https://doi.org/10.1088/0004-637X/728/2/118, arXiv:1012.0609

Malone CM, Zingale M, Nonaka A, Almgren AS, Bell JB (2014) Multidimensional modeling of Type I X-ray bursts. II. Two-dimensional convection in a mixed H/He accretor. Astrophys J 788:115. https://doi.org/10.1088/0004-637X/788/2/115, arXiv:1404.6286

Maraschi L, Cavaliere A (1977) X-ray bursts of nuclear origin? In: X-ray binaries and compact objects, pp 127–128

Matteucci F, Raiteri CM, Busson M, Gallino R, Gratton R (1993) Constraints on the nucleosynthesis of CU and Zn from models of chemical evolution of the Galaxy. Astron Astrophys 272:421–429

Mazurek TJ, Wheeler JC (1980) Thermonuclear explosions in stars. Fundam Cosm Phys 5:193–286

McClintock JE, Canizares CR, Cominsky L, Li FK, Lewin WHG, van Paradijs J, Grindlay JE (1979) A 3-s delay in an optical burst from X-ray burst source MXB1735-44. Nature 279:47–49. https://doi.org/10.1038/279047a0

McCully C, Jha SW, Foley RJ, Bildsten L, Fong WF, Kirshner RP, Marion GH, Riess AG, Stritzinger MD (2014) A luminous, blue progenitor system for the Type Iax supernova 2012Z. Nature 512:54–56. https://doi.org/10.1038/nature13615, arXiv:1408.1089

Messenger S, Keller LP, Stadermann FJ, Walker RM, Zinner E (2003) Samples of stars beyond the solar system: silicate grains in interplanetary dust. Science 300:105–108. https://doi.org/10.1126/science.1080576

Metzger BD (2017) Kilonovae. Living reviews in relativity 20:3. https://doi.org/10.1007/s41114-017-0006-z, arXiv:1610.09381

Metzger BD, Piro AL, Quataert E (2009) Nickel-rich outflows from accretion discs formed by the accretion-induced collapse of white dwarfs. Mon Not R Astron Soc 396:1659–1664. https://doi.org/10.1111/j.1365-2966.2009.14909.x, arXiv:0812.3656

Meyer BS, Zinner E (2006) Nucleosynthesis. The University of Arizona space science series, Tucson, pp 69–108

Milne PA, The L, Leising MD (2001) Late light curves of Type Ia supernovae. Astrophys J 559:1019–1031. https://doi.org/10.1086/322352, arXiv:astro-ph/0104185

Minkowski R (1941) Spectra of supernovae. Publ Astron Soc Pac 53:224–225. https://doi.org/10.1086/125315

Mochkovitch R (1994) An introduction to the physics of Type II supernova explosions. In: Latal H, Schweiger W (eds) Matter under extreme conditions. Lecture notes in physics, vol 440. Springer, Berlin, p 49

Mochkovitch R, Hernanz M, Isern J, Martin X (1993) Gamma-ray bursts as collimated jets from neutron star/black hole mergers. Nature 361:236–238. https://doi.org/10.1038/361236a0

Molaro P, Izzo L, Mason E, Bonifacio P, Della Valle M (2016) Highly enriched ^{7}Be in the ejecta of Nova Sagittarii 2015 No. 2 (V5668 Sgr) and the Galactic ^{7}Li origin. Mon Not R Astron Soc 463:L117–L121. https://doi.org/10.1093/mnrasl/slw169, arXiv:1609.07297

Moll R, Woosley SE (2013) Multi-dimensional models for double detonation in sub-Chandrasekhar mass white dwarfs. Astrophys J 774:137. https://doi.org/10.1088/0004-637X/774/2/137, arXiv:1303.0324

Mostefaoui S, Hoppe P (2004) Discovery of abundant in situ silicate and spinel grains from red giant stars in a primitive meteorite. Astrophys J 613:L149–L152. https://doi.org/10.1086/424842

Mukai K, Nelson T, Sokoloski J, Chomiuk L, Finzell T, Linford J, Weston J, Rupen M, Mioduszewski A (2017) NuSTAR observations of Fermi-detected novae, V339 Delphini and V5668 Sagitarii. In: Ness JU, Migliari S (eds) The X-ray Universe 2017, p 152

Narayan R, Paczynski B, Piran T (1992) Gamma-ray bursts as the death throes of massive binary stars. Astrophys J 395:L83–L86. https://doi.org/10.1086/186493, arXiv:astro-ph/9204001

Ness JU (2012) High-resolution spectroscopy and high-density monitoring in X-rays of novae. Bull Astron Soc India 40:353. arXiv:1209.2153

Ness JU, Schwarz GJ, Retter A, Starrfield S, Schmitt JHMM, Gehrels N, Burrows D, Osborne JP (2007) Swift X-ray observations of classical novae. Astrophys J 663:505–515. https://doi.org/10.1086/518084, arXiv:astro-ph/0703286

Nguyen AN, Zinner E (2004) Discovery of ancient silicate stardust in a meteorite. Science 303:1496–1499. https://doi.org/10.1126/science.1094389

Niemeyer JC, Woosley SE (1997) The thermonuclear explosion of Chandrasekhar mass white dwarfs. Astrophys J 475:740–753. https://doi.org/10.1086/303544, arXiv:astro-ph/9607032

Nittler LR (1997) Presolar oxide grains in meteorites. In: Bernatowicz TJ, Zinner v (eds) American Institute of Physics conference series, vol 402, pp 59–82. https://doi.org/10.1063/1.53320

Nittler LR, Ciesla F (2016) Astrophysics with extraterrestrial materials. Annu Rev Astron Astrophys 54:53–93. https://doi.org/10.1146/annurev-astro-082214-122505

Nittler LR, Hoppe P (2005) Are presolar silicon carbide grains from novae actually from supernovae? Astrophys J 631:L89–L92. https://doi.org/10.1086/497029

Nittler LR, O'D Alexander CM, Gao X, Walker RM, Zinner EK (1994) Interstellar oxide grains from the Tieschitz ordinary chondrite. Nature 370:443–446. https://doi.org/10.1038/370443a0

Nittler LR, Hoppe P, Alexander CMO, Amari S, Eberhardt P, Gao X, Lewis RS, Strebel R, Walker RM, Zinner E (1995) Silicon nitride from supernovae. Astrophys J 453:L25–L28. https://doi.org/10.1086/309743

Nittler LR, Alexander CMO, Gao X, Walker RM, Zinner E (1997) Stellar sapphires: the properties and origins of presolar AL 2O 3 in meteorites. Astrophys J 483:475–495. https://doi.org/10.1086/304234

Nittler LR, Alexander CMO, Nguyen AN (2006) Extreme 13C and 15N enrichments in a Murchison presolar SiC grain. Meteorit Planet Sci Suppl 41:5316

Nofar I, Shaviv G, Starrfield S (1991) The formation of Al-26 in nova explosions. Astrophys J 369:440–450. https://doi.org/10.1086/169772

Nomoto K (1982) Accreting white dwarf models for Type 1 supernovae. II – Off-center detonation supernovae. Astrophys J 257:780–792. https://doi.org/10.1086/160031

Nomoto K, Kondo Y (1991) Conditions for accretion-induced collapse of white dwarfs. Astrophys J 367:L19–L22. https://doi.org/10.1086/185922

Nomoto K, Sugimoto D, Neo S (1976) Carbon deflagration supernova, an alternative to carbon detonation. Astrophys Space Sci 39:L37–L42. https://doi.org/10.1007/BF00648354

Nomoto K, Thielemann F, Yokoi K (1984) Accreting white dwarf models of Type I supernovae. III – carbon deflagration supernovae. Astrophys J 286:644–658. https://doi.org/10.1086/162639

Ore A, Powell JL (1949) Three-photon annihilation of an electron-positron pair. Phys Rev 75:1696–1699. https://doi.org/10.1103/PhysRev.75.1696

Orio M, Covington J, Ögelman H (2001) X-ray emission from classical and recurrent novae observed with ROSAT. Astron Astrophys 373:542–554. https://doi.org/10.1051/0004-6361:20010537, arXiv:astro-ph/0104219

Orio M, Rana V, Page KL, Sokoloski J, Harrison F (2015) A NuSTAR observation of the fast symbiotic nova V745 Sco in outburst. Mon Not R Astron Soc 448:L35–L39. https://doi.org/10.1093/mnrasl/slu195, arXiv:1412.2088

Osborne JP (2015) Getting to know classical novae with Swift. J High Energy Astrophys 7:117–125. https://doi.org/10.1016, arXiv:1507.02153

Ott U, Begemann F (1990) Discovery of s-process barium in the Murchison meteorite. Astrophys J 353:L57–L60. https://doi.org/10.1086/185707

Page D, Cumming A (2005) Superbursts from strange stars. Astrophys J 635:L157–L160. https://doi.org/10.1086/499520, arXiv:astro-ph/0508444

Papish O, Perets HB (2016) Supernovae from direct collisions of white dwarfs and the role of helium shell ignition. Astrophys J 822:19. https://doi.org/10.3847/0004-637X/822/1/19, arXiv:1502.03453

Parikh A, José J, Moreno F, Iliadis C (2008) The effects of variations in nuclear processes on Type I X-ray burst nucleosynthesis. Astrophys J Suppl Ser 178:110–136. https://doi.org/10.1086/589879, arXiv:0802.2819

Payne Gaposchkin CH (1957) The galactic novae. Interscience Publishers, Amsterdam

Perlmutter S, Gabi S, Goldhaber G, Goobar A, Groom DE, Hook IM, Kim AG, Kim MY, Lee JC, Pain R, Pennypacker CR, Small IA, Ellis RS, McMahon RG, Boyle BJ, Bunclark PS, Carter D, Irwin MJ, Glazebrook K, Newberg HJM, Filippenko AV, Matheson T, Dopita M, Couch WJ, The Supernova Cosmology Project (1997) Measurements of the cosmological parameters omega and lambda from the first seven supernovae at $Z \geq 0.35$. Astrophys J 483:565–581. https://doi.org/10.1086/304265, arXiv:astro-ph/9608192

Phillips MM (1993) The absolute magnitudes of Type IA supernovae. Astrophys J 413:L105–L108. https://doi.org/10.1086/186970

Phillips MM, Wells LA, Suntzeff NB, Hamuy M, Leibundgut B, Kirshner RP, Foltz CB (1992) SN 1991T – further evidence of the heterogeneous nature of Type IA supernovae. Astron J 103:1632–1637. https://doi.org/10.1086/116177

Piersanti L, Gagliardi S, Iben I Jr, Tornambé A (2003a) Carbon-oxygen white dwarf accreting CO-rich matter. II. Self-regulating accretion process up to the explosive stage. Astrophys J 598:1229–1238. https://doi.org/10.1086/378952

Piersanti L, Gagliardi S, Iben I Jr, Tornambé A (2003b) Carbon-oxygen white dwarfs accreting CO-rich matter. I. A comparison between rotating and nonrotating models. Astrophys J 583:885–901. https://doi.org/10.1086/345444, arXiv:astro-ph/0210624

Pietsch W, Haberl F (2005) XMM-Newton detection of Type I X-ray bursts in M 31. Astron Astrophys 430:L45–L48. https://doi.org/10.1051/0004-6361:200400128, arXiv:astro-ph/0412373

Pignatari M, Wiescher M, Timmes FX, de Boer RJ, Thielemann FK, Fryer C, Heger A, Herwig F, Hirschi R (2013) Production of carbon-rich presolar grains from massive stars. Astrophys J 767:L22. yhttps://doi.org/10.1088/2041-8205/767/2/L22, arXiv:1303.3374

Pinto PA, Eastman RG, Rogers T (2001) A test for the nature of the Type IA supernova explosion mechanism. Astrophys J 551:231–243. https://doi.org/10.1086/320059, arXiv:astro-ph/0008330

Politano M, Starrfield S, Truran JW, Weiss A, Sparks WM (1995) Hydrodynamic studies of accretion onto massive white dwarfs: ONeMg-enriched nova outbursts. I. Dependence on white dwarf mass. Astrophys J 448:807–821. https://doi.org/10.1086/176009

Pope SB (2000) Turbulent flows. Cambridge University press, Cambridge

Pozdnyakov LA, Sobol IM, Syunyaev RA (1983) Comptonization and the shaping of X-ray source spectra – Monte Carlo calculations. Astrophys Space Phys Rev 2:189–331

Prantzos N, Diehl R (1996) Radioactive 26Al in the galaxy: observations versus theory. Phys Rep 267:1–69. https://doi.org/10.1016/0370-1573(95)00055-0

Prialnik D, Kovetz A (1984) The effect of diffusion on prenova evolution – CNO-enriched envelopes. Astrophys J 281:367–374. https://doi.org/10.1086/162107

Prialnik D, Kovetz A (1995) An extended grid of multicycle nova evolution models. Astrophys J 445:789–810. https://doi.org/10.1086/175741

Prialnik D, Shara MM, Shaviv G (1978) The evolution of a slow nova model with a Z = 0.03 envelope from pre-explosion to extinction. Astron Astrophys 62:339–348

Prialnik D, Shara MM, Shaviv G (1979) The evolution of a fast nova model with a Z = 0.03 envelope from pre-explosion to decline. Astron Astrophys 72:192–203

Pskovskii IP (1977) Light curves, color curves, and expansion velocity of Type I supernovae as functions of the rate of brightness decline. Sov Astron 21:675–682

Rakowski CE, Badenes C, Gaensler BM, Gelfand JD, Hughes JP, Slane PO (2006) Can ejecta-dominated supernova remnants be typed from their X-ray spectra? The case of G337.2-0.7. Astrophys J 646:982–1000. https://doi.org/10.1086/505018, arXiv:astro-ph/0604246

Raskin C, Timmes FX, Scannapieco E, Diehl S, Fryer C (2009) On Type Ia supernovae from the collisions of two white dwarfs. Mon Not R Astron Soc 399:L156–L159. https://doi.org/10.1111/j.1745-3933.2009.00743.x, arXiv:0907.3915

Rauch T, Suleimanov V, Werner K (2008) Absorption features in the spectra of X-ray bursting neutron stars. Astron Astrophys 490:1127–1134. https://doi.org/10.1051/0004-6361:200810129, arXiv:0809.2170

Riess AG, Press WH, Kirshner RP (1996) A precise distance indicator: Type IA supernova multicolor light-curve shapes. Astrophys J 473:88–109. https://doi.org/10.1086/178129, arXiv:astro-ph/9604143

Riess AG, Filippenko AV, Li W, Treffers RR, Schmidt BP, Qiu Y, Hu J, Armstrong M, Faranda C, Thouvenot E, Buil C (1999) The rise time of nearby Type IA supernovae. Astron J 118:2675–2688. https://doi.org/10.1086/301143, arXiv:astro-ph/9907037

Ritossa C, Garcia-Berro E, Iben I Jr (1996) On the evolution of stars that form electron-degenerate cores processed by carbon burning. II. Isotope abundances and thermal pulses in a 10 M_{sun}

model with an ONe core and applications to long-period variables, classical novae, and accretion-induced collapse. Astrophys J 460:489–505. https://doi.org/10.1086/176987
Romano D, Matteucci F, Molaro P, Bonifacio P (1999) The galactic lithium evolution revisited. Astron Astrophys 352:117–128. arXiv:astro-ph/9910151
Röpke FK, Hillebrandt W, Niemeyer JC, Woosley SE (2006) Multi-spot ignition in Type Ia supernova models. Astron Astrophys 448:1–14. https://doi.org/10.1051/0004-6361:20053926, arXiv:astro-ph/0510474
Rosswog S, Kasen D, Guillochon J, Ramirez-Ruiz E (2009a) Collisions of white dwarfs as a new progenitor channel for Type Ia supernovae. Astrophys J 705:L128–L132. https://doi.org/10.1088/0004-637X/705/2/L128, arXiv:0907.3196
Rosswog S, Ramirez-Ruiz E, Hix WR (2009b) Tidal disruption and ignition of white dwarfs by moderately massive black holes. Astrophys J 695:404–419. https://doi.org/10.1088/0004-637X/695/1/404, arXiv:0808.2143
Ruiz-Lapuente P, Cappellaro E, Turatto M, Gouiffes C, Danziger IJ, Della Valle M, Lucy LB (1992) Modeling the iron-dominated spectra of the Type IA supernova SN 1991T at premaximum. Astrophys J 387:L33–L36. https://doi.org/10.1086/186299
Salaris M, Dominguez I, Garcia-Berro E, Hernanz M, Isern J, Mochkovitch R (1997) The cooling of CO white dwarfs: influence of the internal chemical distribution. Astrophys J 486:413–419. https://doi.org/10.1086/304483, arXiv:astro-ph/9704038
Schatz H, Aprahamian A, Goerres J, Wiescher M, Rauscher T, Rembges JF, Thielemann F, Pfeiffer B, Moeller P, Kratz K, Herndl H, Brown BA, Rebel H (1998) rp-Process nucleosynthesis at extreme temperature and density conditions. Phys Rep 294:167–264. https://doi.org/10.1016/S0370-1573(97)00048-3
Schatz H, Bildsten L, Cumming A, Wiescher M (1999) The rapid proton process ashes from stable nuclear burning on an accreting neutron star. Astrophys J 524:1014–1029. https://doi.org/10.1086/307837, arXiv:astro-ph/9905274
Schatz H, Aprahamian A, Barnard V, Bildsten L, Cumming A, Ouellette M, Rauscher T, Thielemann F, Wiescher M (2001) End point of the rp process on accreting neutron stars. Phys Rev Lett 86:3471–3474. https://doi.org/10.1103/PhysRevLett.86.3471, arXiv:astro-ph/0102418
Schmidt W, Niemeyer JC (2006) Thermonuclear supernova simulations with stochastic ignition. Astron Astrophys 446:627–633. https://doi.org/10.1051/0004-6361:20054145, arXiv:astro-ph/0510427
Schmidt W, Niemeyer JC, Hillebrandt W, Röpke FK (2006) A localised subgrid scale model for fluid dynamical simulations in astrophysics. II. Application to Type Ia supernovae. Astron Astrophys 450:283–294. https://doi.org/10.1051/0004-6361:20053618, arXiv:astro-ph/0601500
Schwarz GJ, Ness JU, Osborne JP, Page KL, Evans PA, Beardmore AP, Walter FM, Helton LA, Woodward CE, Bode M, Starrfield S, Drake JJ (2011) Swift X-ray observations of classical novae. II. The super soft source sample. Astrophys J Suppl Ser 197:31. https://doi.org/10.1088/0067-0049/197/2/31, arXiv:1110.6224
Shafter AW (1997) On the nova rate in the galaxy. Astrophys J 487:226–236. https://doi.org/10.1086/304609
Shafter AW (2002) The galactic nova rate. In: Hernanz M, José J (eds) Classical nova explosions. American Institute of Physics conference series, vol 637, pp 462–471. https://doi.org/10.1063/1.1518246
Shafter AW, Ciardullo R, Pritchet CJ (2000) Novae in external galaxies: M51, M87, and M101. Astrophys J 530:193–206. https://doi.org/10.1086/308349
Shara MM (1982) Localized thermonuclear runaways and volcanoes on degenerate dwarf stars. Astrophys J 261:649–660. https://doi.org/10.1086/160376
Shara MM, Doyle T, Lauer TR, Zurek D, Baltz EA, Kovetz A, Madrid JP, Mikołajewska J, Neill JD, Prialnik D, Welch DL, Yaron O (2017) A Hubble space telescope survey for novae in M87. II. Snuffing out the maximum magnitude-rate of decline relation for novae as a non-standard candle, and a prediction of the existence of ultrafast novae. Astrophys J 839:109. https://doi.org/10.3847/1538-4357/aa65cd, arXiv:1702.05788

Shen KJ, Moore K (2014) The initiation and propagation of helium detonations in white dwarf envelopes. Astrophys J 797:46. https://doi.org/10.1088/0004-637X/797/1/46, arXiv:1409.3568

Shen KJ, Kasen D, Weinberg NN, Bildsten L, Scannapieco E (2010) Thermonuclear .Ia supernovae from helium shell detonations: explosion models and observables. Astrophys J 715:767–774. https://doi.org/10.1088/0004-637X/715/2/767, arXiv:1002.2258

Shore SN (2007) Astrophysical hydrodynamics: an introduction. Wiley, Hoboken

Shore SN, Schwarz G, Bond HE, Downes RA, Starrfield S, Evans A, Gehrz RD, Hauschildt PH, Krautter J, Woodward CE (2003) The early ultraviolet evolution of the ONeMg Nova V382 Velorum 1999. Astron J 125:1507–1518. https://doi.org/10.1086/367803, arXiv:astro-ph/0301415

Silverman JM, Foley RJ, Filippenko AV, Ganeshalingam M, Barth AJ, Chornock R, Griffith CV, Kong JJ, Lee N, Leonard DC, Matheson T, Miller EG, Steele TN, Barris BJ, Bloom JS, Cobb BE, Coil AL, Desroches LB, Gates EL, Ho LC, Jha SW, Kandrashoff MT, Li W, Mandel KS, Modjaz M, Moore MR, Mostardi RE, Papenkova MS, Park S, Perley DA, Poznanski D, Reuter CA, Scala J, Serduke FJD, Shields JC, Swift BJ, Tonry JL, Van Dyk SD, Wang X, Wong DS (2012) Berkeley Supernova Ia Program – I. Observations, data reduction and spectroscopic sample of 582 low-redshift Type Ia supernovae. Mon Not R Astron Soc 425:1789–1818. https://doi.org/10.1111/j.1365-2966.2012.21270.x, arXiv:1202.2128

Sim SA, Sauer DN, Röpke FK, Hillebrandt W (2007) Light curves for off-centre ignition models of Type Ia supernovae. Mon Not R Astron Soc 378:2–12. https://doi.org/10.1111/j.1365-2966.2007.11795.x, arXiv:astro-ph/0703764

Sim SA, Röpke FK, Hillebrandt W, Kromer M, Pakmor R, Fink M, Ruiter AJ, Seitenzahl IR (2010) Detonations in sub-Chandrasekhar-mass C+O white dwarfs. Astrophys J 714:L52–L57. https://doi.org/10.1088/2041-8205/714/1/L52, arXiv:1003.2917

Sim SA, Fink M, Kromer M, Röpke FK, Ruiter AJ, Hillebrandt W (2012) 2D simulations of the double-detonation model for thermonuclear transients from low-mass carbon-oxygen white dwarfs. Mon Not R Astron Soc 420:3003–3016. https://doi.org/10.1111/j.1365-2966.2011.20162.x, arXiv:1111.2117

Simonenko VA, Gryaznykh DA, Litvinenko IA, Lykov VA, Shushlebin AN (2012a) Mechanism of thermonuclear burning propagation in a helium layer on a neutron star surface: a refined model with heat conduction and subgrid turbulence. Astron Lett 38:305–320. https://doi.org/10.1134/S1063773712050088

Simonenko VA, Gryaznykh DA, Litvinenko IA, Lykov VA, Shushlebin AN (2012b) Mechanism of thermonuclear burning propagation in a helium layer on a neutron star surface: a simplified adiabatic model. Astron Lett 38:231–237. https://doi.org/10.1134/S1063773712040056

Smale AP (2001) A second intense burst with photospheric radius expansion from X2127+119 in M15. Astrophys J 562:957–962. https://doi.org/10.1086/323517, arXiv:astro-ph/0107593

Starrfield S (1989) Thermonuclear processes and the classical nova outburst. In: Classical novae. Wiley, Chichester, pp 39–60

Starrfield S, Truran JW, Sparks WM (1978a) CNO abundances and hydrodynamic studies of the Nova outburst. V – 1.00-solar-mass models with small mass envelopes. Astrophys J 226:186–202. https://doi.org/10.1086/156598

Starrfield S, Truran JW, Sparks WM, Arnould M (1978b) On Li-7 production in nova explosions. Astrophys J 222:600–603. https://doi.org/10.1086/156175

Starrfield S, Shore SN, Sparks WM, Sonneborn G, Truran JW, Politano M (1992) A prediction of the gamma-ray flux from Nova Herculis 1991. Astrophys J 391:L71–L74. https://doi.org/10.1086/186401

Starrfield S, Truran JW, Politano M, Sparks WM, Nofar I, Shaviv G (1993) ^{22}Na and ^{26}Al production in nova outbursts. Phys Rep 227:223–234. https://doi.org/10.1016/0370-1573(93)90067-N

Starrfield S, Truran JW, Wiescher MC, Sparks WM (1998) Evolutionary sequences for Nova V1974 Cygni using new nuclear reaction rates and opacities. Mon Not R Astron Soc 296:502–522. https://doi.org/10.1046/j.1365-8711.1998.01312.x

Starrfield S, Iliadis C, Hix WR, Timmes FX, Sparks WM (2009) The effects of the pep nuclear reaction and other improvements in the nuclear reaction rate library on simulations of the classical nova outburst. Astrophys J 692:1532–1542. https://doi.org/10.1088/0004-637X/692/2/1532, arXiv:0811.0197

Starrfield S, Iliadis C, Hix WR (2016) The thermonuclear runaway and the classical nova outburst. Publ Astron Soc Pac 128(5):051001. https://doi.org/10.1088/1538-3873/128/963/051001, arXiv:1605.04294

Stritzinger M, Leibundgut B, Walch S, Contardo G (2006) Constraints on the progenitor systems of Type Ia supernovae. Astron Astrophys 450:241–251. https://doi.org/10.1051/0004-6361:20053652, arXiv:astro-ph/0506415

Strohmayer T, Bildsten L (2006) New views of thermonuclear bursts. Cambridge University Press, Cambridge, pp 113–156

Strohmayer TE, Brown EF (2002) A remarkable 3 hour thermonuclear burst from 4U 1820-30. Astrophys J 566:1045–1059. https://doi.org/10.1086/338337, arXiv:astro-ph/0108420

Strohmayer TE, Zhang W, Swank JH, Smale A, Titarchuk L, Day C, Lee U (1996) Millisecond X-ray variability from an accreting neutron star system. Astrophys J 469:L9–L12. https://doi.org/10.1086/310261

Strope RJ, Schaefer BE, Henden AA (2010) Catalog of 93 nova light curves: classification and properties. Astron J 140:34–62. https://doi.org/10.1088/0004-6256/140/1/34, arXiv:1004.3698

Swank JH, Becker RH, Boldt EA, Holt SS, Pravdo SH, Serlemitsos PJ (1977) Spectral evolution of a long X-ray burst. Astrophys J 212:L73–L76. https://doi.org/10.1086/182378

Taam RE (1980) X-ray bursts from thermonuclear runaways on accreting neutron stars. Astrophys J 241:358–366. https://doi.org/10.1086/158348

Taam RE, Picklum RE (1978) Nuclear fusion and carbon flashes on neutron stars. Astrophys J 224:210–216. https://doi.org/10.1086/156367

Tajitsu A, Sadakane K, Naito H, Arai A, Aoki W (2015) Explosive lithium production in the classical nova V339 Del (Nova Delphini 2013). Nature 518:381–384. https://doi.org/10.1038/nature14161, arXiv:1502.05598

Tajitsu A, Sadakane K, Naito H, Arai A, Kawakita H, Aoki W (2016) The ^{7}Be II resonance lines in two classical novae V5668 Sgr and V2944 Oph. Astrophys J 818:191. https://doi.org/10.3847/0004-637X/818/2/191, arXiv:1601.05168

Takei D, Tsujimoto M, Kitamoto S, Ness JU, Drake JJ, Takahashi H, Mukai K (2009) Suzaku detection of superhard X-ray emission from the classical nova V2491 Cygni. Astrophys J 697:L54–L57. https://doi.org/10.1088/0004-637X/697/1/L54, arXiv:0904.1693

Tammann GA (1970) On the frequency of supernovae as a function of the integral properties of inter mediate and late type spiral galaxies. Astron Astrophys 8:458–475

Tan WP, Fisker JL, Görres J, Couder M, Wiescher M (2007) O15(α,γ)Ne19 breakout reaction and impact on X-ray bursts. Phys Rev Lett 98(24):242503/1–242503/4. https://doi.org/10.1103/PhysRevLett.98.242503, arXiv:0706.2349

Tang M, Anders E (1988) Interstellar silicon carbide – how much older than the solar system? Astrophys J 335:L31–L34. https://doi.org/10.1086/185333

Tanvir NR, Levan AJ, Fruchter AS, Hjorth J, Hounsell RA, Wiersema K, Tunnicliffe RL (2013) A 'kilonova' associated with the short-duration γ-ray burst GRB 130603B. Nature 500:547–549. https://doi.org/10.1038/nature12505, arXiv:1306.4971

Tatischeff V, Hernanz M (2007) Evidence for nonlinear diffusive shock acceleration of cosmic rays in the 2006 outburst of the recurrent nova RS Ophiuchi. Astrophys J 663:L101–L104. https://doi.org/10.1086/520049, arXiv:0705.4422

Tatischeff V, Tavani M, von Ballmoos P, Hanlon L, Oberlack U, Aboudan A, Argan A, Bernard D, Brogna A, Bulgarelli A, Bykov A, Campana R, Caraveo P, Cardillo M, Coppi P, De Angelis A, Diehl R, Donnarumma I, Fioretti V, Giuliani A, Grenier I, Grove JE, Hamadache C, Hartmann D, Hernanz M, Isern J, Kanbach G, Kiener J, Knödlseder J, Labanti C, Laurent P, Limousin O, Longo F, Marisaldi M, McBreen S, McEnery JE, Mereghetti S, Mirabel F, Morselli A, Nakazawa K, Peyré J, Piano G, Pittori C, Sabatini S, Stawarz L, Thompson DJ, Ulyanov A, Walter R, Wu X, Zdziarski A, Zoglauer A (2016) The e-ASTROGAM gamma-ray space

mission. In: Space telescopes and instrumentation 2016: ultraviolet to gamma ray, Proc SPIE, vol 9905, p 99052N. https://doi.org/10.1117/12.2231601, arXiv:1608.03739
Taubenberger S (2017) The extremes of thermonuclear supernovae. arXiv:1703.00528
Thielemann F, Brachwitz F, Freiburghaus C, Kolbe E, Martinez-Pinedo G, Rauscher T, Rembges F, Hix WR, Liebendörfer M, Mezzacappa A, Kratz K, Pfeiffer B, Langanke K, Nomoto K, Rosswog S, Schatz H, Wiescher W (2001) Element synthesis in stars. Prog Part Nucl Phys 46:5–22. https://doi.org/10.1016/S0146-6410(01)00103-X, arXiv:astro-ph/0101476
Thielemann F, Brachwitz F, Höflich P, Martinez-Pinedo G, Nomoto K (2004) The physics of Type Ia supernovae. New Astron Rev 48:605–610. https://doi.org/10.1016/j.newar.2003.12.038
Timmes FX, Woosley SE (1992) The conductive propagation of nuclear flames. I – degenerate C + O and O + NE + MG white dwarfs. Astrophys J 396:649–667. https://doi.org/10.1086/171746
Truran JW, Arnett WD, Cameron AGW (1967) Nucleosynthesis in supernova shock waves. Can J Phys 45:2315–2332
van den Bergh S, Tammann GA (1991) Galactic and extragalactic supernova rates. Annu Rev Astron Astrophys 29:363–407. https://doi.org/10.1146/annurev.aa.29.090191.002051
van Horn HM, Hansen CJ (1974) A model for the transient X-ray sources. Astrophys J 191:479–482. https://doi.org/10.1086/152987
van Paradijs J, McClintock JE (1995) Optical and ultraviolet observations of X-ray binaries. In: Lewin WHG, van Paradijs J, van den Heuvel EPJ (eds) X-ray binaries. Cambridge University Press, Cambridge, pp 58–125
Villarreal AR, Strohmayer TE (2004) Discovery of the neutron star spin frequency in EXO 0748-676. Astrophys J 614:L121–L124. https://doi.org/10.1086/425737, arXiv:astro-ph/0409384
Vurm I, Metzger BD (2018) High-energy emission from nonrelativistic radiative shocks: application to Gamma-ray novae. Astrophys J 852:62. https://doi.org/10.3847/1538-4357/aa9c4a
Waki I, Inoue H, Koyama K, Matsuoka M, Murakami T, Ogawara Y, Ohashi T, Tanaka Y, Hayakawa S, Tawara Y, Miyamoto S, Tsunemi H, Kondo I (1984) Discovery of absorption lines in X-ray burst spectra from X1636-536. Publ Astron Soc Jpn 36:819–830
Walker MF (1954) Nova DQ Herculis (1934): an eclipsing binary with very short period. Publ Astron Soc Pac 66:230. https://doi.org/10.1086/126703
Ward RA, Fowler WA (1980) Thermalization of long-lived nuclear isomeric states under stellar conditions. Astrophys J 238:266–286. https://doi.org/10.1086/157983
Warner B (1995) Cataclysmic variable stars. Cambridge University Press, Cambridge
Webbink RF (1984) Double white dwarfs as progenitors of R Coronae Borealis stars and Type I supernovae. Astrophys J 277:355–360. https://doi.org/10.1086/161701
Weinberg NN, Bildsten L (2007) Carbon detonation and shock-triggered helium burning in neutron star superbursts. Astrophys J 670:1291–1300. https://doi.org/10.1086/522111, arXiv:0706.3062
Weinberg NN, Bildsten L, Schatz H (2006) Exposing the nuclear burning ashes of radius expansion Type I X-ray bursts. Astrophys J 639:1018–1032. https://doi.org/10.1086/499426, arXiv:astro-ph/0511247
Weiss A, Truran JW (1990) Na-22 and Al-26 production and nucleosynthesis in novae explosions. Astron Astrophys 238:178–186
Wheeler JC, Harkness RP (1990) Type I supernovae. Rep Prog Phys 53:1467–1557. https://doi.org/10.1088/0034-4885/53/12/001
Whelan J, Iben I Jr (1973) Binaries and supernovae of Type I. Astrophys J 186:1007–1014. https://doi.org/10.1086/152565
White NE, Nagase F, Parmar AN (1995) The properties of X-ray binaries. In: Lewin WHG, van Paradijs J, van den Heuvel EPJ (eds) X-ray binaries. Cambridge University Press, Cambridge, pp 1–57
Wiescher M, Gorres J, Thielemann F, Ritter H (1986) Explosive hydrogen burning in novae. Astron Astrophys 160:56–72
Wijnands R (2001) Recurrent very long Type I X-ray bursts in the low-mass X-Ray binary 4U 1636-53. Astrophys J 554:L59–L62. https://doi.org/10.1086/320922, arXiv:astro-ph/0103125

Williams RE (1992) The formation of novae spectra. Astron J 104:725–733. https://doi.org/10.1086/116268

Woosley SE (1986) Nucleosynthesis and stellar evolution. In: Audouze J, Chiosi C, Woosley SE (eds) Saas-fee advanced course 16: nucleosynthesis and chemical evolution. Geneva Observatory, Geneva, p 1

Woosley SE, Taam RE (1976) Gamma-ray bursts from thermonuclear explosions on neutron stars. Nature 263:101–103. https://doi.org/10.1038/263101a0

Woosley SE, Weaver TA (1986) The physics of supernova explosions. Annu Rev Astron Astrophys 24:205–253. https://doi.org/10.1146/annurev.aa.24.090186.001225

Woosley SE, Heger A, Cumming A, Hoffman RD, Pruet J, Rauscher T, Fisker JL, Schatz H, Brown BA, Wiescher M (2004) Models for Type I X-ray bursts with improved nuclear physics. Astrophys J Suppl Ser 151:75–102. https://doi.org/10.1086/381533, arXiv:astro-ph/0307425

Yaron O, Prialnik D, Shara MM, Kovetz A (2005) An extended grid of nova models. II. The parameter space of nova outbursts. Astrophys J 623:398–410. https://doi.org/10.1086/428435, arXiv:astro-ph/0503143

Yoon S, Langer N (2005) On the evolution of rapidly rotating massive white dwarfs towards supernovae or collapses. Astron Astrophys 435:967–985. https://doi.org/10.1051/0004-6361:20042542, arXiv:astro-ph/0502133

Zheng W, Shivvers I, Filippenko AV, Itagaki K, Clubb KI, Fox OD, Graham ML, Kelly PL, Mauerhan JC (2014) Estimating the first-light time of the Type Ia supernova 2014J in M82. Astrophys J 783:L24. https://doi.org/10.1088/2041-8205/783/1/L24, arXiv:1401.7968

Zingale M, Dursi LJ (2007) Propagation of the first flames in Type Ia supernovae. Astrophys J 656:333–346. https://doi.org/10.1086/510306, arXiv:astro-ph/0610297

Zingale M, Timmes FX, Fryxell B, Lamb DQ, Olson K, Calder AC, Dursi LJ, Ricker P, Rosner R, MacNeice P, Tufo HM (2001) Helium detonations on neutron stars. Astrophys J Suppl Ser 133:195–220. https://doi.org/10.1086/319182

Zingale M, Malone CM, Nonaka A, Almgren AS, Bell JB (2015) Comparisons of two- and three-dimensional convection in Type I X-ray bursts. Astrophys J 807:60. https://doi.org/10.1088/0004-637X/807/1/60, arXiv:1410.5796

Zinner EK (2005) Presolar grains. Elsevier, Oxford, pp 17–39

Zwicky F (1938) On collapsed neutron stars. Astrophys J 88:522–525. https://doi.org/10.1086/144003

Part III
Special Places to Observe Cosmic Isotopes

The preceding chapters presented what happens in individual source objects. We started from a theoretical understanding of these objects, and pointed out where observational data, in particular those related to radioactive decays, contributed significantly. Now we make the connection to the more complex astrophysical sites for which we have measurements of radioactivities. Here theory has a more difficult task, as observations cover a very broad range, and the interplay of different objects and of processes acting on different time scales adds complexity. The Sun and the solar system are unique places for which we actually hold material samples and posses detailed historic knowledge. We would like to link the origins of life on Earth to our cosmic roots through broader and more general astrophysical concepts, but we must bear in mind that special circumstances can distort our special vantage point. Interstellar gas, on the other hand, is the reservoir where the specific sources discussed in preceding chapters contribute their nucleosynthesis ejecta to the medium out of which they were formed originally, and continue being formed. We observe this gas in its current state, but the sources also bear a record of its past evolution. Therefore, this chapter refers to the complex astrophysics of chemical evolution, that will be addressed in a separate chapter in Part IV of the book.

Chapter 6
The Early Solar System

Maurizio Busso

This chapter reviews the early history of the solar system from radioactive nuclei of very different half-lives, which were recognized to have been present alive in pristine solids. Such radioactivities open a unique window on the evolution of the solar nebula and provide tools for understanding the crucial events that determined and accompanied the formation of the Sun. The understanding of the astrophysical origin of these unstable isotopes is still not complete and leaves puzzles and questions to our nucleosynthesis and stellar evolution models. We need to consider following aspects, among others: (1) The determination of an age for solar system bodies, as it emerged especially from the application of radioactive dating from long-lived isotopes. (2) A synthetic account of the measurements that proved the presence of shorter-lived radioactive nuclei (especially those of half-life lower than about 100 Myr) in the Early Solar System (hereafter ESS). (3) An explanation of their existence in terms of nuclear processes. We often only have incomplete and/or qualitative views of such complex processes. Additionally, these may have occurred at a galactic scale (providing a *galactic inheritance*), or at the level of the molecular cloud in which the Sun was formed, or also locally, either through a single, late stellar contamination affecting the pre-collapse solar nebula or through high-energy phenomena induced by the same early sun, in its flares or in the bombardment of pristine solids with the energetic particles of its intense winds. We cannot give a complete analysis of the many branches of research on the early phases of our Solar System. Rather, we focus on those more directly connected with the general theme of this book, as the study of the early solar system promoted and opened new views on the importance of radioactive isotopes as tools for the study of stellar and galactic nuclear astrophysics. For reviews of aspects such as geochemistry and cosmochemistry, see e.g. Arnould and Prantzos (1999), Busso et al. (1999), Kratz et al. (2004), McKeegan and Davis (2005), Goswami et al. (2005), Wasserburg et al.

M. Busso (✉)
Università di Perugia and Sezione INFN di Perugia, Perugia, Italy

R. Diehl et al. (eds.), *Astrophysics with Radioactive Isotopes*, Astrophysics and Space Science Library 453, https://doi.org/10.1007/978-3-319-91929-4_6

(2006), Wadhwa et al. (2007), Huss et al. (2009), Davies et al. (2014) and Davis and McKeegan (2014), for stellar condensation in clusters, and nuclear enrichment from ejecta in sequential star formation see Adams et al. (2014) and Pfalzner et al. (2015).

6.1 The Age of the Solar System

In the second half of the nineteenth century, Helmholtz and Kelvin independently noticed that the Sun would survive no more than about 20 Million years (the *Kelvin-Helmholtz time scale*), if evolving only under the opposite effects of gravity and thermodynamics. Hints on the fact that this estimate was too short first came from Charles Darwin and his disciples, as this datum appeared to be by far insufficient to permit the biological and geological evolution they were discovering.

The first role played by radioactivities in the history of our understanding of the solar system was then the solution of this dilemma concerning its age. This solution became available rather soon, less than two decades after the same discovery of radioactivity as a physical phenomenon.

6.1.1 The Beginnings

In 1895 Wilhelm Conrad Röntgen showed the amazing imaging properties of the energetic radiation emanating (sometimes) from matter; the X-rays. Soon after Becquerel made the fortuitous discovery that uranium produced X-rays leaving traces on photographic plates. Further researches on these "beams" were carried out by Pierre and Marie Curie. In 1898 they showed that pitchblende (the major uranium ore, mainly made of uranium oxide, UO_2), emitted X-rays 300 times stronger than those expected from U-rich compounds. This required the presence of another X-ray emitter, a nucleus that was called polonium (from Marie's native land). Then they discovered radium, inventing the term "radio-activity", and showed that the X-ray emission could be quantitatively predicted as a function of time.

It was then Rutherford who unambiguously showed how radioactivity were an intrinsic property of certain atoms, linked to their intimate structure; and Arthur Holmes, in 1911, presented the first systematic attempt at dating rocks, based on their content of U and Pb. The ages of the rocks derived from radioactivity were scattered from 0.64 to more than 1.4 Gyr, in sharp contrast with all non-nuclear estimates. The decades to follow would prove that Holmes was right: the age of the solar system has now been suggested to be 4.566 ± 0.001 Gyr, see e.g. Allègre et al. (1995).

Rutherford (1929) also demonstrated that the ratio $^{235}U/^{238}U$ (0.007 presently) was probably as high as 0.3 at the Sun's formation, and concluded that an extrapolation backward would yield the production ratio in the extra-solar environment where nuclei had been produced. Hence, the use of radioactivity for dating not only planetary rocks and the Sun, but even the environment from where the solar system was formed has ultimately its roots in Rutherford's work, as well. He also correctly estimated that U decay was already going on 4–5 Gyrs ago, although at his epoch the age of the Sun was still largely debated. Most of the tools for dating samples with radioactive nuclei had been, at this point, made available.

Rutherford's and Holmes' suggestion of using radioactivities as clocks capable of dating ancient samples remains as one of the most important scientific results of the early twentieth century. Depending on the isotope used and its lifetime, the technique is good for measuring very different ages: from those of historical and archaeological objects to those of materials in the ESS and in old stars of the Galaxy. These tools are generally known under the name of "radio-chronology" or, in geological and astronomical contexts, "nuclear cosmo-chronology" (Clayton 1988). For a detailed history of the Earth's age estimates see Dalrymple (1991); for an account of more modern efforts, after Holmes' work, see Chap. 2 in this volume.

6.1.2 Long-Lived Nuclei for Solar System Dating

Estimates of the solar-system age are now based on the measured abundances of long-lived[1] radioactive (parent) nuclei and of their stable daughters in pristine meteorites. This kind of radiometric dating can be performed through modern upgrades of the mass spectrometer, from tiny samples of ancient solid materials, making use of unstable isotopes preserving today a residual abundance. This implies that their lifetime must be very long; and, actually, the shortest-lived nucleus fulfilling this requirement is ^{235}U ($\tau = 1.015 \times 10^9$ years).

Various parent-daughter couples can be used, exploiting the existing channels for natural decay (schematically summarized in Table 6.1). General references where these techniques are discussed in an astrophysical context are numerous: see e.g. Chap. 2, and Clayton (1988), Pagel (1997), Cowley (1995) and Dalrymple (1991).

Deriving an age requires to take into account not only the presence of the radioactive parent (P), but also the original amount of the stable daughter (D) at the beginning of the time interval (see Fig. 6.1). In general, when there is a non-radiogenic isotope of the daughter element in the mineral, it can be used as a reference to compute abundance ratios (in this case this reference nucleus is often indicated as R).

Suppose a series of rocks are characterized, at formation, by the values A, B, $\ldots E$ of the ratio N_P/N_R; they all obviously have the same *initial* $(N_D/N_R)_0 = D_0$.

[1]In this Chapter, the terms *long-lived* and *short-lived* are used as related to the existence, or not, of daughter isotopes from the radioactive decay in the material sample. For short-lived radioactivities, only the daughter isotopes are present, while for long-lived, both parent and daughter isotopes are found in the sample.

Table 6.1 A summary of the weak interactions (decay modes) at atomic mass A and charge Z

Decay mode	Particles involved	Daughter nucleus
β^--decay	Emission of an e^- and $\bar{\nu}$	(A, Z+1)
β^+-decay	Emission of an e^+ and a ν	(A, Z−1)
e^--capture	e^- capture, ν emission	(A, Z−1)
Double β^--decay	Emission of two e^- and two $\bar{\nu}$s	(A, Z+2)
Double e^--capture	Capture of two e^-, two ν emissions	(A, Z−2)
e^--capture, e^+-emission	Capture of an e^-, emission of an e^+ and two νs	(A, Z−2)
Double e^+ emission	Emission of two e^+ and two νs	(A, Z−2)
α-Decay	An α particle (A=4, Z=2) is emitted	(A−4, Z−2)
Proton emission	A proton is ejected	(A−1, Z−1)
Neutron emission	A neutron is emitted	(A−1, Z)
Spontaneous fission	Two or more smaller nuclei emitted	
Cluster decay	Emission of a nucleus of A1, Z1	(A−A1, Z−Z1) + (A1, Z1)
γ-decay	Photon emission from excited states	(A, Z)

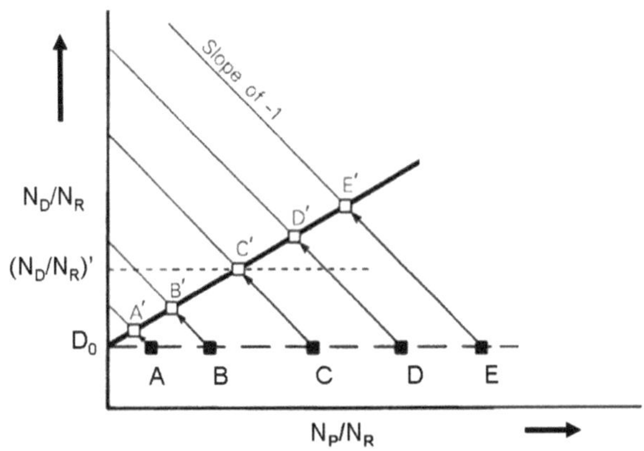

Fig. 6.1 The scheme for the construction of an isochron in the decay of an unstable parent to its daughter. See text for comments

After a time interval t, the content of P will be diminished in each sample, so that, for example:

$$(N_P/N_R)_t^A = (N_P/N_R)_0^A \exp(-\lambda t)$$

where λ is the decay constant ($\lambda = 0.6931/t_{1/2}$). As the product λt is the same for all rocks, the new contents of P (abscissas of the measured points A', B', ...) are proportional to the original ones and the measurement results fall on a straight line (*isochrone*). The value of N_D/N_R is the sum of the *radiogenic contribution* and of the *initial value* D_0. It is straightforward that:

$$(N_D/N_R)_t = D_0 + (N_P/N_R)_t[\exp(\lambda t) - 1] \tag{6.1}$$

When measuring rocks with the same age Δt, the isochrone passing through the points $A^{'}, B^{'}, \ldots$ will intersect the ordinate axis at a value providing the initial isotopic ratio of the daughter, D_0. Least square minimization then allows to determine both the age and the initial D_0 ratio. For an unstable nucleus U decaying to a stable isotope D, with a half-life $t_{1/2}$, we derive that:

$$\Delta t = \frac{1}{\lambda} \times ln\left[1 + \frac{\Delta D}{\Delta U}\right]. \tag{6.2}$$

An example of such an age determination, involving ^{87}Rb (P), ^{87}Sr (D) and ^{86}Sr (R), can be found in Anderson et al. (2007).

Several such parent-daughter couples are available for isotope geology. Examples are the ^{40}K – ^{40}Ar pair (^{40}K producing ^{40}Ar through β^+-decay and e^--captures, with a half-life of 1.25 Gyr), and the ^{147}Sm – ^{143}Nd pair (the first α-decays to the second with a half life of 10.6 Gyr).

Alternatively, the series of α and β decays of uranium (and sometimes transuranic nuclei) can be exploited. Perhaps the most commonly-used pair is in this case U-Pb, actually made of two separate chains, one leading from ^{238}U to ^{206}Pb, with a half-life of 4.5 Gyr, the second leading from ^{235}U to ^{207}Pb, with a half-life of 0.7 Gyr. The two processes are shown in Fig. 6.2. Recently, cautions have been advanced on the accuracy of the U-Pb datation, due to the discovery of live ^{247}Cm in the ESS (Brennecka 2010). This nucleus decays to ^{235}U, with a half-life of 15.6 Myr, so that the initial inventory of U must be corrected for this effect in order to avoid inaccuracies in the solar system age, which can be as high as 5 Myr. (See also Sect. 2.2.4. for early solar system radioactivity studies).

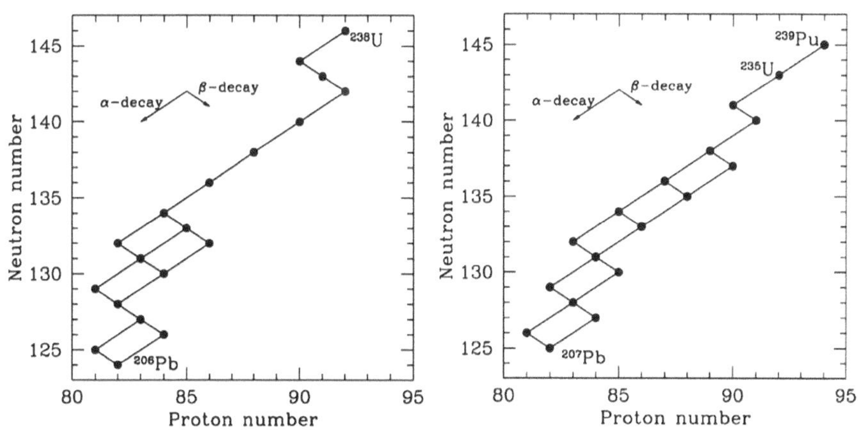

Fig. 6.2 The decay chains of ^{238}U and ^{235}U, commonly used for dating rocks. The U-Pb method is applied to the mineral zircon ($ZrSiO_4$), containing some U (because it can chemically substitute Zr) and strongly rejecting lead

6.2 Short-Lived Radioactive Nuclei in the ESS

Research on the early solar system in recent decades pursued the following main questions:

1. Can we learn details on how the Solar System was formed, through radioactivities of moderate lifetime?
2. Can we use any understanding obtained on the origins of our own astronomical "home" to infer which processes are expected to control star and planet formation elsewhere?
3. Was our star typical in its formation phases, or alternatively was it somehow peculiar and influenced by rare or unique events?
4. Can we use the lessons on the early solar system for advances in our knowledge of nuclear physics, meteoritics, stellar evolution and the chemical/nuclear evolution of the Galaxy?

We think that questions 1 and 4 and their pursuit produced most of the scientific profits. Despite many uncertainties, it is clear that the present view of the formation of our star and planetary system owes crucial advancements to the study of radioactivities in ancient solids.

To achieve this, we needed better nuclear half-lives and nuclear level schemes of several key nuclei; we also discovered that accounting for the abundances in pristine meteorites of different heavy nuclei (such as ^{129}I and ^{182}Hf) required us to re-think basic nucleosynthesis processes, such as that of fast neutron captures. On the other hand, these studies and revisions led us to a better understanding of apparently very different astrophysical sites, such as the early low-mass and very metal-poor stars, their photospheres preserving the signatures of the first nucleosynthesis events, as well as later-born stars, sampling each their current mix of nucleosynthesis over more than 13 Gyr, in the complex isotopic and elemental composition of gas in our Galaxy. For questions 2 and 3, the situation is still more confused; but direct observation of other planetary systems in recent years, such as from the Kepler space-borne observatory, have been very stimulating.

The original measurements revealing the presence of radioactive species in the ESS, with mean lifetimes lower than about 100 Myr (generally *in this context* referred to as *short-lived radioactivities*, or SLR) have by now been recognized as one of the most important scientific achievements in the second half of the twentieth century. Identification of the decay product allows us to draw a plot similar to Fig. 6.1, if we replace the parent nucleus P (that no longer exists) in the abscissa by another stable isotope of the same element (sometimes called the *substitute*, or *index* nucleus), which supposedly entered into the sample together with P[2] (Busso et al. 1999; Wasserburg et al. 2006). In this way, after a careful demonstration that the excess in the stable daughter's abundance is quantitatively correlated with

[2]The main formal change with respect to the appearance of Fig. 6.1 is in the tracks with slope −1, which in the new plot would be vertical, as both nuclei of the abscissa would be stable.

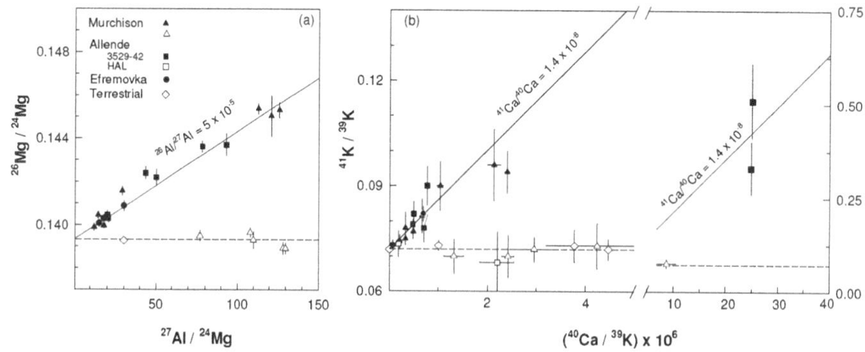

Fig. 6.3 Panel (**a**) shows the Al-Mg data from CAIs in different meteorites; filled symbols indicate the original claim of canonical value of ^{26}Al in ESS at the level ^{26}Al/^{27}Al = 5×10^{-5}. Panel (**b**) displays in a similar way the Ca-K data from the same samples. The first published data, and the line showing $[^{41}\text{Ca}/^{40}\text{Ca}]_0 = 1.4 \times 10^{-8}$ were due to Srinivasan et al. (1994, 1996). Other data and the demonstration of the correlation with Al are from Sahijpal et al. (1998). This figure is reproduced from these last authors and from Wasserburg et al. (2006). Copyright Nature Publishing Group

the chemical properties of the parent element (using the index isotope), one can reconstruct isochrones and derive the initial concentration of P in the sample. In a diagram similar to Fig. 6.1, and using the pair ^{26}Al – ^{26}Mg as an example, we use as abscissa the ratio ^{27}Al/^{24}Mg (^{27}Al is then the index nucleus) and on the ordinate we plot ^{26}Mg/^{24}Mg. This can be done, for example, for a mineral intrinsically poor in Mg, where the presence of ^{26}Mg can be ascribed largely to the ^{26}Al-decay. A description of the procedure can be found in Lee et al. (1976). Figure 6.3 shows the technique applied to ^{26}Al and ^{41}Ca, as obtained by Sahijpal et al. (1998). These authors showed that ^{26}Al and ^{41}Ca were either both present and correlated, or both absent from the samples, thus demonstrating their likely common origin. As we shall see, due to the very short half-life of ^{41}Ca, the correlation of Fig. 6.3 is particularly difficult to explain.

The above studies revealed an unexpected complexity of the physical processes that affected the early phases of the solar system. It became clear that stellar nucleosynthesis, galactic chemical evolution and the energetic processes in the winds of the early Sun all contributed, and a single explanation alone would lead to more or less obvious conflicts with ESS radio-isotope measurements (see also Chaps. 2 and 7). These insights began from the work by Reynolds (1960), who was the first to show that a now extinct nucleus, ^{129}I (τ = 23 Myr), had existed alive in the ESS. This was demonstrated analyzing the concentration of its decay product, ^{129}Xe, in meteorites (Jeffery and Reynolds 1961). It was then suggested by Wasserburg et al. (1960) and by Cameron (1960) that the abundance of ^{129}I could be attributed to the continuous production of *r*-process nuclei in the galaxy, provided the solar nebula had been isolated for about 10^8 years.

Shortly after the discovery of iodine, the presence of another relatively long-lived nucleus, ^{244}Pu (τ = 115 Myr), was inferred from excesses in planetary differentiates (PD) of neutron-rich Xe isotopes (Rowe and Kuroda 1965). It was then demonstrated that the enrichment in Xe isotopes correlated with excess fission tracks in meteoritic materials rich in both U, Th, and rare earth elements, showing that fission of heavy nuclei had occurred in situ (Wasserburg et al. 1969; Stoerzer and Pellas 1977).

More systematic measurements, extending to nuclei of shorter lifetime and to very early ESS constituents, became possible after 1969, also thanks to the fall of two extremely primitive meteorites: one in the Mexican village Allende and the second in the Australian village Murchison. They are rich in very refractory aggregates, called *Calcium & Aluminium Inclusions*, or CAIs (MacPherson et al. 2003), which are believed to be among the first-condensing solids in the solar nebula.

Measurements of Xe isotopes on CAIs in Allende (Podosek and Lewis 1972) soon confirmed both the presence of ^{129}I and the in situ fission of actinides. Even more important was the discovery of ^{26}Mg excesses that, thanks to the technique illustrated above in Figs. 6.1 and 6.3, were proven to be the relics of the earlier presence of ^{26}Al (Lee et al. 1976, 1977). The mean life of ^{26}Al (τ = 1.03 Myr) is much shorter than for iodine, and its presence seemed to confirm older suggestions (Urey 1955; Urey and Donn 1956), where it had been anticipated that ^{26}Al decay would be a very likely origin for the early heating of planetesimals, as other internal energy sources appeared insufficient to melt them. Searches for ^{26}Al had been done before the fall of Allende (Schramm and Wasserburg 1970), but they could not find the nucleus in solid solar-system samples. After the works by Lee et al. (1976, 1977), several studies then confirmed the presence of ^{26}Al at a high isotopic ratio $^{26}\mathrm{Al}/^{27}\mathrm{Al} \simeq 5 \times 10^{-5}$. These repeated confirmations led to the indication of the mentioned ratio as being the *canonical* one of the ESS. Later, however, the existence of values higher than it (sometimes referred to as "supra canonical") was also reported (Young et al. 2005; Simon et al. 2005; Liu et al. 2005). Recently, these last indications, based mainly on micro-analytical techniques, like e.g. laser ablation MC-ICPMS (multiple collector-inductively coupled plasma mass spectrometry), received some further experimental support (Simon and Young 2011). A possible interpretation was that bulk CAI fragments may record the cessation of thermal processing rather than the true initial formation of CAI themselves (Simon and Young 2011). If this were the case, then initial ^{26}Al/^{27}Al ratios different from the canonical one would be possible, and values as high as 6.5×10^{-5} were actually derived for some of the earliest solar materials (Krot et al. 2012; Makide et al. 2013; Kerekgyarto et al. 2015).

Contrary to the above interpretation, precision measurements made by Jacobsen et al. (2008) re-confirmed the original idea of a universal abundance of ^{26}Al in the very early solar system, fixing the canonical ratio to $^{26}\mathrm{Al}/^{27}\mathrm{Al} = 5.23(\pm 0.13) \times 10^{-5}$, by comparing Al-Mg and U-Pb isotopic systematics. According to these authors the supra-canonical values have to be attributed to post-crystallization inter-mineral redistributions of Mg isotopes within individual inclusions. The existence of a canonical value and the attribution of discrepancies to local disturbances was subsequently confirmed on independent bases (Jacobsen et al. 2013; Kuffmeier et al. 2016). In what follows I shall therefore accept, despite the remaining uncertainties, that a canonical ESS concentration of ^{26}Al close to $^{26}\mathrm{Al}/^{27}\mathrm{Al} = 5.23 \times 10^{-5}$ did in fact exist.

The above-mentioned discoveries of isotopic anomalies, induced by the in-situ decay of radioactive species, were soon followed by several others. The *p*-nucleus ^{146}Sm (τ = 148 Myr) was identified in meteoritic materials more evolved than CAIs, showing the signs of ongoing planetary differentiation: it was identified from its decay product, ^{142}Nd (Lugmair and Marti 1977). Then the abundance of ^{107}Pd (τ = 9.4 Myr) was established, again in planetary differentiates (Kelley and Wasserburg 1978). The fact that planetary cores could contain a relatively short-lived nucleus like ^{107}Pd was a proof that they formed very quickly in the history of the solar system.

Excesses of ^{205}Tl were also inferred by Chen and Wasserburg (1987). They would indicate the presence of ^{205}Pb (τ = 22 Myr). Nielsen et al. (2006) found evidence of a correlation between the ^{205}Tl excess in iron meteorites and ^{204}Pb. This would imply the in-situ decay of ^{205}Pb, at the level $^{205}\mathrm{Pb}/^{204}\mathrm{Pb} = 1\text{–}2 \times 10^{-4}$. Values even larger than that were inferred by Baker et al. (2010). We notice that ^{205}Pb is a shielded ("*s*-process only") *n*-capture nucleus, whose presence would certify that the ESS was somehow influenced by recent *s*-process nucleosynthesis events, either from high mass stars (Busso and Gallino 1985) or from low-intermediate mass red giants (Gallino et al. 1998); see also later, in Sect. 6.5.2.

The presence of ^{53}Mn (τ = 5.3 Myr) in CAIs, through a correlation of ^{53}Cr with Mn, was suggested by Birck and Allègre (1985, 1988). This was later confirmed by measurements of abundant ^{53}Mn in PDs, made first by Lugmair et al. (1992) and subsequently by Lugmair and Shukolyukov (1998) and Hutcheon et al. (1998). ^{53}Mn is very effectively produced in supernovae (SNe II or SNIa), and is also a possible outcome of spallation processes.

Tungsten isotopic anomalies discovered in early planetary materials, with deficiencies in ^{182}W, and their correlation with hafnium in chondritic meteorites demonstrated the presence of ^{182}Hf (τ = 13 Myr), see Lee and Halliday (1995, 1996) and Harper and Jacobsen (1996). This nucleus is traditionally ascribed to the *r*-process and can indeed by fully explained by it. A revision of the level scheme for ^{181}Hf subsequently led Lugaro et al. (2014) to shed doubts on this interpretation; however this is based so far on a single measurement and waits for confirmations (see Sect. 6.3).

The presence of ^{60}Fe (τ =3.75 Myr) was early discovered by Shukolyukov and Lugmair (1993a,b). Subsequently ^{60}Ni excesses were shown to be correlated with Fe/Ni ratios in meteoritic chondrites and the ^{60}Fe concentration was found to be rather high in the ESS (Mostefaoui et al. 2003, 2004; Tachibana and Huss 2003). This nucleus is a product of neutron captures at a relatively high neutron density, and can be synthesized both in massive stars exploding as supernovae and in intermediate-mass stars in their final evolutionary stages, but not in spallation processes (Lee et al. 1998a). Estimates of the initial ^{60}Fe content of the ESS have oscillated noticeably, ranging from a minimum of ^{60}Fe/^{56}Fe = 10^{-8} to a maximum some 200s times higher (Wadhwa et al. 2007). A discussion of some of the many revisions can be found in Tachibana et al. (2006), Gounelle and Meibom (2010), Huss et al. (2009), Moynier et al. (2011) and Tang and Dauphas (2012). Presently, the estimates seem to settle around the lowest values suggested. A recent analysis indicates an upper limit of 3.7×10^{-7} (Donohue et al. 2017) and the recent review by Davis and McKeegan (2014) points to even lower values, at the level of 10^{-8}. A remarkable suggestion was due to Bizzarro et al. (2007), according to which deficits in ^{60}Ni would exist in early meteorites as compared to the planets, and would indicate a contamination in ^{60}Fe that occurred shortly after, not before, the solar nebula formation. This idea seems now to be the result of experimental uncertainties and has not since been confirmed (Dauphas et al. 2008a,b).

^{10}Be (τ =2.3 Myr) was shown to be present in the ESS by McKeegan et al. (2000); as it is not produced by stellar nucleosynthesis, its presence certifies either the in situ proton bombardment of small solids, or some contributions from galactic cosmic rays. Further measurements fix the initial ^{10}Be/^{9}Be ratio at (0.4–1) $\times 10^{-3}$ (Liu et al. 2005).

The chemical and isotopic evidence for the existence of the radionuclide ^{135}Cs (with an half-life of 2.3 Myr) in the ESS was reported by Hidaka and Yoneda (2013). The estimated ratio of $^{135}\mathrm{Cs}/^{133}\mathrm{Cs} = (6.8 \pm 1.9) \times 10^{-4}$ was very large and the authors suggested remobilization of Cs, including ^{135}Cs, in the chondrules of the meteorite parent body; the value should therefore be seen as an upper limit.

Also very-short lived nuclei, with lifetimes below 1 Myr, were discovered in the last decade of the twentieth century. The presence of ^{41}Ca (τ = 0.15 Myr) was ascertained by Srinivasan et al. (1994, 1996). The initial ratio derived was ^{41}Ca/^{40}Ca = 1.5×10^{-8}, a datum that was later suggested to be a lower limit (Gounelle et al. 2006). ^{41}Ca is abundantly produced from neutron captures in stellar environments, but the short time scale for its decay (actually an e^- capture, likely increasing its efficiency in the high e^--density environment of stars) poses hard constraints on any astrophysical scenario for its formation. At the abundances observed, it might be produced by proton bombardment, as well. Only its correlation with ^{26}Al, shown in Fig. 6.3 (Sahijpal et al. 1998) and the evidence that ^{26}Al must be exotic (Fitoussi et al. 2008), suggest also for ^{41}Ca a stellar origin, despite the enormous difficulties related to its very short lifetime.

Hints on the presence of alive ^{36}Cl (τ=0.43 Myr) in the ESS were early found by Goebel et al. (1982), through measurements of ^{36}Ar abundances in Allende samples (including CAIs). The large shifts found in ^{36}Ar/^{38}Ar correlated with Cl abundances, and would imply an initial ratio ^{36}Cl/^{35}Cl $\simeq 2\times 10^{-8}$ (U. Ott, personal communication). A more recent study by Murty et al. (1997) also found ^{36}Ar shifts that were attributed to ^{36}Cl. Subsequent works (Lin et al. 2004; Leshin et al. 2004) clearly demonstrated a correlation of ^{36}S/^{34}S with ^{35}Cl/^{34}S in late-formed halogen-rich phases in CAIs (^{36}S is another decay product of ^{36}Cl). Sulphur anomalies were shown to be uncorrelated with ^{26}Al. These results demonstrate the presence of ^{36}Cl in the ESS with a high concentration (^{36}Cl/^{35}Cl $\simeq 10^{-4}$). Stellar sources would not be capable of explaining ^{36}Cl at such a high abundances, and it would require to be attributed to bombardment of solids by the early solar wind, while ^{26}Al would in this case come from stars. These results were subsequently strengthened by a work from Hsu et al. (2006), where the decoupling of chlorine and aluminium was unambiguously and clearly demonstrated. Further evidence have then been added (Jacobsen et al. 2009; Matzel et al. 2010), confirming that the use of ^{36}Cl as a chronometer for ESS events is unfeasible and that chlorine was added by phenomena internal to the system, occurred well after the injection of ^{26}Al.

The wealth of new measurements on ESS samples has now become impressive. The above account is certainly incomplete, but should at least focus our attention on the fact that no simple explanation for the origin of the complex nucleosynthesis pattern revealed by SLR can be invoked. It is evident that, although solid materials formed in a very short lapse of time (from a fraction of a Myr to a few Myrs), they maintain the records of several phenomena, from a blend of different stellar nucleosynthesis processes, to solar wind bombardment and possibly also to galactic cosmic ray spallation. Table 6.2 summarizes the status of SLR in the ESS with lifetimes τ_R lower than 25 Myr. The *p*-process nucleus ^{146}Sm is also shown, as an example of a longer-lived isotope. Table 6.2 is largely based on the review by Davis and McKeegan (2014), except for the uncertainty still left on ^{60}Fe, for which a general consensus seems not to have been obtained yet. For notations, see later Eqs. (6.3) and (6.4).

Table 6.2 The abundances of Short-Lived Nuclei with lifetime below 25 Myr in the ESS (the longer-lived ^{146}Sm is also shown for comparison)

Rad.	Stable	τ_R (Myr)	$[N^R/N^S]_{\mathrm{Meas.}}$
^{10}Be	^{9}Be	2.0	$(8.8 \pm 0.6) \times 10^{-4}$
^{26}Al	^{27}Al	1.03	$(5.23 \pm 0.13) \times 10^{-5}$
^{36}Cl	^{35}Cl	0.43	1.8×10^{-5}
^{41}Ca	^{40}Ca	0.15	4×10^{-9}
^{53}Mn	^{55}Mn	5.3	$(6.7 \pm 0.56) \times 10^{-6}$
^{60}Fe	^{56}Fe	3.75	10^{-8}–10^{-6}
^{107}Pd	^{108}Pd	9.4	$(5.9 \pm 2.2) \times 10^{-5}$
^{129}I	^{127}I	23	1.0×10^{-4}
^{135}Cs	^{133}Cs	3.3	4.8×10^{-4}
^{182}Hf	^{180}Hf	12.8	$(9.81 \pm 0.41) \times 10^{-5}$
^{205}Pb	^{204}Pb	22	10^{-3}
^{247}Cm	^{232}Th	23	$(1.1$–$2.4) \times 10^{-3}$
^{146}Sm	^{144}Sm	148	1.0×10^{-2}

For general references see Wasserburg et al. (2006) and Davis and McKeegan (2014). Among recently revised abundances, for ^{26}Al I adopt the value given by Jacobsen et al. (2008); for ^{60}Fe see Donohue et al. (2017); for ^{107}Pd see Schönbächler et al. (2008); for ^{135}Cs see Hidaka and Yoneda (2013); for ^{182}Hf see Burkhardt et al. (2008); for ^{205}Pb see Nielsen et al. (2006); for ^{247}Cm, see Brennecka et al. (2010)

6.3 The Galactic Inheritance

For most radioactive nuclei with sufficiently long mean life ($\tau \geq 10$ Myr) the ESS abundance was probably inherited from the equilibrium established in the ISM by galactic evolution (see also Chaps. 7 and 2). For illustration purposes, we use a rather simple model, representing the Galaxy as a *closed box* with *instantaneous recycling*, evolving for a time interval T. For this case, if we assume that the observation of a decay product (the daughter nucleus D) descends from the equilibrium ratio N^R/N^S established in the Galaxy for its parent radioactive nucleus R, referred to a stable reference isotope S produced in the same process, then Schramm et al. (1970) derived this equilibrium ratio to be:

$$N^R(T)/N^S(T) \simeq P^R \cdot p(T)\tau_P/P^S < p > T. \tag{6.3}$$

Here $P^S < p >$ is the stellar production rate of the stable reference nucleus, expressed as the product of an invariant stellar production factor P^S and of the average over a time interval T of a variable scaling factor $p(t)$. P^R is the stellar production factor for the parent radioactive isotope and $p(T)$ is the scaling factor for it at time T. If the ESS is separated from the ISM for a time interval Δ_1 before forming the first solid condensates, then the abundance ratio in the oldest meteoritic material will be decreased with respect to the ISM equilibrium by $\exp(-\Delta_1/\tau_R)$. This old material is usually identified with CAIs. Any younger solid body, formed at a time $t = \Delta_2$ after the CAI's condensation, will have an abundance ratio further reduced by $\exp(-\Delta_2/\tau_R)$. In this framework, Wasserburg et al. (1996) demonstrated that ^{244}Pu (an actinide nucleus produced by the r-process and with a relatively long mean life) and several other nuclei have ESS abundances compatible with uniform production in the Galaxy over about 10^{10} years. On this subject, see also Cameron (1993), Cameron et al. (1995) and Meyer and Clayton (2000).

We already stated that the solar nebula was probably isolated from the main events of galactic nucleosynthesis for a time Δ_1 of the order of 10^7 years (see Sect. 6.4), i.e. the time required by a cloud core to evolve. For reproducing the ESS amounts of ^{182}Hf through r-processes, this delay was found by Wasserburg et al. (1996) to be adequate. If the r-process has a unique source, however, that scheme would overestimate the concentration of ^{129}I by a large amount, so that the above authors assumed that this isotope was synthesized by a different supernova type, presenting an r-process distribution substantially different from the typical producer of heavy r-nuclei and whose last explosion would have occurred long before the one accounting for these last.

These suggestions stimulated a lively debate and a series of studies, both theoretical and observational, on the ensuing multi-modal nature of the r-process.

We refer to Chap. 4 in this book for an in-depth presentation of the r-process researches both in core-collapse supernovae and in colliding neutron stars (NSM, see e.g. Baiotti and Rezzolla 2017, and references therein), including the recent discussions after the observation of some evidence for n-capture nucleosynthesis in the kilonova associated with the gravitational wave event GW170817 (Abbott et al. 2017).

Here we recall simply that for the production of neutron-rich SLR heavier than Fe, of interest here, at least three distinct mechanisms must be at play (see also Vescovi et al. 2018).

1. An explosive process involving nuclei from Fe up to A $\simeq$ 100, in which the number of neutrons per heavy seed is very low ($\lesssim$ 1), so that the nucleosynthesis is still dominated by charged particle interactions.
2. A *weak* r-process, involving nuclei heavier than those quoted before and reaching (with decreasing efficiency) the rising wing of the main r-process peak, at A $\simeq$ 120–130. This process would produce iodine at the level of a few percent as compared to typical, purely r-process nuclei. It was ascertained to occur (and to be rather frequent) in old stars, such as those observed by Honda et al. (2006).
3. A *main* r-process mechanism, accounting for the most typical r-nuclei, such as Eu. Also this process was active in old stars of our Galaxy and seems to produce a pattern of nuclei heavier than A $\simeq$ 130 looking very stable and universal, mimicking the abundance ratios in the Sun. Observations of low metallicity stars with this imprinting were discussed by Sneden et al. (2008); the producers of this *main* type of the r-processing should be more efficient in yielding neutron-capture nuclei.

In the above scheme, the conundrum of the ratio I/Hf may be explained by the fact that iodine might derive mainly by processes of the type (2) above, producing it at low efficiency. Hf would instead originate from the more efficient processes of type (3). Then the origin of these two elements would be decoupled, and the low abundance of ^{129}I with respect to ^{182}Hf would derive by its limited yields in the respective explosive producers. The values of $< p >$ to be used for I and Hf in Eq. (6.3) would then differ by a large factor (10–20) and the entire problem would essentially disappear, with the two elements being both synthesized by the r-process, albeit by its different components and with different enhancement factors in the ejecta. This may seem ad-hoc, but recent studies of heavy-element nucleosynthesis and chemical evolution point in similar directions (Wehmeyer et al. 2017).

Further complexities may apply to the synthesis of heavier nuclei. Hints arise from the presence of ^{247}Cm in early solids, with an abundance ratio $(^{247}\text{Cm}/^{235}\text{U})_{ESS} = 1 - 2.4 \times 10^{-4}$ (Brennecka et al. 2010). This has two main implications. First, it shows that the ancient inventory of ^{235}U might have been modified by actinide decays and should be corrected accordingly, to avoid errors in the U-Pb datation system (see Sect. 6.1). Second, the low inferred ^{247}Cm abundance would be another proof that fast neutron-capture processes must be more complicated than assumed in the simplest possible uniform production models, adopting *standard* r-process production factors from a unique mechanism. This then would confirm that different r-process sources are characterized by different production factors $< p >$, thus making the hypotheses made for ^{129}I more plausible.

Revised scenarios of the r-process in which ^{129}I and ^{182}Hf find their explanation in changes of the production factors $< p >$ of Eq. (6.3) imply that iodine comes from essentially the same sources as ^{107}Pd, which would be produced by r-processing more efficiently than imagined in Wasserburg et al. (2006). Since ^{107}Pd can be synthesized also by the s-process in AGB stars and (partly) in He- and C-burning phases of massive stars, Wasserburg et al. (2006) found natural to assume that a late contamination event, induced by the close-by passage of an AGB star or a Supernova would have produced ^{107}Pd together with ^{26}Al, ^{41}Ca, ^{60}Fe, and possibly ^{205}Pb (Wasserburg et al. 1994, 2006). Presently, a revision seems to be needed: the required late production of ^{107}Pd would be now minimal, as its abundance might come almost entirely from the chemical evolution of the Galaxy, together with ^{129}I and ^{182}Hf.

As if not complicated enough, a revision by Bondarenko et al. (2002) of the level scheme for ^{181}Hf did not confirm the presence of some excited levels previously accepted, especially the one at 68 keV, which had previously induced to assume an increased decay of ^{181}Hf to ^{181}Ta in stellar conditions. On this basis Lugaro et al. (2014) proposed that ^{182}Hf might be produced efficiently also in s-process conditions, as neutron captures on ^{181}Hf would have a better chance to compete with β-decays. In view of the scheme discussed so far, in which ^{182}Hf can be completely accounted for by the r-process, this matter needs to be further verified by new measurements, as it would lead to an excessive production for this isotope. This caution seems also wise in view of the fact that the idea of multiple irradiations outlined in Lugaro et al. (2014) does not succeed in explaining all the SLR: ^{26}Al is actually strongly underproduced. These authors accepted the idea that heterogeneous ^{26}Al values existed in the ESS, indicating an original state with low Al and high Hf concentrations, due to prior stellar pollution, followed by in-situ irradiation of pristine solids from the early Sun to produce most of the solar ^{26}Al. This possibility seems however to be excluded today by revisions both of the homogeneity of ^{26}Al abundance (Jacobsen et al. 2013) and of the irradiation model itself: see later Sect. 6.6 (Sossi et al. 2017).

Summing up the complex evolution of the ideas outlined in this section, we summarize in Table 6.3 a general view of the synthesis of ESS radioactivities that might come from the galactic evolution. In Table 6.3, the production factors in the original astrophysical environments are taken from the hypotheses by Wasserburg et al. (2006); however, they do not differ much (within a factor of 2) from those of more recent computations. The production ratio ^{129}I/^{127}I is tentatively multiplied by a reduction factor of the order of ten to take into account the previous discussion of points from (1) to (3). If the numbers in Table 6.3 are not wrong by orders of magnitude, a late contamination by a close-by star should actually account only for ^{26}Al, ^{41}Ca, ^{135}Cs, and possibly some ^{60}Fe.

Table 6.3 A summary of the production of SLR from the galactic inheritance

Rad.	Stable	τ_R (Myr)	$[P^R/P^S]_{CE}$	$[N^R/N^S]_0$	$[N^R/N^S]_{10}$	$[N^R/N^S]_{20}$	$[N^R/N^S]_{Meas.}$
^{26}Al	^{27}Al	1.03	5.4×10^{-3}	5.6×10^{-7}	–	–	$(5.23 \pm 0.13) \times 10^{-5}$
^{41}Ca	^{40}Ca	0.15	1.4×10^{-3}	2.2×10^{-8}	–	–	4×10^{-9}
^{53}Mn	^{55}Mn	5.3	0.189	1.0×10^{-4}	1.5×10^{-5}	2.2×10^{-6}	10^{-6}
^{60}Fe	^{56}Fe	3.75	2.3×10^{-3}	8.0×10^{-7}	5.6×10^{-8}	3.8×10^{-9}	10^{-8}–10^{-6}
^{107}Pd	^{108}Pd	9.4	0.66	6.2×10^{-4}	2.1×10^{-4}	7.4×10^{-5}	$(5.9 \pm 2.2) \times 10^{-5}$
^{129}I	^{127}I	23	1.30×0.1[a]	3.0×10^{-4}	1.9×10^{-4}	1.3×10^{-4}	10^{-4}
^{135}Cs	^{133}Cs	3.3	0.724	2.1×10^{-4}	1.0×10^{-5}	4.9×10^{-7}	4.8×10^{-4}
^{182}Hf	^{180}Hf	12.8	0.346	4.5×10^{-4}	2.1×10^{-4}	9.4×10^{-5}	$(9.81 \pm 0.41) \times 10^{-5}$
^{247}Cm	^{235}U	23	3.95	8.9×10^{-3}	5.8×10^{-3}	3.7×10^{-3}	$(1.1$–$2.4) \times 10^{-3}$
^{205}Pb	^{204}Pb	22	1.05	2.3×10^{-3}	1.5×10^{-3}	9.3×10^{-4}	10^{-3}
^{146}Sm	^{144}Sm	148	0.675	9.9×10^{-3}	9.2×10^{-3}	8.6×10^{-3}	10^{-2}

Nuclei with lifetime $\tau_R \lesssim 25$ Myr, expected to be produced in massive stars, are shown. In the last two rows the *s*-process nucleus ^{205}Pb and the longer-lived *p*-nucleus ^{146}Sm are also added. Columns 6 and 7 show the predicted abundance ratios after a delay time of 10 and 20 Myr, respectively

[a]This multiplying factor would result if the production of iodine occurs at low efficiency in the *weak* *r*-process, with a production factor $< p >$ at least ten times smaller than what would be given by the *main* *r*-process (which instead produces ^{135}Cs and ^{182}Hf). This possibility would require that the *main* *r*-process does not extend below $A \simeq 130$ (see the text for details)

6.4 Expected Conditions in the ESS and Its Environment

The presence, in the ESS, of short-lived radioactive nuclei, some of which exclusively produced in stars, calls for a detailed study of the mutual relations between the forming Sun and the sources of galactic nucleosynthesis. In order to do this one needs to know the general scheme of low-mass star formation, then inquiring whether some special condition was necessary, for the Sun to carry live nuclei of short mean-life. The conditions prevailing in star-forming regions have started to be known, at least in part, over the last 30 years, mainly thanks to the advances in the techniques for infrared-, millimeter- and radio-astronomy. Accounts can be found in Lada and Lada (2003) and in Zuckerman and Song (2004). On the other hand, more recent observations of nearby cold clouds now provides insight into the processes affecting the local ISM near the Sun (Frisch et al. 2011; McComas et al. 2015). The evidence deriving from the quoted works tells us that some 70–90% of stars are born in clusters or multiple systems, while the remaining part undergoes more isolated processes of slow accretion. In the first phases not only gravity, but also galactic magnetic fields, velocity fields of the interstellar medium (ISM), and possibly the extra-pressure exerted by supernovae and other triggering phenomena can affect the proto-stellar clouds (Boss 2005). In most cases, and especially when stars are born through the formation of isolated cloud cores or globules, the initial phases are characterized by a slow accretion of cool materials thanks to plasma processes, namely the diffusion of positive and negative particles at virtually the same rate, due to interactions via the electric fields. This mechanism, called *ambipolar diffusion* allows weakly ionized or neutral gas to separate from the galactic plasma and to accumulate until the minimum mass necessary for gravitational contraction (the so-called *Jeans mass*) is reached. Then the collapse starts from inside out and a protostar and a disk are formed: see e.g. Shu et al. (1987a,b) and references therein.

In the above context, a lively scientific debate has examined extensively the various possible sources for the production of short lived nuclei: see e.g. Davies et al. (2014) for a recent review. The processes considered included the continuous synthesis in the Galaxy, briefly summarized in the previous Section and discussed in Chapter 7 of this book and in Schramm et al. (1970). They also took into account the local pollution by a nearby star, early introduced by Cameron and Truran (1977) for a massive supernova, and, for a few species, the production by spallation processes in the winds of the forming Sun, suggested by Shu et al. (2001).

The scenario invoking a nearby stellar source was in particular reanalysed repeatedly. This was done, e.g., by G.J. Wasserburg and co-workers (Wasserburg et al. 1994, 2006, 2017) and by Trigo-Rodríguez et al. (2009), for low- and intermediate-mass stars and by various authors (see e.g. Ouellette et al. 2007a; Boss et al. 2010; Boss and Keiser 2010; Tatischeff et al. 2010) for supernovae and/or Wolf-Rayet stars. Large uncertainties in the interpretations still remain, and are due to the insufficiently known details of nuclear physics in stars (weak interactions in particular) and to the complex physics of magnetic winds in star-forming regions. On the other hand, new insight has been provided by recent estimates for densities and lifetimes of stellar clusters (Dukes and Krumholz 2012). The presence of a dying star extremely close (1–2 pc) to a star forming cloud was found to be very unlikely (Williams and Gaidos 2007a; Gounelle and Meynet 2012a). While the time scales imposed by short-lived radio-nuclei remained a precious tool for inferring the origins of the solar system (McKeegan and Davis 2005), those works displayed a general picture in which the formation of these isotopes might occur in a complex variety of different stellar sources. In this case, as an example, the short-lived ^{26}Al could be produced by a lately-occurred contamination (see e.g. Tatischeff et al. 2010), in an environment previously enriched with ^{60}Fe from different massive stars (Gounelle et al. 2009). For very short-lived nuclei like ^{26}Al and ^{60}Fe diverse molecular clouds were recently analysed by Kuffmeier et al. (2016) showing a variety of behaviors (see also Chap. 7). On this basis they also confirmed that very anomalous ^{26}Al abundances like those found in peculiar condensates like the so-called FUN inclusions (see Chap. 1) do not need a separate explanation by exogenic nucleosynthesis sources.

The idea of a previously polluted and homogenized cloud might actually be applied also to ^{53}Mn. Indeed, despite the fact that it has a rather short lifetime ($\tau = 5.3$ Myr), its production in supernovae can be so huge that the requirement, for it, of a very late single addition, like for ^{26}Al, is questionable (Wasserburg et al. 2006): in Table 6.3 we actually showed that it can be fully explained by the chemical evolution of the Galaxy prior to the solar formation. These issues will be commented later, in Sect. 6.4.2.

6.4.1 Processes in Star-Forming Clouds and Protostellar Disks

The various phases that lead a forming star to reach its final structure and burn hydrogen on the Main Sequence correspond to different classes of objects seen in molecular clouds at long wavelengths. Starting from a proto-stellar cloud (often indicated as being of class -I), one has first the formation of a condensed core with a surrounding envelope (class 0). This core then evolves, with the growth of the central condensation, through classes I, II and III, characterized by a varying spectral energy distribution at mid-infrared wavelengths, informative of the amount of cold circumstellar dust. Class I objects have large mid-infrared excesses and are optically invisible, Class II sources have infrared fluxes decreasing for increasing wavelength, as the percentage of dust is reduced to $\sim 0.01\,M_{\odot}$. In the total emission, a flattened disk becomes gradually more important than the outside fading envelope and the central star becomes visible in the form of a variable object, of the T Tauri group (a star approaching the Main Sequence). Class III sources, then, have essentially no remaining mid-infrared excess from the original envelope, being *naked* T Tauri stars, with disks whose masses decrease in time, although they are partially preserved up to the Main Sequence phase. The whole duration of the above processes is highly uncertain theoretically and also strongly dependent on the total mass. For solar-like stars the whole pre-Main Sequence evolution may last for about 20 Myr, half of which spent in the form of a pre-collapse cloud core. Then the formation of the central protostar is fast (virtually in free-fall conditions), and the T Tauri star may need another 10 Myr to approach the main Sequence and start the fusion of hydrogen into helium.

The above mentioned time scales suggest that the proto-Sun must have been isolated from the global nucleosynthesis processes that control the nuclear evolution of the Galaxy as a whole for a time of 1–2 $\times$ 10^7 years. Such a long quiescent time would imply large dilution factors and long free-decay periods for any radioactive nucleus created in a galactic evolution framework. Roughly speaking, one might expect that nuclei with mean lives longer than about 5–10 Myr (^{244}Pu, ^{146}Sm, ^{247}Cm, ^{129}I, ^{205}Pb, ^{182}Hf, ^{107}Pd) might have survived the above-mentioned pre-stellar phases. The same may be true for ^{53}Mn ($\tau = 5.3$ Myr), due to its enormous production in CCSNe. On the contrary, shorter-lived nuclei ($\tau \lesssim 5$ Myr), would require a "local" production.

With the term "local" one may actually mean a wealth of different phenomena, from the repeated episodes of supernova and massive star contributions to a molecular cloud in which one or more clusters were formed, to a last-minute contamination from a single star, dying very closely to the ESS both in time and space, and up to endogenic phenomena producing radioactive nuclei in the solar nebula itself. Very short-lived nuclei will have to be attributed to one of these last processes.

A very late, local production of radioactive nuclei, had been proposed by Cameron and Truran (1977), who suggested that a supernova might have been responsible both for introducing unstable nuclei in the ESS and for triggering the cloud contraction forming the Sun. An alternative assumption was that in the ESS, a combination of magnetized winds and particle bombardments from the early Sun might have induced spallation processes on early solids, breaking heavy nuclei into lighter unstable isotopes (Shu et al. 1997). This idea descends from the observation of strong bipolar emissions from star forming clouds, showing that matter is lost along the rotational axis through magnetic winds (Shu et al. 1987a,b; Boss 2005). The physical origins of these winds have been identified in three main phenomena, namely: (1) a coronal wind, originating directly from the star; (2) a disk wind, starting from the surface of the accretion disk over a wide range of distances from the central star (from less than one to more than a hundred astronomical units), which is probably the main source of mass loss in T Tauri variables (Hartmann and Kenyon 1996); and (3) a wind driven by the interaction of the stellar corona with the inner edge of the accretion disk, launched at a distance of a few stellar radii from the center. This last is the so-called "X wind" (Shu et al. 1997), illustrated in Fig. 6.4, and is the type of magnetic interaction that might be important also for the explanation of SLR in the solar nebula. In such a case, its role would be that of lifting early solids from the symmetry plane of the disk, exposing them to high temperatures and to fluxes of energetic particles from the Sun. The irradiated dust would then fall back on the disk itself at large distances from the internal regions where it had been originally created, preserving the record of new radioactive species produced by solar spallation (Shu et al. 1997; Boss 2005). Confirmations

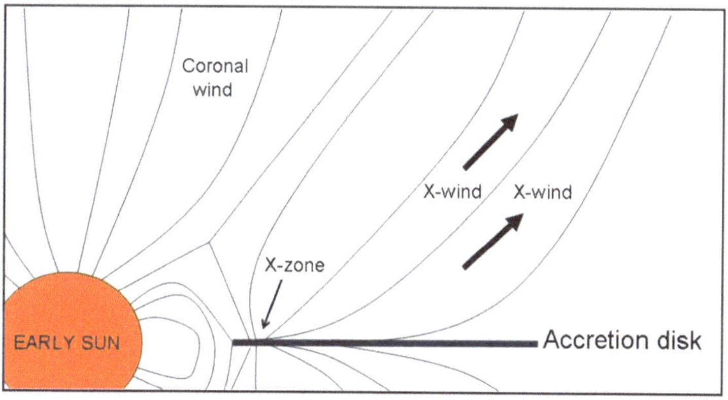

Fig. 6.4 A schematic view of the X-wind model

of these lifting and transporting processes recently came from the first results of the STARDUST mission (Brownlee et al. 2006).

6.4.2 Local Contamination, I: An Individual Star?

Since the original proposal by Cameron and Truran (1977), the idea of a close stellar encounter for explaining the presence in the ESS of radioactive nuclei with short mean-life has evolved into different branches, and both massive and low/intermediate-mass stars have been considered as possible causes (Podosek and Nichols 1997; Wasserburg et al. 1998; Busso et al. 1999; Meyer and Clayton 2000; Wasserburg et al. 2006; Gounelle et al. 2006; Takigawa et al. 2008; Huss et al. 2009).

Various objections to the supernova trigger model were advanced, with the gradual increase of our knowledge of star formation (Evans 1999). Although some evidence of star birth triggered by massive star ejecta was known to exist (Kothes et al. 2001; Palla and Stahler 2000; Zinnecker 2002), the required distance did not seem to match the needs for having live SLR in the ESS. Actually, the distance of a single SN explosion contaminating the solar nebula can be estimated from the dilution required to obtain in the ESS the observed abundances of short-lived nuclei, starting with nucleosynthesis yields predicted in SN ejecta. Useful distances then are from a fraction of a pc to a few pc, depending on the models; see also Cameron et al. (1995). However, both the evolution SN remnants and the observational indications on star formation triggering suggested that new stars could form only when the shock fronts have slowed down to a few km/sec or less (Preibisch and Zinnecker 1999). Such slow motions may require that the expanding SN shells have traveled much larger distances, of the order of tens of pc: see e.g. Scheffler and Elsaesser (1987, chapter 6). This would imply a dilution in excess by a factor of at least 10 as compared to the requirements by Cameron et al. (1995). A single SN trigger like the one originally suggested (Cameron and Truran 1977), if it ever occurred, must therefore be a very rare or unique event.

For some radioactive nuclei, then, an alternative was suggested by Wasserburg et al. (1994, 2006) and by Busso et al. (1999, 2003). This idea foresees the close-by passage of an AGB star in its final evolutionary stages, producing ^{26}Al, ^{41}Ca, ^{60}Fe, ^{107}Pd and *s*-process radioactivities like ^{135}Cs and ^{205}Pb. For updates on the production of some radioactive nuclei in AGB stars of intermediate mass see Trigo-Rodríguez et al. (2009) and Wasserburg et al. (2017). However, the close encounter with an AGB star is hard to motivate on statistical grounds. In this case we have not to face the risk of cloud disruption (as for fast supernova winds), but we require the simultaneous occurrence of two rare phenomena (an isolated, low-mass, cloud-core collapse and a planetary nebula ejection by a relatively old star passing there just by chance). Critical remarks on this idea, pointing out that such an event must be extremely unlikely, were advanced by Kastner and Myers (1994). Although one has to warn that some of these remarks were subsequently criticized (Trigo-Rodríguez et al. 2009) on the basis of upgrades in our knowledge of the Initial Mass Function and of the very poor statistics available at the time of the analysis by Kastner and Myers (1994), it is clear that invoking a chance encounter is a very special requirement that can be accepted only in the lack of any explanation from more likely pictures of the ESS contamination with radioactive nuclei.

6.4.3 Local Contamination, II: Sequential Episodes in a Molecular Cloud?

Alternatives to the chance encounter require to consider the possibility of contamination from supernovae and/or Wolf-Rayet stars physically connected to the Sun by being members of a same cluster or complex. Suggestions of this kind were advanced by Adams and Laughlin (2001), Hester and Desch (2005), Williams and Gaidos (2007b) and Smith and Brooks (2007). In such a case the contaminated solar nebula may have already evolved forming a proto-stellar disc (Adams and Laughlin 2001; Ouellette et al. 2005, 2007b). A critical analysis of the above scenario was performed by Gounelle and Meibom (2010). For the contamination of a contracting nebula having already developed the protostar/disc structure these authors estimated the useful distance of a SN source to be very small (at the level of 1 pc scale). This corresponds to less than 1 Myr in time. It was also argued that massive-star evolution timescales are too long for them to explode as supernovae with the due timeliness to contaminate forming stars in a same cluster. As a consequence, in these works the probability of contamination from a supernova belonging to the same cluster turns out to be very small (of the order of 10^{-3}). Similarly low probabilities were found by the above authors in the case of a more distant supernova, for a cloud still in its initial evolutionary stages, those of a molecular cloud core, a situation that would mimic the original idea by Cameron and Truran (1977).

In the framework of several attempts made at overcoming the difficulties related to an unlikely stellar encounter, the hypothesis was also advanced that the concentration of ^{26}Al in the oldest solar system solids could be due to its presence at high abundance in the parent molecular cloud, where it would have been contributed by several very massive stars evolving to the Wolf-Rayet stage (Gaidos et al. 2009b). Although this idea is interesting in itself, in its first formulation ad hoc hypotheses had to be advanced to avoid overproduction (by a up to a factor of 10) of ^{60}Fe from the supernovae expected to have occurred in the cloud.

At a first sight, the situation looks embarrassing. A few radioactive isotopes alive in the ESS seem to require necessarily a stellar production very close in time and space to the formation of the Sun. However, realistic (although highly speculative) models of how such a production might have occurred, always lead to unlikely conditions. It is not comfortable, for science, to call for special or unique conditions, occurred only for our Sun. Somehow, this seems to shed doubts on the accepted principle, according to which our position in the Universe has nothing special.

Several works addressed in the past years the above difficulties, reconsidering on new grounds the reasonable view of a pollution of the ESS from massive stars born sequentially in the same environment from which the Sun was formed. This started from the evidence recently accumulated that the solar nebula did actually condense within a stellar cluster (Pfalzner et al. 2015). Out of the two main forms currently assumed by this type of star formation in our Galaxy (starbursts or OB associations), numerical N-body simulations suggested that only the second one would be a viable way for generating the known solar system structure (Pfalzner 2013). Indeed, in the other case the central density during the first several million years would be so high to hamper the survival of any cluster member having the shape of a central protostar with a surrounding disc. A cluster of the second variety, with a sufficient number of objects ($N \geq 4000$) would instead provide a reasonable chance of hosting massive enough stars ($M \geq 20$–$25\ M_{\odot}$) to allow some of the radioactive nuclei in the ESS to be explained rather naturally as a product of the parent cluster itself (Lee et al. 1998b; Adams 2010). The above consideration of the likely environment, as well as the mentioned suggestions, sometimes advanced, for heterogeneous distributions of some SLR, then yielded a number of more sophisticated studies on the origins of the short-lived species, like those by Young (2014, 2016). It was also inferred that a high initial ^{26}Al concentration like the one of the solar system might not be extremely rare in star forming regions (Jura et al. 2013; Gounelle 2015). This might have resulted from the accumulation of this nuclide in a dense shell created by a nearby massive star wind. Within less than 1 Myr this shell might induce the formation of new stars enriched in ^{26}Al (see Pfalzner et al. 2015, and their Figure 1 in particular). The parent star should not have produced much ^{60}Fe, and be a non-exploding Wolf-Rayed star. Its products would be added to a cloud already enriched in ^{60}Fe by previous supernovae: the various revisions of the initial ^{60}Fe concentration, now pointing to rather low values (see Table 6.2) were of considerable help for this idea. An important contribution to this general view came from the work by Gounelle and Meynet (2012b), based on stellar physics models by Meynet et al. (2008) and on a star formation scheme discussed in Hennebelle et al. (2009). More recently, a noticeable contribution in this field was also due to Dwarkadas et al. (2017), who also suggested that a late WR wind may act as a trigger for the solar system formation and explain both the high abundance of ^{26}Al and the low concentration of ^{60}Fe.

Several words of caution are nevertheless necessary for the models outlined in this Section. One has first of all to remember that all studies involving massive star nucleosynthesis suffer for the large uncertainties involved in crucial cross sections, especially the one for the $^{12}C(\alpha, \gamma)^{16}O$ reaction. Careful analyses of nucleosynthesis yields for ^{26}Al and ^{60}Fe as done by, e.g., Tur et al. (2009) conclude with serious cautions about the uncertain nature of several such predictions. Moreover, one has to mention that to limit these uncertainties, the use of stellar models fully accounting for explosive nucleosynthesis would be mandatory, while this is very rarely the case in the scenarios illustrated in this section.

Even independently from the above warnings, some strong problems exist that look very difficult to avoid. In particular, the ESS contained not only ^{26}Al and ^{60}Fe, which may find an explanation in the scenarios here discussed. The early Sun contained also the other SLR discussed in this chapter; and they do need to be addressed. As an example, in all the models for the sequential contamination of the cloud from which the Sun formed, previous CCSNe should have necessarily produced at least the other short-lived isotopes not explained by Galactic evolution, i.e. ^{41}Ca and ^{135}Cs (see discussion of Table 6.3). This crucial point is never considered in the works we are speaking of. Much more important and problematic is the fact that, if these two nuclei were indeed produced by some events involving massive stars in a cloud or cloud complex, then also several other SLR would have been synthesized there, including some (like ^{53}Mn) whose abundance is already well explained by the previous Galactic evolution (see Table 6.3 and the relative discussion). Then these models, if analyzed in depth, seem to imply, almost necessarily, some excesses with respect to ESS measurements for crucial nuclei. Last, but certainly not least, if those SN materials had to be diluted adequately to account for the ESS measurements of ^{41}Ca and ^{135}Cs, then they would certainly predict excesses on stable isotopes at the level of several percent (see later comments on this point in Sect. 6.5.1). However, as already mentioned, widespread anomalies on stable elements in early meteorites are today excluded at levels in excess of a few per mil, except for some limited cases discussed and explained by Wasserburg et al. (2015).

6.5 Arguments Left for a Single Close Stellar Encounter?

Let us consider those nuclei that cannot be ascribed to normal galactic evolution, isotopes with mean life lower than a few Myr (maybe with the exclusion of ^{53}Mn): These would decay during the delay time required either by a dense cloud core to form one or a few new stars or by its parent giant molecular cloud to evolve and disrupt: typical evolutionary time scales of these last are indeed of a few 10^7 years

(Murray 2011). In principle, different nucleosynthesis phenomena, occurring inside or outside the giant molecular cloud itself, might have stopped influencing the solar nebula at different moments (see e.g. Lugaro et al. 2014). Since for some SLR (e.g. ^{60}Fe) in situ spallation processes cannot play a role, such nearby nucleosynthesis events seem a plausible suggestion.

In the specific case of very short-lived nuclei (e.g. ^{26}Al), a close encounter or very late pollution would permit the nucleus to be present in sufficient concentration at the moment of the formation of the first solids (this moment will be indicated here below by Δ_1). In such a case, the relations that short-lived nuclei should obey at t = Δ_1 are (Wasserburg et al. 1994, 1998):

$$\alpha_{\Delta_1}^{R,S} = \left(\frac{N^R}{N^S}\right)_{\Delta_1} \simeq q^R(w) \cdot \frac{N^R(w)}{N^S(w)} \cdot d \cdot e^{-\frac{\Delta_1}{\tau_R}} \tag{6.4}$$

where $\alpha_{\Delta_1}^{R,S}$ is the abundance ratio (radioactive nucleus to stable reference) established in the solar nebula at t = Δ_1, $q^S(w)$ is the production factor of the stable nucleus S in the stellar wind, $N^R(w)/N^S(w)$ is the abundance ratio (radioactive to stable) in the wind, d is the dilution factor at t = 0 and τ_R is the mean life of the nucleus R. Abundances in PDs (t = Δ_2) can then be found by considering a further exponential decay for the nuclei of interest. According to recent results from absolute dating of ESS samples (Zinner and Göpel 2002; Amelin et al. 2005) the values of Δ_2 should not exceed few Myr.

6.5.1 *Short-Lived Nuclei: A Late Supernova Origin?*

Two decades ago Wasserburg et al. (1998), from the assessment of SN yields by Woosley and Weaver (1995), indicated a fundamental constraint that must be satisfied by a SN explosion (at a time t = Δ_1 before the formation of CAIs), providing short-lived nuclei in the ESS: The admixture of SN ejecta with the material of the forming solar nebula should be compatible with the general knowledge of isotopic anomalies on stable elements in meteorites. The main doubt that was raised in that paper is that a close SN would introduce several anomalies over the distribution of stable isotopes and mainly on typical SN products like oxygen and α-rich nuclei, at levels that should be within the present experimental possibilities and nevertheless were not observed. These problems were underlined and repeated in great detail by Wasserburg et al. (2006), considering also models by Rauscher et al. (2002) and by Limongi and Chieffi (2003). A remarkable anomaly, known since several years, concerns ^{16}O (Clayton 1973; Clayton and Nittler 2004). However, it is not associated to shifts in other oxygen isotopes (Thiemens and Heidenreich 1983; Mauersberger et al. 1999) and it is now attributed to chemical processes which do not involve any nucleosynthesis input. In this context, Young et al. (2009) recently noted that the ^{18}O/^{17}O ratio is anomalous in the solar system as compared to the ISM. This anomaly was interpreted as due to the addition of SN-polluted material to the solar nebula, from various recent SNe in a picture of sequential star formation. Notice however that this anomaly with respect to the present ISM, if confirmed, can certainly be a signature of ongoing nucleosynthesis, but not necessarily of a solar contamination by a late SN. Addition in the solar neighborhood, in the last 4.5 Gyr, of materials richer in ^{17}O than in ^{18}O, as expected from low mass stars experiencing extra-mixing (Palmerini et al. 2011a) would induce the same difference, which should in this case be interpreted in terms of the continuing production of ^{17}O (and destruction of ^{18}O) after the Sun's formation (see also Nollett et al. 2003; Palmerini et al. 2009, 2011a). This would underline the relevance of low mass star nucleosynthesis in changing the chemical composition of objects younger than the Sun. This is a characteristic of recent galactic chemical evolution that is now inferred from other contexts (D'Orazi et al. 2009; Maiorca et al. 2011, 2012). Similar conclusions (although not involving extra mixing) were presented by Gaidos et al. (2009a), who also pointed out how this scenario could explain the gradual decrease in the ^{12}C/^{13}C ratio of the Galaxy.

Supernova nucleosynthesis models have been significantly improved since the time of the analysis by Wasserburg et al. (1998); nevertheless the uncertainties concerning the yields (as discussed by e.g. Tur et al. 2009), remain very high. A lively discussion, both in modelling massive star explosions and nucleosynthesis and in analyzing the possible pollution of the solar nebula from a supernova took place (Chieffi and Limongi 2004; Meyer 2005; Limongi and Chieffi 2006; Nomoto et al. 2006; Huss et al. 2007; Woosley and Heger 2007; Takigawa et al. 2008; Kuroda et al. 2008; Huss et al. 2009). These studies and the general constraints from ESS abundances (e.g. avoiding the overproduction of ^{53}Mn and of ^{60}Fe as compared to ^{26}Al, guaranteeing enough production for very short-lived nuclei like ^{41}Ca, etc.) focused the attention on SN models including fallback of a consistent amount of matter and including internal mixing among a previously-layered composition (Joggerst et al. 2009; Takigawa et al. 2008). In particular, Takigawa et al. (2008) showed that virtually all SN models not modified by these effects would overproduce ^{53}Mn by large amounts, and most models would also overproduce ^{60}Fe. In contrast, a *faint supernova*, including fallback and mixing, more likely is able to reproduce the observed distribution of SLR. The mass cut below which material has to fall back and the level of mixing among the stellar layers may then be varied as free parameters to reproduce the measurements in the ESS. We recall that the first idea of a modified supernova model with a high mass cut (to avoid ^{53}Mn overproduction) was suggested by the Clemson group (Meyer and Clayton 2000). The models by Takigawa et al. (2008) showed an impressively good accord between the ejecta of a fall-back SN around 30 $M_{\odot}$ and the ESS record of ^{26}Al, ^{41}Ca, ^{53}Mn and ^{60}Fe. A slight deficit in ^{53}Mn from this model seems not a real problem, in view of the abundant production of this isotope in the continuous galactic nucleosynthesis processes.

Comparisons between the studies by Limongi and Chieffi (2006) and by Woosley and Heger (2007) also clarify that accounting for the galactic inventory of ^{26}Al and ^{60}Fe (for any reasonable value of their abundance ratio in the range indicated by measurements) is possible in principle, provided the proper choices are made for parameters such as the mass cut, the mass loss rate and the most crucial reaction cross sections. In this respect, the stronger mass loss rates by Langer et al. (1989) seem to be preferable to the choice by Nugis and Lamers (2000). The reproduction of the observed ^{26}Al/^{60}Fe in the Galaxy remains an important point to consider.

The idea of a late supernova production of short-lived nuclei is still the subject of a controversial debate. Among the reasons that make SNe appealing candidates are: (1) SNe are in general the site where most nucleosynthesis processes occur. (2) A close-by SN might explain the SLR abundances in the ESS in the same framework in which the equilibrium abundances of radioactive nuclei in the Galaxy are explained. (3) Despite the difficulties we mentioned, star formation in the vicinity of SNe and possibly triggered by them is an appealing and still non-excluded possibility. (4) Faint supernovae, with internal mixing and limited or no contribution to iron do exist, somehow justifying the speculative part of the models required by SLR in the early solar nebula.

However, serious concerns arise by considering other relevant issues, such as: (1) The fact that the probability of a close-by SN encounter has been shown to be extremely small (Gounelle and Meibom 2010), possibly smaller than for an AGB star (Kastner and Myers 1994; Trigo-Rodríguez et al. 2009). (2) The need for ad-hoc choices of parameters, like e.g. the mixing extension, the mass cut, the rate of mass loss in any SN model suitable to explain SLR and the fact that results strongly depend on very uncertain choices for cross sections (Tur et al. 2009). (3) The lack of any answer to the risk of introducing unwanted anomalies on stable isotopes. Widespread anomalies in the abundances of stable nuclei of early meteorites are in general at the ϵ (part per ten thousand) or δ (part per mil) level, except for few cases discussed in Wasserburg et al. (2015). See on this point also the discussion in Vescovi et al. (2018). As an example, the best model by Takigawa et al. (2008), from a 30 $M_\odot$ star, ejects almost 7 $M_\odot$ of processed material, where at least C, O, Ne, and Mg should be highly enhanced (maybe by factors close to 100, as other SN nucleosynthesis calculations suggest). At the adopted dilution factor of 4×10^{-4} this should induce anomalies at percent level on major elements, which are not observed. The same effects are introduced by the recent model by Banerjee et al. (2016): see e.g their Table 3.

On this problem of predicting unobserved anomalies on stable isotopes one has to underline again that a proper estimate of these effects requires reference to massive star calculations including a full account of explosive nucleosynthesis, something that is not frequently done in works on the ESS abundances. As an example of what comes our from such calculations, we refer to Fig. 6.5. There we show results from the recent series of models for CCSN by Chieffi and Limongi (2013), using as a typical example cases computed with a high initial rate for the equatorial rotation velocity (300 km/s). The models are computed with the FRANEC code and include a detailed hydrodynamical estimate for the effect of explosive nucleosynthesis ensuing from the collapse and bounce of the stellar layers external to the central neutron star. The upper panel in Fig. 6.5 shows the results for SLR, after the ESS abundances of ^{26}Al and ^{41}Ca are fitted and used to determine the proper values

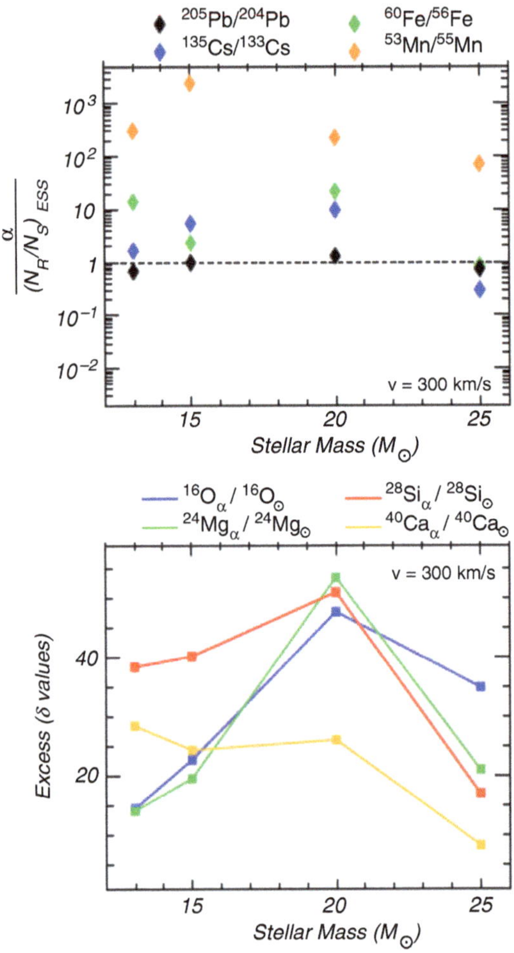

Fig. 6.5 Upper panel: predictions for abundance ratios involving SLR of high atomic mass in the ESS, as derived from the CCSN models by Chieffi and Limongi (2013), including rotation and explosive nucleosynthesis. The ordinate shows the ratio between the prediction and the actual measurement of any abundance ratio: values close to unity would therefore mean agreement with the data. The stellar mass is indicated in abscissa. For each model mass the parameters d and Δ_1 of Eq. (6.4) have been preliminarily fixed so that the ejecta reproduce the ESS ratios for the lower atomic mass radioactivities (i.e. ^{26}Al/^{27}Al and ^{41}Ca/^{40}Ca). In the best case of a 25 $M_{\odot}$ and if we adopt the highest possible value for the ^{60}Fe content of the ESS, then only ^{53}Mn is largely discrepant: but its abundance ratio might be accommodated if the model had a higher mass cut, ejecting less ^{53}Mn. For such a compromise the dilution factor d is about 10^{-4} and the time delay is $\Delta_1 \simeq 1.5$ Myr. Lower panel: Excesses on some stable isotopes implied by the same models of the previous panel and using the dilution factors d derived from them. Any possible coherent picture inferred by the previous plot through some fine-tuning of parameters (e.g. the mass cut for the mentioned 25 $M_{\odot}$ model) is spoiled by the large shifts introduced on stable nuclei shown in this second panel, which are at odds with meteoritic measurements

of the free parameters d and Δ_1 in Eq. (6.4). In the ordinate of the plot the ratio between the prediction α and the measured abundance ratio for each SLR considered is shown. A general problem of all the results would be a large excess for ^{60}Fe if we adopt for the ESS the low values now generally accepted. To avoid this, and for pure illustration purposes, we adopted here an initial solar system ratio ^{60}Fe/^{56}Fe = 10^{-6}. With this (now unlikely) choice, the panel shows that a 25 $M_\odot$ model would permit an apparently good picture for the other nuclei: not only ^{60}Fe, but also ^{135}Cs and ^{205}Pb would have predictions close to the measurements (the ratios in the ordinate are not far from 1). The only remaining exception is ^{53}Mn, which has a large excess; that might be accounted for if the model had a large mass cut, as suggested by Meyer and Clayton (2000). However, in the lower panel of Fig. 6.5 we see that also this tentative solution poses large unsolved problems. There we show a partial record of the excesses on stable nuclei, in terms of δ values, implied by the same computations used above. All the values shown exceed the percent level and unfortunately force one to say that this scenario is not acceptable, because present measurements in early meteorites exclude the existence of similar anomalies above a few per mil (see discussion in Section 2 of Vescovi et al. 2018).

Summing up on this point, the present knowledge of CCSN nucleosynthesis seems for the moment to exclude that an event of this kind may be at the origin of the ESS record of radioactivities. This is so firstly because it looks very difficult to account self-consistently for all SLR inside the nucleosynthesis processes of massive stars, without introducing very special, ad hoc sets of parameterizations. Even more important, unwanted excesses on stable nuclei would emerge, at levels not permitted by the present accuracy of meteoritic measurements.

Frustrating consequences also emerge, from inspection of Fig. 6.5, for the attempts at invoking a cluster-like evolution or a sequential contamination of the pre-solar molecular cloud (see Sect. 6.4.3). Even the mentioned scenarios attributing ^{26}Al to the winds of a late, non-exploding WR star, one obviously needs to explain also the other SLR, attributing them to previous supernovae. If a pre-solar molecular cloud was similar to those we know today (with masses between 10^5 and 3×10^6 $M_\odot$), then one can easily compute the required dilution for any CCSN material ejected. Let's suppose, as an example, that a particular SN ejected 10 $M_\odot$ of processed matter. If the cloud mass was at the lowest limit of what was indicated above, then we would have a dilution $d \simeq 10^{-4}$, very close to the one found in Fig. 6.5 for the 25 $M_\odot$ case. Then, within a factor of a few, we might find that particular SN to provide an acceptable prediction for a few SLR; but the limited dilution of its material would induce also the large anomalies on stable nuclei mentioned for the second panel of Fig. 6.5. The model would then need to be excluded for the same reasons. If instead the cloud mass were close to the maximum estimate, then most SLR would not be explained. Exploring intermediate cases would require a careful analysis, including ad-hoc parameters and a full treatment for explosive nucleosynthesis and for the hydrodynamics of the nebula pollution.

6.5.2 Contributions from a Nearby AGB Star?

The possibility of a close-by contamination by an AGB star was extensively explored by Wasserburg et al. (1994, 1995, 2006) and by Trigo-Rodríguez et al. (2009). These last authors also demonstrated that several previous criticisms to the AGB contamination had to be reconsidered, or looked at with caution.

The most appealing property of AGB models is that they include a limited number of free parameters (i.e. parameters that are not constrained otherwise). In general the main points are the necessity to provide enough ^{60}Fe (something that is now favored by the mentioned downward revisions of the ESS ^{60}Fe abundance) and enough ^{26}Al, as compared to heavier radioactive nuclei easily synthesized by slow neutron captures processes (like e.g. ^{107}Pd or ^{135}Cs). Both problems seem to be in principle affordable if the mass is sufficiently high ($M \geq 5 - 6$ M$_\odot$) to induce high temperatures in the thermal pulses (hence high neutron densities, favoring ^{60}Fe production) and high temperatures at the base of the convective envelope, allowing proton captures (producing ^{26}Al) to occur there in the so-called Hot Bottom Burning (HBB) process. This was indeed the idea explored by Trigo-Rodríguez et al. (2009).

For lower masses, Wasserburg et al. (2006) suggested that ^{26}Al might be produced rather efficiently by extra mixing mechanisms, while high enough neutron densities to produce ^{60}Fe at the high concentrations assumed at that time were shown to be present in stars of relatively low metallicity ($Z = Z_\odot/3$). It was also inferred that the effectiveness of extra mixing could be calibrated through the reproduction of the abundances in pre-solar grains of AGB origin (see Chaps. 3 and 10).

In general, however, both low and intermediate mass star models will suffer for the need of rather free choices for mass loss rates, which only recently have started to be quantitatively determined, thanks to extensive sky surveys in the infrared by space-based instruments (Guandalini et al. 2006; Busso et al. 2007a; Guandalini and Busso 2008; Guandalini and Cristallo 2013).

With the above warnings, it was shown by Wasserburg et al. (2006) and by Trigo-Rodríguez et al. (2009) that models for a 3 M$_\odot$ and a 6.5 M$_\odot$ star were acceptable to reproduce the ESS measurements, if they could avoid hosting a ^{13}C pocket (i.e. considering only their neutron captures occurring inside convective thermal pulses, induced by the ^{22}Ne(α,n)^{25}Mg source). On this basis, ESS measurements for ^{26}Al, ^{41}Ca, ^{60}Fe, and ^{107}Pd, could be reproduced, with dilution factors of a few $\times 10^{-3}$. These results, obtained about one decade ago, are shown in Table 6.4; note that estimates for ^{135}Cs and ^{205}Pb are available only for the Wasserburg et al. (2006) work. In the 3 M$_\odot$ model ^{26}Al would come from the mentioned extra-mixing processes, while in the 6.5 M$_\odot$ star it would come from HBB. Both models tried to reproduce the ESS abundance for ^{60}Fe at the high level then assumed; this estimate has now been reduced by roughly a factor between 10 and 100.

We must underline that AGB stars of intermediate mass would imprint in the solar nebula only marginal anomalies in stable nuclei. The most critical effect is probably a 1% shift in ^{17}O for a star of 6.5 M$_\odot$, see Trigo-Rodríguez et al. (2009). A

Table 6.4 Previous predictions for short lived nuclei from AGB stars according to: (a) Wasserburg et al. (2006), and (b) Trigo-Rodríguez et al. (2009)

Rad. R	Stable S	$3.0\ M_\odot$, $Z = Z_\odot/3$, $d = 0.004$ (a)		$6.7\ M_\odot$, $Z = Z_\odot$, $d = 0.003$ (b)		Meas.
		$(N^R/N^S)_{\Delta_1}$	$(N^R/N^S)_{\Delta_2}$	$(N^R/N^S)_{\Delta_1}$	$(N^R/N^S)_{\Delta_2}$	at $t = \Delta_i$
		$\Delta_1 = 0.53$	$\Delta_2 = 6.5$	$\Delta_1 = 0.53$	$(\Delta_2) = 6.0$	
^{26}Al	^{27}Al	5.0×10^{-5}	8.5×10^{-8}	3.2×10^{-5}	9.8×10^{-8}	5.23×10^{-5}
^{41}Ca	^{40}Ca	1.5×10^{-8}	–	1.5×10^{-8}	–	$\geq 1.5 \times 10^{-8}$
^{60}Fe	^{56}Fe	2.1×10^{-6}	1.0×10^{-7}	2.6×10^{-6}	1.7×10^{-7}	$10^{-8} - 10^{-6}$
^{107}Pd	^{108}Pd	3.8×10^{-5}	2.0×10^{-5}	3.8×10^{-5}	2.0×10^{-5}	$5.9(\pm 2.2) \times 10^{-5}$
^{135}Cs	^{133}Cs	3.6×10^{-5}	3.5×10^{-6}	n.a.	n.a.	$1\text{–}2 \times 10^{-4}$
^{205}Pb	^{204}Pb	$\leq 3.4 \times 10^{-4}$	$\leq 2.5 \times 10^{-4}$	n.a.	n.a.	$1\text{–}2 \times 10^{-4}$

The time intervals Δ_1 and Δ_2 are in million years lower-mass star would introduce no measurable changes on stable isotopes, except perhaps for the (unmeasurable) addition of some carbon. In view of the depressing results shown previously in Fig. 6.5 for massive stars, This is a special merit of AGB models. If one also considers that heavy s-process nuclei like ^{205}Pb are typical products of AGB nucleosynthesis, it is understandable that the above authors suggested some caution before excluding an intermediate-mass star as a polluting source, at least until the massive star models could be shown to permit similar global consistencies.

As mentioned, doubts on an AGB pollution of the ESS were raised from the very low (a mere 1% or less) probability of a close encounter (Kastner and Myers 1994) (see also Sect. 6.4.2). For them, Trigo-Rodríguez et al. (2009) presented several reasons to alleviate these criticisms; and moreover Gounelle and Meibom (2010) indicated that chance encounter probabilities for individual massive stars are at least as small as for AGBs. If a close encounter with a star producing radioactive nuclei ever occurred, this must have been necessarily an unusual event, maybe a unique occurrence, independently on the stellar mass. Relative statistics then do not help discriminating between these possibilities.

An important point for all AGB star models would be the easy production of the s-only isotope ^{205}Pb. However, here the predictions are particularly uncertain due to the need of having a good description for the complex coupled weak interactions linking ^{205}Pb to ^{205}Tl, which becomes unstable at high temperature. On this point and in general on many issues involving electron-captures and β-decays along the s-process path, new efforts by theoretical and (when possible) experimental nuclear physics are needed. In particular, treatments similar to the one performed by Simonucci et al. (2013) and Palmerini et al. (2011b) for ^{7}Be seem to be quite urgent.

Very recently new model efforts on evolved intermediate-mass stars have shed more light on the nucleosynthesis expected by AGB phases. This unfortunately introduces serious doubts on their chances to be at the origin of SLR in the ESS. On the basis of AGB calculations including HBB, performed with new reaction rates and following previous work by Lugaro et al. (2014) and Wasserburg et al. (2017) showed that their models could not be reconciled with the solar-system record of SLR. The most massive of the stars in their calculations ($M \geq 6\,M_{\odot}$) produced too much ^{26}Al with respect to s-process nuclei; on the contrary, the lower masses had the opposite problem. Their suggestion then was that maybe a solution might be found in a very thin mass interval at the separation of those explored (i.e. for $5 - 5.5\,M_{\odot}$). These suggestions descended by the scheme they adopted for s-processing: they included a ^{13}C pocket below about $5\,M_{\odot}$, but not for more massive models. The extensions of the assumed ^{13}C reservoirs (where they were included) were in any case rather small as compared to recent findings by Trippella et al. (2016).

In the framework of recent computations for AGB stars in which extra-mixing mechanisms and the formation of the neutron source ^{13}C are no longer parameterized, but derive self-consistently from simulations of the magneto-hydrodynamics (MHD) of magnetically-active stars (Nucci and Busso 2014) the situation for the production of the neutron source ^{13}C and also for the deep mixing mechanisms yielding ^{26}Al above the H-burning shell appears today rather different. For the approaches that led to this scenario see e.g. Busso et al. (2007b), Nordhaus et al. (2008) and Denissenkov et al. (2009). The descending scheme for mixing mechanisms on the AGB was then computed from a full solution of MHD equations in the typical geometry of an AGB star and was recently shown to satisfy quite a number of important constraints deriving from both the solar system and the pre-solar grain isotopic admixtures of trace elements (Trippella et al. 2016; Palmerini et al. 2017, 2018). Within these models a ^{13}C source is essentially always formed and is considerably larger than considered by Wasserburg et al. (2017). In a recent analysis by Vescovi et al. (2018) the results of our approach and those discussed by Cristallo et al. (2011, 2015), where deep mixing was modelled on the basis of opacity-induced phenomena, were shown to be almost coincident for the inventory of SLR. The considerations we discuss here therefore apply equally well to the revisions of the AGB models performed by our group and by Cristallo et al. (2015).

We illustrate the emerging situation with reference to the model of a $5M_{\odot}$ star. The evolution was computed with a pure Schwarzschild criterion for convection and any deep-mixing or overshoot from the envelope convection was attributed to magnetic buoyancy effects. These would occur in quasi-ideal MHD conditions, allowing them to be much faster than any previously suggested diffusive mechanism for mass circulation, like e.g. thermohaline mixing (see e.g. Eggleton et al. 2006, 2008).

The model presented synthesizes ^{26}Al in a mass circulation induced by magnetic buoyancy and the same mechanism also yields the penetrations of protons into the He-rich layers at TDU, generating there a ^{13}C reservoir where subsequently protons are produced.

If we apply the above scheme, it necessarily produces a ^{13}C pocket also for masses well above 2–3 $M_{\odot}$, at odds with the assumptions originally done, on parameterized approaches, by Wasserburg et al. (2006). In the pocket, s-processing would produce heavy radioactive nuclei in quantities that seem now to overproduce inevitably ^{107}Pd with respect to ^{26}Al, jeopardizing the chances of AGB stars of intermediate mass to offer the previously inferred coherent picture. Preliminary results from this new approach are shown in Table 6.5.

The basic point is that, since the calculations are no longer parameterized, one has no freedom of choice. Once the model is verified on other constraints (as done e.g. in Palmerini et al. 2017, 2018) its results must be accepted as they are: and they seem to say that the possibility for an AGB star without HBB to reproduce the record of early SLR is now over. This, integrated with the mentioned results by Wasserburg et al. (2017) seem at the present moment to put in serious question the AGB scenarios discussed in the last two decades. One has also to consider that a 5 $M_{\odot}$ star has a typical evolutionary timescale of 10^8 years, roughly a factor of 3 more than the age of a possible molecular cloud being the parent of both this contaminating star and the Sun. The need for a very rare or unique event would remain: it could be accepted only for the appeal of previous models, explaining a

Table 6.5 Recent predictions for short lived nuclei from a revised model of a 5 $M_\odot$ AGB star

Rad.	Ref.	τ_R	N^R/N^S	q^S	$\alpha^{R,S}$	$[N^R/N^S]_{Meas.}$
^{26}Al	^{27}Al	1.03	3.65×10^{-3}	1.002	5.23×10^{-5}	$(5.23 \pm 0.13) \times 10^{-5}$
^{41}Ca	^{40}Ca	0.15	3.57×10^{-5}	0.996	4.00×10^{-9}	4×10^{-9}
^{60}Fe	^{56}Fe	3.75	4.36×10^{-4}	0.995	1.13×10^{-5}	10^{-8}–10^{-6}
^{107}Pd	^{108}Pd	9.4	2.24×10^{-2}	1.139	7.61×10^{-4}	$(5.9 \pm 2.2) \times 10^{-5}$
^{135}Cs	^{133}Cs	3.3	3.09×10^{-2}	1.011	7.60×10^{-4}	4.8×10^{-4}
^{182}Hf	^{180}Hf	12.8	3.66×10^{-4}	1.026	1.15×10^{-5}	$(9.81 \pm 0.41) \times 10^{-5}$
^{205}Pb	^{204}Pb	22	4.42×10^{-2}	1.038	1.44×10^{-3}	10^{-3}

The parameters adopted are: [Fe/H] = 0; dilution $d = 3.27 \times 10^{-2}$–; delay time $\Delta = 0.85$ Myr. As done previously for massive stars, the parameters d and Δ : 1 are fixed by reproducing the ESS abundances of ^{26}Al and ^{41}Ca (see text for discussions)

wide range of ESS radioactive isotopes together; but this possibility appears to be no longer there, when parameterizations are abandoned. Concerning more massive AGB stars, like those analyzed by Wasserburg et al. (2017), one has to remember that their results for HBB (hence for the ratio of ^{26}Al to ^{107}Pd or^{135}Cs, produced ny n-captures) are still strongly model dependent, This still offers a small window of possibilities, which will be mentioned in the conclusions.

6.6 Short-Lived Nuclei: Production Inside the ESS

For a few shorter-lived species, especially ^{10}Be, ^{36}Cl, ^{41}Ca, ^{53}Mn, alternative models for the formation in spallation reactions, from the bombardment of fast particles coming from the magnetically-active early Sun were proposed; see e.g. Shu et al. (1997) and Gounelle et al. (2001). The same magnetic phenomena occurring in most stars, at least during their Main Sequence but probably also in their subsequent Red Giant phases (Andrews et al. 1988), provide a site where nucleosynthesis of radioactive isotopes occurs: this has been shown for very short-lived nuclei (Tatischeff et al. 2006) but might be true also for longer-lived species of interest for the ESS (Palmerini and Busso 2008).

6.6.1 Radioactivities from the Bombardment of Early Solids

This mechanism takes advantage of the already described X-wind scenario (Lee et al. 1998a), lifting the recently formed CAIs from the disk plane and transporting them to large distances. In such models the hypothesis was also advanced that CAIs and chondrules might be produced in those same winds. If this were the case, then the high abundance measured in CAIs for ^{26}Al (^{26}Al/^{27}Al $= 5 \times 10^{-5}$) could be suspected to indicate solar irradiation rather than a uniform stellar contamination.

A thorough discussion of the spallation mechanisms in the ESS was presented by Gounelle et al. (2006), on the basis of nuclear parameters for spallation processes that were standard at that epoch. On the basis of models for the structure of CAIs and for the flux of irradiating particles, the authors first determined the conditions for the production of ^{7}Be, whose very short mean life makes it a discriminating nucleus. They suggested that, for a flux of particles from the Sun of the order of $F \simeq 2 \times 10^{10}\,\mathrm{cm}^{-2}\,\mathrm{s}^{-1}$, both ^{7}Be and ^{10}Be could be produced at the observed levels within the uncertainties. A byproduct of the adopted fluxes was also a noticeable contribution to ^{26}Al, ^{36}Cl and ^{53}Mn, whose abundances in the ESS were therefore suggested to come from proton bombardment, leaving space to nearby stars mainly to produce ^{60}Fe and heavier SLR.

After the work by Gounelle et al. (2006), it was shown that the cross sections for some crucial spallation processes, in particular ^{24}Mg(^{3}He,p)^{26}Al, needed important revisions (Fitoussi et al. 2008). This led to the conclusion that ^{26}Al cannot be produced at sufficient levels by spallation processes in the early solar nebula, so that its synthesis seems now to require unambiguously nucleosynthesis processes in stars, at odds with suggestions by Lugaro et al. (2014). The clear decoupling between ^{26}Al and ^{36}Cl (Hsu et al. 2006) in early meteorites then suggests that ^{36}Cl, instead, was formed in the X-wind scenario, together with Be-isotopes. Further limits to the level of production of radioactive nuclei from solid particle bombardment were put, after the first results of the STARDUST mission (Brownlee et al. 2006), by new models using improved cross sections for the reactions leading to ^{7}Be, ^{10}Be, ^{26}Al, ^{36}Cl, ^{41}Ca, and ^{53}Mn (Duprat and Tatischeff 2008). According to these new findings, the role of non-thermal nucleosynthesis in the ESS might actually be limited to the production of ^{10}Be and some ^{41}Ca.

A very recent study by Sossi et al. (2017) reconsidered the effects of the irradiation of pristine solids from the intense cosmic ray fluxes originated in the atmosphere of the early Sun (mainly in flares). By correlating the evidence provided by ^{10}Be with that from ^{50}V, these authors succeeded in showing that early solar system dust was probably exposed to a bombardment induced by gradual flares for a very short period of time (maybe lower than 300 years) and at a distance from the central Sun compatible with the mentioned X-wind scenario. Such a process might contribute to nuclei like ^{41}Ca and ^{53}Mn but again not to ^{26}Al.

It seems therefore that the solar-wind model could indeed contribute to the inventory of SLR, but only for a limited set of them, confirming the need for a variety of sources. In particular, ^{41}Ca might be produced in this way, while for both ^{26}Al and ^{60}Fe a stellar origin was confirmed. This would however leave the correlation between ^{26}Al and ^{41}Ca show in Fig. 6.3 as an unanswered mystery.

6.6.2 Solar Activity and the Production of ^{7}Be

Nuclear interactions occurring in solar explosions are revealed by their induced prompt emission of γ-ray lines; these are due to the de-excitation of nuclei excited by reactions occurring between flare-accelerated particles and the solar atmospheric material. The first observations of these solar γ-ray lines were obtained by the experiment GRS (Gamma Ray Spectrometer), on board of the OSO-7 space-borne observatory (Chupp et al. 1973). Subsequently, the evidence for spallation-induced reactions in flares grew thanks to the measurements of various instruments: the Solar Maximum Mission (Share and Murphy 1995), the Compone Gamma-Ray Observatory (Share et al. 1997) and the Ramaty High Energy Solar Spectroscopic Imager (RHESSI), see e.g. Lin et al. (2003).

In recent years, it was suggested that also delayed X- and γ-ray emission might occur from solar flares, thanks to the production of short-lived radioactive nuclei, whose subsequent decay would be accompanied by emission lines (Ramaty and Mandzhavidze 2000; Kozlovsky et al. 2002). In particular, detailed predictions of line emissions form such decays in solar active regions, after the occurrence of intense flares, were presented by Tatischeff et al. (2006, 2007). These authors estimated the cross sections for the formation of several radio-nuclei from interactions of protons and ^{3,4}He particles with isotopes of elements up to Ni. Then, on the basis of a thick-target model, they provided expected yields for 25 radio-nuclei with half-life in the range from 10 min (^{13}N) to 77 days (^{56}Co). Fluxes in γ-ray lines interesting for a possible detection from future experiments were found for ^{52}Mn (1434 keV), ^{60}Cu (1332, 1792 keV), ^{34}Cl (2127 keV), ^{24}Na (1369, 2754 keV) and ^{55}Co (931.1 keV).

The above findings are important for explaining the γ-ray fluxes of solar activity phenomena and for providing information on details of the solar plasma physics. On the contrary, spallation processes on solar and stellar atmospheres are not expected to be relevant for galactic nucleosynthesis, as production of nuclei by similar processes occurring in Galactic Cosmic Rays (GCR) would dominate by several orders of magnitude (Tatischeff and Thibaud 2008). Concerning the SLR discussed in this Chapter, the works mentioned in this subsection might be of interest for the early inventory of very short-lived nuclei, like ^{7}Be. In fact, it has been found that this isotope might have been present alive in CAIs (Chaussidon et al. 2003, 2006). Its production must be essentially contemporaneous to the same CAI formation, due to its very short half-life (53 days), and its (uncertain) ESS abundance was originally indicated as being roughly compatible with X-wind models (Gounelle et al. 2003; Chaussidon et al. 2006). Recent revisions suggest, instead, that its production in that environment should be largely insufficient (Duprat and Tatischeff 2008). The formation of ^{7}Be in solar flares might therefore be considered as a promising alternative possibility to the irradiation of solids. It has also been noticed that production of ^{7}Be (hence of its daughter ^{7}Li) might explain the unexpected detection of Li in several M-type dwarfs, although it would probably be insufficient to account for the trends of the Li abundance in open clusters (Tatischeff et al. 2008).

The above studies represent a remarkable bridge linking the models for the in-situ production of SLR to those for their stellar synthesis. Indeed, some recent models for the deep-mixing phenomena occurring in evolved red giants and accounting, among other things, for the production of abundant ^{26}Al and Li (Palmerini and Busso 2008; Palmerini et al. 2009; Guandalini et al. 2009) consider stellar magnetic fields as the main engine for the transport of proton-capture ashes to the stellar envelope, following ideas by Busso et al. (2007b) and detailed models by Nucci and Busso (2014), Trippella et al. (2016), and Palmerini et al. (2017, 2018). Those models account for the production and destruction of Li in red giants through mechanisms of magnetic buoyancy occurring at rates respectively faster or slower than that for ^{7}Be decay. Such models, however, did not include, so far, the possible contributions from spallation processes in the transported material itself, which is moving relativistically along toroidal flux tubes and their Ω-shaped instabilities. A check of the relevance of stellar spallation processes for explaining the formation of super-Li rich stars, and for the evolution of the Li abundance in the Sun seems to be really necessary now.

6.7 Lessons from the Early Solar System

The general scenarios explored so far in order to account for the presence of short-lived radioactive nuclei in the ESS is far from satisfactory and still quite confused. We can summarize the indications emerged from the many efforts dedicated to this field in the past years in the following points.

- Decay of radioactive isotopes of long lifetime (longer than 1 Gyr) offers us a number of tools for estimating the global age of the Earth and of the other solid bodies orbiting around our star. They tell us that the solar system was formed when the Galaxy was already quite mature, having spent 2/3 of its present age. The resulting age of the solar system is close to 4.566 Gyr.
- The ESS abundances of shorter-lived isotopes, with half lives from 10–20 Myr to a few hundred Myr, can be used for reconstructing the history of nucleosynthesis in the solar neighborhood, as their initial concentration in the solar nebula is probably compatible with the equilibrium abundances established by SN explosions and nuclear decay in the local interstellar medium, during several cycles of molecular cloud aggregation and destruction, occurred before the formation of the specific one from which the solar system and many more stars were born. From these nuclei it was inferred that the r-process has a multi-modal nature. This suggestion, which in the last 20 years produced considerable insight on both solar system formation and the abundance patterns observed in old stars, has stimulated further studies that are now helping in disentangling the various nucleosynthesis contributions displayed by population II stars. For sure this has ascertained that some of the lighter nuclei previously attributed to the r process, might be produced instead in a series of explosive phenomena involving mainly charged-particle interactions; these may affect isotopes up $A \simeq 100$–120.
- The original event suggested for explaining the SLR, that of a nearby passage of a CCSN is now excluded, at least in its simplest form, because any dilution of its ejecta suitable to account for a few SLR would inevitably introduce excessive contributions of stable isotopes, thus predicting in the solar nebula isotopic and

elemental anomalies at the level of a few percent at least, which are today excluded by the accuracy achieved in measurements on early meteorites.

- The solar nebula was probably formed inside a complex stellar aggregation, belonging to a parent molecular cloud. This is expected to have undergone a series of massive star nucleosynthesis episodes, some of which terminating with a SN explosion, some with more quiet Wolf-Rayet-type winds; these last were suggested to be responsible for the late addition of ^{26}Al and perhaps ^{60}Fe. This idea would however require to reconstruct which SN contributions created a previous homogeneous composition in the parent cloud over a time scale of 2–3 $\times$ 10^7 years, before a last contamination, possibly by WR winds, added the shortest lived nuclei. This verification should be done carefully, to ascertain that the mechanism does not suffer for the same problems of excesses on stable species mentioned for a single nearby CCSN. Probably this difficult task requires better nuclear inputs than available today, especially for explosive phases.
- There are signs that the forming Sun was really affected by a last episode of stellar nucleosynthesis rather close in time and space, adding the radioactivity with lifetimes below $\tau_R \lesssim 5$ Myr. This event had a low probability of occurrence from a purely statistical point of view; this difficulty might however be mitigated if it was not a chance encounter, but rather a last nucleosynthesis episode in the lifetime of the parent molecular cloud from which the Sun formed.
- Attribution of this last event to an intermediate-mass AGB star, which was a subject of lively debates in the past, seems now excluded by more quantitative and non-parametric models. At masses below about 6–7 M$_\odot$ the inevitable presence of a ^{13}C pocket seems to produce too high abundances of n-rich radioactivities with respect to ^{26}Al; at higher masses the reverse may be true, thanks to HBB nucleosynthesis. This is however a subject that must be re-examined, as the models are still uncertain and very model dependent. There is therefore still some space for revisions; this might also be so for slightly higher masses, below the lowest limit of those forming CCSN, i.e. the so-called super-AGB stars, probable parent of O-Ne white dwarfs.
- Subsequently to its formation, the Sun itself, in its fully convective, pre-Main Sequence phase, must have added new nuclei (^{7}Be, ^{10}Be, ^{36}Cl and perhaps some ^{41}Ca), through high-energy processes occurring either directly in coronal flares, or in the interactions of fast solar particles with early solids that were forming in the innermost regions of an accretion disk.

Despite the many uncertainties, the wealth of the short-lived or very-short-lived radioactivities discussed in this Chapter remains an invaluable source of information on the timing of the first events occurred in the solar nebula and has also remarkably contributed to the understanding of nucleosynthesis processes generating very heavy nuclei in the Galaxy.

References

Abbott BP, Abbott R, Abbott TD, Acernese F, Ackley K, Adams C, Adams T, Addesso P, Adhikari RX, Adya VB et al (2017) GW170608: observation of a 19 solar-mass binary black hole coalescence. Astrophys J 851:L35. https://doi.org/10.3847/2041-8213/aa9f0c, arXiv:1711.05578

Adams FC (2010) The birth environment of the solar system. Annu Rev Astron Astrophys 48:47–85. https://doi.org/10.1146/annurev-astro-081309-130830

Adams FC, Laughlin G (2001) Constraints on the birth aggregate of the solar system. Icarus 150:151–162

Adams FC, Fatuzzo M, Holden L (2014) Distributions of short-lived radioactive nuclei produced by young embedded star clusters. Astrophys J 789:86. https://doi.org/10.1088/0004-637X/789/1/86

Allègre CJ, Manhès G, Göpel C (1995) The age of the earth. Geochim Cosmochim Acta 59:2445–1456

Amelin Y, Ghosh A, Rotenberg E (2005) Unraveling the evolution of chondrite parent asteroids by precise U-Pb dating and thermal modeling. Geochim Cosmochim Acta 69:505–518. https://doi.org/10.1016/j.gca.2004.05.047

Anderson FS, Whitaker TJ, Young D, Peterson B (2007) Rb-Sr dating using LDRIMS. In: Chronology of meteorites and the early solar system, pp 22–23

Andrews AD, Rodono M, Linsky JL, Brown A, Butler CJ, Catalano S, Scaltriti F, Busso M, Nha I, Oh JY, Henry MCD, Hopkins JL, Landis HJ, Engelbrektson S (1988) Rotational modulation and flares on RS CVn and BY DRA stars. Astron Astrophys 204:177–192

Arnould M, Prantzos N (1999) Cosmic radioactivities. New Astron 4:283–301. https://doi.org/10.1016/S1384-1076(99)00016-0, arXiv:astro-ph/9907275

Baiotti L, Rezzolla L (2017) Binary neutron star mergers: a review of Einstein's richest laboratory. Rep Prog Phys 80(9):096, 901. https://doi.org/10.1088/1361-6633/aa67bb

Baker RGA, Schönbächler M, Rehkämper M, Williams HM, Halliday AN (2010) The thallium isotope composition of carbonaceous chondrites. New evidence for live ^{205}Pb in the early solar system. Earth Planet Sci Lett 291:39–47. https://doi.org/10.1016/j.epsl.2009.12.044

Banerjee P, Qian YZ, Heger A, Haxton WC (2016) Evidence from stable isotopes and ^{10}Be for solar system formation triggered by a low-mass supernova. Nat Commun 7:13639. https://doi.org/10.1038/ncomms13639, arXiv:1611.07162

Birck J, Allègre CJ (1985) Evidence for the presence of Mn-53 in the early solar system. Geophys Res Lett 12:745–748. https://doi.org/10.1029/GL012i011p00745

Birck J, Allègre CJ (1988) Manganese-chromium isotope systematics and the development of the early solar system. Nature 331:579–584. https://doi.org/10.1038/331579a0

Bizzarro M, Ulfbeck D, Trinquier A, Thrane K, Connelly JN, Meyer BS (2007) Evidence for a Late Supernova Injection of ^{60}Fe into the protoplanetary disk. Science 316:1178–. https://doi.org/10.1126/science.1141040

Bondarenko V, Berzins J, Prokofjevs P, Simonova L, von Egidy T, Honzátko J, Tomandl I, Alexa P, Wirth HF, Köster U, Eisermann Y, Metz A, Graw G, Hertenberger R, Rubacek L (2002) Interplay of quasiparticle and phonon excitations in ^{181}Hf observed through (n,γ) and (d−− >p) reactions. Nucl Phys A 709:3–59

Boss AP (2005) The solar nebula. Elsevier, New York, p 63

Boss AP, Keiser SA (2010) Who pulled the trigger: a supernova or an asymptotic giant branch star? Astrophys J 717:L1–L5. https://doi.org/10.1088/2041-8205/717/1/L1

Boss AP, Keiser SA, Ipatov SI, Myhill EA, Vanhala HAT (2010) Triggering collapse of the presolar dense cloud core and injecting short-lived radioisotopes with a shock wave. I. Varied shock speeds. Astrophys J 708:1268–1280. https://doi.org/10.1088/0004-637X/708/2/1268

Brennecka GA (2010) A complication in determining the precise age of the solar system. Tech rep, Arizona State University

Brennecka GA, Weyer S, Wadhwa M, Janney PE, Zipfel J, Anbar AD (2010) $^{238}U/^{235}U$ variations in meteorites: extant ^{247}Cm and implications for Pb-Pb dating. Science 327:449–. https://doi.org/10.1126/science.1180871

Brownlee D, Tsou P, Aléon J, Alexander CMO, Araki T et al (2006) Comet 81P/Wild 2 under a microscope. Science 314:1711. https://doi.org/10.1126/science.1135840

Burkhardt C, Kleine T, Bourdon B, Palme H, Zipfel J, Friedrich JM, Ebel DS (2008) Hf W mineral isochron for Ca,Al-rich inclusions: age of the solar system and the timing of core formation in planetesimals. Geochim Cosmochim Acta 72:6177–6197. https://doi.org/10.1016/j.gca.2008.10.023

Busso M, Gallino R (1985) The production of neutron-rich isotopes during He burning in massive stars. Astron Astrophys 151:205–214

Busso M, Gallino R, Wasserburg GJ (1999) Nucleosynthesis in asymptotic giant branch stars: relevance for galactic enrichment and solar system formation. Annu Rev Astron Astrophys 37:239–309. https://doi.org/10.1146/annurev.astro.37.1.239

Busso M, Gallino R, Wasserburg GJ (2003) Short-lived nuclei in the early solar system: a low mass stellar source? Publ Astron Soc Aust 20:356–370. https://doi.org/10.1071/AS03035

Busso M, Guandalini R, Persi P, Corcione L, Ferrari-Toniolo M (2007a) Mid-infrared photometry of mass-losing asymptotic giant branch stars. Astron J 133:2310–2319. https://doi.org/10.1086/512612, arXiv:astro-ph/0701501

Busso M, Wasserburg GJ, Nollett KM, Calandra A (2007b) Can extra mixing in RGB and AGB stars be attributed to magnetic mechanisms? Astrophys J 671:802–810. https://doi.org/10.1086/522616, arXiv:0708.2949

Cameron AGW (1960) New neutron sources of possible astrophysical importance. Astron J 65:485. https://doi.org/10.1086/108085

Cameron AGW (1993) Nucleosynthesis and star formation. In: Levy EH, Lunine JI (eds) Protostars and planets III, pp 47–73

Cameron AGW, Truran JW (1977) The supernova trigger for formation of the solar system. Icarus 30:447–461. https://doi.org/10.1016/0019-1035(77)90101-4

Cameron AGW, Hoeflich P, Myers PC, Clayton DD (1995) Massive supernovae, orion gamma rays, and the formation of the solar system. Astrophys J 447:L53. https://doi.org/10.1086/309554

Chaussidon M, Robert F, Russel SS, Gounelle M, Ash RD (2003) B and Mg isotopic variations in Leoville MRS-06 type B1 CAI:origin of ^{10}Be and ^{26}Al. In: EGS - AGU - EUG joint assembly, p 9292

Chaussidon M, Robert F, McKeegan KD (2006) Reply to the comment by Desch and Ouellette on Li and B isotopic variations in an Allende CAI: evidence for the in situ decay of short-lived ^{10}Be and for the possible presence of the short-lived nuclide ^{7}Be in the early solar system. Geochim Cosmochim Acta 70:5433–5436. https://doi.org/10.1016/j.gca.2006.08.042

Chen JH, Wasserburg GJ (1987) A search for evidence of extinct lead 205 in iron meteorites. In: Lunar and planetary institute science conference abstracts, vol 18, p 165

Chieffi A, Limongi M (2004) Explosive yields of massive stars from $Z = 0$ to $Z = Z_{solar}$. Astrophys J 608:405–410. https://doi.org/10.1086/392523, arXiv:astro-ph/0402625

Chieffi A, Limongi M (2013) Pre-supernova evolution of rotating solar metallicity stars in the mass range 13-120 $M_{\odot}$ and their explosive yields. Astrophys J 764:21. https://doi.org/10.1088/0004-637X/764/1/21

Chupp EL, Forrest DJ, Higbie PR, Suri AN, Tsai C, Dunphy PP (1973) Solar gamma ray lines observed during the solar activity of August 2 to August 11, 1972. Nature 241:333–335. https://doi.org/10.1038/241333a0

Clayton RN (1973) Oxygen isotopic composition of the Luna 20 soil. Geochim Cosmochim Acta 37:811–813. https://doi.org/10.1016/0016-7037(73)90177-4

Clayton DD (1988) Nuclear cosmochronology within analytic models of the chemical evolution of the solar neighbourhood. Mon Not R Astron Soc 234:1–36

Clayton DD, Nittler LR (2004) Astrophysics with presolar stardust. Annu Rev Astron Astrophys 42:39–78. https://doi.org/10.1146/annurev.astro.42.053102.134022

Cowley CR (1995) An introduction to cosmochemistry. Cambridge University Press, Cambridge

Cristallo S, Piersanti L, Straniero O, Gallino R, Domínguez I, Abia C, Di Rico G, Quintini M, Bisterzo S (2011) Evolution, nucleosynthesis, and yields of low-mass asymptotic giant branch stars at different metallicities. II. The FRUITY database. Astrophys J Suppl 197:17. https://doi.org/10.1088/0067-0049/197/2/17, arXiv:1109.1176

Cristallo S, Straniero O, Piersanti L, Gobrecht D (2015) Evolution, nucleosynthesis, and yields of AGB stars at different metallicities. III. Intermediate-mass models, revised low-mass models, and the ph-FRUITY interface. Astrophys J Suppl 219:40. https://doi.org/10.1088/0067-0049/219/2/40, arXiv:1507.07338

Dalrymple GB (1991) The age of the earth. Stanford University Press, Palo Alto

Dauphas N, Cook DL, Sacarabany A, Fröhlich C, Davis AM, Wadhwa M, Pourmand A, Rauscher T, Gallino R (2008a) ^{60}Fe in the cosmic blender. Geochim Cosmochim Acta 72:200

Dauphas N, Cook DL, Sacarabany A, Fröhlich C, Davis AM, Wadhwa M, Pourmand A, Rauscher T, Gallino R (2008b) Iron 60 evidence for early injection and efficient mixing of stellar debris in the protosolar nebula. Astrophys J 686:560–569. https://doi.org/10.1086/589959, 0805.2607

Davies MB, Adams FC, Armitage P, Chambers J, Ford E, Morbidelli A, Raymond SN, Veras D (2014) The long-term dynamical evolution of planetary systems. In: Protostars and planets VI, pp 787–808

Davis AM, McKeegan KD (2014) Short-lived radionuclides and early solar system chronology, pp 361–395

Denissenkov PA, Pinsonneault M, MacGregor KB (2009) Magneto-thermohaline mixing in red giants. Astrophys J 696:1823–1833. https://doi.org/10.1088/0004-637X/696/2/1823, arXiv:0806.4346

Donohue PH, Huss GR, Nagashima K, Telus M (2017) Live(?) 60Fe during aqueous alteration of chondrite parent bodies: evidence from UOCs and CV chondrites. In: Lunar and planetary science conference, vol 48, pp 2307

D'Orazi V, Magrini L, Randich S, Galli D, Busso M, Sestito P (2009) Enhanced production of barium in low-mass stars: evidence from open clusters. Astrophys J 693:L31–L34. https://doi.org/10.1088/0004-637X/693/1/L31

Dukes D, Krumholz MR (2012) Was the sun born in a massive cluster? Astrophys J 754:56. https://doi.org/10.1088/0004-637X/754/1/56

Duprat J, Tatischeff V (2008) On non-thermal nucleosynthesis of short-lived radionuclei in the early solar system. New Astron Rev 52:463–466. https://doi.org/10.1016/j.newar.2008.06.016

Dwarkadas VV, Dauphas N, Meyer B, Boyajian P, Bojazi M (2017) Triggered star formation inside the shell of a Wolf-Rayet bubble as the origin of the solar system. Astrophys J 851:147. https://doi.org/10.3847/1538-4357/aa992e, 1712.10053

Eggleton PP, Dearborn DSP, Lattanzio JC (2006) Deep mixing of ^{3}He: reconciling big bang and stellar nucleosynthesis. Science 314:1580–. https://doi.org/10.1126/science.1133065, arXiv:astro-ph/0611039

Eggleton PP, Dearborn DSP, Lattanzio JC (2008) Compulsory deep mixing of ^{3}He and CNO isotopes in the envelopes of low-mass red giants. Astrophys J 677:581–592. https://doi.org/10.1086/529024

Evans NJ II (1999) Physical conditions in regions of star formation. Annu Rev Astron Astrophys 37:311–362. https://doi.org/10.1146/annurev.astro.37.1.311, arXiv:astro-ph/9905050

Fitoussi C, Duprat J, Tatischeff V, Kiener J, Naulin F, Raisbeck G, Assunção M, Bourgeois C, Chabot M, Coc A, Engrand C, Gounelle M, Hammache F, Lefebvre A, Porquet M, Scarpaci J, de Séréville N, Thibaud J, Yiou F (2008) Measurement of the ^{24}Mg(^{3}He,p)^{26}Al cross section: implication for Al26 production in the early solar system. Phys Rev C 78(4):044613. https://doi.org/10.1103/PhysRevC.78.044613

Frisch PC, Redfield S, Slavin JD (2011) The Interstellar Medium Surrounding the Sun. Annu Rev Astron Astrophys 49:237–279. https://doi.org/10.1146/annurev-astro-081710-102613

Gaidos E, Krot AN, Huss GR (2009a) On the oxygen isotopic composition of the solar system. Astrophys J 705:L163–L167. https://doi.org/10.1088/0004-637X/705/2/L163, arXiv:0909.3589

Gaidos E, Krot AN, Williams JP, Raymond SN (2009b) ^{26}Al and the formation of the solar system from a molecular cloud contaminated by Wolf-Rayet winds. Astrophys J 696:1854–1863. https://doi.org/10.1088/0004-637X/696/2/1854, arXiv:0901.3364

Gallino R, Arlandini C, Busso M, Lugaro M, Travaglio C, Straniero O, Chieffi A, Limongi M (1998) Evolution and nucleosynthesis in low-mass asymptotic giant branch stars. II. Neutron capture and the s-process. Astrophys J 497:388. https://doi.org/10.1086/305437

Goebel R, Begemann F, Ott U (1982) On neutron-induced and other noble gases in Allende inclusions. Geochim Cosmochim Acta 46:1777–1792. https://doi.org/10.1016/0016-7037(82)90117-X

Goswami JN, Marhas KK, Chaussidon M, Gounelle M, Meyer BS (2005) Origin of short-lived radionuclides in the early solar system. In: Krot AN, Scott ERD, Reipurth B (eds) Chondrites and the protoplanetary disk. Astronomical Society of the Pacific conference series, vol 341, p 485

Gounelle M (2015) The abundance of ^{26}Al-rich planetary systems in the galaxy. Astron Astrophys 582:A26. https://doi.org/10.1051/0004-6361/201526174

Gounelle M, Meibom A (2010) The origin of ^{60}Fe and other short-lived radionuclides in the early solar system. In: Montmerle T, Ehrenreich D, Lagrange A-M (eds) EAS publications series, vol 41, pp 301–311. https://doi.org/10.1051/eas/1041021

Gounelle M, Meynet G (2012a) Solar system genealogy revealed by extinct short-lived radionuclides in meteorites. Astron Astrophys 545:A4. https://doi.org/10.1051/0004-6361/201219031

Gounelle M, Meynet G (2012b) Solar system genealogy revealed by extinct short-lived radionuclides in meteorites. Astron Astrophys 545:A4

Gounelle M, Shu FH, Shang H, Glassgold AE, Rehm KE, Lee T (2001) Extinct radioactivities and protosolar cosmic rays: self-shielding and light elements. Astrophys J 548:1051–1070. https://doi.org/10.1086/319019

Gounelle M, Shang S, Glassgold AE, Shu FH, Rehm EK, Lee T (2003) Early solar system irradiation and beryllium-7 synthesis. In: Mackwell S, Stansbery E (eds) Lunar and planetary institute science conference abstracts, vol 34, p 1833

Gounelle M, Shu FH, Shang H, Glassgold AE, Rehm KE, Lee T (2006) The irradiation origin of beryllium radioisotopes and other short-lived radionuclides. Astrophys J 640:1163–1170. https://doi.org/10.1086/500309, arXiv:astro-ph/0512517

Gounelle M, Meibom A, Hennebelle P, Inutsuka Si (2009) Supernova propagation and cloud enrichment: a new model for the origin of ^{60}Fe in the early solar system. Astrophys J 694:L1–L5. https://doi.org/10.1088/0004-637X/694/1/L1

Guandalini R, Busso M (2008) Infrared photometry and evolution of mass-losing AGB stars. II. Luminosity and colors of MS and S stars. Astron Astrophys 488:675–684. https://doi.org/10.1051/0004-6361:200809932, arXiv:0806.4591

Guandalini R, Cristallo S (2013) Luminosities of carbon-rich asymptotic giant branch stars in the Milky Way. Astron Astrophys 555:A120. https://doi.org/10.1051/0004-6361/201321225

Guandalini R, Busso M, Ciprini S, Silvestro G, Persi P (2006) Infrared photometry and evolution of mass-losing AGB stars. I. Carbon stars revisited. Astron Astrophys 445:1069–1080. https://doi.org/10.1051/0004-6361:20053208, arXiv:astro-ph/0509739

Guandalini R, Palmerini S, Busso M, Uttenthaler S (2009) Extra-mixing in luminous cool red giants: hints from evolved stars with and without Li. Publ Astron Soc Aust 26:168–175. https://doi.org/10.1071/AS08063, arXiv:0905.4458

Harper CL, Jacobsen SB (1996) Evidence for ^{182}Hf in the early solar system and constraints on the timescale of terrestrial accretion and core formation. Geochim Cosmochim Acta 60:1131–1153. https://doi.org/10.1016/0016-7037(96)00027-0

Hartmann L, Kenyon SJ (1996) The FU orionis phenomenon. Annu Rev Astron Astrophys 34:207–240. https://doi.org/10.1146/annurev.astro.34.1.207

Hennebelle P, Mac Low MM, Vazquez-Semadeni E (2009) Diffuse interstellar medium and the formation of molecular clouds. Cambridge University Press, Cambridge, p 205

Hester JJ, Desch SJ (2005) Understanding our origins: star formation in HII region environments. In: Krot AN, Scott ERD, Reipurth B (eds) Chondrites and the protoplanetary disk. Astronomical Society of the Pacific conference series, vol 341, p 107

Hidaka H, Yoneda S (2013) Radioactive Cs capture in the early solar system. Sci Rep 3:1330. https://doi.org/10.1038/srep01330

Honda S, Aoki W, Ishimaru Y, Wanajo S, Ryan SG (2006) Neutron-capture elements in the very metal poor star HD 122563. Astrophys J 643:1180–1189. https://doi.org/10.1086/503195, arXiv:astro-ph/0602107

Hsu W, Guan Y, Leshin LA, Ushikubo T, Wasserburg GJ (2006) A late episode of irradiation in the early solar system: evidence from extinct ^{36}Cl and ^{26}Al in meteorites. Astrophys J 640:525–529. https://doi.org/10.1086/500043

Huss GR, Goswami JN, Meyer BS, Sahijpal S, Wasserburg GJ (2007) Stellar sources of short-lived radionuclides. In: Chronology of meteorites and the early solar system, pp 71–72

Huss GR, Meyer BS, Srinivasan G, Goswami JN, Sahijpal S (2009) Stellar sources of the short-lived radionuclides in the early solar system. Geochim Cosmochim Acta 73:4922–4945

Hutcheon ID, Krot AN, Keil K, Phinney DL, Scott ERD (1998) ^{53}Mn-^{53}Cr dating of Fayalite formation in the CV3 chondrite Mokoia: evidence for asteroidal alteration. Science 282:1865. https://doi.org/10.1126/science.282.5395.1865

Jacobsen B, Yin Qz, Moynier F, Amelin Y, Krot AN, Nagashima K, Hutcheon ID, Palme H (2008) ^{26}Al- ^{26}Mg and ^{207}Pb- ^{206}Pb systematics of Allende CAIs: canonical solar initial ^{26}Al/ ^{27}Al ratio reinstated. Earth Planet Sci Lett 272:353–364

Jacobsen B, Matzel JEP, Hutcheon ID, Ramon E, Krot AN, Ishii HA, Nagashima K, Yin Q (2009) The ^{36}Cl-^{36}S systematics of Wadalite from the Allende meteorite. In: Lunar and Planetary Institute Science conference abstracts, vol 40, p 2553

Jacobsen B, Wasserburg GJ, McKeegan KD, Hutcheon ID, Krot AN, Yin QZ, Matzel JE (2013) Resetting and disturbance to the Al-Mg system in Allende type B CAIs. In: Lunar and planetary science conference. Lunar and Planetary Institute Technical Report, vol 44, p 2941

Jeffery PM, Reynolds JH (1961) Origin of excess Xe^{129} in stone meteorites. J Geophys Res 66:3582–3583. https://doi.org/10.1029/JZ066i010p03582

Joggerst CC, Woosley SE, Heger A (2009) Mixing in zero- and solar-metallicity supernovae. Astrophys J 693:1780–1802. https://doi.org/10.1088/0004-637X/693/2/1780, arXiv:0810.5142

Jura M, Xu S, Young ED (2013) ^{26}Al in the early solar system: not so unusual after all. Astrophys J 775:L41. https://doi.org/10.1088/2041-8205/775/2/L41, 1308.6325

Kastner JH, Myers PC (1994) An observational estimate of the probability of encounters between mass-losing evolved stars and molecular clouds. Astrophys J 421:605–615. https://doi.org/10.1086/173676

Kelley WR, Wasserburg GJ (1978) Evidence for the existence of Pd-107 in the early solar system. Geophys Res Lett 5:1079–1082. https://doi.org/10.1029/GL005i012p01079

Kerekgyarto AG, Jeffcoat CR, Lapen TJ, Andreasen R, Righter M, Ross DK, Simon JI (2015) Supra-canonical initial 26Al/27Al from a reprocessed Allende CAI. In: Lunar and planetary science conference, vol 46, pp 2918

Kothes R, Uyaniker B, Pineault S (2001) The supernova remnant G106.3+2.7 and its pulsar-wind nebula: relics of triggered star formation in a complex environment. Astrophys J 560:236–243. https://doi.org/10.1086/322511, arXiv:astro-ph/0106270

Kozlovsky B, Murphy RJ, Ramaty R (2002) Nuclear deexcitation gamma-ray lines from accelerated particle interactions. Astrophys J Suppl 141:523–541. https://doi.org/10.1086/340545

Kratz K, Pfeiffer B, Cowan JJ, Sneden C (2004) r-process chronometers. New Astron Rev 48:105–108. https://doi.org/10.1016/j.newar.2003.11.014

Krot AN, Makide K, Nagashima K, Huss GR, Ogliore RC, Ciesla FJ, Yang L, Hellebrand E, Gaidos E (2012) Heterogeneous distribution of ^{26}Al at the birth of the solar system: evidence from refractory grains and inclusions. Meteorit Planet Sci 47:1948–1979. https://doi.org/10.1111/maps.12008

Kuffmeier M, Frostholm Mogensen T, Haugbølle T, Bizzarro M, Nordlund Å (2016) Tracking the distribution of 26Al and 60Fe during the early phases of star and disk evolution. Astrophys J 826:22. https://doi.org/10.3847/0004-637X/826/1/22, arXiv:1605.05008
Kuroda T, Wanajo S, Nomoto K (2008) The r-process in supersonic neutrino-driven winds: the role of the wind termination shock. Astrophys J 672:1068–1078. https://doi.org/10.1086/523795
Lada CJ, Lada EA (2003) Embedded clusters in molecular clouds. Annu Rev Astron Astrophys 41:57–115. https://doi.org/10.1146/annurev.astro.41.011802.094844, arXiv:astro-ph/0301540
Langer N, El Eid MF, Baraffe I (1989) Blue supergiant supernova progenitors. Astron Astrophys 224:L17–L20
Lee D, Halliday AN (1995) Hafnium-tungsten chronometry and the timing of terrestrial core formation. Nature 378:771–774. https://doi.org/10.1038/378771a0
Lee D, Halliday AN (1996) Hf-W isotopic evidence for rapid accretion and differentiation in the early solar system. Science 274:1876–1879. https://doi.org/10.1126/science.274.5294.1876
Lee T, Papanastassiou DA, Wasserburg GJ (1976) Demonstration of ^{26}Mg excess in Allende and evidence for ^{26}Al. Geophys Res Lett 3:109–109. https://doi.org/10.1029/GL003i002p00109
Lee T, Papanastassiou DA, Wasserburg GJ (1977) Aluminum-26 in the early solar system - fossil or fuel. Astrophys J 211:L107–L110. https://doi.org/10.1086/182351
Lee T, Shu FH, Shang H, Glassgold AE, Rehm KE (1998a) Protostellar cosmic rays and extinct radioactivities in meteorites. Astrophys J 506:898–912. https://doi.org/10.1086/306284
Lee T, Shu FH, Shang H, Glassgold AE, Rehm KE (1998b) Protostellar cosmic rays and extinct radioactivities in meteorites. Astrophys J 506:898–912. https://doi.org/10.1086/306284
Leshin LA, Guan Y, Lin Y (2004) Implications of meteoritic ^{36}Cl abundance for the origin of short-lived radionuclides in the early solar system. In: Krot A, Scott E, Keil K, Reipurth B (eds) Workshop on chondrites and the protoplanetary disk, p 9084
Limongi M, Chieffi A (2003) Evolution, explosion, and nucleosynthesis of core-collapse supernovae. Astrophys J 592:404–433. https://doi.org/10.1086/375703, arXiv:astro-ph/0304185
Limongi M, Chieffi A (2006) The nucleosynthesis of ^{26}Al and ^{60}Fe in solar metallicity stars extending in mass from 11 to 120 M_{solar}: the hydrostatic and explosive contributions. Astrophys J 647:483–500. https://doi.org/10.1086/505164, arXiv:astro-ph/0604297
Lin RP, Krucker S, Hurford GJ, Smith DM, Hudson HS, Holman GD, Schwartz RA, Dennis BR, Share GH, Murphy RJ, Emslie AG, Johns-Krull C, Vilmer N (2003) RHESSI observations of particle acceleration and energy release in an intense solar gamma-ray line flare. Astrophys J 595:L69–L76. https://doi.org/10.1086/378932
Lin Y, Guan Y, Leshin LA, Ouyang Z, Wang D (2004) Evidence for live ^{36}Cl in Ca-Al-rich inclusions from the Ningqiang carbonaceous chondrite. In: Mackwell S, Stansbery E (eds) Lunar and Planetary Institute science conference abstracts, vol 35, p 2084
Liu M, Iizuka Y, McKeegan KD, Tonui EK, Young ED (2005) Supra-canonical ^{26}Al/^{27}Al ratios in an unaltered Allende CAI. In: Mackwell S, Stansbery E (eds) 36th annual lunar and planetary science conference. Lunar and Planetary Institute science conference abstracts, vol 36, p 2079
Lugaro M, Heger A, Osrin D, Goriely S, Zuber K, Karakas AI, Gibson BK, Doherty CL, Lattanzio JC, Ott U (2014) Stellar origin of the ^{182}Hf cosmochronometer and the presolar history of solar system matter. Science 345:650–653. https://doi.org/10.1126/science.1253338
Lugmair GW, Marti K (1977) Sm-Nd-Pu timepieces in the Angra DOS Reis meteorite. Earth Planet Sci Lett 35:273–284. https://doi.org/10.1016/0012-821X(77)90131-5
Lugmair GW, Shukolyukov A (1998) Early solar system timescales according to ^{53}Mn-^{53}Cr systematics. Geochim Cosmochim Acta 62:2863–2886. https://doi.org/10.1016/S0016-7037(98)00189-6
Lugmair GW, MacIsaac C, Shukolyukov A (1992) The ^{53}Mn-^{53}Cr isotope system and early planetary evolution. In: Lunar and planetary institute science conference abstracts, vol 23, p 823
MacPherson GJ, Huss GR, Davis AM (2003) Extinct ^{10}Be in type A calcium-aluminum-rich inclusions from CV chondrites. Geochim Cosmochim Acta 67:3165–3179. https://doi.org/10.1016/S0016-7037(02)01298-X

Maiorca E, Randich S, Busso M, Magrini L, Palmerini S (2011) s-processing in the galactic disk. I. Super-solar abundances of Y, Zr, La, and Ce in young open clusters. Astrophys J 736:120. https://doi.org/10.1088/0004-637X/736/2/120

Maiorca E, Magrini L, Busso M, Randich S, Palmerini S, Trippella O (2012) News on the s process from young open clusters. Astrophys J 747:53. https://doi.org/10.1088/0004-637X/747/1/53

Makide K, Nagashima K, Krot AN, Huss GR, Hutcheon ID, Hellebrand E, Petaev MI (2013) Heterogeneous distribution of ^{26}Al at the birth of the solar system: evidence from corundum-bearing refractory inclusions in carbonaceous chondrites. Geochim Cosmochim Acta 110:190–215. https://doi.org/10.1016/j.gca.2013.01.028

Matzel JEP, Jacobsen B, Hutcheon ID, Krot AN, Nagashima K, Yin Q, Ramon E, Weber PK, Wasserburg GJ (2010) Distribution and origin of ^{36}Cl in Allende CAIs. In: Lunar and Planetary Institute science conference abstracts, vol 41, p 2631

Mauersberger K, Erbacher B, Krankowsky D, Gunther J, Nickel R (1999) Ozone isotope enrichment: isotopomer-specific rate coefficients. Science 283:370. https://doi.org/10.1126/science.283.5400.370

McComas DJ, Bzowski M, Fuselier SA, Frisch PC, Galli A, Izmodenov VV, Katushkina OA, Kubiak MA, Lee MA, Leonard TW, Möbius E, Park J, Schwadron NA, Sokół JM, Swaczyna P, Wood BE, Wurz P (2015) Local interstellar medium: six years of direct sampling by IBEX. Astrophys J Suppl 220:22. https://doi.org/10.1088/0067-0049/220/2/22

McKeegan KD, Davis AM (2005) Early solar system chronology. Elsevier, New York, p 431

McKeegan KD, Chaussidon M, Robert F (2000) Incorporation of short-lived ^{10}Be in a calcium-aluminum-rich inclusion from the Allende meteorite. Science 289:1334–1337. https://doi.org/10.1126/science.289.5483.1334

Meyer BS (2005) Synthesis of short-lived radioactivities in a massive star. In: Krot AN, Scott ERD, Reipurth B (eds) Chondrites and the protoplanetary disk. Astronomical Society of the Pacific conference series, vol 341, p 515

Meyer BS, Clayton DD (2000) Short-lived radioactivities and the birth of the sun. Space Sci Rev 92:133–152. https://doi.org/10.1023/A:1005282825778

Meynet G, Ekström S, Maeder A, Hirschi R, Georgy C, Beffa C (2008) Developments in physics of massive stars. In: Bresolin F, Crowther PA, Puls J (eds) Massive stars as cosmic engines. IAU symposium, vol 250, pp 147–160

Mostefaoui S, Lugmair GW, Hoppe P, El Goresy A (2003) Evidence for live iron-60 in Semarkona and Chervony Kut: a NanoSIMS study. In: Mackwell S, Stansbery E (eds) Lunar and Planetary Institute science conference abstracts, vol 34, p 1585

Mostefaoui S, Lugmair GW, Hoppe P, El Goresy A (2004) Evidence for live ^{60}Fe in meteorites. New Astron Rev 48:155–159. https://doi.org/10.1016/j.newar.2003.11.022

Moynier F, Blichert-Toft J, Wang K, Herzog GF, Albarede F (2011) The elusive ^{60}Fe in the solar nebula. Astrophys J 741:71. https://doi.org/10.1088/0004-637X/741/2/71

Murray N (2011) Star formation efficiencies and lifetimes of giant molecular clouds in the milky way. Astrophys J 729:133. 1007.3270

Murty SVS, Goswami JN, Shukolyukov YA (1997) Excess ^{36}Ar in the Efremovka meteorite: a strong hint for the presence of ^{36}Cl in the early solar system. Astrophys J 475:L65+. https://doi.org/10.1086/310449

Nielsen SG, Rehkämper M, Halliday AN (2006) Large thallium isotopic variations in iron meteorites and evidence for lead-205 in the early solar system. Geochim Cosmochim Acta 70:2643–2657. https://doi.org/10.1016/j.gca.2006.02.012

Nollett KM, Busso M, Wasserburg GJ (2003) Cool bottom processes on the thermally pulsing asymptotic giant branch and the isotopic composition of circumstellar dust grains. Astrophys J 582:1036–1058. https://doi.org/10.1086/344817, arXiv:astro-ph/0211271

Nomoto K, Tominaga N, Umeda H, Kobayashi C, Maeda K (2006) Nucleosynthesis yields of core-collapse supernovae and hypernovae, and galactic chemical evolution. Nucl Phys A 777:424–458. https://doi.org/10.1016/j.nuclphysa.2006.05.008, arXiv:astro-ph/0605725

Nordhaus J, Busso M, Wasserburg GJ, Blackman EG, Palmerini S (2008) Magnetic mixing in red giant and asymptotic giant branch stars. Astrophys J 684:L29–L32. https://doi.org/10.1086/591963, arXiv:0806.3933
Nucci MC, Busso M (2014) Magnetohydrodynamics and deep mixing in evolved stars. I. Two- and three-dimensional analytical models for the asymptotic giant branch. Astrophys J 787:141. https://doi.org/10.1088/0004-637X/787/2/141
Nugis T, Lamers HJGLM (2000) Mass-loss rates of Wolf-Rayet stars as a function of stellar parameters. Astron Astrophys 360:227–244
Ouellette N, Desch SJ, Hester JJ, Leshin LA (2005) A nearby supernova injected short-lived radionuclides into our protoplanetary disk. In: Krot AN, Scott ERD, Reipurth B (eds) Chondrites and the protoplanetary disk. Astronomical Society of the Pacific conference series, vol 341, p 527
Ouellette N, Desch SJ, Hester JJ (2007a) Interaction of supernova ejecta with nearby protoplanetary disks. Astrophys J 662:1268–1281. https://doi.org/10.1086/518102
Ouellette N, Desch SJ, Hester JJ (2007b) Interaction of supernova ejecta with nearby protoplanetary disks. Astrophys J 662:1268–1281. https://doi.org/10.1086/518102, arXiv:0704.1652
Pagel BEJ (1997) Nucleosynthesis and chemical evolution of galaxies. Cambridge University Press, Cambridge
Palla F, Stahler SW (2000) Accelerating star formation in clusters and associations. Astrophys J 540:255–270. https://doi.org/10.1086/309312
Palmerini S, Busso M (2008) ^{26}Al production from magnetically induced extramixing in AGB stars. New Astron Rev 52:412–415. https://doi.org/10.1016/j.newar.2008.05.003, arXiv:0806.2733
Palmerini S, Busso M, Maiorca E, Guandalini R (2009) Nucleosynthesis of light-element isotopes in evolved stars experiencing extended mixing. Publ Astron Soc Aust 26:161–167. https://doi.org/10.1071/AS08040, arXiv:0905.4365
Palmerini S, La Cognata M, Cristallo S, Busso M (2011a) Deep mixing in evolved stars. I. The effect of reaction rate revisions from C to Al. Astrophys J 729:3. https://doi.org/10.1088/0004-637X/729/1/3
Palmerini S, Cristallo S, Busso M, Abia C, Uttenthaler S, Gialanella L, Maiorca E (2011b) Deep mixing in evolved stars. II. Interpreting Li abundances in red giant branch and asymptotic giant branch stars. Astrophys J 741:26. https://doi.org/10.1088/0004-637X/741/1/26
Palmerini S, Trippella O, Busso M (2017) A deep mixing solution to the aluminum and oxygen isotope puzzles in pre-solar grains. Mon Not R Astron Soc 467:1193–1201. https://doi.org/10.1093/mnras/stx137
Palmerini S, Trippella O, Busso M, Vescovi D, Petrelli M, Zucchini A, Frondini F (2018) s-processing from MHD-induced mixing and isotopic abundances in presolar SiC grains. Geochim Cosmochim Acta 221:21–36. https://doi.org/10.1016/j.gca.2017.05.030, arXiv:1711.03039
Pfalzner S (2013) Early evolution of the birth cluster of the solar system. Astron Astrophys 549:A82. https://doi.org/10.1051/0004-6361/201218792
Pfalzner S, Davies MB, Gounelle M, Johansen A, Münker C, Lacerda P, Portegies Zwart S, Testi L, Trieloff M, Veras D (2015) The formation of the solar system. Phys Scr 90(6):068001
Podosek FA, Lewis RS (1972) ^{129}I and ^{244}Pu abundances in white inclusions of the Allende meteorite. Earth Planet Sci Lett 15:101. https://doi.org/10.1016/0012-821X(72)90048-9
Podosek FA, Nichols RH (1997) Short-lived radionuclides in the solar nebula. In: Bernatowicz TJ, Zinner E (eds) American Institute of Physics conference series, vol 402, pp 617–647. https://doi.org/10.1063/1.53321
Preibisch T, Zinnecker H (1999) The history of low-mass star formation in the upper Scorpius OB association. Astron J 117:2381–2397. https://doi.org/10.1086/300842
Ramaty R, Mandzhavidze N (2000) Gamma-rays from solar flares. In: Martens PCH, Tsuruta S, Weber MA (eds) Highly energetic physical processes and mechanisms for emission from astrophysical plasmas. IAU Symposium, vol 195, p 123

Rauscher T, Heger A, Hoffman RD, Woosley SE (2002) Nucleosynthesis in massive stars with improved nuclear and stellar physics. Astrophys J 576:323–348. https://doi.org/10.1086/341728, arXiv:astro-ph/0112478

Reynolds JH (1960) Isotopic composition of primordial xenon. Phys Rev Lett 4:351–354. https://doi.org/10.1103/PhysRevLett.4.351

Rowe MW, Kuroda PK (1965) Fissiogenic xenon from the Pasamonte meteorite. J Geophys Res 70:709–714. https://doi.org/10.1029/JZ070i003p00709

Rutherford E (1929) Origin of actinium and age of the earth. Nature 123:313–314

Sahijpal S, Goswami JN, Davis AM, Lewis RS, Grossman L (1998) A stellar origin for the short-lived nuclides in the early solar system. Nature 391:559. https://doi.org/10.1038/35325

Scheffler H, Elsaesser H (1987) Physics of the galaxy and interstellar matter. Springer, Berlin

Schönbächler M, Carlson RW, Horan MF, Mock TD, Hauri EH (2008) Silver isotope variations in chondrites: volatile depletion and the initial ^{107}Pd abundance of the solar system. Geochim Cosmochim Acta 72:5330–5341. https://doi.org/10.1016/j.gca.2008.07.032

Schramm DN, Wasserburg GJ (1970) Nucleochronologies and the mean age of the elements. Astrophys J 162:57. https://doi.org/10.1086/150634

Schramm DN, Tera F, Wasserburg GJ (1970) The isotopic abundance of ^{26}Mg and limits on ^{26}Al in the early solar system. Earth Planet Sci Lett 10:44–59. https://doi.org/10.1016/0012-821X(70)90063-4

Share GH, Murphy RJ (1995) Gamma-ray measurements of flare-to-flare variations in ambient solar abundances. Astrophys J 452:933. https://doi.org/10.1086/176360

Share GH, Murphy RJ, Ryan J (1997) Solar and stellar gamma ray observations with COMPONE. In: Dermer CD, Strickman MS, Kurfess JD (eds) Proceedings of the fourth Compton symposium. American Institute of Physics conference series, vol 410, pp 17–36. https://doi.org/10.1063/1.54117

Shu FH, Adams FC, Lizano S (1987a) Star formation in molecular clouds - observation and theory. Annu Rev Astron Astrophys 25:23–81. https://doi.org/10.1146/annurev.aa.25.090187.000323

Shu FH, Lizano S, Adams FC (1987b) Star formation in molecular cloud cores. In: Peimbert M, Jugaku J (eds) Star forming regions. IAU symposium, vol 115, pp 417–433

Shu FH, Shang H, Glassgold AE, Lee T (1997) X-rays and fluctuating X-winds from protostars. Science 277:1475–1479. https://doi.org/10.1126/science.277.5331.1475

Shu FH, Shang H, Gounelle M, Glassgold AE, Lee T (2001) The origin of chondrules and refractory inclusions in chondritic meteorites. Astrophys J 548:1029–1050. https://doi.org/10.1086/319018

Shukolyukov A, Lugmair GW (1993a) Fe-60 in eucrites. Earth Planet Sci Lett 119:159–166. https://doi.org/10.1016/0012-821X(93)90013-Y

Shukolyukov A, Lugmair GW (1993b) Live iron-60 in the early solar system. Science 259:1138–1142. https://doi.org/10.1126/science.259.5098.1138

Simon JI, Young ED (2011) Resetting, errorchrons and the meaning of canonical CAI initial ^{26}Al/^{27}Al values. Earth Planet Sci Lett 304:468–482. https://doi.org/10.1016/j.epsl.2011.02.023

Simon JI, Young ED, Russell SS, Tonui EK, Dyl KA, Manning CE (2005) A short timescale for changing oxygen fugacity in the solar nebula revealed by high-resolution ^{26}Al- ^{26}Mg dating of CAI rims. Earth Planet Sci Lett 238:272–283. https://doi.org/10.1016/j.epsl.2005.08.004

Simonucci S, Taioli S, Palmerini S, Busso M (2013) Theoretical estimates of stellar e^- captures. I. The half-life of ^{7}Be in evolved stars. Astrophys J 764:118. https://doi.org/10.1088/0004-637X/764/2/118

Smith N, Brooks KJ (2007) A census of the carina nebula - II. Energy budget and global properties of the nebulosity. Mon Not R Astron Soc 379:1279–1292. https://doi.org/10.1111/j.1365-2966.2007.12021.x, arXiv:0705.3053

Sneden C, Cowan JJ, Gallino R (2008) Neutron-capture elements in the early galaxy. Annu Rev Astron Astrophys 46:241–288. https://doi.org/10.1146/annurev.astro.46.060407.145207

Sossi PA, Moynier F, Chaussidon M, Villeneuve J, Kato C, Gounelle M (2017) Early solar system irradiation quantified by linked vanadium and beryllium isotope variations in meteorites. Nat Astron 1:0055

Srinivasan G, Ulyanov AA, Goswami JN (1994) Ca-41 in the early solar system. Astrophys J 431:L67–L70. https://doi.org/10.1086/187474
Srinivasan G, Sahijpal S, Ulyanov AA, Goswami JN (1996) Ion microprobe studies of Efremovka CAIs: II. Potassium isotope composition and ^{41}Ca in the early solar system. Geochim Cosmochim Acta 60:1823–1835. https://doi.org/10.1016/0016-7037(96)00054-3
Stoerzer D, Pellas P (1977) Angra DOS Reis - Plutonium distribution and cooling history. Earth Planet Sci Lett 35:285–293. https://doi.org/10.1016/0012-821X(77)90132-7
Tachibana S, Huss GR (2003) The initial abundance of ^{60}Fe in the solar system. Astrophys J 588:L41–L44. https://doi.org/10.1086/375362
Tachibana S, Huss GR, Kita NT, Shimoda G, Morishita Y (2006) ^{60}Fe in chondrites: debris from a nearby supernova in the early solar system? Astrophys J 639:L87–L90. https://doi.org/10.1086/503201
Takigawa A, Miki J, Tachibana S, Huss GR, Tominaga N, Umeda H, Nomoto K (2008) Injection of short-lived radionuclides into the early solar system from a faint supernova with mixing fallback. Astrophys J 688:1382–1387. https://doi.org/10.1086/592184, arXiv:0808.1441
Tang H, Dauphas N (2012) Abundance, distribution, and origin of ^{60}Fe in the solar protoplanetary disk. Earth Planet Sci Lett 359:248–263. https://doi.org/10.1016/j.epsl.2012.10.011
Tatischeff V, Thibaud J (2008) Li production by stellar flares in young open clusters. New Astron Rev 52:423–426. https://doi.org/10.1016/j.newar.2008.06.013
Tatischeff V, Kozlovsky B, Kiener J, Murphy RJ (2006) Delayed X- and gamma-ray line emission from solar flare radioactivity. Astrophys J Suppl 165:606–617. https://doi.org/10.1086/505112, arXiv:astro-ph/0604325
Tatischeff V, Kozlovsky B, Kiener J, Murphy RJ (2007) Radioactive line emission from solar flares. In: ESA special publication, vol 622, p 135
Tatischeff V, Thibaud J, Ribas I (2008) Nucleosynthesis in stellar flares. ArXiv e-prints 0801.1777
Tatischeff V, Duprat J, de Séréville N (2010) A runaway Wolf-Rayet star as the origin of ^{26}Al in the early solar system. Astrophys J 714:L26–L30. https://doi.org/10.1088/2041-8205/714/1/L26
Thiemens MH, Heidenreich JE III (1983) The mass-independent fractionation of oxygen - a novel isotope effect and its possible cosmochemical implications. Science 219:1073–1075. https://doi.org/10.1126/science.219.4588.1073
Trigo-Rodríguez JM, García-Hernández DA, Lugaro M, Karakas AI, van Raai M, García Lario P, Manchado A (2009) The role of massive AGB stars in the early solar system composition. Meteorit Planet Sci 44:627–641. arXiv:0812.4358
Trippella O, Busso M, Palmerini S, Maiorca E, Nucci MC (2016) s-processing in AGB stars revisited. II. Enhanced 13C production through MHD-induced mixing. Astrophys J 818:125. https://doi.org/10.3847/0004-637X/818/2/125
Tur C, Heger A, Austin SM (2009) Production of 26Al, 44Ti, and 60Fe in core-collapse supernovae: sensitivity to the rates of the triple alpha and 12C(a,g)16O reactions. ArXiv e-prints 0908.4283
Urey HC (1955) The cosmic abundances of potassium, uranium, and thorium and the heat balances of the earth, the moon, and mars. Proc Natl Acad Sci 41:127–144. https://doi.org/10.1073/pnas.41.3.127
Urey HC, Donn B (1956) Chemical heating for meteorites. Astrophys J 124:307. https://doi.org/10.1086/146223
Vescovi D, Busso M, Palmerini S, Trippella S O anc Cristallo, Piersanti L, Chieffi A, Limongi M, Hoppe P, Kratz KL (2018, in press) On the origin of the early solar system radioactivities. Astrophys J
Wadhwa M, Amelin Y, Davis AM, Lugmair GW, Meyer B, Gounelle M, Desch SJ (2007) From dust to planetesimals: implications for the solar protoplanetary disk from short-lived radionuclides. In: Protostars and planets V, pp 835–848
Wasserburg GJ, Fowler WA, Hoyle F (1960) Duration of nucleosynthesis. Phys Rev Lett 4:112–114. https://doi.org/10.1103/PhysRevLett.4.112
Wasserburg GJ, Huneke JC, Burnett DS (1969) Correlations between fission tracks and fission type xenon in meteoritic whitlockite. J Geophys Res 74:4221–4232. https://doi.org/10.1029/JB074i017p04221

Wasserburg GJ, Busso M, Gallino R, Raiteri CM (1994) Asymptotic giant branch stars as a source of short-lived radioactive nuclei in the solar nebula. Astrophys J 424:412–428. https://doi.org/10.1086/173899

Wasserburg GJ, Gallino R, Busso M, Goswami JN, Raiteri CM (1995) Injection of freshly synthesized Ca-41 in the early solar nebula by an asymptotic giant branch star. Astrophys J 440:L101–L104. https://doi.org/10.1086/187771

Wasserburg GJ, Busso M, Gallino R (1996) Abundances of actinides and short-lived nonactinides in the interstellar medium: diverse supernova sources for the r-processes. Astrophys J 466:L109+. https://doi.org/10.1086/310177

Wasserburg GJ, Gallino R, Busso M (1998) A test of the supernova trigger hypothesis with 60Fe and 26Al. Astrophys J 500:L189+. https://doi.org/10.1086/311414

Wasserburg GJ, Busso M, Gallino R, Nollett KM (2006) Short-lived nuclei in the early solar system: possible AGB sources. Nucl Phys A 777:5–69. https://doi.org/10.1016/j.nuclphysa.2005.07.015, arXiv:astro-ph/0602551

Wasserburg GJ, Trippella O, Busso M (2015) Isotope anomalies in the Fe-group elements in meteorites and connections to nucleosynthesis in AGB stars. Astrophys J 805:7. https://doi.org/10.1088/0004-637X/805/1/7, arXiv:1503.02256

Wasserburg GJ, Karakas AI, Lugaro M (2017) Intermediate-mass asymptotic giant branch stars and sources of ^{26}Al, ^{60}Fe, ^{107}Pd, and ^{182}Hf in the solar system. Astrophys J 836:126. https://doi.org/10.3847/1538-4357/836/1/126

Wehmeyer B, Pignatari M, Thielemann FK (2017) Inhomogeneous chemical evolution of r-process elements in the galactic halo. In: Kubono S, Kajino T, Nishimura S, Isobe T, Nagataki S, Shima T, Takeda Y (eds) 14th international symposium on nuclei in the cosmos (NIC2016), p 020201. https://doi.org/10.7566/JPSCP.14.020201

Williams JP, Gaidos E (2007a) On the likelihood of supernova enrichment of protoplanetary disks. Astrophys J 663:L33–L36. https://doi.org/10.1086/519972

Williams JP, Gaidos E (2007b) On the likelihood of supernova enrichment of protoplanetary disks. Astrophys J 663:L33–L36. https://doi.org/10.1086/519972, arXiv:0705.3459

Woosley SE, Heger A (2007) Nucleosynthesis and remnants in massive stars of solar metallicity. Phys Rep 442:269–283. https://doi.org/10.1016/j.physrep.2007.02.009, arXiv:astro-ph/0702176

Woosley SE, Weaver TA (1995) The evolution and explosion of massive stars. II. Explosive hydrodynamics and nucleosynthesis. Astrophys J Suppl 101:181. https://doi.org/10.1086/192237

Young ED (2014) Inheritance of solar short- and long-lived radionuclides from molecular clouds and the unexceptional nature of the solar system. Earth Planet Sci Lett 392:16–27. https://doi.org/10.1016/j.epsl.2014.02.014

Young ED (2016) Bayes theorem and early solar short-lived radionuclides: the case for an unexceptional origin for the solar system. Astrophys J 826:129

Young ED, Simon JI, Galy A, Russell SS, Tonui E, Lovera O (2005) Supra-canonical ^{26}Al/^{27}Al and the residence time of CAIs in the solar protoplanetary disk. Science 308:223–227. https://doi.org/10.1126/science.1108140

Young ED, Gounelle M, Smith RL, Morris MR, Pontoppidan KM (2009) The oxygen isotopic composition of the solar system in a galactic context: new results for CO in young stellar objects and implications for the birth environment of the solar system. In: Lunar and Planetary Institute science conference abstracts, vol 40, p 1967

Zinnecker H (2002) From local star formation to global star formation. Astrophys Space Sci 281:147–157. https://doi.org/10.1023/A:1019532503571

Zinner E, Göpel C (2002) Aluminum-26 in H4 chondrites: implications for its production and its usefulness as a fine-scale chronometer for early solar system events. Meteorit Planet Sci 37:1001–1013

Zuckerman B, Song I (2004) Young stars near the sun. Annu Rev Astron Astrophys 42:685–721. https://doi.org/10.1146/annurev.astro.42.053102.134111

Chapter 7
Distributed Radioactivities

Roland Diehl, Dieter H. Hartmann, and Nikos Prantzos

Radioactive nuclei freshly produced in compact sources of cosmic nucleosynthesis are ejected into the source surroundings. Then, their decay may be directly observable through characteristic gamma-ray lines, or they may be transported directly to a terrestrial detector. Also, condensation into dust and solids may occur, and then freeze the composition of gas at the dust formation site, which can be inferred from such stardust as found in meteorites. The intrinsic clocks provided by the radioactive decay of each isotope then provide a tool to investigate the characteristics of transport away from the sources, and through interstellar medium, and in the solidification and planet formation phase of our solar system around its earliest history. In this Chapter we discuss what radioactive isotopes such as ^{26}Al, ^{60}Fe, ^{129}I, ^{59}Ni, but also very heavy element isotopes such as ^{244}Pu, or positrons from β^+ decay can teach us about sources and transport of radioactive nuclei.

7.1 Radioactivities in the Interstellar Medium

Radioactive isotopes are ejected into the surroundings of their sources, and become observable through their gamma-ray line emission once having left dense production sites where not even gamma-rays may escape. Thus, radioactivity sources of

R. Diehl (✉)
Max Planck Institut für extraterrestrische Physik, Garching, Germany
e-mail: rod@mpe.mpg.de

D. H. Hartmann
Clemson University, Clemson, SC, USA
e-mail: hdieter@g.clemson.edu

N. Prantzos
Institut d'Astrophysique, Paris, France
e-mail: prantzos@iap.fr

R. Diehl et al. (eds.), *Astrophysics with Radioactive Isotopes*, Astrophysics and Space Science Library 453, https://doi.org/10.1007/978-3-319-91929-4_7

gamma-ray lines are *diffuse* by this very nature. A second path for us as observers arises when dusts and solids form in interstellar space, and include radioactive nuclei: The composition of such solids as frozen in at their formation is altered from radioactive decay, and recognition of daughter isotopes thus tells us about the radioactivity in the surroundings of the dust/solid formation site.

In this Chapter, we discuss such radioactivities as seen or inferred in interstellar space, tracing them back to their sources, and extracting what they may tell us about the links between sources and interstellar medium. Complementing the discussions of isotopes which originate from single sources (the short-lived isotopes) in Chaps. 3–5, here we address the ejections of sources as accumulated in interstellar space. The radioactive lifetime of each of these isotopes is related to its specific role in astrophysical research (see Table 1.1 in the Introduction of this book).

When the lifetime of a radioactive nucleus is not very large compared to the timescale of its propagation away from its source into interstellar space, we may not be able to distinguish between a point source and a diffuse source. This depends also on the resolving power of the observing instrument, i.e. diffuse emission is seen if typical distances—projected on the sky—between point sources are larger than the instrument's resolution. Our definition of *diffuse* here attempts to focus on the physical connection between the source and what we observe: When the observed source is identical to the object of our study, we call this a *point source*. Examples are novae, supernovae, and type-I X-ray bursts, and are addressed in Chaps. 4 and 5. When the observed source includes a transport process between the production site of radioactivity and its observational manifestation, we call this a *diffuse source or emission*.

Key isotopes and their specific astrophysical implications are summarised here first:

- ^{56}Ni and ^{44}Ti isotopes decay as the supernova that produced them explodes and expands into a supernova remnant. The emerging gamma rays from the ^{56}Ni decay chain, with radioactive lifetimes of 8.8 days (^{56}Co* lines) and 111 days (^{56}Fe* lines), primarily tell us about how the supernova unfolds and dilutes after explosion. This information is convolved with the known radioactive clock of the decays. A supernova becomes transparent to gamma rays within a few months. Late emission from ^{56}Co and ^{44}Ti decay, therefore, tells us about the total amount of radioactivity produced in the source, and about ejecta kinematics and spatial distribution. Comparing to other radiation measured from the remnant, the amount of radioactive energy deposit and the remnant leakage of positrons from β^+ decay can be inferred. All of these are properties related more to the source rather than to circum- or interstellar medium. Therefore, the sources of these isotopes and their science are not considered 'diffuse', and are therefore discussed in preceding Chaps. 4 and 5. Also, short-lived actinides and lanthanides, which are inferred to power light from neutron star collisions in the form of macro-novae (also called kilonovae; see Chap. 4), belong into this characteristic class of radioactivities that radiate from an expanding explosive nucleosynthesis source. They are mentioned here to illustrate the key aspect of astronomy with radioactive nuclei: an emission with known characteristics and energy output, leaving behind material that was not present originally.

- ^{26}Al and ^{60}Fe have radioactive decay times of 1.04 and 3.8 My, respectively. Therefore, they may safely assumed to have left their production sites. The dispersal of nucleosynthesis ejecta from a single source can be traced over a time scale of millions of years, from observing their decay or daughter isotopes. This is approximately the time scale expected to characterise the return of stellar gas through an explosive event into the ambient interstellar gas in the source surroundings. One of the questions addressed would be how far out in space away from the source ejecta may travel within a few My. Another question would be if the dispersal shows any preferred directions, such as possibly escape flows from the galactic-plane (*chimneys*). Yet another aspect is the state of circum-source, hot interstellar medium with cavities of age 1 My or younger, yet older than the 0.1 My scale which can be addressed through observations of gas in supernova remnants by observations of atomic transitions in partly-ionized plasma. Measuring the width of radioactive-decay gamma-ray lines encodes turbulence and bulk flow through Doppler broadening and shifts of the gamma-ray lines, respectively. These long-lived isotopes ^{26}Al and ^{60}Fe are neither part of the *light-element* family (originating from primordial or cosmic-ray spallation nucleosynthesis) nor of the *heavy-element* family (originating from probably explosive sites with extreme densities of matter, energy, and neutrons). Rather, ^{26}Al and ^{60}Fe most likely originate from (normal and frequent/omnipresent) stellar sources, hence from nuclear reactions inside stars with contributions from their terminal explosions. But the explosions mainly help to eject stellar matter into interstellar space, while their contribution to production probably is sub-dominant (see Chaps. 3 and 4). Thirty years of gamma-ray and meteoritic measurements carried the celestial signal from ^{26}Al decay through a history from *establishing proof of past and current nucleosynthesis* in the Galaxy to *identifying ^{26}Al origin* predominantly from massive stars into an *astronomical tool* for the study of massive-star/ISM interactions in nearby stellar groups.
- Isotopes such as ^{129}I, ^{146}Sm, ^{182}Hf, and ^{244}Pu are parent radioactivities inferred to have existed once in the early solid material that formed the Earth. They are rather long-lived with decay lifetimes of 23–115 Myrs, but *extinct* and only evident through their characteristic decay products. The isotopic composition of solids of different formation times thus reveal characteristic anomalies, found through precision mass spectroscopy in terrestrial laboratories. The inferred existence of those radioactivities thus tell us about pre-solar nucleosynthesis in the surroundings accessible to interstellar transport during the radioactive lifetime of each isotope. Early-solar system astrophysics involving homogeneity and effectiveness of mixing before first meteorites were formed can be addressed by 'shortlived' (0.1-few My) radioactivities, including the above-mentioned ^{26}Al and ^{60}Fe, but also ^{10}Be, ^{36}Cl, ^{41}Ca, and ^{53}Mn. Probes on a somewhat longer time scale are ^{129}I, ^{146}Sm, ^{182}Hf, and ^{244}Pu. This application of radioactivity focuses on the early solar system, hence is dealt with in detail in the book's Chap. 6 dedicated to the Solar System.

- The even more long-lived (>100 My) radioactivities of ^{187}Re, ^{235}U, ^{238}U, ^{232}Th connect us to nucleosynthesis at cosmic time scales. They provide age dating nucleosynthesis since the formation of the Galaxy, and also dating of the age of the solar system. Therefore you find more details in book Chaps. 2 and 6.
- ^{59}Ni is an example of a useful subclass of radioactive isotopes, which can decay only through electron capture. Therefore, its decay implies an environment of incomplete if any ionization, i.e., an environmental temperature below 10^5 K. Cosmic ray nuclei are characterised by the other extreme, they are relativistic and always fully ionized. Therefore, the observed abundance of ^{59}Ni in cosmic ray nuclei directly encodes the delay time between nucleosynthesis of this ^{59}Ni ($\tau_{decay} \sim 10^5$ years) and its acceleration to cosmic-ray energies (and hence full ionization) (see Sect. 7.4). ^{57}Co and ^{56}Ni are more-shortlived electron-capture only radio-isotopes, similarly useful to constrain the times between their ionization and recombination. Here, the property of radioactivity (the clock) combines with (otherwise usefully-unimportant, but here diagnostic impact) of the atomic shell. More sophisticated analysis is required to understand the cosmic-ray abundances of unstable isotopes which include electron capture as a main decay mechanism, such as ^{7}Be, ^{49}V, ^{51}Cr, and ^{55}Fe: Their abundances result from how much of their trajectory had involved low energies with chances to attach electrons. Hence they address the topic of cosmic-ray re-acceleration and/or successive cosmic-ray acceleration by different shock regions along their journey.
- In direct detections of cosmic rays also radioactive species and their abundances can be measured. The observed abundance, e.g. as seen for ^{60}Fe, can be combined with known production cross sections for spallation reactions to determine the distance to the source, that is, the effectively-traversed path length of cosmic rays producing such abundance from collisions with ambient interstellar gas. Cosmic-ray isotopes ^{10}Be, ^{26}Al, ^{36}Cl, and ^{54}Mn have thus been used to constrain the time spent in the Galaxy to $\sim 10^7$ years, while the more-shortlived ^{60}Fe was attributed to supernova explosions rather nearby and recent (see Sect. 7.4).
- Positrons produced in various radioactive decays add interesting aspects. Their propagation properties as charged particles are reminiscent of cosmic rays: they will be directed by the morphology of the magnetic field, and will lose energy in collisions with ambient gas. Their annihilation provides a characteristic gamma-ray signature, thus making positrons a rather special astronomical messenger. Annihilation is favoured in a rather narrow energy range between $\sim$7 and 200 eV (corresponding to temperatures in the 1000 K regime). This makes positron annihilation gamma rays special probes of this phase of the interstellar medium. On the other hand, positrons from radioactive decays may deposit radioactive energy efficiently in expanding envelopes from stellar explosions, and their abundance in interstellar space must reflect a significant contribution from nucleosynthesis from a variety of sources, and throughout the Galaxy. This yields information which complements other observables of the same radioactive decays (e.g. bolometric broad-band emission, or ^{26}Al-decay gamma-rays). Even

though there are other candidate sources of positrons, the research with positrons and their annihilation is included in this Chapter (see Sect. 7.5).

In the following, we discuss candidate sources with their astrophysical and nuclear-reaction aspects, then their observations are presented, and we close the treatment with a summary of the lessons learned and the remaining questions.

7.2 ^{26}Al

7.2.1 *Nuclear Reactions, Candidate Sources, and Observability*

^{26}Al is produced by proton capture on ^{25}Mg, the ^{25}Mg (p,γ) ^{26}Al reaction that is characteristic for hydrogen burning (Iliadis et al. 2011). Pre-existing ^{25}Mg is processed in this way, although more-abundant ^{24}Mg could also be transferred into ^{25}Mg by proton capture and successive β decay. The amount of ^{26}Al thus depends on pre-existing Mg, and scales with the metallicity of the star as it was formed. In principle, such hydrogen-burning environments may further process ^{26}Al by proton capture into ^{27}Si; but the rising Coulomb repulsion makes this destruction channel rather negligible (see experimental constraints and discussion by Parikh et al. 2014), compared to the neutron capture reactions that may destroy ^{26}Al through ^{26}Al(n, p)^{26}Mg and ^{26}Al(n, α)^{23}Na. The ^{27}Al$(p,\alpha)^{24}$Na reaction will close a nuclear-reaction cycle against further nuclear reaction flow towards heavier isotopes. This avoids loss of nucleons away from locking them into the cycle around ^{26}Al and its neighbouring isotopes, as long as the leakage reaction from that cycle ^{27}Al$(p,\gamma)^{28}$Si proceeds at a low rate. Within this reaction cycle, shown in Fig. 7.1, an equilibrium abundance of ^{26}Al would be established, balancing production and destruction reaction flows. Due to its relative stability (lifetime against β decay is 1.04×10^6 years), the ^{26}Al abundance in such an equilibrium is rather high and comparable to the Mg isotopes. In explosive environments, only the charged-particle and photon reactions are dominant, as the dynamics of the explosion disfavours the slower β decay reactions.

In sufficiently-hot environments, ^{26}Al may be thermally excited to its higher levels at excitation energies of 228, 417, and 1058 keV with spin states 0+, 3+, and 1+, respectively. Although only the high-energy tail of the thermal distribution will make such excitations possible, the daughter spin states of ^{26}Mg (2+ for the excited state at 1809 keV, 0+ for the ground state) allow much higher transition rates than the highly-forbidden transition from ^{26}Al's ground state with spin 5+ to any of the two Mg states. Therefore, in hot environments, the excited states of ^{26}Al must be included in a nuclear reaction network, and lead to a much larger destruction of ground state ^{26}Al; ^{26m}Al decays in just 0.109 s. Thermal excitation with subsequent rapid decay of ^{26}Al^m through this energy level of ^{26}Al at 228 keV excitation energy thus becomes relevant at T $\geq$ 0.1 10^9 K, adding destruction channels to the otherwise dominant neutron capture reactions. For ^{26}Al, a lifetime

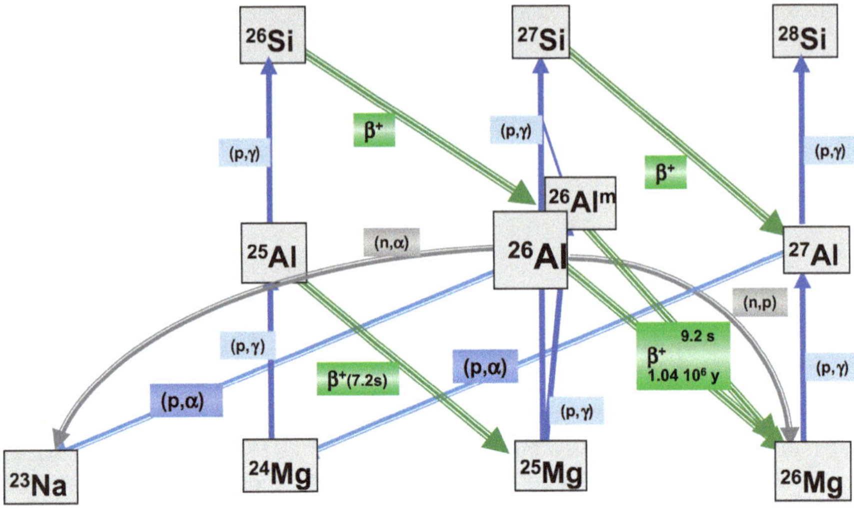

Fig. 7.1 The Na-Al-Mg reaction cycle, characterised by proton capture reactions and β decays, mainly, and closed by a (p,α) reaction (after Prantzos and Diehl (1996), Fig. 3.1)

of 1.042×10^6 years (or, equivalently, a half-life of 722,000 years) is obtained from averaging over four independent measurements; the value recommended in the *Table of Isotopes* is (1.034 ± 0.046) My, Endt 1990).

In principle, all hydrogen-burning environments are expected to process pre-existing ^{25}Mg into ^{26}Al. Depending on these other reactions occurring in those environments, the yield of a particular source can vary substantially. Mixing processes among regions containing Mg seeds and hydrogen, as well as rapid convective transport away from the hot zones and thus avoiding destruction reactions, all have important impact on the final ^{26}Al yields (see Chap. 3). Here, 'yield' defines the amount of ^{26}Al that emerges from the source. Therefore, core hydrogen burning in stars less massive than several $M_\odot$ is unimportant, as any ^{26}Al produced during main sequence hydrogen burning will decay before any interior material could be ejected in a giant phase (see Chap. 3). But for stars more massive than about 3 $M_\odot$, shell hydrogen burning further out becomes interesting. Modeling convective overturn may implement the *Schwarzschild criterion* where buoyancy of a homogeneous gas mixture is compared to restoring forces from pressure gradients (gravity and thermal), or the *Ledoux criterion* where the net effect of all restoring forces is evaluated, including chemical-diffusion gradients. More complexity may be added by chemical-diffusion gradient, which leads to *thermohaline mixing* or *semiconvection*, depending on the direction of dominance of between chemical-diffusion gradient and thermal gradient forces. Stellar rotation brings in exchanges between local and global angular momenta inside a star, adding other origins of mixing processes. This has been shown to be an important ingredient, significantly enhancing mixing and transport of fuel and ashes into and away from the nuclear reaction regions within a star. As a result, typically the inclusion of stellar rotation enhances the amounts of ejected ^{26}Al (Palacios et al. 2005). Also in the Ne-O layers, where ^{25}Mg nuclei are abundant from Ne-burning reactions, and protons may be produced by (X, p) reactions, ^{26}Al may be produced through hydrogen burning. Stellar structure instabilities are characteristic of those more-massive stars in their evolution along the asymptotic giant branch (AGB), as Luminous Blue Variables (LBV), and more-intense mass ejections during those evolutionary phases help to make them candidate sources of interstellar ^{26}Al. An ambient-medium abundance of ^{25}Mg of log(Mg) = 6.6 (where log(H) = 12) (Asplund et al. 2009, solar abundance assumed for the current Galaxy) would thus be available for ^{26}Al synthesis in more-massive stars; for hotter environments, more abundant ^{24}Mg (abundance of log(Mg) = 7.49) may provide ^{26}Al fuel as well, as it is processed into ^{25}Mg through p capture reaction and intermediate β-decay.

Also in stellar explosions, novae (Chap. 5) and core-collapse supernovae (Chap. 4), explosive nucleosynthesis of ^{26}Al may occur, as stellar plasma is heated by the explosion shock wave to temperatures where protons are released and may undergo nuclear reactions on ambient ^{25}Mg nuclei. The intense flux of neutrinos from the collapsed stellar core may produce additional ^{26}Al depending on the poorly known average neutrino energy (Woosley and Weaver 1995)—a process that has not been included in many ^{26}Al source models yet. These neutrino reactions on infalling material in the region close to the proto-neutron star result in a *neutrino process* , releasing nucleons (protons, neutrons, α particles) from heavier nuclei, and thus enabling ^{26}Al synthesis on Mg again. It has been estimated that this could enhance ^{26}Al yields by 20–30% (Sieverding et al. 2017).

In interstellar space, nuclear reactions occur from cosmic ray collisions with ambient gas. Cosmic rays predominantly have a nucleonic component, with protons and nuclei as heavy as iron, while electrons are a small contribution at the percent level. If energetic nuclei collide with interstellar gas, they will be broken up and fragment into lighter nuclei such as ^{26}Al. Similarly, heavy nuclei in interstellar gas will be broken up by collisions with any component of energetic cosmic rays. Such spallation reactions therefore may produce ^{26}Al—either from heavy-nuclei cosmic rays being broken up or from ambient-gas nuclei being broken up to produce ^{26}Al with its 13 nucleons only. The cross sections for such reactions are rather well known, and thus, yields have been estimated for plausible cosmic-ray intensities and interstellar medium parameters (Ramaty et al. 1979). From this, the cosmic-ray produced ^{26}Al appears to be several orders of magnitude below detectability by gamma ray telescopes.

The amount of ^{26}Al ejected into the interstellar medium depends on many ingredients of the process from its production until release from the star. Typical values are 10^{-5} to 10^{-4} $M_\odot$ (see Fig. 4.20). Several factors within stellar models play a significant role:

Theoretical predictions ^{26}Al yields from models of massive stars and their supernovae have been provided by Woosley and Weaver (1995), Timmes et al. (1995), Thielemann et al. (1996) and Limongi and Chieffi (2006b, the 'first-generation' models), the Wolf-Rayet contributions which crucially depend on mass loss properties were discussed in specific detail by Palacios et al. (2005), later the aspects of stellar rotation and lessons from 3D considerations were included (Meynet et al. 2006; Ekström et al. 2012; Chieffi and Limongi 2013). Extended sets of model yields for stars of masses up to 120 $M_\odot$ are available from Ekström et al. (2012), Chieffi and Limongi (2013), Pignatari et al. (2016) and Sukhbold et al. (2016) (see also Chap. 4). For the less massive AGB stars, yields from Meynet et al. (1997, first generation) with a recent detailed treatment by Karakas and Lattanzio (2014) (see also Chap. 3). Nova contributions to ^{26}Al are addressed by Jose and Hernanz (1998) and Gehrz et al. (1998) (see also Chap. 5).

It turned out that, from all these candidate sources, massive stars such as Wolf-Rayet and possibly the high-mass fraction of AGB stars, and the core-collapse supernovae that mark the end points of their evolution, very likely dominate above the other candidate sources in producing interstellar ^{26}Al (see the detailed discussion of observational results below, and considerations on astrophysical source models as presented in Chaps. 3–5).

A supernova explosion will obviously disperse ejecta in the interstellar medium, including ^{26}Al as the synthesis occurs in outer regions (the O/Ne shell), far above the inner supernova core where some material may fall back onto the remnant star or black hole. ^{26}Al produced hydrostatically in stars is ejected only in two possible ways: Either the supernova explosion occurs sufficiently soon after its production so that it still has not decayed, or else convective transport within the star will bring it up to the surface and a sufficiently-strong wind will carry it away from the star. Two types of stars fulfill the latter conditions: In asymptotic-giant branch stars (AGB), ashes from hydrogen-shell burning are brought to the surface, the convective shell may be a *hot-bottom*-producer of ^{26}Al (see Chap. 3). Alternatively, core hydrogen-burning ashes from the Helium shell are revealed in the strong winds of the massive-stars' Wolf-Rayet phase, in a later stage of this phase after the hydrogen envelope has been stripped (see Chap. 4).

Once transported away from the site of its production and of any further nuclear reactions, ^{26}Al will β-decay with its 1.04 Myr-lifetime (see Fig. 1.3 in the Introduction of this book). ^{26}Al decay produces a characteristic gamma-ray line signature. From above-mentioned spin states and selection rules, decay is inhibited, leading to its relative stability with τ 1.04 $\times$ 10^6 years. In 99.76% of all decays, a gamma-ray photon of energy 1808.63 keV ($\pm$0.07 keV uncertainty) is emitted. 82.1% of decays are β^+-decays which also emit a positron, the remainder of decays occur through capture of an electron from the atomic shell. Photon yields at 1130 (2.4% of decays) and 2938 keV (0.24 of decays) are weak by comparison with the 1808.63 keV line, and the photon yield from positron annihilation ($\simeq$1.6 photons of 511 keV, depending on annihilation conditions) is more indirect, as the positron's lifetime depends on gas and magnetic-field conditions and can reach the order of 10^5 years (Jean et al. 2009).

7.2.2 Observations of ^{26}Al

The first report of a gamma-ray line from a radioactive nucleus of cosmic origin arose from the HEAO-C instrument's 1979/1980 measurements, from a satellite mission carrying a Ge detector which located a line at 1809 keV and plausibly attributed it to decay of live ^{26}Al in the Galaxy's interstellar medium (Mahoney et al. 1984). ^{26}Al radioactive decay within 1 My requires a source within the past million years, which is a rather brief recent time span in the Galaxy's history which spans 12 Gy or more. Therefore this is proof of currently-ongoing nucleosynthesis.

Independently, meteoritic studies had shown (Schramm et al. 1970; Lee et al. 1976; Wasserburg et al. 1977) an excess of the abundance of ^{26}Mg in Al-rich inclusions, which pointed to an earlier existence of the ^{26}Al parent isotope in the aluminium minerals at the time this material had condensed (see discussion in MacPherson et al. 1995, for consolidated meteoritic results). It turned out that ^{26}Al is among the most informative nucleocosmochonometers to study the formation of first solids in the young, pre-solar, nebula (Srinivasan and Chaussidon 2013, see also Chapter 6). The origin of ^{26}Al in the early solar nebula remained unclear, and a subject of (see Chap. 6). With improvements of methods to study also different components within meteorites, in particular refractory SiC grains, ^{26}Al traces were also found in inclusions in meteorites which are clearly not of solar-system origin, and called 'stardust' (Nittler et al. 2008). Their origins could be related to an origin in AGB stars, from isotope ratios in C, N, Si, and Al (see Chaps. 3 and 10).

The COMPTEL sky survey, accumulated over 9 years, then provided a sky image in the ^{26}Al gamma-ray line. This reflects a map of nucleosynthesis activity in our current Galaxy.[1] The resulting ^{26}Al map showed structured ^{26}Al emission, extended along the plane of the Galaxy (Plüschke et al. 2001a; Knödlseder et al. 1999; Oberlack et al. 1996; Diehl et al. 1995) (see Fig. 7.2), in broad agreement with earlier expectations of ^{26}Al being produced throughout the Galaxy and mostly from massive stars and their supernovae (Prantzos 1993; Lingenfelter and Ramaty 1978).

COMPTEL's detectors lacked the spectral resolution required for line identification and spectroscopic studies, with $\simeq$200 keV instrumental resolution, compared to $\simeq$3 keV for Ge detectors, at the energy of the ^{26}Al line. A 1995 balloon experiment also carrying high-resolution Ge detectors had provided an indication that the ^{26}Al line was significantly broadened to 6.4 keV (Naya et al. 1996). This implied kinematic Doppler broadening of astrophysical origin of 540 km s^{-1}. Considering the 1.04 × 10^6 years decay time of ^{26}Al, such a large line width would naively translate into kpc-sized cavities around ^{26}Al sources or major fractions of ^{26}Al to be condensed on grains (Chen et al. 1997; Sturner and Naya 1999). The INTEGRAL space observatory was launched in 2002. With its Ge-detector based spectrometer SPI, it provided high-quality spectroscopic results (Fig. 7.3), adding a wealth of

[1]Exposure time is million years, while the Galaxy's age is 10–13 Gyrs.

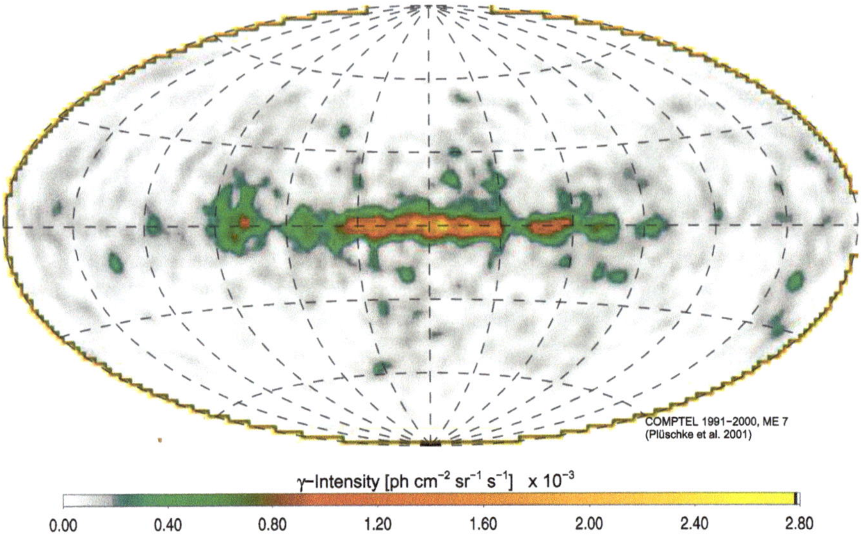

Fig. 7.2 The ^{26}Al sky as seen with the COMPTEL telescope. This image was obtained from measurements taken 1991–2000, and using a maximum-entropy regularization together with likelihood to iteratively fit a best image to the measured photons (from Plüschke et al. 2001)

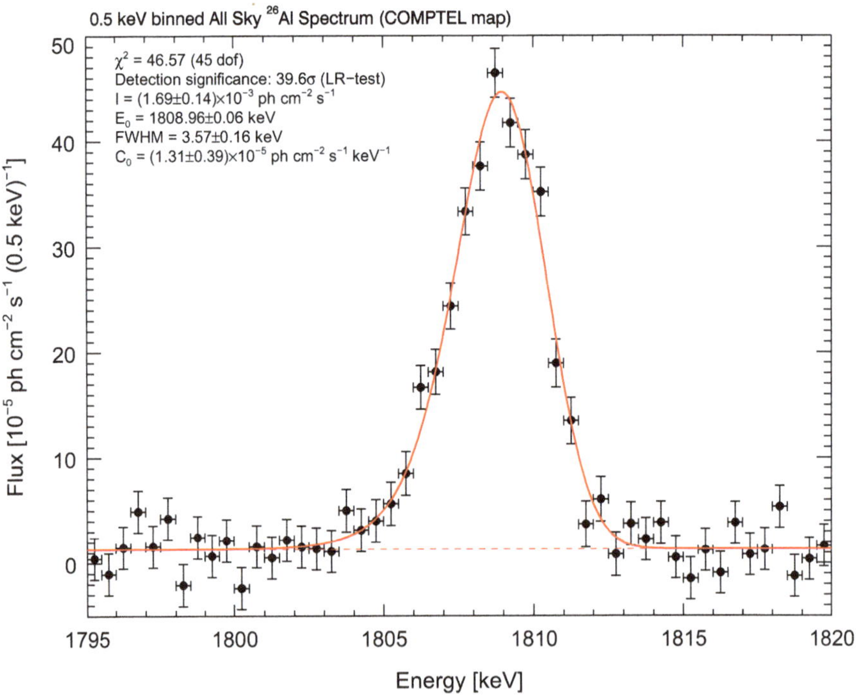

Fig. 7.3 The ^{26}Al line as seen with INTEGRAL's high-resolution spectrometer SPI and 13 years of measurements integrated (Siegert 2017)

detail from data accumulated over its more than 15-year long mission. This allowed not only higher precision study of the Galaxy-wide ^{26}Al aspects (Diehl et al. 2006a,b), but also the detailed test of our understanding of massive-star activity in specific and well-constrained massive star groups such as in Orion or Scorpius-Centaurus. Kinematic constraints from the ^{26}Al line width and centroid (Kretschmer et al. 2013), and multi-messenger studies using the stellar census and information on atomic and molecular has from radio data as well as hot plasma from X ray emission (Voss et al. 2009, 2010; Krause et al. 2013, 2014), all have led to a kind of ^{26}Al astronomy as part of the studies of stellar feedback and massive-star nucleosynthesis (e.g., Krause et al. 2015).

^{26}Al gamma-rays had established proof of current nucleosynthesis in our Galaxy. With better observations, this is now turned into an astronomical window of its own. The brightness of Galactic ^{26}Al emission allows measurement constraints not only from the integrated Galaxy's signal, but also for specific regions. A dream of cosmic nucleosynthesis studies through radioactivity gamma-ray lines had come true: Astrophysical models of massive stars and their evolution could be confronted with observations of source aggregates in our Galaxy to learn about validity of such models.

7.2.3 *^{26}Al Throughout the Galaxy*

7.2.3.1 Large-Scale Aspects

The distribution of ^{26}Al emission on the sky and within the Galaxy hold the clues to the ^{26}Al sources. Already in HEAO-C data (Mahoney et al. 1984), it had been shown that an extended emission region in the inner Galaxy is plausible, rather than assuming a point-like source or an assembly of such The COMPTEL image (Fig. 7.2) confirmed and firmly established the extended nature of ^{26}Al gamma-ray emission along the entire disk of the Galaxy.

The image shown in Fig. 7.2 is one of many possible representations of the celestial distribution of emission that are consistent with measurements. For an instrument with a rather broad instrumental imaging-response function from its Compton scattering measurement principle, and a large instrumental background, the process of image extraction, which is constrained by statistical fluctuations, also must include constraints from prior knowledge, in order to direct the otherwise under-determined imaging process. In the image shown, the two criteria are maximum likelihood (i.e., quality of the data fit to the image), and maximum image entropy, which prefers an image with least new information content. One of the alternative choices is to build up the image successively from largest to smaller scales, at each stage estimating statistical noise if the image were true already, and suppressing its effects in the iterative addition of next-finer spatial information; this obtains a smooth image, by design (Fig. 7.4, top). Many images would be consistent with the measured data, and careful evaluation of priors, biases, and systematics are important to judge the constraints on astrophysics.

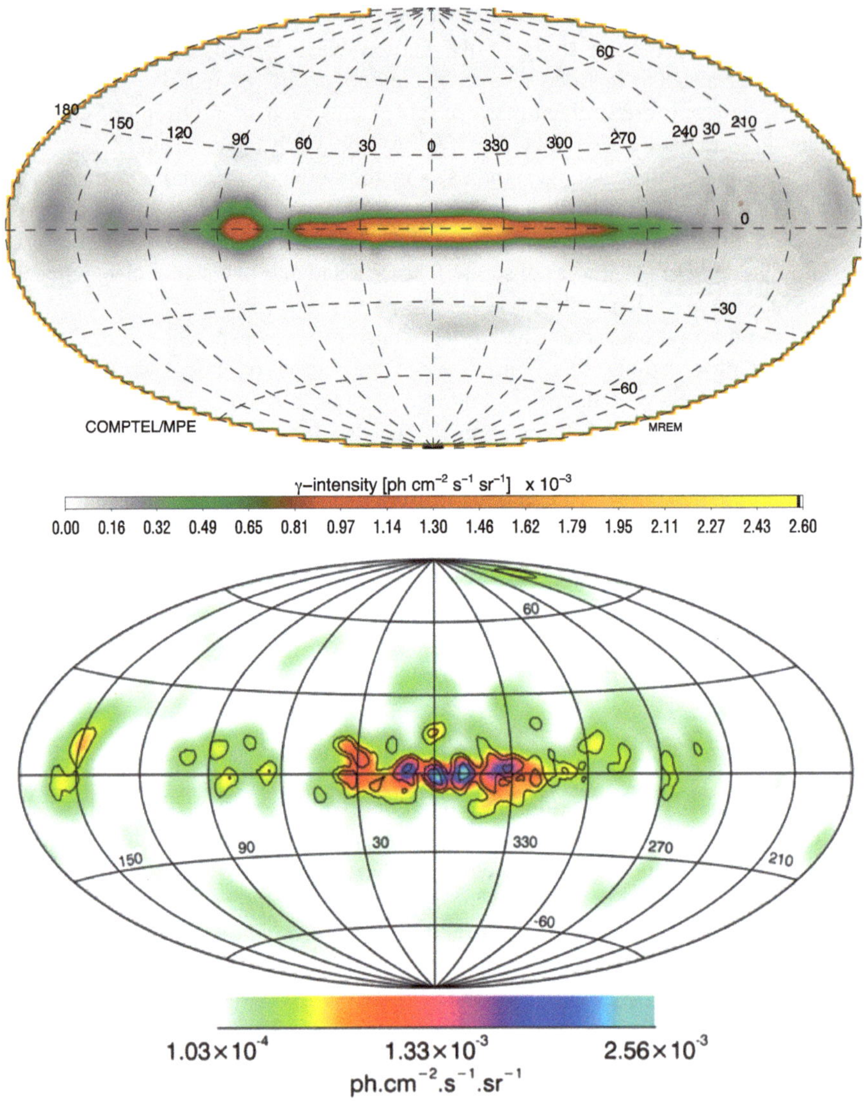

Fig. 7.4 *Top:* The ^{26}Al sky as deconvolved from the same data as shown in Fig. 7.2, now using a multi-resolution expectation maximization based on likelihood and wavelet-domain hierarchical noise filtering (Knödlseder et al. 1999). *Bottom:* The ^{26}Al sky as deconvolved from INTEGRAL's sky survey from 10 years of data with the SPI imaging spectrometer, and a maximum-likelihood imaging method (Bouchet et al. 2015)

Following the science principle of reproducibility of results, it is helpful to also have entirely-different instruments measuring the same celestial information. The spectrometer SPI on ESA's INTEGRAL space observatory features Ge detectors with high spectral resolution, and obtains its imaging information from a coded mask shadow onto its hexagonal 19-detector camera. Although the sky exposure during the INTEGRAL mission was largely emphasised on the inner Galaxy, an all-sky image had been constructed from 10 years of data (Bouchet et al. 2015), shown in Fig. 7.4 (lower graph). This image also is potentially susceptible to systematics and uncertainties from the instrument, yet those will be different here. The total background count rate of this kind of instrument is much higher than for the Compton telescopes, and its imaging principle is more straightforward with a single interaction and detector type. The images show striking similarities, and several features of the ^{26}Al image shown in Fig. 7.2 reappear. In view of the different ways those images were obtained, we consider as confirmed that there is diffuse emission all along the plane of the galaxy, with emission peaks or hot spots reminiscent of known massive star groups.

The irregular distribution of ^{26}Al emission all along the plane of the Galaxy provided a main argument for the idea that massive stars dominate the production of ^{26}Al (Prantzos and Diehl 1996). Massive stars preferentially form in clusters; nearby massive-star regions appear prominent in ^{26}Al emission (e.g. the Cygnus region appears in all three images).

A Galaxy-wide interpretation of the ^{26}Al gamma-ray measurements needs to resolve the distance uncertainty when assigning a measured flux along a line of sight to source intensities. The possibility of localised regions which may efficiently produce ^{26}Al needs a proper account. Since the massive star census in the Galaxy is well known from astronomical measurements in thermal emission from those stars only out to distances of a few kpc, and many regions of the Galaxy are occulted for direct measurements, one is left with some uncertainty about their Galaxy-wide distribution. Probably, the *molecular ring* around the center of our Galaxy at a radial distance of 3–4 kpc from the center is a prominent birth site for massive stars, as are molecular clouds swept up along the Galaxy's spiral arms. The Doppler shift systematics for the ^{26}Al line as measured with INTEGRAL/SPI suggests that much or most of the ^{26}Al seen towards the inner Galaxy directions is taking part in large-scale Galactic rotation. Therefore, this ^{26}Al emission originates from sources at kpc distances reaching to the Galaxy's center and beyond, and thus are not local or foreground sources.

Supported by this, the total amount of ^{26}Al in the Galaxy can be estimated from the measured gamma ray flux, using as a plausible assumption for the distances to the emission regions a galaxy-wide distribution. This has been done by several authors, based on COMPTEL and on INTEGRAL data, and using as models smooth double-exponential disk models, or models including more structure such as spiral arms (such as Robin et al. 2003; Taylor and Cordes 1993). With such methods, an ^{26}Al amount between 1.5 and 3 $M_\odot$ is found, depending on data and models used; INTEGRAL/SPI data were used to obtain (2.8±0.8) $M_\odot$ of ^{26}Al (Diehl et al. 2006a) in a study comparing many alternative views of massive star activity in the Galaxy. A systematic limitation of any galaxy-wide parameter determination is potential local bias. Gamma rays are at an advantage here due to their penetrating nature, reaching us also from sources of the distant regions of our Galaxy. However, there may be sources in the vicinity of our vantage point in the solar system, 8 kpc from the

Galaxy's center. The Cygnus region is prominent as an ^{26}Al source, and, although with member associations at distances between 800 and 1500 pc (Plüschke et al. 2001b). The Gould Belt has been recognised as a region of local stellar associations arranged in an elliptical-belt-like structure(Olano 1982; Pöppel 1997; Pöppel et al. 2010; Maíz-Apellániz 2004; Perrot and Grenier 2003). Several member associations of the Gould Belt appear to be aligned with ^{26}Al emission peaks. In particular, the most-nearby association of Scorpius-Centaurus at merely 120–150 pc distance was detected as an ^{26}Al source with INTEGRAL (Diehl et al. 2010), supporting indications from COMPTEL. Accounting for such foreground sources leads to a reduction of the Galaxy-wide ^{26}Al content. Values range from 1.7 ± 0.2 M$_\odot$ (Martin et al. 2009) to 2.0 ± 0.3 M$_\odot$ (Diehl et al. 2010; Diehl 2016), depending on estimates of residual systematics uncertainty.

Such ^{26}Al mass determination based on observations sampling the *entire* Galaxy now allows a comparison with what would be expected from their candidate sources on theoretical grounds. Early predictions for supernovae from core collapse of massive stars were 2.1 ± 1.1 M$_\odot$ (Timmes et al. 1995); the earlier Wolf-Rayet phase was estimated to contribute a galactic total of 0.5 M$_\odot$ with a factor $\sim$3 uncertainty (see Prantzos and Diehl 1996, and references therein). Uncertainties are due to the incomplete census of WR stars in our Galaxy (van der Hucht 2001) and metallicity dependence uncertainties. Recent models have assessed a Galactic contribution from WR stars of $\sim$0.6–1.4 M$_\odot$ (Palacios et al. 2005), accounting for new insights on impacts of stellar rotation and WR winds (see also Chap. 4). Models for these two candidate sources were best established, models for novae and AGB stars were uncertain, by comparison. Estimates for novae ranged from 0.1 to 5 M$_\odot$ with large uncertainties, mainly from a lack of progenitor knowledge, and from lacking still an also quantitatively-realistic nova model. About 0.2 M$_\odot$ of Galactic ^{26}Al may be due to classical novae, with again a factor $\sim$3 uncertainty (see Jose and Hernanz 1998, for a description of nova nucleosynthesis, but Chapter 5 presents an updated discussion in detail). Although AGB stars were once thought to be possibly important ^{26}Al sources (e.g. Forestini and Charbonnel 1997), their contribution is presently estimated (Karakas and Lattanzio 2014) to be below 10% of that of massive stars (see Chap. 3). Note that both AGB stars and novae are clearly identified as ^{26}Al producers from interstellar-grain inclusions in meteorites (see Clayton and Nittler 2004, for a review, and Chapters 3,5). Both these two types of sources are copious dust producers, more so than supernovae or WR stars; interstellar grain samples are biased towards measuring dust-producing sources. But in summary, none of the ^{26}Al sources discussed so far can be ruled out, based on such global, galaxy-wide considerations.

We now conclude at least that massive stars are the *dominating* contributors. Then we may use their theoretical yield estimates per mass and integrate over the mass distribution function to compare with the observed ^{26}Al amount. The normalization factor in such comparison is the Galactic supernova rate (see discussion in Chap. 11). ^{26}Al measurements interpreted within this framework yield a core-collapse supernova rate; values of 1.9 ± 1.1 SN century^{-1} (Diehl et al. 2006a), or, correspondingly reduced from foreground sources, 1.3 ± 0.4 SN century (Diehl 2016) have been estimated.

The line width of the ^{26}Al line as seen from the Galaxy reflects Doppler broadening from the large-scale rotation within the Galaxy, but also from turbulent ^{26}Al nuclei motions in the presumably-hot interstellar gas phase that ^{26}Al may still reside in at the time of its decay, after its ejection from the source now typically traveling through interstellar space for $\simeq$1 My. A balloon experiment based report of a rather broad ^{26}Al line with 5.4 keV broadening beyond instrumental resolution of the Ge detector (Naya et al. 1996), the My-averaged velocity broadening correspondingly would be about 500 km s^{-1}. Over such long time scale, either ^{26}Al would have to reside in highly-turbulent regions of the ISM, or cavities surrounding Al sources would be kpc-sized, or ^{26}Al would have to condense on grains early-on so that its coasting through lower-velocity ISM at original ejecta velocities would be possible (Chen et al. 1997). As shown above, the kinematic signature from large-scale Galactic rotation has been detected with sufficient exposure (see discussion in Diehl 2013, for how sufficient detail had been accumulated over the years of the INTEGRAL mission to enable such a measurement). An astrophysical broadening of the all-sky integrated ^{26}Al line was discovered also with INTEGRAL. The broadening value of 1.4 keV (±0.3 keV) is much smaller than what had been reported from the above balloon instrument data, and corresponds to 175 km s^{-1} (±45 km s^{-1}) in velocity space.

The bulk velocities seen in the Doppler shifts of the ^{26}Al line as viewed towards different Galactic longitudes (Fig. 7.5) turn out to be on the order of several 100 km s^{-1}. If interpreted as large scale Galactic rotation on Keplerian orbits around

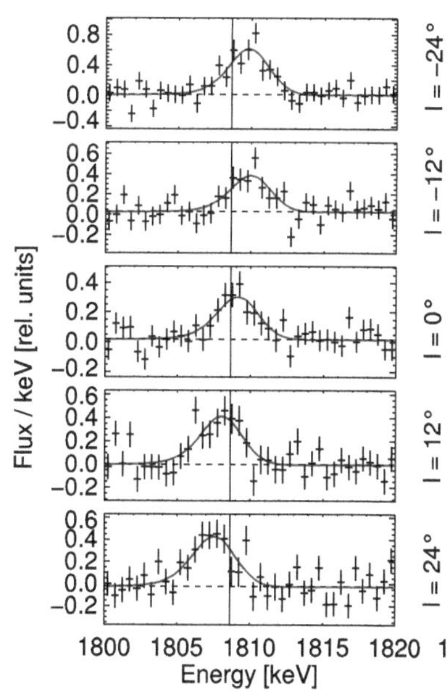

Fig. 7.5 ^{26}Al line variation with Galactic longitude (Diehl et al. 2006a). This shift of the line centroid reflects the kinematics of ^{26}Al towards the inner Galaxy in INTEGRAL/SPI measurements (Kretschmer et al. 2013)

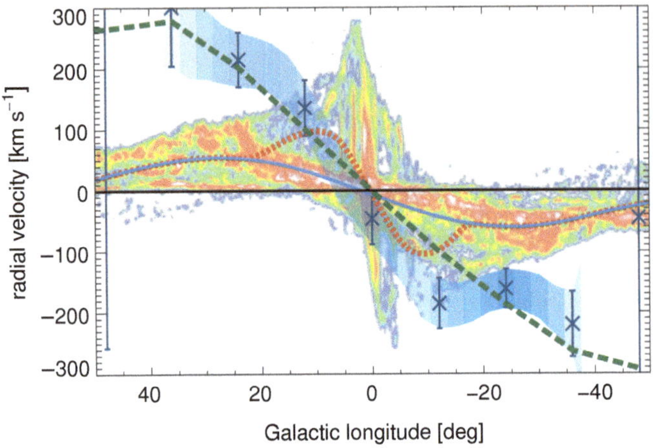

Fig. 7.6 Kinematics of ^{26}Al towards the inner Galaxy, from INTEGRAL/SPI measurements. This longitude-velocity diagram for hot ISM as traced through ^{26}Al in the inner Galaxy shows the trend from the Galaxy's large-scale rotation. The underlying color plots show the corresponding kinematics from molecular gas as traced through CO data. The ^{26}Al traced hot gas shows systematically higher velocities by about 200 km s^{-1} in the direction of Galactic rotation (Kretschmer et al. 2013)

the central regions of the Galaxy, this should be consistent for different objects within the Galaxy. Figure 7.6 compares the velocities determined from ^{26}Al and thus representing hot interstellar medium around sources of nucleosynthesis, with measurements of bulk velocities of molecular clouds as traced by carbon monoxide (Dame et al. 2001). Evidently, the velocities measured from ^{26}Al kinematics exceed typical Galactic rotation significantly, by about 200 km s^{-1}. Something must be different for hot gas near nucleosynthesis sources; but such generally more rapid rotation is not possible, and radioactive decay probably is part of the explanation.

When we consider that ^{26}Al will stream into the medium surrounding its sources while decaying, it is clear that the velocity measurement from ^{26}Al will be biased towards the conditions in the vicinity of the source, and ^{26}Al at locations more distant from the source will have decayed to ^{26}Mg and not emit gamma rays. Ejection from the sources plausibly should be isotropic. Typical ejection velocities are on the order of a few 1000 km s^{-1}, both for corec-collapse supernovae and for Wolf-Rayet stellar winds. But streaming away from its sources, if there would be an asymmetry of circum-source medium density such that in the direction of Galactic rotation more ^{26}Al would be streaming freely than in directions opposing Galactic rotation, the measurements could be understood. Based on such ideas, an asymmetry scenarion was proposed: Massive stars inside the Galaxy's co-rotation radius would be formed within spiral arms, but travel towards the leading edges of spiral arms while evolving into their Wolf Rayet phases, and in any case before they explode as supernovae. Then, the ejection of nucleosynthesis ejecta would occur in a region with a density gradient, higher density surroundings from the spiral arm

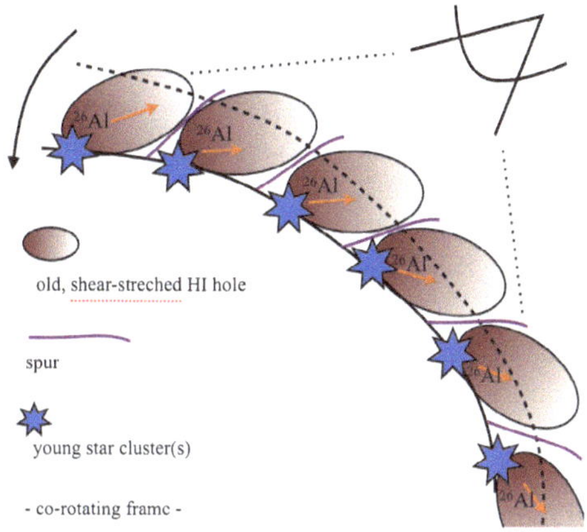

Fig. 7.7 Scenario for asymmetric surroundings of ^{26}Al sources. At time of nucleosynthesis product ejection, massive star groups could be located at the leading edges of spiral arms, thus presenting more material moving away from spiral arms at higher velocities than moving 'backwards' into higher density regions (Krause et al. 2015)

being located preferentially behind the massive stars, as seen from the perspective of an object at large-scale Keplerian rotation. So, ejecta would be decelerated if streaming towards the spiral arms, while streaming at higher velocities would be allowed into the inter-arm regions. Such a scenario (Krause et al. 2015), placing the candidate ^{26}Al sources along inner spiral arms and preferentially onto their inner ends approaching the Galaxy's bar, is illustrated in Fig. 7.7, and results in a longitude-velocity trend as shown by the green-dashed line in Fig. 7.6. In these calculations, the pitch angle of a logarithmic spiral structure model has been fitted (and obtains values in agreement with common results inferred otherwise), and an ejecta velocity of 200 km s^{-1} had been assumed, which corresponds to the typical expected sound velocity in superbubbles.

The significant enhancement of apparent velocities as observed from ^{26}Al therefore suggests that sources of ^{26}Al may be typically surrounded by interstellar cavities. These could have been created by previous stellar generations, or, alternatively, by the winds of the most-massive stars of a massive-star group, which evolve on shortest time scales and thus enter their Wolf-Rayet phase with strong winds only about 3 My after their formation. Superbubbles as typical surroundings of massive stars with ages of few to tens of My then presents a new scenario for how new nuclei may be recycled into the general flow of cosmic gas as part of cosmic chemical evolution. The sizes of such cavities plausibly extend up to kpc (Krause et al. 2015). Ejecta may thus be returned into ambient interstellar medium only as the walls of such superbubbles fragment and dissolve, on time scales beyond 10^7 years. ^{26}Al has shed new light onto the evolution of enriched gas on such time scales, which are difficult to constrain through observations otherwise.

Excess ^{26}Mg has been found in meteorites and associated with prior existence of ^{26}Al within these samples. Within meteoritic samples, one finds characteristic condensations, the CAI chondrules, which are associated with the first solids condensing in the early solar system which are preserved in their original composition.

These show a rather well-determined isotopic ratio ^{26}Al/^{27}Al of 4.5×10^{-5} (MacPherson et al. 1995, 2010), and are shown in the upper graph of Fig. 7.8 (see details in Chap. 6). Much smaller particles of stardust are also found as inclusions in meteorites, and these show a large spread of ^{26}Al/^{27}Al with generally higher values up to 1. Figure 7.8 (below) shows a more-recent representation of such stardust results. These are understood to originate from the immediate vicinity of sources of nucleosynthesis, where copious ^{26}Al production can lead to such high values, in particular for AGB stars (see Chap. 3). It is now interesting to see if, at the larger scale of the Galaxy which is sampled by the gamma ray measurements, the isotopic ratio for Al would be similar to values found for solar system material. Estimating the ^{27}Al amount from the Galaxy's gas mass estimates and solar abundance, and using the ^{26}Al mass derived from the gamma ray data as discussed above, one obtains a present-day ratio ^{26}Al/^{27}Al of $(6\pm3)\times10^{-6}$; the uncertainty of the current-ISM ratio was estimated from the gamma-ray flux uncertainty and uncertain ^{27}Al abundance. The result shown in Fig. 7.8 (top) is consistent with the conjecture (discussed in detail in Chap. 6) that the solar system is particularly enriched in ^{26}Al from some nearby nucleosynthesis.

7.2.4 *^{26}Al from Specific Regions*

The deep exposures of regions along the Galactic plane provided by the long-duration missions of COMPTEL and INTEGRAL were sufficient to identify and discriminate emission from specific source regions within the Galaxy (Fig. 7.9). In the following, the most-prominent regions are discussed in more detail (see also Chap. 11 and Sect. 11.1.2.4).

7.2.4.1 ^{26}Al in the Cygnus Region

The ^{26}Al gamma-ray images show emission from the Cygnus region at Galactic longitude near 80° as the clearest and most-prominent feature beyond the bright ridge in the inner Galaxy (see Figs. 7.2 and 7.4). From 13 years of INTEGRAL observations, the ^{26}Al line is seen at 11σ significance and an intensity of $9.3 \pm 1.8)\times10^{-5}\,\mathrm{ph\,cm^{-2}\,s^{-1}}$ (Fig. 7.10, from Siegert 2017).

Along the line of sight towards Cygnus, there are six prominent OB associations at distances ranging from 0.7 to 2.5 kpc (Plüschke et al. 2002), plus about a dozen open clusters, some associated to those OB associations (Fig. 7.11). Their ages range from 2.5 to 7.5 My.

It appears that the Cygnus OB2 association dominates by far the stellar census of this *Cygnus complex*; possibly, the associations Cyg OB1, 2, and 9 are related to OB2 and may originate from the same parental molecular cloud (Knödlseder et al. 2002). Cyg OB2 may even be considered the most prominent case in our Galaxy of

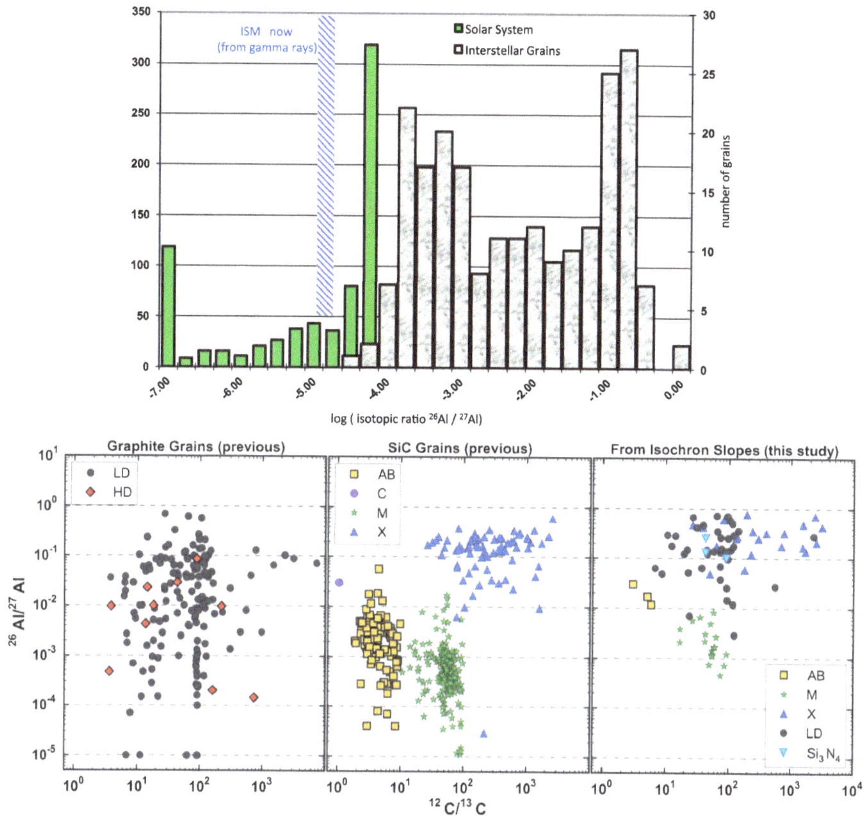

Fig. 7.8 *Top:* Isotopic ratios of radioactive versus stable Al, ^{26}Al/^{27}Al, are determined rather accurately for meteoritic components. Stardust particles (grey) show high values, while solids associated with the early solar system condensations (green) yield a limiting value of 4.5×10^{-5}. The global galactic value inferred from the gamma-ray data (blue hatched) lies somewhat below this, consistent with the early solar system's special enrichment discussed in detail in Chap. 6. *Bottom:* Stardust sample isotopic signatures. In three-isotope graphs, the characteristics of the dust formation site are imprinted, and compositional patterns of ^{26}Al enrichments show that high ratios of ^{26}Al/^{27}Al up to 1 were characteristic in the formation sites of this stardust. This is orders of magnitude above the ratio of ^{26}Al/^{27}Al that is estimated for the Galaxy's interstellar gas as a whole, from the ^{26}Al gamma ray line observations. Grains of type AB are thought to originate from AGB stars, while grain types X and LD are attributed to a supernova origin (after Groopman et al. 2015)

extremely-rich superclusters, which appear prominent in other galaxies but are hard to recognize within our own Galaxy; about 120 stars in the high-mass range (20–120 $M_{\odot}$) have been identified to relate to Cyg OB2; the other associations typically are ten times smaller. The age and distance of Cyg OB2 is 2.5 My and 1.57 kpc, respectively.

Other prominent objects towards this line of sight include the Cygnus Loop supernova remnant, and a diffuse and extended structure of X-ray emitting (hot) gas

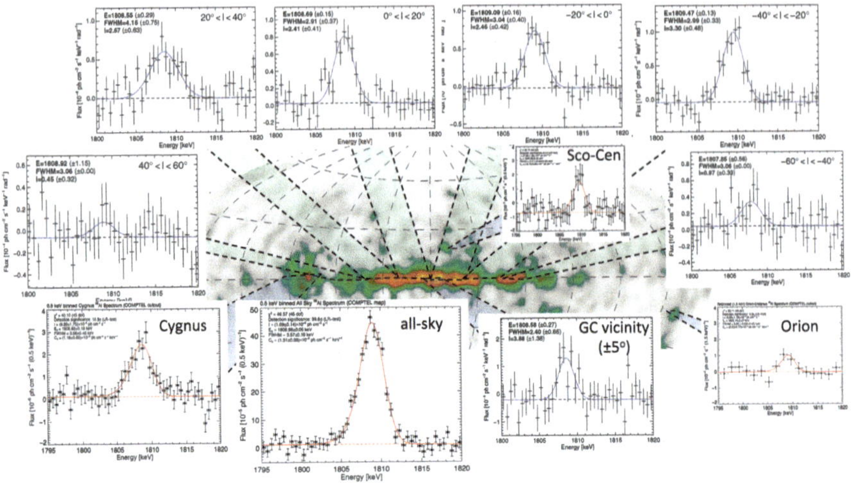

Fig. 7.9 The ^{26}Al line is measured separately for different emission regions along the plane of the Galaxy (from 5 years of INTEGRAL/SPI observations, Wang et al. 2009, and updates from internal reports MPE)

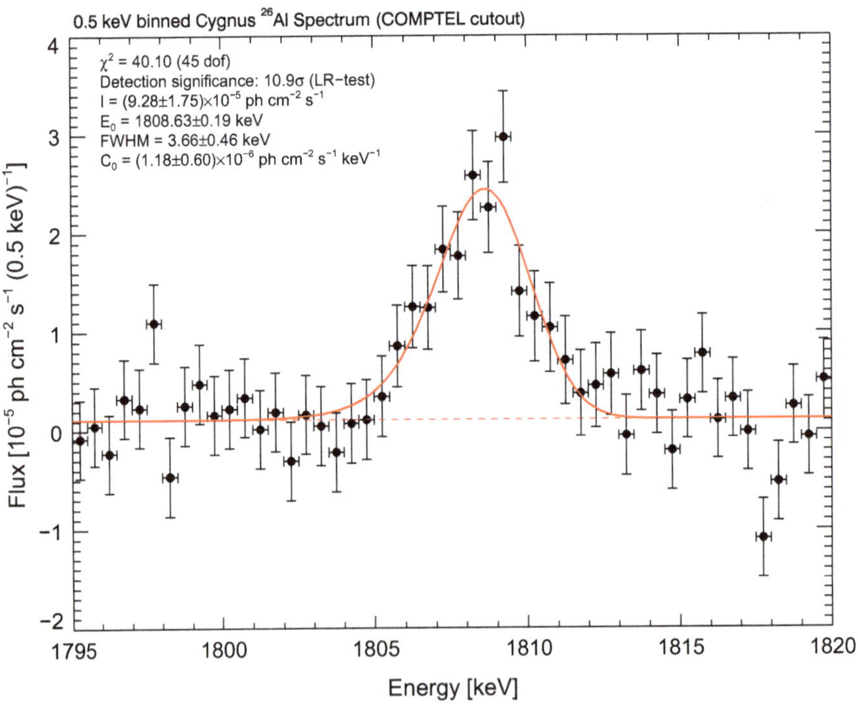

Fig. 7.10 Gamma-ray spectrum with the ^{26}Al line as measured with INTEGRAL/SPI towards Cygnus (see also Fig. 7.11; from Siegert 2017)

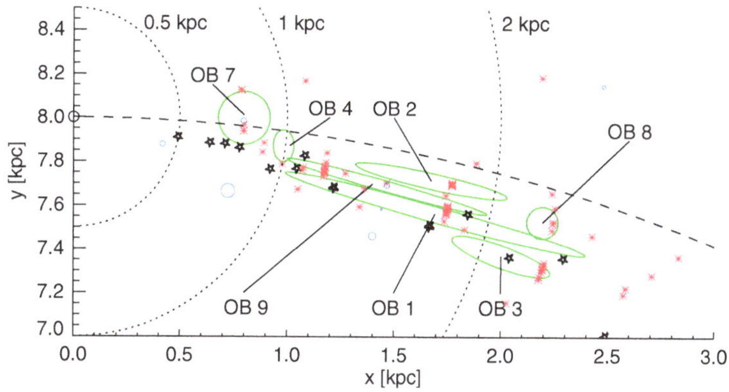

Fig. 7.11 Top view of the Galactic plane, showing the position of the Sun and prominent objects towards Cygnus. The OB associations along the line of sight towards the Cygnus region cover a rather large range of distances (ellipses illustrate distance uncertainties; Plüschke et al. 2002). The Cygnus OB2 group is by far the richest group of stars, and probably dominates ^{26}Al production

called the Cygnus Superbubble. The Cygnus Loop is a young supernova remnant with an estimated age of 10,000 years, and relatively nearby at 540 pc distance (Blair et al. 2005), attributed to a $\sim$12 $M_\odot$ progenitor star. By itself, it is a candidate ^{26}Al source, yet not more prominent than other massive stars if viewed at the characteristic ^{26}Al time scale of 1 My. Its proximity and age makes it appear as a bright X-ray and radio source; its ^{26}Al contribution appears unrelated to the Cygnus complex, however. The Cygnus Superbubble (Cash et al. 1980) is potentially more interesting for ^{26}Al studies, as it is much more extended and thus may be related to the collective effects of many past supernovae and/or massive-star winds. It has been re-assessed from radio observations as being most likely a superposition of many smaller hot-gas filaments at a range of distances (Uyanıker et al. 2001). The current picture is that within the Cygnus region, star formation has been ongoing for more than 10 My, and has led to a significant population of field and dispersed stars, in addition to the now-observed relatively-young OB associations. The large variations of visual magnitude extinction over small angular scales also supports a picture where the interstellar medium in the Cygnus complex is very heterogeneous and filamentary, with hot cavities bounded by dense remains of the parental molecular clouds (Lozinskaya et al. 2002; Comerón et al. 2008). Thus, one difficulty is to constrain the stellar population of a specific OB association (such as Cyg OB2) from all O stars seen towards this sightline, and to avoid inclusions of indirectly or even unrelated stars in such stellar budget (as discussed in detail by Comerón et al. 2008).

The total ^{26}Al gamma-ray brightness towards the Cygnus direction in the longitude interval [70°, 96°] has been associated to the specific sources in Cygnus with $\sim 6\times10^{-5}$ ph cm^{-2} s^{-1} (from the line of sight towards the Cygnus OB associations, accounting for a large-scale Galactic-disk background), and the contribution from the *Cygnus complex* alone was estimated as $\sim 3.9 \times 10^{-5}$ ph cm^{-2} s^{-1} (Martin et al. 2009).

Because of this young age of the dominant source region of Cyg OB2, stellar evolution even for the most-massive stars should still not be completed, and contributions from core-collapse supernovae to ^{26}Al production should be small or absent. Wolf-Rayet-wind ejected ^{26}Al from hydrostatic nucleosynthesis may be assumed to dominate, currently originating from Cyg OB2 stars. In that case ^{26}Al gamma-rays from the Cygnus region potentially could disentangle the different ^{26}Al production phases and regions within the same massive stars: In galactic-averaged analysis, one assumes a *steady state* situation of ^{26}Al decay and production, such that the complete age range of stars is represented and contributes to ^{26}Al production with its time-averaged numbers of stars per age interval and their characteristic ^{26}Al ejection from either process (hydrostatic, or late-shell burning plus explosive; Limongi and Chieffi (see 2006b, and Chapter 4)).

From a first comparison of measured versus expected ^{26}Al emission, as estimated from the number of massive stars in the region, ^{26}Al emission seemed surprisingly bright. But a revision of the stellar census using IR data of the 2MASS survey brought population synthesis estimates into closer agreement with expectations (Knödlseder et al. 2002; Martin et al. 2009, 2010). One can use *population synthesis* to account for time dependent ^{26}Al ejections from a çoeval group of massive stars. In such an approach, the stellar evolution model results for different initial masses of a star are evaluated and accumulated, for a group of stars with a specified total stellar mass and individual mass values and numbers drawn from an initial-mass distribution function (for details of the method see Voss et al. 2009). Martin et al. (2009) applied this to the stellar groups of the Cygnus complex. The resulting time dependent ejection of ^{26}Al is shown in Fig. 7.12, and compared to the observed ^{26}Al intensity. It appears that the measurement and prediction are in agreement, within uncertainties. Then, indeed, the observed ^{26}Al should be attributed to Wolf Rayet wind ejections, with contributions from core-collapse supernovae just beginning or expected in the near future. This seems consistent with the absence of signs of core collapse supernovae such as radio- or X-ray supernova remnants and pulsars (see discussion by Martin et al. 2010). Martin et al. (2010) point out, however, that there is some uncertainty about the metallicity in the Cygnus region, which could be lower than solar by a factor of a few. Then, the Wolf Rayet winds would likely be weaker, shifting the time of most ^{26}Al ejection towards a peak driven by supernovae, and at a later time. The consistency shown in Fig. 7.12 is suggestive that this is not the case.

For a young and active region of massive-star action, one may plausibly assume that the interstellar medium would be peculiar and probably more dynamic than in a large-scale average. With the Fermi satellite, evidence for cosmic ray produced high-energy gamma ray emission has been found at GeV energies (Tibaldo and Grenier 2013), confirming such turbulent interstellar medium with likely production of the shock fronts that are plausible cosmic-ray accelerators. With the fine spectroscopic resolution of the INTEGRAL measurements, therefore constraints

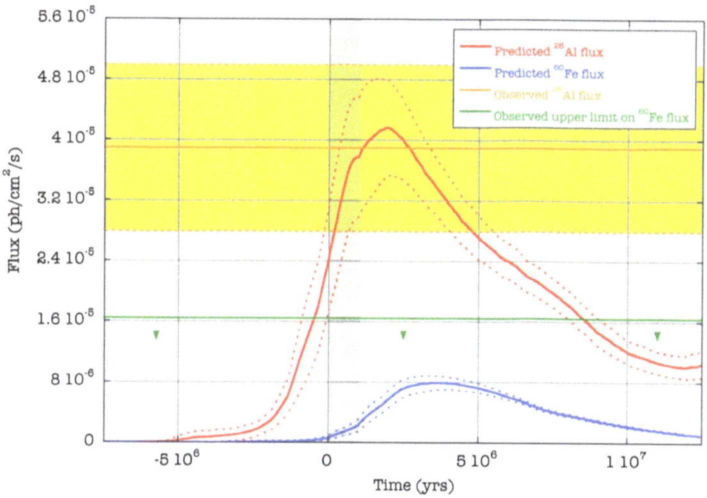

Fig. 7.12 The time history of ^{26}Al production in the Cygnus complex, as compared to the gamma-ray observations. Expectations from such populations synthesis are on the low side of observed ^{26}Al gamma-rays, in particular if a lower metallicity is adopted for the Cygnus region (solar metallicity is assumed, 0.04). The horizontal shaded area presents the range given by the ^{26}Al gamma-ray data, the dashed lines bracket the uncertainty range of predictions from recent massive-star models through population synthesis, and the vertical shaded area indicates the current time (from adopted cluster ages; the impact of stellar rotation on age estimates determines the width of the shaded area) (Figure adapted from Martin et al. 2010)

for a broadened ^{26}Al gamma-ray line would be interesting. As shown in Fig. 7.10, broadening is modest but present, at a level comparable to the Galactic-averaged value (see Fig. 7.3), i.e. at or below ~200 km s^{-1}).

7.2.4.2 ^{26}Al in the Orion Region

The Orion region is the most-nearby region of massive stars, at a distance of ~450 pc (Bally 2008; Genzel and Stutzki 1989). Its location towards the outer Galaxy and at Galactic latitudes around 20° is favorable, as potential confusion from other Galactic sources is negligible. The groups of massive stars, and in particular the Orion Nebula Cluster of stars, have been studied extensively, and are considered the prototype laboratory for astrophysical studies of *normal* massive-star activity. The dominating group of massive stars is the Orion OB1 association (Brown et al. 1994) with three major subgroups of different ages, one oldest subgroup *a* at 8–12 My, and two possibly coeval subgroups *b* (5–8 My) and *c* (2–6 My); subgroup *d* is the smallest and youngest at 1 My or below (see re-assessment of the stellar census by Voss et al. 2010). Subgroup c hold most massive stars, about 45 in the mass range 4–120 M$_{\odot}$. These groups are located on the near side of the Orion A and B molecular

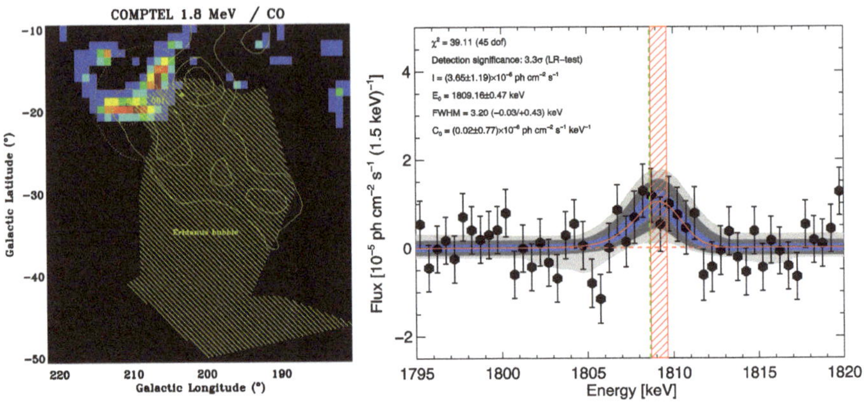

Fig. 7.13 The ^{26}Al gamma-ray signal seen by COMPTEL (left) and INTEGRAL/SPI (right) towards Orion. The gamma-ray intensity map contours from COMPTEL measurements are inconsistent with a concentrated source, and suggest extended emission away from the Orion molecular clouds (color pixels in the map, from CO) and the OB1 association subgroups (circles). The location of the interstellar cavity of Eridanus is indicated (hatched). From Diehl et al. (2003). The spectrum measured with SPI's Ge detectors shows a clear detection of the ^{26}Al line, at instrumental line width, with an indicated bulk motion blue shift (from Siegert and Diehl 2017)

clouds, which extend from 320 to 500 pc distance away from the Sun, and span a region of ~120 pc perpendicular to our viewing direction.

With the COMPTEL imaging telescope, only faint hints for emission in the wider Orion region were noticed, at low surface brightness and apparently only at the level of typical noise (Fig. 7.13). Upon a closer inspection, a clear line at 1.8 MeV could be seen (which is typically not seen for other low-brightness regions in the COMPTEL map), and a model fit suggested a significant (5σ) detection of ^{26}Al from Orion (Diehl 2002)). INTEGRAL observations could confirm this signal, though marginally at 3σ, at an intensity of $(3.65 \pm 1.2) \times 10^{-5}$ ph cm^{-2} s^{-1} (Fig. 7.13, from Siegert 2017).

Earlier X-ray studies and HI maps had revealed that a huge interstellar cavity was extending from the Orion molecular clouds towards the Sun, banana-shaped and extending over almost 300 pc (Burrows et al. 1993) (see sketch in Fig. 7.14). The oldest, and most-nearby, OB1 subgroup *a* plausibly may have created the Eridanus cavity on the near side of the Orion molecular clouds, triggering subsequent star formation (the later and more-distant OB1 subgroups) into the molecular cloud. The scenario illustrated in Fig. 7.14 plausibly explains the offset of ^{26}Al gamma-rays from their sources, as well as the indicated blue shift: fresh ejecta propagate further into the cavity, and ^{26}Al decays along with its flow. Population synthesis for the OB1 subgroups and their expected ejections of kinetic energy and ^{26}Al support this scenario (Fig. 7.15): Energy for blowing the Eridanus cavity had been available, and ^{26}Al production appears ongoing almost steadily, from OB1 subgroups. Hydrodynamical simulations of such massive-star feedback resulted

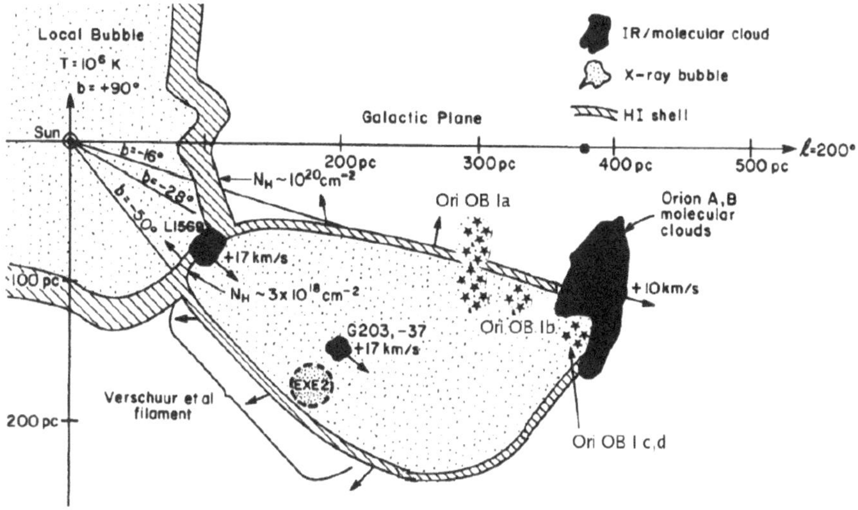

Fig. 7.14 A sketch of the region between us and the molecular clouds in Orion at about 450 pc distance, with the OB1 stellar association and its subgroups on the near side of the clouds, and the Eridanus cavity extending from the clouds towards the Sun. A scenario of ^{26}Al distribution from ejecta of the Orion OB1 association, blown into the Eridanus cavity, is indicated. Adapted from Burrows et al. (1993), see Fierlinger et al. (2016)

in predictions of gas morphology and X-ray emission from the Eridanus cavity, which are consistent with observational constraints from HI and diffuse X-ray emission measurements (Fierlinger et al. 2016; Krause et al. 2013, 2014). Thus, the Orion region appears to resemble one of the massive-star clusters in an asymmetric geometry that was suggested above (Fig. 7.7) to explain the velocity offset of ^{26}Al versus large-scale Galactic rotation (Fig. 7.15).

7.2.4.3 ^{26}Al in the Sco-Cen Region

The most-nearby groups of stars with members massive enough to significantly shape their surrounding medium is the stellar association of Scorpius-Centaurus and its subgroups, at a distance of about 100–150 pc (de Geus 1992; de Zeeuw et al. 1999; Preibisch and Zinnecker 1999).

Its location in the sky is up to 20° above the plane of the Galaxy, quite extended though, due to its proximity; the member stars are recognised through their coherent motion, as shown in Fig. 7.16 in the righthand graph.

The COMPTEL ^{26}Al image shows some hints for emission that may be associated to Sco-Cen, only recognisable due to its location above the Galactic plane. INTEGRAL observations obtained a very deep exposure of the region of the inner Galaxy, so that ^{26}Al emission from Sco-Cen could be detected from an emission region of about 50 degrees^2 centered at $(l, b) = 350°, 20°)$ (Diehl et al. 2010). The

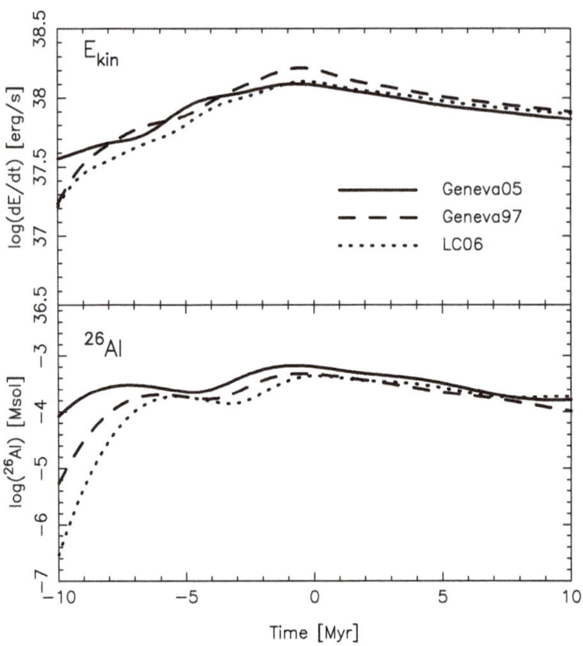

Fig. 7.15 The predicted time dependence of ejections of kinetic energy (above) and ^{26}Al (below), from population synthesis of the Orion OB1 association and its subgroups (Voss et al. 2010)

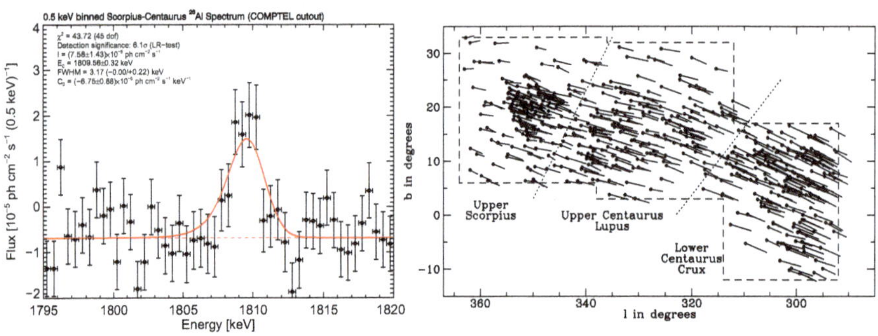

Fig. 7.16 The ^{26}Al signal disentangled from the Scorpius-Centaurus region with INTEGRAL (left; Diehl et al. 2010; Siegert 2017), and an illustration of the stars as tracked by Hipparcos in this nearby region (right, de Zeeuw et al. 1999)

spectrum from a similar region from 13 years of data is shown in Fig. 7.16 (left) (from Siegert 2017). The intensity in the ^{26}Al line of $(7.6 \pm 1.4) \times 10^{-5}$ ph cm^{-2} s^{-1} can plausibly be associated to massive star outputs from the Upper Sco subgroup (Diehl et al. 2010); the population synthesis estimate predicts 7×10^{-5} ph cm^{-2} s^{-1} from the stellar census, which should be rather complete for this nearby association.

Signs of past supernova activity from these stars are imprinted on the morphology of the nearby interstellar medium, which could be mapped in quite some detail from absorption line measurements towards nearby stars (Frisch 1995; Lallement 2007). Estimates are that the most massive star in Upper Sco presumably had $\sim 50\,M_\odot$ and thus may have exploded as a supernova about 1.5 Myr ago, and the pulsar PSR J1932+1059 may be its compact remnant (Hoogerwerf et al. 2000; Chatterjee et al. 2004). Several loops/shells reminiscent of supernova remnants have been identified (de Geus 1992). The Local Bubble is the most-nearby of such a cavity, surrounding the Sun (Breitschwerdt et al. 1998), as seen in X-ray emission from hot gas in its interior. Its origin has been related to subgroups associated with the Sco-Cen association (Breitschwerdt and de Avillez 2006; Fuchs et al. 2009; Welsh and Shelton 2009). Also, the supernova origin of live radioactivity of ^{60}Fe that was found in oceanfloor and lunar material (Wallner et al. 2016) is attributed to an origin related to Sco-Cen (Breitschwerdt et al. 2016) (see details in Chaps. 4 and 6).

At least three subgroups of different ages can be distinguished among the Sco-Cen stars (de Geus et al. 1989; de Zeeuw et al. 1999): The Upper Centaurus-Lupus (UCL), Lower Centaurus Crux (LCC), and Upper Centaurus (USco) groups, as shown in Fig. 7.16 in the righthand graph. Their ages are 17, 15, and 5 My, respectively, where typical age uncertainties are 1–2 My (de Geus et al. 1989; Slesnick et al. 2008; Pecaut and Mamajek 2016). Stellar subgroups of different ages would result from a star forming region within a giant molecular cloud if the environmental effects of massive-star action of a first generation of stars (specifically shocks from winds and supernovae) would interact with nearby dense interstellar medium, in a scenario of *propagating* or *triggered star formation*. Then later-generation ejecta would find the ISM pre-shaped by previous stellar generations. Such a scenario was proposed (de Geus et al. 1989; Preibisch and Zinnecker 1999) based on the different subgroups of the Scorpius-Centaurus association and the stellar groups surrounding it (e.g. Preibisch and Mamajek 2008; Fernández et al. 2008). It is illustrated in Fig. 7.17 (upper set of graphs).

Indications of recent star formation have been found in the L1688 cloud as part of the ρ Oph molecular cloud, and may have been triggered by the winds and supernovae causing the ^{26}Al we observe. The young ρ Oph stars then could be interpreted as the latest signs of propagating star formation originally initiated from the oldest Sco-Cen subgroup in Upper Centaurus Lupus (Wilking et al. 2008). Also the Lupus cloud shows recent star forming activity (Gaczkowski et al. 2015, 2017).

However, there is tension for a triggering scenario as originally proposed by Preibisch and Zinnecker (1999). Krause et al. (2018) combine data from the entire Sco-Cen region and wavelengths from radio to X- and gamma-rays, and rather propose that a concerted fireworks of star forming activity in different regions of an original, parental giant molecular cloud would be a more plausible description. They argue that the triggering of secondary star formation is more complex, and often arises from converging shells of expanding bubbles and superbubbles. They term this process "surround and squash" (see Fig. 7.17 lower graph) to illustrate how feedback from massive stars destroys the parental cloud little by little.

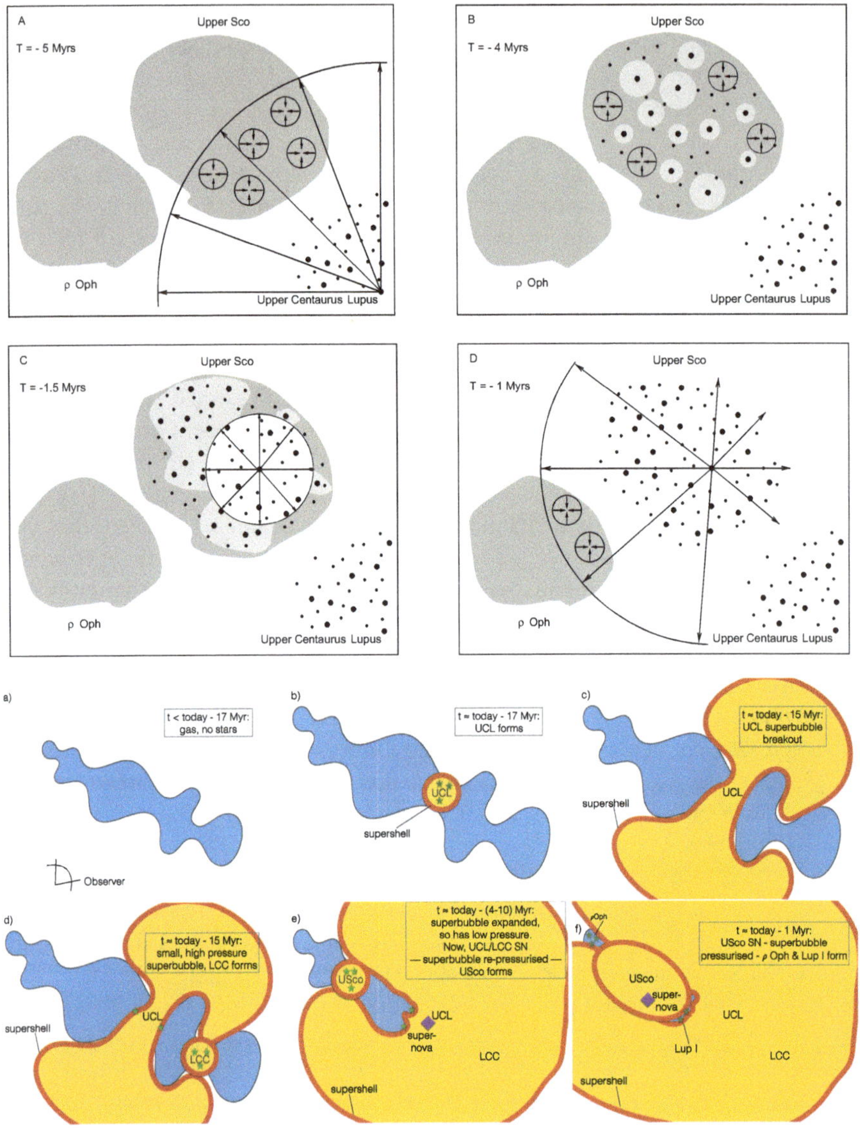

Fig. 7.17 *Top:* The objects in the Sco-Cen region, as they could have evolved in a scenario of *triggered star formation* (from Preibisch and Zinnecker 1999). *Bottom:* A scenario of successive erosion of molecular clouds by *propagating star formation* (from Krause et al. 2018)

7.3 ^{60}Fe

7.3.1 Nuclear Reactions, Candidate Sources, and Observability

^{60}Fe is a prominent neutron-rich isotope that can be produced from the abundant stable Fe isotopes if irradiated with neutrons (Fig. 7.18). The characteristic reaction sequence shown in Fig. 7.18 suggest that ^{60}Fe is produced by an s process of slow neutron capture reaction flow, allowing for (some!) intermediate β decay. The reaction path thus remains close to the valley of stable isotopes, as is characteristic for the s process. Note that the reaction domain of the r process (see Chap. 4) is far from the regimes of instability where neutrons cannot be bound easily. With Fe as seed elements being abundant, ^{60}Fe appears to be an excellent astrophysical probe of cosmic s-process environments.

Nuclear reaction uncertainties in ^{60}Fe production, therefore, are the neutron capture cross sections of ^{59}Fe and ^{60}Fe, for the production and destruction of ^{60}Fe, and their β decay rates, reducing ^{60}Fe yields. It is difficult to set up proper experiments for neutron capture, as its lifetime against β decay is 64 days only. Moreover, the resonant neutron capture, which probably dominates at stellar energies, also has a contribution from direct capture. Therefore, different measurements for different aspects and reaction channels involved have to be combined with theoretical estimates for reactions that may occur through excited stages of the nucleus (see Chap. 9).

The β decay itself had been theoretically estimated by FFN, later revised substantially by proper account of the Gamov-Teller transitions... Recently, Li et al. (2016) again revised the rate for ^{59}Fe, to now fall a factor 5 above Langanke+'s values, yet orders of magnitude below FFN's values; this shows that considerable uncertainty remains for the weak reactions involved in ^{60}Fe production.

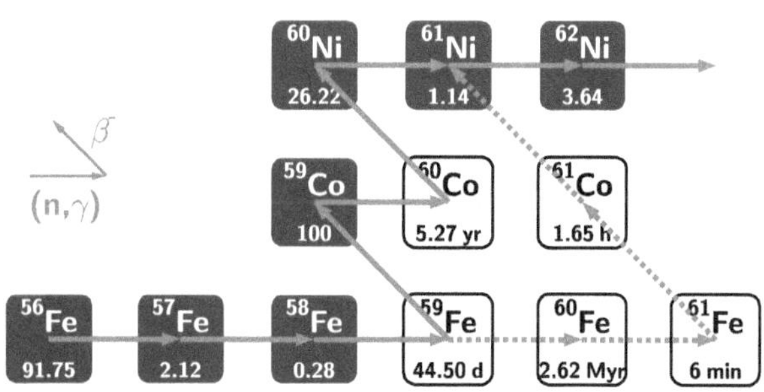

Fig. 7.18 The ^{60}Fe production through successive neutron capture reactions, with competing β decays indicated (from Heftrich et al. 2015)

The astrophysical setting of ^{60}Fe production is the existence of a neutron source in an environment where heavy elements such as Fe still may exist. These appear to be the shell burning stages of massive stars. Chapters 3 and 4 discuss these regions inside stars more thoroughly. Similar to ^{26}Al, ^{60}Fe also is a rather long-lived isotope, with a radioactive lifetime of 3.8×10^6 years (Rugel et al. (2009); corresponding half-life is 2.62 My). And, again, this lifetime is also short compared to a normal star's evolution, and on the order of the lifetime of the more-massive stars only. Hence, again, the ejection from the production site into interstellar space appears to be an issue. More-massive stars which undergo a supernova explosion soon after shell burning has been activated appear as most plausible sources.

Candidate production environments of ^{60}Fe are the zones of helium burning inside the more-massive stars. Here, the ^{22}Ne(α,n)^{25}Mg reaction liberates neutrons as helium undergoes this reaction with pre-existing neon that is convectively mixed into the reaction zone (Limongi and Chieffi 2006a,b). Later shell burning stages of more-massive stars may also produce some ^{60}Fe, as α particle release reactions again produce helium for this neutron-producing reaction. But, at those higher temperatures, ^{60}Fe production is less efficient, as it now must compete with increased rates of the β decays of ^{59}Fe and ^{60}Fe, as well as with photo-destruction reactions that begin to destroy heavier nuclei such as neon. Hence, the production of ^{60}Fe is a delicate process, requiring several conditions to be favourable. Conversely, if produced, the abundance of ^{60}Fe may reveal details about the conditions in the reaction site, that cannot be measured otherwise.

Contrary to the case of ^{26}Al, ^{60}Fe is expected to be ejected *only* by the SN explosion and not by the stellar wind, since it is buried to deep (up to the Ne-O shell) for ejection into interstellar space during any pre-supernova wind phase. It had been suggested, therefore, that detection of ^{60}Fe in the Galaxy would help to decide whether WR stars or core-collapse SN are the major sources of observed ^{26}Al. Stated more precisely, any massive-star region which is not in a steady state of production versus decay of these radioactivities from continued star formation, would show a variation of the γ-ray flux ratio from ^{26}Al and ^{60}Fe with age (see Voss et al. 2009, for model predictions) (see following section).

Also in rare thermonuclear explosions, ^{60}Fe could be produced (Woosley and Weaver 1994). If ignition occurs at highest densities, nuclear burning is expected to produce several neutron rich species, among them ^{60}Fe, which could be produced in substantial amounts. The necessary high central ignition densities will probably rarely be obtained, if the accretion is not exceptionally slow so that the white dwarf can remain cold up to reaching the Chandrasekhar mass limit. Accretion induced collapse, however, is a competing pathway for the evolution in such binary systems.

Upon its radioactive decay, ^{60}Fe emits gamma-rays from β^- decay. Through this radiation signature, it is suitable for remote studies of cosmic nucleosynthesis. Gamma-ray energies are 1173 and 1332 keV from cascade transitions in the final ^{60}Ni daughter nucleus of the β-decay in 99.85 and 99.98% of decays, respectively.

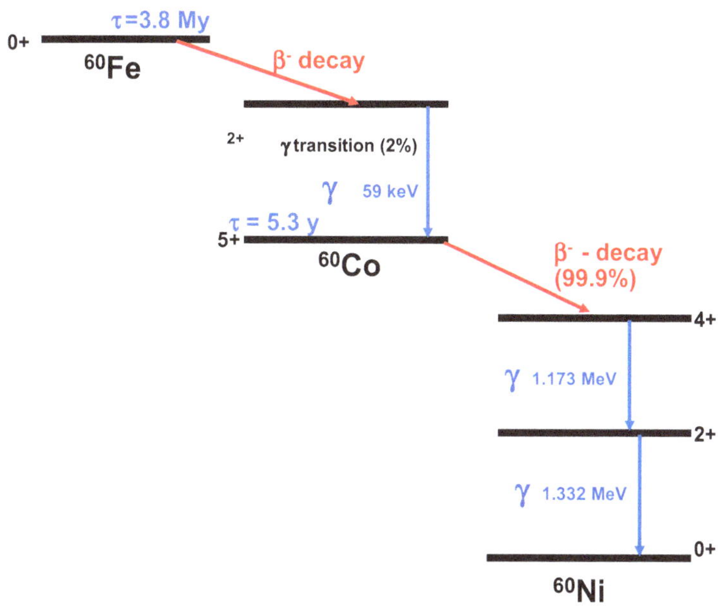

Fig. 7.19 The ^{60}Fe isotope decays with a radioactive lifetime of 3.8 My through ^{60}Co to ^{60}Ni. Note that per decay, two gamma-ray photons are obtained

A photon at 59 keV energy accompanies this ^{60}Fe–^{60}Co–^{60}Ni-decay chain, from a transition in the primary daughter nucleus ^{60}Co, is seen in 2% of ^{60}Fe-decays only (most de-excitations occur through internal conversion), and thus relatively unimportant. The decay chain of ^{60}Fe (see Fig. 7.19) involves intermediate ^{60}Co and produces two gamma-rays at 1332 and 1173 keV, respectively, which are well-known from laboratory work through the ^{60}Co calibration sources.

7.3.2 Observations Throughout the Galaxy

^{60}Fe gamma-rays are hard to detect with current telescope sensitivities. RHESSI reported a marginal signal (2.6 σ for the combined ^{60}Fe lines at 1.173 and 1.332 MeV) (Smith 2004) from the inner Galaxy, at the 10%-level of ^{26}Al brightness; SPI aboard INTEGRAL obtains a similarly low value, around 10%, also at the 3σ-level from early analysis (Harris et al. 2005), but more convincingly the ^{60}Fe lines were seen at 5σ significance with 4 years of data (Wang et al. 2007)[2] (see Fig. 7.20), and reported a ^{60}Fe to ^{26}Al brightness ratio of 14 ($\pm$6)%. Efforts

[2]Instrumental background from ^{60}Co produced locally in the satellite by cosmic rays may be a remaining concern.Wang et al. (2007) have accounted for such contribution within instrumental

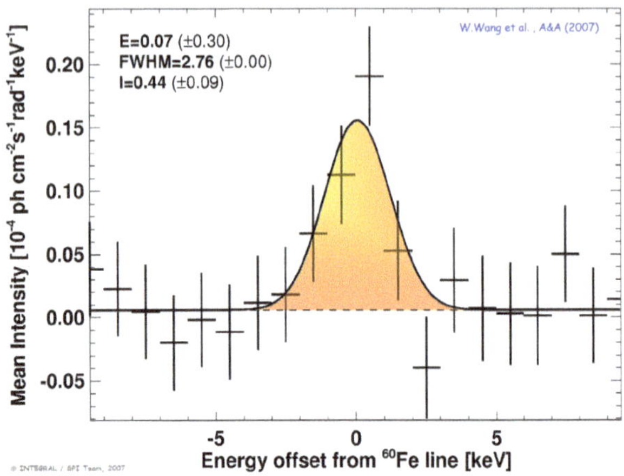

Fig. 7.20 The ^{60}Fe signal (both gamma-ray lines superimposed at their laboratory energies of 1173 and 1332 keV) as observed with INTEGRAL/SPI from the Galaxy (Wang et al. 2007)

are made to improve upon this measurement with deeper exposure collected with INTEGRAL. However, it has been shown that ^{60}Fe detectability suffers from a specific problem: The isotope ^{60}Co apparently is built up within the INTEGRAL spacecraft, background thus growing linearly with time (Diehl et al. 2018). The important test if indeed the spatial distribution of ^{60}Fe emission is identical to that of ^{26}Al thus remains a challenge. Nevertheless, obviously ^{60}Fe γ-ray intensity from the inner Galaxy remains substantially below the brightness from ^{26}Al.

Massive stars are the most plausible sources of both ^{60}Fe and ^{26}Al, but their production occurs at quite different regions and burning episodes in those same stars. The determination of the ratio of their yields $r = Y_{60Fe}/Y_{26Al}$ should be a very sensitive global diagnostic of the validity of massive-star nucleosynthesis models (e.g. Woosley and Heger 2007, and references therein). In steady-state approximation of current galactic nucleosynthesis (i.e. the galactic average synthesis rate of ^{26}Al and ^{60}Fe equals its decay rate), this production ratio and the total gamma-ray flux ratio accessible to gamma-ray telescopes are related through

$$\frac{I(^{60}Fe)}{I(^{26}Al)} = \frac{\dot{M}(^{60}Fe)}{\dot{M}(^{26}Al)} \cdot \frac{26}{60} \cdot 2 \tag{7.1}$$

where $\dot{M}(^{60}Fe)$ is the total Galactic production in $M_{\odot}$ year^{-1} (similarly for ^{26}Al). The mass of each isotope maintained in steady state throughout the Galaxy then is

$$\langle M(^{60}Fe)\rangle = \dot{M}(^{60}Fe) \cdot \tau(^{60}Fe) \tag{7.2}$$

knowledge. Note that insufficient subtraction of instrumental contributions would make the Galactic ^{60}Fe contribution even smaller.

For determination of the integrated production rate, the yield in ^{60}Fe per star of initial mass M is weighted with the number of stars of mass M, the *initial mass function*, summing over the mass range for massive stars $[M_{low}, M_{up}]$. Normalisation with the star formation rate cancels when the ratio for the ^{60}Fe and ^{26}Al isotopes is determined. This illustrates that much uncertainty related to unknowns of Galactic nucleosynthesis cancel for the isotope ratio ^{60}Fe/^{26}Al, and a measurement of the production ratio for all massive stars is obtained, which only depends on nucleosynthetic yields per star of mass M and the integration over the mass distribution of stars.[3]

Current models predict a gamma-ray flux ratio around 16% (Timmes et al. 1995; Woosley and Heger 2007) or ∼18 (±4)% (Limongi and Chieffi 2006b).[4]

Though apparently observations and predictions agree, there are uncertainties in both (gamma-ray telescope data analysis; nuclear-physics issues; massive-star shell burning). In several simulations, the nucleosynthesis imprint for the ^{60}Fe/^{26}Al ratio has been studied on a larger scale beyond individual sources. If the evolution of an entire star forming region is traced over its evolutionary time scale of order 10^7 years, apparently the ratio observed from gamma rays for the current milky way galaxy is compatible with expectations, for a more-evolved situation where also supernovae from the less-massive of the massive-star population contribute with enhanced ^{26}Al contributions—the earlier evolution with most-massive stars dominating produces ratios up to factors 2–3 higher (Kuffmeier et al. 2016, and references therein). But also, large fluctuations on a local scale of order of a star forming region occur, as well as fluctuations in time. Both are of order of a factor 10. This is relevant for the early solar system, discussed in Chap. 6, and for observations of specific regions: A determination of the ^{60}Fe to ^{26}Al ratio not only for integrated observations of the Galaxy, but more locally for specific regions of massive star nucleosynthesis would be revealing, with respect to the ^{60}Fe sources (see discussion above, Sect. 7.2.4.1).

7.3.3 *Observations of ^{60}Fe in Solar-System Material*

Live ^{60}Fe had been found more than 10 years ago in oceancrust material from the deep pacific ocean (Knie et al. 2004). This initiated a new 'astronomy' of radioactivity, the search for ejecta from nucleosynthesis sources near the solar system as captured directly. Radioactivity is the process that allows us to distinguish

[3]The mass range commonly assumed is 8–120 $M_\odot$; but both mass limits are subject to some uncertainty (see discussion in Zinnecker and Yorke 2007). Limongi and Chieffi (2006b) discuss in detail the impact of varying the slope of the mass distribution function, and the upper mass limit for the integrated range of masses.

[4]A period of irritation occurred between 2002 and 2005, as nuclear cross sections and models were updated, and seemed to predict much higher ratios up to 1; see discussion in Woosley and Heger (2007) and Prantzos (2004).

such ejecta samples from the bulk matter of the solar system bodies themselves. ^{60}Fe is located in a domain of atomic nuclei, where the prospects of finding such ejecta tracers are maximised: On one hand, normal solar system matter is made radioactive through energetic interactions with cosmic rays, through spallation reactions (see also Sect. 7.4). Spallation reactions thus break up nuclei, and due to the local abundance maximum around iron group elements, all nuclei lighter than the iron group elements are potentially created from such cosmic ray interactions. As discussed in Sect. 7.4, specific and widely used examples are ^{10}Be and ^{53}Mn, but also ^{26}Al is produced in significant abundance by cosmic ray interactions, for example in the Earth atmosphere. The positive side of such production is that one can measure cosmic ray intensities as well as atmospheric and terrestrial (e.g. subsurface water reservoirs) using radioactive isotopes. But detecting cosmogenic ^{26}Al on Earth is a challenge, due to the high cosmic ray production. Not so for ^{60}Fe, because the abundance of candidate nuclei which could be fragmented by cosmic-ray interactions to produce ^{60}Fe are much smaller, and, additionally, the production of a neutron rich daughter isotope in such a spallation reaction is small. Therefore, ^{60}Fe is a good candidate for studies of terrestrial or lunar samples of nucleosynthesis ejecta, as confirmed by the oceancrust detection of Knie et al. (2004).

Following this exciting discovery, ^{60}Fe as a nucleus was re-investigated, and its radioactive lifetime was found significantly longer in such improved measurements (Rugel et al. 2009; Ostdiek et al. 2017). This was important, as the inference of a nucleosynthetic origin of the terrestrial ^{60}Fe sample implies to calculate the probability that ejecta from an explosion at some recent time can reach the oceanfloor before the radioactive decay wipes out its information.

One might be concerned that a single sample of oceancrust, through carefully selected to minimise anthropogenic contaminations, could be a single outlier. Analysis of a sediment (Fitoussi et al. 2008) had been less significant, which seemed puzzling. Note that the Accelerator Mass Spectrometry method is one of the most sensitive isotopic abundance detection methods, with a sensitivity exceeding 10^{-17} (Korschinek and Kutschera 2015; Kutschera 2013). More than 10 years later, a broad collection of oceancrust and sediment probes from three different regions, the Indian, Pacific, and Atlantic ocean, covering an age range from current to 11 My ago, had been analysed for ^{60}Fe (Wallner et al. 2016). These consistently showed ^{60}Fe signals in an age range 1.5–3.5 My ago, and a possible second marginal signal around 8 My ago. The sediment data here have a much better time resolution around 10,000 years. From the analysis of lunar surface material obtained from the Apollo mission, also a significant signal from ^{60}Fe could be obtained (Fimiani et al. 2016). Here, the turnover of surface material ('gardening') prevents an age determination of the material; but the mere detection of ^{60}Fe was an important, independent, confirmation. Microfossils may help to enhance iron content locally, and were exploited by Ludwig et al. (2016) for yet another independent ^{60}Fe detection on a terrestrial material probe.

The infall period of ^{60}Fe in the composite data from all samples clearly is resolved, that is, it appears to be extended and therefore not likely due to a single supernova ejecta wave crossing the solar system. It would have been tempting to 'date' a nearby supernova explosion in this way. But reality appears to be more complex. Interstellar transport of ejecta from supernova explosions is not easy to model, and ^{26}Al kinematic information from superbubbles as discussed above (see Fig. 7.7 and discussion thereof) may help to understand some of such transport on larger scales. But even on scales of the Local Bubble, the dynamical evolution of the nearby morphology of cavities and walls is complex. Breitschwerdt et al. (2016) showed through 3D hydrodynamical modeling of the nearby interstellar medium, that both, a supernova from a nearby stellar group occurring within the solar cavity, as well as the passing of the cavity wall of this Local Bubble across the solar system, could be explanations for the signal. The extent in time that is observed may originate from multiple supernova ejections, or from ejecta being reflected on the nearby cavity wall (see also Schulreich et al. 2017). This is illustrated in Fig. 7.21, where data are shown as compared to each of these scenarios.

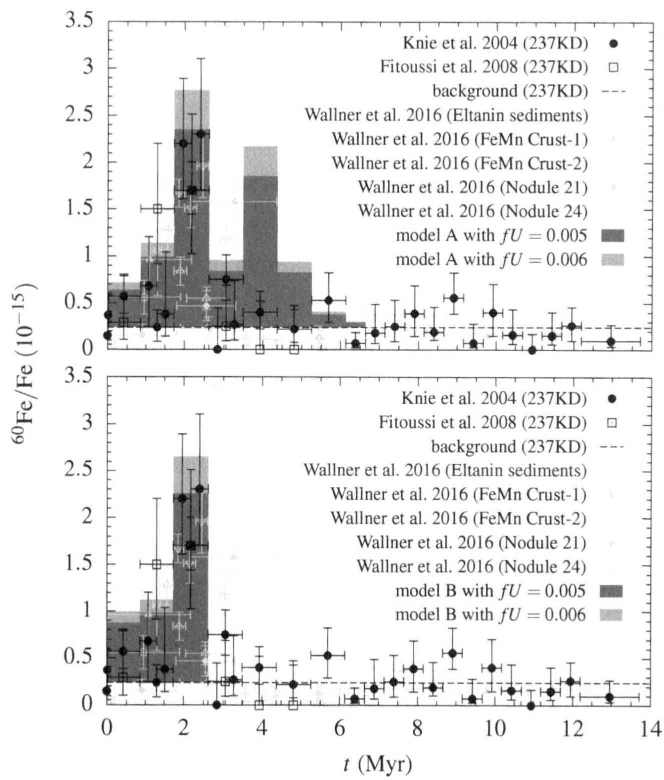

Fig. 7.21 The ^{60}Fe signal from several oceancrust and sediment analyses, as dated in time from age dating of the corresponding layer with cosmic-ray produced radioactive clocks ^{10}Be, ^{26}Al. The cases of ejections from supernovae and transport within the Local Bubble *(above)*, and from passing of the cavity wall across the solar system *(below)* are shown (from Schulreich et al. 2017)

7.4 Radioactivities in Cosmic Rays

Along with stellar grains, Galactic Cosmic Rays (GCR) provide a sample of matter from outside the Solar system. Despite almost a century of active research, the physics of GCR (concerning their sources, acceleration and propagation in the Galaxy) is not yet thoroughly understood. In particular, key questions regarding GCR physics are related to the timescales of various processes (acceleration in one event or in a series of events, confinement in the Galaxy, etc.). Radionuclides unstable to $\beta^{\pm}$ decay or e-capture, with laboratory lifetimes close to the timescales of interest for GCR studies, provide important probes of the aforementioned processes.

7.4.1 Sources, Acceleration, and Propagation of Cosmic Rays

From the previous section one may infer that several steps are involved between the production of the GCR nuclides in stellar interiors and their detection near Earth: (1) stellar nucleosynthesis, (2) ejection by stellar winds and explosions, (3) elemental fractionation, (4) acceleration of primary GCR nuclides, by shocks due to SN and winds of massive stars, (5) propagation through the ISM of the Galaxy, (6) modulation at the heliospheric boundary and (7) detection of arriving GCR. In particular, GCR transport through the ISM has been studied with models of varying sophistication, which account for a large number of astrophysical observables (see the comprehensive review of Strong et al. (2007) and references therein).

To describe the composition data, less sophisticated models are sufficient, like e.g. the *leaky-box* model. In that model, GCR are assumed to fill homogeneously a cylindrical box (the Galactic disk) and their intensity in the ISM is assumed to be in a steady state (equilibrium), between several production and destruction processes. The former involve acceleration in GCR sources and production *in-flight* through fragmentation of heavier nuclides, while the latter include either physical losses from the *leaky box* (*escape* from the Galaxy) or losses in energy space: *fragmentation*, *ionization losses* and *radioactive decay*. Most of the physical parameters describing these processes are well known, although some spallation cross sections still suffer from considerable uncertainties. The many intricacies of GCR transport are encoded in a simple parameter, the *escape length* Λ_{esc} (in $\mathrm{g\,cm^{-2}}$): it represents the average column density traversed by GCR nuclei before escaping the Galactic leaky box.

The abundance ratio of a secondary to a primary nuclide depends essentially on Λ_{esc}. Observations of $\frac{LiBeB}{CNO}$ (and more recently of $\frac{ScTiV}{Fe}$) in arriving GCR, interpreted in this framework, suggest a mean escape length Λ_{esc} ~7 $\mathrm{g\,cm^{-2}}$. In fact, the observed GCR secondary/primary ratios display some energy dependence, which translates into an energy dependent $\Lambda_{esc}(E)$, going through a maximum at E ~1 GeV/nucleon and decreasing both at higher and lower energies. The observed

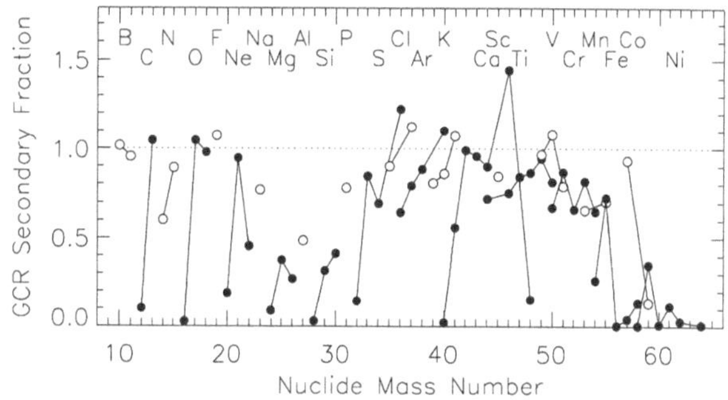

Fig. 7.22 Secondary fractions in arriving GCR for isotopes from B to Ni, after the analysis of ACE/CRIS measurements. Nuclides with filled (open) symbols correspond to isotopes with even (odd) Z, while solid lines connect isotopes of the same element. *From* Wiedenbeck et al. (2007)

energy dependence of that key phenomenological parameter can be interpreted in the framework of more sophisticated GCR transport models and provides valuable insight into the physics of GCR transport (role of turbulent diffusion, convection by a Galactic wind, re-acceleration, etc.); those same models can be used to infer the injection spectra of GCR at the source (see Jones et al. 2001; Strong et al. 2007).

Once the key parameters of the leaky-box model are adjusted to reproduce the key secondary/primary ratios, the same formalism may be used in order to evaluate the secondary fractions (produced by fragmentation in-flight) of all GCR nuclides. Those fractions depend critically on the relevant spallation cross-sections (well known in most cases) and appear in Fig. 7.22. Fractions close to 1 imply an almost purely secondary nature while fractions close to 0 characterize primary nuclides (like, e.g. ^{12}C, ^{16}O, ^{24}Mg, ^{56}Fe etc.). The former are very sensitive to the adopted Λ_{esc}, contrary to the latter (Wiedenbeck et al. 2007).

The *source abundances of primary GCR* (GCRS) derived that way display both similarities and differences when compared to the solar ones (Fig. 7.23). It was recognised quite early on, that the observed pattern of GCRS/Solar abundances shows some correlation with the First Ionisation Potential (FIP): elements of high FIP are depleted in GCRS relatively to low FIP ones. Since the latter are more easily ionized than the former, it is conceivable that they are also more easy to pre-accelerate (in some suitable environment) before SN shocks decerebrate them to relativistic GCR energies (Meyer 1985).Although similar FIP-dependent abundance patterns are observed in the solar corona, this idea did not evolve in a fully self-consistent model for the origin of GCR (Fig. 7.23).

Another, long standing, idea attracted considerable attention in the past 15 years. It invokes chemical volatility to explain fractionation in the GCR sources. Of course, volatility is somewhat related to FIP: high FIP elements (e.g. He, N, Ne, Ar) are, in general, volatile, where low FIP elements are, in general, refractories (e.g. Mg, Si, Fe etc.), and condense readily into dust grains. Meyer et al. (1997) and Ellison et al. (1997) suggested that grains are efficiently accelerated in SN shocks, because of their high mass/charge ratio; subsequent sputtering of atoms from these fast grains creates a *pool* of *suprathermal* ions, that are further accelerated to relativistic GCR energies from SN shocks. Building on those ideas, Meyer et al. (1997) proposed a model explaining quantitatively the inferred pattern of GCRS/Solar abundances.

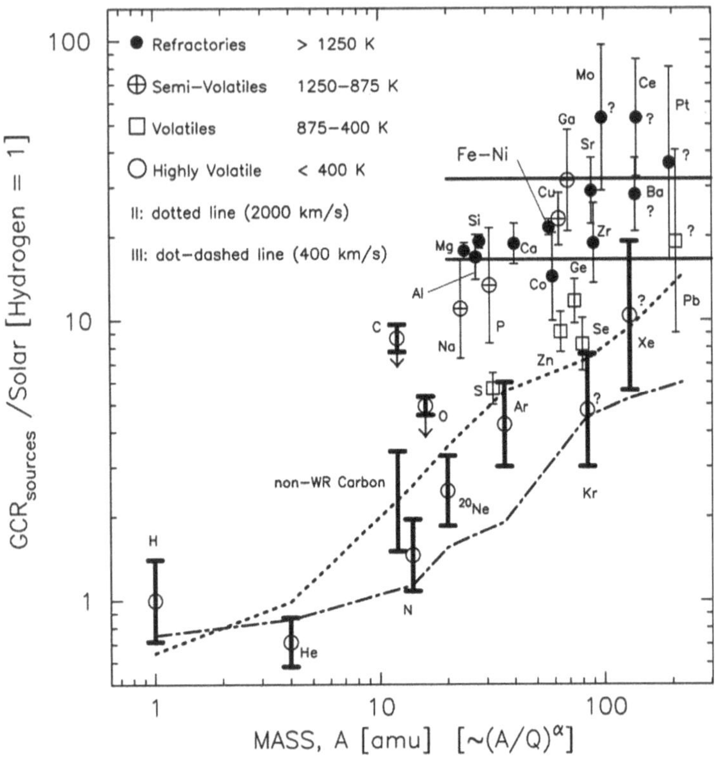

Fig. 7.23 GCR source abundance relative to solar abundance vs. atomic mass number. All values are measured relative to GCR hydrogen at a given energy per nucleon. The elements are divided, on the basis of condensation temperature, into refractory, semivolatile, volatile, and highly volatile groups. The refractories are essentially completely locked in grains in the ISM, while the highly volatile ones are gaseous. The arrows on carbon and oxygen indicate that these elements have an additional source from ^{22}Ne-C-O enriched Wolf-Rayet wind material. Model predictions for the abundances of volatile elements from a high Mach number shock model are shown with a dotted line, and those for a lower Mach number model with a dot-dashed line. The horizontal solid lines on the right side of the plot are limits on predicted abundance of iron and other refractory elements. The label on the abscissa [$\sim$(A/Q)$^{\alpha}$, where α is some unspecified constant] is a reminder that, for most ionization models, A/Q is a roughly monotonically increasing function of mass. The abundances of Kr, Xe, Mo, Ba, Ce, Pt, and Pb relative to Fe may contain systematic errors that are difficult to evaluate (indicated with a question mark) (from Ellison et al. 1997)

Despite its success, the model relies on several assumptions which lack empirical support (notice, however, that interplanetary dust particles display traces of grain acceleration, as reported by Westphal and Bradley (2004).

In view of the complexities introduced by fractionation, it appears safer to rely either on refractory primary elements (unaffected by fractionation) or to ratios of isotopes belonging to the same element (since fractionation effects cancel out), in order to probe the source composition of GCR. The ACE/CRIS observations allowed to determine the primary GCRS abundances of 19 refractory nuclides, belonging to Mg, Al, Si, Ca, Fe and Ni. As reviewed in Wiedenbeck et al. (2007) those nuclides have a solar composition (within uncertainties), with the possible exception of ^{58}Fe. Since more than half of solar Fe come from a *long-lived* source (SNIa, Goswami and Prantzos 2000), the GCRS composition has to originate from an extremely well mixed sample of the ISM (or of solar type stars).

Contrary to the refractory nuclides, the volatile ^{22}Ne/^{20}Ne ratio in the GCR source has long been known to exceed its solar value by a factor of $\sim$5. This excess of ^{22}Ne is attributed to the contribution of Wolf-Rayet winds to GCRs, as originally suggested by Casse and Paul (1982) and quantitatively elaborated in subsequent studies (e.g. Prantzos et al. 1985).

It is not clear at present whether the totality of the GCRS composition originates from the same site (where the nucleosynthetic products of many stars and SN are well mixed) or whether it originates as a diversified sample of nuclei (produced from a variety of sources distributed all over the Galaxy), which are mixed well after their acceleration by SN shocks, during their 10^7 years travel through the Milky Way. The latter case corresponds to the, now defunct, FIP-based paradigm of GCRS composition (Meyer 1985). The former possibility is now embodied in the *super-bubble* paradigm, advocated in Higdon et al. (1998) and further supported by Binns et al. (2005) on the basis of the ^{22}Ne/^{20}Ne analysis. GCR (both refractories produced by sputtered grains and volatiles) are accelerated—with higher efficiency than in the normal ISM—from the thermal pool of ions present in a super-bubble, which is created and chemically enriched by the action of dozens or hundreds of stars and SN. In that model, SN explosions have to occur $>10^5$ years apart from each other, in order to allow for *e*-capture decay of ^{59}Ni, which is absent in arriving GCR (see next section).

The super-bubble paradigm has been criticized on several grounds in Prantzos (2012), who showed that a superbubble, resulting from explosions of a representative sample of massive stars, is expected to have a solar ^{22}Ne/^{20}Ne ratio and cannot explain the excess observed in GCRs.

7.4.1.1 Radioactive Isotopes in Cosmic Rays

Radioactive nuclides present in GCR may provide key information about various timescales concerning the acceleration and propagation of cosmic rays in the Galaxy. Primary nuclides unstable against $\beta^{\pm}$ and α decays or spontaneous fission, are sensitive to the timescale between nucleosynthesis and arrival on Earth; however, U and Th have such large lifetimes that they are essentially stable during the propagation of GCR in the Galaxy and they cannot be used as probes of that timescale. Primary nuclides unstable against *e*-capture only, are sensitive to the timescale from their nucleosynthesis to acceleration, since after acceleration to GCR energies they can hardly capture an orbital electron and they propagate essentially without decay; ^{57}Co and, especially, ^{59}Ni belong to this category. Secondaries unstable against $\beta^{\pm}$ and α decays or spontaneous fission can probe the residence timescale of GCR in the Galaxy, provided their half-lives are comparable to

Table 7.1 Decay modes and half-lives of key secondary radioactivities in GCR (from Yanasak et al. 2001)

Isotope	Decay mode	Mode half-life (year)
^{10}B	β^-	1.51×10^6
^{14}C	β^-	5.70×10^3
^{26}Al	β^+	8.73×10^5
	e-Capture	8.45×10^6
^{36}Cl	β^-	3.07×10^5
	e-Capture	1.59×10^7
^{54}Mn	β^-	$(6.3 \pm 1.7)\times 10^5$
	e-Capture	0.8548

that timescale; a handful of nuclides belong to this class (see Table 7.1). The GCR residence time, combined with the mean path length (obtained from the secondary/primary ratio) and the particle velocity, allows one to evaluate the mean density of the ISM traversed by GCR. Finally, secondary nuclides unstable against e-capture only, can be used as probes of re-acceleration effects, because e-capture is more likely at low energies. The next two subsections summarize the excellent review of Mewaldt et al. (2001) on radioactivities in GCR.

7.4.2 Observations of Cosmic Rays

7.4.2.1 Spectra and Composition

Spectra of Galactic cosmic rays as measured in Earth's vicinity have the same overall shape for all chemical elements. Above a few GeV/nucleon, and up to 10^6 GeV/nucleon, the GCR intensity $I(E)$ (in cm^{-2} s^{-1} st^{-1} GeV^{-1}) is described by a practically featureless power-law as a function of kinetic energy E

$$I(E) = 1.8\,E^{\alpha} \tag{7.3}$$

with slope $\alpha = -2.7$. At $E \sim 10^6$ GeV/nucleon (the *knee*), the GCR spectrum steepens to $\alpha = -3$. At E<1 GeV/nucleon, the GCR spectrum progressively flattens (with α becoming even positive) and its intensity varies, in a way which anti-correlates with the solar activity (*solar modulation*); the solar wind prevents the lowest energy GCR from entering the heliosphere. Correcting for that effect (*demodulating*) allows one to infer, through the time-averaged locally observed spectra, the *true* spectrum of GCR as they propagate in the ISM (Fig. 7.24). This allows one to evaluate the kinetic energy density of GCR ϵ_{GCR} as

$$\epsilon_{GCR} = 4\pi \int \frac{E_k}{v}\, I(E)\, dE \tag{7.4}$$

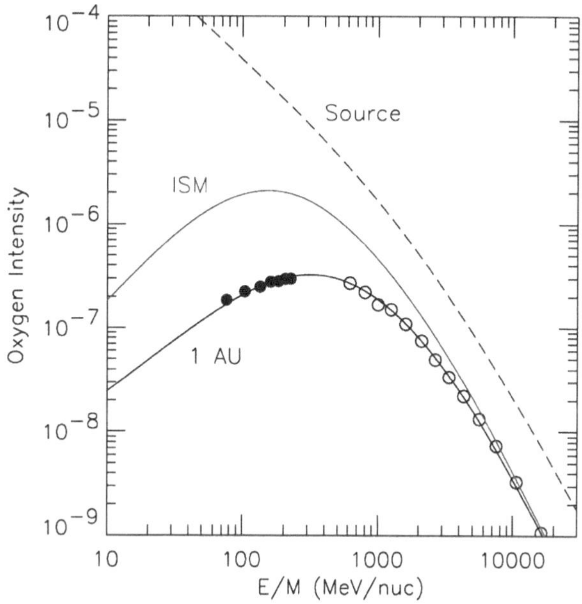

Fig. 7.24 Low energy GCR spectra of oxygen, observed near Earth (*bottom*, in the ISM (demodulated for the solar wind, *middle*) and in the GCR source (theoretically inferred, *top*) (from Wiedenbeck et al. 2007)

where υ is the velocity of a particle with kinetic energy E. The inferred local GCR energy density is $\epsilon_{GCR} \sim 1\,\text{eV}\,\text{cm}^{-3}$. It is comparable to the local energy densities of starlight (ϵ_{Light}), of ISM gas thermal motion ($\epsilon_{Gas} \sim n\kappa T$) and of the magnetic field ($\epsilon_B \sim \frac{B^2}{8\pi}$), i.e. GCR constitute an important dynamical agent in the Galaxy.

As seen above, GCR are expected to be confined for a timescale $\tau_{Conf} \sim 10^7$ years in the Galactic disk (before escaping to the halo and then to the intergalactic space), i.e. in a roughly cylindrical volume V of radius $R \sim 15\,\text{kpc}$ and height $h \sim 4\,\text{kpc}$. Assuming that ϵ_{GCR} is the same throughout that volume, one finds that the total GCR power is $P_{GCR} = \epsilon V/\tau_{Conf} \sim 2 \times 10^{41}\,\text{erg}\,\text{s}^{-1}$. Taking into account the expected supernova frequencies in the Milky Way ($f_{GalSN} \sim 2\,\text{century}^{-1}$) and assuming a kinetic energy of $E_{SN} = 1.5\ 10^{51}\,\text{erg}$ for each SN, one sees that $P_{GCR} \sim 0.1\ P_{GalSN}$ (where $P_{GalSN} = f_{GalSN} E_{SN} \sim 1.5 \times 10^{42}\,\text{erg}\,\text{s}^{-1}$). Those numbers suggest that supernova remnants accelerate cosmic rays in the Galaxy with an efficiency of the order of 10%.

The composition of cosmic rays arriving at Earth is best measured in the low energy range (0.1–1 GeV/nucleon), because in that region GCR intensity is highest and experimental techniques for particle identification have better resolution than at higher energies. In Fig. 7.25, the arriving GCR composition is compared to the solar system one. Both samples of matter are dominated by the elements H, He, C, O, Ne, Mg, Si and Fe. The overall similarity of the two abundance curves suggests that they have similar nucleosynthetic origin, i.e. from massive stars and supernovae (for nuclei in the C-Fe peak range). Volatile elements (He, O, Ne, S and Ar) are somewhat depleted relative to refractory Si, perhaps as a result of some fractionation process, favouring the acceleration of more refractory elements (see next section).

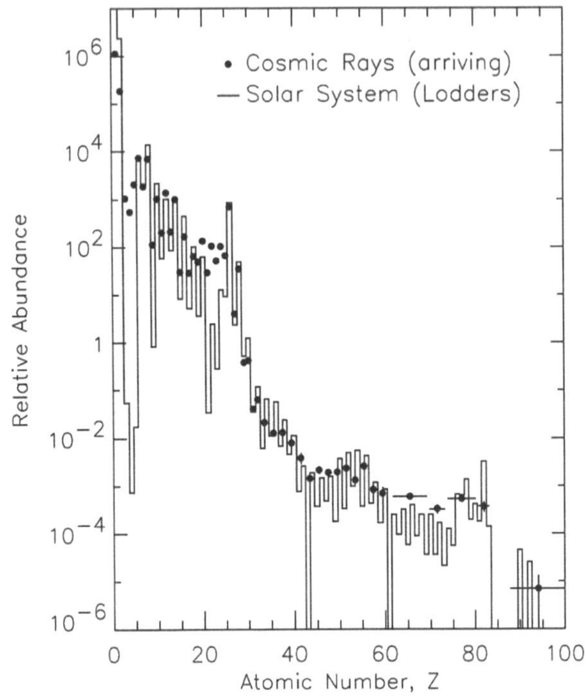

Fig. 7.25 Elemental abundances in arriving GCR (*dots*), compared to the solar system ones (*histogram*) (from Wiedenbeck et al. 2007)

The most striking difference between GCR and solar abundances is the GCR overabundance of some nuclides which are relatively rare in the solar system composition. The atomic numbers of those rare nuclides are only a few units below those of the nuclides dominating the abundance curves. This is the case of e.g. Li, Be and B with respect to the slightly heavier C, N and O nuclei ($(\frac{LiBeB}{CNO})_{solar} \sim 10^{-6}$, while $(\frac{LiBeB}{CNO})_{GCR} \sim 10^{-1}$) or of Sc, Ti and V nuclei with respect to Fe peak nuclei (see Fig. 7.25). The relative overabundance of those rare nuclei in GCR has long been recognised to be the result of their *secondary* nature: they are produced by the nuclear fragmentation (*spallation*) of their heavy progenitors (*primaries*), during the GCR propagation in the ISM.

The properties of GCR secondaries, both stable and radioactive, provide important information on the physics of Galactic cosmic rays.

7.4.2.2 Radioactive Clocks and Cosmic-Ray Timescales

Determination of the confinement (or residence or escape) timescale τ_{Conf} of GCR in the Galaxy is a key issue, because τ_{Conf} determines the power required to sustain the energy density of GCR (see Sect. 7.4.2). All four nuclides of Table 7.1 have been measured in arriving GCR after the Ulusses and ACE results and the situation is summarized in Fig. 7.26. Confinement times are obtained in the framework of a

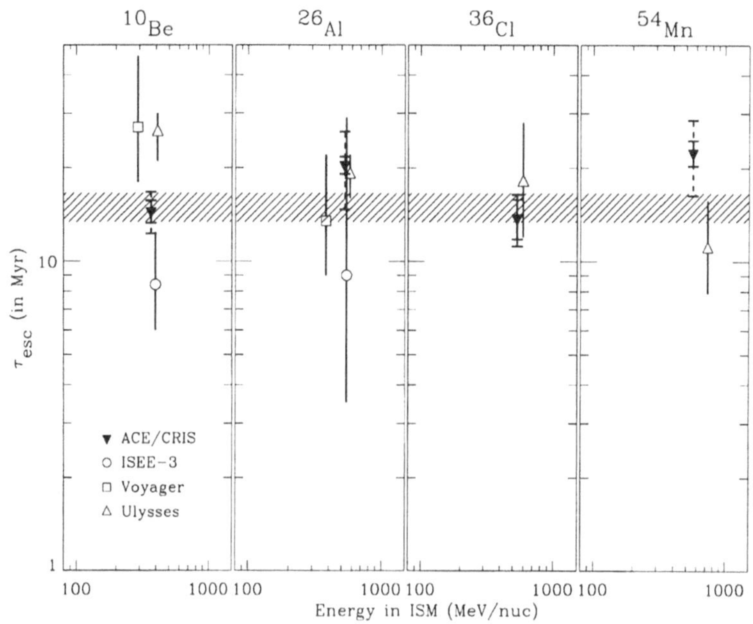

Fig. 7.26 Confinement times obtained by ACE/CRIS and previous experiments. Uncertainties shown with solid error bars are 1 standard deviation statistical. The average value of the confinement time, $\tau_{esc} = 15.0 \pm 1.6$ Myr, indicated by the CRIS data for the four clock isotopes is shown as a hatched band (From Yanasak et al. 2001)

Leaky box model with energy dependent escape length and the average value τ_{Conf} is found to be 15.0 ± 1.6 Myr. In the Leaky box model, $\tau_{Conf} = \Lambda_{esc}/(\upsilon\rho)$, where $\upsilon = \beta c$ is the GCR velocity (at <1 GeV/nucleon, the effects of solar modulation have to be accounted for in the calculation of β). This allows one to evaluate the average ISM density traversed by GCR as $n = \rho/m_p$ =0.36 H atoms cm^{-3} (Yanasak et al. 2001). This is a factor of ~3 lower than the canonical value of the local ISM density (~1 H atom cm^{-3}).

In the Leaky box model, where GCR intensities and ISM densities are uniform in the GCR propagation volume (and in time), such measurements can probe only the average density of the confinement region. The above result implies that GCR spend a large fraction of their confinement time in a volume of smaller average density than the one of the local gas, i.e. in the Galactic halo. In more realistic models, involving e.g. diffusion (see Strong et al. 2007), the aforementioned radioactivities probe a volume of radius R which is limited by their mean life τ, such that $R \sim (\gamma D \tau)^{1/2}$, where D is the spatial diffusion coefficient and $\gamma = (1 - \beta^2)^{-1/2}$. In that scheme, at 1 GeV/nucleon, ^{10}Be probes regions out to ~400 pc (i.e. beyond the gaseous layer), while ^{14}C probes the immediate vicinity of the solar system; however, its expected signal is lower than the background due to ^{14}C produced inside the ACE/CRIS instrument. Notice that uncertainties in the derived ISM densities are dominated by uncertainties in fragmentation cross-sections, rather than by measurement uncertainties (Yanasak et al. 2001).

Finally, in diffusion models of GCR propagation, the expected surviving fraction of secondary radioactivities depends on the assumed diffusion coefficient D. Comparing the ACE/CRIS measurements for ^{10}Be, ^{26}Al and ^{36}Cl at 400 MeV/nucleon with theoretical predictions of Ptuskin and Soutoul (1998), values of $D \sim 2 \times 10^{28}\,\mathrm{cm^2\,s^{-1}}$ are found (see Fig. 9 in Mewaldt et al. 2001).

Radionuclides unstable to e-capture, if accelerated to >500 MeV/nucleon, become fully stripped of orbital electrons, and cannot decay. This suggests that, if $f = X/D_X$ (the abundance fraction of such a nuclide X relative to its decay product D_X) arriving on Earth >>1, then the timescale Δt between the nucleosynthesis and acceleration of X is smaller than its decay timescale τ_X by e-capture; inversely, an arriving fraction $f = X/D_X$ <<1 would imply that $\Delta t > \tau_X$. Three such radionuclides were identified by Soutoul et al. (1978) as potentially important in that respect: ^{56}Ni, ^{57}Co and ^{59}Ni, with half-lives against e-capture from 6 days to 7.6×10^4 years. It is well known now that ^{56}Ni is the dominant product of explosive nucleosynthesis in supernovae, however, GCR composition is dominated by its decay product ^{56}Fe; thus, Δt is much larger than 6 days.

The most interesting case is the one of ^{59}Ni, which probes rather large values of the timescale Δt. Nucleosynthesis calculations (Woosley and Weaver 1995) suggest that ~2/3 of the A = 59 isobars is synthesized as ^{59}Ni The analysis of ACE/CRIS data showed that very little (if any at all) ^{59}Ni exists in arriving GCR, while its daughter nuclide ^{59}Co is present at the level of ^{59}Co/^{60}Ni~0.25 (Fig. 7.27). An absence of ^{59}Ni from GCR implies that $\Delta t > 10^5$ years.

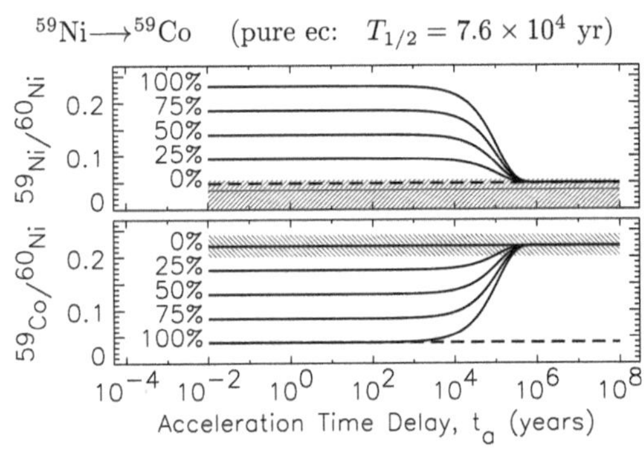

Fig. 7.27 Calculated abundances at Earth of ^{59}Ni/^{60}Ni (upper panel) and ^{59}Co/^{60}Ni (lower panel) are shown as a function of the time delay between nucleosynthesis and GCR acceleration. Calculated abundances are a combination of a secondary component (*dashed lines*) produced by nuclear fragmentation during transport and a surviving primary component. The different curves correspond to different assumed fractional contributions of ^{59}Ni in the primary A=59 material, as indicated by the labels on the curves. The time dependencies are the result of the exponential decay of the primary ^{59}Ni into ^{59}Co as the result of the electron-capture decay of ^{59}Ni before acceleration. The hatched regions indicate the abundances measured with CRIS, including 1 σ uncertainties. *From* Wiedenbeck et al. (1999)

Such a long timescale between nucleosynthesis and GCR acceleration would have important implications for our understanding of GCR composition. It would imply that SN cannot accelerate their own ejecta: even in a rarefied medium, the SN ejecta are completely diluted in (and mixed up with) the interstellar gas on timescales as long as 105 years.

There is a loop-hole to the above argument: while e-capture is essentially impossible at sufficiently high energies (>500 MeV/nucleon), it becomes possible at lower energies. The average energies of ACE/CRIS measurements (~300 MeV/nucleon) correspond to ~450 MeV/nucleon outside the heliosphere. If ^{59}Ni would not be directly accelerated to high energies after its ejected from the supernova source, it might capture an electron and decay before it became a cosmic-ray component. In that case, its absence in arriving GCR provides no information on a delay between its nucleosynthesis and acceleration.

Recent updates of supernova yields, including the effects of stellar rotation, however obtain (Neronov and Meynet 2016) a much lower abundance of ^{59}Ni compared to ^{59}Co, compared to the earlier supernova nucleosynthesis calculations (Woosley and Weaver 1995). This then reduces the abundances that would be expected in cosmic rays even excluding any decay prior to acceleration. Therefore the above-discussed exploitation of the decay properties of ^{59}Ni may have been premature.

There is an important issue of GCR, which is re-acceleration: do GCR get their high energies in a single SN blast wave, or in a series of SN shock waves (*distributed acceleration*)? Secondary radionuclides, unstable against e-capture, with short enough timescales, can be used as probes of such processes: these include ^{7}Be, ^{37}Ar, ^{44}Ti, ^{49}V, ^{55}Fe, with laboratory lifetimes extending up to 67 years. Comparing the measured abundance ratios of those nuclides to their stable daughter nuclei in various energy bands, one may, in principle, constrain the extent of re-acceleration. Unfortunately, the analysis of observations from various experiments (*Ulysses, Voyager, ACE*) has produced contradicting results up to now. However, as stressed in Mewaldt et al. (2001), measurements of e-capture radionuclides in GCR are still in their infancy. The same holds for models to interpret the data, which suffer from uncertainties related to cross-sections (for fragmentation, e-capture and loss as a function of energy), the history of GCR acceleration and re-acceleration, inhomogeneities of the ISM, etc. Improvements in both measurements and interpretation will contribute to a much better understanding of the physics of GCR propagation (Fig. 7.28).

Spallation of cosmic rays during their journey has been shown above to produce a number of interesting radioactive species. However, the discovery of ^{60}Fe in cosmic rays (Binns et al. 2016; Fig. 7.28) came as a surprise, because this cannot originate from such spallation.

The detection of ^{60}Fe thus must be understood from a different viewpoint: As Binns et al. (2016) argue, its detection presents an upper limit on the time it takes to accelerate cosmic rays. ^{60}Fe obviously survived a journey from its production through acceleration to detection, which is compatible with its 3.6 Myr radioactive lifetime, that, in this case, is independent of the atomic electron shell population.

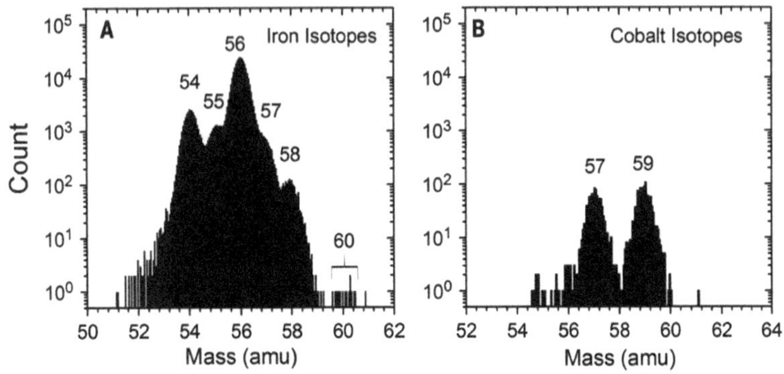

Fig. 7.28 Composition of iron and cobalt isotopes in cosmic rays near Earth (as observed with ACE/CRIS Binns et al. 2016)

Binns et al. (2016) argue that only less than 10% of their detected ^{60}Fe isotopes could originate from heavier nuclei being fragmented and hence be 'secondary', and most of the nuclei therefore are considered 'primary'. They use then a simple diffusive propagation model to argue for an origin within 620 pc of the Sun, making supernovae from the Scorpius-Centaurus association potential sources. The propagation of ejecta from source to detector obviously remains a major scientific issue today.

It is interesting that, from a different viewing point, Kachelrieß et al. (2015) infer from the anisotropy of the cosmic ray spectral signatures that a nearby supernova explosion a few Myrs ago is suggested, to provide peculiar conditions for cosmic ray acceleration in the nearby interstellar medium. Taken together, this is reminiscent of the scenario discussed above for the Scorpius Centaurus association and its history of the past 10–20 Myrs (Sect. 7.2.4.3).

7.5 Positrons and Their Annihilation

One of the most direct and unambiguous manifestations of radioactivity is the emission of positrons from the β^+ decay of unstable nuclei. Such nuclei are located on the proton rich side of the valley of nuclear stability in the isotope chart. This is the case, e.g., for ^{26}Al, ^{44}Ti or ^{56}Ni and ^{56}Co, produced in massive stars and supernova explosions, also for ^{13}N, ^{14,15}O, ^{17}F, and ^{22}Na produced in novae. If the lifetime of the unstable nucleus is larger than the timescale τ_T that may be characteristic for the *trapping* of the positrons within the expanding explosive production site, then a substantial fraction of the released positrons may escape from that site of nucleosynthesis and find its way into the interstellar medium. Most of the positrons from the decay of ^{44}Ti ($\tau \sim 89$ years) and ^{26}Al ($\tau \sim$1 Myr), and ^{22}Na ($\tau \sim$3.8 years) are therefore expected to be released within the interstellar medium, while the fate of those released by short-lived ^{56}Ni ($\tau \sim$8.8 days) and

^{56}Co (τ ~77 days) is not clear. Note that, for example, the late bolometric emission from supernovae that is seen after 100–1000 days mostly is energised by the energy deposits from positrons within the expanding supernova.

But there is a variety of other candidate sources of positrons, either through pair plasma production in high-density and high-energy environments, or from high-energy particle collisions, or even dark matter annihilation or decay.

Positrons have been observed directly in particle detectors, but more commonly their annihilation is seen, through a characteristic gamma-ray signature that involves a unique line at 511 keV. First measured from the Sun in solar flares (Chupp et al. 1973), the gamma-ray signals of positron annihilation were observed from our Galaxy's central region (Leventhal et al. 1978; Purcell et al. 1993; Knödlseder et al. 2005; Siegert et al. 2016b) and also from transient sources (Sunyaev et al. 1992; Bouchet et al. 1991; Siegert et al. 2016a). By now, imaging and spectroscopic results have been established, which make cosmic positrons to be one of the major current puzzles in astrophysics. Prantzos et al. (2011) have reviewed this topic in great detail; we summarise this here, with some updates since their review. This Section describes the processes relating positrons across all energies to their annihilation gamma-ray signature, with an aim to learn about the contributions from radioactivities.

7.5.1 Candidate Sources and Expected Observables

Positrons can be produced either by radioactivity in stars, novae and supernovae, or through various high energy processes in compact objects (production of positrons by cosmic rays, which are accelerated by supernovae, belongs to the latter class); processes involving dark matter (DM) annihilation, decay or de-excitation is a third possibility.

Among the various astrophysical sources of positrons proposed so far, the only one known certainly[5] to release e^+ in the ISM is β^+ radioactivity of ^{26}Al; the observed intensity of its characteristic 1.8 MeV emission in the Galaxy corresponds to $\sim$ 3–4 $\times$ 10^{42} e^+ s^{-1}. A similar amount is expected from the decay of ^{44}Ti, on the grounds of nucleosynthesis arguments (it is the parent nucleus of stable ^{44}Ca). Both radionuclides are produced mostly in massive stars and their positrons should be released along the Galactic plane, as traced by the 1.8 MeV emission; they could thus account for the observed disk 511 keV emission.

Radioactivity of ^{56}Co from SNIa was traditionally considered to be the major e^+ producer in the Galaxy. Both the typical ^{56}Ni yield of a SNIa and the Galactic SNIa rate are rather well constrained, resulting in 5 $\times$ 10^{44} e^+ s^{-1} produced *inside* SNIa. If only f_{esc} ~4% of them escape the supernova to annihilate in the ISM, the

[5] That is, by proof from observations: ^{26}Al gamma rays are observed, and ^{26}Al decay is known to emit positrons.

observed total e^+ annihilation rate can be readily explained. However, observations of two SNIa, interpreted in the framework of 1-D (stratified) models, suggest that the positron escape fraction is negligible *at late times*. On the other hand, both observations of early spectra and 3-D models of SNIa suggest that a sizeable fraction of ^{56}Ni is found at high velocity (close to the surface), making—perhaps—easier the escape of ^{56}Co positrons. SNIa remain a serious candidate, with a potential Galactic yield of 2×10^{43} e^+ s^{-1}. But the expected spatial distribution of SNIa in the Galaxy would predict a much smaller bulge-to-disk ratio than the one of the observed 511 keV profile.

Several candidate sources will likely produce positrons through various high-energy processes: pulsars, millisecond pulsars, magnetars, accreting binary systems such as microquasars, and the central supermassive black hole (SMBH) in the Milky Way. There is little observational evidence that either of those sources produces positrons, however, and the e^+ yields evaluated by various authors are often (intentionally) optimistic rather than typical values. Among those sources, the most promising appear to be accreting binaries and the massive black hole at the Galactic centre. Binaries are observed as LMXRB X-ray sources; the microquasar variant of that class of sources with its plasma jets are plausible positron sources. In one case, a flaring microquasar has been shown to emit a gamma-ray signal that suggests the presence and ejection of high-energy positrons (see below). Because of the current low activity of the central SMBH it has to be assumed that the source was much more active in the past, thus dropping the assumption of *steady state* between e^+ production and annihilation, which is adopted in all other cases.

Positrons originating from radioactive decay are released with typical energies as they are characteristic of the difference between nuclear energy levels, i.e. of the order of ~MeV. With their rest mass energy of 511 keV, these positrons will thus have relativistic velocities. The alternative source processes for positrons fall into two classes, with respect to positron energies: If pair plasma is involved, and the source is compact and at rather high density, back-reactions between charged particles and photons (also virtual photons representing a high magnetic field) will limit production processes by the pair creation threshold to energies around MeV. In the case of relativistic jets, this may be boosted by relativistic motion. If the production region is less extreme, the positron energy is not limited from such considerations, and depends on the primary particle energies. In pulsar magnetospheres, therefore, GeV to TeV energies can be reached. For secondary processes involving dark matter, the mass of the dark matter particle sets an upper limit; commonly, masses of GeV and above are considered plausible.

7.5.1.1 Signatures of Positron Annihilation

Once in the interstellar medium, the positrons slow down through the various processes of Inverse-Compton scattering, synchrotron emission, Coulomb collisions, ionization losses, and finally they can annihilate when at eV energies (Bussard et al. 1979). Annihilation may occur either directly with electrons, or after formation of

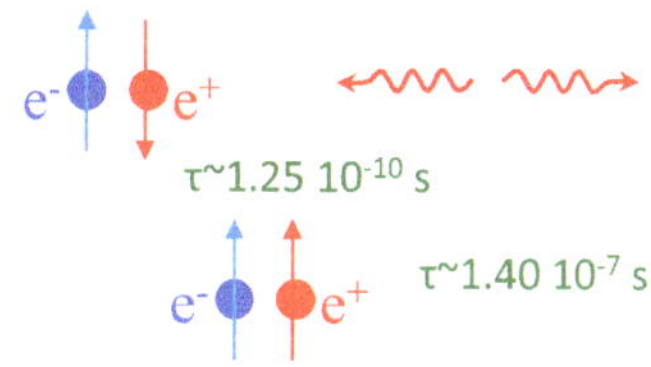

Fig. 7.29 The two configurations of the positronium atom, as an intermediate stage before annihilation of a positron with an electron

a *positronium* atom (with the positron substituting the positively-charged atomic nucleus); momentum conservation sets constraints. In both cases, the characteristic γ-ray line at 511 keV is emitted, whereas annihilation from a positronium also produces a γ-ray continuum below 511 keV.

The annihilation of a positron with an electron releases a total (rest-mass) energy of 1022 keV in the form of two or more photons. Direct annihilation of a e^--e^+ pair at rest produces two photons of 511 keV each. The situation is more complex in the case of positronium (Ps). Positronium has two basic states, depending on the relative orientations of the spins of the electron and the positron (see Fig. 7.29). The singlet state has antiparallel spins, total spin S = 0, is denoted as $^1S_0$ and is known as *para-positronium* (p-Ps). The triplet state has parallel spins, total spin S = 1, is denoted as $^3S_1$ and is known as *ortho-positronium* (o-Ps). From the (2S + 1) spin degeneracy, it follows that Ps will be formed 1/4 of the time in the p-Ps state and 3/4 of the time in the o-Ps state. The energy difference between the two spin states (*hyperfine splitting*) is 8.4×10^{-4} eV. Transitions between these states similar to the spin-flip transition in hydrogen, which produces the astrophysically-important 21 cm line of HI, are unimportant due to the short Ps lifetimes.

Spin and momentum conservation control the release of annihilation energy in the form of photons. Para-positronium annihilation releases two photons of 511 keV each in opposite directions (as in the case of direct e^--e^+ annihilation). Ortho-positronium annihilation requires a final state with more than two photons from spin conservation; momentum conservation distributes the total energy of 1022 keV among three photons producing a continuum of energies up to 511 keV (Fig. 7.30). The corresponding lifetimes before annihilation (in vacuum) are 1.2×10^{-10} s for para-Ps and 1.4×10^{-7} s for ortho-Ps (see Fig. 7.29).

If a fraction f_{Ps} of the positrons annihilate via positronium formation, then the 3-photon γ-ray continuum of ortho-positronium will have an integrated intensity of

$$I_{3\gamma} \propto \frac{3}{4} 3 f_{Ps} \tag{7.5}$$

The remaining fraction $1 - f_{Ps}$ will annihilate directly to 2 photons of 511 keV each. We should add to this the 2-photon contribution of the para-Positronium state, so that the 2-photon (511 keV line) intensity will be:

$$I_{2\gamma} \propto 2(1 - f_{Ps}) + \frac{1}{4} 2 f_{Ps} = 2 - 1.5 f_{Ps} \tag{7.6}$$

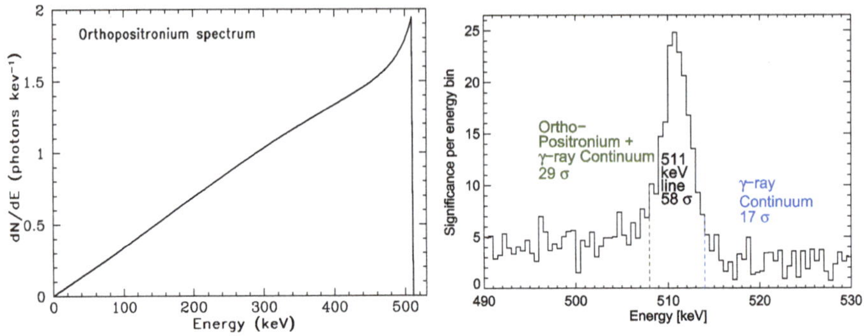

Fig. 7.30 *Left:* Spectrum of ortho-positronium annihilation with the three-photon continuum (from Ore and Powell 1949). *Right:* Spectrum of the annihilation emission measured by SPI, with the 511 keV line and the 3-photon continuum from annihilation through positronium. An underlying Galactic continuum emission is seen as well (from Siegert et al. 2016b)

This expected spectrum is shown in Fig. 7.30. By measuring the intensities of the 511 keV line and of the Ps continuum one can then derive the positronium fraction

$$f_{Ps} = \frac{8\, I_{3\gamma}/I_{2\gamma}}{9 + 6 I_{3\gamma}/I_{2\gamma}} \qquad (7.7)$$

This quantity offers a valuable diagnostic of the physical conditions of the ISM where positrons annihilate, as these impact on the positronium formation efficiency.

7.5.1.2 Issues of Positron Transport in the Interstellar Medium

The propagation and transport of positrons has important implications for the interpretation of Galactic 511 keV emission with respect to positron origins (first raised by Prantzos 2006a). Positrons injected from various sources interact with the surrounding medium either through collisions (collisional transport) or through interaction with plasma waves (collisionless transport). If only limited by collisions with ambient gas, MeV particles can propagate large distances—of the order of 10 kpc/*n* (*n* is the local density in cm^{-3}). These distances would be shorter by a factor 0.75 if the tangled components of the magnetic field are added (Jean et al. 2009; Alexis et al. 2014). Such distances are much larger than the typical sizes of structures in the hot and warm phases of the ISM.

Collisionless transport is governed by wave-particle interactions (Bykov and Treumann 2011). In the case of a magnetised plasma, positrons spiral along the magnetic field lines. The gyroradius of a positron with Lorentz factor γ is $r_g \sim 1.7 \times 10^9\, B_{\mu G}^{-1} (\gamma^2 - 1)^{1/2}$ cm, where the local mean magnetic field $B_{\mu G}$ is expressed in μG. In a magnetised, turbulent, plasma, the most efficient of collisionless processes is scattering off magnetic fluctuations of size $r_B \simeq r_g$, which induce *resonant* pitch angle scattering of positrons (e.g. Kulsrud 2005, and references therein), or *non-resonant* interactions with fluctuations on scales just above r_g (see e.g. Toptygin 1985; Ragot 2006). The resonant interactions involve either the particle gyromotion around the mean magnetic field (*cyclotrotron resonance*) or the parallel motion of the particle along the field line (*Cherenkov resonance*). The Larmor radius of the resonant positron being small—of the order of $r_g \simeq 10^9$cm and the positron-gyromotion polarisation being left-handed only MHD waves (Alfvén or fast magneto-sonic) can fulfil the resonance conditions. Non-resonant compressible perturbations may also affect the positron mean free path, as it seems to be the case for sub-MeV electrons in the solar wind, where there are strong indications for MeV electron re-acceleration (Ragot 1999). Moreover, *adiabatic deceleration* of positrons in jets or expanding shells (for example in SN remnants) results in positron cooling, even without Coulomb collisions; this occurs if the positron mean free path, which is dominated by e^+ scattering by waves, is shorter than the typical scale of bulk plasma motion.

In summary, wave-particle interactions, both resonant and adiabatic non-resonant, could result in particle deceleration, but also *re-acceleration*, depending strongly on the local conditions. The transport of energetic (>GeV) Galactic cosmic rays is driven by such collisionless processes. The case of MeV positrons is not clear, as the level and scale of interstellar turbulence is not well understood.

For low-energy positrons, Alexis et al. (2014) performed a detailed Monte Carlo study of propagation in the various phases of the ISM and found that propagation over distances larger than ~1 kpc through the Galaxy seems rather unlikely (neglecting, however, any re-acceleration). This implies that the morphology of the 511 keV emission does not necessarily reflect the morphology of the underlying e^+ source distribution, as propagation up to kpc distances appears possible. As an example, positrons from SNIa are expected to be released away from dense gas and in the hot and rarefied ionised medium, since the scale height of SNIa is considerably larger than the scale height of the cool, dense gas in the Galactic disk (see Fig. 11.15). Another example is the release of massive star nucleosynthesis ejecta with preference in large superbubbles surrounding somewhat-evolved massive star groups (see Sect. 7.2.3). The e^+ propagation distances could then be quite large, possibly with channels open towards the Galactic halo ('chimneys'), thus allowing e^+ from the disk to annihilate far away from their sources (perhaps in the halo, where a low surface brightness emission should be expected). We shall discuss further those issues in the last section of this chapter.

7.5.2 *Observations of Positrons*

The 511 keV emission of interstellar e^+ annihilation was first detected from the general direction of the Galactic centre in the early 1970s, by balloon borne experiments of low energy resolution (Johnson et al. 1972). It was unambiguously identified a few years later with high resolution Ge detectors (Leventhal et al. 1978).

It is the first and most intense γ-ray line originating from outside the solar system that was ever detected. Its flux on Earth ($\sim 10^{-3}$ cm^{-2} s^{-1}), if combined with the distance to the Galactic center ($\sim$8 kpc) (assuming that annihilation occurs in the inner Galaxy), implies the annihilation of $\sim 2 \times 10^{43}$ e$^+$ s^{-1} (see below), releasing a power of $\sim 10^{37}$ erg s^{-1} or $\sim 10^4$ L$_\odot$ in γ-rays. Assuming a steady state, i.e. equality between production and annihilation rates of positrons, one should then seek for a source able to provide $\sim 2\times 10^{43}$ e$^+$ s^{-1}. If the activity of that site were maintained at such level during the $\sim 10^{10}$ years of the Galaxy's lifetime, a total amount of positrons equivalent to $\sim$3 M$_\odot$ would have been annihilated.

Imaging the Galaxy in annihilation γ-rays was considered to be the exclusive way to identify the cosmic e$^+$ sources (assuming that the spatial morphology of the γ-ray emission reflects the spatial distribution of the sources, i.e. that positrons annihilate close to their production sites). Because of the difficulties of imaging in the MeV region, progress was extremely slow in that field: Only in the 1990s the first hints on the spatial distribution of the 511 keV emission were obtained by the OSSE instrument aboard the Compton Gamma Ray Observatory (CGRO, Cheng et al. (1997); Purcell et al. (1997); see Chap. 10). The first reliable imaging of the 511 keV emission was obtained by the SPI coded-mask imaging instrument aboard ESA's INTEGRAL Gamma Ray Observatory (see Chap. 10). Around 0.5 MeV, SPI has a spatial resolution of 3° (FWHM) and a spectral resolution of $\sim$2.1 keV (FWHM, at 0.5 MeV). This allows for imaging, but also a spatially resolved fine spectroscopy of the signal (including the underlying continuum emission).

7.5.2.1 Imaging Studies

The first all-sky maps of positron annihilation gamma rays, which have been presented from 1 year of observations by Knödlseder et al. (2005) for the 511 keV line and for the positronium continuum Weidenspointner et al. (2006b), respectively (see Fig. 7.31), showed that the emission is strongly concentrated in the inner Galaxy, and much weaker brightness is seen from the Galactic disk (Weidenspointner et al. 2008). These two maps are identical within their uncertainties, which suggests that the positronium fraction does not vary much over the different emission regions across the sky.

The images show a remarkable predominance of the spheroidal component from the inner Galaxy. The earlier imaging study with OSSE had suggested a clear emission component from the Galactic disk, but this seemed to be completely absent in the first images made with SPI (see Fig. 7.31 top). Also model fitting indicated only a marginal signal from the Galactic disk, and a bulge-to-disk flux ratio $>$ 1 had been suggested (Knödlseder et al. 2005).

This emission morphology is strikingly-different than the ones seen in any other wavelength for radiation from the Galaxy. Such strong dominance of the Galactic bulge, unseen in any other wavelength, stimulated above-mentioned *unconventional* models involving dark matter.

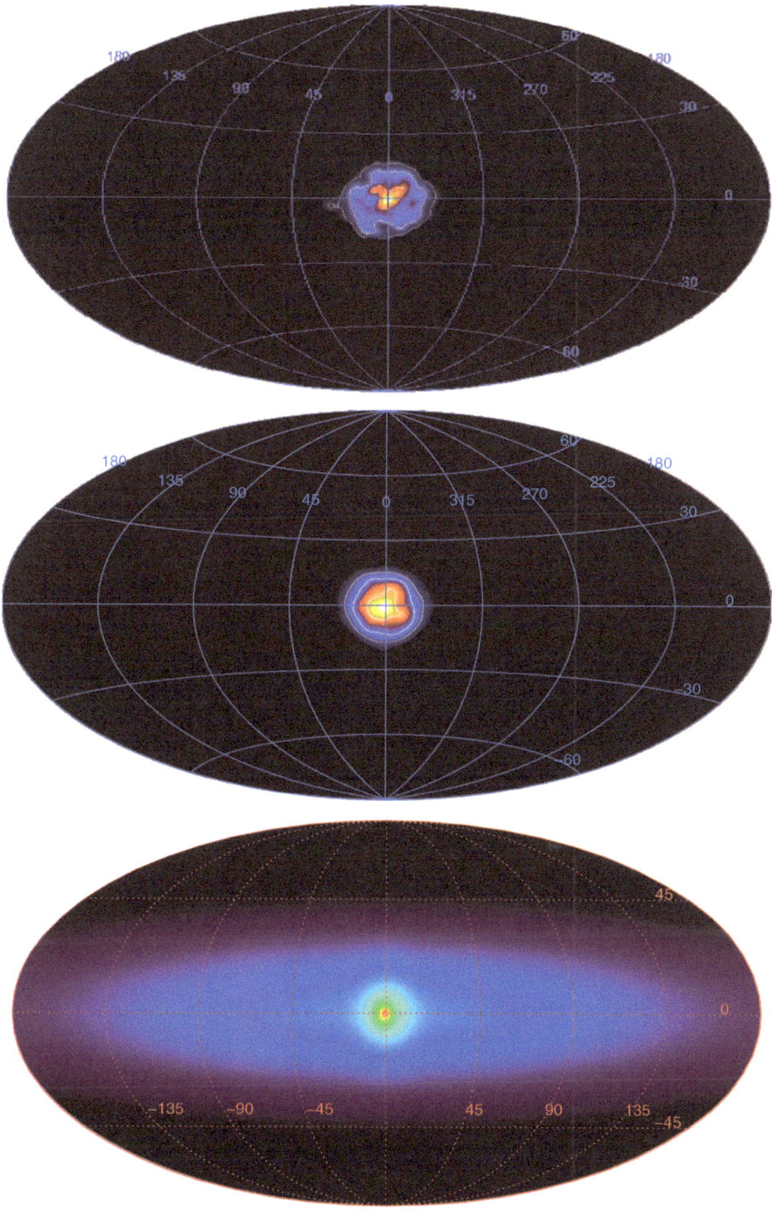

Fig. 7.31 Images of positron annihilation gamma rays as derived from INTEGRAL/SPI data. *Top:* First image from Richardson-Lucy deconvolution, the 511 keV line (Knödlseder et al. 2005). *Middle:* Image from Richardson-Lucy deconvolution, the positronium 3-photon continuum (Weidenspointner et al. 2006b). *Bottom:* Image from model fit optimisations, the 511 keV line (Siegert et al. 2016b; Skinner et al. 2014)

The detailed quantitative characterisation of the different components of 511 keV emission requires parameterising these in the form of (necessarily idealised) spatial emission models fitted to the data. Such decomposition is not unique, both the spheroid and the disk may have faint extensions contributing substantially to their total γ-ray emissivities. In the early years of INTEGRAL/SPI analyses, thin to moderately extended disk models had been tested. With more exposure, the disk emission became clear, and also why it had been difficult to detect it: In a parameter study from 13 years of observations, Siegert et al. (2016b) showed that the disk component appears to have a low surface brightness, although as a whole being as intense as the emission from the inner Galaxy. The bulge-to-disk flux ratio derived from these deeper observations now falls well below the values beyond 1 that stimulated the above discussions of exotic origins, and is determined as (0.58 ± 0.13) (Siegert et al. 2016b). The disk component of annihilation gamma rays seems quite extended, up to kpc in latitude. This suggests that positrons may fill a much larger volume than previously thought, and may annihilate as they leave the gaseous disk of the Galaxy towards the halo.

The bulge emission is best described by combining a narrow and a broad source region that can be described as a 2-dimensional Gaussian, with widths (FWHM, projected onto the sky) of 5–6° and 20°, respectively. The rather thick disk of vertical extent 25° (FWHM projected on the sky) also may not only be bright in the inner quadrants, as suggested by the large extent in latest model fits of almost 150° (FWHM) (Siegert et al. 2016b). But morphological modelling remains uncertain, even a very significant total-sky signal (see Fig. 7.30) becomes marginal as it is split up on the sky, in particular as the bulge region dominates so clearly in surface brightness.

There was considerable excitement about an apparent asymmetry of the emission with respect to the direction towards our Galaxy's centre. The flux asymmetry for fourth versus first galactic quadrant as reported by Weidenspointner et al. (2008) seemed like an invitation to look for positron sources with a similar asymmetric distribution in the Galaxy. Low-mass X-ray binaries appeared suggestive, although it remained unclear how those should emit large amounts of positrons. Their extreme variants, when the compact component would be a stellar-mass black hole, are called 'microquasars', and are known to eject plasma jets, as seen in radio emission. Although microquasars have been among the candidate sources, only few of those are identified in our Galaxy and appear throughout the disk; but from observational biases due to jet aspects, uncertainties are large. Currently, there is a consensus that the bright bulge component is not centred in the Galaxy, but rather offset by 1.2° towards the fourth quadrant. Although this is a conclusion that had first been suggested also by imaging deconvolution (Bouchet et al. 2010), and is different from an asymmetric disk component, this had stimulated ideas that the Galaxy altogether and including its candidate positron sources as possibly dominated by radioactivities was indeed not entirely symmetric nor centred at Galactic coordinates (0°,0°) (Higdon et al. 2009) (Fig. 7.32).

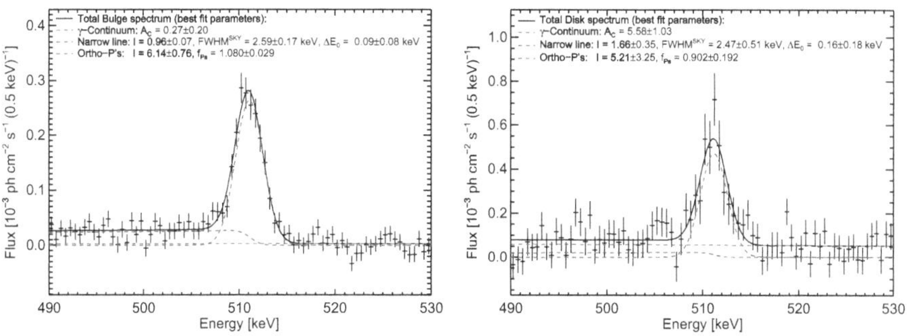

Fig. 7.32 Spectra of the annihilation emission from bulge (left) and disk (right) (Siegert et al. 2016b). The different annihilation conditions are indicated from the different intensities of the 3-photon continuum

7.5.2.2 Spectroscopy

The excellent spectral resolution of SPI allowed for the first time to study the spectrum of the 511 keV emission in great detail and for different regions (Figs. 7.32 and 7.33). The spectra of the Galactic spheroidal emission were analysed by Churazov et al. (2005) and Jean et al. (2006), based on the first year of SPI data. The line displays no significant overall spectral shift, i.e. it appears at the expected energy $E = 511$ keV within 0.05 keV or less (Siegert et al. 2016b) and it is composed of two spectral components (assumed, to first order, to be represented by Gaussians): a narrow line with a width of FWHM $= 2 \pm 0.08$ keV and a broad component with a width of FWHM $= 10 \pm 0.5$ keV (Fig. 7.34 and Table 7.2). The width of the broad line is in agreement with the broadening expected from positronium annihilation via charge exchange with hydrogen atoms. The narrow line component contains $\sim 2/3$ of the total annihilation line flux while the broad one makes up the remaining $\sim 1/3$ of the flux. Table 7.2 summarises the results of the spectral analysis of the Galactic 511 keV emission after the first year of SPI data.

SPI also clearly detected the ortho-positronium continuum with an intensity that corresponds to a positronium fraction of $f_{Ps} = 100 \pm 10\%$ for bulge and for disk component (Siegert et al. 2016b) (see Eq. 7.7). This confirms earlier measurements obtained by SPI and other instruments for the bulge (Jean et al. 2006; Kinzer et al. 1996; Harris et al. 1998, (97±2 %) for SPI, for OSSE (97±3 %) and for TGRS (94± 4 %)). This suggests that practically all annihilations occur through the intermediate step of the positronium atom (Fig. 7.33).

The shape of the annihilation line and the relative intensity of the ortho-positronium continuum are closely related to the physical conditions such as density, temperature and chemical abundances of the interstellar medium in which positrons annihilate. Figure 7.34 shows that the 511 keV line is rather narrow, with some broadening indicated in the wings. The results of the spectral analysis of the bulge emission suggest that positrons annihilate mostly in a warm medium (T$\sim 10^4$ K)

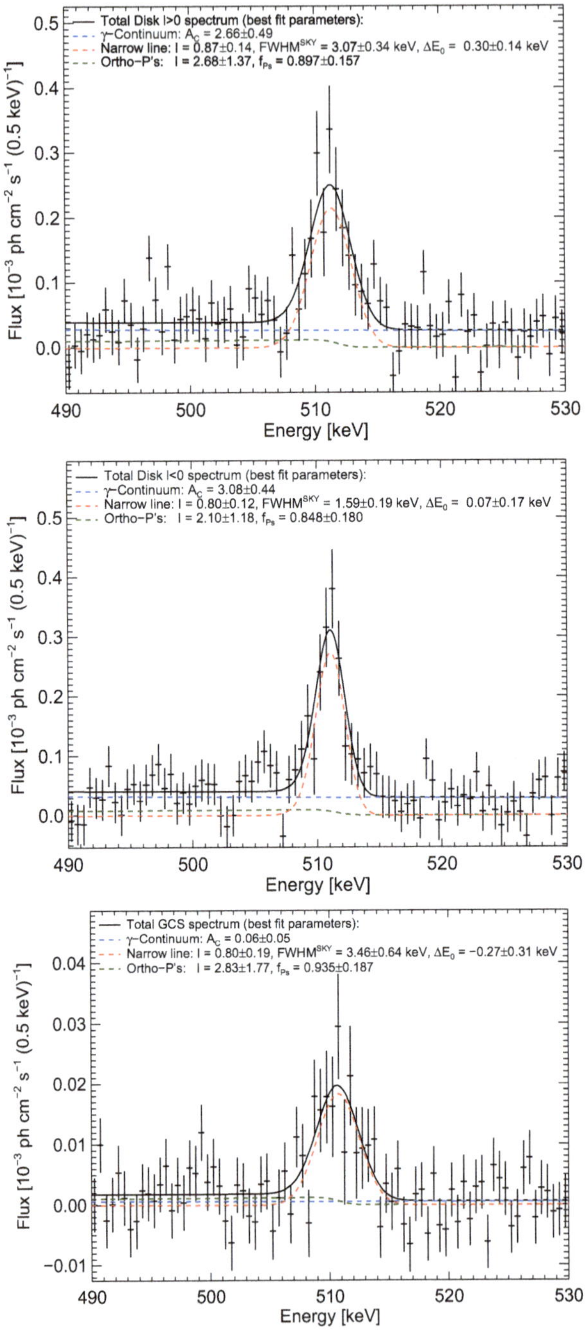

Fig. 7.33 Spectra of the different aspects of annihilation spectroscopy. The left and right (first and fourth) quadrants of the Galaxy show very similar spectra, although intensity differs slightly. The line for the point source located in the vicinity of Sgr A* may have a slightly broader line (from Siegert et al. 2016b)

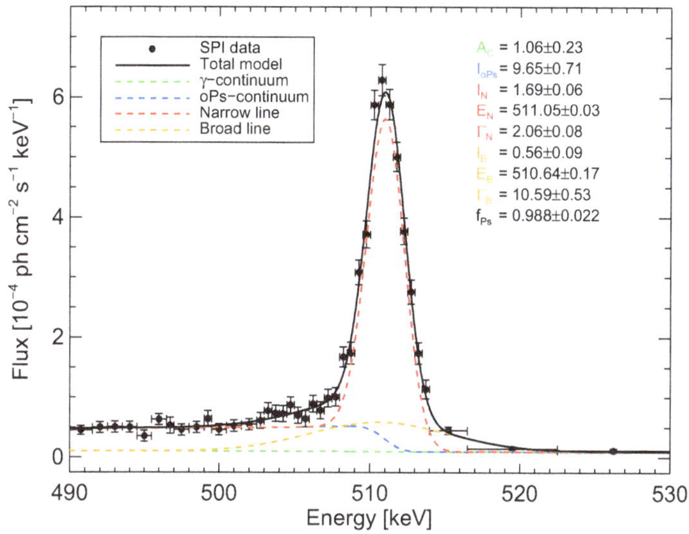

Fig. 7.34 Best fit of the spectrum measured by SPI using different components of the annihilation spectrum from the ISM and the Galactic continuum (Siegert et al. 2016b)

Table 7.2 Results of spectral analysis of Galactic 511 keV emission from the entire sky region exposed by INTEGRAL after 15 years

Parameter	Measured value
E_n (keV)	511.05 ± 0.03
I_n (10^{-3} s^{-1} cm^{-2})	1.69 ± 0.06
Γ_n (keV)	2.06 ± 0.08
I_b (10^{-3} s^{-1} cm^{-2})	0.56 ± 0.09
Γ_b (keV)	10.59 ± 0.53
$I_{3\gamma}$ (10^{-3} s^{-1} cm^{-2})	9.65 ± 0.71
A_c (10^{-6} s^{-1} cm^{-2} keV^{-1})	1.06 ± 0.23

Identified components are narrow and broad lines at 511 keV, and underlying continuum from ortho-positronium and Galactic diffuse emission. E_n the centroid of the narrow line, $I_{n/b}$ and $\Gamma_{n/b}$ are the flux and width (FWHM) of the narrow and broad lines, respectively. $I_{3\gamma}$ is the flux of the ortho-positronium continuum and A_c is the amplitude of the Galactic continuum at 511 keV (From Siegert et al. 2016b)

with a non-negligible ionised fraction: >1% (according to Churazov et al. 2005), and up to 50% (according to Jean et al. 2006). The latter adopted spectral models appropriate for e^+ annihilation in the different ISM phases, and adjusted the phase fractions f_i (with i = {molecular, cold, warm neutral, warm ionised, hot}) so as to obtain the best fit to the spectral details measured by SPI. Since the bulge is dominated by hot gas (see Chap. 11), one may ask how positrons end up annihilating in sub-dominant (by volume) phases of the ISM. This immediately suggests that positron propagation may be crucial in understanding the 511 keV emission. Furthermore, the small width of the 511 keV line, along with the high positronium fraction, both suggest that positrons annihilate at very low energies (<10 keV).

7.5.2.3 Relevant Observations at MeV Energies

The observed γ-ray emission from the decay of ^{26}Al in the Galaxy (see Sect. 7.2) implies that ^{26}Al provides an important contribution to the Galactic amount of positrons. The detected flux translates into a decay rate of ^{26}Al which depends slightly on the adopted 3D distribution of ^{26}Al in the Galaxy. The most recent analysis of SPI data results in a rate of $\dot{N}_{26} = 4.3\ 10^{42}$ s^{-1} or 2.7 M$_\odot$/Myr (Wang et al. 2009). Assuming a steady state, i.e. equality between production and decay rates, this is also the present production rate of ^{26}Al in the Galaxy (Sect. 7.2).

Being predominantly a β^+-emitter (with a branching ratio of $f_{e^+,26} = 82\%$, see Fig. 1.3 in Chap. 1) ^{26}Al is itself a source of positrons. The corresponding Galactic e$^+$ production rate is $\dot{N}_{e^+,26} = f_{e^+,26}\dot{N}_{26} \sim 3.5 \times 10^{42}$ s^{-1}. This constitutes a significant contribution to the total Galactic e$^+$ production rate: 17% of the total e$^+$ annihilation rate and almost half of the (thick) disk in the baseline model, or 10% of the total and 70% of the thin disk in the Halo+thin disk model (Weidenspointner et al. 2006a). We shall see that positrons from this and other β^+-decaying nuclei could plausibly explain the disk emissivity, while the bulge emissivity remains a challenge for our understanding.

Important complementary information on the energies of the annihilating positrons is obtained from the analysis of the observed continuum emission at somewhat higher energies (above 511 keV and into the MeV region) The reason is that positrons in several e$^+$ candidate sources are typically emitted at relativistic energies, in some cases even far above 1 MeV. They behave essentially like relativistic electrons of cosmic rays, producing bremsstrahlung and inverse-Compton emission while slowing down to thermal energies (eV) of the interstellar medium, where they eventually annihilate. But positrons may also annihilate *in flight* while still having relativistic energies, giving rise to a unique γ-ray continuum signature at energies above 511 keV (as the centre-of-mass energy is transferred to annihilation photons, systematically increasing and broadening their typical energies in observer's frame). The shape and amplitude of this γ-ray emission depend on the injection spectrum of positrons and the corresponding total *in-flight* annihilation rate. For positrons injected at low energies (of the order of ~MeV, such as those released by radioactivity), the amplitude of the in-flight annihilation continuum above 1 MeV is quite small, while for sources injecting positrons at much higher energy (such as cosmic-ray positrons from pion decay), the annihilation γ-ray spectrum would extend up to GeV energies and include a considerable γ-ray flux. The high energy γ-ray continuum above 1 MeV therefore constrains the energy and the annihilation rate of relativistic positrons, when all other sources of such high energy emission are properly accounted for.

Diffuse Galactic continuum emission has been well-measured at least in the inner part of the Galactic disk (longitudes $-30° < l < 30°$) in the hard-X-ray through γ-ray regime by INTEGRAL, OSSE, COMPTEL, and EGRET Bouchet et al. (2008). It is mostly due to various interactions of cosmic rays propagating with interstellar gas, which also produce secondary positrons (Strong et al. 2007). In view of the uncertainties still affecting the propagation parameters, soe (albeit very little) room is still left for a contribution of in-flight e^+ annihilation to the MeV continuum.

The constraints to the injection energy of positrons have been pointed out a long time ago by Agaronyan and Atoyan (1981). They showed that the positrons which are responsible for the Galactic 511 keV line cannot be produced in a steady state by the decay of the π^+ created in proton-proton collisions (between cosmic rays and the ISM) or else the in-flight annihilation emission should have been detected. A similar argument was used by Beacom and Yüksel (2006) and Sizun et al. (2006) to constrain the mass of the candidate dark matter particle which could be the source of positrons in the Galactic spheroid. If such particles produce positrons (in their decay or annihilation) at a rate which corresponds to the observed 511 keV emission, then their mass should be less than a few MeV, otherwise the kinetic energy of the created positrons would have been sufficiently high to produce a measurable γ-ray continuum emission in the 1–30 MeV range (Fig. 7.35). The same argument allows one to constrain the initial kinetic energy of positrons and thus to eliminate several classes of candidate sources, like e.g. pulsars, ms pulsars, magnetars, cosmic rays etc., as major positron producers, due to their high positron injection energies.

An interesting opportunity occurred in 2015 when a nearby microquasar, called V404 Cygni, went into a spectacular outburst, and made it the brightest source on the gamma ray sky for about 2 weeks, at hard-X-ray brightness exceeding the Crab

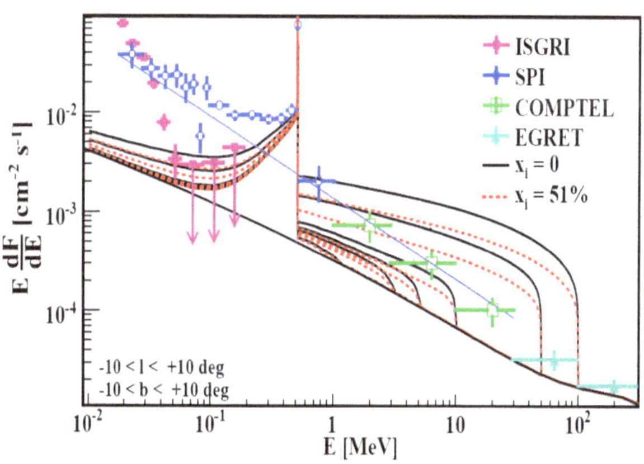

Fig. 7.35 Spectrum of the inner Galaxy as measured by various instruments, compared to various theoretical estimates made under the assumption that positrons are injected at high energy: the four pairs of curves result from positrons injected at 100, 30, 10 and 3 MeV (from top to bottom) and correspond to positrons propagating in neutral (*solid*) or 50% ionised (*dotted*) media (*from* Sizun et al. 2006). This constrains the injected positron energy (or, equivalently, the mass of decaying/annihilating dark matter particles) to a few MeV

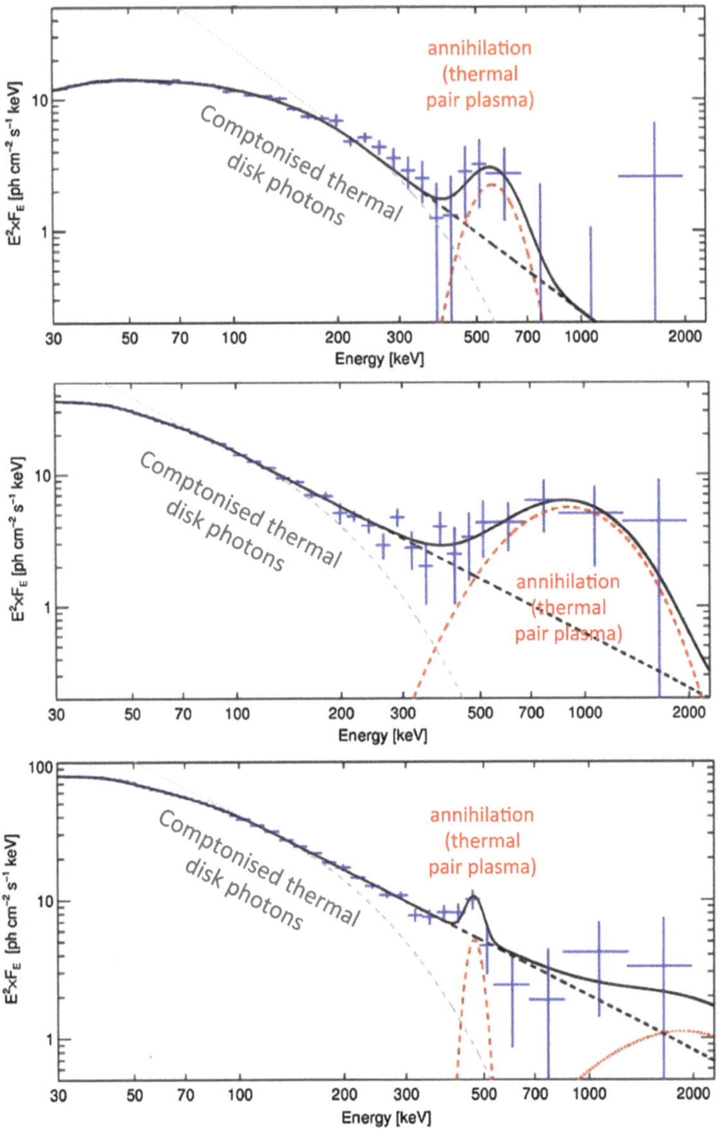

Fig. 7.36 Signature from pair plasma positron annihilation, as observed during the flaring episode June 2015 of microquasar V404 Cyg (Siegert et al. 2016a)

by a factor of 40. INTEGRAL observations were fortunate to catch transient spectral features at high energy, in addition to the bright hard X-ray emission that had been expected from such a flaring episode. After some study, the high-energy spectral features shown in Fig. 7.36 plausibly and best associated with positron annihilation of a relativistic plasma (Siegert et al. 2016a).

7.5.3 Assessment of Candidate Positron Sources

Each of the candidate positron sources should be critically discussed the light of the observational constraints in all different aspects as presented in the previous section. Here we use three main criteria: (1) the total e^+ annihilation rate ($\sim 5 \times 10^{43}\,s^{-1}$), (2) the typical energy of the injected positrons, or the equivalent mass of annihilating DM particles (<3–7 MeV) and (perhaps, most significantly) (3) the morphology of the 511 keV emission (parameterised by a bulge/disk ratio B/D∼1, higher than in all other wavelengths.[6]

Positron Production Rate Assuming a steady state regime, the e^+ annihilation rate has to be equal to the *average* e^+*production rate during the lifetime of* e^+*in the ISM*. The only source definitely known to provide substantial amounts of e^+ at a well constrained rate is the radioactive decay of ^{26}Al: $0.4 \times 10^{43}\,e^+\,s^{-1}$. The decay of ^{44}Ti probably provides another $0.3 \times 10^{43}\,e^+\,s^{-1}$. GCRs probably provide $0.1 \times 10^{43}\,e^+\,s^{-1}$. Nova models (as constrained against several observables such as ejecta abundances, velocities etc.) may provide a e^+ yield from ^{22}Na decay not be much below the reported value of $10^{41}\,e^+\,s^{-1}$. The e^+ of all other candidate sources is entirely speculative at present. The values reported in Table 7.3 for the other candidate positron sources should be considered as optimistic rather than typical ones. Indeed, no useful observational constraints exist up to now on the e^+ yields of hypernovae/GRBs, pulsars, ms pulsars, magnetars, microquasars, the SMBH at the Galactic center, or dark matter annihilation. SNIa remain an intriguing, but serious candidate, with a potential Galactic yield of $2 \times 10^{43}\,e^+\,s^{-1}$ (assuming an escape fraction of 4%).

Positron Energy Radioactive decay produces positrons of $E \leq 1$ MeV, naturally fulfilling the observational constraint on continuum γ rays from in flight annihilation. The same applies to pair creation through $\gamma - \gamma$ collisions in the inner accretion disk or at the base of the jets of LMXRBs, microquasars and the SMBH at the Galactic centre. Conversely, pair creation involving very high energy photons, as in e.g. pulsars or magnetars, will produce positrons of too high energy. The same holds for energetic p-p collisions in Galactic cosmic rays or in the baryonic jets of LMXRBs, microquasars and the Galactic MBH. Those processes produce e^+ of >30 MeV, thus may be discarded as major e^+ sources in the Milky Way. Also, that same constraint limits the mass of putative decaying or annihilating DM particles to <10 MeV, while it does not constrain the mass of de-exciting DM particles.

Source Morphology None of the e^+ sources reproduces the large bulge-to-disk ratio∼1 ratio inferred from SPI data. The best-established e^+ sources, β^+-decay from ^{26}Al and ^{44}Ti produced in massive stars, yield a bulge-to-disk ratio≤0.2, as derived from the observed distribution of the 1.8 MeV ^{26}Al line (Fig. 7.37). Such

---

[6]We do no include in this list of constraints the intriguing disk asymmetry reported byWeidenspointner et al. (2008), since it has not been subsequently confirmed, e.g. Bouchet et al. (2008).

Table 7.3 Properties of candidate positron sources in the Milky Way (adapted from Prantzos et al. 2011)

Source	Process	$E(e^+)$[a] (MeV)	e^+ rate[b] $\dot{N}_{e^+}(10^{43}\ s^{-1})$	Bulge[c]/Disk[c] B/D	Comments
Massive stars: ^{26}Al	β^+	~1	0.4	<0.2	$\dot{N}$, B/D: Observationally inferred
Supernovae: ^{44}Ti	β^+	~1	0.3	<0.2	$\dot{N}$: Robust estimate
SNIa: ^{56}Ni	β^+	~1	2	<0.5	Assuming $f_{e^+,esc} = 0.04$
SN91bg-like: ^{44}Ti	β^+	1	4	~ 0.5	Assuming 0.03 $M_\odot$ of ^{44}Ti per SN
Novae	β^+	~1	*0.02*	<0.5	Insufficient e^+ production
Hypern./GRB: ^{56}Ni	β^+	~1	?	<0.2	Improbable in inner MW
Cosmic rays	p-p	~30	0.1	<0.2	Too high e^+ energy
LMXRBs	$\gamma - \gamma$	~1	*2*	<0.5	Assuming L_{e^+} ~0.01 $L_{obs,X}$
Microquasars	$\gamma - \gamma$	~ 1	*1*	<0.5	e^+ load of jets uncertain
Pulsars	$\gamma - \gamma$	>30	*0.5*	<0.2	Too high e^+ energy
ms pulsars	$\gamma - \gamma$	>30	*0.15*	<0.5	Too high e^+ energy
Magnetars	$\gamma - \gamma$	>30	*0.16*	<0.2	Too high e^+ energy
Central black hole	p-p	High	?		
	$\gamma - \gamma$	1	?		Requires e^+ diffusion to ~1 kpc
Dark matter	Annih.	1 (?)	?		Light scalar particle, cuspy DM profile
	Deexcit.	1	?		Only cuspy DM profiles allowed
	Decay	1	?		Ruled out for all DM profiles
Observational constraints		<7	2	1.4	

[a]Typical values are given
[b]e^+ rates: in roman: observationally deduced or reasonable estimates; in italic: speculative (and rather closer to upper limits)
[c]Sources are simply classified as belonging to either young ($B/D < 0.2$) or old(<0.5) stellar populations

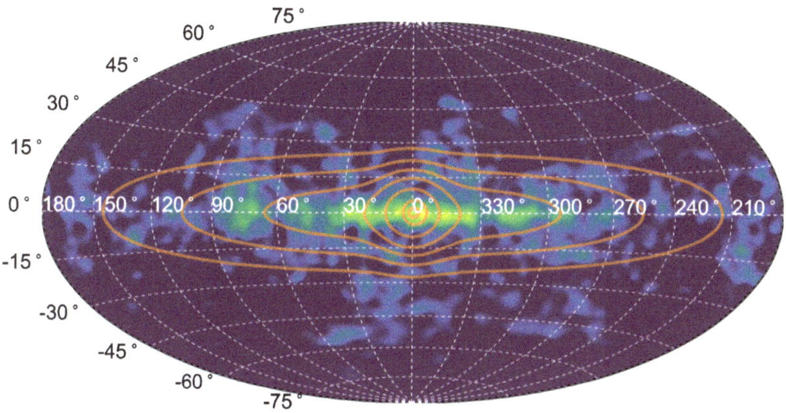

Fig. 7.37 Map of Galactic 511 keV emission detected with SPI/INTEGRAL (contours) superimposed onto the Galactic emission map at 1.8 MeV from ^{26}Al decay (color) (Siegert et al. 2016c)

a distribution reflects essentially the corresponding present star formation rates in the bulge and the disk. On the other hand, an older stellar population, reflecting the time-integrated rather than the present-day star formation, is expected to have a larger bulge/disk ratio (due to the inside-out formation of the Milky Way). Binaries involving low mass stars, such as SNIa, novae and LMXRBs, are expected to have a steeper longitude profile, with a maximal bulge-to-disk ratio$\leq$0.5 (see Prantzos et al. 2011, for a review of the expected profiles of the various candidate sources in the Galaxy).

The morphology of the observed 511 keV emission provides also some interesting constraints in the case of dark matter particles as positron sources (under the assumption of negligible e^+ propagation) (as analysed in Ascasibar et al. 2006): (1) Particle candidates with velocity dependent cross section are excluded as the main source of 511 keV emission, (2) Decaying dark matter cannot be the main source of low energy positrons, because the resulting flux profile is too flat, compared to SPI data. Notice that this latter feature is a generic property of all models involving decaying particles, where the positron production (and annihilation) rate is proportional to the DM density profile: even *cuspy* profiles, such as the *Navarro-Frenk-White* profile do not provide a γ-ray flux profile sufficiently peaked towards the inner Galaxy. Annihilating or de-exciting DM produces positrons at a rate proportional to the square of the DM density profile and leads to a much more peaked γ-ray profile. Light scalar annihilating particles remain a possible candidate, provided the dark matter halo is at least as cuspy as the Navarro-Frenk-White profile with $\gamma \sim 1$; however, astrophysical evidence favors flatter DM halo profiles.

A stacking analysis for 39 dwarf spheroidal galaxies nearby (Siegert et al. 2016c), finally, yields a strong argument against dark matter origin of any kind being the dominant source of Galactic positrons. Those galaxies are expected to be *dominated* in their mass by dark matter, as apposed to relatively little baryonic matter. The

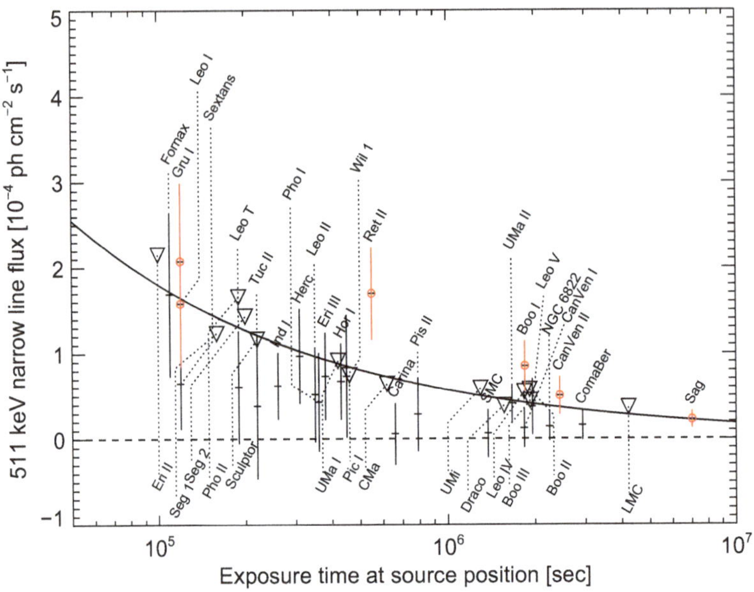

Fig. 7.38 Limits of positron annihilation signal in the 511 keV line from 39 nearby dwarf spheroidal galaxies. The continuous line shows the expected trend if dark matter and distance would contribute to a small signal contribution (Siegert et al. 2016c)

stacking analysis does not find a significant positron annihilation signal, with the exception of one galaxy that also is peculiar for other reasons, Ret II. But the limits obtained for all other dSph galaxies were expected to correlate at least somewhat with their distance and dark matter content, while even an opposite trend was found (see Fig. 7.38). Scaling with matter content, the stacking limit is incompatible with a dark-matter dominated positron annihilation luminosity for the Galaxy, and hence argues that dark matter contributions, if existing, must be small.

The main features of all these candidate e^+ sources are summarised in Table 7.3. The e^+ production rates of all those sources are extremely uncertain (except those of ^{26}Al, ^{44}Ti and GCRs) and the values listed above should be considered as optimistic rather than typical ones. Only in the case of novae may the estimated production value be used to eliminate those sources as important e^+ producers. Source morphology and high energy of produced positrons appear to exclude pulsars, magnetars and Galactic cosmic rays as major contributors to the observed 511 keV emission from the bulge. Source morphology alone would exclude CCSN, hypernovae and GRBs. The high energy of positrons disfavours ms pulsars, as well as p-p collisions from any source (micro-quasars, LMXRB jets, the central SMBH).

A rare sub-class of SNIa, named after their 'prototype' event, SN1991bg, has been recently suggested as the main source of Galactic positrons (Crocker et al. 2017). That class represents $\sim$15% of all SNIa and they are several times less luminous than the average SNIa. Theoretical ? and still uncertain ? models find that their explosion may produce up to a few 0.01 $M_\odot$ of ^{44}Ti, providing enough positrons to explain the observed 511 keV emission *and* its Galactic distribution: indeed, assuming that the Delay Time Distribution (DTD) of those objects is different than the one of standard SNIa (peaking several Gy after the formation of the progenitor stars, in contrast to the DTD $\propto time^{-1}$ of SNIa), one finds that the early enhanced star formation in the bulge may produce today few SNIa but enough SNIbg to provide a large B/D ratio. That scenario might also explain the paucity of Galactic sources of ^{44}Ti: that radionuclide is the progenitor of stable ^{44}Ca and, if the source of the solar abundance of the latter is CCSN of low ^{44}Ti yield (as usually assumed), one should expect several SN remnants glowing in the ^{44}Ti γ-ray lines to be seen by INTEGRAL, whereas only Cas A is currently detected.

The possibility of explaining at one stroke both the Galactic 511 keV emission and the paucity of ^{44}Ti sources makes the idea of Crocker et al. (2017) appealing. However, two key ingredients of the model, namely the ^{44}Ti yields and the evolution of the rate of SNIbg-type supernovae should be substantiated by further studies (including 3D models of supernova nucleosynthesis) before concluding.

If positrons annihilate near their sources, one has to conclude that (1) either poorly understood class of sources (such as SNIbg-like objects) dominates e^+ production, or that (2) positrons are produced by a combination of the sources of Table 7.3, e.g. (a) ^{26}Al +^{44}Ti for the disk and dark matter for the bulge, or (b) ^{26}Al +^{44}Ti + LMXRBs(or microquasars) for the disk *and* the bulge plus a contribution from the central SMBH for the inner bulge, or (c) some other combination.

In order to alleviate the morphology problem, it has been suggested that positron transport might help. Prantzos (2006b) suggested that *if* the magnetic field of the Milky Way halo has a strong poloidal component, then some positrons escaping the disk may be channeled into the bulge and annihilate there, enhancing the bulge/disk e^+ annihilation ratio. In that case, positrons from SNIa may suffice to explain *quantitatively both* the total observed e^+ annihilation rate ($\sim 2\times 10^{43}$ e^+ s^{-1}) *and* the corresponding bulge-to-disk ratio, *provided* that the escaping e^+ fraction from SNIa is $\sim$3–4%. However, observations of external spirals suggest rather an X-shaped halo field in which case it would be difficult for disk positrons to find their way into the bulge. Still, the issue is of considerable interest to urge a better assessment of the poorly known global configuration of the Galactic magnetic field.

In the same framework of "outside-in" positron transport, Higdon et al. (2009) suggested that positron propagation through the Galaxy may be all that is needed for understanding not only the spatial morphology of the 511 keV emission, but also its spectral properties. They assumed that radioactivity (from ^{26}Al, ^{44}Ti and, mostly, from ^{56}Co) is the sole e^+ source in the Galaxy and they considered (1) a fairly detailed description of the various phases of the ISM and (2) a particular phenomenological model of collisionless scattering of MeV positrons by turbulent fluctuations of the ISM, allowing to transport positrons from the inner 3 kpc into the bulge region.

The aforementioned ideas were put in test through detailed numerical simulations of positron transport, either with Monte Carlo methods (Alexis et al. 2014) or with cosmic ray propagation codes such as GALPROP (Martin et al. 2012). In both cases, it was found that although positrons may travel up to a few kpc from their birth place, the bulk of them is annihilated rather close to their birth places, making it difficult to reproduce the observed high B/D ratio; re-acceleration of positrons, not considered in those studies, might help in that respect.

Finally, the idea of an "inside-out" propagation of positrons was explored, in order to investigate the possibility of positrons produced by the activity of a central Galactic source (the super-massive black hole of SgA) and diffusing throughout the bulge. The spectral signature of the 511 keV emission, suggesting that positrons annihilate mostly in the warm ISM, provides a powerful constraint in that case. The Monte Carlo study of Jean et al. (2009) investigated collisional transport in the ISM of the bulge and found the diffusion length of positrons to exceed typical size scales of the warm ISM, where they are thought to annihilate. On the other hand, Panther et al. (2018) investigated the transport of positrons coupled to the turbulent, magnetized plasma outflowing from the inner Galaxy (as evidenced from infra-red and γ-ray observations). They found that although positrons may indeed be advected to scales of $\sim$2 kpc and fill the bulge, they would annihilate mostly in a hot, ionised plasma, while observations point to a warm ISM. That study concerns a steady plasma and positron outflow, while Alexis et al. (2014) argued that a burst of activity in the galactic center 1–10 My ago could make positrons annihilate in a warm environment, in agreement with observations.

In summary, more than 40 years after its discovery, the origin of positrons annihilating in the Galaxy remains unknown. Progress in the field will require advances in several directions:

1. *Observations of 511 keV emission*: what is the true spatial distribution of the emission? how far the spheroid and disk extend? are there yet undetected regions of low surface brightness? is the disk emission asymmetric indeed? how do the 1.8 MeV and 511 keV disk emissions compare to each other? A much deeper exposure of the Galaxy and a better understanding of the backgrounds will be required to tackle those issues. Even if INTEGRAL's mission is extended into 2020ies, it appears questionable that it will be able to provide the answers; even worse, no other mission in this energy range is scheduled at present.
2. *Physics of e^+ sources*: what is the e^+ escaping fraction in SNIa ? what is the SNIa rate in the inner (star forming) and in the outer (inactive) bulge? what are the e^+ yields, activity timescales, and spatial distribution in the inner bulge of LMXRBs or microquasars? how can the past level of activity of the central massive black hole be reliably monitored?
3. *Positron propagation:* what is the large scale configuration of the Galactic magnetic field? what are the properties of interstellar plasma turbulence and how they affect the positron transport? what are the dominant propagation modes of positrons and what the role of re-acceleration might be?

The many faces of the Galactic 511 keV emission make this problem one of the most intriguing problems in high energy astrophysics today and for the years to come.

References

Agaronyan FA, Atoyan AM (1981) Sov Astron Lett 7:395
Alexis A, Jean P, Martin P, Ferrière K (2014) Astron Astrophys 564:A108
Ascasibar Y, Jean P, Bœhm C, Knödlseder J (2006) Mon Not R Astron Soc 368:1695
Asplund M, Grevesse N, Sauval AJ, Scott P (2009) Annu Rev Astron Astrophys 47:481
Bally J (2008) Overview of the orion complex. In: Reipurth B (ed) ASP monograph publications. Handbook of star forming regions, Valencia, vol I. ASP, p 459
Beacom JF, Yüksel H (2006) Phys Rev Lett 97:071102
Binns WR, Wiedenbeck ME, Arnould M et al (2005) Astrophys J 634:351
Binns WR, Israel MH, Christian ER et al (2016) Science 352:677
Blair WP, Sankrit R, Raymond JC (2005) Astron J 129:2268
Bouchet L, Mandrou P, Roques J-P et al (1991) Astrophys J 383:L45+
Bouchet L, Jourdain E, Roques J-P et al (2008) Astrophys J 679:1315
Bouchet L, Roques JP, Jourdain E (2010) Astrophys J 720:1772
Bouchet L, Jourdain E, Roques J-P (2015) Astrophys J 801:142
Breitschwerdt D, de Avillez MA (2006) Astron Astrophys 452:L1
Breitschwerdt D, Freyberg MJ, Truemper J (eds) (1998) Lecture notes in physics. The local bubble and beyond, vol 506. Springer, Berlin
Breitschwerdt D, Feige J, Schulreich MM et al (2016) Nature 532:73
Brown AGA, de Geus EJ, de Zeeuw PT (1994) Astron Astrophys 289:101
Burrows DN, Singh KP, Nousek JA, Garmire GP, Good J (1993) Astrophys J 406:97
Bussard RW, Ramaty R, Drachman RJ (1979) Astrophys J 228:928
Bykov AM, Treumann RA (2011) Astron Astrophys Rev 19:42
Cash W, Charles P, Bowyer S et al (1980) Astrophys J 238:L71
Casse M, Paul JA (1982) Astrophys J 258:860
Chatterjee S, Cordes JM, Vlemmings WHT et al (2004) Astrophys J 604:339
Chen W, Diehl R, Gehrels N et al (1997) In: Winkler C, Courvoisier TJL, Durouchoux P (eds) The transparent universe. ESA special publication, vol 382, pp 105–+
Cheng LX, Leventhal M, Smith DM (1997) Astrophys J 481:L43+
Chieffi A, Limongi M (2013) Astrophys J 764:21
Chupp EL, Forrest DJ, Higbie PR et al (1973) Nature 241:333
Churazov E, Sunyaev R, Sazonov S, Revnivtsev M, Varshalovich D (2005) Mon Not R Astron Soc 357:1377+
Clayton DD, Nittler LR (2004) Annu Rev Astron Astrophys 42:39
Comerón F, Pasquali A, Figueras F, Torra J (2008) Astron Astrophys 486:453
Crocker RM, Ruiter AJ, Seitenzahl IR et al (2017) Nat Astron 1:0135
Dame TM, Hartmann D, Thaddeus P (2001) Astrophys J 547:792
de Geus EJ (1992) Astron Astrophys 262:258
de Geus EJ, de Zeeuw PT, Lub J (1989) Astron Astrophys 216:44
de Zeeuw PT, Hoogerwerf R, de Bruijne JHJ, Brown AGA, Blaauw A (1999) Astron J 117:354
Diehl R (2002) New Astron Rev 46:547
Diehl R (2013) Astron Rev 8:4
Diehl R (2016) J Phys Conf Ser 703:012001
Diehl R, Dupraz C, Bennett K et al (1995) Astron Astrophys 298:445
Diehl R, Kretschmer K, Plüschke S, Cerviño M, Hartmann DH (2003) In: van der Hucht K, Herrero A, Esteban C (eds) IAU symposium. A massive star odyssey: from main sequence to supernova, vol 212, pp 706–+
Diehl R, Halloin H, Kretschmer K et al (2006a) Nature 439:45
Diehl R, Halloin H, Kretschmer K et al (2006b) Astron Astrophys 449:1025
Diehl R, Lang MG, Martin P et al (2010) Astron Astrophys 522:A51+
Diehl R, Siegert T, Greiner J, Krause M, Kretschmer K, Lang M, Pleintinger M, Strong AW, Weinberger C, Zhang X (2018) Astron Astrophys 611:A12. https://doi.org/10.1051/0004-6361/201731815

Ekström S, Georgy C, Eggenberger P et al (2012) Astron Astrophys 537:A146
Ellison DC, Drury LO, Meyer J (1997) Astrophys J 487:197
Endt PM (1990) Nucl Phys A 521:1
Fernández D, Figueras F, Torra J (2008) Astron Astrophys 480:735
Fierlinger KM, Burkert A, Ntormousi E et al. (2016) Mon Not R Astron Soc 456:710
Fimiani L, Cook DL, Faestermann T et al. (2016) Phys Rev Lett 116:151104
Fitoussi C, Raisbeck GM, Knie K et al. (2008) Phys Rev Lett 101:121101
Forestini M, Charbonnel C (1997) Astron Astrophys Suppl Ser 123:241
Frisch PC (1995) Space Sci Rev 72:499
Fuchs B, Breitschwerdt D, de Avillez MA, Dettbarn C (2009) Space Sci Rev 143:437
Gaczkowski B, Preibisch T, Stanke T et al. (2015) Astron Astrophys 584:A36
Gaczkowski B, Roccatagliata V, Flaischlen S et al. (2017) Astron Astrophys 608:A102
Gehrz RD, Truran JW, Williams RE, Starrfield S (1998) Publ Astron Soc Pac 110:3
Genzel R, Stutzki J (1989) Annu Rev Astron Astrophys 27:41
Goswami A, Prantzos N (2000) Bull Astron Soc India 28:305
Groopman E, Zinner E, Amari S et al (2015) Astrophys J 809:31
Harris MJ, Teegarden BJ, Cline TL et al (1998) Astrophys J 501:L55+
Harris MJ, Knödlseder J, Jean P et al (2005) Astron Astrophys 433:L49
Heftrich T, Bichler M, Dressler R et al (2015) Phys Rev C 92:015806
Higdon JC, Lingenfelter RE, Ramaty R (1998) Astrophys J 509:L33
Higdon JC, Lingenfelter RE, Rothschild RE (2009) Astrophys J 698:350
Hoogerwerf R, de Bruijne JHJ, de Zeeuw PT (2000) Astrophys J 544:L133
Iliadis C, Champagne A, Chieffi A, Limongi M (2011) Astrophys J Suppl 193:16
Jean P, Knödlseder J, Gillard W et al (2006) Astron Astrophys 445:579
Jean P, Gillard W, Marcowith A, Ferrière K (2009) Astron Astrophys 508:1099
Johnson WN, Harnden FR, Haymes RC (1972) Astrophys J 172:L1+
Jones FC, Lukasiak A, Ptuskin VS, Webber WR (2001) Adv Space Res 27:737
Jose J, Hernanz M (1998) Astrophys J 494:680
Kachelrieß M, Neronov A, Semikoz DV (2015) Phys Rev Lett 115:181103
Karakas AI, Lattanzio JC (2014) Publ Astron Soc Aust 31:e030
Kinzer RL, Purcell WR, Johnson WN et al (1996) Astron Astrophys Suppl Ser 120:C317+
Knie K, Korschinek G, Faestermann T et al (2004) Phys Rev Lett 93:171103
Knödlseder J, Dixon D, Bennett K et al (1999) Astron Astrophys 345:813
Knödlseder J, Cerviño M, Le Duigou J-M et al (2002) Astron Astrophys 390:945
Knödlseder J, Jean P, Lonjou V et al (2005) Astron Astrophys 441:513
Korschinek G, Kutschera W (2015) Nucl Phys A 944:190
Krause M, Fierlinger K, Diehl R et al (2013) Astron Astrophys 550:A49
Krause M, Diehl R, Böhringer H, Freyberg M, Lubos D (2014) Astron Astrophys 566:A94
Krause MGH, Diehl R, Bagetakos Y et al (2015) Astron Astrophys 578:A113
Krause M et al. (2018) Astron Astrophys (in press)
Kretschmer K, Diehl R, Krause M et al (2013) Astron Astrophys 559:A99
Kuffmeier M, Frostholm Mogensen T, Haugbølle T, Bizzarro M, Nordlund Å (2016) Astrophys J 826:22
Kulsrud RM (2005) Plasma physics for astrophysics. Princeton University Press, Princeton
Kutschera W (2013) In: European physical journal web of conferences, vol 63, pp 03001
Lallement R (2007) Space Sci Rev 130:341
Lee T, Papanastassiou DA, Wasserburg GJ (1976) Bull Am Astron Soc 8:457
Leventhal M, MacCallum CJ, Stang PD (1978) Astrophys J 225:11+
Limongi M, Chieffi A (2006a) New Astron Rev 50:474
Limongi M, Chieffi A (2006b) Astrophys J 647:483
Li KA, Lam YH, Qi C, Tang XD, Zhang NT (2016) Phys Rev C 94(6):065807. https://doi.org/10.1103/PhysRevC.94.065807
Lingenfelter RE, Ramaty R (1978) Phys Today 31:40
Lozinskaya TA, Pravdikova VV, Finoguenov AV (2002) Astron Lett 28:223

Ludwig P, Bishop S, Egli R et al (2016) Proc Natl Acad Sci 113:9232
MacPherson GJ, Davis AM, Zinner EK (1995) Meteoritics 30:365
MacPherson GJ, Bullock ES, Janney PE et al (2010) Astrophys J 711:L117
Mahoney WA, Ling JC, Wheaton WA, Jacobson AS (1984) Astrophys J 286:578
Maíz-Apellániz J (2004) In: Alfaro EJ, Pérez E, Franco J (eds) How does the galaxy work?. Astrophysics and space science library, vol 315, pp 231
Martin P, Knödlseder J, Diehl R, Meynet G (2009) Astron Astrophys 506:703
Martin P, Knödlseder J, Meynet G, Diehl R (2010) Astron Astrophys 511:A86+
Martin P, Strong AW, Jean P, Alexis A, Diehl R (2012) Astron Astrophys 543:A3
Mewaldt RA, Yanasak NE, Wiedenbeck ME et al (2001) Space Sci Rev 99:27
Meyer J (1985) Astrophys J Suppl 57:151
Meyer J, Drury LO, Ellison DC (1997) Astrophys J 487:182
Meynet G, Arnould M, Prantzos N, Paulus G (1997) Astron Astrophys 320:460
Meynet G, Hirschi R, Maeder A, Ekström S (2006) In: Montmerle T, Kahane C (eds) EAS publications series, vol 19, pp 85–100
Naya JE, Barthelmy SD, Bartlett LM et al (1996) Nature 384:44
Neronov A, Meynet G (2016) Astron Astrophys 588:A86
Nittler LR, Alexander CMO, Gallino R et al (2008) Astrophys J 682:1450
Oberlack U, Bennett K, Bloemen H et al (1996) Astron Astrophys Suppl Ser 120:C311
Olano CA (1982) Astron Astrophys 112:195
Ore A, Powell JL (1949) Phys Rev 75:1696
Ostdiek KM, Anderson TS, Bauder WK et al (2017) Phys Rev C 95:055809
Palacios A, Meynet G, Vuissoz C et al (2005) Astron Astrophys 429:613
Panther FH, Crocker RM, Birnboim Y, Seitenzahl IR, Ruiter AJ (2018) Mon Not R Astron Soc 474:L17
Parikh A, José J, Karakas A, Ruiz C, Wimmer K (2014) Phys Rev C 90:038801
Pecaut MJ, Mamajek EE (2016) Mon Not R Astron Soc 461:794
Perrot CA, Grenier IA (2003) Astron Astrophys 404:519
Pignatari M, Herwig F, Hirschi R et al (2016) Astrophys J Suppl 225:24
Plüschke S, Diehl R, Schönfelder V et al (2001a) In: Gimenez A, Reglero V, Winkler C (eds) Exploring the gamma-ray universe. ESA special publication, vol 459, pp 55–58
Plüschke S, Kretschmer K, Diehl R, Hartmann DH, Oberlack UG (2001b) In: Gimenez A, Reglero V, Winkler C (eds) Exploring the gamma-ray universe. ESA special publication, vol 459, pp 91–95
Plüschke S, Cerviño M, Diehl R et al (2002) New Astron Rev 46:535
Pöppel W (1997) Fundam Cosm Phys 18:1
Pöppel WGL, Bajaja E, Arnal EM, Morras R (2010) Astron Astrophys 512:A83+
Prantzos N (1993) Astron Astrophys Suppl Ser 97:119
Prantzos N (2004) Astron Astrophys 420:1033
Prantzos N (2006a) αp 449:869
Prantzos N (2006b) Astron Astrophys 449:869
Prantzos N (2012) Astron Astrophys 538:A80
Prantzos N, Diehl R (1996) Phys Rep 267:1
Prantzos N, Arnould M, Arcoragi JP, Casse M (1985) In: Jones FC (ed) International cosmic ray conference, vol 3, pp 167–170
Prantzos N, Boehm C, Bykov AM et al (2011) Rev Mod Phys 83:1001
Preibisch T, Zinnecker H (1999) Astron J 117:2381
Preibisch T, Mamajek E (2008) In: Reipurth B (ed) The nearest OB association: Scorpius-Centaurus (Sco OB2), pp 235
Preibisch T, Brown AGA, Bridges T, Guenther E, Zinnecker H (2002) Astron J 124:404
Ptuskin VS, Soutoul A (1998) Space Sci Rev 86:225
Purcell WR, Grabelsky DA, Ulmer MP et al (1993) Astrophys J 413:L85
Purcell WR, Cheng L-X, Dixon DD (1997) Astrophys J 491:725+
Ragot BR (1999) Astrophys J 518:974

Ragot BR (2006) Astrophys J 642:1163
Ramaty R, Kozlovsky B, Lingenfelter RE (1979) Astrophys J Suppl 40:487
Robin AC, Reylé C, Derrière S, Picaud S (2003) Astron Astrophys 409:523
Rugel G, Faestermann T, Knie K et al (2009) Phys Rev Lett 103:072502
Schramm DN, Tera F, Wasserburg GJ (1970) Earth Planet Sci Lett 10:44
Schulreich MM, Breitschwerdt D, Feige J, Dettbarn C (2017) Astron Astrophys 604:A81
Siegert T (2017) PhD thesis, TU Munich
Siegert T, Diehl R (2017) In: Kubono S, Kajino T, Nishimura S, Isobe T, Nagataki S, Shima T, Takeda Y (eds) 14th international symposium on nuclei in the cosmos (NIC2016), pp 020305
Siegert T, Diehl R, Greiner J et al (2016a) Nature 531:341
Siegert T, Diehl R, Khachatryan G et al (2016b) Astron Astrophys 586:A84
Siegert T, Diehl R, Vincent AC et al (2016c) ArXiv e-prints 1608.00393
Sieverding A, Martínez-Pinedo G, Langanke K, Heger A (2017) In: Kubono S, Kajino T, Nishimura S, Isobe T, Nagataki S, Shima T, Takeda Y (eds) 14th international symposium on nuclei in the cosmos (NIC2016), pp 020701
Sizun P, Cassé M, Schanne S (2006) Phys Rev D 74:063514
Skinner G, Diehl R, Zhang X-L, Jean P (2014) Proc Sci INT2014:054
Slesnick CL, Hillenbrand LA, Carpenter JM (2008) Astrophys J 688:377
Smith DM (2004) In: Schoenfelder V, Lichti G, Winkler C (eds) 5th integral workshop on the integral universe. ESA special publication, vol 552, pp 45–+
Soutoul A, Casse M, Juliusson E (1978) Astrophys J 219:753
Srinivasan G, Chaussidon M (2013) Earth Planet Sci Lett 374:11
Strong AW, Moskalenko IV, Ptuskin VS (2007) Annu Rev Nucl Part Sci 57:285–327. http://dx.doi.org/10.1146/annurev.nucl.57.090506.123011
Sturner SJ, Naya JE (1999) Astrophys J 526:200
Sukhbold T, Ertl T, Woosley SE, Brown JM, Janka H-T (2016) Astrophys J 821:38
Sunyaev R, Churazov E, Gilfanov M et al (1992) Astrophys J 389:L75+
Taylor JH, Cordes JM (1993) Astrophys J 411:674
Thielemann F-K, Nomoto K, Hashimoto M-A (1996) Astrophys J 460:408
Tibaldo L, Grenier IA (2013) Nucl Phys B Proc Suppl 239:70
Timmes FX, Woosley SE, Hartmann DH et al (1995) Astrophys J 449:204
Toptygin IN (1985) Cosmic rays in interplanetary magnetic fields. D. Reidel Publishing Co., Dordrecht
Uyanıker B, Fürst E, Reich W, Aschenbach B, Wielebinski R (2001) Astron Astrophys 371:675
van der Hucht KA (2001) New Astron Rev 45:135
Voss R, Diehl R, Hartmann DH et al (2009) Astron Astrophys 504:531
Voss R, Diehl R, Vink JS, Hartmann DH (2010) Astron Astrophys 520:A51+
Wallner A, Feige J, Kinoshita N et al (2016) Nature 532:69
Wang W, Harris MJ, Diehl R et al (2007) Astron Astrophys 469:1005
Wang W, Lang MG, Diehl R et al (2009) Astron Astrophys 496:713
Wasserburg GJ, Lee T, Papanastassiou DA (1977) Meteoritics 12:377
Weidenspointner G, Shrader CR, Knödlseder J (2006a) Astron Astrophys 450:1013+
Weidenspointner G, Shrader CR, Knödlseder J et al (2006b) Astron Astrophys 450:1013
Weidenspointner G, Skinner G, Jean P et al (2008) Nature 451:159
Welsh BY, Shelton RL (2009) Astrophys Space Sci 323:1
Westphal AJ, Bradley JP (2004) Astrophys J 617:1131
Wiedenbeck ME, Binns WR, Christian ER et al (1999) Astrophys J 523:L61
Wiedenbeck ME, Binns WR, Cummings AC et al (2007) In: von Steiger R, Gloeckler G, Mason GM (eds) An overview of the origin of galactic cosmic rays as inferred from observations of heavy ion composition and spectra. Springer, New York, pp 415–+
Wilking BA, Gagné M, Allen LE (2008) In: Reipurth B (ed) Star formation in the ρ Ophiuchi molecular cloud, pp 351
Woosley SE, Heger A (2007) Phys Rep 442:269

Woosley SE, Weaver TA (1994) Astrophys J 423:371
Woosley SE, Weaver TA (1995) Astrophys J Suppl 101:181
Yanasak NE, Wiedenbeck ME, Mewaldt RA et al (2001) Astrophys J 563:768
Zinnecker H, Yorke HW (2007) Annu Rev Astron Astrophys 45:481

Part IV
Tools for the Study of Radioactivities in Astrophysics

At the foundations of science lie the instruments (detectors) which enable us to record data that encode information about objects under study. Also important are the methods used to organize our concepts and notions about these objects in the forms of astrophysical theories or phenomenological descriptions. These *tools* are mostly based on general physical or mathematical principles, but with increasing sophistication one often requires the development of specialized methods that may not be common in other areas of astronomy. Here we address in three chapters how theorists construct models of astrophysical objects and processes, exemplified by the core-collapse of massive stars as a key source of cosmic radioactivity. We further present the astrophysics of nuclear reactions and discuss them from the laboratory and theory point of view. Then, we present two specific classes of instruments used to measure cosmic radioactivities, one employed in terrestrial laboratories for the study of interstellar dust grains, another operating in space for the study of electromagnetic radiation from cosmic radioactive decays. Thereafter, we present the approach to descriptions of the evolution of cosmic gas composition, which often is called 'chemical evolution', in a separate chapter. A final chapter lists specific nuclei with a discussion of their nucleosynthesis origins, thus adding a different perspective on compositional evolution as ordered by nuclear properties.

Chapter 8
Computer-Modeling of Stars

Matthias Liebendörfer

A human being experiences his immediate environment on the scale of meters, seconds and grams. These are also the natural scales of his actions. Thus, as soon as he starts to explore the laws of physics, he can easily move around masses at the scale of grams, objects on the scale of meters and perform experiments on the scale of seconds. On these scales, the experimentator has full control on the setup of an experiment and direct access to all degrees of freedom during the evolution of the experiment. This direct access is lost in experiments that explore the physics on scales that are many orders of magnitude smaller. The experimentator still has full control on the setup, for example, by putting a specific target into a properly designed accelerator beam. But the measurements are then limited to the far field, where only a superposition of the effects of the microscopic physics becomes detectable. The large number of degrees of freedom that may be present in the microscopic physics must be explored by clever variations of the experimental setup. Most astronomical observations are obviously also taken from the far field, because the distance to the observed source is so much larger than the length scale of the source. Hence, many degrees of freedom of the dynamics on the length scale of the source are only indirectly accessible for the observer. Moreover, it is not possible to efficiently manipulate and prepare matter outside the solar system in order to produce systematic variations in the setup as in terrestrial experiments.

In this situation, a computer model is a particularly useful tool. It implements known input physics and evolves sophisticated coupled systems in space-time in order to reproduce the observed far field response of a corresponding event in nature. In contrast to the real event, the data of the computer model is fully accessible and permits to study all degrees of freedom in detail. Of course there is no guarantee that the results represent an accurate image of the actual astrophysical source. Only

M. Liebendörfer (✉)
Universität Basel, Basel, Switzerland
e-mail: matthias.liebendoerfer@unibas.ch

R. Diehl et al. (eds.), *Astrophysics with Radioactive Isotopes*, Astrophysics and Space Science Library 453, https://doi.org/10.1007/978-3-319-91929-4_8

the real experiment is capable of probing the laws of physics realised in nature, while the computer model is bound to elaborate the properties of the theory it is built on. The strength of the computer simulation is that it demonstrates and extends the predictive power of the underlying theory. It is therefore essential to check the models for predictions that could be in reach of a targeted future observation, so that a specific model can be supported or falsified. The prediction of the production of radioactive nuclei that can be observed from far distances is one possibility.

Before an astrophysical model is constructed, it is of primary importance to perform order of magnitude estimates and to identify relevant and irrelevant processes. The presence of equilibrium conditions can further reduce the degrees of freedom. This information is then used to compose a sequence of relevant processes that form a scenario. In a second step, one substitutes the more complicated processes by simple approximate laws (for example power laws) that allow an analytical investigation of the most important aspects of the model. In stellar structure, for example, order of magnitude arguments show that the matter is in thermodynamical equilibrium. Hence, an equation of state can be defined that expresses the gas pressure, p, as a function of the local density, ρ, temperature and composition. In some regimes, the microscopic physics information in the equation of state is well approximated by an equation of state of the form $p = \kappa\rho^{1+1/n}$. Here, κ is a constant and n a parameter called the polytropic index. If this approximation holds, the equations of stellar structure reduce to the Lane-Emden equation (e.g. Chandrasekhar 1967) which has analytical solutions for specific integer values of n. Scientific computing provides the tools to specify additional solutions for non-integer values of n. An almost arbitrary numerical precision can be obtained because all physical uncertainties remain hidden by the power law of the equation of state and the reduction of the star to spherical symmetry. In this example, the scenario and the simplification of the input physics have to be pre-defined manually before scientific computing can be used to determine specific solutions of a simple model.

It is numerically more challenging to let a computer model contribute at the stage of the order of magnitude estimates and the elaboration of the scenario. Before the irrelevant processes are excluded, the model is rich and complicated. Instead of judging processes in each fluid cell by hand, the finite differencing can be set up in a manner that filters out the irrelevant processes automatically and establishes equilibria accurately. A well-known example of this technique is provided by the numerical propagation of a shock front: Only a minority of computer codes resolves the microscopic width of the shock. Most codes just ensure the accurate conservation of mass, momentum and energy across the shock front to obtain the shock propagation speed and the thermodynamic conditions on both sides of the shock. Here, it is the numerical algorithm that dynamically performs the simplification of the model while the microscopic viscosity of the fluid does not enter the calculation whether it is implemented or not. Hence, one may enable a rich set of input physics and design the finite differencing such that fundamental laws of physics are fulfilled under all possible conditions. Then, the complexity of the model is limited by the scale on which unresolvable small-scale structures are dissipated in space and time. This scale depends on the numerical algorithm and the mesh topology rather than the investigated physics. Because the behaviour at the resolution threshold can greatly vary from numerical algorithm and grid to the next, we believe that it is important for the reliability of astrophysical models to be investigated with several different numerical tools that are simple and efficient enough to be broadly used.

Computer models of stellar evolution and explosion have a long tradition and have mostly been carried out in spherical symmetry (e.g. Henyey et al. 1959; Woosley et al. 2002; Colgate and White 1966; Bethe 1990). The restriction to spherical symmetry is a consequence of the fact that the hydrodynamic time scale is many orders of magnitude smaller than the lifetime of a star. Traditional stellar evolution codes assumed that the matter is in hydrostatic equilibrium and thus close to spherical. If one implements hydrodynamic equilibrium that adjusts slowly on the time scale of the thermodynamic evolution of the star, sufficiently large time steps can be taken to numerically cover the lifetime of the star. The crucial effect of mixing by convection and rotation on the evolution has been included by effective mixing parameters (e.g. Ludwig et al. 1999) and the inclusion of an effective evolution of angular momentum in centrifugally deformed layers (Meynet and Maeder 1997; Langer et al. 1997) that can still be labelled and evolved by a unique radial parameter. Several computer codes have been built to follow the nucleosynthesis through the different stages of stellar evolution. The application of traditional simulation codes relied on instructed users that learned to handle and extend the codes within the different research groups. Publications therefore mostly focus on the extensions of the numerical methods and the properties of the new stellar models that resulted (Nomoto and Hashimoto 1988; Maeder and Meynet 1989; Woosley and Weaver 1995). It is a more recent phenomenon that the different codes obtain a name and an identity as software package (Limongi et al. 2000; Rauscher et al. 2002; Young and Arnett 2005; Eggenberger et al. 2008; Dotter and Paxton 2009).

As soon as the iron core becomes gravitationally unstable at the end of the evolution of massive stars, it starts to collapse on the dynamical time scale. This marks the end of the applicability of stellar evolution codes that usually are not designed to describe dense matter with enhanced weak interactions. For a range of stellar masses, which for solar metallicity is believed to be within $\sim$8–40 $M_\odot$, this last phase of stellar evolution leads to the supernova explosion, which may eject matter into the interstellar medium after its processing by explosive nucleosynthesis. This phase is modelled by 'supernova codes' that traditionally have also assumed spherical symmetry—but for a different reason than the stellar evolution codes. The growing importance of general relativistic corrections around the dense neutron star at the center of the collapse and the large impact of neutrino physics and transport on the dynamics of the explosion makes it computationally very challenging to treat this phase in multidimensional simulations.

Hence, for both stellar evolution and stellar core collapse, realistic three-dimensional turnover and important fluid instabilities were for a long time only simulated in a phenomenological manner. A very simple estimate shows that this unsatisfactory situation is about to change: In order to sample a dynamically interesting profile of a physical quantity, one needs a minimum resolution of ~100 data points. If one requests that the evolution of these data points should not touch the scale on which unresolvable features are numerically dissipated, one needs another order of magnitude, i.e. a resolution ~1000 points in one dimension. For a three-dimensional simulation, this then leads to $\sim 1000^3 = 10^9$ fluid cells. Assuming that each cell holds at least 10 real variables, this means that a three-dimensional astrophysical simulation has to evolve about 80 Gigabyte of data at each time step. As this level of resolution is now achievable on high-performance computer clusters, the link between input physics and astrophysical observables receives a new quality in the transition from generic computer models in spherical or axi-symmetry to detailed three-dimensional models. This transition has already occurred in astrophysical fields that are dominated by hydrodynamics (e.g. cosmology) while it will occur over the next years in the fields that have to couple the hydrodynamics to radiative transfer. In stellar evolution, it becomes possible to perform dynamical global stellar models in three dimensions over short time intervals so that the parameterised long-term models can be gauged (Brun et al. 2004; Meakin and Arnett 2007). Simulations of thermonuclear explosions that lead to the display of a type Ia supernova have also entered the three-dimensional area (Fryxell et al. 2000; Zingale et al. 2005; Röpke et al. 2007). The ignition and detailed propagation of the burning front in the 3D geometry is the crucial ingredient of type Ia supernova simulations. Likewise, the simulations of stellar core collapse are ready to make the transition to three dimensions in spite of GR effects and the difficulties with multidimensional neutrino transport, because there is a global symmetry center that initially permits some simplifying approximations, because the neutrino spectra are continuous and smooth, and because the dynamical time scale and the neutrino propagation time scale differ by less than an order of magnitude (a more detailed discussion follows below).

In addition to these more application-specific code developments, there are several well-documented and publicly available general purpose multidimensional hydrodynamics codes. As starting points we might mention the following ones: GADGET is a cosmological N-body code based on the Smoothed Particle Hydrodynamics method and parallelised using the Message Passing Interface (MPI). It is a Lagrangian approach with a hierarchical tree for the non-local evaluation of Newtonian gravity (Springel et al. 2001). VH-1 is a multidimensional hydrodynamics code based on the Lagrangian remap version of the Piecewise Parabolic Method (PPM) (Blondin and Lufkin 1993) that has been further developed and parallelised using MPI. ZEUS-2D (Stone and Norman 1992a,b; Stone et al. 1992) and ZEUS-3D are widely used grid-based hydrodynamics codes for which a MPI-parallel ZEUS-MP version exists as well. It offers the choice of different advection schemes on a fixed or moving orthogonal Eulerian mesh in a covariant description. While these more traditional approaches distribute in the form of a software package that includes options to switch on or off, recent open source projects try to provide the codes in the form of a generic framework that can host a variety of different modules implementing different techniques. In this category we could mention the FLASH code (e.g. Calder et al. 2002) with its main focus on the coupling of adaptive mesh stellar hydrodynamics to nuclear burning and the WHISKY code (Giacomazzo and Rezzolla 2007) as a recent general relativistic hydrodynamics code based on the CACTUS environment.

For the accurate prediction of yields of radioactive nuclei in astrophysical scenarios it would be necessary to perform accurate models of stellar evolution and the different types of stellar explosions. The involved physics has been discussed in Chaps. 4 and 9. Here we try to give some additional information about the technical tools that enable these simulations. Rather than continuing the discussion for all of these domains, we select the problem of stellar core collapse and explosion, where we think that the technical uncertainties are largest. Nevertheless, many of the discussed considerations and difficulties of current multidimensional models are quite general and also apply to the other astrophysical applications.

8.1 Models of Core-Collapse Supernovae

8.1.1 Basic Physical Description

The general scenario of a core-collapse supernova is described in Chap. 4. Supernova matter is described as a self-gravitating compressible fluid with negligible viscosity. The compactness of the gravitationally bound matter is sufficient to make general relativistic effects important. With respect to the electromagnetic interaction, the fluid is assumed to be locally charge neutral, but it is allowed to couple to dynamical magnetic fields as described by the equations of ideal magneto-hydrodynamics. With respect to the strong interaction, nuclear statistical equilibrium can be assumed in a large part of the computational domain. In this case, it is sufficient to specify the triple (n_b, T, Y_e) in order to uniquely describe the local thermodynamic state and the nuclear composition. The three independent variables are the baryon density n_b, the temperature T, and the electron fraction $Y_e = (n_e - n_{e^+})/n_b$, where n_e is the electron density and n_{e^+} the positron density. The weak interaction rates are not guaranteed to be in equilibrium. In fact, in dense regions of the computational domain the reaction rates are orders of magnitude faster than the dynamical time scale, while in other parts the weak reaction rates are negligible. Among the key reactions are the capture of electrons, e, on protons, p, and the capture of positrons, e^+, on neutrons, n. These reactions produce electron flavour neutrinos, ν_e, and antineutrinos, $\bar{\nu}_e$, and influence the structure and dynamics of the gravitationally bound fluid by changing its temperature and electron fraction. The inverse reactions must be considered as well:

$$\begin{aligned} e + p &\rightleftharpoons n + \nu_e \\ e^+ + n &\rightleftharpoons p + \bar{\nu}_e. \end{aligned} \tag{8.1}$$

At low density, the neutrinos have a long mean free path and escape from the star. At high density, the neutrinos are trapped due to a short mean free path. Being fermions, they can block the reactions (8.1) such that the effective rates of electron and positron capture are determined by the ability of neutrinos to diffuse out of the high-density regime. As the weak interaction rates increase with the square of the neutrino energy, one has to solve an energy-dependent radiative transfer problem. The radiative transfer problem is difficult to solve because the semi-transparent regime can be subject to turbulence and asymmetric convective turnover. Moreover, one has to consider the additional emission of μ flavour neutrinos and τ flavour neutrinos and their antineutrinos.

8.1.2 Basic Mathematical Description

With the ingredients summarised above, the mathematical description of the problem consists of an advection problem of the compressible fluid that is coupled to a spectral radiative transfer problem for the neutrinos. In principle, the description should be given in full general relativity. However, the general relativistic problem is only solved in rare cases (Liebendörfer et al. 2004; Ott 2009). Most multi-dimensional supernova models with spectral neutrino transport use a non-relativistic framework with a modified gravitational potential that makes up for the most important general relativistic effects (Marek et al. 2006). The radiative transfer part is usually kept accurate to order v/c, where v is the velocity of the fluid and c the velocity of light.

Using Cartesian space coordinates x_i and time t in the laboratory frame, the hydrodynamic part can be written as

$$\frac{\partial \rho}{\partial t} + \frac{\partial}{\partial x_j}\left(\rho v_j\right) = 0 \tag{8.2}$$

$$\frac{\partial}{\partial t}\left(\rho v_i\right) + \frac{\partial}{\partial x_j}\left(v_i \rho v_j - b_i b_j\right) + \frac{\partial P}{\partial x_i} + \rho \frac{\partial \phi}{\partial x_i} = \rho \dot{v}_i \tag{8.3}$$

$$\frac{\partial E}{\partial t} + \frac{\partial}{\partial x_j}\left[(E+P)\, v_j - b_i v_i b_j\right] + \rho v_i \frac{\partial \phi}{\partial x_i} = \rho \dot{e} \tag{8.4}$$

$$\frac{\partial}{\partial t}\left(\rho Y_e\right) + \frac{\partial}{\partial x_j}\left(Y_e \rho v_j\right) = \rho \dot{Y}_e . \tag{8.5}$$

The mass density $\rho = m_b n_b$ relates to the baryon density n_b by a constant conversion factor m_b that represents an average mass per baryon. Einsteins convention is used to sum expressions with equal indices over the different spatial directions denoted by indices $i = 1 \ldots 3$. All equations are written in conservative form, i.e.

in the general form

$$\frac{\partial U}{\partial t} + \frac{\partial F_j}{\partial x_j} = S, \tag{8.6}$$

where U is the density of a conserved quantity, $\mathbf{F}$ the flux density of this quantity, and S a local source density. The straight-forward finite differencing of a conservation equation guarantees for a fluid cell with volume dV that the change of the conserved quantity in the cell, UdV, is exactly given by the in/out-flow of the quantity through the cell-boundary and the source or sink of the quantity integrated over the cell volume, SdV. Equation (8.2) expresses the conservation of baryon number $\rho dV = n_b m_b dV$. The conservation of momentum $\rho v_i dV$ is described in Eq. (8.3) and the conservation of energy $EdV = \left(\rho e + \frac{1}{2}\rho v^2 + \frac{1}{2}b^2\right) dV$ in Eq. (8.4). Equation (8.5) describes the conservation of the lepton number $\rho Y_e dV$. The pressure $P(\rho, T, Y_e)$ and the specific internal energy $e(\rho, T, Y_e)$ are provided by an equation of state (EoS). The right hand sides of Eqs. (8.3)–(8.5) contain source terms. The terms $\rho \dot{v}_i$, $\rho \dot{e}$ and $\rho \dot{Y}_e$ stand for the exchange rates of momentum, energy, and lepton number, respectively, with the neutrino radiation field. The gravitational potential ϕ, finally, is determined by the Poisson equation

$$\nabla^2 \phi = 4\pi\rho. \tag{8.7}$$

If a magnetic field $\mathbf{B} = \sqrt{4\pi}\mathbf{b}$ is present, the momentum and energy equations (8.3) and (8.4) obtain contributions from the magnetic stress as indicated. The time-evolution of the magnetic field is described by

$$\frac{\partial \mathbf{b}}{\partial t} - \nabla \times (\mathbf{v} \times \mathbf{b}) = 0, \tag{8.8}$$

which guarantees that the magnetic field stays divergence free. Note that the nine Eqs. (8.2)–(8.8) determine the nine unknowns ρ, T, Y_e, v_i, b_i. Only the three source terms in Eqs. (8.3)–(8.5) remain open. They are determined by the neutrino transport equation that evaluates the weak interactions and the exchange of momentum, energy and lepton number between the fluid and the neutrino radiation field.

For each constituent of the fluid that cannot move with respect to the baryons, it is sufficient to assume thermal equilibrium and to specify an abundance. Constituents like the neutrinos that are not always in thermal equilibrium and that are able to propagate with respect to the fluid are more difficult to describe. The distribution function $f\left(t, \mathbf{x}, \omega, \frac{\mathbf{p}}{p}\right)$ additionally accepts the neutrino energy ω and the direction of propagation $\frac{\mathbf{p}}{p}$. The neutrino mass is neglected in the following kinetic treatment of the neutrinos, $(|\mathbf{p}|\, c = pc = \omega)$. The neutrino abundance is given by the integration of the distribution function over all neutrino energies ω and all propagation directions Ω,

$$Y_\nu(t, \mathbf{x}) = \frac{4\pi}{n_b (hc)^3} \int f\left(t, \mathbf{x}, \omega, \frac{\mathbf{p}}{p}\right) \omega^2 d\omega d\Omega, \tag{8.9}$$

whereas the time evolution of the neutrino distribution function is described by the Boltzmann equation

$$\frac{\partial f}{\partial t} + \frac{\partial f}{\partial \mathbf{x}} \cdot \frac{\mathbf{p}}{m} + \frac{\partial f}{\partial \mathbf{p}} \cdot \mathbf{K} = \dot{f}_{\text{collision}}. \tag{8.10}$$

The first term is the time derivation of the neutrino distribution function, which we would like to calculate. The second term describes the advection of neutrinos in and out of the fluid cell. In contrast to the analogous second term in Eq. (8.6), it depends on the particle velocity $\mathbf{p}/m$, which for the massless neutrinos becomes the light speed c instead of the fluid velocity. The third term describes the action of a force field $\mathbf{K}$ on the propagating neutrinos. The right hand side of Eq. (8.10) is given by the rate of local weak interactions between the neutrinos and the fluid. If this collision term is very large (i.e. if the mean free path of the neutrinos is very small), one can show that the appropriate integration over energy and angle of Eq. (8.10) will lead to an equation like Eq. (8.5) for the time evolution of the neutrino abundance in Eq. (8.9). The full system of equations that needs to be solved is now given by Eqs. (8.2)–(8.5), (8.7) and (8.8), coupled to Eq. (8.10) by the interaction rates on the right hand sides of the equations. The collision integral should at least implement the weak interactions listed in Bruenn (1985).

8.1.3 Scales and Challenges

The large *density range* covered by a supernova model is typical for astrophysical applications. After the gravitational collapse on a time scale of ~100 ms (milliseconds), the central density in the compact remnant can reach up to $\rho_c \sim 10^{15}$ g/cm^3. At the outer boundary of the computational domain (at a radius of ~10,000 km in the progenitor star), densities of order 10^5 g/cm^3 are typical. In the case of a successful explosion, the density between the compact remnant and the ejecta may drop to order 1 g/cm^3. The neutrinos decouple from matter at densities $\sim 10^{11}$–10^{12} g/cm^3, depending on the neutrino energy and type. The location where the energy-dependent optical depth is of order unity is called the *neutrinosphere* (in analogy to the photosphere). The neutrino-opaque region inside the neutrinospheres forms the proto-neutron star. During the simulation, the size of the *proto-neutron star* shrinks from an initial radius ~70 km to ~20 km due to the accumulation of mass and the emission of neutrinos. The proto-neutron star has an initial mass of ~1.2 $M_\odot$ and accretes infalling outer layers until the supernova explosion is launched or the maximum stable mass is exceeded so that a black hole is formed.

In order to resolve the steep density profile at the edge of the proto-neutron star, a *resolution* of $\Delta x < 300$ m is desirable to investigate the proto-neutron star itself. A resolution of $\Delta x \sim 1$ km might just be sufficient if one is only interested in the explosion on the surface of the proto-neutron star. The sound speed in the compact remnant is $c_s \sim 0.3c$, which leads to *time steps* $\Delta t < 10^{-5}$ s if the algorithm has to obey the CFL condition $\Delta t < \Delta x / c_s$. On the one hand, this is a reasonable time step compared to the dynamical time scale $(G\rho_c)^{-\frac{1}{2}} \sim 0.2$ ms of the proto-neutron star. On the other hand, the post-bounce phase between the collapse and the explosion is thought to extend on a time scale of order seconds. The assumed neutrino diffusion time scale is consistent with the detection of neutrinos over several seconds from supernova SN1987A in the Magellanic cloud. However, even with time steps of order µs, the reaction rates in the neutrino-opaque regime are fast enough to require an *implicit finite differencing* to find the correct equilibrium value.

Since implicit finite differencing is an important and ubiquitous concept for the modelling of stars, we dedicate a paragraph to this powerful technique. Consider two containers with contents I_1 and I_2, respectively, as drawn in Fig. 8.1. The containers are able to exchange content by two channels having fluxes $|f_1| = I_1/\tau$ and $|f_2| = I_2/\tau$ in the direction indicated by the arrows. τ is an arbitrary time scale. With this, the system is described by the coupled system of equations

$$\frac{\partial I_1}{\partial t} = \frac{I_2 - I_1}{\tau}, \qquad \frac{\partial I_2}{\partial t} = -\frac{I_2 - I_1}{\tau}. \tag{8.11}$$

The system is readily solved analytically by investigating the time evolution of the total quantity $I_2 + I_1$ and the content difference $I_2 - I_1$. Now, imagine a computer code that solves the finite differenced system

$$\begin{aligned} I_1^{n+1} &= I_1^n + \frac{I_2^{n+k} - I_1^{n+k}}{\tau} \Delta t \\ I_2^{n+1} &= I_2^n - \frac{I_2^{n+k} - I_1^{n+k}}{\tau} \Delta t. \end{aligned} \tag{8.12}$$

The upper index indicates the time step number: I^0 is the initial container content, I^1 the value calculated after the first time step Δt, I^2 the value after the second time step, and so forth. If one sets $k = 0$ in Eq. (8.12) one obtains Euler forward

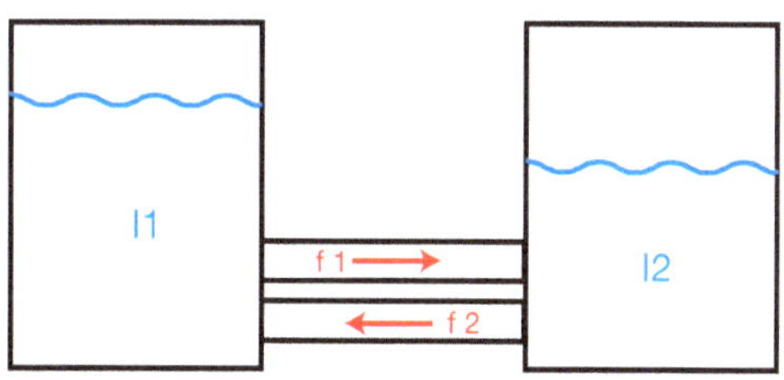

Fig. 8.1 Two containers with content I_1 and I_2 exchanging a conserved quantity on a characteristic time scale τ

finite differencing or explicit finite differencing. In this case, the values of I^{n+1} are directly expressed by the values of I^n of the previous time step as described in Eq. (8.12). Now let's start from the initial values $I_1^0 = 0.6$ and $I_2^0 = 0.4$. Thinking through few time steps shows clearly that a reasonable solution can only be expected when the time step Δt is much smaller than the characteristic time scale τ. If it is chosen larger, the solution diverges! This cannot happen with Euler backward or implicit finite differencing. For implicit finite differencing one sets $k = 1$ in Eq. (8.12). Because I^{n+1} now appears on both sides of Eq. (8.12) one has to solve the system of equations first for I^{n+1} and obtains

$$\begin{aligned} I_2^{n+1} + I_1^{n+1} &= I_2^n + I_1^n \\ I_2^{n+1} - I_1^{n+1} &= \left(I_2^n - I_1^n\right)\left(1 + \frac{2\Delta t}{\tau}\right)^{-1} \end{aligned} \tag{8.13}$$

One can now convince oneself that Eq. (8.13) is equivalent to the explicit version of Eq. (8.12) for very small time steps $\Delta t \ll \tau$, while the equilibrium configuration is obtained without difficulty for very large time steps $\Delta t \gg \tau$. Hence, an implicitly finite differenced equation puts automatically all processes into equilibrium that have a faster characteristic time scale than the time step. This technique is crucial for the successful evolution of the weak interaction rates, where many different characteristic time scales couple across the divers thermodynamic conditions found in a supernova. However, this stability comes at the cost of solving the system for I^{n+1}, which may quickly become a limiting factor for the memory and CPU-time of the simulation.

Even if the *computational domain* has to extend to 10,000 km radius in order to capture most of the infalling layers during the post-bounce evolution, the initially interesting dynamics is confined to a sphere of about 500 km radius (see e.g. Marek and Janka 2009). This still gives a ratio of neutrino-opaque to neutrino-transparent volume of $50^3/500^3 = 10^{-3}$. At the first glance it is therefore surprising that most investigators choose a *comoving frame* description of the neutrino transport which is excessively inefficient in the neutrino-transparent regimes that constitute 99.9% of the computational domain. The reason is that the diffusive limit of the approach has to be very accurate in order to resolve the *small neutrino flux* H at high neutrino density J in the core, where ratios of $\frac{H}{cJ} \sim 10^{-4}$ can prevail. While the diffusion equation has no difficulty to resolve arbitrarily small fluxes (the flux in the co-moving frame naturally scales with the mean free path), it seems close to impossible for a Boltzmann equation that is discretised in the laboratory frame to obtain equally accurate results.

With the neutrino-driven (Marek and Janka 2009), the magneto-rotational (Bisnovatyi-Kogan et al. 2008), the acoustic (Burrows et al. 2007) and the phase-transition induced (Sagert et al. 2009) explosion mechanisms, several supernova explosion mechanisms are discussed as cause of the energetic stellar explosions. The difficulty to pinpoint a specific mechanism is not due to the lack of available energy to explode the star. The gravitational binding energy released during the collapse of the inner stellar core to the proto-neutron star amounts to several 10^{53} erg, i.e. much more than the $\sim 10^{51}$ erg kinetic energy of a typical supernova explosion. The major part of the released energy is first stored as thermal or rotational energy in the proto-neutron star and then emitted from there by the emission of neutrinos or build-up of magnetic fields. The difficulty is to reliably *quantify the energy transfer* that occurs from the small coupling of the large energy reservoir in the proto-neutron star to its surface layers that have densities of 10^{10}–10^{8} g/cm^3 with much lower characteristic energies. In the neutrino-driven explosion mechanism the energy is transferred by the neutrinos. The energy deposition rate of the neutrinos depends on their spectra and on the propagation angles in the layers that are close to the energy-dependent neutrinospheres. In the magneto-rotational explosion mechanism the rotational energy from the collapse is transferred via its coupling to magnetic fields, in the (magneto-)acoustic explosion mechanism the energy is transferred by (magneto-)acoustic waves, and in the phase-transition induced mechanism the energy is transferred by a shock that emerges from the proto-neutron star.

A further difficulty is that there are *fluid instabilities* like the standing accretion shock instability (SASI; Foglizzo et al. 2007) and large scale convective turnover from neutrino heating (Herant et al. 1994) that lead to flow patterns with large regions of outflow alongside of narrow downstreams that affect the efficiency of the energy transfer. Typical dynamical time scales in this less dense regime are of order of several 10 ms. Finally, the effects of *magnetic fields* in combination with rotation have been discussed over decades, but are only starting to be quantitatively explored by the advent of three-dimensional simulations that are based on the ideal magneto-hydrodynamic (MHD) equations. In combination with differential rotation, the magneto-rotational instability (MRI) might develop on the length scale of meters, which cannot be resolved in global simulations. The magnetic fields might grow over a longer post-bounce time scale and lead to deformations of the fluid flow in the layers above the proto-neutron star that are crucial for the energy transfer.

8.1.4 Solution Strategies

8.1.4.1 Spherically Symmetric Models

An important milestone has been reached by the traditional spherically symmetric approach. In this case, the degrees of freedom of the distribution function reduce to $f(t, r, \omega, \mu)$, where r is the radial coordinate, $\mu = \cos\theta$ and θ is the angle between the propagation direction of the neutrino and the radial direction. For example, $\mu = 1$ means 'radially outward' propagation, $\mu = 0$ means 'tangential' propagation and $\mu = -1$ means 'radially inward' propagation. At first, one might be surprised that the neutrino distribution function has an angular dependence in spite of the assumption of spherical symmetry. This fact is understandable in the following way: Let's assume that the abundance of neutrinos is spherically symmetric. If all neutrinos propagate radially outward, the configuration is obviously spherically symmetric. But if all neutrinos propagate tangentially to a sphere around the origin and isotropically within the surface of the sphere, the configuration is also spherically symmetric. Hence, there are different spherically symmetric configurations that can be distinguished by the angle of the neutrino propagation with respect to the radial direction. The neutrino number density, for example, is then given by the integration of the neutrino distribution function over energy and the propagation angle

$$n_\nu(t,r) = \frac{4\pi}{(hc)^3}\frac{1}{2}\int_{-1}^{+1}\int_0^\infty f(t,r,\omega,\mu)\,\omega^2 d\omega d\mu. \tag{8.14}$$

The time evolution of the neutrino distribution function is described by the Boltzmann transport equation (Lindquist 1966; Castor 1972; Mezzacappa and Bruenn 1993; Liebendörfer et al. 2004)

$$\begin{aligned}
&\frac{df}{cdt} + \mu\frac{\partial f}{\partial r} + \left[\mu\left(\frac{d\ln\rho}{cdt} + \frac{3v}{cr}\right) + \frac{1}{r}\right]\left(1-\mu^2\right)\frac{\partial f}{\partial \mu} \\
&+ \left[\mu^2\left(\frac{d\ln\rho}{cdt} + \frac{3v}{cr}\right) - \frac{v}{cr}\right]\omega\frac{\partial f}{\partial \omega} = j\,(1-f) - \chi f \\
&+ \frac{1}{c\,(hc)^3}\left[(1-f)\int R^{\text{in}} f'\omega'^2 d\omega d\mu' - f\int R^{\text{out}}\left(1-f'\right)\omega'^2 d\omega d\mu'\right].
\end{aligned} \tag{8.15}$$

Why does Eq. (8.15) look so complicated with respect to Eq. (8.10)? The transport terms on the left hand side of the Boltzmann equation would be equally simple as in Eq. (8.10) if the momentum phase space of the neutrino distribution function was described in the *laboratory frame*. However, most supernova modellers choose a description in the *co-moving frame* because there are very sensitive regimes where the neutrinos have a small mean free path. There, the neutrinos diffuse with respect to the turbulent fluid and not with respect to the inertial frame. Moreover, the computationally expensive reaction rates on the right hand side of the Boltzmann equation are difficult to calculate and tabulate in the laboratory frame, where the emission angles depend on the fluid velocity. In the co-moving frame, the propagation terms naturally vanish with increasing opacity and the weak interaction rates can establish an accurate thermal and weak equilibrium between the fluid and the trapped neutrinos. On the other hand, the neutrinos propagate between fluid cells with relative velocities so that their energy and propagation direction must be corrected for Doppler frequency shift and angular aberration. These velocity-dependent terms are responsible for the complicated appearance of Eq. (8.15). The collision integral on the right hand side of Eq. (8.15) describes weak interactions between the neutrinos and the fluid. Examples are the isotropic neutrino emissivity $j\,(\rho, T, Y_e, \omega)$ and absorptivity $\chi\,(\rho, T, Y_e, \omega)$ for the reactions (8.1) and the collision integral for the scattering of neutrinos with the nucleons or electrons in the fluid. The latter depends on scattering kernels that describe inward scattering, $R^{\text{in}}(\rho, T, Y_e, \omega, \mu, \omega', \mu')$, and outward scattering, $R^{\text{out}}(\rho, T, Y_e, \omega, \mu, \omega', \mu')$. Due to the fermionic nature of the neutrinos, the appropriate blocking factors $(1 - f)$ must be applied. The solution of Eq. (8.15) determines the collision integral and therefore the exchange rates $\rho\dot{\mathbf{v}}$, $\rho\dot{e}$ and $\rho\dot{Y}_e$ that couple to Eqs. (8.3)–(8.5). From the emission and absorption terms alone one gets for example (Mezzacappa and Bruenn 1993)

$$\rho\dot{v} = \frac{2\pi}{(hc)^3}\int (j+\chi)\, f\omega^3 d\omega\mu d\mu$$

$$\rho\dot{e} = \frac{2\pi c}{(hc)^3}\int [j-(j+\chi)\, f]\,\omega^3 d\omega d\mu$$

$$\rho\dot{Y}_e = \frac{2\pi c m_b}{(hc)^3}\int [j-(j+\chi)\, f]\,\omega^2 d\omega d\mu.$$

Successful comparisons of the results have been performed with alternative radiative transfer algorithms like the Variable Eddington Factor (VEF) method (Rampp and Janka 2002; Liebendörfer et al. 2005) and the Multi-Group Flux-Limited Diffusion (MGFLD) approximation (Bruenn et al. 2001; Liebendörfer et al. 2004). However, models in spherical symmetry cannot treat the fluid instabilities and mixing that turn out to be crucial for the onset of the explosion and the further evolution of the expanding ejecta.

8.1.4.2 Axisymmetric Models

In spherically symmetric models, only very peculiar progenitor stars (Kitaura et al. 2006) or exotic input physics (Sagert et al. 2009) can explain explosions. It is well-known that fluid instabilities between the proto-neutron star and the standing accretion shock increase the absorption of neutrino energy behind the shock (Herant et al. 1994). Indeed, more progenitor stars are found to explode in axisymmetric models. However, the results of different groups have not yet converged: Models from the Florida/Oak Ridge group predict explosions ~300 ms post-bounce (Messer et al. 2008), models from the Garching group predict weak explosions ~600 ms (Marek and Janka 2009), and models from the Arizona group predict explosion ~1200 ms (Burrows et al. 2007) based on a different explosion mechanism.

Models in axi-symmetry have not been performed without approximations in the neutrino transport part. In attempts to solve a direct discretisation of the analogue of Eq. (8.15) (Livne et al. 2004), one had to drop the complicated but important velocity-dependent observer corrections, so that the first law of thermodynamics is violated for the trapped neutrinos. Implementations of the VEF method were more successful, but so far only in combination with the ray-by-ray approximation (Marek and Janka 2009). The ray-by-ray approach treats the neutrino transport in separate angular wedges with periodic boundary conditions at their interfaces to the neighbour wedges. Lateral transport is only supported in the neutrino-opaque regime by a term that advects trapped neutrinos with the fluid. The MGFLD approximation has been implemented in true axi-symmetry (Dessart et al. 2006) and in ray-by-ray approaches (Bruenn et al. 2009). The VEF and MGFLD approximations have both the important advantage that they are based on the solution of moments equations of the radiation field that can naturally resolve small neutrino fluxes in the diffusion limit. While both methods share the drawback that the implicit solution of moments equations is very inefficient in the neutrino-transparent regime, a traditional flux limiter in the MGFLD approach additionally looses the information about propagation directions outside the neutrinospheres (Ott et al. 2008). We believe that the VEF approach can successfully cope with most of the challenges listed in Sect. 8.1.3. However, because a single run with a well-tested and comprehensive VEF approach in two-dimensional models still takes a wallclock time of 6–12 months. For the time being, it appears uneconomical to use this otherwise favourable approach in global three-dimensional models.

8.1.4.3 Three-Dimensional Models

Even if axi-symmetry supports crucial fluid instabilities in all these models, it still severely restricts the degrees of freedom of the fluid. Patterns of the accretion flow that include accretion tubes or funnels instead of cones and the three-dimensional spreading of non-axisymmetric modes cannot be investigated in axi-symmetry. Also the Lorentz force, that acts on the fluid when magnetic fields are present, has an intrinsically three-dimensional nature. In search of a neutrino transport approximation that can be used as soon as possible to explore the rich physics of three-dimensional dynamical phenomena, we step back from the ideal that the numerical algorithm or method has to be uniform across the whole computational domain. There is no 'best algorithm' without a clear specification what problem the algorithm has to solve. In global astrophysical models, many different problems have to be solved in different parts of the computational domain and hence, it is unlikely that there is *one* best algorithm to do this *everywhere*. In three-dimensional supernova models one should exploit that a lot of computational cost could be avoided by adaptive algorithms. A numerically well-resolved model that is based on good physical approximations is much easier to control and safer to interpret than a very comprehensive model that is numerically under-resolved or too inefficient to be sufficiently checked before the production phase. In this spirit, we start with a list of the most important technical requirements that characterise a useful supernova model:

1. The hydrodynamics algorithm must prevent that errors from the calculation of the gravitational potential propagate into the specific internal energy of the fluid and it must conserve baryon number, energy and momentum in order to be able to handle shocks and dissipate unresolved turbulence.

2. Reaction rates must be time-implicitly finite differenced and obey detailed balance in order to be able to find thermodynamic equilibrium where appropriate.
3. In the diffusion limit, mechanical work acting on the neutrino gas must change the neutrino temperature consistently with the first law of thermodynamics and the *discretised* equations must deliver diffusive fluxes that scale with the mean free path to arbitrary small numbers.
4. The neutrino spectrum must be retained when neutrinos are transported between fluid cells of different thermodynamic state and the flux factor in the transparent regime should rather depend on the distance to the neutrinosphere of appropriate neutrino energy than on irrelevant local conditions of the fluid.
5. In order to explore the large parameter space of possible initial configurations, the code must be capable to run over thousands of dynamical time scales within reasonable time.

In order to implement these requirements we developed the Isotropic Diffusion Source Approximation (IDSA) (Liebendörfer et al. 2009). It attempts to implement an 'adaptive algorithm' in the sense that the separation into hydrodynamics and radiative transfer is not based on particle species as in Eqs. (8.2)–(8.5) and (8.15), but on the local opacity. One particle species is allowed to have a component that evolves in the hydrodynamic limit, while another component of the same particle species is treated by radiative transfer. The distribution function of the neutrinos is decomposed into a trapped neutrino component and a streaming neutrino component. The thermodynamics of trapped neutrinos is included in the standard hydrodynamics scheme that evolves the background matter. The neutrino losses due to diffusion are described by an additional source term, the 'diffusion source', that converts trapped neutrinos into streaming neutrinos. The obtained streaming neutrino emission rates are then integrated over space to obtain the neutrino flux. Finally, the neutrino flux is converted to a local streaming neutrino abundance Eq. (8.9) that interacts with matter and feeds back into the evaluation of the diffusion source for the next time step. The conversion from the neutrino flux to the neutrino abundance is performed by a geometric estimate of the flux factor based on the location of the neutrinospheres.

An 'adaptive algorithm' can significantly boost the efficiency of a simulation because good approximations can be used locally, even if they do not hold in the full computational regime. For example, the diffusion limit can be used in the computational domain of large optical depth and ray-tracing can be used in the computational domain with negligible optical depth. The challenge of this approach is the implementation of a smooth and reasonably consistent transition between the different regimes of approximation. The 3D results shown in Fig. 4.14 have been produced by implementing the IDSA in the magnetohydrodynamics code FISH (Käppeli et al. 2011).

The theoretical understanding of the supernova explosion mechanism is crucial for the understanding of the stellar life cycle, the feedback of internal energy to the interstellar gas in star-forming regions, and the enrichment of the Galaxy with heavy elements. Supernovae are active in all observational windows and emit a broad spectrum of electro-magnetic waves, neutrinos, cosmic rays and gravitational waves. A quantitative understanding of the supernova explosion mechanism may grant observational access to matter under extreme conditions where new physics could be discovered. The knowledge of the input physics is continuously improving by terrestrial experiments and theoretical developments. Similarly, the data base of astronomical observations is growing due to the increased sensitivity of observational instruments and improved statistics. The computational model as link between the input physics and the astronomical observable is most productive if the uncertainties in the model are smaller than the uncertainties in the input physics or the observables. The main difficulty in quantifying the uncertainties in the understanding of the explosion mechanism is that the interaction between the neutrinos and matter constitute a 3D dynamical radiative transfer problem that requires the solution of a spectral transport equation for the neutrinos in a general relativistically curved environment where turbulence and mixing spans many orders of magnitude in density and time scales.

8.2 Models of Ejecta for Nucleosynthetic Yield Prediction

As discussed in Chaps. 4 and 9, the nucleosynthetic yields are determined by (1) the dynamics of the ejection process, (2) the initial composition and thermodynamic conditions of the ejecta and (3) the weak interactions between the ejecta and the neutrinos that are emitted from the neutrino spheres. In the remainder of this section we will discuss these ingredients and their uncertainties based on the supernova models outlined above.

8.2.1 The Dynamics of the Ejection Process

The described uncertainties about the supernova explosion mechanism and the discussed technical difficulties to constrain the uncertainties by accurate three-dimensional computer models are responsible for the largest part of the uncertainties of the core-collapse supernova nucleosynthesis yields. Fortunately, not every detail of the multidimensional computer model impacts the ejection dynamics with similar significance. The first and perhaps most significant unknown is the delay time $\sim$ 500 ms between the core bounce and the onset of the explosion. If this delay time is long, the innermost stellar layers that surround the collapsing iron core will accrete onto the proto-neutron star and settle there before the explosion. The shock wave of the explosion then runs through the outer stellar layers performing explosive nucleosynthesis. The amount of mass that remains bound (including the proto-neutron star) is called the *mass cut*. It determines which stellar layers form the innermost ejecta and has therefore a significant impact on the ejected supernova yields. Since it determines the total amount of ejected ^{56}Ni, the mass cut can be inferred from the observed mass of ^{56}Ni in the ejecta (Thielemann et al. 1996). A very long delay time between the core bounce and the supernova explosion increases the chance that the proto-neutron star accretes more than its maximum stable mass so that the proto-neutron star collapses to a black hole instead of driving an explosion. Thus, the unknown delay time determines in this way also the mass range of progenitor stars that will eject nucleosynthesis yields.

The second important unknown in current supernova models is the explosion energy. The otherwise most plausible neutrino-driven supernova mechanism (Bethe and Wilson 1985) shows a tendency of producing too low explosion energies throughout the literature. This is mostly due to a negative feedback between the accretion rate and the neutrino heating: Whenever the explosion starts, the accretion rate is reduced. This leads to a significant decrease of the neutrinos that are emitted from the compressed accreted matter. This decrease in the neutrino luminosity from the neutrinospheres reduces the neutrino heating behind the shock and hinders the explosion to pick up as much energy as the observation of supernovae suggests. This problem is most severe in spherically symmetric approaches where accretion and explosion cannot coexist at the same time. In multidimensional models, it is plausible that narrow accreting downward streams maintain a high accretion luminosity while regions in between them are already expanding toward the explosion (Herant et al. 1994). Multidimensional simulations of the post-explosion phase show strong mixing by the Rayleigh-Taylor instabilities that occur when the cold outer stellar layers are accelerated from below by the hot ejecta (Kifonidis et al. 2003; Hammer et al. 2010). However, this late-time mixing determines rather the geometrical distribution of the yields than the nucleosynthesis itself.

8.2.2 The Thermodynamic Conditions of the Ejecta

The nucleosynthesis is mainly influenced by the initial composition and the time evolution of the thermodynamic conditions in the ejecta. Traditional yield predictions adopt the composition of the progenitor star as initial condition of the ejecta, infer the mass cut from the observed ^{56}Ni mass and set artificially a kinetic piston or thermal energy (Aufderheide et al. 1991) into a hydrodynamical model in order to produce timelines of thermodynamic conditions in the ejecta that can be post-processed with large nuclear reaction networks (see Chap. 9). Individual fluid elements from different stellar layers are post-processed in this way until the reactions freeze out. The resulting yields are then obtained by collecting the abundances from the treated fluid elements. This approach is very clever, because the above-mentioned two big uncertainties of supernova models with respect to the mass cut and the explosion energy are circumvented by observational information. In a spherically symmetric model, the temperature of the matter at the time it is passed by the shock can even be estimated analytically from Chap. 4, Eq. (4.7). If one determines the thermodynamical conditions of the ejecta more accurately from a hydrodynamical simulation, one faces the problem that the thermodynamic evolution of the ejecta depends itself on the energy budget of the nuclear reactions that one would like to evaluate after the fact in a post-processing step. Fortunately, it is a good approximation to run the hydrodynamics together with an efficient small network of selected (mostly α-)nuclei (Hix et al. 1998) that correctly deliver the overall energy budget of the nucleosynthesis to the thermodynamical evolution.

8.2.3 The Exposure of the Ejecta to Neutrinos

The large reaction networks that are used to determine the nucleosynthesis trace the evolution of the abundances of thousands of nuclei. Hence, they keep an accurate track of the composition—with one exception: the neutrinos. Because the neutrinos are not trapped in the ejecta, their abundance is not determined by local conditions, but by the distance to the neutrinospheres and the conditions at the neutrinospheres. The influence of the neutrinos on the nucleosynthesis is a third factor of uncertainty in the yield prediction that relates to the underlying supernova model. The exchange of neutrons into protons and vice versa by the reactions in Eq. (8.1) depends on the ratio of the neutrino to anti-neutrino abundances and their spectra (they are not in thermal equilibrium with the ejecta). These neutrino properties are determined at the neutrinospheres close to the surface of the proto-neutron star.

While the mass cut and the explosion energy could be gauged by the observation, it is not possible to reliably derive the influence of neutrinos from observations. The only detection of neutrinos from supernova 1987A was statistically too scarce to pinpoint the neutrino energies. Fortunately the neutrino energies and spectra are less difficult to predict from the above-described supernova models than the multidimensional explosion dynamics: The neutrinos decouple from the matter at the surface of the proto-neutron star where the geometry is still reasonably spherically symmetric. Hence, the spherical models described in Sect. 8.1.4.1 can be used to derive the neutrino abundances as function of radius and time in general relativity with sophisticated input physics. In this way, the neutrino luminosities can be chosen consistently with the progenitor model that serve for the ejection model. This is important because the neutrino luminosities and spectra can strongly vary between progenitor models of different main sequence mass. A compilation of luminosities and mean energies is given in Liebendörfer et al. (2003). As discussed in Chap. 4, yield predictions have shown that the neutrinos have a significant impact on the nucleosynthesis of the innermost ejecta, where the neutrino flux and target density in the ejecta is largest (Woosley et al. 1990; Fröhlich et al. 2006b; Pruet et al. 2006).

How can the three uncertainties *dynamics*, *thermodynamic conditions* and *neutrino influence* further be reduced? In the mean time, the neutrino luminosities are also available from accurate axisymmetric supernova models (Marek and Janka 2009). The main difference to the spherical case is that the multidimensional accretion flow permits time-variable fast downflows that contribute a time-variable accretion luminosity component to the total neutrino luminosities. Additionally, it is now interesting to not only use the neutrinos from the supernova model, but also the dynamics of the ejection process and the consistent thermodynamic condition of the ejecta. However, because there are only very few supernova models that explode naturally (Hoffman et al. 2008; Wanajo et al. 2009), one has to produce series of explosion models by artificially adding external energy to the neutrino heating in a parameterised form (Fröhlich et al. 2006a). As a next step, the axisymmetric supernova models discussed in Sect. 8.1.4.2 should be refined and brought to mutual agreement and convergence between the different groups so that the three-dimensional models with spectral neutrino transport that are constructed now along the lines of Sect. 8.1.4.3 find again a reliable reference. Then the new generation of three-dimensional models will lead to new insight: If they produce explosions that are consistent with the observations, they can serve as models to predict the nucleosynthesis yields in much more detail. If, on the other hand, they still do not produce sufficiently energetic explosions, one has for the first time proven missing physics in the models and can start a new quest for the supernova mechanism.

References

Aufderheide MB, Baron E, Thielemann F (1991) Shock waves and nucleosynthesis in type II supernovae. Astrophys J 370:630–642. https://doi.org/10.1086/169849

Bethe HA (1990) Supernova mechanisms. Rev Mod Phys 62:801–866. https://doi.org/10.1103/RevModPhys.62.801

Bethe HA, Wilson JR (1985) Revival of a stalled supernova shock by neutrino heating. Astrophys J 295:14–23. https://doi.org/10.1086/163343

Bisnovatyi-Kogan GS, Moiseenko SG, Ardelyan NV (2008) Different magneto-rotational supernovae. Astron Rep 52:997–1008. https://doi.org/10.1134/S1063772908120056

Blondin JM, Lufkin EA (1993) The piecewise-parabolic method in curvilinear coordinates. Astrophys J Suppl 88:589–594. https://doi.org/10.1086/191834

Bruenn SW (1985) Stellar core collapse - numerical model and infall epoch. Astrophys J Suppl 58:771–841. https://doi.org/10.1086/191056

Bruenn SW, De Nisco KR, Mezzacappa A (2001) General relativistic effects in the core collapse supernova mechanism. Astrophys J 560:326–338. https://doi.org/10.1086/322319, arXiv:astro-ph/0101400

Bruenn SW, Mezzacappa A, Hix WR, Blondin JM, Marronetti P, Messer OEB, Dirk CJ, Yoshida S (2009) 2D and 3D core-collapse supernovae simulation results obtained with the CHIMERA code. J Phys Conf Ser 180(1):012018. https://doi.org/10.1088/1742--6596/180/1/012018

Brun AS, Miesch MS, Toomre J (2004) Global-scale turbulent convection and magnetic dynamo action in the solar envelope. Astrophys J 614:1073–1098. https://doi.org/10.1086/423835, arXiv:astro-ph/0610073

Burrows A, Livne E, Dessart L, Ott CD, Murphy J (2007) Features of the acoustic mechanism of core-collapse supernova explosions. Astrophys J 655:416–433. https://doi.org/10.1086/509773, arXiv:astro-ph/0610175

Calder AC, Fryxell B, Plewa T, Rosner R, Dursi LJ, Weirs VG, Dupont T, Robey HF, Kane JO, Remington BA, Drake RP, Dimonte G, Zingale M, Timmes FX, Olson K, Ricker P, MacNeice P, Tufo HM (2002) On validating an astrophysical simulation code. Astrophys J Suppl 143:201–229. https://doi.org/10.1086/342267, arXiv:astro-ph/0206251

Castor JI (1972) Radiative transfer in spherically symmetric flows. Astrophys J 178:779–792. https://doi.org/10.1086/151834
Chandrasekhar S (1967) An introduction to the study of stellar structure. Dover, New York
Colgate SA, White RH (1966) The hydrodynamic behavior of supernovae explosions. Astrophys J 143:626. https://doi.org/10.1086/148549
Dessart L, Burrows A, Ott CD, Livne E, Yoon S, Langer N (2006) Multidimensional simulations of the accretion-induced collapse of white dwarfs to neutron stars. Astrophys J 644:1063–1084. https://doi.org/10.1086/503626, arXiv:astro-ph/0601603
Dotter A, Paxton B (2009) Evolutionary implications of the new triple-α nuclear reaction rate for low mass stars. Astron Astrophys 507:1617–1619. https://doi.org/10.1051/0004-6361/200912998, 0905.2397
Eggenberger P, Meynet G, Maeder A, Hirschi R, Charbonnel C, Talon S, Ekström S (2008) The Geneva stellar evolution code. Astrophys Space Sci 316:43–54. https://doi.org/10.1007/s10509-007-9511-y
Foglizzo T, Galletti P, Scheck L, Janka H (2007) Instability of a stalled accretion shock: evidence for the advective-acoustic cycle. Astrophys J 654:1006–1021. https://doi.org/10.1086/509612, arXiv:astro-ph/0606640
Fröhlich C, Hauser P, Liebendörfer M, Martínez-Pinedo G, Thielemann F, Bravo E, Zinner NT, Hix WR, Langanke K, Mezzacappa A, Nomoto K (2006a) Composition of the innermost core-collapse supernova ejecta. Astrophys J 637:415–426. https://doi.org/10.1086/498224, arXiv:astro-ph/0410208
Fröhlich C, Martínez-Pinedo G, Liebendörfer M, Thielemann F, Bravo E, Hix WR, Langanke K, Zinner NT (2006b) Neutrino-induced nucleosynthesis of A>64 nuclei: the νp process. Phys Rev Lett 96(14):142502. https://doi.org/10.1103/PhysRevLett.96.142502, arXiv:astro-ph/0511376
Fryxell B, Olson K, Ricker P, Timmes FX, Zingale M, Lamb DQ, MacNeice P, Rosner R, Truran JW, Tufo H (2000) FLASH: an adaptive mesh hydrodynamics code for modeling astrophysical thermonuclear flashes. Astrophys J Suppl 131:273–334. https://doi.org/10.1086/317361
Giacomazzo B, Rezzolla L (2007) WhiskyMHD: a new numerical code for general relativistic magnetohydrodynamics. Classical Quantum Gravity 24:235. https://doi.org/10.1088/0264-9381/24/12/S16, arXiv:gr-qc/0701109
Hammer NJ, Janka HT, Müller E (2010) Three-dimensional simulations of mixing instabilities in supernova explosions. Astrophys J 714:1371–1385. https://doi.org/10.1088/0004-637X/714/2/1371, 0908.3474
Henyey LG, Wilets L, Böhm KH, Lelevier R, Levee RD (1959) A method for atomic computation of stellar evolution. Astrophys J 129:628. https://doi.org/10.1086/146661
Herant M, Benz W, Hix WR, Fryer CL, Colgate SA (1994) Inside the supernova: a powerful convective engine. Astrophys J 435:339–361. https://doi.org/10.1086/174817, arXiv:astro-ph/9404024
Hix WR, Khokhlov AM, Wheeler JC, Thielemann F (1998) The quasi-equilibrium-reduced alpha-network. Astrophys J 503:332. https://doi.org/10.1086/305968, arXiv:astro-ph/9805095
Hoffman RD, Müller B, Janka H (2008) Nucleosynthesis in O-Ne-Mg supernovae. Astrophys J 676:L127–L130. https://doi.org/10.1086/587621, 0712.4257
Käppeli R, Whitehouse SC, Scheidegger S, Pen UL, Liebendörfer M (2011) FISH: a three-dimensional parallel magnetohydrodynamics code for astrophysical applications. Astrophys J Suppl 195:20. https://doi.org/10.1088/0067-0049/195/2/20, 0910.2854
Kifonidis K, Plewa T, Janka H, Müller E (2003) Non-spherical core collapse supernovae. I. Neutrino-driven convection, Rayleigh-Taylor instabilities, and the formation and propagation of metal clumps. Astron Astrophys 408:621–649. https://doi.org/10.1051/0004-6361:20030863, arXiv:astro-ph/0302239
Kitaura FS, Janka H, Hillebrandt W (2006) Explosions of O-Ne-Mg cores, the Crab supernova, and subluminous type II-P supernovae. Astron Astrophys 450:345–350. https://doi.org/10.1051/0004-6361:20054703, arXiv:astro-ph/0512065
Langer N, Fliegner J, Heger A, Woosley SE (1997) Nucleosynthesis in rotating massive stars. Nucl Phys A 621:457–466. https://doi.org/10.1016/S0375-9474(97)00290-X

Liebendörfer M, Mezzacappa A, Messer OEB, Martinez-Pinedo G, Hix WR, Thielemann F (2003) The neutrino signal in stellar core collapse and postbounce evolution. Nucl Phys A 719:144. https://doi.org/10.1016/S0375-9474(03)00984-9, arXiv:astro-ph/0211329

Liebendörfer M, Messer OEB, Mezzacappa A, Bruenn SW, Cardall CY, Thielemann F (2004) A finite difference representation of neutrino radiation hydrodynamics in spherically symmetric general relativistic spacetime. Astrophys J Suppl 150:263–316. https://doi.org/10.1086/380191, arXiv:astro-ph/0207036

Liebendörfer M, Rampp M, Janka H, Mezzacappa A (2005) Supernova simulations with Boltzmann neutrino transport: a comparison of methods. Astrophys J 620:840–860. https://doi.org/10.1086/427203, arXiv:astro-ph/0310662

Liebendörfer M, Whitehouse SC, Fischer T (2009) The isotropic diffusion source approximation for supernova neutrino transport. Astrophys J 698:1174–1190. https://doi.org/10.1088/0004-637X/698/2/1174, 0711.2929

Limongi M, Straniero O, Chieffi A (2000) Massive stars in the range 13-25 $M_{\odot}$: evolution and nucleosynthesis. II. The solar metallicity models. Astrophys J Suppl 129:625–664. https://doi.org/10.1086/313424, arXiv:astro-ph/0003401

Lindquist RW (1966) Relativistic transport theory. Ann Phys 37:487–518. https://doi.org/10.1016/0003-4916(66)90207-7

Livne E, Burrows A, Walder R, Lichtenstadt I, Thompson TA (2004) Two-dimensional, time-dependent, multigroup, multiangle radiation hydrodynamics test simulation in the core-collapse supernova context. Astrophys J 609:277–287. https://doi.org/10.1086/421012, arXiv:astro-ph/0312633

Ludwig H, Freytag B, Steffen M (1999) A calibration of the mixing-length for solar-type stars based on hydrodynamical simulations. I. Methodical aspects and results for solar metallicity. Astron Astrophys 346:111–124. arXiv:astro-ph/9811179

Maeder A, Meynet G (1989) Grids of evolutionary models from 0.85 to 120 solar masses - observational tests and the mass limits. Astron Astrophys 210:155–173

Marek A, Janka H (2009) Delayed neutrino-driven supernova explosions aided by the standing accretion-shock instability. Astrophys J 694:664–696. https://doi.org/10.1088/0004-637X/694/1/664, 0708.3372

Marek A, Dimmelmeier H, Janka H, Müller E, Buras R (2006) Exploring the relativistic regime with Newtonian hydrodynamics: an improved effective gravitational potential for supernova simulations. Astron Astrophys 445:273–289. https://doi.org/10.1051/0004-6361:20052840, arXiv:astro-ph/0502161

Meakin CA, Arnett D (2007) Turbulent convection in stellar interiors. I. Hydrodynamic simulation. Astrophys J 667:448–475. https://doi.org/10.1086/520318, arXiv:astro-ph/0611315

Messer OEB, Bruenn SW, Blondin JM, Hix WR, Mezzacappa A (2008) Multidimensional, multiphysics simulations of core-collapse supernovae. J Phys Conf Ser 125(1):012010. https://doi.org/10.1088/1742-6596/125/1/012010

Meynet G, Maeder A (1997) Stellar evolution with rotation. I. The computational method and the inhibiting effect of the μ-gradient. Astron Astrophys 321:465–476

Mezzacappa A, Bruenn SW (1993) Stellar core collapse - a Boltzmann treatment of neutrino-electron scattering. Astrophys J 410:740–760. https://doi.org/10.1086/172791

Nomoto K, Hashimoto M (1988) Presupernova evolution of massive stars. Phys Rep 163:13–36

Ott CD (2009) Topical review: the gravitational-wave signature of core-collapse supernovae. Class Quantum Gravity 26(6):063001. https://doi.org/10.1088/0264-9381/26/6/063001, 0809.0695

Ott CD, Burrows A, Dessart L, Livne E (2008) Two-dimensional multiangle, multigroup neutrino radiation-hydrodynamic simulations of postbounce supernova cores. Astrophys J 685:1069–1088. https://doi.org/10.1086/591440, 0804.0239

Pruet J, Hoffman RD, Woosley SE, Janka H, Buras R (2006) Nucleosynthesis in early supernova winds. II. The role of neutrinos. Astrophys J 644:1028–1039. https://doi.org/10.1086/503891, arXiv:astro-ph/0511194

Rampp M, Janka H (2002) Radiation hydrodynamics with neutrinos. Variable Eddington factor method for core-collapse supernova simulations. Astron Astrophys 396:361–392. https://doi.org/10.1051/0004-6361:20021398, arXiv:astro-ph/0203101

Rauscher T, Heger A, Hoffman RD, Woosley SE (2002) Nucleosynthesis in massive stars with improved nuclear and stellar physics. Astrophys J 576:323–348. https://doi.org/10.1086/341728, arXiv:astro-ph/0112478

Röpke FK, Hillebrandt W, Schmidt W, Niemeyer JC, Blinnikov SI, Mazzali PA (2007) A three-dimensional deflagration model for type Ia supernovae compared with observations. Astrophys J 668:1132–1139. https://doi.org/10.1086/521347, 0707.1024

Sagert I, Fischer T, Hempel M, Pagliara G, Schaffner-Bielich J, Mezzacappa A, Thielemann F, Liebendörfer M (2009) Signals of the QCD phase transition in core-collapse supernovae. Phys Rev Lett 102(8):081101. https://doi.org/10.1103/PhysRevLett.102.081101, 0809.4225

Springel V, Yoshida N, White SDM (2001) GADGET: a code for collisionless and gasdynamical cosmological simulations. New Astron 6:79–117. https://doi.org/10.1016/S1384-1076(01)00042-2, arXiv:astro-ph/0003162

Stone JM, Norman ML (1992a) ZEUS-2D: A radiation magnetohydrodynamics code for astrophysical flows in two space dimensions. I - The hydrodynamic algorithms and tests. Astrophys J Suppl 80:753–790, https://doi.org/10.1086/191680

Stone JM, Norman ML (1992b) ZEUS-2D: A radiation magnetohydrodynamics code for astrophysical flows in two space dimensions. II. The magnetohydrodynamic algorithms and tests. Astrophys J Suppl 80:791. https://doi.org/10.1086/191681

Stone JM, Mihalas D, Norman ML (1992) ZEUS-2D: a radiation magnetohydrodynamics code for astrophysical flows in two space dimensions. III - the radiation hydrodynamic algorithms and tests. Astrophys J Suppl 80:819–845. https://doi.org/10.1086/191682

Thielemann F, Nomoto K, Hashimoto M (1996) Core-collapse supernovae and their ejecta. Astrophys J 460:408. https://doi.org/10.1086/176980

Wanajo S, Nomoto K, Janka H, Kitaura FS, Müller B (2009) Nucleosynthesis in electron capture supernovae of asymptotic giant branch stars. Astrophys J 695:208–220. https://doi.org/10.1088/0004-637X/695/1/208, 0810.3999

Woosley SE, Weaver TA (1995) The evolution and explosion of massive stars. II. Explosive hydrodynamics and nucleosynthesis. Astrophys J Suppl 101:181. https://doi.org/10.1086/192237

Woosley SE, Hartmann DH, Hoffman RD, Haxton WC (1990) The nu-process. Astrophys J 356:272–301. https://doi.org/10.1086/168839

Woosley SE, Heger A, Weaver TA (2002) The evolution and explosion of massive stars. Rev Mod Phys 74:1015–1071. https://doi.org/10.1103/RevModPhys.74.1015

Young PA, Arnett D (2005) Observational tests and predictive stellar evolution. II. Nonstandard models. Astrophys J 618:908–918. https://doi.org/10.1086/426131, arXiv:astro-ph/0409658

Zingale M, Woosley SE, Rendleman CA, Day MS, Bell JB (2005) Three-dimensional numerical simulations of Rayleigh-Taylor unstable flames in type Ia supernovae. Astrophys J 632:1021–1034. https://doi.org/10.1086/433164, arXiv:astro-ph/0501655

Chapter 9
Nuclear Reactions

Michael Wiescher and Thomas Rauscher

Nuclear reaction rates determine the abundances of isotopes in stellar burning processes. A multitude of reactions regulate the reaction flow pattern, which is described in terms of reaction network simulations. The reaction rates depend on laboratory experiments supplemented by nuclear reaction and structure theory. In this chapter we will discus the experimental approach as well as the theoretical tools for obtaining the stellar reaction rates.

A detailed analysis of a reaction is only possible for a few selected cases, which will be highlighted in this section. The bulk of nuclear reaction processes is however described in terms of a statistical model approach, which relies on global nuclear structure and reaction parameters such as level density and mass and barrier penetration, respectively. For light nuclei, the *statistical Ansatz* becomes less reliable because the specific structure of the compound nucleus plays an increasingly important role. In recent years, attempts have been made to use statistical methods for predicting these parameters, such as single particle, α or γ strengths distributions to reduce the structure related uncertainties which are associated with the predictions of reaction rates.

We will discuss a variety of experimental facilities and techniques used in the field, this includes low energy stable beam experiments, measurements at radioactive beam accelerators, and neutron beam facilities.

M. Wiescher (✉)
University of Notre Dame, Notre Dame, IN, USA
e-mail: Michael.C.Wiescher.1@nd.edu

T. Rauscher
Universität Basel, Basel, Switzerland

University of Hertfordshire, Hatfield, UK
e-mail: Thomas.Rauscher@unibas.ch

R. Diehl et al. (eds.), *Astrophysics with Radioactive Isotopes*, Astrophysics and Space Science Library 453, https://doi.org/10.1007/978-3-319-91929-4_9

9.1 Nuclear Reactions in Astrophysical Environments

Nuclear reactions are the engine of stellar evolution and have been dictating the emerging abundance pattern in the chemical evolution of the universe. They also determine the overall production of the long-lived radioactive isotopes in a variety of nucleosynthesis processes which characterise the distribution of radioactivities our galaxy. A detailed understanding of the characteristic production and depletion rates within the framework of the different nucleosynthesis processes is crucial for reliable model predictions and the interpretation of the observed abundances.

There are several experimental and theoretical challenges in obtaining stellar reaction rates. The interaction energies in stellar environments extend from basically zero projectile energy up to only several MeV. This is especially challenging for the measurement of the relevant reaction cross sections which can be extremely small, especially for reactions with charged projectiles. This also makes theoretical predictions extremely difficult because several quantum mechanisms may contribute to a reaction and simple approximations may only be of limited use.

Another challenge arises from the fact that nuclear burning at high temperature takes place far away from stability and therefore produces very short-lived isotopes which subsequently decay to long-lived and stable species. Current experimental technology can only access a fraction of the nuclei associated with such processes and is still limited in obtaining detailed information on their properties. The possibility to measure cross sections of reactions involving such highly unstable nuclei is even more limited currently. Thus, investigations of nucleosynthesis in high-temperature environments largely rely on theoretical models, which not only have to treat the reaction mechanisms properly but also are required to predict nuclear properties far from stability.

Moreover, depending on the actual plasma conditions, reactions in an astrophysical plasma may proceed fundamentally differently from those in the laboratory. This is due to two effects. On one hand, laboratory experiments involve nuclei in atomic or molecular configurations whereas nuclear burning in stellar environments involves fully ionised nuclei immersed in a cloud of free electrons (and photons). The Coulomb charge of a nucleus is partially shielded by the surrounding electrons but this shielding (or screening) effect will be different for an atom or molecule and a plasma because of the different electron distribution and kinetics. While theoretical cross sections always imply bare nuclei, the values have to be appropriately converted (also based on a theoretical treatment of different screening mechanisms) for comparison to low-energy laboratory cross sections and for application in astrophysical plasmas. Additionally, the quantum mechanically and geometrically different electron distribution in a plasma directly affects electron capture reactions. For example, nuclei such as ^{7}Be or ^{44}Ti, decaying by capturing an electron from the K-shell of the atom under laboratory conditions will not be able to do this in a stellar plasma. Instead, electron capture inside a star involves capturing a free electron from the plasma, which is more unlikely and therefore the terrestrial half-life can be shorter than the one in a stellar environment (Iliadis 2007; Johnson et al. 1992).

Finally, due to the high photon and matter densities in astrophysical environments, nuclei very quickly reach thermal equilibrium with their surroundings by excitation and de-excitation via photons and by collisions. In most cases, this happens on a shorter timescale than that of nuclear transformations (one of the exceptions being isomeric states). Consequently, the nuclei involved in the reactions occur not only in their ground states, as in the laboratory, but also their excited states which are populated with a probability involving the Boltzmann factor. Reactions on nuclei in excited states are mostly treated theoretically. The population of excited nuclear states does not only depend on the plasma temperature but also on the structure of a nucleus. Nuclei with isolated levels at low excitation energies of only a few keV will exhibit pronounced thermal population already at low plasma temperature that can participate in the reaction process. This is especially important for modern s-process studies, which require high accuracy knowledge of neutron capture rates (Rauscher et al. 2011; Rauscher 2012a). Thermal effects are also important in decays and neutrino reactions because the available phase space of the reaction products is altered, leading to a modification of the rate. For example, electron capture rates in stellar core collapse are enhanced at temperatures $T > 1.5\,\mathrm{MeV}$ because of the unblocking of low-lying neutron states by thermal excitation (Cooperstein and Wambach 1984).

9.1.1 Reaction Networks and Thermonuclear Reaction Rates

The change of abundances Y with time due to nuclear processes is traced by coupled differential equations. To be fully solvable, the number of equations N must equal the number of involved nuclei acting as reaction partners, and thus an equation matrix of size N^2 has to be solved. Such a set of coupled equations is called *reaction network*. It can generally be written as

$$\dot{Y}_i = \frac{1}{\rho N_\mathrm{A}}\dot{n}_i = \frac{1}{\rho N_\mathrm{A}}\left\{\sum_j {}^1_iP_j\;{}_i\lambda_j + \sum_j {}^2_iP_j\;{}_ir_j + \sum_j {}^3_iP_j\;{}_i\hat{r}_j + \ldots\right\} \quad , \tag{9.1}$$

where $1 \leq i \leq N$ numbers the nucleus, ${}_i\lambda_j$ is the jth rate for destruction or creation of the ith nucleus without a nuclear projectile involved (this includes spontaneous decay, lepton capture, photo-disintegration), and ${}_ir_j$ is the rate of the jth reaction involving a nuclear projectile and creating or destroying nucleus i. Similarly, we have three-body reactions where nucleus i is produced or destroyed together with two other (or similar) nuclei. Reactions with more participants (denoted by ... above) are unlikely to occur at astrophysical conditions and are usually neglected. The quantities ${}^1_iP_j$, ${}^2_iP_j$, and ${}^3_iP_{jk}$ are positive or negative integer numbers specifying the amount of nuclei i produced or destroyed, respectively, in the given process. As shown below, the rates λ, r, and $\hat{r}$ contain the abundances of the interacting nuclei. Rates of type λ depend on one abundance (or number density), rates r depend on the abundances of two species, and rates $\hat{r}$ on three.

Using abundances Y instead of number densities $n = Y\rho N_{\mathrm{A}}$ (where ρ is the plasma density) has the advantage that a change in the number of nuclei in a given volume due to density fluctuations is factored out and only changes by nuclear processes are considered. Using abundance changes, the total energy generation rate per mass due to nuclear reactions can easily be expressed as

$$\dot{\epsilon} = -\sum_i \dot{Y}_i N_A M_i c^2 \quad , \tag{9.2}$$

with the rest masses $M_i c^2$ of the participating nuclei.

The rates ${}_i\lambda_j$ appearing in the first term of Eq. (9.1) are reactions per time and volume, and only contain the abundance Y_j. For example, ${}_i\lambda_j$ is simply $n_j L_j = Y_j \rho N_{\mathrm{A}} L_j$ for β-decays. The factor $L_j = (\ln 2)/{}^jT_{1/2}$ is the usual decay constant (with the unit 1/time) and is related to the half-life ${}^jT_{1/2}$ of the decaying nucleus j. It has to be noted that some decays depend on the plasma temperature and thus L_j is not always constant, even for decays.

Two-body rates r include the abundances of two interacting particles or nuclei. In general, target and projectile follow specific thermal momentum distributions dn_1 and dn_2 in an astrophysical plasma. With the resulting relative velocities $\mathbf{v}_1 - \mathbf{v}_2$, the number of reactions per volume and time, is given by

$$r_{12} = \int \hat{\sigma}(|\mathbf{v}_1 - \mathbf{v}_2|)|\mathbf{v}_1 - \mathbf{v}_2| dn_1 dn_2 \quad , \tag{9.3}$$

and involves the reaction cross section $\hat{\sigma}$ as a function of velocity/energy, the relative velocity $\mathbf{v}_1 - \mathbf{v}_2$ and the thermodynamic distributions of target and projectile dn_1 and dn_2. The evaluation of this integral depends on the type of particles (fermions, bosons) and distributions which are involved.

However, many two-body reactions can be simplified and effectively expressed similarly to one-body reactions, only depending on one abundance (or number density). If reaction partner 2 is a photon, the relative velocity is always c and the quantities in the integral do not depend on dn_1. This simplifies the rate expression to

$$\lambda_1 = L_\gamma(T) n_1 \quad , \tag{9.4}$$

where $L_\gamma(T)$ stems from an integration over a Planck distribution for photons of temperature T. This is similar to the decay rates introduced earlier and therefore we replaced r by λ in our notation and can include this type of reaction in the first term of Eq. (9.1). A similar procedure is used for electron captures by protons and nuclei. Because the electron is about 2000 times less massive than a nucleon, the velocity of the nucleus is negligible in the center-of-mass system in comparison to the electron velocity ($|\mathbf{v}_{\text{nucleus}} - \mathbf{v}_{\text{electron}}| \approx |\mathbf{v}_{\text{electron}}|$). The electron capture cross section has to be integrated over a Fermi distribution of electrons. The electron capture rates are a function of T and $n_e = Y_e \rho N_A$, the electron number density. In a neutral, completely ionized plasma, the electron abundance Y_e is equal to the total proton abundance $Y_e = \sum_i Z_i Y_i$ and thus

$$\lambda_{\text{nucleus,ec}} = L_{\text{ec}}(T, \rho Y_e) n_{\text{nucleus}} \quad . \tag{9.5}$$

Again, we have effectively a rate per target L (with unit 1/time) similar to the treatment of decays earlier and a rate per volume including the number density of only one nucleus. We denote the latter by λ and use it in the first term of Eq. (9.1). This treatment can be applied also to the capture of positrons, being in thermal equilibrium with photons, electrons, and nuclei. Furthermore, at high densities ($\rho > 10^{12}\,\text{g}\,\text{cm}^{-3}$) the size of the neutrino scattering cross section on nucleons, nuclei, and electrons ensures that enough scattering events occur to lead to a continuous neutrino energy distribution. Then also the inverse process to electron capture (neutrino capture) can occur as well as other processes like, e.g., inelastic scattering, leaving a nucleus in an excited state which can emit nucleons and α particles. Such reactions can be expressed similarly to photon and electron captures, integrating over the corresponding neutrino distribution.

In the following, we focus on the case of two interacting nuclei or nucleons as these reactions will be extensively discussed in Sects. 9.3 and 9.4. This will result in an actual two-body rate r to be used in the second term of Eq. (9.1). Here, we mention in passing that Eq. (9.3) can be generalised to 3 and more interacting nuclear species by integrating over the appropriate number of distributions, leading to rates $\hat{r}$ and higher order terms in Eq. (9.1).

Turning our attention back to two-body reactions, we note that the velocity distributions can be replaced by energy distributions. Furthermore, it can be shown that the two distributions in Eq. (9.3) can be replaced by a single one in the center-of-mass system. This time the resulting expression describes a rate r including two abundances (or number densities) and showing up in the second term of Eq. (9.1). The rate r is defined as an interaction of two reaction partners with an energy distribution $\phi(E)$ according to the plasma temperature T and a reaction cross section $\sigma^x(E)$ specifying the probability of the reaction on the excited state x of the target nucleus (Fowler 1974):

$$r = \frac{n_1 n_2}{1 + \delta_{12}} \frac{2J_x + 1}{G(T)} \sum_x \int_{E_x}^{\infty} \sigma^x (E - E_x) e^{-\frac{E_x}{kT}} \phi(E - E_x)\, dE \quad . \tag{9.6}$$

The factor $1/(1 + \delta_{12})$ with the Kronecker symbol δ is introduced to avoid double counting. The nuclear cross section is defined as in standard scattering theory by

$$\sigma = \frac{\text{number of reactions target}^{-1}\ \text{sec}^{-1}}{\text{flux of incoming projectiles}} \quad . \tag{9.7}$$

However, in an astrophysical plasma, nuclei quickly (on the timescale of nuclear reactions and scattering) reach thermal equilibrium with all plasma components. This allows thermal excitation of nuclei, following a Boltzmann population law $e^{-\frac{E_x}{kT}}$, and requiring to account for reactions on the excited states in the sum running over all excited states x of the target (for simplicity, here we assume the projectile, i.e. the second reaction partner, does not have excited states) with spin J_x and excitation energy E_x. The quantity $G(T)$ is the partition function of the nucleus. Cross sections $\sigma = \sigma^{x=0}$ measured in terrestrial laboratories do not include such thermal effects. At the large temperatures reached in explosive burning, thermal enhancement can lead to a considerable deviation from the ground-state cross section. Because of the exponential dependence of the population factors on temperature and E_x, a lower importance of reactions on excited states is expected. It was found, however, that stellar rates may considerably differ from rates determined from $\sigma^{x=0}$ already for the comparatively low temperatures encountered in the s-process, especially for deformed nuclei in the rare-earth region (Rauscher et al. 2011; Rauscher 2012a).

To assess the importance of thermally excited states, the *ground-state contribution to the stellar rate* can be used. It is defined as (Rauscher 2012a, 2011)

$$X_0 = \frac{1}{G(T)}\frac{r_0}{r} \quad , \tag{9.8}$$

where r_0 is the rate calculated from the ground-state cross section $\sigma^{x=0}$ and r is the stellar rate calculated by accounting for all thermally excited states. Extended tables of these ground-state contributions to the stellar rate can be found in Rauscher (2012a,b). It should be noted that the stellar enhancement factor (SEF) previously often used in literature does not correctly represent the thermal enhancement because the factor $1/G(T)$ was erroneously omitted.

Nuclei in an astrophysical plasma obey a Maxwell-Boltzmann (MB) distribution $\phi(E) = \phi_{\mathrm{MB}}$ and we obtain finally (Rauscher 2011):

$$r = \frac{n_1 n_2}{1+\delta_{12}} < \sigma v >^* \tag{9.9}$$

$$< \sigma v >^* = (\frac{8}{\mu\pi})^{1/2}(kT)^{-3/2}\frac{1}{G(T)} \times \int_0^\infty \left\{ \sum_x \frac{g_x}{g_0}\sigma^x(E-E_x)\,(E-E_x) \right\} e^{-\frac{E}{kT}}\, dE \quad . \tag{9.10}$$

Here, μ denotes the reduced mass of the two-particle system and $< \sigma v >^*$ is the stellar reaction rate per particle pair or *stellar reactivity*, using the spin J_x of the labelled state ($x = 0$ being the ground state) in $g_x = 2J_x + 1$.

As mentioned above the charge of the reaction partners can be screened. For most astrophysical conditions this can be included by introducing a screening factor $f_{\rm screen}$, modifying the above rate for bare nuclei (Iliadis 2007; Salpeter and Van Horn 1969)

$$r^{\rm scr} = f_{\rm screen} r \quad . \tag{9.11}$$

The screening factor is derived from the plasma conditions of the specific stellar environment. At high densities and low temperatures screening factors can enhance reactions by many orders of magnitude and lead to *pycno-nuclear ignition* (Yakovlev et al. 2006). However, note that the above factorisation is not valid for vanishing temperatures when nuclei are trapped in a Coulomb lattice. The screening factor in astrophysical plasmas depends on phenomenological models and is not experimentally confirmed yet. New experiments at laser plasma facilities are in preparation to test directly the theoretical predictions on the impact of plasma screening (Cerjan et al. 2017)

Forward and reverse rates are related. Applying the well-known reciprocity theorem for nuclear transitions (Blatt and Weisskopf 1991) and further assuming that the reaction partners in the entrance channel a as well as the reaction products in the exit channel b are instantaneously thermalised (the *detailed balance* principle), the relation (Holmes et al. 1976; Iliadis 2007)

$$< \sigma^* v >^{b \to a} = \frac{1 + \delta_{b_1 b_2}}{1 + \delta_{a_1 a_2}} \frac{G_{a_1} G_{a_2}}{G_{b_1} G_{b_2}} \left(\frac{\mu_a}{\mu_b} \right)^{3/2} e^{-\frac{Q}{kT}} < \sigma^* v >^{a \to b} \quad , \tag{9.12}$$

relating the *stellar* reverse rate to the *stellar* forward rate. The latter has the reaction Q-value Q. For captures (forward channel a) and photo-disintegrations (reverse channel b), Eq. (9.12) transforms to

$$L_\gamma = \frac{1}{1 + \delta_{a_1 a_2}} \frac{G_{a_1} G_{a_2}}{G_b} \left(\frac{\mu_a kT}{2\pi \hbar^2} \right)^{3/2} e^{-\frac{Q}{kT}} < \sigma^* v >^{\rm capture} \quad . \tag{9.13}$$

These expressions will not be valid anymore if any of the involved rates was derived from a laboratory cross section. They also imply that the detailed balance assumption is valid. Detailed balance can be violated in nuclei with long-lived isomeric states which are not populated or depopulated during regular reaction timescales. For these cases, reactions to separate final states have to be calculated and the (de)population of these states by photon transitions have to be followed explicitly (Ward and Fowler 1980). Important examples for such nuclei are ^{26}Al and ^{180}Ta (Rauscher et al. 2002).

9.1.2 Reaction Equilibria

It is not always necessary to solve a full reaction network (Eq. (9.1)) including all the rates. On one hand, simplifications can often be made by omitting slow reactions which will not contribute significantly during the timescale of the astrophysical event. These are, for example, charged-particle reactions on heavy targets in hydrostatic stellar burning. On the other hand, high temperature can establish reaction equilibria. When both forward and reverse reactions become sufficiently fast to reach equilibrium with abundances set at equilibrium values. The equilibrium abundances of nuclei can be derived by using the relations (9.12) and (9.13) in the network equation (9.1) and assuming $\dot{Y} = 0$. Somewhat depending on the density, for $T > 4$–5 GK all reactions are in a full *nuclear statistical equilibrium* (NSE) and the abundances are given by

$$Y_i = G_i \left(\rho N_\mathrm{A}\right)^{A_i - 1} \frac{A_i^{3/2}}{2_i^A} \left(\frac{2\pi \hbar^2}{m_\mathrm{u} kT}\right)^{\frac{3}{2}(A_i - 1)} e^{\frac{B_i}{kT}} Y_\mathrm{n}^{N_i} Y_\mathrm{p}^{Z_i} \quad , \tag{9.14}$$

$$\sum_i A_i Y_i = 1 \sum_i Z_i Y_i = Y_\mathrm{e} \tag{9.15}$$

with Z_i, N_i, A_i, and B_i being the charge, neutron number, mass number, and the binding energy of the nucleus i, respectively, the atomic mass unit m_u, and the abundances of free neutrons Y_n, free protons Y_p, and free electrons Y_e. Here, it is assumed that reactions via the strong and electromagnetic interactions are in equilibrium while the weak interaction is not. Therefore, Y_e can still be time-dependent and thus also the resulting NSE abundances Y_i.

At $T < 4$ GK and/or low densities only some reactions may be in equilibrium while others are too slow. This gives rise to the so-called *quasi-statistical equilibrium* (QSE) where only groups of nuclei are equilibrated and those groups are connected by slower reactions which are not in equilibrium (Hix and Thielemann 1999). Abundance ratios within a QSE group can be determined by application of Eq. (9.14) while the connecting, slow reaction determines the amount of matter in each group relative to the other groups at a given time. QSE occurs in low temperature, low density Si-burning and in O-burning of stars. Often, the slowest rate falling out of equilibrium first is that of the strongly density-dependent triple-α reaction.

A special case of a QSE is the *waiting-point approximation*, often used in r-process calculations (Cowan et al. 1991; Arnould et al. 2007). There, the network is reduced to neutron capture reactions and their reverse reactions, and β^--decay (with possible release of neutrons). Assuming equilibrated capture and photodisintegration, QSE within an isotopic chain is obtained and the relative abundances are given by

$$\frac{Y_{i+1}}{Y_i} = n_{\mathrm{n}} \frac{G_{i+1}}{2G_i} \left(\frac{A_i + 1}{A_i}\right)^{3/2} \left(\frac{2\pi\hbar^2}{m_{\mathrm{u}}kT}\right)^{3/2} e^{\frac{Q_{\mathrm{ncap}}}{kT}} \quad . \tag{9.16}$$

The neutron number density is denoted by n_{n} and the neutron capture Q-value is given by the neutron separation energy in nucleus $i+1$: $Q_{\mathrm{ncap}} = S_{\mathrm{n},i+1}$. The indices i are ordered by increasing neutron number. The β^--decays are much slower and not in equilibrium. Synthesis of the next element is delayed until the decay of the produced isotopes. Typically, only one or two nuclides have significant abundances in such an isotopic QSE chain, hence the name waiting-point approximation.

The advantage of using equilibria is that the rates—and thus the cross sections—do not have to be known explicitly. The resulting abundances are completely determined by basic nuclear properties and the conditions in the astrophysical environment.

9.2 Relevant Energy Range of Astrophysical Cross Sections

In the general calculation of the reaction rate according to Eq. (9.10) the nuclear cross section has to be known. Although the integration limits in Eq. (9.10) run from Zero to Infinity, significant contributions to the integral only come from a comparatively narrow energy range. This is due to the shape of the MB distribution, showing a peak around the energy $E_{\mathrm{MB}} = kT$ and quickly approaching very small values both towards $E = 0$ and $E \gg kT$. For a slowly varying cross section (as found, e.g., in non-resonant neutron-induced reactions), the relevant energy range is simply given by the peak of the distribution, $E_0 = E_{\mathrm{MB}}$ and its width $\Delta_0 = \Delta_{\mathrm{MB}}$. For partial waves higher than s-waves, the additional centrifugal barrier introduces a stronger energy dependence in the cross section and shifts the relevant range to slightly higher energy, i.e. $E_0 \approx 0.172 T_9(\ell + 1/2)$ MeV and $\Delta_0 \approx 0.194 T_9 \sqrt{\ell + 1/2}$ for partial waves $\ell > 0$ (Rauscher et al. 1997; Wagoner 1969). Charged-particle cross sections exhibit a strong energy dependence at energies close to and below the Coulomb barrier. They decrease by many orders of magnitude towards lower energy. Using the astrophysical S-factor

$$S(E) = \sigma E e^{2\pi\eta} \quad , \tag{9.17}$$

with η being the Sommerfeld parameter describing the barrier penetrability, most of the Coulomb suppression is taken out and $S(E)$ is easier to handle because it is varying less with energy than σ. Inserting definition (9.17) into Eq. (9.10) shows that the penetration factor causes a significant shift of the relevant energy range towards higher energy. The resulting energy window (the *Gamow window*, given by the *Gamow peak* appearing when folding the charged particle cross section with the

MB distribution) can be approximated by (Iliadis 2007; Rauscher et al. 1997)

$$E_0 = 0.12204 \left(Z_1^2 Z_2^2 \mu\right)^{1/3} T_9^{2/3} \,\text{MeV} \tag{9.18}$$

$$\Delta_0 = 4\sqrt{\frac{E_0 kT}{3}} = 0.23682 \left(Z_1^2 Z_2^2 \mu\right)^{1/6} T_9^{5/6} \,\text{MeV} \quad . \tag{9.19}$$

Here, T_9 is the plasma temperature in GK. The idea of a single, relevant energy window is only viable for non-resonant cross sections or reactions with broad resonances. Strong, narrow resonances lead to fragmentation of the peak and split it up in several small energy ranges around the resonance energies, with decreasing weight towards higher energy.

It is important to note that Eq. (9.18) is not always valid. It is based on the assumption that the energy dependence of the cross section is mainly determined by the penetration of the projectile through the Coulomb barrier. However, the dependence is dominated by the one of the smallest width in the entrance or exit channel for resonant reactions or smallest *averaged* width in the case of Hauser-Feshbach compound reactions (see Sect. 9.3.2). This smallest width can also be the one of the exit channel, leading to a different maximum in the contribution to the reaction rate integral than estimated from Eq. (9.18). This is often the case in capture reactions when $\Gamma_{\text{projectile}} \gg \Gamma_\gamma$ (Iliadis 2007). Because of the weak energy dependence of the γ-width, there would not be a Gamow window. Effectively, however, the Gamow window is shifted to energies where $\Gamma_{\text{projectile}}$ (which is strongly energy dependent) becomes smaller than Γ_γ. Since reaction rates at higher temperature are determined by cross sections at higher energy, the discrepancy between Eq. (9.18) and the true maximum of the integrand is more pronounced at high temperature than at low temperature. Therefore, the relevant energy range for reactions in explosive burning should be derived by a proper inspection of the product of the (predicted) cross sections and the MB distribution. For other charged particle captures in astrophysics, often $\Gamma_{\text{projectile}} \ll \Gamma_\gamma$ due to the low interaction energy implied by $E_{\text{MB}} = kT = T_9/11.6045\,\text{MeV}$, unless for light nuclei (with low Coulomb barrier). Regarding neutron captures, although $\Gamma_{\text{n}} \gg \Gamma_\gamma$ will apply in most cases (unless very close to the reaction threshold), the shape of the integrand is mostly determined by the shape of the MB distribution and obviously not by any Coulomb penetration. Therefore, the relevant energy window for neutrons can still be estimated from the MB distribution as shown above.

A numerical investigation of the reaction rate integrals has been published in Rauscher (2010b). This reference also includes extended tables of astrophysically relevant energy windows for reactions with neutrons, protons, and α-particles.

9.3 Nuclear Reaction Models

Having identified the relevant energy range, the cross sections have to be predicted by reaction models or determined experimentally. As previously mentioned, often measurements for astrophysics prove difficult due to small cross sections or/and unstable nuclei involved. However, even if a measurement is feasible, the resulting cross section has to be corrected for effects of electron screening and thermal excitation of the target via theoretical models before being used to compute an astrophysical reaction rate.

Here, we provide a brief overview of approaches to predict low-energy cross sections of reactions involving the strong force. Decays and other reactions via the weak force are important but cannot be discussed due to limited space. The reader is referred to other sources, e.g. Möller et al. (2003), Fuller et al. (1982), Vogel (2006) and references therein. We also do not cover fission reactions which are important in extremely neutron-rich explosive environments where a r-process could occur and reach the region of fissionable nuclei. Current predictions of fission barriers, however, carry large uncertainties. For details, see e.g. Cowan et al. (1991), Arnould et al. (2007), Panov et al. (2005, 2010), Goriely et al. (2009) and references therein. We also only discuss reactions between a nucleus and a nucleon or an α-particle as the majority of reactions in astrophysics is of that type. There are a number of interesting fusion reactions such as $^{12}C+^{12}C$, and $^{16}O+^{16}O$ during the late phases of stellar burning or in explosive stellar environments, but these reactions only facilitate the energy production and contribute little to the production of long-lived radioactivities and are therefore omitted here.

The interaction of a particle with a nucleus can excite few or many degrees of freedom, i.e. transfer energy to few (or none) or to many of the nucleons constituting the target nucleus. In nature, all interaction types are, in principle, possible but often only one will be dominating at a given interaction energy but with gradual transitions from one type to the other within certain energy intervals. For theory, it is simpler to consider extreme, idealised cases. Interdependence and interference effects between different reaction mechanisms, even if in principle understood, are very difficult to predict and especially so for the required large number of reactions with unstable nuclei required in astrophysics. In the following we introduce a selection of relevant reaction mechanisms considered in literature. The number of degrees of freedom which can be excited depends on the number of states or levels present in the system formed by projectile and target (Descouvemont and Rauscher 2006). Therefore, it is helpful to distinguish between compound systems with low and high level densities.

9.3.1 Resonance and Potential Models

Low level-density systems exhibit no or only few, isolated resonances in the relevant energy range. These involve mostly light nuclei which have few, widely spaced excited states within several tens of MeV above the ground state and therefore also show only few resonances even when the separation energy of the projectile from the compound system is large. A similar situation also occurs for heavier nuclei with closed shells or heavier nuclei far off stability and close to the drip lines, where the projectile separation energy becomes very low (e.g. in neutron capture on extremely neutron-rich nuclei), and in consequence the compound system is formed at very low relative energy.

In principle, isolated resonances can be included by the *Breit-Wigner formula* (Blatt and Weisskopf 1991)

$$\sigma^x = \frac{\omega^2}{4\pi} \frac{2J+1}{(2J_x+1)(2J_{\mathrm{proj}}+1)} \frac{\Gamma_a^x \Gamma_b}{(E-E_{\mathrm{res}})^2 + \frac{\Gamma_{\mathrm{tot}}^2}{4}} , \tag{9.20}$$

where J and E_{res} refer to the spin and energy of the resonance, ω is the de Broglie wavelength, and Γ_{tot} is the total resonance width, including the entrance and exit widths Γ_a^x and Γ_b plus all other open channels. Note that the widths are energy dependent. For a narrow resonance, inserting the above in Eq. (9.10) yields

$$N_{\mathrm{A}} \langle \sigma v \rangle = 1.54 \times 10^{11} \frac{1}{(\mu T_9)^{3/2}} \frac{2J+1}{(2J_x+1)(2J_{\mathrm{proj}}+1)} \frac{\Gamma_a^x \Gamma_b}{\Gamma_{\mathrm{tot}}} e^{-\frac{11.6045 E_{\mathrm{res}}}{T_9}} . \tag{9.21}$$

This gives the reactivity in units of cm^3 s^{-1} mole^{-1} when the widths and the resonance energy E_{res} are inserted in units of MeV (Iliadis 2007). (Note that the above equations do not involve stellar cross sections. For a true stellar cross section and rate, a thermally weighted sum of target states has to be used, according to Eq. (9.6).) However, tails of resonances with the same J may interfere and there may also be interference with a direct component (see below). Therefore, additional interference terms may have to be added (see, e.g., Rauscher and Raimann 1996). Furthermore, location of the resonance and the widths have to be predicted from nuclear structure. Currently, this is not possible from first principles (except for the lightest nuclei) with the accuracy needed in applications. Therefore, this information usually has to come from experiments. The contribution of possible resonance or bound states near the particle threshold energy has always been a major challenge. While the resonance energy is frequently known by transfer reaction studies, information on the level parameters such as spin parity, and more so the partial widths Γ_a^x, Γ_b, and Γ_{tot} is frequently not known. For the estimating a reaction rate often "standard" recommended strengths parameters were adopted that typically varied between 1 and 10% of a single particle or α cluster strength component. The uncertainty in this assumption can be very large because these configurations are not well known and depend on the specific nuclear structure of threshold levels. A statistical model approach was developed to reduce the uncertainty in these often arbitrary assignments by describing the single particle or α cluster configuration of the experimentally observed states by a Porter-Thomas distribution, adopting the average values taken from the distribution as typical parameters for calculating an averaged resonance strength with an uncertainty extracted from the range of the Porter-Thomas distribution (Iliadis et al. 2015). The claim is that this approach allows to assign a uncertainty range to the predicted resonance strength, it remains unclear if such a approach is suitable. There are indications that the assumption of a Porter-Thomas distributions is not valid for threshold states (Volya et al. 2015) since this does not account for effects that are largely associated with the specific shape of the nucleus near threshold or break-up conditions. The validity of the assumption of a Porter-Thomas distribution for the reduced width of threshold states has to be confirmed experimentally to reduce the possibility of large systematic uncertainties not considered in this approach.

Instead of Breit-Wigner formulas and interference terms, often the *R-matrix method* (Lane and Thomas 1958) is used. It is applied to parameterise experimentally known cross sections with as few parameters as possible, implicitly accounting for resonances and their interference. The R-matrix is a phenomenological approach that traditionally has been used can be used to fit the experimental cross section data and extract the observed level and interference parameters in that particular reaction channel. The R-matrix had therefore limited predictive power. This has changed when the R-matrix was used for fitting multiple reaction channels of one compound nucleus in a parallel mode (Azuma et al. 2010). This not only limited the number of free fitting parameters enormously but also improves the predictive power of the approach and allowed to reliably extrapolate the cross section prediction of for a single reaction channel from the experimentally observed range to the Gamow range of stellar burning since nuclear structure information about other resonance levels and non-resonant reaction contributions are being extracted from the fit to the other reaction channels. The power of this approach was recently demonstrated on the classical example of the $^{12}C(\alpha, \gamma^{16}O)$ reaction, where the multichannel R-matrix analysis led to a substantially reduced uncertainty in the predictions for the stellar reaction rate (deBoer et al. 2017).

In addition to possible resonance contributions a direct capture process can occur. These are fast, one-step processes in which a captured particle directly enters the final state. Typical reaction timescales of direct processes are of the order of 10^{-22} s whereas compound reactions, distributing the energy among a large number of nucleons, take of the order of 10^{-16} s. Direct reactions include transfer processes where a particle exchange takes place between projectile and target nucleus, and capture processes in which the projectile is being fully captured by the target nucleus. These two reaction types can be treated in *ab initio* models, determining the cross sections from wave functions obtained by solving the Schrödinger equation using effective potentials. However, recent developments in the R-matrix formalism include the direct reaction mode with the traditional resonance formalism permitting a consistent fit and extrapolation mode for all modes of reaction contributions (Azuma et al. 2010).

For transfer reactions often the *Distorted Wave Born Approximation* (DWBA) (Satchler 1983; Glendenning 2004) is used, utilising optical potentials to compute the cross sections from the overlap integral of distorted scattering wave functions and the bound state wave function. The DWBA implicitly assumes that elastic scattering is dominant while non-elastic contributions can be treated in the framework of perturbation theory.

On the other hand, capture reactions can be calculated with a simple potential model, which is a first-order approach involving an electromagnetic operator describing the emission of photons due to the dynamics in the movement of electric charges (Descouvemont and Rauscher 2006). In the potential model the differential cross section is proportional to the matrix element defined by the overlap of the final state ϕ_β of the final nucleus and the initial state composed of the target wave function ϕ_α and a (distorted) scattering wave of the projectile χ_α. This can be decomposed into an overlap function S of the target and the final nucleus and a radial integral containing the scattered wave $\chi_\alpha^x(r)$, the bound state wave function of the projectile in the target $\phi_{\mathrm{a+A}}$, and the radial form of the electromagnetic operator $\mathcal{O}_{\mathrm{EM}}$ (Kim et al. 1987)

$$\frac{d\sigma^x}{d\Omega} \propto \left|\langle\phi_\beta|\mathcal{O}_{\mathrm{EM}}|\chi_\alpha\phi_\alpha^x\rangle\right|^2 \propto S\left|\phi_{\mathrm{a+A}}(r)\mathcal{O}_{\mathrm{EM}}(r)\chi_\alpha^x(r)\,dr\right|^2 \quad . \tag{9.22}$$

The wave functions $\phi_{\mathrm{a+A}}(r)$ and $\chi_\alpha^x(r)$ are obtained by solving the radial Schrödinger equation with appropriate effective potentials.

Both approaches, DWBA and potential model, require a renormalisation of the resulting cross section through spectroscopic factors S, describing nuclear structure effects by the overlap of initial and final state of the system. These spectroscopic factors have to be obtained from nuclear structure models or by comparison with experiment (Satchler 1983; Glendenning 2004).

Microscopic reaction models are first principle methods, starting from effective nucleon-nucleon interactions and treating all nucleons in a Hamiltonian with exact anti-symmetrisation of the wave functions. Because of this, no artificial distinction between direct and resonant contributions has to be made. Unfortunately, such reaction models are limited to systems of few nucleons. Although the Quantum Monte Carlo method (Pieper and Wiringa 2001) is promising, it is currently limited to $A \leq 10$ and not applicable to continuum states. Cluster models have been often used for light systems so far (see Descouvemont and Rauscher 2006; Descouvemont 2003, and references therein). They assume that the nucleons are grouped in clusters and use cluster wave functions defined in the shell model and computed with an adapted effective nucleon-nucleon force. The Resonating Group model (RGM) and the Generator Coordinate Method (CGM) are two equivalent implementations differing in the definition of the relative wave function of the clusters (Descouvemont and Rauscher 2006; Descouvemont 2003).

9.3.2 Statistical Model

In systems with high level density $\rho(J, \pi, E)$, individual resonances cannot be resolved anymore and an average over the overlapping resonances can be used instead. Further assuming that the relative phases are randomly distributed, interferences will cancel and a simple sum of Breit-Wigner contributions can be replaced by a level-density weighted sum of averaged widths $\langle \Gamma \rangle$ over all spins J and parities π (Descouvemont and Rauscher 2006; Gadioli and Hodgson 1992)

$$\sigma^x(E) \propto \frac{1}{(2J_x+1)(2J_{\rm proj}+1)} \tag{9.23}$$

$$\times \sum_{J,\pi} [(2J+1)\rho(J,\pi,E_{\rm c}) \tag{9.24}$$

$$\times \left\langle \Gamma^x_{\rm pro}(\{J_x,\pi_x\} \rightarrow \{J,\pi\},E) \right\rangle \tag{9.25}$$

$$\frac{\left\langle \Gamma_{\rm b}\left(\sum_{J_{\rm fin},\pi_{\rm fin},E_{\rm fin}} (\{J,\pi\} \rightarrow \{J_{\rm fin},\pi_{\rm fin}\},E_{\rm fin})\right)\right\rangle}{\langle \Gamma_{\rm tot} \rangle} \tag{9.26}$$

$$\times W(J,\pi,E_{\rm c})] \quad , \tag{9.27}$$

$$E_{\rm c} = E + E_{\rm sep,pro} - E_x \quad , \tag{9.28}$$

$$E_{\rm fin} = E_{\rm c} - E_{\rm sep,fin} - E_{x,\rm fin} \quad . \tag{9.29}$$

This is called the *Hauser-Feshbach* or *statistical model of compound reactions* (Hauser and Feshbach 1952). Width fluctuation corrections W account for non-statistical correlations but are only important close to channel openings (Ericson 1960). The separation energy $E_{\rm sep,pro}$ of the projectile in the compound system determines at which energy $E_{\rm c}$ the compound system is formed. The averaged width of the exit channel $\langle \Gamma_{\rm b} \rangle$ usually includes a sum over energetically possible final states at energy $E_{x,\rm fin}$ or an integral over a level density of the final system when individual states are not known or numerous. For capture, compound and final system are identical. The averaged widths are related to transmission coefficients $\mathcal{T} = 2\pi\rho\,\langle \Gamma \rangle$. The latter are calculated from the solution of a (radial) Schrödinger equation using an optical potential. (It is to be noted that these potentials differ from the ones employed for low-density systems described in Sect. 9.3.1.)

The challenge for nuclear astrophysics lies in the determination of globally applicable descriptions of low-energy optical potentials as well as level densities, masses (determining the separation energies), and spectroscopy (energies, spins, parities) of low-lying excited states, to be applied for a large number of nuclei at and far from stability. For details on the different properties and the remaining open

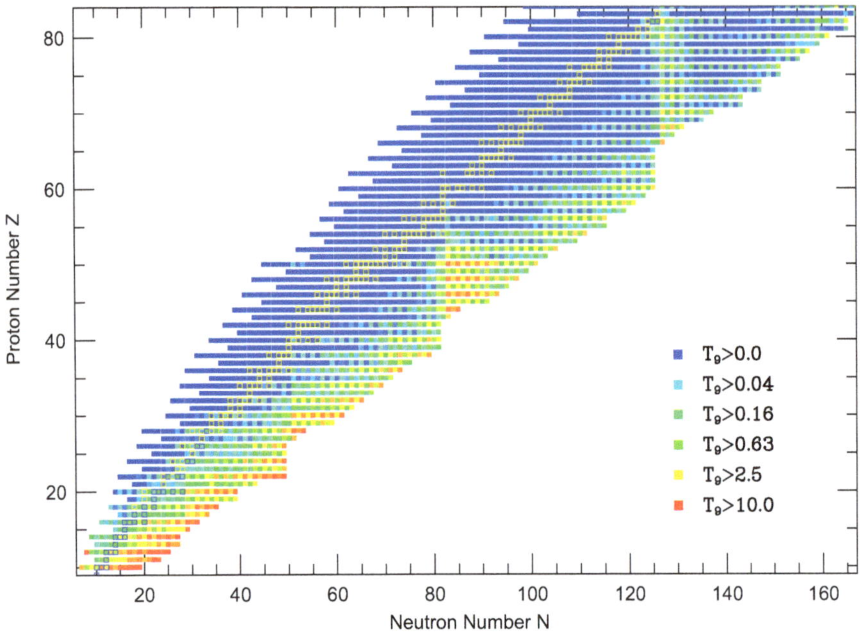

Fig. 9.1 Applicability of the Hauser-Feshbach model to calculate astrophysical reaction rates for neutron-induced reactions: Shown are the stellar temperatures above which the rate can be calculated from Hauser-Feshbach cross sections (reprinted from Rauscher et al. 1997, with permission)

problems in their treatment (see, e.g., Descouvemont and Rauscher 2006; Rauscher et al. 1997; Arnould et al. 2007; Goriely et al. 2008; Rauscher 2010a, and references therein). For a general discussion of the applicability of the statistical model, see Rauscher et al. (1997), Rauscher and Thielemann (2000) (Fig. 9.1).

9.4 Experimental Facilities and Techniques

The experimental determination or verification of nuclear reaction rates requires a large variety of facilities and techniques. This is in particular true if one wants to establish experimentally reaction rates associated with the production of long-lived radioactive isotopes associated with galactic gamma sources. Nuclear astrophysics related experiments include low energy high intensity accelerator measurements with stable beams to study charged particle reactions of relevance for quiescent stellar burning which may possibly lead to the production of ^{7}Be, ^{18}F, ^{22}Na and ^{26}Al. High flux neutron beam studies to explore neutron induced reactions for the weak and main s-process which can be associated with the production of long-lived radioactive isotopes such as ^{60}Fe and ^{98}Tc. Real and virtual photon beams are increasingly used for probing nuclear reactions associated with explosive nucleosynthesis events such as the p-process but can also be used to probe indirectly neutron capture reactions associated with the s-process. Intense radioactive beams are the primary tools for exploring nuclear reactions and decay mechanisms far of stability which are expected to occur in explosive stellar environments and may lead to the production of long-lived radioactive elements such as ^{18}F, ^{26}Al, ^{44}Ti, ^{60}Fe, and ^{56}Ni. The specific origin of the observed radioactivities is frequently a matter of debate and relies strongly on the efficiency of the production mechanism in specific quiescent or explosive stellar environments. The experimental study of such reactions often presents a considerable challenge and requires sophisticated experimental techniques.

9.4.1 Low-Energy Facilities, Underground Techniques

Low energy charged particle measurements belong to the most challenging experiments in nuclear astrophysics. The cross sections need to be measured at the extremely low energies associated with the Gamow range of quiescent stellar burning. This requires determining the cross sections of proton capture reactions for hydrogen burning in main sequence stars at energies well below 100 keV. Measurements for helium burning in red giant stars need to be explored in the 200–500 keV range and heavy ion fusion reactions in subsequent stellar evolution phases need to be measured near 1–2 MeV center of mass energy. The cross sections are extremely low, typically in the pico- to femto-barn range, which requires a long time, in excess of days, to accumulate a statistical relevant amount of reaction yield data. Typical experimental techniques are summarised in the text book literature (Iliadis 2007) and will not be discussed here.

The critical issue with low cross sections is that the yield of the characteristic gamma or particle radiation associated with a reaction is extremely low at stellar energies and often blanked out by environmental background radiation in the detectors. This requires using high efficiency detector material with high resolution to separate the characteristic events from random background events. High beam intensity is desired to increase the event rate, however it may also increase beam induced background on target impurities and is limited by target stability.

The second critical issue is the background rate in the detector. There are typically three different kind of background, cosmic ray induced or cosmogenic background in the detector environment, natural long-lived radioactivity or radiogenic background in the detector material and the surrounding environment, and finally beam induced background on low Z target impurities and beam defining slits or apertures. These background components must be reduced as far as possible to identify reaction events in the spectrum.

Cosmogenic background affects the spectra of radiative capture reactions up to very high energies and makes it difficult to extract weak signals. That background is the most important to remove since most of the stellar reactions have high Q-values and emit high energy γ radiation. Radiogenic background will be strong in an underground environment except for salt mine locations. But the radiogenic γ lines are mainly below 3 MeV and can be shielded locally. Neutron background affects the study of stellar (p,n) or α,n) reactions. Cosmic neutron background again can be reduced in an underground environment, but radiogenic neutron are more difficult to absorb and need special shielding arrangements. Beam induced background depends critically on the target as well as the choice and preparation of the target material. It is difficult to suppress and may require active shielding procedures.

The latter can be done by identifying the event electronically by its particular characteristics such as coincidence conditions in a particular decay sequence, pulse shape or timing conditions and reject the background events which do not fulfil these requirements. This can lead to active background suppression by up to three orders of magnitude (Runkle et al. 2005; Couture et al. 2008). While this clearly helps in many cases a more efficient background reduction is desired.

The high energy cosmic ray induced background can be most successfully suppressed by operating the experiments in a deep underground environment where the cosmic ray flux is heavily reduced. This was demonstrated with the installation of the LUNA accelerator facility at the Gran Sasso deep underground laboratory in Italy. The cosmic ray induced background was successfully removed and several critical reactions of the pp-chains and the CNO cycles were successfully measured in the or near the Gamow energy range (Costantini et al. 2009). As a consequence of this successful operation new underground accelerator facilities are being proposed or planned which would allow covering reactions over a wider energy range than available at LUNA. This is of particular importance for an improved R-matrix analysis and extrapolation. Higher energies are also of great relevance for the underground measurements of α capture reactions and stellar neutron sources in helium burning. In particular it will also improve the chances for pursuing heavy ion fusion reaction studies towards lower energies. There are presently three major initiatives for the construction of new underground accelerator facilities. The CASPAR accelerator at the Sanfield Underground Research facility (SURF) at Homestake mine in South Dakota is optimised for the measurement of α beam induced reactions of interest for stellar during the helium burning and subsequent heavy ion burning phases of stellar evolution. The upgrade of the LUNA facility by adding a higher energy accelerator LUNA-MV has a similar goal of studying primarily α induced reactions but also plans new initiatives towards the measurement of fusion reactions during the late phases of stellar burning. A major effort is the development of the JUNA accelerator laboratory at the Jinping underground laboratory in China. The goal is to have two high intensity accelerators installed to cover the broad range of nuclear reactions that characterises all nucleosynthesis phases and environments associated with quiescent stellar burning. JUNA has the largest depth location of all existing and proposed facilities. A smaller project is the on-going construction of the Felsenkeller facility in Germany at a more modest depth level. Yet it will provide easy access and test opportunities for the community. With these facilities the nuclear astrophysics community hopes to address the new and critical questions

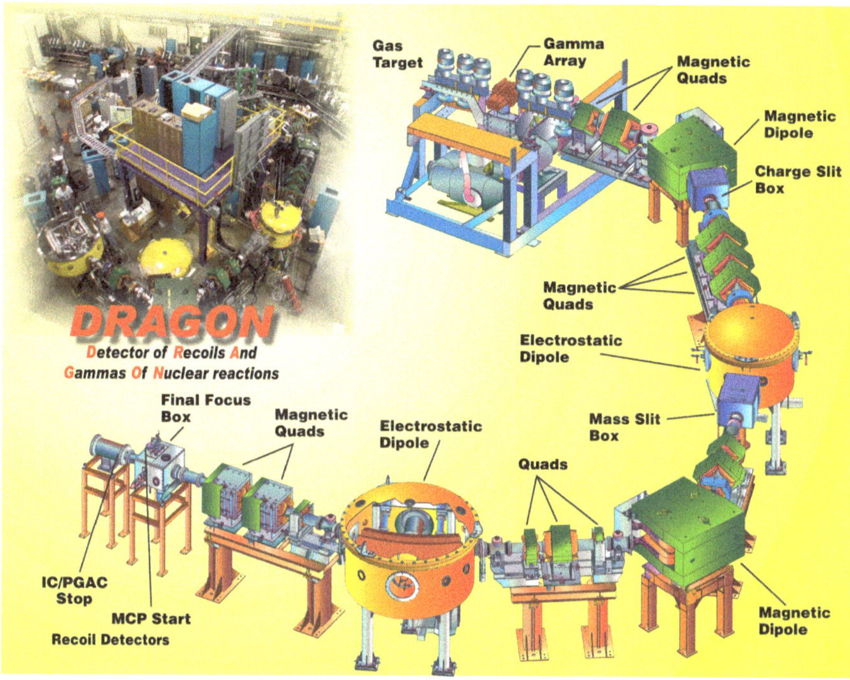

Fig. 9.2 The *Dragon* facility in Vancouver, Canada, is an example of nuclear experiment facilities, now aimed at experiments for astrophysically-relevant reactions: Radioactive isotopes can be selected and accelerated to form projectiles for such reactions of interest

about stellar reaction cross sections and provide the final answer on the nuclear engine of stellar evolution.

It has been demonstrated that alternative inverse kinematics methods are a very powerful tool in reducing the background. They are based on the technique of using a high intensity heavy ion beam on a hydrogen or helium gas target and separate the heavy ion recoil reaction products from the primary beam through a high resolution electromagnetic mass separator system from the primary beam. This method has been demonstrated to be successful at a number of different separator facilities such as DRAGON at TRIUMF, Vancouver (Vockenhuber et al. 2007) (Fig. 9.2) and ERNA at the Ruhr University Bochum (Di Leva et al. 2009).

The detection of the recoiling charged particle has a clear efficiency advantage compare to the gamma detection. The possible detection of the gamma rays in coincidence with the reaction products reduces dramatically the backgrounds. However, there are several experimental challenges associated with using recoil separators. At the low stellar energies, the energy spread and the angular aperture are much larger than the acceptance of any of the cited existing recoil separators. In order to measure an absolute cross section the transmission of the recoils should ideally be 100% or exactly known. It is also necessary to know precisely the charge state distribution of the recoil products. In addition, since the primary beam intensity is typically many orders of magnitude larger than the recoiling reaction products, a large spatial separation between the reaction products and the beam is required, which is difficult to realise for beams with a large energy spread. Therefore, solar fusion reactions are particularly challenging to measure with recoil separators and are typically used for higher energies and for the helium or heavy ion burning reactions.

Dedicated next generation separators for low energy nuclear astrophysics studies with stable ion beams coming on line are the ST.GEORGE facility in Notre Dame (Couder et al. 2008) and the modified and upgraded ERNA facility at CIRCE in Caserta, Italy (Di Leva et al. 2017). Both separators feature large angular and energy acceptance and are equipped with high density gas jet targets that ensure a well defined interaction region.

9.4.2 Laboratory Neutron Sources

Many of the observed or anticipated long lived radioactive isotopes in our galaxy are produced by neutron induced nucleosynthesis in the weak or main s-process taking place in helium and carbon burning stellar environments. This includes ^{41}Ca, ^{60}Fe, ^{63}Ni, but also more massive isotopes such as ^{98}Tc and ^{99}Tc and possibly numerous long lived isomers.

The study of neutron induced stellar reactions leading to the production of such isotopes requires high intensity neutron sources with a well defined energy distribution to determine the reaction cross sections at stellar energies of a few keV. Neutrons in that energy range can be produced in several ways. Nuclear reactions such as ^{7}Li(p,n) or ^{3}H(p,n) with high intensity proton beams provided by low-energy particle accelerators offer the possibility of tailoring the neutron spectrum to the energy range of interest; this has the advantage of low backgrounds. A particularly successful approach is to simulate a quasi-stellar neutron spectra in the laboratory. In bombarding thick metallic lithium targets with protons of 1912 keV, the resulting neutrons exhibit a continuous energy distribution with a high-energy cutoff at $E_n = 106$ keV and a maximum emission angle of 60°. The angle-integrated spectrum corresponds closely to a Maxwell–Boltzmann distribution for $kT = 25$ keV (Ratynski and Käppeler 1988). Hence, the reaction rate measured in such a spectrum yields immediately the proper stellar cross section.

Higher intensities can be achieved via photon production by bombarding heavy-metal targets with typically 50-MeV electron beams from linear accelerators. When these energetic neutrons are slowed down in a moderator, the resulting spectrum contains all energies from thermal energy to nearly the initial energy of the electron beam. Since the astrophysical relevant range corresponds to only a narrow window in this spectrum, background conditions are more complicated and measurements need to be carried out at larger neutron flight paths. In turn, the longer flight paths are advantageous for neutron-resonance spectroscopy with high resolution.

The most intense keV neutron flux is produced by spallation reactions. The LANSCE facility at Los Alamos is particularly suited for neutron TOF work due to the favourable repetition rate of only 12 Hz (Lisowski et al. 1990), and because the accumulation of a number of beam pulses in an external storage ring yields extremely intense neutron bursts. Accordingly, excellent signal-to-background ratios can be achieved. The n_ToF facility at CERN provides high intensity neutron beam pulses with a lower repetition rate of 0.4 Hz (Borcea et al. 2003). This has proved highly advantageous for a large number of experimental neutron capture studies along the s-process path.

The experimental methods for measuring (n,γ) cross sections fall into two groups: TOF techniques based on the detection of the prompt capture γ rays and activation methods.

The TOF techniques can be applied in measurements of most stable nuclei but require a pulsed neutron source to determine the neutron energy via the flight time between target and detector. Capture events in the samples are identified by the prompt γ ray cascade in the product nucleus.

The best signature for the identification of neutron capture events is the total energy of the capture gamma cascade, which corresponds to the binding energy of the captured neutron. Hence, accurate measurements of (n,γ) cross sections require a detector that operates as a calorimeter with good energy resolution and is insensitive to neutron exposure. In the gamma spectrum of such a detector, all capture events would fall in a line at the neutron binding energy (typically between 5 and 10 MeV), well separated from the gamma-ray backgrounds that are inevitable in neutron experiments. Such detectors have been successfully developed at the various laboratories using arrays of 4π BaF_2 scintillator detectors with a large number of independent detector modules (Heil et al. 2001) and have emerged as standard technology for these kinds of measurements.

A completely different approach to determining stellar (n,γ) rates is activation in a quasi-stellar neutron spectrum. Compared with the detection of prompt capture gamma rays, this method offers superior sensitivity, which means that much smaller samples can be investigated. Since it is also selective with respect to various reaction products, samples of natural composition can be studied instead of the expensive enriched samples required by the TOF techniques. However, the activation technique is restricted to cases where neutron capture produces an unstable nucleus, and it yields the stellar rate only for two thermal energies at kT = 25 and 52 keV. This method is however particularly powerful in obtaining cross sections for reactions producing long-lived radioactive materials which can be identified by their particular decay characteristics and signature. This activation technique has been used for a variety of measurements. The technique can be applied to short-lived products with half-lives in the millisecond range and allows for cross section measurements with uncertainties of a few percent.

9.4.3 *Accelerator Mass Spectroscopy*

Classical activation techniques require a characteristic decay signal associated with the decay pattern or the half-life of the produced radioactive isotope. This can be difficult in cases where no characteristic gamma or particle decay pattern exists or where the decay analysis of the β decay signal is prohibited by high background activity. In these cases activation analysis through accelerator mass spectrometry (AMS) offers a powerful tool to measure cross sections through ultra-low isotope-ratio determination. The AMS method was successfully introduced for the study of the neutron-capture cross section of ^{62}Ni(n,γ)^{63}Ni (Nassar et al. 2005), and extended to other neutron and charge-particle-induced reactions, such as ^{25}Mg(p,γ)^{26}Al (Arazi et al. 2006) and ^{40}Ca(α,γ)^{44}Ti (Nassar et al. 2006).

In these cases samples were either irradiated in a neutron spectrum resembling a stellar Maxwell-Boltzmann distribution or by charged particles of well known energies. After the irradiation the samples must be chemically treated to extract the radioactive reaction products. This requires some time and limits AMS activation studies to more longer lived isotopes. Since isotopic and isobaric interferences may represent a major challenge in AMS measurements of irradiated samples, extensive background studies for these isotopes are always necessary prior to the irradiations in order to demonstrate that the required sensitivity can be reached. In AMS, negative ions are extracted from an ion source which have to pass a low energy mass spectrometer prior to entering a tandem accelerator. When passing the stripper, positive ions are produced while within this stripping process molecular isobars are destroyed. One positive charge state is selected with a second (high-energy) mass spectrometer system which is optimised for mass, charge and isobar separation trough possible combination of dipole magnet separators, Wien-filters, and more recently magnetic gas filled separators for improved isobar separation. With such a system the concentration ratio of the radioisotope is determined relative to a stable isotope by measuring the number of radionuclides relative to the current of the isotopic ions in front of the detector, after adjusting the injector magnet, terminal and Wien-filter voltage appropriately. By measuring relative to a standard sample of known isotopic ratio, factors like stripping yields and transmissions mostly cancel.

The difficulties with AMS experiments is in the chemical preparation of the sample and the sufficient separation of the extracted radioactive ions from background events. While AMS is a widely established method with many applications, the analysis of the very limited number of radioactive products from low cross section reactions remains challenging. Systematic studies are necessary to reduce possible uncertainties.

Dedicated AMS facilities with an established nuclear astrophysics program are the Vienna Environmental Research Accelerator (VERA) (Kutschera et al. 1997), the Center for Isotopic Research on Cultural and Environmental Heritage (CIRCE) in Caserta/Italy (Terrasi et al. 2007) or the Munich Tandem accelerator facility (Knie et al. 2000), the Notre Dame tandem accelerator (Robertson et al. 2007), and the Pelletron Tandem machine at the Australian National University (Wallner et al. 2016). The last three of these facilities are optimised for the analysis of more massive radioactive isotopes.

9.4.4 Radioactive Beam Techniques

The development of radioactive accelerated beams for low energy nuclear astrophysics experiments has been one of the large challenges of the field. The experimental study of nuclear reactions and decay processes far of stability is necessary for the understanding of explosive nucleosynthesis processes such as the rp-process in cataclysmic binary systems or the r-process and p-process in the supernova shock front. These processes can in particular contribute to the production of long-lived galactic radioactivity by primary reaction or also by secondary decay processes from the reaction path towards the line of stability.

For the purpose of studying the origin of long-lived radioactive isotopes in astrophysical environments radioactive beams are utilised in two ways, for producing long-lived targets by implantation for subsequent irradiation with neutron, charged particle or possibly intense photon beams or for direct reaction measurement in inverse kinematics on light ion target materials. The later approach requires well defined mono-energetic and intense radioactive beams and a detection system for light or heavy recoil reaction products.

The main challenge in this approach is to produce a sufficiently high intensity of radioactive beams which have to be produced on-line as a secondary reaction product. This requires high cross sections for the production process and high primary beam intensities. A variety of different approaches has been chosen in the past to optimise the production efficiency and maximise the intensity of the radioactive beams. A technique developed for small scale facilities is the selection of specific nuclear reactions tailored for the on-line production of radioactive beams at optimum conditions. The secondary particles can be used for subsequent nuclear reaction studies after blocking and separation from the primary beams (Kolata et al. 1989). The efficient separation of a suitably high intensity beam of radioactive species is the most challenging problem for this approach.

An alternative approach is the ISOL (Isotope Separation On-Line) technique where high energy protons are used to bombard heavy ion targets for producing a large number of radioactive species through spallation processes. These isotopes diffuse out of the target into an ion source for being charged and re-accelerated for secondary beam decay or reaction experiments. The method has been proven to be very powerful over the years but is limited to isotopes with lifetimes appreciably longer that the time necessary for the diffusion transport and ionisation process. This can be different for different elements because of the associated chemical processes between the isotopes and the surrounding environment.

The third approach is based on the use of energetic heavy projectiles bombarding light or target nuclei fragmenting on impact. This fragmentation process generates a cocktail beam of many radioactive species which move forward with high velocity because the initial momentum of the primary particles is maintained. For experiments with a specific secondary particle, it must be selected by fragment separator systems which separate and focus the isotopes by magnetic fields and energy loss characteristics in heavy wedge materials. For nuclear astrophysics related experiments the fast beam particles need to be slowed down by energy loss in gas or solid material and re-accelerated to energies corresponding to the temperatures in the explosive stellar scenarios.

There is a number of laboratories which have focused on nuclear reaction studies with radioactive beams. The first fully operating radioactive beam laboratory based on the ISOL principle was the coupled cyclotron facility at Louvain la Neuve which did a number of successful radioactive beam studies of relevance for investigating the production of ^{18}F in novae (De Sérville et al. 2009). These measurements were complemented by measurements at the HRIBF facility at Oak Ridge using intense ^{18}F beams (Chae et al. 2006). Both facilities produce the radioactive species by nuclear reactions on thin production targets, with the reaction products being transported into an ion source for producing and subsequently accelerating the secondary beam. The intensity is largely limited by target technology and beam transport and re-ionisation efficiency.

The premier ISOL radioactive beam facility is ISAC at TRIUMF Canada. The primary 600 MeV proton beam is provided by the TRIUMF cyclotron. The reaction products are post-accelerated in an RFQ SC LINAC accelerator combination to energies of 0.3–3 MeV/u. ISAC has successfully performed a number of radioactive beam experiments of relevance for explosive hydrogen and helium burning. Most notable a direct study of ^{21}Na(p,γ)^{22}Mg in inverse kinematics to probe the production mechanism of ^{22}Na in Ne nova explosion environments (D'Auria et al. 2004). The facility also runs a successful program with stable beams which was utilized to investigate the production of ^{44}Ti (Vockenhuber et al. 2007). Presently a number of studies associated with the production of the long-lived γ emitter ^{26}Al are being performed.

Other ISOL based radioactive beam facilities such as Spiral facility at GANIL in Caen, France or REX-ISOLDE at CERN have been used to perform interesting experiments for nuclear astrophysics but have been less concerned with the question of nuclear production mechanisms for long lived cosmic gamma emitters.

There have been a number of fast radioactive beam facilities with scientific programs in nuclear astrophysics primarily aimed at the study of nucleosynthesis processes far off stability. However the rapid new developments in fast beam physics promises a number of new experimental opportunities which can provide benefits for studying reactions associated with the production of long-lived gamma emitters in explosive nucleosynthesis events.

There are currently four major fragmentation facilities in the world: GANIL and GSI in Europe, NSCL/MSU in the US and RIKEN in Japan. They are all based on Heavy Ion accelerators which operate in complementary energy domains. Because of the high energy of the fragment products low energy reaction experiments for nuclear astrophysics are not possible but the development of indirect techniques to determine critical reaction or decay parameters has been the primary goal. In the context of long-lived isotopes of astrophysical interest a major contribution was the development of fast beams such as ^{8}B at NSCL/MSU, RIKEN, and GSI for utilising Coulomb dissociation techniques for probing critical reactions such as ^{7}Be(p,γ)^{8}B. The NSCL and RIKEN also successfully developed a ^{44}Ti beam for new measurements of its half-life (Görres et al. 1998). More half-life measurements of long-lived isotopes such as ^{60}Fe are presently underway to re-evaluate these critical parameters.

9.5 Specific Experiments

The complexities of the experiments and the uncertainties in the experimental results affect the reliability of model predictions on the nucleosynthesis of long-lived radioactive species. In particular recent studies of critical nuclear reactions and decay processes exhibit considerable differences to earlier studies which so far have been the reference point for nucleosynthesis simulations and predictions for long-lived radioactive isotope abundances in stellar burning environments. It is therefore important to carefully evaluate the experimental results and clarify possible discrepancies and inconsistencies in the data. This section will discuss the present status of the experimental reaction rates and evaluate future opportunities to improve the existing data base.

9.5.1 Experiments with Stable Beams

Many of the long-lived radioactive gamma emitters in our universe have been produced by radiative capture reactions on stable isotopes. The best known examples are ^{26}Al, which is primarily formed by proton capture on stable ^{25}Mg isotopes, ^{25}Mg(p,γ)^{26}Al, and ^{44}Ti which is most likely produced via alpha capture on stable ^{40}Ca isotopes, ^{40}Ca(α,γ)^{44}Ti. Extensive measurements using in-beam γ spectroscopy techniques have been made for both reactions and have formed the basis for earlier reaction rate compilations.

The low energy reaction cross section of ^{25}Mg(p,γ)^{26}Al is characterised by several resonances with energies between 30 and 400 keV . The reaction rate is directly correlated to the strengths $\omega\gamma$ of the resonances. The strengths for the resonances above 190 keV have been determined from the on-resonance thick target yield in radiative capture measurements (Elix et al. 1979; Iliadis et al. 1990). The strengths of lower energy resonances are estimated on the basis of single particle transfer reaction studies. Of particular importance are three resonances at 90, 130 and 190 keV which determine the reaction rate at temperatures typical for stellar hydrogen burning in AGB stars and nova explosions. Because the low energy radiative capture measurements have been handicapped by cosmic ray induced background, an alternative measurement was done using the AMS technique to analyse the number of ^{26}Al reaction products after irradiation at resonance energies (Arazi et al. 2006). The experiment was successful and confirmed the resonance strengths of the known resonances at 304, 347, and 418 keV resonance energy. However the results indicated a substantially lower strength for the critical resonance at 190 keV. This would reduce the reaction rate by about a factor of five at the temperature range between 0.2 and 1.0 GK. This result introduced a large uncertainty in the reaction rate which motivated a new experimental study at LUNA in the Gran Sasso laboratory using in-beam gamma spectroscopy techniques with a variety of high efficiency and high resolution gamma detector devices. The measurements confirmed earlier gamma spectroscopy studies of the strengths of higher energy resonances (Elix et al. 1979; Iliadis et al. 1990) tabulated in the NACRE compilation (Angulo et al. 1999). These results have been now published (Strieder et al. 2012). Parallel to the gamma spectroscopy measurement, the irradiated samples were analysed for their ^{26}Al content using AMS techniques. The AMS measurements were performed at the CIRCE facilities. Excellent agreement is demonstrated for the resonance at 304 keV, additional experiments are being pursued for lower energy resonances to address the inconsistencies in the strength determination for the 190 keV resonance (Limata et al. 2010)

The ^{40}Ca$(\alpha, \gamma)^{44}$Ti reaction is considered to be one of the major production reactions for ^{44}Ti in supernova shock front nucleosynthesis. The cross section for this radiative capture process has been explored in a number of in-beam gamma spectroscopy studies down to centre of mass energies of 2.5 MeV (Simpson et al. 1971; Cooperman et al. 1977). The cross section is characterised by a large number of resonances and the initial reaction rate determinations were based on an analysis of resonance strengths. Despite the high level density in ^{44}Ti, it was noted that the experimental reaction rate is substantially smaller than the reaction rate based on statistical model Hauser Feshbach predictions (Rauscher et al. 2000). The reaction was studied independently using a thick He-gas cell target and counting the long lived ^{44}Ti reaction products by AMS techniques (Nassar et al. 2005) to determine the integral yield over an energy range of 1.7–4.2 MeV. The extracted reaction rate is substantially higher than the ones discussed in the literature (Rauscher et al. 2000). A more recent study of the reaction using inverse kinematics techniques was performed at the ISAC facility at TRIUMF, Vancouver, separating the ^{44}Ti reaction products on-line with the DRAGON recoil separator. The measurements covered the energy range of 2.3–4.2 MeV (centre of mass) in more than 100 small energy steps. The extracted yield was mostly interpreted as on-resonance resonance thick target yield and translated to a resonance strength. There are large uncertainties associated with this approach, in particular with the determination of the resonance energies, which have not been unequivocally determined in the experiment. In some cases several of the quoted resonances agree with previously identified states, in other cases it needs to be confirmed that the observed yields really correspond to additional resonances and do not originate from tail contributions of resonant yield curves associated with the different states. As far as the resonance levels which have been observed in both studies are concerned the published strengths are comparable to each other. Nevertheless the reaction rate suggested by Vockenhuber et al. (2007) is larger by more than a factor of two than the rates projected on the basis of the in-beam gamma spectroscopy measurements, but it is in agreement with the projections by Rauscher et al. (2000). A new gamma spectroscopy study with thin targets was performed to measure directly the ^{40}Ca(α, γ)^{44}Ti resonances over a comparable energy range of 3.0–4.6 MeV to remove the existing uncertainties in the experimental rates. The results (Robertson et al. 2012) confirm the reaction rate by Vockenhuber et al. (2007). At higher energies the rate agrees with Non-Smoker Hauser-Feshbach predictions but decreases with respect to the Non-Smoker predictions towards lower temperatures since resonances below 2.5 MeV. An independent analysis using the Monte Carlo method summarised in Iliadis et al. (2015) was performed by Pogrebnyak et al. (2013). The projected rate agrees well with the experimental one at temperatures relevant for the explosive conditions of the production site. At lower temperatures the rate agrees well with traditional Non-Smoker predictions.

9.5.2 Experiments with Neutron Beams

A particularly interesting case is the origin of the long-lived gamma emitter ^{60}Fe. Its characteristic γ-radioactivity has been observed with the INTEGRAL gamma ray telescope in supernova remnants near the solar system. These observations are complemented by recent AMS studies which suggest high ^{60}Fe abundance in deep sea ferromanganese sediments (Knie et al. 2004; Wallner et al. 2016). These ^{60}Fe observations have been interpreted as indication for the existence of a recent ($\approx$3 million years) supernova event in the solar system vicinity. A more quantitative interpretation of the time and distance of the proposed supernova event requires a detailed knowledge of the nucleosynthesis history of ^{60}Fe.

The radioactive ^{60}Fe isotope is produced by a sequence of neutron capture reactions of stable iron isotopes such as ^{58}Fe(n,γ)^{59}Fe(n,γ)^{60}Fe, the production rate and final abundance of the long-lived ^{60}Fe depends on the reaction rate of these feeding processes as well as on the rates of the ^{60}Fe(n,γ)^{61} and ^{60}Fe(β^-, ν)^{60}Co depletion processes. No experimental information are available about the associated cross sections except for the neutron capture reaction ^{58}Fe(n,γ)^{59}Fe. Simulations of the ^{60}Fe nucleosynthesis relied entirely on statistical model predictions of the neutron capture rates. Because of the relatively low level density in the associated ^{60}Fe, ^{61}Fe compound nuclei these model predictions are unreliable and need to be tested experimentally. This is underlined by the direct comparison between the experimental cross sections for neutron capture on the stable isotopes ^{56}Fe, ^{57}Fe, and ^{58}Fe which were all measured through neutron activation techniques and theoretical Hauser Feshbach predictions which show considerable discrepancies in particular in the cases of ^{56}Fe(n,γ)^{57}Fe and ^{57}Fe(n,γ)^{58}Fe. For ^{58}Fe(n,γ)^{59}Fe on the other hand , the agreement seems reasonable well but that cannot be extrapolated towards neutron captures on the more neutron rich Fe isotopes which are subject of the here proposed measurements.

Particularly important is the determination of the reaction rate of ^{59}Fe(n,γ)^{60}Fe since it competes directly with the ^{59}Fe β-decay which would by-pass the production of ^{60}Fe. A direct measurement of this critical reaction in the traditional activation or time of flight spectroscopy technique is not feasible because the target is radioactive and only small amounts can be accumulated. These small amounts nevertheless produce a large background activity level, which would prohibit any of the described methods. The cross section for the ground state decay branch of ^{59}Fe(n,γ_0)^{60}Fe has been investigated using inverse ^{60}Fe(γ,n)^{59}Fe Coulomb dissociation techniques (Uberseder et al. 2014). The experiment was performed at the GSI Helmholtz Center in Darmstadt. This experiment only probed the strength of the ^{59}Fe(n,γ)^{60}Fe ground state transition; γ transitions to higher excited states were estimated using Hauser-Feshbach predictions. This introduces a certain model dependence in the extracted reaction rate. A direct measurement of the reaction is presently not possible. But new initiatives have emerged to develop techniques for direct measurement by coupling radioactive beams with high neutron flux facilities. (Reifarth et al. 2017)

9.5.3 *Experiments with Radioactive Beams or Targets*

The depletion processes of long-lived radioactive isotopes includes the natural decay. Simulating this branch requires not only a good knowledge of the laboratory lifetime but also of the nature of the decay process since extreme environmental effects can change the decay rates drastically. In terms of β decay, the decay can be accelerated through the decay of thermally excited states as in the case of ^{26}Al. For decay through electron capture, the decay can be slowed down since the nuclei are completely ionised and the electrons have to be captured from the stellar plasma rather than from the inner K- or L-shell of the atom. This affects in particular the lifetime of ^{44}Ti, which primarily decays by electron capture.

Often the depletion is primarily driven by nuclear reactions, such as ^{22}Na(p,γ)^{23}Mg, ^{26}Al(p,γ)^{27}Si, ^{44}Ti(α,p)^{47}V, or ^{60}Fe(n,γ)^{61}Fe, but also capture reactions on shorter-lived excited configurations of these nuclei are possible, such as ^{26}Al*(p,γ)^{27}Si. There are two possibilities for experimental studies of the reaction cross sections. The first one is based on the production of highly enriched long-lived radioactive targets, which can be prepared through standard chemical target preparation techniques using externally bred radioactive material, or by implantation of radioactive ions at low energy ISOL facilities. The disadvantage of both techniques is that the actual γ measurements have to be performed in a high radiation background environment produced by the sample itself.

Nevertheless, earlier measurements of reactions such as ^{22}Na(p,γ)^{23}Mg (Seuthe et al. 1990) and ^{26}Al(p,γ)^{27}Si (Buchmann et al. 1984) relied entirely on this approach. In both cases a large number of resonances were detected and the resonance strength determined for calculating the reaction rates. The results for $^{22}Na(p,\gamma)^{23}Mg$ were confirmed by new direct measurements using improved target and detection techniques (Stegmüller et al. 1996), resulting in the observation of an additional low energy resonance at lower energies. Complementary spectroscopy techniques such as the study of the β-delayed proton decay of ^{23}Al (Peräjärvi et al. 2000) and the heavy ion reaction induced γ decay of proton unbound states in ^{23}Mg (Jenkins et al. 2004) provided additional nuclear structure information which led to the reduction of uncertainties in the reaction rate.

The situation is similar with $^{26}Al(p,\gamma)^{27}Si$; after the initial study with radioactive targets (Buchmann et al. 1984). A number of transfer experiments (Schmalbrock et al. 1986; Vogelaar et al. 1996) providing complementary information about the threshold levels in ^{27}Si not accessible to direct study by radiative capture measurements lead to an improved reaction rate for ^{26}Al ground state capture. A first direct study of a lower energy resonance was successfully performed in inverse kinematics at the ISAC facility at TRIUMF using the DRAGON recoil separator (Ruiz et al. 2006). The resonance value is substantially smaller than the value quoted before (Vogelaar et al. 1996), which reduced the reaction rate slightly at temperatures anticipated for nova burning conditions.

Not included in the reaction rate calculations are possible contributions of proton capture on the thermally first excited state in ^{26}Al (Runkle et al. 2001). Recently number of indirect measurements have been performed to explore the possible contribution to the total reaction rate of $^{26}Al(p,\gamma)^{27}Si$. Transfer reactions have been used to populate proton unbound states in ^{27}Si measuring the subsequent proton decay to the ground state and the first excited state in ^{26}Al (Deibel et al. 2009). This approach allows to determine the branching and the relative strength of the proton decays for each of the unbound states. This can be used to scale the reaction rate component for the proton capture on the first excited state.

Possible lower energy resonance contributions to the proton capture rates on the ground state (Lotay et al. 2009a) and the excited state of ^{26}Al (Lotay et al. 2009b) have been explored by γ spectroscopy techniques probing the proton unbound state in ^{27}Si through heavy ion fusion evaporation reactions and measuring the γ decay of proton unbound states. This is a particular efficient method to explore the levels near the threshold where proton decay is suppressed by the Coulomb barrier. The measurements provide critical information about spin and parity of the observed states but gives only limited information about the resonance strengths which is primarily determined by the proton decay strength.

The main reaction for the depletion of ^{60}Fe in neutron rich environments is $^{60}Fe(n,\gamma)^{61}Fe$. The reaction rate used for nucleosynthesis simulations was for many years based on theoretical Hauser Feshbach model predictions. Recently an experiment has been performed at the FZK Karlsruhe in Germany to determine the stellar reaction cross section experimentally by neutron activation with the neutron beam resembling a quasi-stellar neutron spectrum (Uberseder et al. 2009). The activated ^{60}Fe sample was prepared from PSI beam stop material. The cross section was determined from the characteristic ^{61}Fe γ activity relative to the amount of ^{60}Fe nuclei in the target material. The latter was determined from the characteristic ^{60}Fe

γ activity of the target sample. based on this the experimental results suggest a cross section which is twice as large as standard Hauser Feshbach predictions suggesting a much more rapid depletion of ^{60}Fe in neutron rich environment than previously anticipated. The estimate of the number of ^{60}Fe nuclei, however relied on adopting a half-life of $T_{1/2} = 1.49$ Gy (Kutschera et al. 1984). Recent work suggested that the half-live is considerably larger model $T_{1/2} = 2.62$ Gy (Rugel et al. 2009). This would translate into a considerably larger amount of ^{60}Fe particles in the sample, suggesting a cross section which would be in fair agreement with the Hauser Feshbach predictions. The new lifetime value has recently been confirmed by two independent measurements at ANU (Wallner 2015) and Notre Dame (Ostdiek et al. 2017).

The direct measurement of the ^{60}Fe(n,γ)^{61}Fe reaction did not provide information about the strengths of the direct and resonant reaction components. To determine the fraction of direct capture in the reaction contribution a ^{60}Fe(d,p) measurement was performed at GANIL using a 27AMeV ^{60}Fe populating the bound states of the ^{61}Fe nucleus (Giron et al. 2017). The single particle strength of the final states were determined and used to calculate the direct reaction component of ^{60}Fe(n,γ)^{61}F, which represents only 2% of the total strength given by Uberseder et al. (2009).

References

Angulo C, Arnould M, Rayet M et al (1999) Nucl Phys A656:3
Arazi A, Faestermann T, Fernandez Niello JO et al (2006) Phys Rev C 74:025802
Arnould M, Goriely S, Takahashi K (2007) Phys Rep 450:97
Azuma RE, Uberseder E, Simpson EC, Brune CR, Costantini H, de Boer RJ, Görres J, Heil M, LeBlanc PJ, Ugalde C, Wiescher M (2010) Phys Rev C 81:04580
Blatt JM, Weisskopf VF (1991) Theoretical nuclear physics. Dover, New York
Borcea C, Cennini P, Dahlfors M et al (2003) Nucl Instrum Methods A 513:524
Buchmann L, Hilgemeier M, Krauss A et al (1984) Nucl Phys A415:93
Cerjan C, Bernstein L, Hopkins L, Bionta R et al (2017) J Phys G-101848:R1
Chae KY, Bardayan DW, Blackmon JC et al (2006) Phys Rev C 74:012801
Cooperman EL, Shapiro MH, Winkler H (1977) Nucl Phys A284:163
Cooperstein J, Wambach J (1984) Nucl Phys A420:591
Costantini H, Formicola A, Imbriani G et al (2009) Rep Prog Phys 72:086301
Couder M, Berg GPA, Görres J et al (2008) Nucl Instrum Methods A 587:35
Couture A, Beard M, Couder M et al (2008) Phys Rev C 77:015802
Cowan JJ, Thielemann F-K, Truran JW (1991) Phys Rep 208:267
D'Auria JM, Azuma RE, Bishop S et al (2004) Phys Rev C 69:065803
deBoer RJ, Görres J, Wiescher M, Azuma RE, Best A, Brune CR, Fields CE, Jones S, Pignatari M, Sayre D, Smith K, Timmes FX, Uberseder E (2017) Rev Mod Phys 89:035007
De Sérv́ille N, Angulo C, Coc A et al (2009) Phys Rev C 79:015801
Deibel CM, Clark JA, Lewis R et al (2009) Phys Rev C 80:035806
Descouvemont P (2003) Theoretical models for nuclear astrophysics. Nova Science Publications, New York
Descouvemont P, Rauscher T (2006) Nucl Phys A777:137
Di Leva A, Gialanella L, Kunz R et al (2009) Phys Rev Lett 102:232502

Di Leva A, Imbriani G, Buompane R, Gialanella L, Best A et al (2017) Phys Rev C 95:045803
Elix K, Becker HW, Buchmann L et al (1979) Z Phys A 293:261
Ericson T (1960) Phys Rev Lett 5:430
Fowler WA (1974) Quart J R Astron Soc 15:82
Fuller GM, Fowler WA, Newman MJ (1982) Astron J 252:715
Gadioli E, Hodgson PE (1992) Pre-equilibrium nuclear reactions. Clarendon Press, Oxford
Giron S, Hammache F, de Séréville N, Roussel P et al (2017) Phys Rev C 95:035806
Glendenning NK (2004) Direct nuclear reactions. World Scientific, Singapore
Goriely S, Hilaire S, Koning AJ (2008) Astron Astrophys 487:767
Goriely S, Hilaire S, Koning AJ, Sin M, Capote R (2009) Phys Rev C 79:024612
Görres J, Meißner J, Schatz H et al (1998) Phys Rev Lett 80:2554
Hauser W, Feshbach H (1952) Phys Rev 87:366
Heil M, Reifarth R, Fowler MM et al (2001) Nucl Instrum Methods A 459:516
Hix WR, Thielemann F-K (1999) Astron J 511:862
Holmes JA, Woosley SE, Fowler WA, Zimmerman BA (1976) At Data Nucl Data Tables 18:305
Iliadis C (2007) Nuclear physics of stars. Wiley-VCH, New York
Iliadis C, Schange T, Rolfs C et al (1990) Nucl Phys A512:509
Iliadis C, Longland R, Coc A, Timmes FX, Champagne AE (2015) J Phys G Nucl Part Phys 42:034007
Jenkins DG, Lister CJ, Janssens RVF et al (2004) Phys Rev Lett 92:031101
Johnson CW, Kolbe E, Koonin SE, Langanke K (1992) Astron J 392:320
Kim KH, Park MH, Kim BT (1987) Phys Rev C 23:363
Knie K, Faestermann T, Korschinek G et al (2000) Nucl Instrum Methods B 172:717
Knie K, Korschinek G, Faestermann T et al (2004) Phys Rev Lett 93:171103
Kolata JJ, Morsad A, Kong XJ et al (1989) Nucl Instrum Methods B 40:503
Kutschera W, Billquist PJ, Frekers D et al (1984) Nucl Instrum Methods B 5:430
Kutschera W, Collon P, Friedmann H et al (1997) Nucl Instrum Methods B 123:47
Lane AM, Thomas RG (1958) Rev Mod Phys 30:257
Limata B, Strieder F, Formicola A, Imbriani G et al (2010) Phys Rev C 82:015801
Lisowski PW, Bowman CD, Russell GJ, Wender SA (1990) Nucl Sci Eng 106:208
Lotay G, Woods PJ, Sewryniak D et al (2009a) Phys Rev Lett 102:162502
Lotay G, Woods PJ, Sewryniak D, Carpenter MP, Janssens RVE, Zhu S (2009b) Phys Rev C 80:055802
Möller P, Pfeiffer B, Kratz K-L (2003) Phys Rev C 67:055802
Nassar H, Paul M, Ahmad I et al (2005) Phys Rev Lett 94:092504
Nassar H, Paul M, Ahmad I et al (2006) Phys Rev Lett 96:041102
Ostdiek KM, Anderson TS, Bauder WK, Bowers MR et al (2017) Phys Rev C 95:055809
Panov IV, Kolbe E, Pfeiffer B, Rauscher T, Kratz K-L, Thielemann F-K (2005) Nucl Phys A747:633
Panov IV, Korneev IY, Rauscher T, Martínez-Pinedo G, Kelić-Heil A, Zinner NT, Thielemann F-K (2010) Astron Astrophys. arXiv:0911.2181
Peräjärvi K, Siiskonen T, Honkanen A, Dendooven P et al (2000) Phys Lett B 492:1
Pieper SC, Wiringa RB (2001) Ann Rev Nucl Part Phys 51:53
Pogrebnyak I, Howard C, Iliadis C, Longland R, Mitchell GE (2013) Phys Rev C 88:015808
Ratynski W, Käppeler F (1988) Phys Rev C 37:595
Rauscher T (2010a) Nucl Phys A834:635c
Rauscher T (2010b) Phys Rev C 81:045807
Rauscher T (2011) Int J Mod Phys E 20:1071
Rauscher T (2012a) Astron J Lett 755:L10
Rauscher T (2012b) Astron J Suppl 201:26
Rauscher T, Raimann G (1996) Phys Rev C 53:2496
Rauscher T, Thielemann F-K (2000) At Data Nucl Data Tables 75:1

Rauscher T, Thielemann F-K, Kratz K-L (1997) Phys Rev C 56:1613
Rauscher T, Thielemann FK, Görres J, Wiescher M (2000) Nucl Phys A675:695
Rauscher T, Heger A, Hoffman RD, Woosley SE (2002) Astron J 576:323
Rauscher T, Mohr P, Dillmann I, Plag R (2011) Astron J 738:143
Reifarth R, Gobel K, Heftrich T, Weigand M, Jurado B, Käppeler F, Litvinov YA (2017) Phys Rev Accel Beams 20:044701
Robertson D, Schmitt C, Collon P et al (2007) Nucl Instrum Methods B 259:669
Robertson D, Görres J, Collon P, Wiescher M, Becker HW (2012) Phys Rev C 85:045810
Rugel G, Faestermann T, Knie K et al (2009) Phys Rev Lett 103:072502
Ruiz C, Parikh A, José J, Buchmann L et al (2006) Phys Rev Lett 96:252501
Runkle RC, Champagne AE, Engel J (2001) Astron J 556:970
Runkle RC, Champagne AE, Angulo C et al (2005) Phys Rev Lett 94:082503
Salpeter EE, Van Horn HM (1969) Astron J 155:183
Satchler GR (1983) Direct nuclear reactions. Clarendon, Oxford
Schmalbrock P, Donoghue TR, Wiescher M, Wijekumar V, Browne CP, Rollefson AA, Rolfs C, Vlieks A (1986) Nucl Phys A457:182
Seuthe S, Rolfs C, Schröder U et al (1990) Nucl Phys A514:471
Simpson JJ, Dixon WR, Storey RS (1971) Phys Rev C 4:443
Stegmüller F, Rolfs C, Schmidt S et al (1996) Nucl Phys A601:168
Strieder F, Limata B, Formicola A, Imbriani G et al (2012) Phys Lett B 707:60
Terrasi F, Rogalla D, De Cesare N et al (2007) Nucl Instrum Methods B 259:14
Uberseder E, Reifarth R, Schumann D et al (2009) Phys Rev Lett 102:151101
Uberseder E, Adachi T, Aumann T et al (2014) Phys Rev Lett 112:211101
Vockenhuber C, Ouellet CO, The L-S et al (2007) Phys Rev C 76:035801
Vogel P (2006) Nucl Phys A777:340
Vogelaar B, Mitchell LW, Kavanagh RW et al (1996) Phys Rev C 53:1945
Volya A, Weidenmüller HA, Zelevinsky V (2015) Phys Rev Lett 115:052501
Wagoner RV (1969) Astron J Suppl 18:247
Wallner A (2015) Phys Rev Lett 114:041101
Wallner A, Fifield L, Tims S et al (2016) Nature 532:69–72
Ward RA, Fowler WA (1980) Astron J 238:266
Yakovlev DG, Gasques LR, Afanasjev AV et al (2006) Phys Rev C 74:035803

Chapter 10
Instruments for Observations of Radioactivities

Gottfried Kanbach and Larry Nittler

This chapter describes key tools used to observe cosmic radioactivity including astronomical methods, laboratory measurements of meteorites and detection of Galactic cosmic rays. Cosmic nucleosynthesis, that is, the creation of new elements including radioactive isotopes, occurs in the most energetic, often explosive, sites in the universe. To observe these targets and processes in the light of high-energy photons, which are emitted in nuclear transitions and particle interactions, sensors for photon energies from around 100 keV to more than 10 MeV have been developed and employed on satellites and balloon platforms, outside the Earth's atmosphere, which is opaque to this radiation. The basic interactions for such photons are the photoelectric effect, Compton scattering, and pair creation. Typical examples for instrument designs are described in the first section of this chapter, followed by a presentation of successful missions since the 1980s (SMM, Compton Gamma-Ray Observatory CGRO), then currently operational missions (INTEGRAL, NuStar, Fermi), and perspectives for future telescopes with advances in technology. The second section addresses radioactivities in meteorite samples, which are generally measured by means of mass spectrometry. The most widely used methods are thermal ionisation (TIMS), multi-collector inductively-coupled-plasma (MC-ICPMS), secondary ion- (SIMS), and resonance ionisation mass spectrometry (RIMS). Parent and daughter nuclides can be measured on a variety of sample sizes, with precision depending on the size of the sample and concentrations of the elements of interest. The ultimate attainable precision is generally limited by the number of atoms in a given sample. New developments

G. Kanbach
Max Planck Institut für extraterrestrische Physik, Garching, Germany
e-mail: gok@mpe.mpg.de

L. Nittler (✉)
Carnegie Institution for Science, Washington, DC, USA
e-mail: lnittler@ciw.edu

R. Diehl et al. (eds.), *Astrophysics with Radioactive Isotopes*, Astrophysics and Space Science Library 453, https://doi.org/10.1007/978-3-319-91929-4_10

in RIMS, accelerator-based SIMS, and laser-assisted atom-probe tomography all hold promise for pushing meteoritic measurements to higher sensitivity and smaller spatial scales. Galactic cosmic rays are addressed in a third section. These are analysed by a variety of instruments from the ground, on high altitude balloons, or on spacecraft. Basic principles are discussed as well as specific experiments, including the Pierre Auger Observatory, the Cosmic Ray Isotope Spectrometer on the ACE spacecraft, TIGER, and PAMELA.

10.1 Astronomical Telescopes

10.1.1 Measuring Radiation from Cosmic Radioactivity

10.1.1.1 General Considerations

Radioactivity is characterised by the emission of particles or photons that accompany the nuclear transformations of unstable isotopes. Direct observations of secondary particles (e.g., $\beta^{\pm}$, or α) are only possible with in-situ measurements in the *local environment* of the solar system and often the detected particles are not very specific as to their parent nuclei. Transitions between energy levels of radioactive and excited nuclei however produce characteristic X- and γ-ray lines that can be detected from astronomical distances. Only one secondary *particle* resulting from radioactive decay, the positron, signals its presence in a characteristic γ-ray line: positrons annihilate with their anti-particle (electrons) and convert the pair's rest mass into a line at 511 keV (see Chap. 7). Nuclear energy levels range from the atomic levels at 10s of keV upwards to energies of 10s of MeV with most important astrophysical lines in the range from about 100 keV to several MeV.

Detection of MeV photons must exploit the dominant interaction processes in this energy range: the photoelectric effect (*'photoeffect'*) and *Compton scattering*. Both these interaction processes and their cross-sections depend on photon energy and target material. The typical energy of transition from dominating photoeffect to Compton scattering is around the rest mass energy of the electron, i.e. 511 keV. For photon energies above the highest atomic binding energy the photoelectric cross section is given by $\sigma \propto Z^n E^{-3}$, where Z is the atomic number (of protons) and n is an index with a value between 4 and 5. At lower photon energies characteristic line-structures appear in the cross section (K-, L-, M-edges, etc.) which indicate the atomic energy levels of the electrons released in an atomic transition. The experimental consequence is that a high-efficiency photoelectric detector should be made of high-Z material, and it will work best below a few 100 keV. Spectroscopy with such detectors however requires careful calibration, because these structures in the cross section will shape the instrumental response and appear in the measurements. At higher energies, incoherent scattering of photons with electrons, (*Thomson* and *Compton scattering*), leads to the release of an energetic electron and a secondary photon, and dominates photon interactions. The maximum of this interaction cross section is around 511 keV (m_ec^2). The high- and low-energy ends of the cross section behavior can be described by asymptotic limits in terms of the *Thomson* cross section

$$\sigma_T = \frac{8\pi}{3}\left(\frac{\alpha\hbar}{m_e c}\right)^2 \simeq 6.652 \times 10^{-25}\ \mathrm{cm}^2 \tag{10.1}$$

and the photon energy in units of the electron rest mass, $\epsilon = h\nu/m_ec^2$. In the *Thomson regime*, $\epsilon \ll 1$, the total cross section is about $\sigma = \sigma_T(1 - 2\epsilon \ldots)$. In the high energy limit (the *Klein Nishina* regime, $\epsilon \gg 1$) the total cross section is a decreasing function of energy:

$$\sigma = \frac{3}{8}\sigma_T\epsilon^{-1}(1 + 2\ln\epsilon) \tag{10.2}$$

Detailed tables of photon cross sections for various elements and compounds are available at http://physics.nist.gov/PhysRefData/Xcom/Text/XCOM.html.

10.1.1.2 Instrument Types

Building an efficient detector for ~MeV gamma radiation however not only demands a good choice of detector material, but also requires that the detectors are sensitive to measure the secondary particles (electrons, positrons) released by the incoming energetic photons. This can be achieved in several ways:

- **Ionisation chambers:** the detector is intrinsically capable of measuring the presence of ionisation and generating an electronic signal. Gas-filled ionisation chambers, proportional counters, or Geiger counters were the original devices to detect high energy radiation. The small amounts of charge generated in the detector volume by a single photon is often amplified by a strong electric field generated from electrodes. This results in acceleration, collisional secondary ionisation and the formation of a break-down cascade, that is more easily detectable. Variations of this principle led to gas-filled drift chambers or spark chambers that also allowed to locate the impact coordinates of primary photons. After semiconductors with sufficient volume had been developed (e.g. Si, Ge, Cadmium Zinc Telluride, CZT, or Cadmium Telluride, CdTe) the principles and concepts of ionisation chambers were transferred to the new solid state detectors. The higher densities and generally higher atomic numbers of solid state detectors led to much higher efficiencies for the conversion of γ-rays and, in the case of cryogenically cooled detectors, also to very fine energy resolution (Ge detectors).
- **Scintillation detectors:** the detector material emits light under the impact of ionisation (*fluorescence, scintillation*). In addition to the scintillator material itself, these detectors require a photon detector to record the emitted light. Commonly used are *organic scintillators* (liquid or solid), scintillating *crystals*

(e.g., NaI, CsI, BGO, La_2Br_3), or *noble gases* (e.g., liquid Xenon). Traditionally the detection of scintillation light was achieved with *photo-multiplier tubes (PMTs)*, which are low noise and very fast detectors. The typical quantum efficiency of a PMT photocathodes however is only around ~20%. More recently the readout of scintillation light is also done with *solid-state photodetectors* such as PIN diodes, Si drift detectors (SDD), and avalanche photo diodes (APD). The advantages of solid state readout are a higher detection efficiency, the small size and the possibility of complex arrangements (pixels, arrays) of Si detectors, in addition to the straightforward interface with modern readout electronics.

All high energy detectors in space are exposed to radiation of energetic particles (cosmic rays, radiation belts) and thus *photons of non-astronomical origin*. The structures of spacecraft and detector alike will become sources of local background under this irradiation. Direct nuclear interactions and the creation of radioactive isotopes may lead to intense levels of background, unless counter measures are taken. Direct and prompt interactions of incident charged particles can be rejected by enclosing the sensitive volumes in veto shield detectors. These *anti-coincidence* detectors are often made of plastic scintillator (e.g., *COMPTEL*, Schönfelder et al. (1993), *EGRET*, Kanbach et al. (1989), *Fermi*, Atwood et al. (2009)) but massive shields of inorganic scintillators, like *CsI (SMM-GRS*, Forrest et al. (1980)) or *BGO (INTEGRAL-SPI*, Vedrenne et al. (2003)) are also employed to better cover the low energy range. There is a trade-off to be made between adding detection capability for undesired events and adding mass which generates such undesired events.

To improve handling of the *intrinsic background* of a gamma-ray telescope, several options are available:

- The structural materials close to the sensitive detectors should be chosen specifically to ensure low activation by the dominant external particle environment (i.e., low cross sections for such reactions). For example, aluminium, which activates to radioactive ^{24}Al, could be replaced by beryllium; hydrazine propellant, which thermalises and captures neutron on protons, emitting 2.2 MeV photons, should be minimised.
- The sensitive parts of a detector could be mounted at a distance from the mass of the spacecraft, e.g., on a boom.
- the detection process should fully exploit the characteristics of gamma-ray interactions to discriminate against background through selections in the measured interaction details. Examples that have been used include *pulse shape discrimination* to separate neutron and γ-ray interactions or specific coincidence trigger requirements.
- The *choice of orbit* of a low-energy gamma-ray telescope should avoid regions of elevated particle background in the Earth's magnetosphere (radiation belts, South Atlantic Anomaly). A low altitude, circular equatorial orbit provides the best environment for such telescopes, since it is also shielded against low energy solar particles.

Even with the above counter measures, the instrumental background of a nuclear line telescope will remain at a significant level and often dominates celestial signals even for strong sources (see Chap. 7). The spectrum of this background extends from several 100 keV up to about 5 MeV and shows a continuum and several strong complexes of nuclear lines. Detailed simulations are essential, and were developed based on the Monte-Carlo principle and detailed high-energy interaction physics software packages such as *GEANT*, e.g., by Weidenspointner et al. (2005). This allows one to estimate the intensity and spectrum of instrumental background, for future and past missions. Simulations and trade-offs for options of telescope design and orbit can thus be investigated.

The usefulness of an astronomical telescope can be expressed as the level of *sensitivity* reachable during typical periods of observation. Sensitivity is given as the minimum source flux that can be detected significantly above the level of background. Generally the instrumental-background signature is spread out, and varies smoothly over the field of view, but it contains the spectral features mentioned above. In a telescope with temporal, angular and energy resolution, the relevant level of background is then the level contained in the temporal, angular and spectral resolution elements which would be populated by the signal from a real source (*point spread function*). The resolution elements of measurement and analysis, i.e., bins in a measured signal parameter, pixels in an image, or $\Delta E/E$ in a spectrum, should therefore correspond to the properties of the astronomical sources in order to optimize the sensitivity. Of course the combination of angular and spectral resolution will lead to best results, but as we discuss below, temporal resolution for transient or variable sources (flares, bursts) may well substitute a lack of angular resolution. We discuss now typical examples of successful instruments, that pioneered nuclear-line astrophysics.

10.1.2 Photon Collectors

SMM-GRS, the γ-ray spectrometer of the *Solar Maximum Mission (SMM)* was an actively shielded multi-crystal scintillation γ-ray spectrometer with a wide field of view, sensitive to γ-rays in the energy range from about 300 keV to 100 MeV (Forrest et al. 1980). Its goal was to measure solar flares and during the mission from 1980 to 1989 a total of 185 flares were detected (Vestrand et al. 1999). The discrimination of signal and background in *SMM-GRS* was based on the time profile of the flares riding on top of a slowly varying background and on a model of the orbital background averaged over locations of similar magnetospheric conditions. *SMM-GRS* was therefore a typical photon collector with little angular resolution. The γ-ray spectrometer, shown in Fig. 10.1, was composed of 7 cylindrical NaI(Tl) crystals with a diameter of 7.6 cm and a height of 7.6 cm. Each of these crystals was viewed by a PMT. The crystals were surrounded on the sides by a 2.5 cm thick CsI annulus. On the rear side a circular CsI crystal with a thickness of 7.6 cm and a diameter of 25 cm was placed. The 7 NaI (Tl) crystals and the back CsI crystal

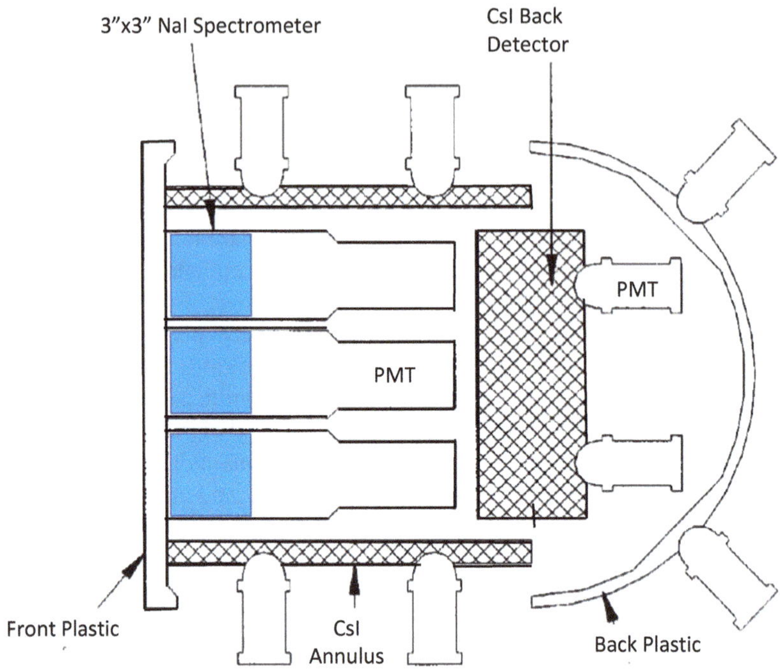

Fig. 10.1 The *Gamma Ray Spectrometer* on the *Solar Maximum Mission* (Forrest et al. 1980). Seven 3" NaI detectors are enclosed in an anti-coincidence shield made of a plastic scintillator (front and back) and of CsI scintillators (sides and back)

formed together the high-energy detector. In order to suppress charged particles the front and rear sides were covered with sheets of plastic scintillators thus shielding the main detector from all sides. The γ-ray spectrometer had a very wide field of view with a diameter of $\simeq 120°$, an energy resolution of 7% at 662 keV and an effective area ranging from 20 to 200 cm^2 as a function of energy.

10.1.3 Imaging Instruments

COMPTEL was the first truly imaging telescope for the MeV range and was in orbit for 9 years from 1991 to 2000 (Schönfelder et al. 1993) aboard NASA's *Compton Gamma-Ray Observatory*. A photomontage of *COMPTEL* is shown in Fig. 10.2.

The upper detector, called D1, consists of 7 modules filled with the liquid scintillator NE 213A. Each module of 28 cm diameter and 8.5 cm thickness is viewed from the sides by 8 photomultiplier tubes. The total geometrical area of D1 is 4300 cm^2. The lower detector, D2, separated from D1 by 1.5 m, consists of 14 modules of NaI(Tl)-scintillator. Each module of 28 cm diameter and 7.5 cm thickness is viewed from below by 7 photomultipliers. The total geometrical area

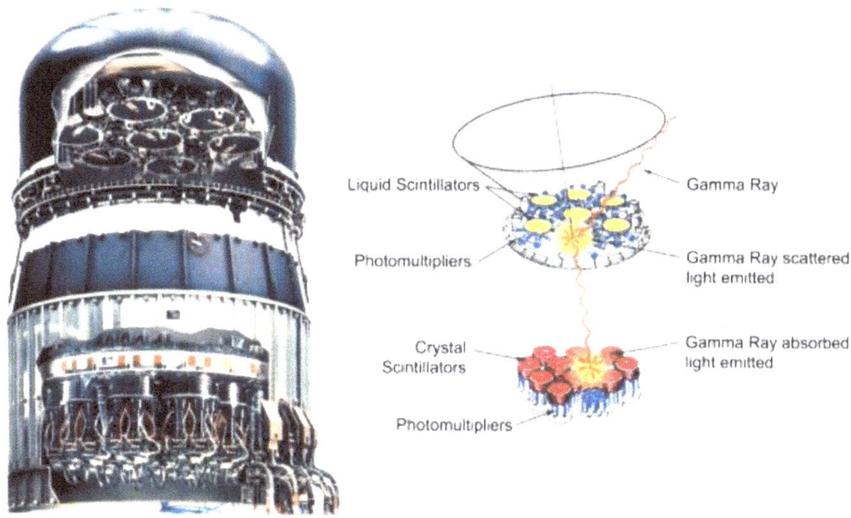

Fig. 10.2 The *COMPTEL* instrument on the *Compton Gamma Ray Observatory CGRO* (Schönfelder et al. 1993). A schematic drawing shows a typical scattering event and the cone with opening angle ϕ which contains the incident photon

of D2 is 8600 cm^2. From the relative pulse heights of the photomultiplier tubes in each D1- and D2-module the locations of the interactions are determined to within about 2 cm. The sum of the photomultiplier signals of each module provides the energy losses E1 and E2 . Both, D1 and D2 are completely surrounded by veto-domes of 1.5 cm thick plastic scintillator to reject charged particles. A valid event trigger is generated by a time-of-flight delayed coincidence between D1 and D2 and the absence of a veto signal. The positions of the interactions in D1 and D2 define the direction of the scattered photon and the Compton scattering formula

$$cos(\phi) = 1 - m_e c^2 (\frac{1}{E2} - \frac{1}{E1 + E2})$$

is used to estimate the opening angle of a cone that contains the incident gamma ray (see schematic in Fig. 10.2)

COMPTEL covers the energy range 0.8–30 MeV. Within its large field-of view of about 1 steradian (64°, FWHM), its angular resolution rages from 1.7° to 4.4° (FWHM; depending on energy). The energy resolution is in the range 5–8% (FWHM), and the effective detection area is in the range 20–30 cm^2 (again depending on energy). The sensitivity of *COMPTEL* is significantly determined by the instrumental background. A substantial suppression is achieved by the combination of the charged particle veto domes, the time-of-flight measurement technique, pulse shape discrimination in D1, Earth horizon angle cuts, and proper event selection cuts. In spite of these significant countermeasures the actual total in-flight instrumental background was higher than expected prior to launch by about a factor of 4. As a consequence, the sensitivity of *COMPTEL* was a factor of two worse than expected. Fortunately, however, this loss in sensitivity could be compensated by the four times longer mission life-time (which originally was planned to be only 2.25 years). The actually achieved point source sensitivity for deep observations ($T = 6 \times 10^6$ s) was 6.3×10^{-5} photons/cm^2 s for continuum emission between 1 and 30 MeV (corresponding to about 3% of the Crab-flux) and 1.6×10^{-5} photons/cm^2 s for line emission at 1.157 and 1.809 MeV (^{44}Ti and ^{26}Al lines).

10.1.4 Current Spectrometry and Imaging: INTEGRAL/SPI, NuSTAR, and COSI

As we have discussed above, angular resolution is an essential requirement to achieve astronomical sensitivity. When the spectrometer instrument for ESA's INTEGRAL mission was designed it was clear that the superb energy resolution of a Ge spectrometer had to be combined with a telescope that provides at least moderate angular resolution. In the Spectrometer for Integral, SPI, this was achieved with a coded-mask system (see Fig. 10.3). The SPI mask, which is mounted 171 cm above the detection plane, features a HURA (for 'Hexagonal Uniform Redundant Array') coded mask pattern with 120° symmetry enclosed within a diameter of 72 cm. Of the 127 individual cells (60 mm side to side) in the mask, 63 are opaque (3 cm thick blocks of Densimet, a tungsten alloy, with an opacity of about 90%) and 64 are transparent. The *shadowgram* projected onto the detector plane is measured with 19 Ge detectors. The reverse-electrode n-type Ge detectors also have a hexagonal shape with a side length of 3.2 cm and a height of 69.42 mm and are operated at a temperature of ~90 K cooled by a Sterling system. The detector is enclosed in a massive anti-coincidence shield of BGO scintillation detectors. The energy range extends from 20 keV to 8 MeV with a typical energy resolution of 2.5 keV at 1.33 MeV. The coded mask system provides an angular resolution of about 2.5° with a positioning accuracy for a strong source of 10 arc min. The fully coded field-of-view has a diameter of 16°. Further details can be found in Vedrenne et al. (2003). ESA's INTEGRAL observatory with its main instruments SPI (spectrometer) and IBIS (imager) was launched on October 17, 2002 and has been operating successfully since then (Diehl et al. 2017). Several astrophysical results from INTEGRAL/SPI observations address radioactivities and are described in Chaps. 4–7 of this book.

The NASA small explorer mission *NuSTAR* (Nuclear Spectroscopic Telescope Array) was launched on June 13, 2012. NuSTAR employs two co-aligned grazing incidence X-ray telescopes with a focal length of about 10 m. These nested Wolter-1 type optics are coated with multilayers allowing effective reflection of X-rays in the energy range 3–79 keV. Each of the two focal plane detectors consists of a 2×2

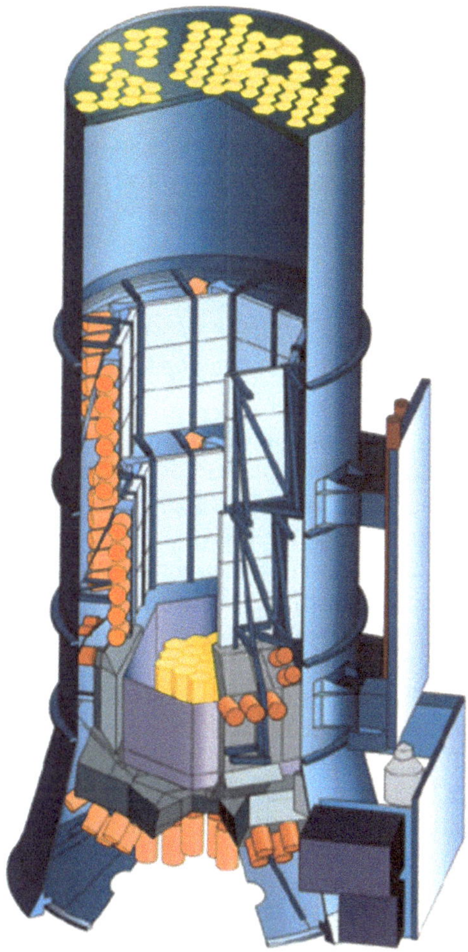

Fig. 10.3 The *Spectrometer Instrument (SPI)* on ESA's *INTEGRAL* mission (Vedrenne et al. 2003). The γ-ray sky is projected through a coded mask onto an array of 19 Ge detectors. Deconvolution of the shadowgram allows an angular resolution of 2.5 deg and the Ge detectors provide an energy resolution $E/\Delta E$ of $\sim$600

array of CdZnTe pixel detectors, each with 32 by 32 pixels, 0.6 mm in size. With a field of view between 10 arcmin (at 10 keV) and 6 arcmin (at 68 keV), NuSTAR achieves an angular resolution of 18 arcsec (FWHM). The prime objectives of this telescope are therefore the observation of galactic and extra-galactic point sources and small scale structures such as young SNe. The effective area of NuSTAR in the range around 70 keV, which is relevant for some low energy radioactivity, is about 60 cm^2. A full description of NuSTAR is given by Harrison et al. (2013). A major result from NuSTAR was the imaging of radioactive debris in the supernova remnant Cas A, where the ^{44}Ti decay lines at 68 and 78 keV were used for locating the radioactive ^{44}Ti and determining its kinematics (see Chap. 4).

The balloon experiment *COSI* (Compton Spectrometer and Imager) is based on Ge detector units that record Compton interactions in the photon energy range 0.2–5 MeV. The compact Compton telescope design consists of 12 cross-strip, high-purity germanium detectors, each of size $8 \times 8 \times 1.5\,\text{cm}^3$ with strip read-out electronics allowing 3-dimensional resolution. The total active detector volume is $972\,\text{cm}^3$. In May 2016 the COSI instrument was flown for 46 days on NASA's new super pressure balloon; launched from Wanaka, New Zealand the balloon circled around the globe one and a half times and landed in Peru. The instrument and preliminary results from this balloon campaign are described in Kierans et al. (2017).

10.1.5 Perspectives of Cosmic Gamma-Ray Imaging and Spectroscopy

In the years since the termination of the *CGRO* mission, many efforts and studies were undertaken to develop the next generation of Compton telescopes. Some projects, such as the *Advanced Compton Telescope ACT* (Boggs et al. 2006), *GRIPS* (Greiner et al. 2009), and *GRI* (Knödlseder 2007) were proposed to the space agencies but were not selected. Other experiments were built as prototypes or balloon payloads. Among those are MEGA (Kanbach et al. 2003), the *Nuclear Compton Telescope NCT* (NCT Collaboration et al. 2007), and *TIGRE* (Zych et al. 2006).

A very compact instrument using a Compton scattering technique—the Soft Gamma Ray Detector, *SGD*, was developed for the Japanese mission *ASTRO-H* (Takahashi et al. 2009, 2004). The Compton camera *SGD* was built with Si and CdTe detectors and was surrounded by an active BGO shield. Its energy range was 40–600 keV. After launch on Feb. 17, 2016, the mission was named *HITOMI*. Unfortunately the satellite was lost due to a malfunction of the attitude control system on March 26, 2016 and only very few useful observations were recovered.

Two gamma-ray telescopes covering the range from about 300 keV up to 10s of GeV are currently (2017) being studied in the U.S. (*AMEGO*) and in Europe (*e-ASTROGAM*). Both concepts combine the capabilities of the major *CGRO* instruments: *COMPTEL* in the Compton range from 300 keV to about 10 MeV and *EGRET* in the pair creation range (>10 MeV) in a large instrument that records both types of interactions. These telescopes are based on modern solid-state detectors and feature large fields of view (>2 sr), angular resolution of $\sim$1–3 deg, and energy resolution of $\sim$10 keV in the range of nuclear lines. Polarisation sensitivity is a further goal of high interest in gamma-ray astronomy.

The European *e-ASTROGAM* mission (De Angelis et al. 2017) features a deep tracker unit made of Si strip detectors that allows imaging of the secondary particles released in Compton scattering or pair creation events. The total energy of the incoming photons is partially measured in the tracker and finally deposited in a massive calorimeter made of CsI scintillators. The gamma-ray telescope is surrounded by an anticoincidence veto shield of plastic scintillator to reject charged background radiation. In Fig. 10.4 the expected sensitivity of the e-ASTROGAM

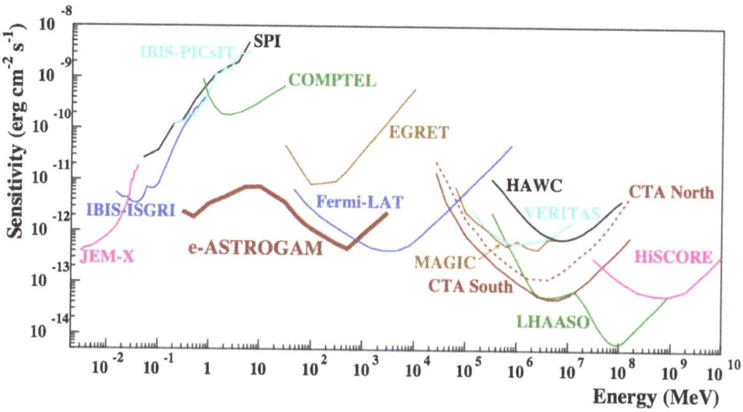

Fig. 10.4 Point source continuum sensitivity of different X- and gamma-ray instruments. The curves for INTEGRAL/JEM-X, IBIS (ISGRI and PICsIT), and SPI are for an observing time Tobs = 1 Ms. The COMPTEL and EGRET sensitivities are given for the time accumulated during the duration of the CGRO mission (about 9 years). The Fermi/LAT sensitivity is for a high Galactic latitude source over 10 years. For MAGIC, VERITAS, and CTA, the sensitivities are given for $T_{obs} = 50$ h. For HAWC $T_{obs} = 5$ years, for LHAASO $T_{obs} = 1$ year, and for HiSCORE $T_{obs} = 1000$ h. The e-ASTROGAM sensitivity is for an effective exposure of 1 year for a source at high Galactic latitude (adapted from De Angelis et al. 2017)

gamma-ray telescope is shown. The range of nuclear lines between ~300 keV up to ~10 MeV is well covered by *e-ASTROGAM* with a sensitivity that exceeds previous coverage by more than an order of magnitude.

The mission *AMEGO* (All-sky Medium Energy Gamma-ray Observatory, McEnery 2017) is under study by a large collaboration in the U.S. The basic design is similar to *e-ASTROGAM*, and the use of pixelated CdTe detectors in the calorimeter enhances the low energy response of *AMEGO* between 200 and 300 keV.

From *COMPTEL* and *INTEGRAL*, we have learned that the sky is rich in phenomena and objects that can be studied around 1 MeV. But it is also true, that with *COMPTEL* we could see only the tip of the iceberg. The achieved sensitivity was still modest. The next telescope will have to have sensitivity better than an order of magnitude. In the studies listed above different concepts of Compton telescopes are presently investigated and tested. Instead of scintillators also other detector materials are considered and tested, like silicon strip detectors, position sensitive germanium detectors, CdTe-detectors, liquid xenon gas detectors, and high pressure gas detectors. In order to achieve the required improvement in sensitivity, the detection efficiency has to be drastically increased and the background has to be reduced considerably. Several of the proposed experiments are capable of achieving this higher level of sensitivity and hopefully will advance gamma-ray astronomy in this largely undeveloped MeV wavelength band. What is needed is a decision by the international space agencies that this window into the universe is worth exploring with a new mission with a sensitivity to go beyond the *tip of the iceberg* results addressed in Chaps. 4–7.

10.2 Analyzing Material Samples from and Within Meteorites

Radioactivities in meteorite samples are in general detected by separating individual atoms from a sample and weighing them with a device known as a mass spectrometer. For long-lived radionuclides, that is those with lifetimes significantly longer than the 4.6 billion year age of the Solar System, the nuclides themselves are detected, along with their decay products. In the case of short-lived isotopes created by cosmic-ray interactions with the meteorites during the time interval between ejection from a parent body and arrival on Earth, the parent isotopes can also be directly measured. For short-lived isotopes that existed in the early Solar System but have since fully decayed (*extinct* isotopes), only the decay products can now be determined. In most cases, stable isotopes of the parent and daughter elements must also be measured in order to derive reliable information. Depending on the sample and the isotopic system, a wide array of instrumentation and techniques can be used to derive information on radioactive elements in extraterrestrial samples. These are the subject of this section.

10.2.1 Measurement Principles and Techniques

When a radionuclide decays within a solid sample, the isotopic compositions of both the parent and daughter elements are changed. By plotting isotopic and/or elemental ratios, one can infer either the age of the sample (time elapsed since the sample solidified to the point that the parent isotope was fixed in the structure) or, for extinct isotopes, the initial abundance of the parent isotope. This is illustrated in Fig. 10.5, which shows an example of *Rb-Sr* dating based on the decay of long-lived ($\tau \sim 7.2 \times 10^{11}$ years) ^{87}Rb to ^{87}Sr. Each plotted point represents a different mineral that obtained a different Rb/Sr elemental ratio upon crystallisation of the rock. Initially all of the minerals had the same ^{87}Sr/^{86}Sr ratio (open circles), but with time, some of the ^{87}Rb decayed, moving the points to the up and left on the plot (filled symbols). The age of the rock can be calculated from the slope of the resulting line; this line is called an *isochron.*

For the case of extinct nuclides, the daughter isotope ratio is plotted versus a stable isotope of the parent element, and the slope of an isochron gives the initial ratio of the extinct nuclide to the stable isotope. For example, to detect extinct ^{26}Al, one plots the measured ^{26}Mg/^{24}Mg ratio versus the ^{27}Al/^{24}Mg ratio and the slope gives the initial ^{26}Al/^{27}Al ratio of the sample (see Chap. 6). Often, one seeks information on a single object with a single parent/daughter element ratio, for example pre-solar grains (Chap. 3). In such cases, isochrons cannot be constructed

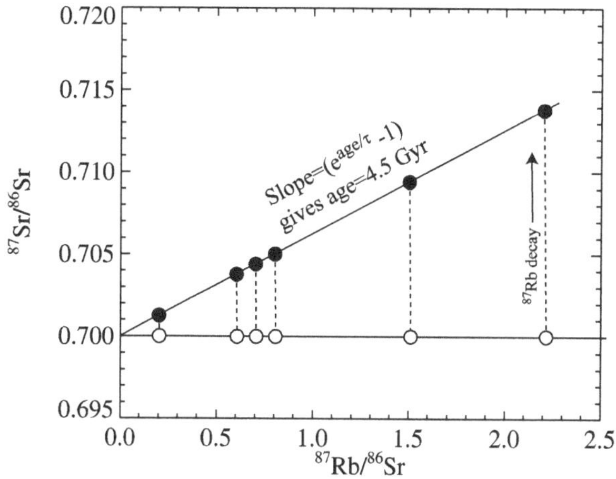

Fig. 10.5 Schematic example of radiometric dating of rocks. Open symbols represent the initial isotopic compositions of a set of minerals that co-formed with varying Rb/Sr ratios. After 4.5 Gyr, some of the long-lived ^{87}Rb has decayed to ^{87}Sr (filled symbols); the slope of the resulting *isochron* gives the age of the rock

and assumptions regarding the initial stable isotopic composition must be made to determine initial abundances of radionuclides. Usually, such assumptions can be made with sufficient confidence for the problem at hand.

Note that there are many complications to radiometric age-dating and determination of extinct isotopes, both arising from the laboratory analyses (see below) and from the samples themselves. For example laboratory contamination can compromise analyses, especially for very-low-abundance elements. Atoms of either the parent and/or daughter elements may have diffused over time into or out of the sample, so that the measured ratios do not reflect the simple isochron behaviour described above. Researchers have developed many techniques for identifying and overcoming such problems.

Mass spectrometers determine the elemental, isotopic or molecular composition of a sample by first converting the atoms or molecules of the sample into a beam of ions and then using electric and/or magnetic fields to disperse the beam in mass. Thus every mass spectrometer requires an ion source, a filter to separate masses and one or more ion detectors. Several examples of different types of mass spectrometers used for meteoritic radioactivity research are detailed in subsequent sections. However, we note here that a key distinction must be made between *bulk* analyses—those that are made on relatively large samples, usually chemically treated to concentrate the elements of interest, and *in situ* analyses—those that sample specific locations within a solid sample. For bulk analyses, the sample may be a single mineral grain a few hundred microns in size up to several grams of a whole meteorite. The sample is first dissolved in pure acids and other solvents and then chemical techniques (e.g., cation exchange columns) are used to purify the element(s) of interest. For *in situ* measurements, no chemical preparation is necessary; polished rock sections or dispersed mineral grains can be directly sampled by the instrumentation, down to sub-micron spatial scales.

Which technique is used for a specific problem depends on the problem itself, but is largely a question of sample size, abundances of the elements of interest and the required analytical precision. Because fundamentally this work involves counting atoms, the ultimate precision depends on the number of atoms that can be measured in a given sample. Since the magnitude of a radiogenic isotope effect depends on the parent/daughter ratio (e.g., Fig. 10.5), the lower this ratio is, the higher a precision is required and thus the larger number of atoms must be counted. In general, the highest precisions can be obtained with bulk techniques, but the attainable precision of modern *in situ* instrumentation has greatly increased in recent years, any many problems can now be addressed at smaller scales than was historically possible.

Most of the mass spectrometers described below use a magnetic-sector design for mass filtering the ion beam, often with an additional electrostatic analyser (ESA) for double-focusing the beam in energy and mass. For example, Fig. 10.6 shows a schematic diagram of the NanoSIMS 50L ion microprobe (Cameca Instruments)

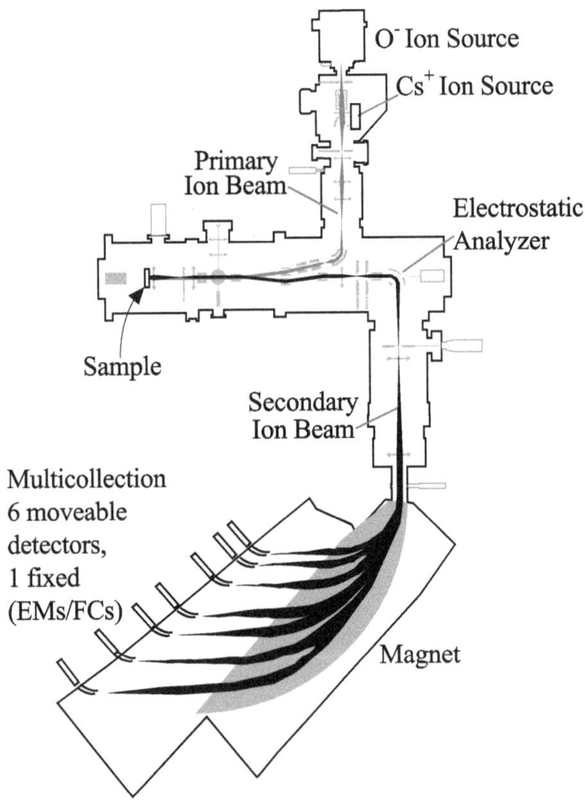

Fig. 10.6 Schematic illustration of Cameca Instruments NanoSIMS 50L ion microprobe

used for many *in situ* measurements of pre-solar grains and other meteoritic components. In this double-focusing mass spectrometer, the ESA is used to disperse the beam of ions from the sample according to kinetic energy and an electromagnet is used to disperse the beam in mass/charge ratio. The ESA and magnet are matched such that ions of a given mass/charge ratio are brought to the same focal point regardless of their kinetic energy.

The most common detectors used for meteoritic mass spectrometry are Faraday Cups (FCs) and electron multipliers (EMs). FCs measure the total charge deposited in a small conductive cup; they have essentially 100% detection efficiency but electronic noise requires minimum ion currents of $>10^4 - 10^5\,s^{-1}$ for accurate measurements. In an EM, a single ion triggers an exponentially increasing cavalcade of electrons, which generates a measurable electronic pulse. EMs are thus used for low-counting-rate situations. However, in general the detection efficiency of an EM is less than unity, varies with element (and isotope) and changes with time as it is bombarded with more and more ions (*aging*). These problems limit the ultimate precision that may be attained by measurements that use EMs, compared to those that use FCs. Some mass spectrometers are built with a single ion detector. In these systems, the current in the magnet is repeatedly changed (*peak-jumping*) in order to focus different isotopes onto the detector. Repeated cycles of peak-jumping are used to build up statistics for the isotopes of interest and calculate ratios. In contrast, in a *multi-collection* system (e.g., Fig. 10.6), the magnet spreads the ion beam out along its focal plane so that different isotopes are measured simultaneously in a set of detectors. Multi-collection both decreases the time required for analyses and improves the accuracy of measurements compared to peak-jumping.

Other types of mass spectrometry, including accelerator mass spectrometry (AMS), gas-source mass spectrometry, and several types of organic mass spectrometry, while used for some meteorite research, are not used commonly for research related to the topics of this book and are not discussed further here. The exception is AMS, used for nuclear reaction rates, and described in Chap. 9, and for analysis of material samples of solar-system probes that hold traces of cosmic radioactivity (see Fig. 10.7; discussed in Chaps. 4 and 7).

10.2.2 Bulk Techniques

The most commonly-used methods for bulk isotopic analysis of meteoritic components are Thermal Ionisation Mass Spectrometry (TIMS) and Multi-Collector Inductively-Coupled Mass Spectrometry (MC-ICPMS). In TIMS, the element to be analysed is chemically separated from the sample, and deposited from aqueous solution onto a metal (commonly W, Re, or Ta) filament. As the filament is heated by an electric current, atoms from the sample boil off and become ionised. The ions then pass through a magnetic mass spectrometer and are counted, usually by a multi-collection system of Faraday Cups. TIMS has the advantages of very high transmission of ions to the detectors as well as producing very stable beams,

Fig. 10.7 A sample of ocean crust from the deep Pacific ocean. Such a crust grows slowly but steadily from material sedating at a few mm per 1000 years. The fraction of radioactive ^{60}Fe nuclei was enough so that with sensitive accelerator mass spectrometry its detection was possible, and lead to the enrichment history diagram shown in the inset, using depth dating with radioactivity of ^{10}Be and ^{53}Mn (Knie et al. 2004)

especially for elements with relatively low ionisation potentials, such as Mg, Cr, Fe, Ni, Sr, Pb and many others. For example, the original discovery of extinct ^{26}Al in the Solar System was made by TIMS measurement of Mg isotopes (Gray 1974; Lee et al. 1976). However, the thermal ionisation processes introduces a mass fractionation, such that lighter atoms more easily get ionised, and the data must be corrected for this. Quantification of concentration ratios of different elements (needed to determine radiometric ages and abundances of extinct isotopes, see above) is accomplished by isotope dilution, where a known amount of artificially-enriched isotopic tracers are added to the sample and measured along with the isotopes of interest.

In bulk MC-ICPMS analysis, drops of a solution of purified element are sprayed into an Ar plasma at very high temperature. The plasma evaporates the droplets and ionises the sample atoms with high efficiency and the ions are then passed through a multi-collector double-focusing mass spectrometer. MC-ICPMS has the advantages of high ionisation efficiency for all elements, including some elements difficult to measure by TIMS (Lee and Halliday 1995), as well as relatively rapid data collection. As in TIMS, mass fractionation effects are significant and must be corrected for, often by measuring standards with known isotopic composition.

10.2.3 In Situ Techniques

In addition to being used for bulk analyses, MC-ICPMS instruments can be equipped with a laser system for *situ* analysis. A UV laser (e.g., 193 nm) is focused into a ∼10–100 μm spot on a sample, ablating material that is then transferred into the Ar plasma. High precisions can be obtained, but this method has the disadvantage that all elements in the sample are transferred to the plasma and ionised, leading to the possibility of unresolved isobaric interferences in mass spectra. Nonetheless, the technique has been successfully applied to studies of extinct radioactivity in meteorites (Young et al. 2005).

The most widely used *in situ* mass spectrometric technique for meteoritic research is Secondary Ion Mass Spectrometry (SIMS, Fig. 10.6). In SIMS, a beam of primary ions is focused onto a solid sample in the presence of a strong electric field. The primary beam sputters atoms from the surface, some of which get ionised and transferred into a mass spectrometer. Either a Cs^+ beam is used to generate negative secondary ions of electronegative elements (e.g., C, O, S, etc.) or an O^- beam is used to generate positive ions of electropositive species (e.g., Mg, Ti etc.). SIMS combines high sensitivity with high spatial resolution to allow isotopic ratios of even minor elements to measured in very small samples and was crucial in the discovery and detailed isotopic characterisation of pre-solar grains. Modern SIMS instruments, e.g., the Cameca NanoSIMS (Fig. 10.6) and ims-1280 ion probes, include very high transmission of secondary ions and multi-collection detector systems. In the case of the NanoSIMS, a resolution of <100 nm can be attained with a Cs^+ source, compared to the best resolution of ∼1 μm with previous instruments. These characteristics allow higher-precision data to be acquired on smaller spatial scales than was ever possible before, allowing, for example, detailed isotopic characterisation of sub-micron pre-solar grains (Nguyen and Zinner 2004; Zinner et al. 2005) as well as high-precision ^{26}Al systematics on small scales in meteoritic inclusions (Kita et al. 2012) and in cometary samples returned by NASA's Stardust mission (Matzel et al. 2010).

An extraordinarily useful feature of SIMS instruments is the ability to produce mass-filtered images of a sample. Two types of SIMS imaging are commonly used in meteoritic research. In *direct imaging,* the instrument behaves as a microscope: the primary beam is defocused on the sample and the ion optics of the instrument transfer an image of the surface through the mass spectrometer onto an imaging detector (e.g., a micro-channel plate or a CCD). The spatial resolution of this technique is set by the ion optics and is typically >500 nm. In *scanning* or *raster* imaging, a focused beam is scanned over the sample with synchronised collection of secondary ions; spatial resolution is set by the size of the primary beam (>50 nm in the NanoSIMS). Both techniques are extremely useful for scanning large numbers of meteoritic grains to search for isotopically anomalous, pre-solar grains of stardust (Nguyen and Zinner 2004; Nittler et al. 1994; Nagashima et al. 2004). Examples of NanoSIMS raster imaging are shown in Fig. 10.8. In the top panels, O isotopic ratio images of an area of a primitive meteorite clearly reveal two sub-micrometer

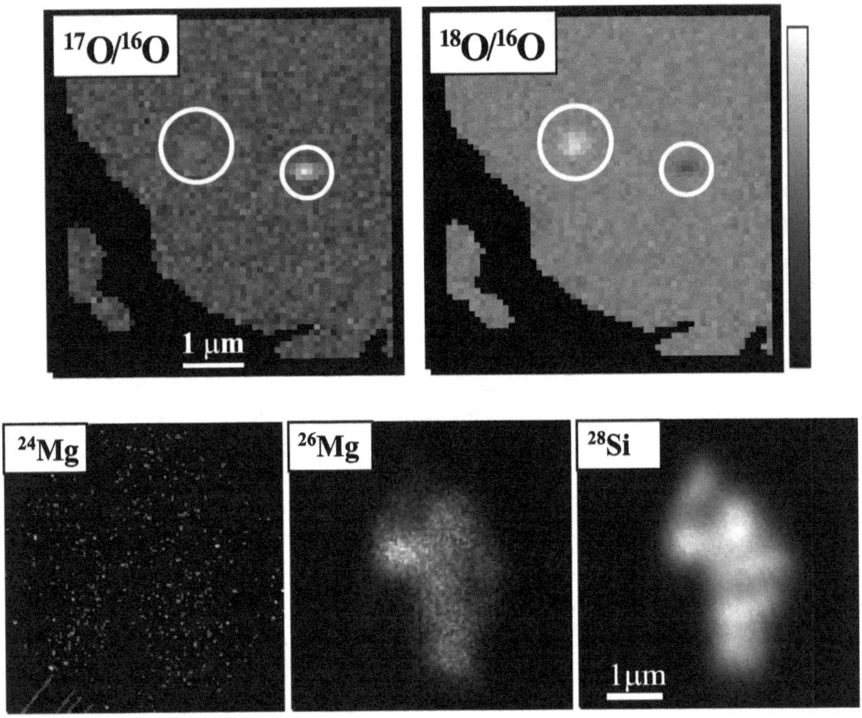

Fig. 10.8 Example NanoSIMS isotopic images of presolar grains. Top Panels: Isotopic images of a 5 × 5 μm^2 area of a primitive meteorite; circles indicate highly anomalous presolar grains surrounded by isotopically normal (solar-composition) material (Nguyen et al. 2010). Bottom panels: Isotopic images of a pre-solar SiC grain from a supernova. Mg in this grain is mono-isotopic ^{26}Mg, due to decay of ^{26}Al (Nittler et al. 2007)

grains with anomalous ^{17}O and ^{18}O abundances (circled). In the bottom panels, a SiC grain believed to have originated from a supernova is revealed to have mono-isotopic ^{26}Mg, from *in situ* decay of ^{26}Al (the inferred ^{26}Al/^{27}Al ratio for this grain is ∼0.5).

Additional *in situ* techniques used in meteoritic radioactivity research include laser-heating noble-gas analyses and Resonance Ionisation Mass Spectrometry (RIMS). In the former, a sample is melted with a laser, releasing atoms of noble gases trapped within. These gases are collected, ionised by bombarding them with electrons and measured with a mass spectrometer. This technique has been successfully applied to He and Ne isotopes in individual pre-solar grains of graphite and SiC, in some cases providing evidence of extinct ^{22}Na (Heck et al. 2007). RIMS uses lasers to ablate material from samples and then to selectively ionise atoms of a specific element; these are measured by means of a time-of-flight mass spectrometer. This technique has extraordinary sensitivity for the selected element and eliminates isobaric interferences that precludes measurement of some isotopes by SIMS. It has provided extremely useful isotopic data for trace elements within single pre-solar grains, including Zr, Mo, Ba, and Ru (Nicolussi et al. 1997; Savina et al. 2004).

A fundamental question for analysis of small samples like pre-solar grains is whether a given isotopic signature might be detectable in a given sample. Because *in situ* isotopic measurements are governed by Poisson statistics, the answer depends on the total number of atoms of the isotope of interest that can be detected, which clearly depends on the specific problem. The number of detectable atoms depends on the abundance of the element in the sample (determined by chemistry), the size of the sample (giving the total number of atoms) and the efficiency of the instrument (giving the fraction of atoms in a sample that can actually be detected and counted). For SIMS, the latter efficiency typically ranges from 10^{-7} to 10^{-2} and depends both on the low ionisation probability of secondary ions during the sputtering process and the efficiency of transmitting ions through the mass spectrometer. As an example, let us consider the detection by SIMS of extinct ^{26}Al in pre-solar SiC grains. A 1-μm SiC grain contains some 5×10^{10} atoms. Assuming an initial ^{26}Al/^{27}Al ratio of 10^{-3}, a typical Al concentration of 1% in pre-solar SiC, and a detection efficiency of 10^{-2}, consuming an entire 1-μm grain would yield some 5000 atoms of radiogenic ^{26}Mg, corresponding to a statistical uncertainty of ∼1.4%. Measuring only a fraction of the grain (to preserve it for additional isotopic analyses, for instance), or smaller grains, or grains with lower initial ^{26}Al contents would correspondingly lower the number of detected atoms and decrease the precision. Moreover, whether the radiogenic signature is observable depends also on the amount of non-radiogenic Mg present in the grain. Clearly, similar considerations must be made for any given problem at hand.

10.2.4 Perspectives for Astronomy with Meteorite Samples

As in many fields, the technology for chemical and isotopic analysis of geological and cosmo-chemical materials is rapidly evolving and improving and this will certainly continue into the future. Recent improvements in the stability of commercial instruments and advances in chemical treatments used to purify samples are now providing for extremely high precision measurements of bulk samples by TIMS and ICP-MS (Brennecka et al. 2010; Budde et al. 2016) and of intermediate-scale (tens of microns) *in situ* analyses by SIMS (Villeneuve et al. 2009). For analysis of tiny samples like pre-solar grains, significant effort is being expended to improve both sensitivity and spatial resolution. While the NanoSIMS has long achieved sub-100 nm spatial resolution with a Cs^+ primary ion source, the duoplasmatron commonly used to generate O^- beams can achieve at best ∼400–500 nm. However, a new RF plasma-based O^- ion source, developed by Oregon Physics, LLC, can match the spatial resolution of the Cs^+ beam and has been installed on several ion microprobes (Matzel et al. 2014; Liu et al. 2018). This source will greatly improve measurements of key radionuclides like ^{26}Al, ^{41}Ca, and ^{60}Fe in ever-smaller meteoritic samples and pre-solar grains.

While SIMS has the advantage of achieving high spatial resolution, it is fundamentally limited by the low ionisation probability for secondary ions during sputtering. In contrast, previous RIMS instruments have proven to have extraordinary sensitivity for certain elements, but limited spatial resolution due to the use of lasers to ablate atoms from samples. A new instrument, named CHILI (CHicago Instrument for Laser Ionisation), has been developed in recent years at the University of Chicago (Stephan et al. 2016) and combines a very high resolution sputtering Ga^{+} ion gun with multiple lasers for resonance ionisation to achieve both sensitivity and spatial resolution beyond what is currently possible. As of this writing, the instrument is generating important data on pre-solar grains and other meteoritical problems with a laser ablation source (Liu et al. 2017; Trappitsch et al. 2018; Stephan et al. 2018; Kodolányi et al. 2018) while development of the scanning Ga sputtering source is ongoing.

Another approach is to use the front end of a commercial SIMS instrument to produce secondary ions which are then further accelerated to 300–1000 keV energies allowing for highly efficient suppression of molecular interferences, and thus allowing for higher sensitivity. The first such instrument, the *MegaSIMS* constructed at the University of California, Los Angeles (McKeegan et al. 2009), was developed specifically for analysis of solar-wind samples returned by NASA's Genesis mission. More recently, a similar instrument called SIMS-SSAMS has been developed at the United States Naval Research Laboratory (Groopman et al. 2017). With its micron-scale spatial resolution and high sensitivity, this instrument has great promise for addressing many problems in meteoritical science.

A fundamentally different technology, atom probe tomography, also holds great promise for isotopic and analysis of small extraterrestrial samples. In an atom probe, the sample is embedded in a very small, extremely sharp tip with a high potential applied to it. Applying a laser to the tip allows for single atoms to be extracted from the tip, one by one, and transmitted to a mass spectrometer. Reconstruction of atom trajectories allows the 3-dimensional reconstruction of the original positions and chemical identity of up to tens of millions of individual atoms. Technical limitations have thus far mostly limited the use of atom probe tomography to materials science applications, but recent attempts to apply the technique to pre-solar grains (Heck et al. 2014) suggest that it is likely to become a very useful tool for cosmochemistry in coming years.

10.3 Detection and Analysis of Cosmic Rays

The discovery of Galactic cosmic rays was inextricably tied to the development of sub-atomic particle physics in the early decades of the twentieth century. In particular, the discoveries in 1912–1914 by Hess and Kolhörster, through high-altitude balloon flights, of a source of ionisation that increases with altitude clearly indicated an extraterrestrial source of high-energy radiation (see historical review in Longair 1992). Through the decades, increasingly more sophisticated instruments and experiments have been developed and deployed to determine the composition and energy spectra of cosmic rays. Because primary cosmic rays cannot penetrate the Earth's thick atmosphere to be detected at the ground, balloon-borne and space-based experiments remain crucial to progress in cosmic ray physics to the present day. However, due their very low flux, ground-based telescopes are still required to detect the highest-energy cosmic rays, through the *air showers* of secondary particles they produce as they traverse the atmosphere. This section describes some of the basic techniques used to detect and characterise GCRs both from the ground and at high altitudes.

10.3.1 Ground-Based Observations

When Galactic cosmic rays enter the Earth's atmosphere, they interact with its atoms and molecules, generating a cascade of secondary particles. For most cosmic rays, these cascades overlap and lead to a relatively constant flux at ground level of secondary particles, dominated by muons. In contrast, air showers are caused by single primary cosmic rays with sufficient energy that their secondary particle cascades are detectable at ground level. Modern ground-based cosmic-ray telescopes are comprised of vast areas of particle and light detectors, designed to detect and characterise as many secondary particles as possible and hence reconstruct the characteristics of the original high-energy cosmic ray. For example, the largest such facility in the world, the Pierre Auger Observatory, is planned to ultimately include sites in Argentina and Colorado; the southern site is operational. It uses two common types of air shower detectors: an array of 1600 water tank detectors spaced out over some 3000 km^2 surrounded by four inward-looking *fluorescence detectors* (FDs) (The Pierre Auger Collaboration et al. 2007). The water tanks detect high-energy particles through the Cherenkov light they emit while passing through the tanks (essentially an electromagnetic sonic boom since the speed of light in the water is smaller than the particle velocities). The FDs consist of telescopes designed to detect UV fluorescent light from interaction of the shower particles with atmospheric nitrogen. Together, the two types of detectors allow reconstruction of the initial energy and direction of the primary cosmic ray. Similar techniques are used in other large air shower experiments, such as HiRes (Abbasi et al. 2004, 2008) and AGASA (Takeda et al. 2003).

10.3.2 High-Altitude and Space-Based Observations

As a charged particle passes through matter, it loses energy through ionisation of atoms and molecules of the material. These ionisation losses can cause significant damage to crystal structures and molecular chains. Many early studies of the composition of cosmic rays took advantage of this radiation damage in certain materials, such as many plastics as well as minerals in meteorites and lunar samples. The damaged areas have much higher chemical reactivity than the undamaged material. Thus, chemical etching of the materials reveals *tracks*, the nature (e.g., size) of which can be empirically related to the identities and energies of the incident particles (Fleisher et al. 1975). This technique was used starting in the late 1960s with balloon-borne plastic detectors to identify elements heavier than Fe, including heavy radioactive elements like U, in the primary GCRs (Blanford et al. 1969).

Because of limitations in the track technique (e.g., relatively poor resolution in charge identification), modern determinations of the composition of Galactic cosmic rays are based on electronic measurements of the total energy and energy loss rate as incident particles pass through various combinations of detectors. The rate of energy loss, −dE/dx, is proportional to the square of the particle's charge and depends also on its velocity as well as properties of the material (the Bethe-Bloch formula, Longair 1992). If both −dE/dx and the total kinetic energy can be independently determined, the charge and mass of the particle can be inferred. In practice, stacks of detectors are used to determine these parameters and additional detectors can be used to refine particle trajectories and exclude backgrounds. Examples of modern cosmic-ray experiments are given below.

Detectors used for cosmic-ray experiments commonly include: *solid-state detectors*, Si or Ge crystals in which the incident radiation releases electron-hole pairs that can be measured as an electrical pulse; *scintillation detectors*, crystals (e.g., NaI or certain plastics) in which light is produced by the incident radiation and detected by photomultiplier tubes; *Cherenkov detectors,* a type of scintillator for which particles above a threshold energy emit detectable light; and *transition radiation detectors,* stratified detectors with different indices of refraction; X-rays are emitted when particles cross the interfaces, with the amount of radiation dependent on the velocity of the particle.

A wealth of data on the isotopic and elemental composition of GCR nuclei has been obtained in the last two decades by the Cosmic Ray Isotope Spectrometer (CRIS, Stone et al. 1998) on the *Advanced Composition Explorer* satellite, launched by NASA in 1997 and still operational as of this writing (Binns et al. 2016). A schematic diagram of the CRIS instrument is shown in Fig. 10.9 (from George et al. 2009). The instrument consists of a position sensitive *hodoscope* detector (HNX,Y) to determine particle trajectories and stacks of solid-state Si detectors (E1–E9) to measure energy and energy loss rate. The hodoscope is constructed of layers of criss-crossed scintillating plastic fibers coupled to image-intensified CCD detectors. As a particle passes through the layers, the patterns of light observed on the CCDs can be inverted to determine the trajectory of the particle through the instrument with high precision. Additional layers of scintillating fibers at the top (TX,TY) serve as a trigger to signal the instrument that a particle has arrived. The arrow indicates the trajectory of a particle that enters the instrument and is finally stopped in the Si layer E9. For this event, the signals measured in detectors E1–E6 provide a measure of the energy loss rate (−dE/dx) whereas the total energy is found from that deposited in all seven detectors.

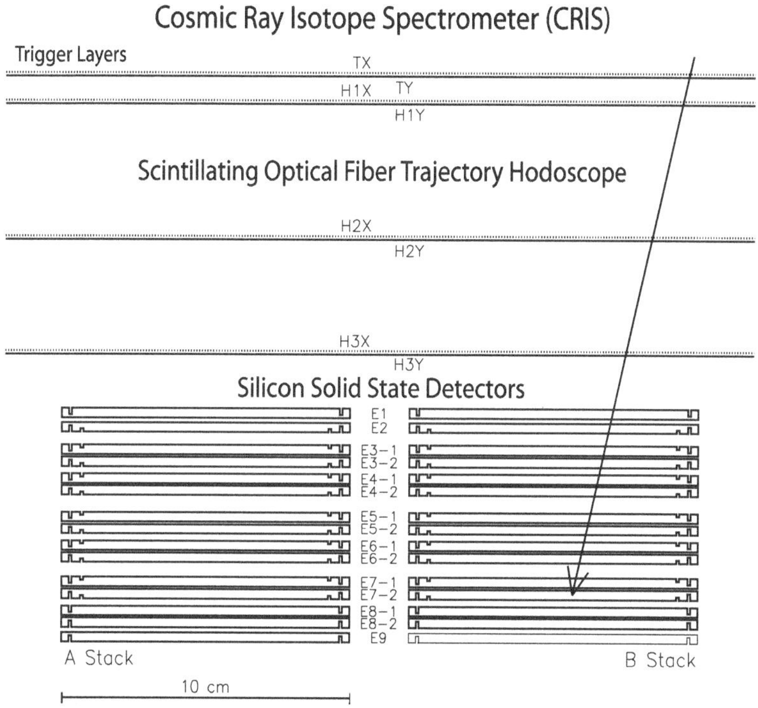

Fig. 10.9 Schematic of CRIS instrument on the *Advanced Composition Explorer* satellite, from George et al. (2009). Arrow indicates the trajectory of a single particle entering the instrument

The process of particle discrimination in CRIS is illustrated in Fig. 10.10 (Stone et al. 1998), showing data obtained with a CRIS Si detector stack from a laboratory calibration experiment. Each data point represents the energy lost in detectors E1 through E3 plotted against the energy deposited in E4 for a single particle traversing the detector stack. Curves corresponding to the distinct isotopes of the indicated elements are clearly seen.

Other modern GCR composition instruments and experiments are designed around similar principles though with widely varying details. For example, the Trans-Iron Galactic Element Recorder (TIGER) experiment, flown on high-altitude balloons in Antarctica in 2001 and 2003, used a combination of scintillation and Cherenkov detectors to determine the composition of GCR with atomic number $26 < Z < 38$ (Rauch et al. 2009). To determine the composition of GCR at higher energies than studied by the experiments described above, the Transition Radiation Array for Cosmic Energetic Radiation (TRACER) instrument uses a combination of Cherenkov, scintillator, gas-filled proportional counter and transition radiation detectors (Ave et al. 2008).

The instruments described above are designed to detect and characterise the composition of GCR nuclei. As discussed in Chap. 7, cosmic-ray positrons also

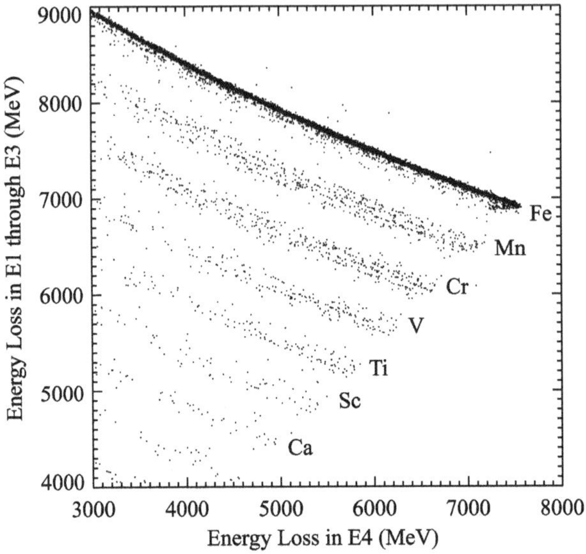

Fig. 10.10 Calibration data for CRIS: plot of energy deposited in Si detectors E1–E3 versus that in detector E4. Particles of a given mass and charge plot along distinct curves, allowing isotope discrimination. Figure taken from Stone et al. (1998)

provide important information regarding the origin and interactions of GCRs in the Galaxy. In 2009 it was reported that the positron fraction of cosmic rays increases with increasing energy, inconsistent with purely secondary sources, based on data from the PAMELA instrument on board the Russian Resurs-DK1 satellite, launched in 2006 (Adriani et al. 2009). PAMELA is broadly similar to other instruments designed to identify positrons (e.g., HEAT, Barwick et al. 1997) and consists of scintillator detectors to measure time of flight and dE/dx for traversing particles and to distinguish upward-traveling electrons from downward-traveling positrons, a magnetic spectrometer to measure the rigidity (momentum per unit charge) of the particles, and an electromagnetic calorimeter consisting of interleaved layers of W absorbers and solid-state Si detectors. The calorimeter allows positrons to be distinguished from protons and anti-protons from electrons in the instrument. More recently, this positron excess has been confirmed and clarified with the Alpha Magnetic Spectrometer that operated on the International Space Station in 2011–2013 (Aguilar et al. 2013; Accardo et al. 2014).

References

Abbasi RU, Abu-Zayyad T, Amann JF et al (2004) Phys Rev Lett 92:151101
Abbasi RU, Abu-Zayyad T, Allen M et al (2008) Phys Rev Lett 100:101101
Accardo L, Aguilar M, Aisa D et al (2014) Phys Rev Lett 113:121101

Adriani O, Barbarino GC, Bazilevskaya GA et al (2009) Nature 458:607
Aguilar M, Alberti G, Alpat B et al (2013) Phys Rev Lett 110:141102
Atwood WB, Abdo AA, Ackermann M et al (2009) Astrophys J 697:1071
Ave M, Boyle PJ, Gahbauer F et al (2008) Astrophys J 678:262
Barwick SW, Beatty JJ, Bhattacharyya A et al (1997) Astrophys J 482:L191
Binns WR, Israel MH, Christian, ER et al (2016) Science 352:677
Blanford GE, Friedlander MW, Klarmann J et al (1969) Phys Rev Lett 23:338
Boggs S, Kurfess J, Ryan J et al (2006) Presented at the Society of Photo-Optical Instrumentation Engineers (SPIE) conference. Society of Photo-Optical Instrumentation Engineers (SPIE) conference series, vol 6266
Brennecka GA, Weyer S, Wadhwa M et al (2010) Science 327:449
Budde G, Burkhardt C, Brennecka GA et al (2016) Earth Planet Sci Lett 454:293
De Angelis A, Tatischeff V, Tavani M et al (2017) Exp Astron 44:25
Diehl R, Siegert T, Greiner J et al (2017) ArXiv e-prints, 1710.10139
Fleisher RL, Price PB, Walker RM (1975) Nuclear tracks in solids: principles and applications. University of California Press, Berkeley
Forrest DJ, Chupp EL, Ryan JM et al (1980) Sol Phys 65:15
George JS, Lave KA, Wiedenbeck ME et al (2009) Astrophys J 698:1666
Gray CM (1974) Nature 251:495
Greiner J, Iyudin A, Kanbach G et al (2009) Exp Astron 23:91
Groopman EE, Grabowski KS, Fahey AJ, Koop L (2017) J Anal At Spectrom 32:2153
Harrison FA, Craig WW, Christensen FE et al (2013) Astrophys J 770:103
Heck PR, Marhas KK, Hoppe P et al (2007) Astrophys J 656:1208
Heck PR, Stadermann FJ, Isheim D et al (2014) Meteorit Planet Sci 49:453
Kanbach G, Bertsch DL, Favale A et al (1989) Space Sci Rev 49:69
Kanbach G, Andritschke R, Bloser PF et al (2003). In: Truemper JE, Tananbaum HD (eds) Presented at the Society of Photo-Optical Instrumentation Engineers (SPIE) conference. Society of Photo-Optical Instrumentation Engineers (SPIE) conference series, vol 4851, pp 1209–1220
Kierans CA, Boggs SE, Chiu J-L et al (2017) ArXiv e-prints, 1701.05558
Kita NT, Ushikubo T, Knight KB et al (2012) Geochim Cosmochim Acta 86:37
Knie K, Korschinek G, Faestermann T et al (2004) Phys Rev Lett 93:171103
Knödlseder J (2007) Adv Space Res 40:1263
Kodolányi J, Stephan T, Trappitsch R et al (2018) Geochim Cosmochim Acta 221:127
Lee D, Halliday AN (1995) Nature 378:771
Lee T, Papanastassiou DA, Wasserburg GJ (1976) Geo Res Lett 3:41
Liu N, Stephan T, Boehnke P et al (2017) Astrophys J 844:L12
Liu M-C, McKeegan KD, Harrison TM, Jarzebinski G, Vltava L (2018) Int J Mass Spectrom 424:1
Longair MS (1992) High energy astrophysics (1992) Vol. 1: Particles, photons and their detection (High energy astrophysics, by MS Longair. Cambridge University Press, Cambridge, pp. 436. ISBN 0521387736
Matzel JEP, Ishii HA, Joswiak D et al (2010) Science 328:483
Matzel JEP, Ishii HA, Joswiak D, Brownlee D, Hutcheon ID (2014) Lunar and Planetary Institute Technical Report. Lunar and planetary science conference, vol 45, p 1645
McEnery JE (2017) AAS/High energy astrophysics division, vol. 16. AAS/High Energy Astrophysics Division, 103.13
McKeegan KD, Kallio AP, Heber V et al (2009) Lunar and Planetary Institute Science conference abstracts. Lunar and Planetary Institute Science conference abstracts, vol 40, p 2494
Nagashima K, Krot AN, Yurimoto H (2004) Nature 428:921
NCT Collaboration, Boggs S, Chang Y (2007) Adv Space Res 40:1281
Nguyen AN, Zinner E (2004) Science 303:1496
Nguyen AN, Nittler LR, Stadermann FJ, Stroud RM, Alexander CMO (2010) Astrophys J 719:166
Nicolussi GK, Davis AM, Pellin MJ et al (1997) Science 277:1281
Nittler LR, Alexander CMO'D, Gao X, Walker RM, Zinner EK (1994) Nature 370:443

Nittler LR, Hoppe P, Stroud RM (2007) Lunar and planetary science conference 38, Abstract #2321

Rauch BF, Link JT, Lodders K et al (2009) Astrophys J 697:2083

Savina MR, Davis AM, Tripa CE et al (2004) Science 303:649

Schönfelder V, Aarts H, Bennett K et al (1993) Astrophys J Suppl 86:657

Stephan T, Trappitsch R, Davis AM et al (2016) Int J Mass Spectrom 407:1

Stephan T, Trappitsch R, Davis AM et al (2018) Geochim Cosmochim Acta 221:109

Stone EC, Cohen CMS, Cook WR et al (1998) Space Sci Rev 86:285

Takahashi T, Awaki A, Dotani T et al (2004). In: Hasinger G, Turner MJL (eds) Proceedings of SPIE. UV and gamma-ray space telescope systems, vol 5488, pp 549–560

Takahashi T, Kelley R, Mitsuda K et al (2009). In: Kawai N, Mihara T, Kohama M, Suzuki M (eds) Astrophysics with all-sky X-ray observations, p 356

Takeda M, Sakaki N, Honda K et al (2003) Astropart Phys 19:447

The Pierre Auger Collaboration, Abraham J, Abreu P et al (2007) Science 318:938

Trappitsch R, Stephan T, Savina MR et al (2018) Geochim Cosmochim Acta 221:87

Vedrenne G, Roques J, Schönfelder V et al (2003) Astron Astrophys 411:L63

Vestrand WT, Share GH, Murphy RJ et al (1999) Astrophys J Suppl 120:409

Villeneuve J, Chaussidon M, Libourel G (2009) Science 325:985

Weidenspointner G, Harris MJ, Sturner S, Teegarden BJ, Ferguson C (2005) Astrophys J Suppl 156:69

Young ED, Simon JI, Galy A et al (2005) Science 308:223

Zinner E, Nittler LR, Hoppe P et al (2005) Geochim Cosmochim Acta 69:4149

Zych AD, O'Neill TJ, Bhattacharya D et al (2006) Presented at the Society of Photo-Optical Instrumentation Engineers (SPIE) conference. Society of Photo-Optical Instrumentation Engineers (SPIE) conference series, vol 6319

Chapter 11
Cosmic Evolution of Isotopic Abundances: Basics

Roland Diehl and Nikos Prantzos

The description of the tempo-spatial evolution of the composition of cosmic gas on galactic scales is called 'galactic chemical evolution'. It combines the knowledge about cosmic sources of nuclei (that is their internal workings and nucleosynthesis yields, and their properties such as frequency of occurrence and spatial distribution), with knowledge about the formation and evolution of these sources in the greater context of a galaxy, as well as transport processes of gas within galaxies. It provides a useful framework, allowing us to interpret the large amount of observational data concerning the chemical composition of stars, galaxies and the interstellar medium.

11.1 Modeling Compositional Evolution of Cosmic Gas

Nucleosynthesis events produce new isotopes, which are mixed with ambient gas to then end up in new generations of stars, which again lead to nucleosynthesis events (see Fig. 11.1). This cycle began from *first stars* (*Population III stars)* created from almost *metal-free primordial* gas, and has since continued to form stars until today.[1]

---

[1] Stars of the Galactic disk with ages comparable to, or much younger than, the Sun (4.5 Gy) are called *Population I*. They are the only stellar population which contains massive (hence short-lived) stars observable today. Stars of the Galactic halo are much older (>10 Gy) and are called *Population II*.

R. Diehl (✉)
Max Planck Institut für extraterrestrische Physik, Garching, Germany
e-mail: rod@mpe.mpg.de

N. Prantzos
Institut d'Astrophysique, Paris, France
e-mail: prantzos@iap.fr

R. Diehl et al. (eds.), *Astrophysics with Radioactive Isotopes*, Astrophysics and Space Science Library 453, https://doi.org/10.1007/978-3-319-91929-4_11

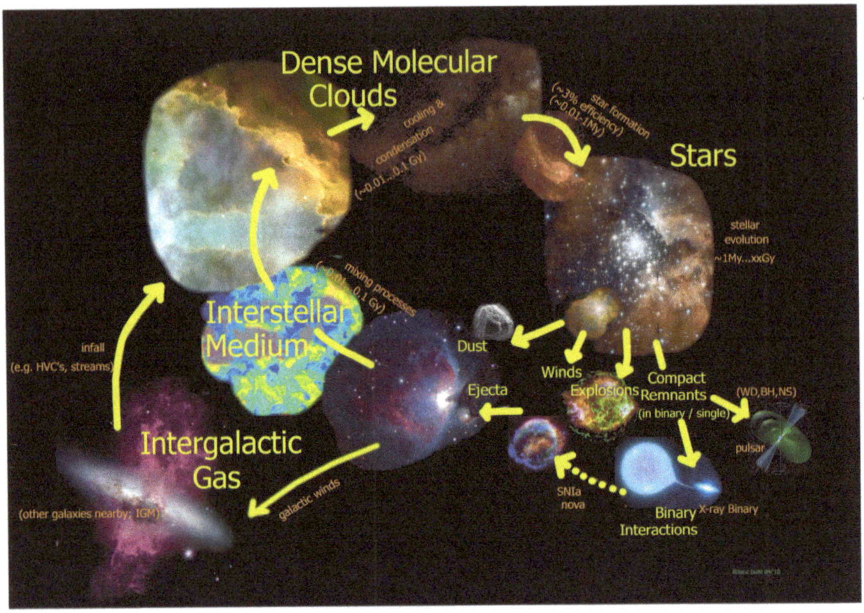

Fig. 11.1 Illustration of the cycle of matter. Stars form from molecular clouds, and eventually return gas enriched with nucleosynthesis products into interstellar space

Star formation, evolution, and nucleosynthesis all vary with changing metal content. It is the challenge of *chemical evolution models* to account for the complex and various astrophysical processes in a suitably-summarizing description to represent the known astronomical constraints. Radioactivities contribute to those constraints, together with the archaeological memory of metal-poor stars in our Galaxy and various measurements of composition and abundances in specific regions and objects throughout the universe.

Analytical treatments of compositional evolution have been proposed and developed 40 years ago, to *relate the elemental-abundance distribution and their evolution in galaxies to the activity of star formation and its history* (Clayton 1968; Cameron and Truran 1971; Truran and Cameron 1971; Audouze and Tinsley 1976; Tinsley 1980, and many others). The physical processes included herein then received more sophisticated treatments, such as allowing for external gas flows, multiple and independent components of a galaxy, and more complex histories of how different stellar components inject their products into the gas cycle (Clayton 1988; Matteucci and Francois 1989; Pagel 1997; Chiappini et al. 1997; Prantzos and Silk 1998; Boissier and Prantzos 1999; François et al. 2004). A specific and useful standard description can be found in Clayton (1985, 1988).

This provides a framework, called Galactic Chemical Evolution (GCE), in which the rich variety of astronomical abundance data can be exploited to obtain a coherent and consistent description. For example, one may investigate which description best represents the observed distribution of stars of different ages in the solar neighborhood. Comparison of the predictions of such a description with observational data and their uncertainties offers clues as to the plausibility of the model and its parameters. Alternatively, one may optimise parameters of the description using the data and parameter fitting algorithms to use measurements within their statistical precision. This also can be used together with probability theory to judge acceptability, or failure, of a particular description.

11.1.1 The Concept and Formalism

In a galaxy, chemical evolution tracks the amounts and composition of the reservoirs of *gas* and *stars* over time. Key concepts are:

(a) *Gas* is *consumed* by the process of star formation
(b) *Stars* evolve and *return gas* enriched with *metals*
(c) *Gas* (and *stars*) may be *lost* from the galaxy
(d) *Gas* and *stars* may be *accreted* from outside the galaxy

These processes are traced through relations among the different components. Mass conservation reads therefore:

$$m = m_{\rm gas} + m_{\rm stars} + m_{\rm infall} + m_{\rm outflow} \tag{11.1}$$

which includes the mass in stars and in gas as well as *infall* and *outflow* terms.The populations of stars may—sometimes usefully—be subdivided into *luminous* (l) and *inert* (c for 'compact remnants') stars:

$$m_{\rm stars} = m_{\rm l} + m_{\rm c} \tag{11.2}$$

Theoretical and/or empirical prescriptions for astrophysical processes can be introduced to obtain a formalism linking different observational quantities:

1. The birth rate of stars is introduced either empirically or through theories of star formation, linking the birth rate to the (atomic, molecular or total) gas content of a galaxy.
2. The theory of stellar evolution allows one to track the stellar population over time, from stellar birth to death and formation of compact stellar remnants.
3. The nucleosynthesis yields of stars in their different evolutionary phases are obtained from stellar nucleosynthesis models and introduced in the model of galactic chemical evolution, leading to progressive enrichment of the gas with metals.
4. The evolving composition of gas is monitored, in principle as a function of time, in practice (since stellar ages are difficult to evaluate) as a function of a "proxy", i.e. an abundant and easy to observe element, like Fe.

5. The, yet poorly understood, dynamics of the gas may affect considerably the overall scheme, i.e. the efficiency of star formation, the distribution of the produced metals in the various gas phases, the preferential ejection of metals from the system, etc.

The observational quantities constraining the models are:

1. Number counts or densities of stars in intervals of intrinsic stellar properties (age, luminosity or metallicity). The results are affected by uncertainties in stellar evolution theory and stellar initial mass function.
2. Abundances of elements or isotopes, in different locations or galaxy components (stars, gas), and for different stellar ages or metallicities.

Within the framework provided by the adopted model of galactic chemical evolution, the various parameters listed above can be adjusted, under observational constraints, in order to end up with a consistent and plausible description of the physical system.

In such a GCE description, a galaxy consists initially of gas of primordial composition, $X_H \sim 0.75$ for H and $X_{He} \sim 0.25$ for ^{4}He, as well as trace amounts of D, ^{3}He and ^{7}Li (abundances are given as *mass fractions* X_i for element or isotope i, with $\Sigma_i X_i = 1$). The gas is progressively turned into stars with a *Star Formation Rate (SFR)* $\Psi(t)$, with the star masses M having a distribution $\Phi(M)$, called the *Initial Mass Function (IMF)*.[2] Depending on its lifetime τ_M, the star of mass M created at time t dies at time $t + \tau_M$ and returns a part of its mass to the interstellar medium (ISM), either through stellar winds (in the case of low mass and intermediate mass stars) or through supernova explosions (in the case of massive stars.[3]) The ejected material is enriched in elements synthesized by nuclear reactions in the stellar interiors, while some fragile isotopes (like D) are absent from its composition. Thus, the ISM is progressively enriched in elements heavier than H, while its D content is reduced. New stellar generations are formed from this ISM, their composition being progressively more enriched in heavy elements, i.e. with an ever increasing *metallicity* Z (where $Z = \Sigma X_i$ for all elements i heavier than He).

In the framework of the simple model of GCE it is assumed that the *stellar ejecta* are immediately and efficiently mixed in the ISM.[4] As a result, the ISM is characterized at every moment by a unique chemical composition $X_i(t)$, which is

[2]In principle, the IMF may depend on time, either explicitly or implicitly (i.e. through a dependence on metallicity, which increases with time); in that case one should adopt a Star Creation Function $C(t, M)$ (making the solution of the GCE equations more difficult). In practice, however, observations indicate that the IMF does not vary with the environment, allowing to separate the variables t and M and adopt $C(t, M) = \Psi(t)\Phi(M)$.

[3]Massive stars also eject part of their mass through a wind, either in the red giant stage (a rather negligible fraction) or in the Wolf-Rayet stage (an important fraction of their mass, in the case of the most massive stars).

[4]This is the so called *Instantaneous Mixing Approximation*, not to be confused with the *Instantaneous Recycling Approximation (IRA)*, to be discussed in Sect. 11.1.3.3.

also the composition of the stars formed at time t. Since the surface composition of stars on the Main Sequence is not affected, in general, by nuclear reactions,[5] observations of stellar abundances reveal, in principle, the composition of the gas of the system at the time when those stars were formed. One may thus recover the chemical history of the system and confront observations to models of GCE.

The GCE scenario sketched in the previous paragraphs can be quantitatively described by a set of integro-differential equations (see Tinsley 1980):

The *evolution of the total mass of the system m(t)* is given by:

$$\frac{dm}{dt} = [\, f - o\,] \tag{11.3}$$

If the system evolves without any input or loss of mass, the right hand member of Eq. (11.3) is equal to zero; this is the so called *Closed Box Model*, the simplest model of GCE. The terms of the second member within brackets are optional and describe *infall* of extragalactic material at a rate $f(t)$ or *outflow* of mass from the system at a rate $o(t)$; both terms will be discussed in Sect. 11.1.3.2.

The *evolution of the mass of the gas* $m_G(t)$ of the system is given by:

$$\frac{dm_G}{dt} = -\Psi + E + [\, f - o\,] \tag{11.4}$$

where $\Psi(t)$ is the Star Formation Rate (SFR) and $E(t)$ is the *Rate of mass ejection by dying stars*, given by:

$$E(t) = \int_{M_t}^{M_U} (M - C_M)\, \Psi(t - \tau_M)\, \Phi(M)\, dM \tag{11.5}$$

where the star of mass M, created at the time $t - \tau_M$, dies at time t (if $\tau_M < t$) and leaves a Compact object (white dwarf, neutron star, black hole) of mass C_M, i.e. it ejects a mass $M - C_M$ in the ISM. The integral in Eq. (11.5) is weighted by the Initial Mass Function of the stars $\Phi(M)$ and runs over all stars heavy enough to die at time t, i.e. the less massive of them has a mass M_t and a lifetime $\tau_M \leq t$. The upper mass limit of the integral M_U is the upper mass limit of the IMF.

Obviously, the *mass of stars* $m_S(t)$ of the system (live + dead) can be derived through:

$$m = m_S + m_G \tag{11.6}$$

[5] An exception to that rule is fragile D, already burned in the Pre-Main Sequence all over the star's mass; Li isotopes are also destroyed, and survive only in the thin convective envelopes of the hottest stars.

The evolution of the chemical composition of the system is described by equations similar to Eqs. (11.4) and (11.5). The *mass of element/isotope i* in the gas is $m_i = m_G X_i$ and its evolution is given by:

$$\frac{d(m_G X_i)}{dt} = -\Psi X_i + E_i + [\, f X_{i,f} - o X_{i,o}\,] \tag{11.7}$$

i.e. star formation at a rate Ψ removes element i from the ISM at a rate ΨX_i, while at the same time stars re-inject in the ISM that element at a rate $E_i(t)$. If infall is assumed, the same element i is added to the system at a rate $f X_{i,f}$, where $X_{i,f}$ is the abundance of nuclide i in the infalling gas (usually, but not necessarily, assumed to be primordial). If outflow takes place, element i is removed from the system at a rate $oX_{i,o}$ where $X_{i,o}$ is the abundance in the outflowing gas; usually, $X_{i,o}$=X_i, i.e. the outflowing gas has the composition of the average ISM, but in some cases it may be assumed that the hot supernova ejecta (rich in metals) leave preferentially the system, in which case $X_{i,o} > X_i$ for metals.

The *rate of ejection of element i by stars* is given by:

$$E_i(t) = \int_{M_t}^{M_U} Y_i(M)\, \Psi(t - \tau_M)\, \Phi(M)\, dM \tag{11.8}$$

where $Y_i(M)$ is the *stellar yield of element/isotope i*, i.e. the mass ejected in the form of that element by the star of mass M. Note that $Y_i(M)$ may depend *implicitly* on time t, if it is metallicity dependent.

The masses involved in the system of Eqs. (11.3)–(11.8) may be either *physical* masses, i.e. m, m_G, m_S etc. are expressed in $M_\odot$ and $\Psi(t)$, $E(t)$, $c(t)$ etc. in $M_\odot\,Gyr^{-1}$, or *reduced* masses (*mass per unit final mass of the system*), in which case m, m_G, m_S etc. have no dimensions and $\Psi(t)$, $E(t)$, $c(t)$ etc. are in Gyr^{-1}. The latter possibility allows to perform calculations for a system of arbitrary mass and normalise the results to the known/assumed present-day mass of that system; note that instead of mass, one may use volume or surface mass densities.

Because of the presence of the term $\Psi(t - \tau_M)$, Eqs. (11.7) and (11.8) have to be solved numerically (except if specific assumptions, like the Instantaneous Recycling Approximation—IRA—are made). The integral (11.8) is evaluated over the stellar masses, properly weighted by the term $\Psi(t - \tau_M)$ corresponding in each mass M. It is explicitly assumed in that case that *all the stellar masses created in a given place, release their ejecta in that same place.*

This assumption does not hold anymore if stars are allowed to travel away from their birth places before dying (see Appendix C in Kubryk et al. (2015b)). In that case, the mass $E_i(t)$ released in a given place of spatial coordinate R and at time t is the sum of the ejecta of stars born in various places R' and times $t - t'$, with different star formation rates $\Psi(t', R')$ for all stellar masses M with lifetimes $\tau_M < t - t'$. Instead of Eq. (11.8), the *isochrone* formalism, concerning instantaneous "bursts" of star formation or Single Stellar Populations (SSP), has to be used then. Equation (11.8) is re-written as

$$E_i(t) = \int_{\tau_{M_U}}^{t} \Psi(t')dt' \left(\frac{Y_i(M)dN}{dt'}\right)_{t-t'} \tag{11.9}$$

where $dN = \Phi(M)dM$ is the number of stars between M and $M + dM$ and $\Psi(t')dt'$ is the mass of stars (in $M_\odot$) created in time interval dt' at time t'. The term $(dN/dt')_{t-t'}$ represents the stellar death rate (by number) at time t of a unit mass of stars born in an instantaneous burst at time $t - t'$. The term $Y_i(M)dN/dt$ represents the corresponding rate of release of element i in $M_\odot$/year.

Expression (11.9) is equivalent to expression (11.8) and it is used in N-body+SPH simulations since it allows one to account for the ejecta released in a given place by "star-particles" produced with different star formation rates in other places. It incorporates naturally the metallicity dependence of the stellar yields and of the stellar lifetimes, both found in the term $Y_i(M, Z)(dN/dt)(Z)$.

The system of integro-differential equations (11.3)–(11.8) can only be solved numerically. Analytical solutions require that some specific assumptions are made, in particular the Instantaneous recycling approximation (IRA, to be discussed in Sect. 11.1.3.3). Its solution requires three types of ingredients:

- *Stellar properties*: stellar lifetimes τ_M, masses of stellar residues C_M and yields $Y_i(M)$; all those quantities can be derived from the theory of stellar evolution and nucleosynthesis and depend (to various degrees) on the initial stellar metallicity Z.
- *Collective stellar properties*: the initial mass function $\Phi(M)$ and the Star Formation Rate Ψ; none of them can be reliably derived from first principles at present, and one has to rely on empirical prescriptions.
- *Gas flows* into and out of the system (infall, inflow, outflow, wind): in simple GCE models these factors are optional, i.e. their introduction depends on the nature of the considered galactic system (e.g. infall for the solar neighborhood or winds for small galaxies). In more physical (e.g. hydrodynamical) models in a cosmological framework, they should stem naturally from the physics of the system.

We discuss those ingredients in the remainder of this chapter.

11.1.2 The Role of Stars

Star formation occurs at a specific efficiency acting on the mass of interstellar gas to produce a presumably universal spectrum of stellar masses (see Krumholz 2014, for a recent review of the physical processes involved). The creation rate of stars, or *stellar birth rate* $B(m, t) = \Psi(t) \cdot \Phi(M)$ links the *star formation rate* $\Psi(t)$ with the *Initial Mass Function* $\Phi(M)$, and explicitly assumes those two key ingredients to be independent (which may not always be true).

11.1.2.1 The Distribution of Stellar Masses

The Initial Mass Function (IMF) cannot be unambiguously calculated at present from first principles and it is mainly derived from observations. Such a derivation is not quite straightforward and important uncertainties still remain, especially in the region of massive stars.

Based on observations of stars in the solar neighborhood, and accounting for various biases (but not for stellar multiplicity), Salpeter (1955) derived the local IMF in the mass range 0.3–10 $M_\odot$ as a power-law function:

$$\Phi(M) = \frac{dN}{dM} = A\,M^{-(1+X)} \tag{11.10}$$

with a slope $X = 1.35$. That slope is deduced from observations covering a large variety of conditions. This "Salpeter IMF" is often used in the entire stellar mass range, from 0.1 to 100 $M_\odot$, especially in studies of the photometric evolution of galaxies. However, it is clear now that there are fewer stars in the low mass range (below 0.5 $M_\odot$) than predicted by the Salpeter slope of $X = 1.35$. As reviewed by Kroupa (2002) (see also Fig. 11.2), a multi-slope power-law IMF may provide a good description, with $X = 0.35$ in the range 0.08–0.5 $M_\odot$. Alternatively, often a log-normal IMF below 1 $M_\odot$ is used (Chabrier 2003, 2005).

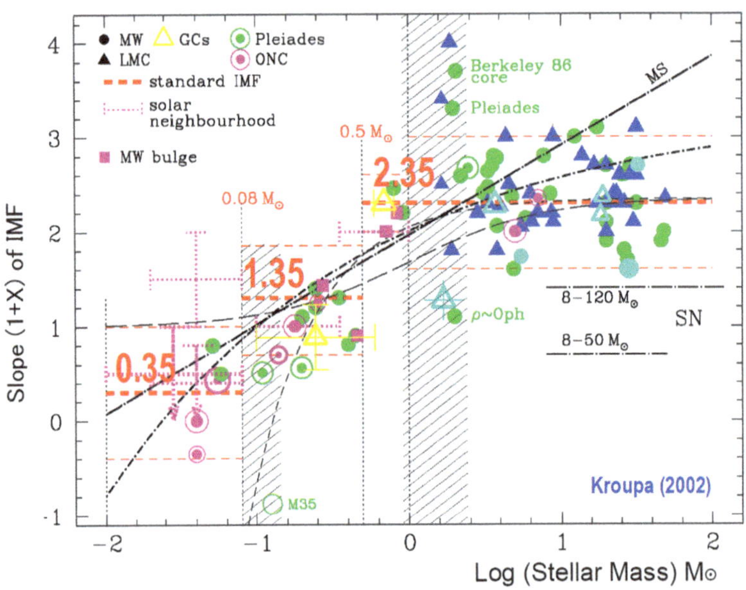

Fig. 11.2 Slope $1+X$ of the IMF (assumed to be described by a multi-power-law form), according to observations in various astrophysical environments; the dashed horizontal lines indicate average values in three selected mass ranges, with $1 + X = 2.35$ being the classical Salpeter value (from Kroupa 2002)

Observations of the IMF in various environments, and in particular young clusters (where dynamical effects are negligible) suggest that a Salpeter slope $X = 1.35$ describes the high-mass range well (see also Fig. 11.2). However, determination of the IMF in young clusters suffers from considerable biases introduced by stellar multiplicity and pre-main sequence evolution. For field stars in the solar neighborhood, Scalo (1986) finds $X = 1.7$, i.e. a much steeper IMF than Salpeter. For GCE purposes, low mass stars are "eternal" and just lock up matter which then is excluded from the recycling in the ISM. Most important for the compositional enrichment is the mass range of high-mass stars with their rapid evolution, that is, the shape of the IMF above 1 $M_\odot$.

Weidner et al. (2011) present a concept, linking a *stellar IMF* as observed in clusters, and possibly controlled by the processes of star formation and feedback, to a *galaxy-integrated IMF* which would apply for a GCE description, i.e. the sum of the action from all clusters, obtaining an IMF which is steeper than the stellar IMF. Although every single cluster is assumed here to have the same stellar IMF (say, $X = 1.35$), the maximum stellar mass $M_{MAX,C}$ in a cluster increases with the total mass of that cluster (Weidner et al. 2010, 2013). Observations also show that small clusters may have $M_{MAX,C}$ as low as a few $M_\odot$, whereas large clusters have $M_{MAX,C}$ up to 150 $M_\odot$. If this were just a statistical effect, the slope of the resulting galaxy-integrated IMF would also be $X = 1.35$. But if there is a physical reason for the observed $M_{MAX,C}$ vs $M_{Cluster}$ relation, then the resulting galaxy-integrated IMF would necessarily be steeper (taking into account the steep decline of the cluster mass function with increasing cluster mass). This concept of a "universal" initial mass function characterising the astrophysical processes, mediated by stellar evolution and observational biases, appears to capture best what we know about the stellar mass distribution (Chabrier et al. 2014; Kroupa et al. 2013).

The IMF is normalised to

$$\Phi(M) = \int_{M_L}^{M_U} \Phi(M)\, M\, dM = 1 \tag{11.11}$$

where M_U is the upper mass limit and M_L the lower mass limit. Typical values are $M_U \sim 100\,M_\odot$ and $M_L \sim 0.1\,M_\odot$, and the results depend little on the exact values (if they remain in the vicinity of the typical ones). A comparison between three normalized IMFs, namely the "reference" one of Salpeter, one proposed by Kroupa (with the Scalo slope at high masses) and one by Chabrier (with the Salpeter slope at high masses) is made in Fig. 11.3.

A useful quantity is the *Return Mass Fraction R*

$$R = \int_{M_T}^{M_U} (M - C_M)\, \Phi(M)\, dM \tag{11.12}$$

i.e. the fraction of the mass of a stellar generation that returns to the ISM. It depends on the IMF as well as on the adopted mass of the stellar remnants C_M. For the three IMFs displayed in Fig. 11.3 one has $R \sim = 0.3$ (Salpeter), 0.34 (Kroupa+Scalo) and

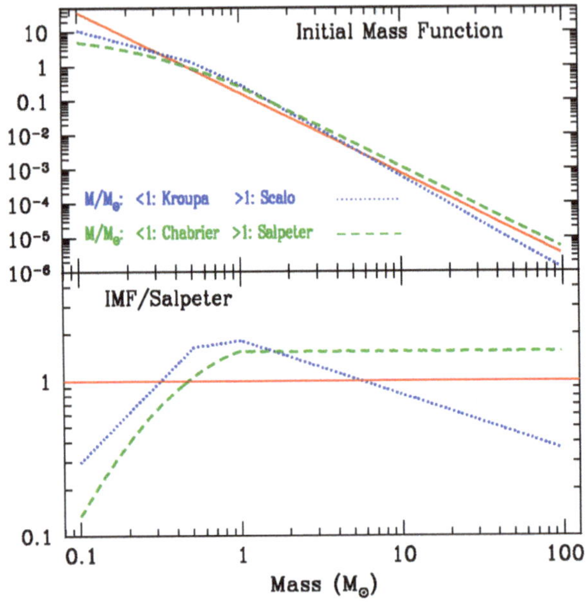

Fig. 11.3 *Top:* Three initial mass functions: *solid curve*: Salpeter (power-law in the whole mass range), *dotted curve*: Kroupa (multi-slope power law for $M<1\,M_{\odot}$) + Scalo ($X=1.7$ for $M>1\,M_{\odot}$), *dashed curve*: Chabrier (log-normal for $M<1\,M_{\odot}$) + Salpeter ($X=1.35$ for $M>1\,M_{\odot}$). *Bottom:* Ratio of the three IMFs to the one of Salpeter

0.38 (Chabrier+Salpeter), respectively, i.e. about 1/3 of the mass gone into stars returns to the ISM.

11.1.2.2 The Rate of Star Formation

Star formation is the main driver of galactic evolution and the most uncertain parameter in GCE studies. Despite decades of intense observational and theoretical investigation (see Elmegreen 2002, and references therein) our understanding of the subject remains frustratingly poor. Observations of various SFR tracers in galaxies provide only relative values, under the assumption that the IMF is the same everywhere (Kennicutt 1998). Moreover, those tracers reveal that star formation occurs in different ways, depending on the type of the galaxy. In spiral disks, star formation occurs mostly inside spiral arms, in a sporadic way. In dwarf, gas rich, galaxies, it occurs in a small number of bursts, separated by long intervals of inactivity. Luminous Infrared galaxies (LIRGS) and starbursts (as well as, most probably, ellipticals in their youth) are characterised by an intense burst of star formation, induced by the interaction (or merging) with another galaxy. Notice that most such tracers concern formation of stars more massive than $\sim 2\,M_{\odot}$; very little information exists for the SFR of low mass stars, even in the Milky Way.

There is no universally-accepted theory to predict large scale star formation in a galaxy, given the various physical ingredients that may affect the SFR (e.g. density and mass of gas and stars, temperature and composition of gas, magnetic fields and frequency of collision between giant molecular clouds, galactic rotation etc.)

Schmidt (1959) suggested that the density of the SFR Ψ is proportional to some power of the density of gas mass m_G:

$$\Psi = \nu\, m_G^N \tag{11.13}$$

This formulation has the merit of reminding us that stars are formed from gas, after all. However, it is not clear whether *volume density* ρ or *surface density* Σ should be used in Eq. (11.13). When comparing data with GCE models for the solar neighborhood, Schmidt (1959) uses surface densities (Σ in $M_\odot/pc^2$). But, when finding "direct evidence for the value of N" in his paper[6] he uses volume densities (ρ in $M_\odot/pc^3$) and finds N=2. Obviously, since $\Sigma = \int_z \rho(z)dz$, one has: $\Sigma^N \neq \int_z \rho(z)^N dz$.

It is not clear then which density should be used in the Schmidt SFR law: Volume density is more "physical" (denser regions collapse more easily) but surface density is more easily measured in galaxies. Furthermore, at first sight, it seems that the density of molecular gas should be used (since stars are formed from molecular gas), and not the total gas density.

Surprisingly enough, Kennicutt (1998) found that, in normal spirals, the surface density of SFR correlates with atomic rather than with molecular gas; this conclusion is based on *average surface densities*, i.e. the total SFR and gas amounts of a galaxy are divided by the corresponding surface area of the disk. In fact, Kennicutt (1998) finds that a fairly good correlation exists between SFR and *total (i.e. atomic + molecular) gas*. This correlations extends over four orders of magnitude in average gas surface density ρ_S and over six orders of magnitude in average SFR surface density Ψ, from normal spirals to active galactic nuclei and starbursts (see Fig. 11.4, left) and can be described as:

$$\Psi \propto \Sigma^{1.4} \tag{11.14}$$

i.e. $N = 1.4$. However, Kennicutt (1998) notes that the same data can be fitted equally well by a different N value, this time involving the *dynamical timescale* $\tau_{dyn} = R/V(R)$, where $V(R)$ is the orbital velocity of the galaxy at the optical radius R:

$$\Psi \propto \frac{\Sigma}{\tau_{dyn}} \tag{11.15}$$

In some cases, it is useful to consider the *efficiency of star formation* ε, i.e. the SFR per unit mass of gas. In the case of a Schmidt law with $N = 1$ one has: $\varepsilon = \nu = const.$, whereas in the case of $N = 2$ one has: $\varepsilon = \nu m_G$.

[6]Schmidt (1959) describes the distributions of gas and young stars perpendicularly to the galactic plane (z direction) in terms of *volume densities* $\rho_{Gas} \propto exp(-z/h_{Gas})$ and $\rho_{Stars} \propto exp(-z/h_{Stars})$ with corresponding scaleheights (obervationally derived) h_{Gas}=78 pc and h_{Stars}=144 pc$\sim$2 h_{Gas}; from that, Schmidt deduces that $\rho_{Stars} \propto \rho^2_{Gas}$, that is N=2.

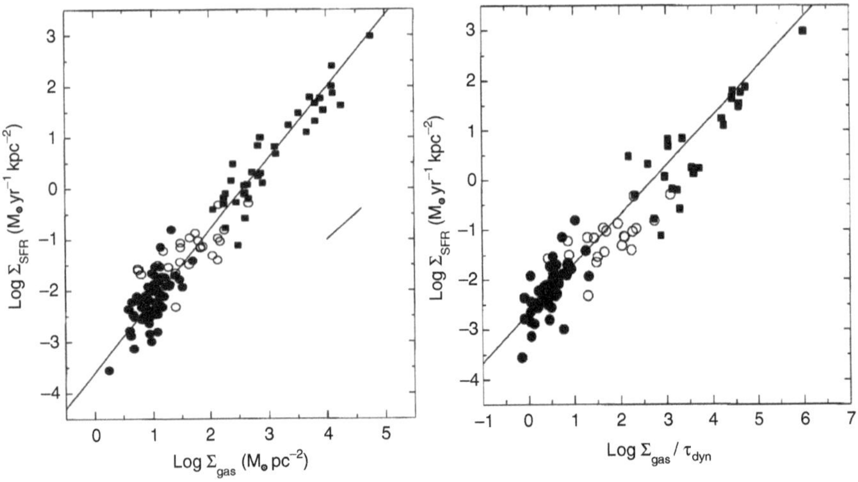

Fig. 11.4 *Left:* Average surface density of star formation rate Ψ (in $M_{\odot}$ year^{-1} pc^{-2}) as a function of average gas (HI+H_2) surface density Σ_G (in $M_{\odot}$ pc^{-2}, in spirals (circles) and starbursts (squares); the solid line corresponds to $\Psi \propto \Sigma^{1.4}$. *Right*: Average surface density of star formation rate Ψ (in $M_{\odot}$ year^{-1} pc^{-2}) as a function of Σ_G/τ_{dyn}, where the dynamical timescale $\tau_{dyn} = R/V$ (for rotational velocity V at radius R); the solid line corresponds to $\Psi \propto \Sigma_G/\tau_{dyn}$ (from Kennicutt 1998)

Determination of absolute values of star formation rates (SFR, in $M_{\odot}$ year^{-1}) constitutes one of the most challenging tasks in modern astrophysics. Each one of the tracers used is sensitive to only some part of the stellar initial mass function and has its own advantages and drawbacks. The derivation of an *absolute* value (i.e. the *calibration* of the SFR profile) remains a delicate enterprise. For that purpose, one needs to know either the total SFR of the MW disk or the local one in the solar neighborhood. A *ball-park* estimate of the former value is obtained by noting that the late spectral type (Sbc) of the MW suggests a slow formation at a relatively steady rate $\langle SFR \rangle$ over the past ΔT ~10 Gyr, leading to $\langle SFR \rangle = M_{disk}/\Delta T$ ~3 $M_{\odot}$ year^{-1}. Most empirical estimates of the present-day total SFR, based on the aforementioned tracers, produce values within a factor of two of the $\langle SFR \rangle$ (Chomiuk and Povich 2011).

The Schmidt/Kennicutt law makes use of the total gaseous amount of the disk. More recent surveys, based on a detailed, sub-kpc scale, observations of large samples of disk galaxies, find that the SFR appears to follow the surface density of molecular gas (H_2), rather than the one of atomic gas (HI) or the total gas surface density; also, the observed radial decline in star formation efficiency is too steep to be reproduced only by increases in the free-fall time or orbital time and no clear indications of a cut-off in the SFR are found.

These results support a SFR depending linearly on the molecular gas surface density, rather than the total one. The H_2 surface density can be obtained by semi-empirical prescriptions for the ratio for the ratio R_{mol}=H_2/HI (Blitz and

Rosolowsky 2006).

$$f_2 = \frac{R_{mol}}{R_{mol} + 1} \tag{11.16}$$

The resulting radial profiles $H_2(R) = f_2(R)\ \Sigma_G(R)$ and $HI(R) = [1 - f_2(R)]\Sigma_G(R)$ compare favorably to the observed ones in the Milky Way and other galaxies (see e.g. Appendix B in Kubryk et al. (2015a)). The corresponding SFR

$$\Psi(R) = \alpha\ f_2(R) \left(\frac{\Sigma_G(R)}{\mathrm{M_\odot/pc^2}}\right) \mathrm{M_\odot/kpc^2/year} \tag{11.17}$$

with coefficient α properly adjusted, reproduces well the "observed" SFR profile of the MW disk.

In the context of Galactic positrons, special attention should be paid to the star formation activity in the central regions of the bulge. The massive star population of the three major star clusters inside the Nuclear Bulge clearly indicate important recent star formation, obviously fed from the gas of the Central Molecular Zone (CMZ). Deep field observations of late-type stars with the *NICMOS/HUBBLE* (Figer et al. 2004) and with *SINFONI/VLT* (Maness et al. 2007) suggest that the star forming activity in that region has proceeded at a relatively steady rate, of the order of a few 10^{-2} M$_\odot$/year, over the past ~10 Gyr.

11.1.2.3 Properties of Stars: Lifetimes, Residues and Yields

Stellar lifetime is a rapidly decreasing function of stellar mass (see Fig. 11.5). Its precise value depends on the various assumptions (about e.g. mixing, mass loss, etc.) adopted in stellar evolution models and, most importantly on stellar metallicity. Indeed, low metallicity stars have lower opacities and are more compact and hot than their high metallicity counterparts; as a result, their lifetimes are shorter (see Fig. 11.5 right). However, in stars with M>2 M$_\odot$, where H burns through the CNO cycles, this is compensated to some degree by the fact that the H-burning rate (proportional to the CNO content) is smaller, making the corresponding lifetime longer; thus, for M>10 M$_\odot$, low metallicity stars live slightly longer than solar metallicity stars. Of course, these results depend strongly on other ingredients, such as stellar rotation (see Chap. 3). In principle, such variations in τ_M should be taken into account in GCE models; in practice, however, the errors introduced by ignoring them are smaller than the other uncertainties of the problem, related e.g. to stellar yields or to the IMF.[7]

[7] Metallicity dependent lifetimes *have to be taken into account* in models of the spectrophotometric evolution of galaxies, where they have a large impact. In galactic chemical evolution calculations, they play an important role in the evolution of s-elements, which are mostly produced by long-lived AGB stars of ~1.5–2 M$_\odot$.

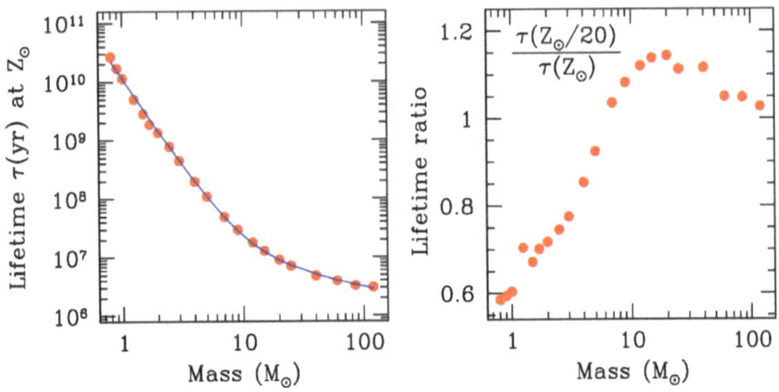

Fig. 11.5 *Left:* Lifetimes of stars of solar metallicity from Schaller et al. (1992) (points), fitted by Eq. (11.18). *Right:* Ratio of stellar lifetimes at metallicity Z = Z⊙/20 to those at Z = Z⊙, from the same reference

The lifetime of a star of mass M (in M⊙) with metallicity Z⊙ can be approximated by:

$$\tau(M) \ = \ 1.13\ 10^{10} M^{-3} \ + \ 0.6\ 10^{8} M^{-0.75} \ + \ 1.2\ 10^{6} \text{ year} \qquad (11.18)$$

This fitting formula is displayed as solid curve in Fig. 11.5 (left). A Z⊙ star of 1 M⊙, like the Sun, is bound to live for 11.4 Gyr, while a 0.8 M⊙ star for ∼23 Gyr; the latter, however, if born with a metallicity Z≤0.05 Z⊙, will live for "only" 13.8 Gyr, i.e. its lifetime is comparable to the age of the Universe (Fig. 11.5 , right). Stars of mass 0.8 M⊙ are thus the lowest mass stars that have ever died since the dawn of time (and the heaviest stars surviving in the oldest globular clusters).

The masses of stellar residues are derived from stellar evolution calculations, confronted to observational constraints (Fig. 11.6). In the regime of Low and Intermediate Mass Stars (LIMS,[8]) i.e. for M/M⊙ ≤8–9, the evolutionary outcome is a white dwarf (WD), the mass of which (in M⊙) is (Weidemann 2000):

$$C_M(WD) \ = \ 0.08\ M \ + \ 0.47 \qquad (M < 8 - 9) \qquad (11.19)$$

Stars more massive than 8–9 M⊙ explode as supernovae (SN), either after electron captures in their O-Ne-Mg core (M≤11 M⊙) or after Fe core collapse (M≥11 M⊙). The nature and mass of the residue depends on the initial mass of the star and on the mass left to it prior to the explosion. It is often claimed that

[8]LIMS are defined as those stars evolving to white dwarfs. However, there is no universal definition for the mass limits characterizing Low and Intermediate Mass stars. The upper limit is usually taken around 8–9 M⊙, although values as low as 6 M⊙ have been suggested (in models with very large convective cores). The limit between Low and Intermediate masses is the one separating stars powered on the Main Sequence by the p-p chains from those powered by the CNO cycle and is ∼1.2–1.7 M⊙, depending on metallicity.

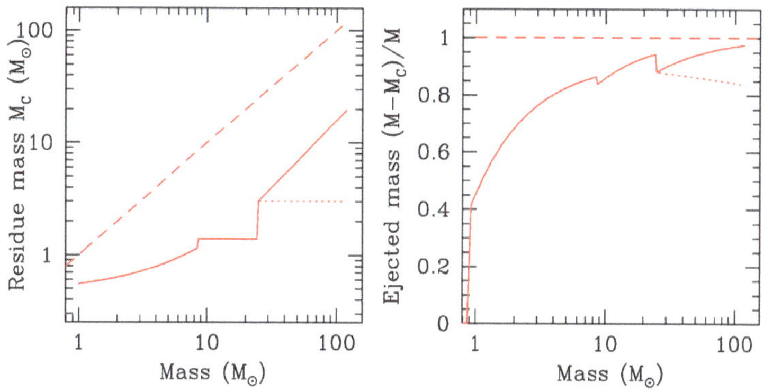

Fig. 11.6 *Left:* Masses of stellar residues as a function of initial stellar mass, for stars of metallicity $Z_\odot$; for massive stars ($M>30\,M_\odot$) the two curves correspond to different assumptions about mass loss, adopted in Limongi and Chieffi (2009) (*solid curve*) and Woosley and Heger (2007) (*dotted curve*), respectively. *Right:* Mass fraction of the ejecta as a function of initial stellar mass; the two curves for $M>30\,M_\odot$ result from the references in the left figure

solar metallicity stars of $M\leq 25\,M_\odot$ leave behind a Neutron Star (NS), while heavier stars leave a black hole (BH). Neutron star masses are well constrained by the observed masses of pulsars in binary systems: $M_{NS} = 1.35 \pm 0.04\,M_\odot$ (Thorsett and Chakrabarty 1999), which is adopted here

$$C_M(NS) = 1.35 \qquad (8-9 < M < 25) \tag{11.20}$$

i.e. C_M is independent of the initial mass M in that case. However, black hole masses are not known observationally as a function of the progenitor mass, while theoretical models are quite uncertain in that respect.

Black hole masses are expected to be larger at low metallicities, where the effects of mass loss are less important. However, the magnitude of the effect could be moderated in models with rotational mixing, which induces mass loss even at very low metallicities. In the last years there has been convergence towards the idea that stars more massive than $25\,M_\odot$ or so actually fail to explode and fully fall back in the remnant. The reasons for this are both observational and theoretical: on the observational side (Pejcha and Prieto 2015) it was found that the kinetic energy of the ejecta in a sample of Type IIP supernovae never exceeds 3 foes (=4.5×10^{51} ergs) while on the theoretical one Sukhbold et al. (2016) find that stellar models more massive than 25–30 $M_\odot$ fail to explode (even if some massive stars—randomly distributed in mass—explode due to a specific overlap of the convective shells in the advanced burning phases) and suggest black hole masses of $\sim 10\,M_\odot$.

The quantities required in Eq. (11.8) are the *stellar yields* $Y_i(M)$, representing the mass ejected in the form of element i by a star of mass M. Those quantities are obviously $Y_i(M) \geq 0$ ($Y_i = 0$ in the case of an isotope totally destroyed in stellar interiors, e.g. deuterium). However, their value is of little help in judging whether star M is an important producer of isotope i (e.g. by knowing that a $20\,M_\odot$ star produces $10^{-3}\,M_\odot$ of Mg or $1\,M_\odot$ of O, one cannot judge whether such a star contributes significantly—if at all—to the galactic enrichment in those elements).

More insight in that respect is obtained through the *net yields* $y_i(M)$, which represent the *newly created mass of nuclide* i from a star, i.e.

$$y_i(M) \;=\; Y_i(M) - \;M_{0,i}(M) \tag{11.21}$$

where $M_{0,i}(M)$ is the mass of nuclide i originally present in the part of the star that is finally ejected:

$$M_{0,i}(M) \;=\; X_{0,i}(M \;-\; C_M) \tag{11.22}$$

and $X_{0,i}$ is the mass fraction of nuclide i in the gas from which the star is formed. Obviously, $y_i(M)$ may be positive, zero or negative, depending on whether star M creates, simply re-ejects or destroys nuclide i. Net yields *are not mandatory* in numerical models of GCE, but *they are used* in analytical models, adopting the Instantaneous recycling approximation (see Sect. 11.8).

Finally, the *production factors* $f_i(M)$ are defined as:

$$f_i(M) \;=\; \frac{Y_i(M)}{M_{0,i}(M)} \tag{11.23}$$

They are useful in the sense that they immediately indicate whether star M is an important producer of nuclide i. For instance, massive stars are the exclusive producers of oxygen, for which $f\ \sim$10 on average (see Fig. 11.7). If such stars produce another nuclide L with, say, $f\ \sim$3 only, they are certainly important contributors, but they cannot account for the solar L/O ratio; another source is then required for nuclide L [*Note:* The example is taken on the case of iron, for which another source is required, beyond massive stars; that source is SNIa (see below)].

Note that the use of production factors as defined in Eq. (11.23) and described in the previous paragraph is interesting only when comparison is made for a star of a given initially metallicity. The properties of the various quantities defined in this section are summarized in Table 11.1. An application of the definitions is given in Fig. 11.7.

Regarding the nucleosynthesis products as a function of stellar mass, it is well established now that massive stars produce the practically all of the nuclides

Table 11.1 Yield definitions

Nuclide i	Yields $Y_i(M)$	Net yields $y_i(M)$	Production factors $f_i(M)$
Created	$>M_{0,i}$	>0	>1
Re-ejected	$=M_{0,i}$	$=0$	$=1$
Destroyed	$<M_{0,i}$	<0	<1

$M_{0,i}$ is defined in Eq. (11.22)

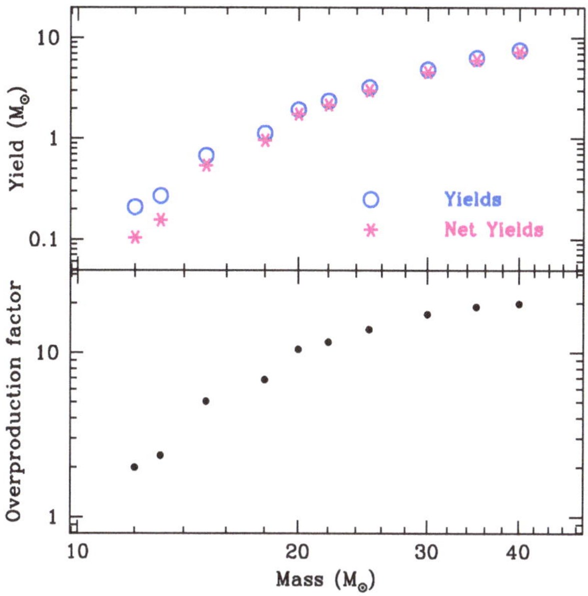

Fig. 11.7 *Top*: Yields (circles) and net yields (asterisks) of oxygen as a function of stellar mass, for stellar models with no mass loss or rotation (from WW95). *Bottom*: Corresponding overproduction factors. Notice that, because of the form of the stellar IMF, average yields (or overproduction factors) correspond to a star of ~25 $M_{\odot}$

between carbon and the Fe-peak, as well as most of the nuclei heavier than iron: the light s- (up to Y), the r- (neutron-rich) and p- (neutron poor, w.r.t. to the nuclear stability valley) nuclides. Oxygen is exclusively produced in massive stars, although its absolute yields are still subject to several uncertainties.

Intermediate mass stars synthesize substantial amounts of several important nuclides, mainly in the Asymptotic Giant Branch (AGB) phase, when H and He burn intermittently in two shells surrounding the inert CO core (see Chap. 3). Nucleosynthesis occurs in those shells, as well as in the bottom of the convective AGB envelope, if it penetrates in regions of high enough temperature (*Hot Bottom Burning* or HBB). Such stars are the main producers of heavy s-nuclei at a galactic level and they synthesize large amounts of ^{4}He, ^{14}N, ^{13}C, ^{17}O, ^{19}F, etc. However, they are not net producers of oxygen, while in some cases, they may even destroy part of their initial O content through HBB. A study of the combined evolution of CNO elements (e.g. of N/O vs O/H) should certainly take into account the role of such stars. The oxygen yields of those stars can be neglected, to a first approximation, but not their H and He ejecta, which contribute to the returned mass in Eq. (11.5).

Iron plays a major role in studies of GCE, because of its high abundance and strong spectral lines. Fe is made in massive star explosions (with fairly uncertain yields, usually taken to be 0.07 $M_{\odot}$, after the case of SN1987A) but also in thermonuclear supernovae (SNIa), where it is produced as radioactive ^{56}Ni. Observations of the peak luminosity of SNIa (powered by the decay of ^{56}Ni) suggest that they produce on average 0.7 $M_\odot$ of ^{56}Fe, the stable product of ^{56}Ni decay; thus, SNIa are major producers of Fe (and Fe-peak nuclides in general). In the case of the Milky Way, this can be seen as follows: The observed frequency of SNIa in external galaxies of the same morphological type (i.e. Sbc/d) is about 5 times smaller than the corresponding frequency of core collapse SN (SNII+SNIb,c), (see, e.g. Mannucci et al. 2005). But core collapse SN produce, on average $\sim$0.1 $M_\odot$ of ^{56}Ni, that is 7 times less than SNIa. Thus, depending on the supernova rate per each type (see Fig. 11.8), SNIa contribute at least as much as massive stars to the production of Fe and Fe-peak nuclides in the Milky Way. In the case of the solar neighborhood, this is corroborated by another observational argument, namely the evolution of the O/Fe ratio: halo stars have a quasi-constant [O/Fe]$\sim$0.5, i.e. three times solar, up to a metallicity of Fe/H]~-1, whereas for higher metallicities (i.e. during the disk phase) that ratio goes smoothly down to 0 (=solar). This implies that $\sim$2/3 of solar Fe are produced by a long-lived source, namely SNIa.

SNIa may have long-lived progenitors, with lifetimes of up to several Gyr; this introduces a substantial delay in their rate of element ejection in the ISM. The evolution of SNIa rate depends on the assumptions made about the progenitor system (see Wang 2018, and references therein). and it obviously cannot be simply proportional to the SFR. For the rate of SNIa a semi-empirical approach is usually adopted: the observational data of extragalactic surveys in the last decay are described well by a power-law in time, of the form $\propto t^{-1}$ (e.g. Maoz and Graur 2017, and references therein). At the earliest times, the DTD is unknown/uncertain, but a cut-off must certainly exist before the formation of the first white dwarfs ($\sim$35–

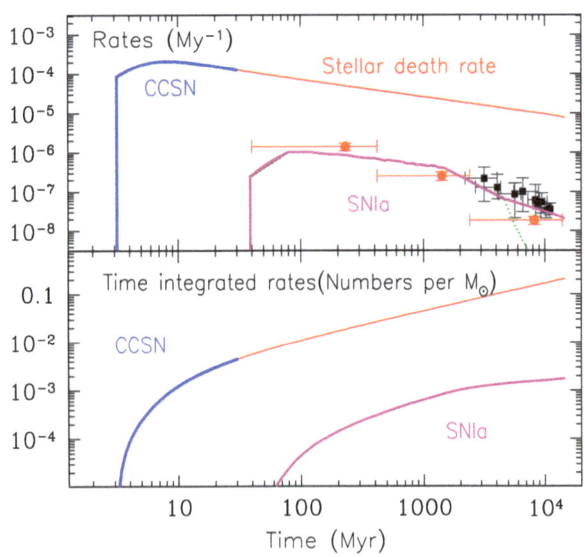

Fig. 11.8 *Top:* Death rates of stars (upper curve), with CCSN appearing in the thick blue part, while SNIa rate appears as magenta curve and is compared to observational data; rates are expressed in units of $M_\odot^{-1}$ My^{-1}. *Bottom:* Time integrated rates of CCSN and SNIa. Adapted from Kubryk et al. (2015b)

40 Myr after the birth of the SSP). The corresponding SNIa rate at time t from all previous SSP is obtained as

$$R_{SNIa}(t) = \int_0^t \Psi(t')DTD(t-t')dt' \tag{11.24}$$

The most widely used set of SNIa yields is probably the one of Iwamoto et al. (1999) for Z = 0 and Z = $Z_\odot$; however, it leads to a systematic overproduction of ^{54}Fe and Ni.

11.1.2.4 Evolving Massive-Star Groups in the Galaxy

All above considerations concern a larger (representative, or averaged) region of a galaxy. But the formation of massive stars occurs in *groups*, from a parental giant molecular cloud coeval groups of hundreds to thousands of massive stars will be born. The stellar content, interstellar-gas enrichment, and dynamical state of restricted regions such as the ones resulting from the evolution of a single giant molecular complex may be different from the galactic average. It may also be more straightforward in its treatment, as approximations in theories may be more valid in this restricted context. What is expected for a group of massive stars, as they evolve from their birth through stellar evolution with terminating core-collapse supernovae?

For such regions with a more limited content of massive stars, a steady-state assumption may not apply, time histories are explicitly evaluated. as an example discussed here, the ^{26}Al content of the interstellar medium around each group evolves with the heterogeneity of ejection given by the evolution of stars of different initial masses. Production of new stars will not continue steadily and will even terminate completely once feedback of energy from those stars will disrupt the parental cloud. The existing massive stars will eventually terminate their stellar evolution in core-collapse supernovae, and any further ^{26}Al production in such region is left to probably less-efficient AGB stars and novae.

Stellar evolution depends on initial mass and metallicity, as analyzed in detail in stellar evolution theory (see Sect. 11.1.2.3 and Chap. 3). The time it takes a star of a given mass to evolve into late phases where significant and enriched material is returned to the interstellar gas varies between less than My for the most massive stars (approximately 100 $M_\odot$) and several Gy for solar-sized stars to enter their giant phase. Substantial enrichment of interstellar gas is probably restricted to stars above masses of 3 $M_\odot$ (see Chap. 3).

The ^{26}Al production though hydrostatic and explosive burnings exhibits large variations of ^{26}Al ejection for a massive star over its evolution. Initially, hydrostatic burning in the core may produce ^{26}Al which is buried and partly decays. But as evolution towards giant phases develops strong stellar winds, the hydrogen envelope is gradually pealed off and interior burning products may be admixed into the wind in the Wolf-Rayet phase of stellar evolution. This typically sets in beyond 3 My for the most-massive stars, which then also are the first to explode as core-collapse supernovae beyond ∼4 My and eject ^{26}Al from late and explosive nuclear burning.

The stellar mass spectrum discussed above (the IMF, Sect. 11.1.2.1) now can be combined with the star formation history and with stellar evolution as a function of initial mass m to determine the *destruction* and *production* rates per isotope in the medium around the group of stars.

For a coeval group of stars, the characteristic ^{26}Al content of surrounding interstellar medium as a function of time after birth is shown in Fig. 11.9 (from Voss et al. 2009; see also Plüeschke et al. 2001; and Cerviño et al. 2000). Shading shows the 68 and 95% variances of ^{26}Al amounts, variations derive from statistical variance due to the limited number of massive stars at each age interval. For this simulation, a group of 100 massive stars in the mass range 8–120 $M_\odot$ were traced, with their initial masses distributed according to the Salpeter mass function.

Figure 11.9 shows that the ^{26}Al content varies by almost an order of magnitude between ∼3 and 20 My after star birth. This may provide a diagnostic for groups of massive stars from their nucleosynthesis: If remaining present-day stars can be counted, the age of the entire population is constrained from ^{26}Al gamma-ray brightness, if age is well-constrained, ^{26}Al gamma-ray emission allows inferences on a possibly embedded population of stars which may have escaped detection. In any case, ^{26}Al measurements provide a consistency check of our understanding of massive-star evolution and -nucleosynthesis.

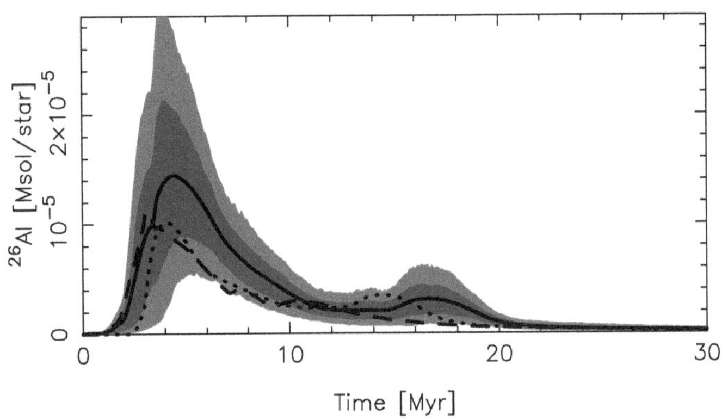

Fig. 11.9 The ^{26}Al content of the interstellar medium around a group of massive stars evolves with the heterogeneity of ejection given by the evolution of stars of different initial masses. This graph shows the simulation for a coeval group of massive stars born at t = 0 with an initial mass function of Salpeter slope (Voss et al. 2009)

11.1.3 The Roles of Gas and Dust

Gas and dust are spread out in interstellar space, subject to gravity from stars and gas of the galaxy components, and to energy injected by stars through radiation, winds, and explosions. The morphology of the interstellar medium is complex on scales below kpc (Fig. 11.10)—understanding of these structures and processes which drive them will be essential for our understanding of the evolution of galaxies. From the immediate surroundings of the Sun, we can map out (to some degree) clouds and hot cavities and their relation to groups of stars (Fig. 11.11), which generally confirms this picture from simulations. Star formation activity and its efficiency is regulated by how energy is transported from the stellar population into the gas (e.g. Jappsen et al. 2005). Turbulent energy and its cascading has been understood to play a major role; the self-gravitation process as estimated through the *Jeans mass* provides crude guidance only. Star formation in Taurus is found to be faster and incompatible with self gravitation only, for example.

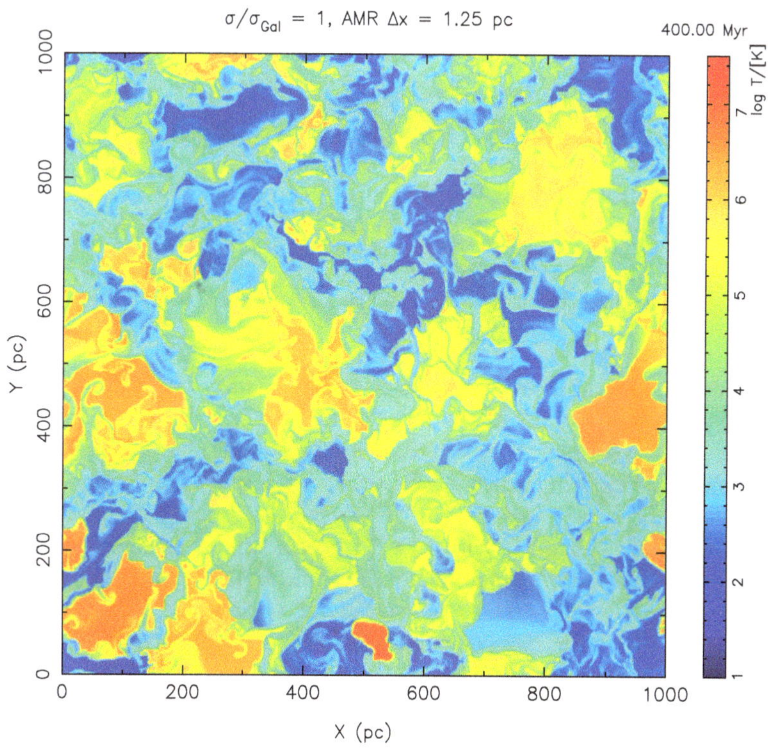

Fig. 11.10 The interstellar medium is dynamic and evolves rapidly, driven by winds and supernova explosion. This graph shows the temperature distribution obtained from hydrodynamical simulations of 300 My of evolution in a cube of dimensions 1 kpc on each side (Breitschwerdt 2004)

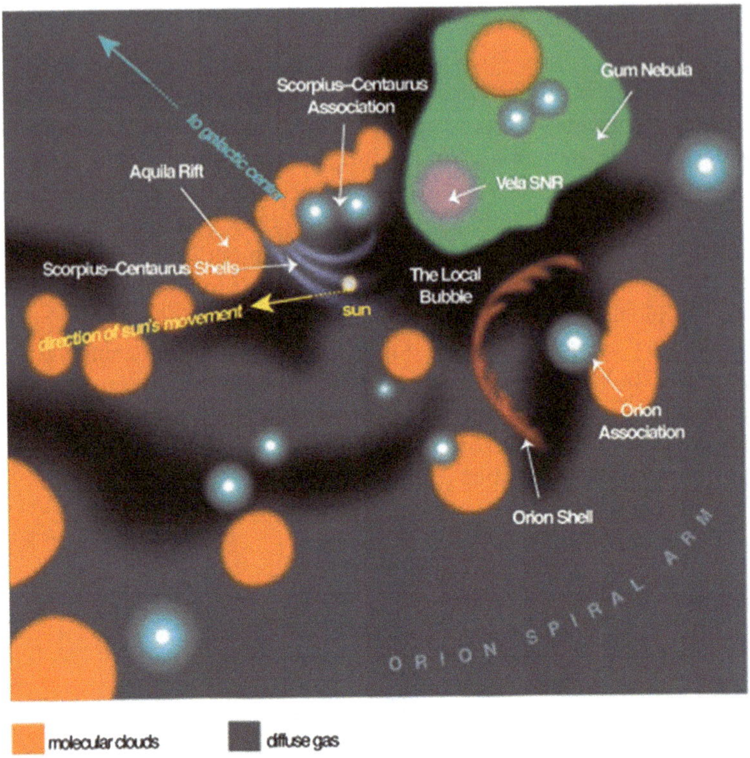

Fig. 11.11 The interstellar medium in the solar vicinity has been mapped in more detail than can be done elsewhere. This figure has been assembled by Frisch et al. (2009), summarizing our current knowledge about prominent clouds and cavities within a few hundred kpc

With the typical time scales of massive-star evolution of about 1–10 My, the current model places origins of massive stars into dense parts of parental giant molecular clouds (see review by McKee and Ostriker 2007; Zinnecker and Yorke 2007). One of the most important questions in models of the star formation process is the release of energy and matter by massive stars and their effects on the surrounding clouds (Vázquez-Semadeni 2015; Zuckerman and Evans 1974). Such feedback on star formation in dense clouds can be either negative (termination of star formation through dispersal of the natal cloud) or positive (triggering of further star formation by compression of cloud material). Most massive-star clusters are believed to be dispersed due to the action of their stars within 10 My (e.g. Pfalzner 2009). ^{26}Al studies from specific regions hosting massive stars thus have a role in improving our understanding of the cycles of star formation and how energy and matter modify the vicinity of their birth sites, thus driving the evolution of galaxies.

Stellar evolution of single stars eventually leads to compact remnant stars (white dwarfs, neutron stars, or black holes, depending on the mass of the star), which locks up a part of stellar gas remaining at the end. Binary systems, however, open channels for re-cycling this locked-up stellar mass into the gas reservoir. Examples are stellar explosions which disrupt the entire star in thermonuclear supernovae (SN Ia; see Chap. 5), or colliding neutron stars and other gamma-ray burst related events (see Chap. 4). Since the interstellar medium environment of different types of stars (low-mass stars, of massive stars, and of the delayed SNIa explosions) all will be different in density and dynamic state (turbulence), corresponding impacts are expected on mixing of gas with different compositions, and on formation of new stars nearby.

The *contribution of new material* into interstellar matter includes two astrophysical processes, the *mass ejection* by stars of different mass during their lifetime, and the *mixing* of ejecta with interstellar gas from which new stars may be formed. The detailed picture of how these processes occur has not been obtained; simulations such as shown in Fig. 11.10 (Breitschwerdt 2001) illustrate typical studies. Their goals are an understanding of distribution and recycling of stellar material and of feedback from massive-star activity.

If complexities of stellar evolution and delayed ejections of new enriched gas are to be avoided, one often assumes *instantaneous recycling*, i.e. the *stellar yield* is employed as a contribution to the gas at the time of star formation in such approximation. For contributions from massive stars this may be not too bad, but for AGB and other giant stars, or even more so for novae or thermonuclear supernovae, delays and extended ejection may be inadequately represented in such treatment. If only the complexities of enrichment of star-forming gas from previous stellar generations are to be avoided, *instantaneous mixing* is assumed as an approximation in all current formulations of chemical evolution. This may be overcome in descriptions of *dynamical* chemical evolution.

Dust formation is rather well modeled in AGB star envelopes (Sedlmayr and Patzer 2004). For more massive stars, this has not been achieved; Wolf-Rayet winds are complex, clumpy, and very energetic. Exploding supernova envelopes are even more dynamic, and dust formation is only beginning to be unexplored.

Interstellar dust is modified in size and composition on its journey through interstellar space (see Jones 2009, for a review). Once created in a 'nucleation', the size of the particle rapidly grows by condensation of interstellar molecules, growing considerable ice mantles. Interstellar shocks, but also the intense radiation near massive stars, can destroy particles, and thus re-processes dust grains through partial or full evaporization of ice mantles. Interstellar shocks enhance grain collisions and may incur sputtering of larger grains into smaller ones. Radiation from dust is a prime tracer for star forming environments, as radiation from stars heats dust particles to higher temperatures than the typical $\sim$10 K in normal interstellar space; thermal emission is observed and studied through infrared telescopes.

11.1.3.1 Scales of Interstellar-Medium Processing

The interstellar medium is a key mediator for the outputs of nucleosynthesis sources, i.e. ejected matter, ionizing radiation, and kinetic energy from winds and explosions. Such impact processes the interstellar medium into phases and states which determine further star formation; this is called *feedback*, and determines the evolution of normal disk galaxies.[9] Turbulence generated by stellar winds and explosions will drive how interstellar gas eventually forms stars, or ceases to form new stars, thus driving galactic evolution on a more fundamental level.[10] Feedback from nucleosynthesis sources occurs throughout a galaxy, and influences its embedded objects. Exactly how matter spreads from nucleosynthesis sites into next-generation stars will determine chemical enrichment over a galaxy's evolution (*mixing*). Major other drivers of galactic evolution are material inflows from extragalactic space through clouds, streams, or mergers, but also a supermassive black hole in a galaxy's center will incur major changes in a galaxy's evolution.

Chemical evolution of the universe at large involves mixing of material at different scales: The early phase of forming a star (before/until planets are being formed), stellar winds and explosions, clusters of co-evolving stars, the disks of typical galaxies, and intergalactic space. We trace matter in its different appearances as plasma (ionized atoms and their electrons), atoms and molecules, and dust particles. We briefly discuss the different spatial scales in more detail:

(a) At the smallest scale, a stellar/planetary formation site evolves from decoupling of its parental interstellar cloud (i.e. no further material exchange with nucleosynthesis events in the vicinity) until the star and its planets have settled and overcome the disk accretion phase with its asteroid collision and jet phases. This phase may have a typical duration of ~My. Issues here are how inhomogeneities in composition across the early solar nebula are smoothed out over the time scales at which chondrites, planetesimals, and planets form. (Chondrites are *early* meteorites, and the most-common meteorites falling on Earth (85%). Their name derives from the term *chondrule*, which are striking spherical inclusions in those rocks. The origin of those is related to melting events in solids of the early solar system, the nature of which is the study of *cosmochemistry* (Cowley 1995). *Carbonaceous chondrites* are 5% of all falling meteors, and are believed to be the earliest solids we know in the solar system.) Inhomogeneities may have been created from (1) the initial decoupling from a triggering event, or from (2) energetic-particle nuclear processing in the jet-wind phase of the newly-forming star. Gaidos et al. (See Chapter 6 and, e.g. 2009, for the case of our Solar system). Radioactive dating is an important tool in such studies.
(b) The fate of the ejecta of a stellar nucleosynthesis event is of concern at the next-larger scale. Stellar winds in late evolutionary stages of stars such as the

[9]In *active galaxies* the central supermassive black hole also plays a role, and even dominates over the impact from massive stars for central regions, and for entire galaxies in late (largely-processed) evolution such as at low redshifts.

[10]In this book we only address the scales and processes within a galaxy, as it is driven by massive stars and can be traced by radioactive material; in this and further sections of this chapter, the broader (cosmic evolution) context is also relevant.

asymptotic giant or *Wolf Rayet* phases, and also explosive events, novae and two kinds of supernovae (according to their different evolutionary tracks) involve different envelope masses, ejection energies, and dynamics. The astronomical display of such injection of fresh nuclei into interstellar space is impressive[11] throughout the early phases of the injection event; however, no real *mixing* with ambient interstellar gas occurs yet at this phase Ejected gas expands into the lower-pressure interstellar medium, but decelerates upon collisional interaction with interstellar atoms, and collisionless interactions with the magnetized plasma. This process is an important ingredient for the acceleration of cosmic rays. Once ejecta velocities have degraded to the velocity range of interstellar gas ($\sim$100–few km s^{-1}), the actual *mixing* process can become efficient. Cooling processes of gas in its different phases are key processes, and also incur characteristic astronomical signatures. (H_α radio emission, C[II] recombination in the IR, or FIR thermal emission of dust are important examples.) Radioactive isotopes are key sources of energy for the astronomical display (supernova light curves), and sensitive tracers of the nucleosynthesis conditions of these events.

(c) Co-evolving stellar groups and clusters provide an astrophysical object in its own right. The combined action of stars, successively reaching their individual wind phases and their terminating supernovae, shape the interstellar environment so that it may vary for each nucleosynthesis event. *Giant HII regions* and *Superbubbles* are the signposts of such 10–100 pc-sized activity, which can be seen even in distant galaxies (Oey et al. 2007). The evolution of disks in galaxies is determined by the processes on this scale: Formation of stars out of Giant Molecular Clouds, as regulated by *feedback* from the massive stars, as it stimulates further star formation, or terminates it, depending on gas dynamics and the stellar population. This is currently the frontier of the studies of cosmic evolution of galaxies (Calzetti and Kennicutt 2009). Cumulative kinetic energy injection may be sufficient to increase size and pressure in a cavity generated in the interstellar medium, such that *blow-out* may occur perpendicularly to the galactic disk, where the pressure of ambient interstellar gas is reduced with respect to the galactic disk midplane. This would then eject gas enriched with fresh nucleosynthesis product into a galaxy's halo region through a *galactic fountain*. Only the fraction of gas below galactic escape velocity would eventually return on some longer time scale ($> 10^7$–10^8 years), possibly as *high-velocity clouds (HVCs)*. Long-lived ($\sim$My) radioactive isotopes contribute with age dating and radioactive tracing of ejecta flows.

(d) In a normal galaxy's disk, large-scale dynamics is set by differential rotation of the disk, and by large-scale regular or stochastic turbulence as it results from star formation and incurred wind and supernova activity (see (c)): This drives

---

[11] AGB stars form colorful planetary nebulae, massive-star winds form gas structures within the *HII-regions* created by the ionizing radiation of the same stars, and thus a similarly-rich variety of colorful filamentary structure from atomic recombination lines results. Supernova remnants are the more violent version of the same processes.

the evolution of a galaxy.[12] As a characteristic time scale for rotation we may adopt the solar orbit around the Galaxy's center of 10^8 years. Other important large-scale kinematics is given by the spiral density waves sweeping through the disk of a galaxy at a characteristic pattern speed, and by the different kinematics towards the central galaxy region with its bulge, where a bar often directs gas and stellar orbits in a more radial trajectory, yet with a bar pattern speed that will differ from Keplerian circular orbits in general. Infalling clouds of gas from the galactic halo, but also gas streams from nearby galaxies and from intergalactic space will add drivers of turbulence in a galaxy's disk at this large-scale. The mixing characteristics of the interstellar medium therefore will, in general, depend on location and history within a galaxy's evolution. Radioactive isotopes are part of the concerted abundance measurement efforts which help to build realistic models of a galaxy's chemical evolution.

(e) On the largest scale, the above gas streams into and away from a galaxy are the mixing agents on the intergalactic scale. Galactic *fountains* thus offer alternative views on above superbubble blow-out, and this may comprise a *galactic wind* ejected from galaxies (observed e.g. in starburst galaxies, see Heckman et al. 1990). Galaxies are part of the cosmic web and appear in coherent groups (and clusters). Hot gas between galaxies in such clusters can be seen in X-ray emission, elemental abundances can be inferred from characteristic recombination lines. Gas clouds between galaxies can also be seen in characteristic absorption lines from distant quasars, constraining elemental abundances in intergalactic space. The estimated budget of atoms heavier than H and He appears incomplete (the *missing metals* issue, Pettini (see 2004, for a global review)), which illustrates that mixing on these intergalactic scales is not understood.

11.1.3.2 Gaseous Flows into and out of a Galaxy

A galaxy is clearly not an isolated system, and it is expected to exchange matter (and energy) with its environment. This is true even for galaxies which are found away from galaxy groups. First of all, most of baryonic matter in the Universe today (and in past epochs) is in the form of gas residing in the intergalactic medium (Fukugita and Peebles 2004) and part of it is slowly accreted by galaxies. Also, small galaxies are often found in the tidal field of larger ones, and their tidal debris (gas and/or stars) may be captured by the latter. In both cases, gaseous matter is accreted by

[12] Feedback from supermassive black holes is small by comparison, but may become significant in AGN phases of galaxies and on the next-larger scale (clusters of galaxies, see (e)). Galaxy interactions and *merging* events are also important agents over cosmic times, their overall significance for cosmic evolution is a subject of many current studies.

galaxies, and in the framework of the simple GCE model this is generically called *infall*.[13]

On the other hand, gas may leave the galaxy, if it gets sufficient (1) kinetic energy or (2) thermal energy and (3) its velocity becomes larger than the escape velocity. Condition (1) may be met in the case of tidal stripping of gas in the field of a neighbour galaxy or in the case of ram pressure from the intergalactic medium. Condition (2) is provided by heating of the interstellar gas from the energy of supernova explosions, especially if collective effects (i.e. a large number of SN in a small volume, leading to a superbubble) become important. Finally, condition (3) is more easily met in the case of small galaxies, with shallow potential wells. Note that, since galaxies (i.e. baryons) are embedded in extended dark matter (non baryonic) haloes, a distinction should be made between gas leaving the galaxy but still remaining trapped in the dark halo and gas leaving even the dark halo. In the former case, gas may return back to the galaxy after "floating" for some time in the dark halo and suffering sufficient cooling. In the framework of the simple GCE model, all those cases are described generically as *outflows*.

The rate of infall or outflow is difficult to calculate from first principles. In the case of infall, this is possible, in principle, for a hydrodynamical model evolving in an appropriate cosmological framework. In the case of outflows, the interaction between stars (SN) and the ISM, known as *feedback*, also requires detailed hydrodynamic modelling. No satisfactory models exist up to now for such complex processes. Incidentally, the treatment of feedback also affects the SFR of the system (by making gas unavailable for star formation, either by heating it or by pushing it out of the system altogether).

In simple GCE models, infall and outflow are treated as free parameters, adjusted as to reproduce observed features of the galaxy systems under study. Such features are the metallicity distributions (MD) of long -lived stars, or the mass-metallicity relationship of external galaxies which provide strong constraints on the history of the systems.

11.1.3.3 Analytical Solutions: Approximating the Recycling of Gas

The system of GCE Eqs. (11.3)–(11.8) can be solved analytically under some conditions, providing useful insights into the evolution of a given galactic system. The key assumption here is the *Instantaneous Recycling Approximation (IRA)*, introduced by Schmidt (1963). Stars are divided in: "eternal" ones (those of low mass, with lifetimes far exceeding the age of the system) and "dead at birth" ones (massive stars, with lifetimes far shorter than the age of the system) for which it is assumed that $\tau_M = 0$. The dividing line between the two classes depends on the

[13] Gaseous flows in the plane of a galactic disk, due e.g. to viscosity, are called *inflows*; for simple GCE models they also constitute a form of infall.

system's age, and for most practical purposes it is put at 1 $M_\odot$, corresponding to an age $T \sim$12 Gyr.

Assuming IRA allows one to replace $\Psi(t - \tau_M)$ in the equations of GCE by $\Psi(t)$ and thus to take out the SFR $\Psi(t)$ from the mass integrals. The equations can then be solved analytically (see, e.g. Tinsley 1980). Solutions involve the *Return fraction* R defined in Eq. (11.12) and the yield p_i of a given nuclide, defined as:

$$p_i = \frac{1}{1-R} \int_{M_T}^{M_U} y_i(M)\, \Phi(M)\, dM \qquad (11.25)$$

The yield p_i is *the newly created amount of nuclide i by a stellar generation, per unit mass of stars blocked into "eternal" objects*: indeed, $y_i(M)$ are the *net yields* of nuclide i and $1 - R$ is the mass blocked in low mass stars and compact objects (for the normalized IMF of Eq. (11.11)). Obviously, the *stellar yields* $Y_i(M)$ and *net yields* $y_i(M)$ are properties of individual stars, while the *yield* p_i is an integrated property of the IMF, as is the return fraction R.

With those definitions, the gas mass fraction X_i of nuclide i in the case of the Closed Box model is:

$$X_i - X_{i,0} = p_i\, ln(\frac{m}{m_G}) = p_i\, ln(\frac{1}{\sigma}) \qquad (11.26)$$

where $\sigma = m_G/m$ is the *gas fraction* and $X_{i,0}$ the initial abundance ($X_{i,0} = 0$ for metals). This is the main result of IRA, relating the chemical enrichment of the gas to the amount of gas left. It is independent on time or on the form of the SFR, and for those reasons it is a powerful tool to study gas flows in the system. It can also be used to derive the metallicity distribution of stars and, thereof, the past history of the system.

If a simple Schmidt law of the form $\Psi = \nu m_G$ is adopted for the SFR (Eq. (11.14)) one can obtain the following solutions for the evolution of gas

$$m_G = m\, e^{-\nu(1-R)t} \qquad (11.27)$$

and for the abundances

$$X_i - X_{i,0} = p_i\, \nu\, (1 - R)\, t \qquad (11.28)$$

which also satisfy the more general solution of Eq. (11.26). Thus, metallicity is roughly proportional to time, a result which is approximately valid even when IRA is relaxed. This property allows one to use stellar metallicity (especially iron, which has many strong and easily identifiable spectral lines) as a proxy for time, since stellar ages are notoriously difficult to evaluate.

IRA turns out to be a surprisingly good approximation for nuclides produced in massive stars, like e.g. oxygen, provided gas fractions stay above $\sim$10%; this is illustrated in Fig. 11.12 for the case of a Closed Box model, with analytical

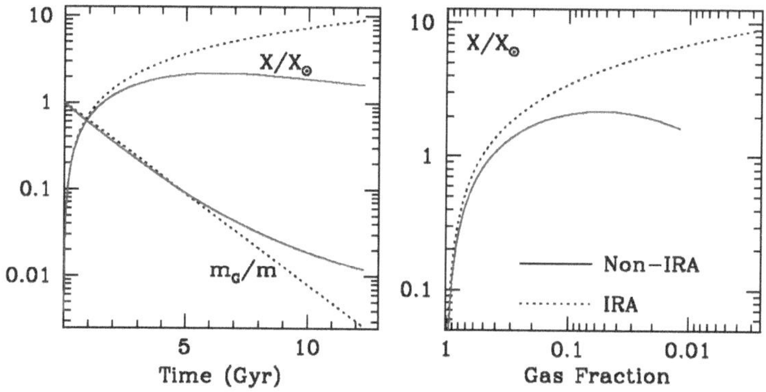

Fig. 11.12 Results of calculations for a Closed Box model with Instantaneous Recycling Approximation (IRA, *dotted curves*) vs Non-IRA (*solid curves*). *Left:* Metallicity (upper curves) and gas fraction (lower curves) as a function of time. *Right:* Metallicity as a function of gas fraction. It is assumed that $\Psi = 1.2\, m_G\, \mathrm{Gyr}^{-1}$

solutions given by (11.27) and (11.28). Analytical solutions assuming IRA can also be obtained in the cases of gaseous flows into or out of the system. However, some specific assumptions have to be made about the form of those flows, and this limits a lot their interest. For instance, in the case of outflow at a rate proportional to the SFR, $o = k\,\Psi$, metallicity evolves as:

$$X_i - X_{i,0} = \frac{p_i}{1+k}\, ln(\frac{1}{\sigma}) \qquad (11.29)$$

i.e. at the same gas fraction metallicity is smaller than in the Closed box or, equivalently, a larger fraction of the system has to turn into stars in order to reach the same metallicity. Equation (11.29) is formally the same as (11.26), with an *effective yield*

$$p_{i,eff} = \frac{p_i}{1+k} \qquad (11.30)$$

in the place of the true yield p_i, and $p_{i,eff} < p_i$. Gas flows always produce *reduced effective yields*, i.e. that metallicity increases most efficiently in the Closed Box model. Other analytical solutions can be obtained in models with specific forms of infall. For instance, in the case of a model with constant SFR and gas mass, driven by an infall of rate $f = \Psi(1 - R)$ and zero metallicity, the resulting solution is:

$$X_i - X_{i,0} = p_i[\,1 - exp(1 - \frac{1}{\sigma})] \qquad (11.31)$$

Equation (11.31) gives similar results to Eq. (11.26) for gas fractions higher than ~0.5; for lower σ values Eq. (11.31) "saturates" (i.e. $X_i \longrightarrow p_i$), while Eq. (11.26) produces values $X_i > p_i$.

Analytical solutions obtained in the framework of IRA can provide important constraints to various quantities of interest in GCE studies. For instance, it is currently admitted that the solar neighborhood did not evolve as a closed box, but with a smoothly declining (or roughly constant) SFR, driven by continuous gas infall (see Sect. 11.3); in other terms, its evolution should be intermediate between the two extreme cases described by the cases of Closed Box and Constant gas mas (but not constant gas fraction!). Assuming that the present day gas fraction is σ_G ~0.24 (see Sect. 11.3.2) and the Oxygen mass fraction is $X_{O,\odot} \sim 7 \times 10^{-3}$, one can solve Eqs. (11.26) and (11.31) to obtain the corresponding "observationally derived" yields, which constrain the oxygen yield in the solar vicinity (see Fig. 11.13).

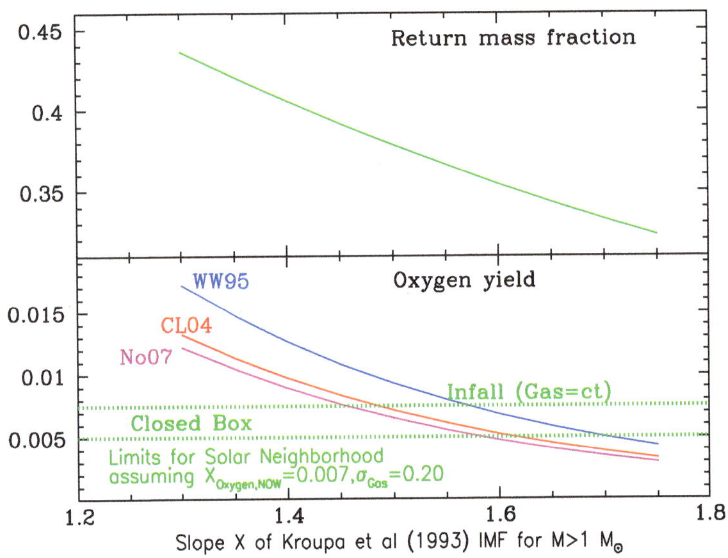

Fig. 11.13 *Top*: Return mass fraction for an IMF given by Kroupa (2002) for stars below 1 $M_\odot$ and by a power-law IMF with a slope X given in the horizontal axis. *Bottom:* Oxygen yields calculated by several groups (Woosley and Weaver 1995; Chieffi and Limongi 2004; Nomoto et al. 2006) and integrated over the IMF, as a function of the slope X of the massive star part of the IMF. The two horizontal dotted lines correspond to empirically determined yields for a closed box model (lower) and evolution at constant gas mass (upper), adapted to match the constraints of the solar neighbourhood (i.e. gas fraction ~0.20 and present-day oxygen mass fraction in the gas of X_O ~7 10^{-3}). These two lines constrain the value of the yield of oxygen

11.2 The Milky Way Galaxy

The galactic distribution of any kind of stellar source of radioactivities is somewhat related to the distribution of stars in the Milky Way. Similarly, the birthrate of any kind of radioactivity source is somewhat related to the Galactic star formation rate. In this section we present a summary of current knowledge about the stellar populations of the Milky Way and their spatial distribution and we discuss the birthrates of stars and supernovae. Long lived radioactivities, such as ^{26}Al and ^{60}Fe, are expected to be thermalised in the ISM; some of their observables should reflect then the ISM properties. Moreover, positrons produced by various processes (including radioactivity), slow down and finally annihilate in the ISM, and the resulting electromagnetic emission also reflects the ISM properties. We present then in this section a brief overview of the various phases of the ISM in the bulge and the disk (including the spiral arms) of the Milky Way.

11.2.1 Stellar Populations

The Milky Way is a typical spiral galaxy, with a total baryonic mass of $\sim 5 \times 10^{10}\,M_\odot$, of which more than 80% is in the form of stars. Stars are found in three main components (Fig. 11.14): the central bulge, the disk and the halo, while the gas is found essentially in the plane of the disk. Because of its low mass, estimated to $4 \times 10^8\,M_\odot$ i.e. less than 1% of the total (Bell et al. 2007), the Galactic halo plays no significant role in the production of distributed radioactivities. The bulge contains $\sim$1/3 of the total mass and an old stellar population (age > 10 Gyr). The dominant component of the Milky Way is the so-called *thin disk*, a rotationally supported structure composed of stars of all ages (0–10 Gyr). A non negligible contribution

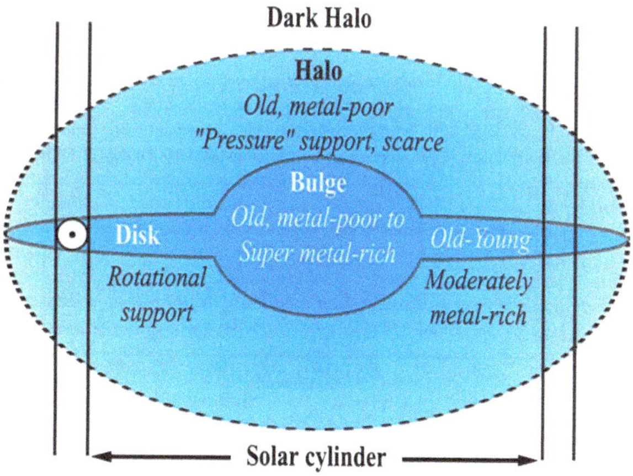

Fig. 11.14 The various components of the Milky Way (bulge, halo, disk) and their main features; a distinction should be made between thin and thick disk (not appearing in the figure). The Solar cylinder is at 8 kpc from the center. Figure adapted from Pagel (1997)

is brought by the *thick disk*, an old (>10 Gyr) and kinematically distinct entity identified by Gilmore and Reid (1983).

11.2.1.1 The Bulge and the Galactic Center

To a first approximation, and by analogy with external galaxies, the bulge of the Milky Way can be considered as spherical, with a density profile either exponential or of Sersic-type ($\rho(r) \propto r^{1/n}$ with $n > 1$). Measurements in the near infrared (NIR), concerning either integrated starlight observations or star counts revealed that the bulge is not spherical, but elongated. Recent models suggest a tri-axial ellipsoid, but its exact shape is difficult to determine (Rattenbury et al. 2007) because of the presence of a Galactic bar. The mass of the bulge lies in the range 1–2 × 10^{10} $M_\odot$ (Robin et al. 2003). By comparing colour-magnitude diagrams of stars in the bulge and in metal-rich globular clusters, Zoccali et al. (2003) find that the populations of the two systems are co-eval, with an age of ~10 Gyr.

The innermost regions of the bulge, within a few hundred pc, are dominated by a distinct, disk-like component, called the Nuclear Bulge which contains about 10% of the bulge stellar population (~1.5 × 10^9 $M_\odot$) within a flattened region of radius 230 ± 20 pc and scaleheight 45 ± 5 pc (Launhardt et al. 2002). It is dominated by three massive stellar clusters (Nuclear Stellar Cluster or NSC in the innermost 5 pc, Arches and Quintuplex), which have a mass distribution substantially flatter than the classical Salpeter IMF Finally, in the center of the Milky Way, at the position of SgrA* source, lies the massive Galactic black hole (MBH) with a total mass of ~4 × 10^6 $M_\odot$ (Gillessen et al. 2008).

11.2.1.2 The Galactic Disk(s)

The Sun is located in the thin disk of the Milky way, at a distance of $R_\odot \sim 8$ kpc from the Galactic center (Groenewegen et al. 2008) and references therein). Furthermore, the Sun is not located exactly on the plane, but at a distance from it $z_\odot \sim 25$ pc, as evaluated from the recent analysis of the Sloan Digital Sky Survey (SDSS) data (Jurić et al. 2008).

In studies of the Milky Way the solar neighborhood plays a pivotal role, since local properties can, in general, be measured with greater accuracy than global ones. The total baryonic surface density of the solar cylinder (which is defined as a cylinder of radius 500 pc centered on Sun's position and extending perpendicularly to the Galactic plane up to several kpc) is estimated to $\Sigma_T = 48.8\, M_\odot\, pc^{-2}$ (Flynn et al. 2006), with ~13 $M_\odot\, pc^{-2}$ belonging to the gas (see Sec. III.B). This falls on the lower end of the dynamical mass surface density estimates from kinematics of stars perpendicularly to the plane) which amount to $\Sigma_D = 50$–$62\, M_\odot\, pc^{-2}$ (Holmberg and Flynn 2004) or 57–66 $M_\odot\, pc^{-2}$ (Bienaymé et al. 2006). Thus, the values for the baryon content of the solar cylinder, summarized in Table 11.2, should

Table 11.2 Properties of the stellar populations of the thin and thick disk[a] (from Prantzos et al. 2011; Kubryk et al. 2015a)

			Thin	Thick
Mass density	$\rho_{0,\odot}$	($M_\odot$ pc^{-3})	4.5×10^{-2}	5.3×10^{-3}
Surface density	$\Sigma_\odot$	($M_\odot$ pc^{-2})	24.5	8.5
Scaleheight	$H_\odot$	(pc)	300	900
Scalelength	L	(pc)	2700	1800
Star mass	M_D	(10^{10} $M_\odot$)	2.3	1.2
$\langle$Age$\rangle_\odot$	$\langle A\rangle_\odot$	(Gyr)	5	10
$\langle$Metallicity$\rangle_\odot$	$\langle$[Fe/H]$\rangle_\odot$	(dex)	−0.1	−0.7

[a]The indice $\odot$ here denotes quantities measured at Galactocentric distance $R_\odot = 8$ kpc. Average quantities are given within $\langle\ \rangle$

be considered rather as lower limits (Flynn et al. 2006): the total stellar surface density could be as high as 40 $M_\odot$ pc^{-2}.

The density profiles of the stellar thin and thick disks can be satisfactorily fit with exponential functions, both in the radial direction and perpendicularly to the Galactic plane. The recent SDSS data analysis of star counts, with no *a priori* assumptions as to the functional form of the density profiles finds exponential disks with scalelengths as displayed in Table 11.2 (from Jurić et al. 2008). The thin and thick disks cannot extend all the way to the Galactic center, since dynamical arguments constrain the spatial co-existence of such rotationally supported structures with the pressure-supported bulge. The exact shape of the *central hole* of the disks is poorly known (see, e.g. Freudenreich (1998), Robin et al. (2003), for parametrizations), but for most practical purposes (i.e. estimate of the total disk mass) the hole can be considered as truly void of disk stars for disk radius $R < 2$ kpc.

The data presented in this section (as summarized in Table 11.2), allow one to estimate the total mass of the thin and thick disks as $M_{D,thin} = 2.3\ 10^{10}\ M_\odot$ and $M_{D,thick} = 1.2\ 10^{10}\ M_\odot$, respectively, in the galactocentric distance range 2–15 kpc. Overall, the disk of the Milky Way is twice as massive as the bulge.

11.2.2 Supernova Rates in the Galaxy

From the theoretical point of view, SN are now classified mainly in *thermonuclear* supernovae (the explosion energy being due to the thermonuclear disruption of a white dwarf accreting matter in a binary system) and *core collapse* supernovae (CCSN, where the energy originates from the gravitational collapse of the iron core of a massive star having exhausted all its nuclear fuel). Thermonuclear supernovae are identified with SNIa (lacking hydrogen in their spectra) and are observed in all types of galaxies: old ellipticals with no current star formation, but also young, star forming, spiral and irregular galaxies. All other supernova types (SNII, SNIb, SNIc) are exclusively observed in the star forming regions of spirals (i.e. inside spiral arms) and irregulars. The degree of mass loss suffered by the massive star prior to the explosion determines the appearance of the core collapse supernova as SNII (little H lost), SNIb (all H and little He lost) or SNIc (all H and most He lost).

No supernovae have been observed in the Galaxy in the past four centuries, and the handful of so called *historical supernovae* offers a very biased estimate of the Galactic SN frequency (Tammann et al. 1994). Some methods used to determine the Galactic SN rate are based exclusively on Galactic data (e.g. counts of SN remnants, the present-day SFR, the present-day mass of ^{26}Al) and are, in principle, able to evaluate solely the core-collapse SN rate ; all those methods suffer from various systematic uncertainties (e.g. shape of the adopted massive star IMF, yields of ^{26}Al etc.) and converge to a value of $R_{CCSN} = 1..2$ per century (Diehl et al. 2006, and references therein).

A most precise value for the Galactic SN rate is probably obtained through counting of SN rates in external galaxies. The work of Mannucci et al. (2005), corrected for various observational biases, offers a valuable database for such an estimate and can be used, along with the stellar masses of the various Galactic components, to derive the Galactic rate of the main SN types (Table 11.3). Note that only SNIa are (perhaps) important producers of positrons, as discussed in Sec. IV.A.4. The spatial distribution of core collapse SN in the MW should obviously follow the one of the SFR (Fig. 11.15). Such an azimuthally averaged surface density masks the fact that CCSN are exclusively concentrated inside spiral arms. The scale height of core collapse SN should be comparable to the scaleheight of the molecular gas, that is, less than 100 pc, and there should be little variation with Galactocentric distance.

Table 11.3 Supernova rates in the Milky Way (from Prantzos and Boissier 2010)

	Stellar mass[a] 10^{10} $M_\odot$	Spectral type	SNIa Specific rate[b] (SNuM)	SNIa Rate (century^{-1})	Core collapse SN Specific rate[b] (SNuM)	Core collapse SN Rate (century^{-1})
Bulge	1.4	E0	0.044	0.062	–	–
Nuclear bulge	0.15	Sbc/d-Irr[c]	0.17–0.77	0.025–0.115	0.86–2.24	0.13–0.33
Thin disk	2.3	Sbc	0.17	0.4	0.86	2
Thick disk	0.5	E0	0.044	0.022	–	
Total bulge	1.5			0.087–0.18		0.13–0.33
Total disk	2.8			0.42		2
Total Milky Way	4.3			0.5–0.6		2.13–2.33
Bulge/disk ratio	<0.5			0.21–0.43		0.06–0.15

SNuM: = 1 SN per 10^{10} $M_\odot$ per century
[a]See Prantzos and Boissier (2010) for references
[b]Mannucci et al. (2005)
[c]Very uncertain, in view of uncertainties in star formation efficiency and slope of IMF (see text)

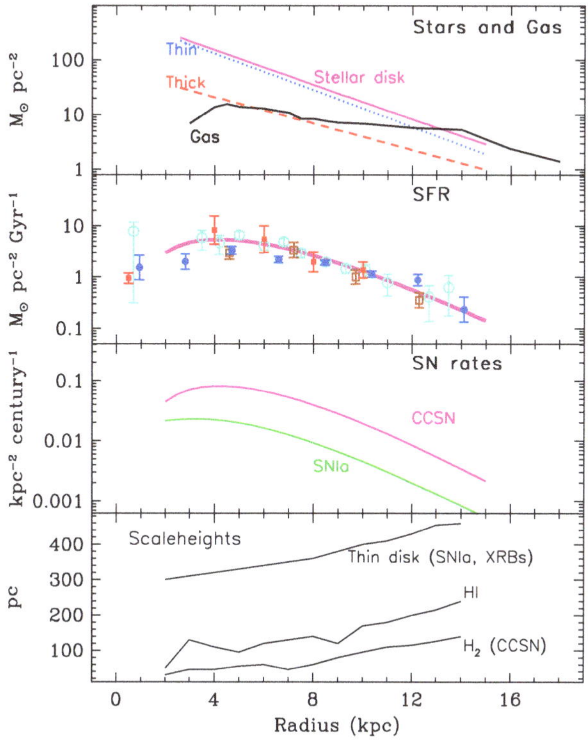

Fig. 11.15 Surface densities of stars+gas, SFR, SN rates and scale heights of gas and stars as a function of Galactocentric distance. Star profiles are from data of Table II, and the gas profile is from Dame (1993). Data for the SFR are from: Lyne et al. (1985) (*open circles*); Case and Bhattacharya (1998) (*filled circles*); McKee and Williams (1997) (*open squares*); Guibert et al. (1978) (*filled squares*). The solid curve is an approximate fit, normalized to 2 $M_\odot$ year^{-1} for the whole Galaxy. The same curve is used for the CCSN rate profile (third panel), normalized to 2 CCSN/century; the SNIa rate profile is normalized to 0.5 SNIa/century (Table 11.3) (from Prantzos et al. 2011)

More difficult is the evaluation of the radial profile of SNIa, since the progenitor white dwarfs may originate from stars of a wide variety of stellar masses (1–8 $M_\odot$) and corresponding lifetimes (10–0.05 Gyr). For the rate of SNIa an empirical approach is usually adopted, taking into account the observational data of extragalactic surveys which are described well by a power-law in time : $R_{SNIa} \propto t^{-1}$ (e.g., Maoz and Graur 2017, and references therein). Taking into account the nature of the SNIa progenitors, it is expected that the distribution of SNIa vertically to the disk plane will follow the corresponding distribution of the thin disk, i.e. with a scale height of 300 pc (an insignificant contribution from the thick disk is also expected).

11.2.3 Interstellar Matter

Interstellar gas is primarily composed of hydrogen, but it also contains helium ($\simeq$10% by number or 28% by mass) and heavier elements, called *metals* ($\simeq$0.12% by number or 1.5% by mass in the solar neighborhood). All the hydrogen, all the helium, and approximately half the metals exist in the form of gas; the other half of the metals is locked up in small solid grains of dust. Overall, gas and dust appear to be spatially well correlated (Boulanger and Perault 1988; Boulanger et al. 1996).

Interstellar gas can be found in molecular, atomic (cold or warm) and ionized (warm or hot) forms. The physical properties of the different gas components in the Galactic disk were reviewed by Ferrière (2001) and are summarized in Table 11.4. The gas properties in the Galactic bulge are less well established, but on the whole, all gas components appear to be hotter and denser in the bulge than in the disk (Ferrière et al. 2007).

Spatially, the molecular gas is confined to discrete clouds, which are roundish, gravitationally bound, and organized hierarchically from large complexes (size $\sim$20–80 pc, mass $\sim 10^5$–$2 \times 10^6\,M_\odot$) down to small clumps (size $\lesssim$0.5 pc, mass $\lesssim 10^3\,M_\odot$) (Goldsmith 1987). The cold atomic gas is confined to more diffuse clouds, which often appear sheet-like or filamentary, cover a wide range of sizes (from a few pc up to $\sim$2 kpc), and have random motions with typical velocities of a few km s^{-1} (Kulkarni and Heiles 1987). The warm and hot components are more widespread and they form the intercloud medium.

The different gas components also differ by their spatial distributions at large scales. The observational situation was reviewed by Ferrière (2001) for the Galactic disk and by Ferrière et al. (2007) for the Galactic bulge. Fig. 11.16 gives the radial variation of the azimuthally-averaged surface densities of H_2, HI, HII and the total gas (accounting for a 28% contribution from He). The distributions of those ISM phases have also different scaleheights, which increase with galactocentric radius (*flaring*), as can be seen in Fig. 11.15; the HII layer (not appearing in that figure) has an even larger scaleheight, of >1 kpc.

The total interstellar masses (including helium and metals) of the three gas components in the Galactic disk are highly uncertain. The masses given below are estimated from references in Fig. 11.16, inside 20 kpc and they are in the range: $\sim(0.9$–$2.5) \times 10^9\,M_\odot$ for the molecular component, $\sim(0.65$–$1.1) \times 10^9\,M_\odot$ for

Table 11.4 Physical properties (typical temperatures, hydrogen densities and ionization fractions) of the different ISM phases in the Galactic disk (Prantzos and Boissier 2010)

Phase		T (K)	n_H (cm^{-3})	x_{ion}
Molecular	(MM)	10–20	10^2–10^6	$\lesssim 10^{-4}$
Cold neutral	(CNM)	20–100	20–100	4×10^{-4}–10^{-3}
Warm neutral	(WNM)	10^3–10^4	0.2–2	0.007–0.05
Warm ionized	(WIM)	$\sim$8000	0.1–0.3	0.6–0.9
Hot ionized	(HIM)	$\sim 10^6$	0.003–0.01	1

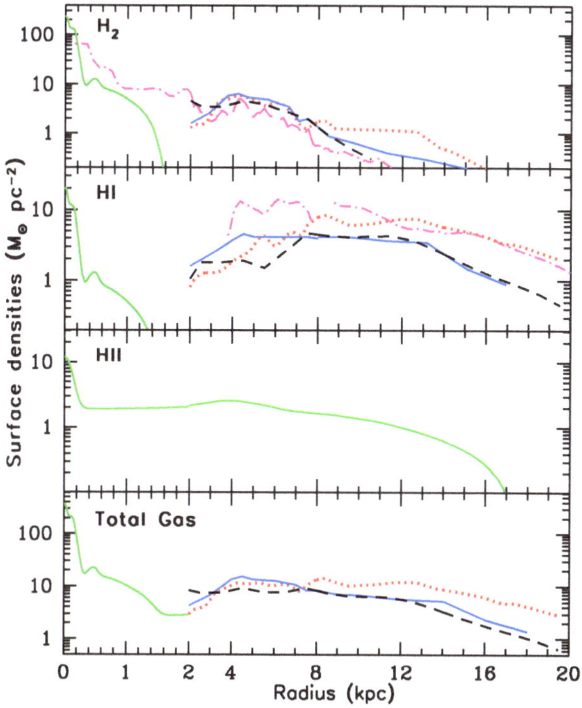

Fig. 11.16 Azimuthally-averaged surface densities of interstellar atomic, molecular and ionized hydrogen as functions of Galactic radius. The total gas (*bottom*) includes a 40% contribution by helium. Notice the change of scale at $R = 2$ kpc. For $R < 2$ kpc (bulge) data are derived by Ferrière et al. (2007), based on a compilation of earlier works: Sawada et al. (2004) for the molecular gas in the Central Molecular Zone, Liszt and Burton (1980) for the neutral gas in the tilted disk, and Cordes and Lazio (2002) for the ionized gas. In all panels, disk data ($R > 2$ kpc) are from: Dame (1993) (*solid*); Olling and Merrifield (2001) (*dotted*); Nakanishi and Sofue (2006) for HI and Nakanishi and Sofue (2003) for H_2, respectively (*dashed*); and Kalberla and Dedes (2008) for HI and Pohl et al. (2008) for H_2, respectively (*dot-dashed*). The curve in the HII panel is from the NE2001 free-electron density model of Cordes and Lazio (2002) (for simplicity, we identified the hydrogen density with the free-electron density, i.e., we neglected the contribution of free electrons originating from helium in the HIM) (from Prantzos and Boissier 2010)

the atomic component, and $\sim 1.5 \times 10^9\,M_\odot$ for the ionized component. The total interstellar mass in the Galaxy is probably comprised between $\sim 0.9 \times 10^{10}\,M_\odot$ and $\sim 1.5 \times 10^{10}\,M_\odot$, representing ~15–25% of the baryonic Galaxy mass or ~25–35% of the total mass of the Galactic disk.

11.2.4 Spiral Arm Structures

The gas distributions of Fig. 11.16 are azimuthally averaged, whereas the Milky Way disk displays a characteristic spiral pattern, concerning mostly molecular gas

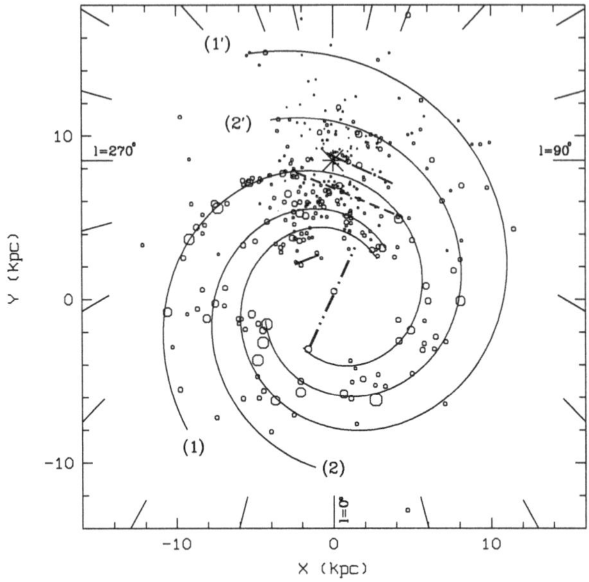

Fig. 11.17 Spiral structure of Milky Way. The Sun position is given by the large star symbol. 1: Sagittarius-Carina arm, 2: Scutum-Crux arm, 1′: Norma-Cygnus arm and 2′: Perseus arm. The local arm feature (long dashed line), the bar (dashed-dot-dot line, from Englmaier and Gerhard (1999)), the expected departure from the logarithmic spiral observed for the Sagittarius-Carina arm (short dashed line) and a feature probably linked to the 3-kpc arm (solid line) are sketched (from Russeil 2003, 2010)

and young stars. Despite several decades of study, neither the number (2 or 4) nor the precise form of the spiral arms is well established yet. A possible description of the spiral arm pattern is provided in Fig. 11.17, which shows deviations from a logarithmic spiral. More recent work (e.g., Pohl et al. 2008; Vallée 2017) suggests a more complex picture: the Milky Way has 4 spiral arms, two of which end inside the corotation radius, while the other two start at the end point of the Galactic bar (∼4 kpc), continue through corotation and branch at 7 kpc into 4 arms, which continue up to 20 kpc.

The spiral pattern of the Milky Way is of key importance for the case of distributed radioactivities produced in massive stars, (such as ^{26}Al and ^{60}Fe), since massive stars are born and die inside spiral arms. Because of their small ejection velocities (a few hundreds of km/s) and ∼Myr lifetimes, radioactive nuclei decay also inside spiral arms, and it is expected that their characteristic γ-ray emission would trace the spiral pattern. Indeed, enhanced emission towards the direction of the spiral arms is detected in the case of ^{26}Al, albeit at a low significance level (see Chap. 7).

The gas distribution in the Galactic bulge is of particular interest in the case of the Galactic positrons, since most of the detected e^+ annihilation emission originates from that region (see Chap. 7). As reviewed by Ferrière (2001), the molecular gas tends to concentrate in the so-called central molecular zone (CMZ), a thin sheet parallel to the Galactic plane, which, on the plane of the sky, extends out to $R \sim 250$ pc at longitudes $l > 0°$ and $R \sim 150$ pc at $l < 0°$ and has a FWHM thickness $\sim$30 pc. Outside the CMZ, the molecular gas is contained in a significantly tilted disk, extending (on the plane of the sky) out to $R \sim 1.3$ kpc on each side of the GC and having a FWHM thickness $\sim$70 pc. The spatial distribution of the atomic gas is arguably similar to that of the molecular gas, with the atomic layer being about three times thicker than the molecular layer. The ionized gas is not confined to either the CMZ or the tilted disk; it appears to fill the entire bulge and to connect with the ionized gas present in the disk.

The dramatic density and temperature contrasts between the different ISM phases as well as the supersonic random motions observed in all of them suggest a highly turbulent state, attributed to the powerful winds and the supernova explosions of massive stars. Interstellar turbulence manifests itself over a huge range of spatial scales, from $\lesssim 10^{10}$ cm up to $\gtrsim 10^{20}$ cm; throughout this range, the power spectrum of the free-electron density in the local ISM is consistent with a Kolmogorov-like power law (Armstrong et al. 1995).

11.2.5 Interstellar Magnetic Fields

The presence of interstellar magnetic fields in our Galaxy was first revealed by the discovery that the light from nearby stars is linearly polarized. This polarization is due to elongated dust grains, which tend to spin about their short axis and orient their spin axis along the interstellar magnetic field; since they preferentially block the component of light parallel to their long axis, the light that passes through is linearly polarized in the direction of the magnetic field. Thus, the direction of linear polarization provides a direct measure of the field direction on the plane of the sky. This technique applied to nearby stars shows that the magnetic field in the interstellar vicinity of the Sun is horizontal, i.e., parallel to the Galactic plane, and that it makes a small angle $\simeq 7°$ to the azimuthal direction (Heiles 1996).

The magnetic field strength in cold, dense regions of interstellar space can be inferred from the Zeeman splitting of the 21-cm line of HI (in atomic clouds) and centimeter lines of OH and other molecules (in molecular clouds). It is found that in atomic clouds, the field strength is typically a few μG, with a slight tendency to increase with increasing density (Troland and Heiles 1986; Heiles and Troland 2005), while in molecular clouds, the field strength increases approximately as the square root of density, from $\sim$10 to $\sim$3000 μG (Crutcher 1999, 2007).

The magnetic field in ionized regions is generally probed with Faraday rotation measures of Galactic pulsars and extragalactic radio sources. An important advantage of pulsars is that their rotation measure divided by their dispersion measure directly yields the electron-density—weighted average value of $B_{\parallel}$ between them and the observer. Faraday rotation studies have provided several interesting pieces of information.

1. The interstellar magnetic field has uniform (or regular) and random (or turbulent) components; near the Sun, the uniform component is $\simeq$1.5 μG and the random component ~5 μG (Rand and Kulkarni 1989).
2. The uniform field is nearly azimuthal in most of the Galactic disk, but it reverses several times along a radial line (Rand and Lyne 1994; Han et al. 1999, 2006; Vallée 2005; Brown et al. 2007). These reversals have often been interpreted as evidence that the uniform field is bisymmetric (azimuthal wavenumber $m = 1$), although an axisymmetric ($m = 0$) field would be expected on theoretical grounds. Recently, Men et al. (2008) showed that neither the axisymmetric nor the bisymmetric picture is consistent with the existing pulsar rotation measures, and they concluded that the uniform field must have a more complex pattern.
3. At low $|Z|$ the uniform field is roughly symmetric in Z, while at high $|Z|$ it is roughly antisymmetric(symmetric) in Z inside(outside) the solar circle (Han et al. 1997, 1999).
4. The uniform field increases toward the GC, from $\simeq$1.5 μG near the Sun to $\gtrsim$3 μG at $R = 3$ kpc (Han et al. 2006); this increase corresponds to an exponential scale length $\lesssim$7.2 kpc. In addition, the uniform field decreases away from the midplane, albeit at a very uncertain rate—for reference, the exponential scale height inferred from the rotation measures of extragalactic sources is ~1.4 kpc (Inoue and Tabara 1981).

The observed antisymmetric pattern inside the solar circle, combined with the detection of vertical magnetic fields near the GC (see Sect. III.D.2) led Han et al. (1997) to suggest that an axisymmetric dynamo mode with odd vertical parity prevails in the thick disk or halo of the inner Galaxy. While this suggestion is certainly reasonable and cannot be excluded at present, one may not jump to the conclusion that the uniform magnetic field in the Galactic halo and near the GC is simply a dipole sheared out in the azimuthal direction by the large-scale differential rotation. Especially as numerical simulations of galactic dynamos do not support this kind of geometry, nor do observations of external edge-on galaxies, which generally reveal X-shaped field patterns (Beck 2008). However, in view of the importance of the magnetic field configuration for positron propagation (Sec. VI) this issue should be further scrutinized.

A more global method to map out the interstellar magnetic field rests on the observed Galactic synchrotron emission. Relying on the synchrotron map of Beuermann et al. (1985) and assuming equipartition between magnetic fields and cosmic rays, Ferriere (1998) found that the total magnetic field has a local value $\simeq$5.1 μG, a radial scale length $\simeq$12 kpc, and a vertical scale height near the Sun $\simeq$4.5 kpc. The local value of the total field is consistent with that inferred from Faraday rotation measures (which applies to ionized regions). In contrast, the scale length and (even more so) the scale height are significantly larger than those obtained from rotation measures for the uniform field in ionized regions. The discrepancy can probably be explained partly by the magnetic field having a variable degree of regularity and partly by the ionized medium occupying only a modest fraction of the interstellar volume.

The properties of the turbulent magnetic field are not well established. Rand and Kulkarni (1989) provided a first rough estimate for the typical spatial scale of magnetic fluctuations, ~55 pc, although they recognized that the turbulent field cannot be characterized by a single scale. Later, Minter and Spangler (1996) presented a careful derivation of the power spectrum of magnetic fluctuations over the spatial range ~ (0.01–100) pc; they obtained a Kolmogorov spectrum below ~4 pc and a flatter spectrum consistent with 2D turbulence above this scale. In a complementary study, Han et al. (2004) examined magnetic fluctuations at larger scales, ranging from ~0.5 to 15 kpc; at these scales, they found a nearly flat magnetic spectrum, with a 1D power-law index ~−0.37 (Fig. 11.18).

The properties of the turbulent Galactic magnetic field are poorly understood at present. However, its local and overall configurations are extremely important for understanding positron propagation in the Milky Way (see Chap. 7 on positrons).

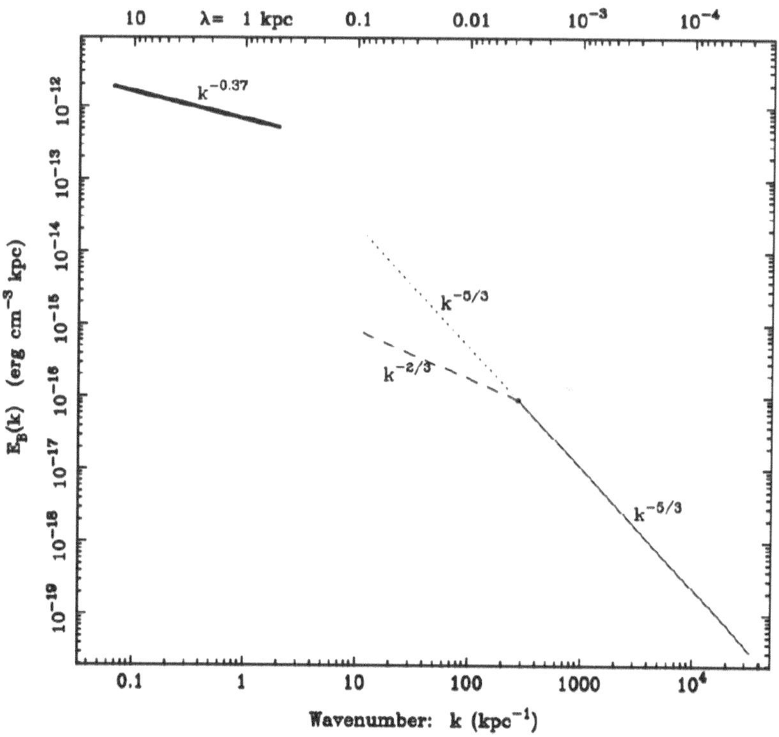

Fig. 11.18 Composite magnetic energy spectrum in our Galaxy. The *thick solid* line is the large-scale spectrum. The *thin solid* and *dashed/dotted* lines give the Kolmogorov and two-dimensional turbulence spectra, respectively, inferred from the Minter and Spangler (1996) study. The two-dimensional turbulence spectrum is uncertain; it probably lies between the *dashed* [$E_B(k) \propto k^{-2/3}$] and *dotted* [$E_B(k) \propto k^{-5/3}$] lines (from Han et al. 2004)

11.2.6 Dark Matter

A large body of observational data on the extragalactic Universe suggests that its mass is dominated by non-baryonic dark matter.[14] In the presently widely accepted "standard" cosmological model (ΛCDM, for Cold Dark Matter with cosmological constant Λ) dark matter accounts for a fraction Ω_{DM} ~24% of the overall matter/energy budget of the Universe, baryons for ~4% and dark energy—or cosmological constant—for the remaining ~72%.

The presence of dark matter in spirals is deduced from the fact that their rotation curves beyond a radius of ~3 scalelengths do not fall off as rapidly as expected from their baryonic content. In the case of the Milky Way, the rotation curve is poorly determined beyond the Sun's location ($R_\odot$ = 8 kpc). It is then assumed, rather than directly inferred from observations, that the MW is found inside a dark matter halo with a density profile $\rho_{DM}(r)$ similar to those found in numerical simulations of structure formation in a ΛCDM universe (e.g. Navarro et al. 1996). In the absence of baryons such simulations predict approximately universal density profiles $\rho_{DM}(r) \propto r^{-k}$, with k being itself a positive function of radius r: $k(r) \propto r^s$ ("Einasto profile").

The shape of the dark matter density profile in the inner Galaxy is obviously crucial for the corresponding profile of the putative electromagnetic radiation emitted from dark matter decay, annihilation or de-excitation. Since dark mater is sub-dominant in the inner Galaxy (see Fig. 11.19) , observations of the rotation curve cannot help determine its density profile. Analysing observations of the optical depth of the inner Galaxy to microlensing events (which are affected only by the baryonic mater) Binney and Evans (2001) find $k \sim 0.3$. On the other hand, rotation curves of smaller than the Milky Way external galaxies (which are dominated by dark matter) systematically suggest flat profiles (Gentile et al. 2007; Spano et al. 2008) with $k \sim 0$, such as those obtained in the case of an "isothermal" dark halo. A useful parametrization of the density profiles is

$$\rho(r) = \frac{\rho_0(r_0)}{(r/r_0)^\gamma \, [1 + (r/r_0)^\alpha]^{(\beta-\gamma)/\alpha}}, \tag{11.32}$$

where ρ_0 and r_0 are, respectively, the characteristic mass/energy density and radius of the halo and α, β, γ are parameters with values (found, either from simulations or from observations), reported in Table 11.5.

The shape of the dark halo profile may deviate from spherical symmetry. A triaxial shape arises naturally from the fact that gravitational collapse of the halo starts first (and proceeds more rapidly) in one direction. However, other processes may subsequently erase it (e.g. gas cooling, Kazantzidis et al. 2004).

Structure formation in the ΛCDM model leads to a hierarchy of dark haloes embedded within the main halo of a galaxy. Since smaller galaxies are more dark

[14] Alternatively, the theory of general relativity may fail at small accelerations.

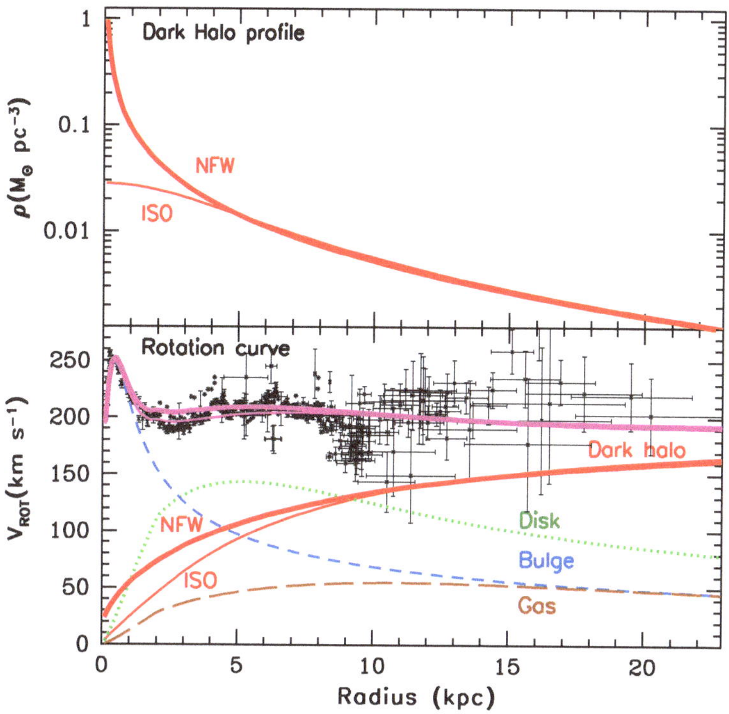

Fig. 11.19 Dark matter density profile (*top*) and rotational velocity (*bottom*) of the Milky Way; the various components (bulge, stellar disk, gas and dark halo) contributing to the latter are also indicated. In both panels *thick* and *thin solid* curves correspond to NFW and isothermal ("ISO") dark halo profiles, respectively. Data points are from Sofue et al. (2008)

Table 11.5 Models for the Milky Way dark halo profile

DM profile	ISO	BE	NFW	M99
α	2	1	1	1.5
β	2	3	3	3
γ	0	0.3	1	1.5
r_0 (kpc)	5	10	20	30
ρ_0 ($M_\odot$ pc^{-3})	5×10^{-2}	7×10^{-2}	1.0×10^{-2}	1.7×10^{-3}
ρ_0 (GeV cm^{-3})	1.89	2.65	0.38	0.065

matter dominated than larger ones, the strongest signal from dark matter annihilation may not arise from the main halo, but from satellite galaxies. This important issue has been extensively studied recently. Analyzing one of the largest "Milky Way size" simulations so far, Springel et al. (2008) find that the most intense emission is expected to arise from the main halo. See Chap. 7, positron annihilation.

11.3 Applications and the Solar Neighborhood

11.3.1 Chemical Evolution of the Local Disk

Obviously, the larger the number of observables, the more the framework provided by GCE can be useful in constraining the history of the system. The Milky Way and, in particular, the *Solar cylinder*, is the best observed system today (for obvious reasons).

The Solar cylinder may be defined as a cylindrical region of radius 0.5 kpc, perpendicular to the galactic plane and centered at Sun's position (at 8 kpc from the Galactic center, see Fig. 11.14). The interstellar gas is located near the plane and it is usually assumed to be chemically homogeneous. On the other hand, three stellar populations coexist, to various extents, in that region; they are distinguished by their kinematic and chemical properties:

- The *thin disk*: young (age $\sim$ 5 Gyr on average), rotationally supported, metal rich ([Fe/H]$\sim$$-$0.1 on average), with small vertical velocity dispersion (σ_W $<$25 km/s) and scaleheight (h $\sim$250 pc on average, but only h $\sim$100 pc for the young stars, i.e. comparable to the scale height of the gaseous layer), dominating the total mass (75–80%).
- The *thick disk*: older than the thin disk ($\sim$10 Gyr), with greater σ_W($>$35 km/s) and more extended (scale height h $\sim$1 kpc), rotating, moderately metal poor ([Fe/H]$\sim$$-$0.7) and contributing $<$20% to the total surface density.
- The *galactic halo*: old (age $\sim$ 12–13 Gy), "pressure" supported, metal poor ([Fe/H]$<$$-$1) with little or no rotation, large velocity dispersion perpendicularly to the disk and small contribution ($<$5%) to the overall surface density.

It is tempting to assume, in the framework of the old, *monolithic collapse* scenario for the Milky Way's evolution (Eggen et al. 1962), a temporal continuity in the formation of those three components, i.e. halo $\longrightarrow$ thick disk $\longrightarrow$ thin disk.[15] Such a continuity is more difficult to establish in the modern framework of hierarchical galaxy formation: the three components may be totally uncorrelated, e.g. the thick disk may have been formed from tidal debris of satellite galaxies (Villalobos and Helmi 2008), while the thin disk from a slow accretion processes In the following we shall study the case of the local thin disk. In fact, if thick disk stars were mostly accreted from merging/disrupted satellites, that population cannot be considered as belonging to the early phase of the local disk. Also, in view of the large scaleheight of the thick disk ($>$1000 pc) and of the elliptical orbits of its stars (vs. circular orbits for those of the thin disk) one may wonder how large the local "chemical box" could be and still be considered as a single system with well defined evolution.

[15] In fact, it is difficult to account for the chemical and kinematical properties of all three galactic components in the monolithic collapse scenario.

11.3.2 Observables

In the case of the Solar neighborhood, the number of available observational data is larger than for any other galactic system, allowing one to constrain strongly (albeit not in a unique way) the history of the system. Those data are (see, e.g. Boissier and Prantzos 1999)

- The current surface densities of gas ($\Sigma_G \sim 12\,M_\odot$ pc^{-2}), live stars ($\Sigma_* \sim 30\,M_\odot$ pc^{-2}), stellar residues ($\Sigma_C \sim 5\,M_\odot$ pc^{-2}) and total amount of baryonic matter ($\Sigma_T \sim 47\,M_\odot$ pc^{-2}), as well as the current star formation rate ($\Psi_0 = 2$–$5\,M_\odot$ pc^{-2} Gyr^{-1}); the corresponding gas fraction is $\sigma \sim 0.22$.
- The elemental and isotopic abundances at solar birth ($X_{i,\odot}$, i.e. it is assumed that the Sun's composition is typical of the ISM 4.5 Gyr ago) and today ($X_{i,Now}$). Those two compositions are quite similar, suggesting little chemical evolution in the past 4.5 Gyr. Of particular importance in that respect is the abundance of deuterium, a fragile primordial isotope destroyed 100% in stellar ejecta, but continuously reintroduced in the system in the case of infall of primordial gas. The evolution of the D abundance is a tracer of the amount of astration plus infall of the system.
- The stellar age-metallicity relationship (AMR), traced by the Fe abundance of long-lived stars, [Fe/H]=$f(t)$ or $log(Z) = f(t)$, Sect. 11.3.3.1.
- The metallicity distribution (MD) of long-lived G-type stars $\frac{dn}{d[Fe/H]}$ or $\frac{dn}{d(logZ)}$, showing that few of them were formed at [Fe/H]< -0.7 ($X_{Fe} \sim 0.2\ X_{Fe,\odot}$). In view of its importance, the MD is discussed in more detail in the following Section.
- The oxygen vs. iron (O-Fe) relationship, interpreted in terms of a delayed (after $\sim$1 Gyr) enrichment of the ISM with products of SNIa. Since Fe/O at low metallicities (where Fe and O are produced only by massive stars) is $\sim$const.$\leq$0.5 (Fe/O)$_\odot$, and since massive stars are the unique source of oxygen, this implies that the complementary source of Fe, SNIa, contribute more than half of the solar Fe (see also Fig. 11.8, upper right panel).

Among those constraints, the age-metallicity relationship and the G-dwarf metallicity distribution are the most important ones. In principle, by combining them, one may derive straightforwardly the SFR history (dn/dt) of the solar neighborhood through:

$$\frac{dn}{dt} = \frac{dn}{d(logZ)} \frac{d(logZ)}{dt} \tag{11.33}$$

In practice, however, this is impossible, because of the sensitivity of the result to the slope of the adopted AMR: a small variation in the form of the AMR produces a dramatic effect on the resulting SFR history. For that reason, the local SFR history is reconstructed only indirectly: models of the GCE of the Solar neighbourhood are developed, which must satisfy all the aforementioned observational constraints.

Before presenting such models, we discuss in some more detail below two of the key observables, namely the AMR and the MD.

11.3.3 The Local Metallicity Distribution in Stars

In the framework of the Closed box model of GCE with IRA, the MD of long-lived stars can be derived as follows. In order to reach a given metallicity Z, a certain amount of stars has to be created, given by (Eq. (11.6)): $m_S = m - m_G$, or (by normalising to the total mass m):

$$n = 1 - \sigma \tag{11.34}$$

where $\sigma = m_G/m$ is the gas fraction and $n = m_S/m$ the star fraction. For a system with a final metallicity Z_1 and star fraction $n_1 = 1 - \sigma_1$, the *cumulative metallicity distribution* (CMD), i.e. the number of stars with metallicity lower than Z as a function of Z, is given by

$$\frac{n(< Z)}{n_1} = \frac{1-\sigma}{1-\sigma_1} \tag{11.35}$$

By using the fundamental result of IRA in the case of the Closed Box, namely $\sigma = exp(-Z/p)$ (Eq. (11.26)) and taking the derivatives of Eq. (11.35), one obtains the *differential metallicity distribution* (DMD):

$$\frac{d(n/n_1)}{d(logZ)} = \frac{ln(10)}{1 - exp(-Z_1/p)} \frac{Z}{p} e^{-Z/p} \tag{11.36}$$

i.e. the number of stars per logarithmic metallicity interval as a function of metallicity Z. This relation has a maximum for $Z = p$, i.e. when the metallicity is equal to the yield (both metallicity and yield can be expressed in units of the solar abundance of the corresponding element). It is important to note that *the MD is independent of the SFR* (at least in the framework of IRA) and for that reason it provides a very powerful constrain on models of GCE.

The local MD obtained from the largest yet survey of the solar neighborhood (Nordström et al. 2004) appears in Fig. 11.20 (histogram in bottom left panel). It peaks at [Fe/H] = −0.1, suggesting an effective yield of p_{Fe} ~0.8 $Z_{\odot}$. The GKS MD can be fitted relatively well by a Gaussian of width $\sigma[Fe/H] = 0.19$ dex (dotted curve in Fig. 11.20 bottom left). Before interpreting it in the framework of some model, corrections should be applied to account for the different scaleheights of the involved stellar populations, which are attributed to the dynamical heating of the stellar disk. These scaleheights are reflected in the larger velocity dispersion σ_W (vertically to the disk) with decreasing metallicity (top left panel of Fig. 11.20): the older (and more metal poor) stars, being scattered for a longer time, obtain

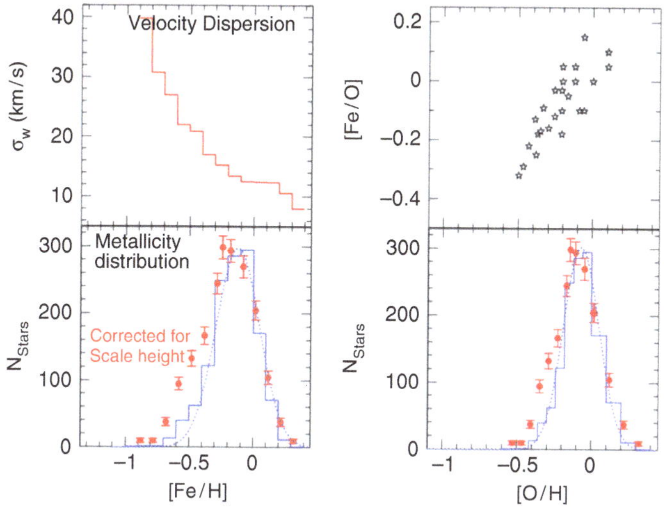

Fig. 11.20 Observations of stars in the solar neighbourhood, as a function of [Fe/H] (*left*) and [O/H] (*right*). *Top left:* Vertical velocity dispersion σ_W; *Bottom left:* Metallicity distribution from Nordström et al. (2004) (*blue histogram*), fitted by a Gaussian of σ([Fe/H]) = 0.19 dex (*dotted curve*) and corrected for velocity dispersion (*points*, from Holmberg et al. (2007)). *Top right:* [Fe/O] vs [O/H] from Edvardsson et al. (1993a). *Bottom right:* as in bottom left, but as a function of [O/H]; note the narrower distribution (fitted with a Gaussian of width σ([O/H]) = 0.13 dex). Model results for the solar cylinder should be compared to the scale-height corrected MD

higher velocities and are found, on average, at larger distances from the plane. This implies that a volume limited local sample misses proportionally more of those low metallicity and old stars; therefore, appropriate corrections should be made before comparison with models. The corresponding corrected MD appears as data points in the lower left panel of Fig. 11.20. Models of the solar neighborhood are usually made for the solar cylinder (i.e. all quantities are expressed per unit surface density) and their results should be compared to the *corrected MD*. One may also use the observed Fe vs O relationship (appearing in the top right panel of Fig. 11.20) to convert the MD as a function of [O/H] (bottom right panel in Fig. 11.20): it is even narrower than as a function of [Fe/H], with a width of $\sigma[O/H] = 0.12$ dex.

Equation (11.36) for the Closed box model with initial metallicity $Z_0 = 0$ appears in Fig. 11.21 (solid curve), where it is compared to (corrected for scaleheight) data for the local disk. It predicts many more stars at low metallicities than observed, a problem known as the "G-dwarf problem".[16]

[16]G-type stars are bright enough for a reasonably complete sample to be constructed and long-lived enough to survive since the earliest days of the disk; the same problem is encountered if F- or K-type stars are used.

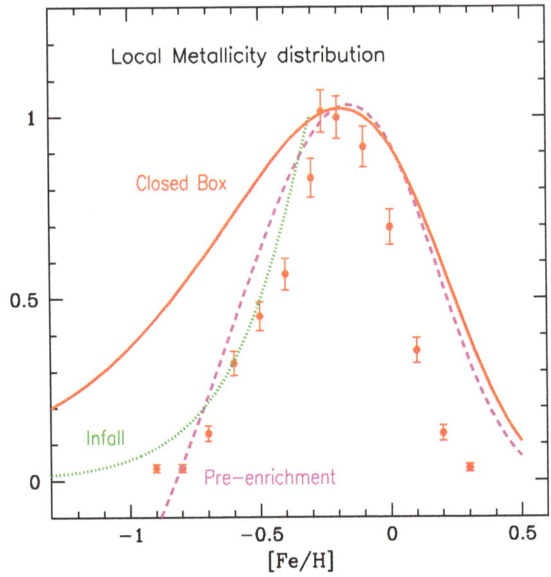

Fig. 11.21 Metallicity distribution of long-lived stars for three models: Closed Box, exponentially decreasing Infall (with a timescale of 7 Gyr) and Pre-enrichment (with $X_0 = 0.08 X_{\odot}$ for Fe), respectively. They are compared to the GKS data for the Solar neighborhood (corrected for scaleheight as in Holmberg et al. (2007))

Two of the main solutions proposed for the G-dwarf problem appear also on Fig. 11.21. According to the first one, the disk started with an initial metallicity $Z_0 \sim 0.1\,Z_{\odot}$ (*pre-enrichment*). In that case, all metallicities in Eq. (11.36) are replaced by $Z - Z_0$ and the resulting curve fits relatively well the data. The main drawback of that hypothesis is that it is hard to justify the origin of such a large pre-enrichment: it is true that the Galactic halo, which preceded disk formation, reached a maximum metallicity of $\sim 0.1\,Z_{\odot}$ (for Fe); but its average (stellar) metallicity is $\sim 0.03\,Z_{\odot}$ (for Fe) and its total mass ($\sim 4 \times 10^8\,M_{\odot}$) is almost 100 times smaller than the one of the disk ($\sim 3.5 \times 10^{10}\,M_{\odot}$). There is simply not enough mass and metals produced in the halo to justify pre-enrichment of the disk to such a high level. Moreover, the halo has a low specific angular momentum (contrary to the disk) and material escaping it should be accreted rather by the bulge, not the disk.

The second hypothesis is that the disk did not evolve as a Closed box, but was gradually built from *infall* of metal free (or metal poor) material. In the Closed box, all the gas of the system is available from the very beginning; a large stellar activity is then required to enrich all that gas to, say, $0.1\,Z_{\odot}$, and correspondingly many small and long-lived stars are formed at low Z. In the case of infall, only a small amount of gas exists early on; it takes then a small number of SN to enrich it to $0.1\,Z_{\odot}$, and correspondingly few low mass stars are formed at low Z.

Infall appears then as an elegant solution to the local G-dwarf problem, especially in view of the fact that gas accretion to galaxies is expected to be a common phenomenon in the Universe. The rate of the infall is not precisely determined by the data of the local disk. An exponentially decreasing infall rate $f(t) = A\, exp(-t/\tau)$

with a long characteristic time-scale of $\tau \sim$7–10 Gyr provides a reasonably good fit to the data and, in view of its simplicity, is often used in models of the solar neighborhood (see next section). However, other forms may do as well and even better; this is the case, for instance, of a Gaussian (as a function of time) infall rate, with a maximum prior to solar system formation (see, e.g. Prantzos and Silk 1998).

The infall rate $f(t)$ must obviously be normalized such as:

$$\int_0^T f(t)dt = \Sigma_T \tag{11.37}$$

where Σ_T is the total surface density in the Solar neighbourhood and T is the age of the Galactic disk.

11.3.3.1 The Local Age-Metallicity Relationship

Stellar ages are much harder to evaluate than stellar metallicities, and the form of the local AMR has varied considerably over the years. The seminal work of Edvardsson et al. (1993a) on 189 F-dwarfs established a clear trend of decreasing metallicity with age, albeit with substantial scatter (Fig. 11.11). A similar trend is obtained in Rocha-Pinto et al. (2000), where ages are evaluated from the chromospheric activity of stars. Such a trend is compatible with (and predicted by) all simple models of local GCE, either closed or open (i.e. with infall) models. It should be noted, however, that the adopted selection criteria in the paper of Edvardsson et al. (1993a) introduced a bias against old metal-rich and young metal-poor stars. Also, considerable systematic errors affect estimates based on chromospheric activity.

The large Geneva-Kopenhagen survey (GKS) of Nordström et al. (2004), concerning ~14,000 F and G stars with 3-D kinematic information (but less precise spectroscopy than the Edvardsson et al. (1993b) study), provides a radically different picture: the volume limited subsample of 462 stars with "well-defined" ages within 40 pc displays a flat AMR (an average metallicity of [Fe/H]~−0.2 at all ages) with a very large scatter; similar results are obtained for a larger volume sample (Fig. 11.20). Accounting for the fact that the oldest stars have the largest age uncertainties does not modify the flatness of the AMR. Such a trend is also obtained in Soubiran and Girard (2005), with spectroscopic metallicities of a large sample of red giant stars. Figure 11.20 suggest that the various studies are in reasonable agreement for ages <8 Gyr, and start diverging at larger ages.

The question of the intrinsic dispersion of the local AMR is also of utmost importance. Rocha-Pinto et al. (2000) find little dispersion (compatible with observational errors), whereas Feltzing et al. (2001) and Nordström et al. (2004) find quite large dispersion at all ages. Even if the dispersion turns out to be real, it is not clear whether it is due to inhomogeneous evolution of the solar vicinity, or to the scattering of stars from regions of different metallicities (mostly inner disk) to the solar neighbourhood. The former possibility seems improbable: indeed, observations of abundances in the local gas, as well as in the young stars in the nearby Orion region, suggest a quite uniform chemical composition *today* with very small dispersion (compared to observational uncertainties) and an average metallicity equal to solar. Thus, the latter possibility, namely the scatter of stars from other Galactic regions—having different evolutionary histories—into the solar neighbourhood, seems more plausible. One should recall that simple GCE models provide a unique AMR (assumed to represent an average trend) and cannot account for dispersion.

11.3.4 A Brief History of the Solar Neighbourhood

The observed properties of the local disk "dictate" the parameters of simple GCE models that may be built for that system. The results of such a model are displayed in Fig. 11.22, as a function of time (left panels) and of metallicity [Fe/H] (right panels). The various parameters of the model are adjusted as follows:

- The total amount of infalling matter is normalized to the local surface density through Eq. (3.5).
- The time-scale of the adopted exponentially decreasing infall rate is sufficiently long ($\tau = 7$–8 Gyr) as to reproduce the (scaleheight-corrected) metallicity distribution.[17]
- The coefficient ν of the SFR $\Psi = \nu\, m_G$ (Eq. (11.13)) is adjusted as to leave the system at $T = 12$ Gyr with a gas fraction $\sigma \sim 0.2$, as observed.

With those parameter adjustments, it remains to be seen whether the other observables of the system are reproduced. In Fig. 11.22 it is seen that the current model SFR is well within observational uncertainties. Note that the derived SFR history is rather flat, around an average value of $\sim 3.8\, M_{\odot}/pc^2/year$ (alternatively, it could be fitted with a broad Gaussian).

The resulting age-metallicity relationship (left bottom panel in Fig. 11.22) fits the data approximately, but its should be stressed that uncertainties in stellar ages are fairly large, and in particularly, at large ages; also, dispersion in the age-metallicity

[17]In fact, the data can be corrected for scaleheight effects only in the framework of a model of the vertical dynamical equilibrium of the local disk, which *requires* as input the "weight" of each stellar population as a function of its metallicity, i.e. a chemo-dynamical model of GCE is required. Thus, the best approach consists, not in comparing the GCE model with the corrected data, but in (a) calculating the local vertical structure by using the GCE data (surface density vs metallicity) as input, (b) correcting the GCE model metallicity distribution (which refers to the total surface area of the local cylinder), by taking into account the vertical disk structure calculated in (a), and finally (c) comparing the corrected MD of the GCE model to the original (local volume) data. Such an approach is used e.g. Sommer-Larsen and Dolgov (2001).

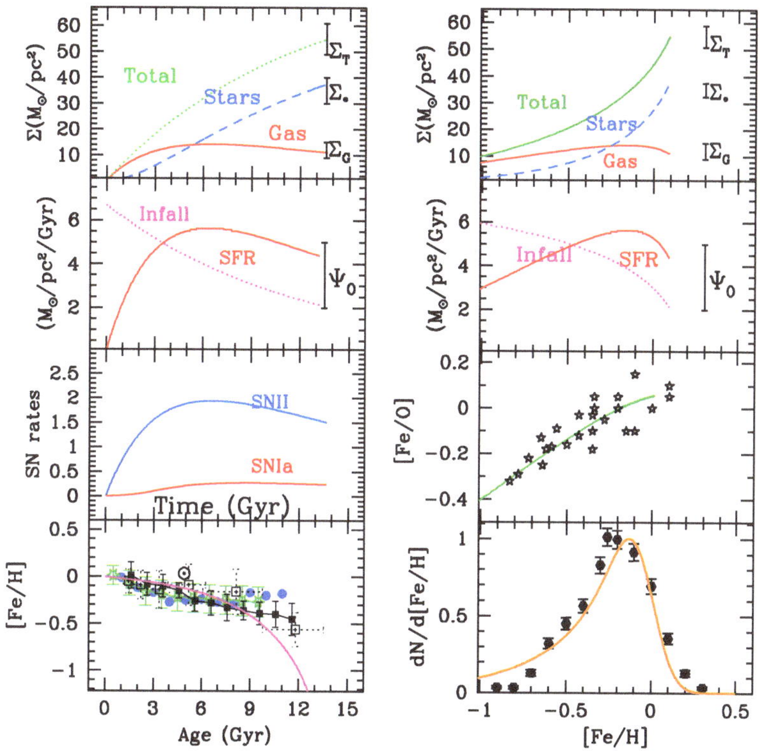

Fig. 11.22 History of the Solar neighborhood, according to a GCE model with infall, constrained by various observables (updated from Boissier and Prantzos 1999). *Left:* Results are plotted as a function of time (or age, for the bottom panel). Data for the present-day local disk are displayed with vertical bars. Data for the age-metallicity relation are from various sources. *Right:* Results are plotted as a function of metallicity [Fe/H]. See text for comments on the various curves

relation appears to be quite large (see previous section). Thus, this observable may be of little use as a constraint for the local GCE.

The rise of Fe/O (right panel of Fig. 11.22) is due to delayed contribution of Fe by SNIa. The resulting SNIa rate as a function of time appears on the left panel.

Finally, the local GCE model, combined with the adopted stellar yields, should also reproduce the pre-solar composition, well established after meteoritic and photospheric measurements (e.g., Lodders 2003). The results of such a comparison (Fig. 11.23) show that all elements and almost all isotopes are nicely co-produced (within a factor of 2 from their pre-solar value), with key elements such as O and Fe being very well reproduced. Taking into account the large abundance variations between O and Sc (a factor of 10^6) this agreement should be considered as a triumph for stellar nucleosynthesis models. That result is quite encouraging, since it shows that the adopted IMF, SFR and stellar yields reproduce very well the solar system composition, but only under the assumption that this composition *is typical of the local ISM 4.5 Gyr ago*. If this is not the case (as the GKS data seem to

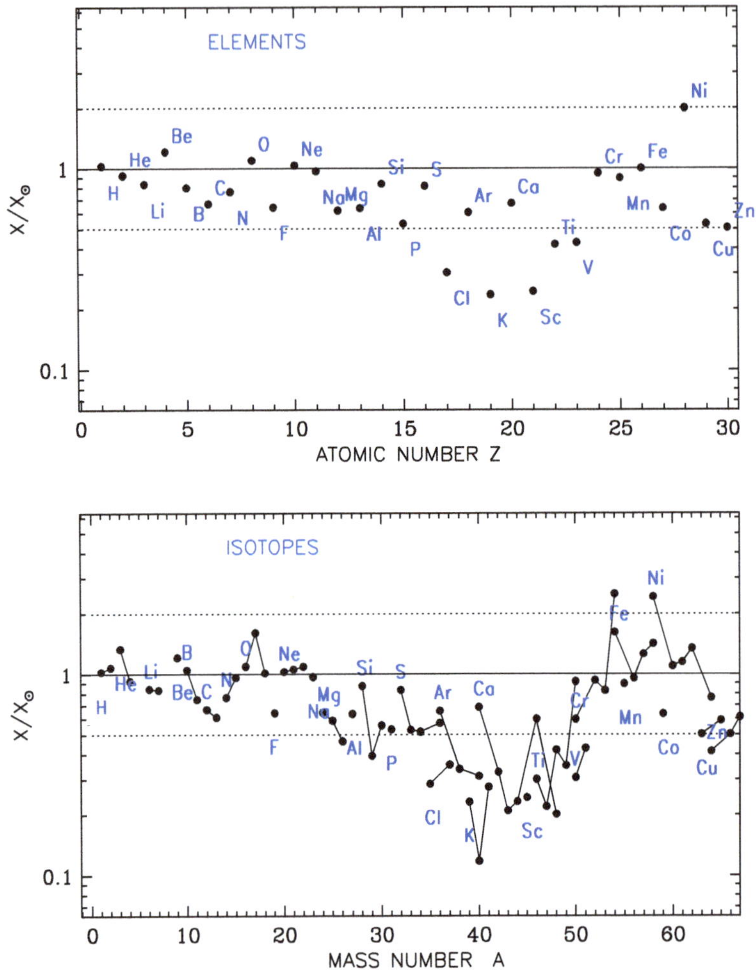

Fig. 11.23 Average composition of 4.5 Gyr old stars in the solar neighbourhood, elemental (top) and isotopic (bottom) compared to the observed solar composition. Most elements and isotopes are co-produced within a factor of two of their solar value. The corresponding values for the local ISM 4.5 Gyr ago are ~0.1–0.15 dex lower for all elements but H and He (from Kubryk et al. 2015a)

indicate), i.e. if the Sun is an outlier (either because it was formed in a higher than average metallicity local cloud, or because it was born in the inner Galactic regions and radially migrated outwards), simple GCE models cannot be used to explain simultaneously local and solar data.

Indeed, in the 2000s, observations revealed the inadequacies of this simple model, regarding several observables in the solar neighborhood. In particular, despite the difficulties in estimating stellar ages, the early age-metallicity relation is flatter than the theoretical one. Moreover, there is considerable dispersion of metallicity at any age, much larger than the one in the local gas (Cartledge et al. 2006); this is impossible to reproduce with 1-zone models where gas is instantaneously and completely mixed. As for the metallicity distribution, one-zone models cannot simultaneously reproduce the metallicity of the local gas and young stars ($\sim Z_\odot$ today) and the most metal-rich old stars (with metallicities $\sim 2\,Z_\odot$). Neither can they explain the observed presence of both old and young stars at all metallicities (Casagrande et al. 2011) in the solar neighborhood.

On the other hand, Sellwood and Binney (2002) showed the potential importance of the dynamical disk evolution to its chemical one. The orbit of a test particle (star) in the potential of a galactic disk is commonly described, to first order approximation, as the superposition of a main circular motion (defining the *guiding radius*), and harmonic oscillations called epicycles. A commonly adopted designation is "blurring" for the radial oscillations around the guiding radius and "churning" for the modifications of the guiding radius. Churning may occur through resonant interactions of the star with non-axisymmetric structures of the gravitational potential (spirals, bar), causing changes in the angular momentum of the stars. The process conserves the overall distribution of angular momentum and does not add random motion, that is, it does not "heat" the disk radially. In contrast, blurring conserves the angular momentum of individual stars but it heats the disk in the radial direction (the epicyclic radius increases with time).

Although the extent of churning cannot be evaluated from fist principles up to now, its is commonly found in numerical N-body simulations (Roškar et al. 2008; Kubryk et al. 2015a) and its effects on the chemical evolution have been studied in the past decade both with N-body and semi-analytical models (Roškar et al. 2008; Schönrich and Binney 2009; Loebman et al. 2011; Minchev et al. 2013; Grand et al. 2014; Kubryk et al. 2015a). The main effect is that churning induces a radial mixing of stellar populations, displacing stars from the inner disk to the outer one (and vice versa, albeit to a smaller extent). This concerns both "passive players" of chemical evolution (low-mass stars keeping on their surfaces the composition of the their birth place which may be far from their current one) and "active players", namely long-lived sources of some elements (SNIa for Fe, AGBs for s-process) which may release their products far away from their birth place (Kubryk et al. 2013).

These developments introduce considerable complexity in studies of galactic chemical evolution and the final picture is far from being clear yet. At the same time, they offer interesting solutions to important observational problems, like the dispersion in the local age-metallicity relation, the double-branch sequence of [α/Fe] vs. [Fe/H] observed in the solar neighborhood, etc. Most importantly, these developments suggest that the evolution of the solar neighborhood is related to a large extent to the evolution of the whole MW disk and cannot be studied in isolation.

References

Armstrong JW, Rickett BJ, Spangler SR (1995) Electron density power spectrum in the local interstellar medium. Astrophys J 443:209–221. https://doi.org/10.1086/175515

Audouze J, Tinsley BM (1976) Chemical evolution of galaxies. Annu Rev Astron Astrophys 14:43–79. https://doi.org/10.1146/annurev.aa.14.090176.000355

Beck R (2008) Galactic and extragalactic magnetic fields. In: American Institute of Physics conference series, vol 1085, pp 83–96. https://doi.org/10.1063/1.3076806

Bell EF, Zucker DB, Belokurov V, Sharma S, Johnston KV, Bullock JS, Hogg DW, Jahnke K, de Jong JTA, Beers TC, Evans NW, Grebel EK, Ivezic Z, Koposov SE, Rix HW, Schneider DP, Steinmetz M, Zolotov A (2007) The accretion origin of the milky way's stellar halo. ArXiv e-prints 706, 0706.0004

Beuermann K, Kanbach G, Berkhuijsen EM (1985) Radio structure of the galaxy - thick disk and thin disk at 408 MHz. Astron Astrophys 153:17–34

Bienaymé O, Soubiran C, Mishenina TV, Kovtyukh VV, Siebert A (2006) Vertical distribution of galactic disk stars. Astron Astrophys. arXiv:astro-ph/0510431

Binney JJ, Evans NW (2001) Cuspy dark matter haloes and the Galaxy. Mon Not R Astron Soc 327:L27–L31. https://doi.org/10.1046/j.1365-8711.2001.04968.x, astro-ph/0108505

Blitz L, Rosolowsky E (2006) The role of pressure in GMC formation II: the H_2-pressure relation. Astrophys J 650:933–944. https://doi.org/10.1086/505417, astro-ph/0605035

Boissier S, Prantzos N (1999) Chemo-spectral evolution of the milky way and of spiral disks. Astrophys Space Sci 265:409–410. https://doi.org/10.1023/A:1002181826858

Boulanger F, Perault M (1988) Diffuse infrared emission from the galaxy. I - solar neighborhood. Astrophys J 330:964–985. https://doi.org/10.1086/166526

Boulanger F, Abergel A, Bernard JP, Burton WB, Desert FX, Hartmann D, Lagache G, Puget JL (1996) The dust/gas correlation at high Galactic latitude. Astron Astrophys 312:256–262

Breitschwerdt D (2001) Modeling the local interstellar medium. Astrophys Space Sci 276:163–176

Breitschwerdt D (2004) Self-consistent modelling of the interstellar medium. Astrophys Space Sci 289:489–498. https://doi.org/10.1023/B:ASTR.0000014982.31688.bf, arXiv:astro-ph/0303237

Brown JC, Haverkorn M, Gaensler BM, Taylor AR, Bizunok NS, McClure-Griffiths NM, Dickey JM, Green AJ (2007) Rotation measures of extragalactic sources behind the southern galactic plane: new insights into the large-scale magnetic field of the inner milky way. Astrophys J 663:258–266. https://doi.org/10.1086/518499, arXiv:0704.0458

Calzetti D, Kennicutt RC (2009) The new frontier: galactic-scale star formation. Publ Astron Soc Pac 121:937–941. https://doi.org/10.1086/605617, 0907.0203

Cameron AGW, Truran JW (1971) The chemical evolution of the galaxy. J Roy Soc Can 65:1

Cartledge SIB, Lauroesch JT, Meyer DM, Sofia UJ (2006) The homogeneity of interstellar elemental abundances in the galactic disk. Astrophys J 641:327–346. https://doi.org/10.1086/500297, astro-ph/0512312

Casagrande L, Schönrich R, Asplund M, Cassisi S, Ramírez I, Meléndez J, Bensby T, Feltzing S (2011) New constraints on the chemical evolution of the solar neighbourhood and Galactic disc(s). Improved astrophysical parameters for the Geneva-Copenhagen Survey. Astron Astrophys 530:A138. https://doi.org/10.1051/0004-6361/201016276, 1103.4651

Case GL, Bhattacharya D (1998) A new sigma -d relation and its application to the galactic supernova remnant distribution. Astrophys J 504:761, arXiv:astro-ph/9807162

Cerviño M, Knödlseder J, Schaerer D, von Ballmoos P, Meynet G (2000) Gamma-ray line emission from OB associations and young open clusters. I. Evolutionary synthesis models. Astron Astrophys 363:970–983. arXiv:astro-ph/0010283

Chabrier G (2003) Galactic stellar and substellar initial mass function. Publ Astron Soc Pac 115:763–795. https://doi.org/10.1086/376392, astro-ph/0304382

Chabrier G (2005) The initial mass function: from salpeter 1955 to 2005. In: Corbelli E, Palla F, Zinnecker H (eds) The initial mass function 50 years later, astrophysics and space science library, vol 327, p 41. https://doi.org/10.1007/978-1-4020-3407-7_5, astro-ph/0409465
Chabrier G, Hennebelle P, Charlot S (2014) Variations of the stellar initial mass function in the progenitors of massive early-type galaxies and in extreme starburst environments. Astrophys J 796:75. https://doi.org/10.1088/0004-637X/796/2/75, 1409.8466
Chiappini C, Matteucci F, Gratton R (1997) The chemical evolution of the galaxy: the two-infall model. Astrophys J 477:765–780. https://doi.org/10.1086/303726, astro-ph/9609199
Chieffi A, Limongi M (2004) Explosive yields of massive stars from $Z = 0$ to $Z = Z_{\odot}$. Astrophys J 608:405–410. https://doi.org/10.1086/392523, arXiv:astro-ph/0402625
Chomiuk L, Povich MS (2011) Toward a unification of star formation rate determinations in the milky way and other galaxies. Astron J 142:197. https://doi.org/10.1088/0004-6256/142/6/197, 1110.4105
Clayton DD (1968) Principles of stellar evolution and nucleosynthesis. McGraw-Hill, New York
Clayton DD (1985) Galactic chemical evolution and nucleocosmochronology: a standard model. In: Arnett WD, Truran JW (eds) Nucleosynthesis: challenges and new developments. University of Chicago Press, Chicago, p 65
Clayton DD (1988) Nuclear cosmochronology within analytic models of the chemical evolution of the solar neighbourhood. Mon Not R Astron Soc 234:1–36. https://doi.org/10.1093/mnras/234.1.1
Cordes JM, Lazio TJW (2002) NE2001.I. A new model for the galactic distribution of free electrons and its fluctuations. ArXiv Astrophysics e-prints. astro-ph/0207156
Cowley C (1995) Introduction to cosmochemistry. Cambridge University Press, Cambridge
Crutcher RM (1999) Magnetic fields in molecular clouds: observations confront theory. Astrophys J 520:706–713. https://doi.org/10.1086/307483
Crutcher RM (2007) Magnetic fields in molecular clouds. EAS Publications Ser 23:37–54. https://doi.org/10.1051/eas:2007004
Dame TM (1993) The distribution of neutral gas in the milky way. In: Holt SS, Verter F (eds) Back to the galaxy. American Institute of Physics conference series, vol 278, pp 267–278
Diehl R, Halloin H, Kretschmer K, Lichti GG, Schönfelder V, Strong AW, von Kienlin A, Wang W, Jean P, Knödlseder J, Roques JP, Weidenspointner G, Schanne S, Hartmann DH, Winkler C, Wunderer C (2006) Radioactive ^{26}al from massive stars in the galaxy. Nature 439:45–47. arXiv:astro-ph/0601015
Edvardsson B, Andersen J, Gustafsson B, Lambert DL, Nissen PE, Tomkin J (1993a) The chemical evolution of the galactic disk - part one - analysis and results. Astron Astrophys 275:101
Edvardsson B, Andersen J, Gustafsson B, Lambert DL, Nissen PE, Tomkin J (1993b) The chemical evolution of the galactic disk - part two - observational data. Astron Astrophys Suppl 102:603
Eggen OJ, Lynden-Bell D, Sandage AR (1962) Evidence from the motions of old stars that the Galaxy collapsed. Astrophys J 136:748. https://doi.org/10.1086/147433
Elmegreen BG (2002) Star formation from large to small scales. Astrophys Space Sci 281:83–95
Englmaier P, Gerhard O (1999) Gas dynamics and large-scale morphology of the Milky Way galaxy. Mon Not R Astron Soc 304:512–534
Feltzing S, Holmberg J, Hurley JR (2001) The solar neighbourhood age-metallicity relation - does it exist? Astron Astrophys 377:911–924. https://doi.org/10.1051/0004-6361:20011119, astro-ph/0108191
Ferriere K (1998) Global model of the interstellar medium in our Galaxy with new constraints on the hot gas component. Astrophys J 497:759. https://doi.org/10.1086/305469
Ferrière KM (2001) The interstellar environment of our galaxy. Rev Mod Phys 73:1031–1066. arXiv:astro-ph/0106359
Ferrière K, Gillard W, Jean P (2007) Spatial distribution of interstellar gas in the innermost 3 kpc of our galaxy. Astron Astrophys 467:611–627. https://doi.org/10.1051/0004-6361:20066992
Figer DF, Rich RM, Kim SS, Morris M, Serabyn E (2004) An extended star formation history for the galactic center from hubble space telescope nicmos observations. Astrophys J 601:319–339. arXiv:astro-ph/0309757

Flynn C, Holmberg J, Portinari L, Fuchs B, Jahreiß H (2006) On the mass-to-light ratio of the local galactic disc and the optical luminosity of the galaxy. Mon Not R Astron Soc 372:1149–1160. arXiv:astro-ph/0608193

François P, Matteucci F, Cayrel R, Spite M, Spite F, Chiappini C (2004) The evolution of the milky way from its earliest phases: constraints on stellar nucleosynthesis. Astron Astrophys 421:613–621. https://doi.org/10.1051/0004-6361:20034140, astro-ph/0401499

Freudenreich HT (1998) A cobe model of the galactic bar and disk. Astrophys J 492:495, arXiv: astro-ph/9707340

Frisch PC, Bzowski M, Grün E, Izmodenov V, Krüger H, Linsky JL, McComas DJ, Möbius E, Redfield S, Schwadron N, Shelton R, Slavin JD, Wood BE (2009) The galactic environment of the sun: interstellar material inside and outside of the heliosphere. Space Sci Rev 146:235–273. https://doi.org/10.1007/s11214-009-9502-0

Fukugita M, Peebles PJE (2004) The cosmic energy inventory. Astrophys J 616:643–668. https://doi.org/10.1086/425155, astro-ph/0406095

Gaidos E, Krot AN, Williams JP, Raymond SN (2009) ^{26}Al and the formation of the solar system from a molecular cloud contaminated by Wolf-Rayet winds. Astrophys J 696:1854–1863. https://doi.org/10.1088/0004-637X/696/2/1854, 0901.3364

Gentile G, Tonini C, Salucci P (2007) Λcdm halo density profiles: where do actual halos converge to nfw ones? Astron Astrophys 467:925–931. arXiv:astro-ph/0701550

Gillessen S, Eisenhauer F, Trippe S, Alexander T, Genzel R, Martins F, Ott T (2008) Monitoring stellar orbits around the massive black hole in the galactic center. ArXiv e-prints 0810.4674

Gilmore G, Reid N (1983) New light on faint stars. iii - galactic structure towards the south pole and the galactic thick disc. Mon Not R Astron Soc 202:1025–1047

Goldsmith PF (1987) Molecular clouds - an overview. In: Hollenbach DJ, Thronson HA Jr (eds) Interstellar processes, astrophysics and space science library, vol 134, pp 51–70

Grand RJJ, Kawata D, Cropper M (2014) Orbits of radial migrators and non-migrators around a spiral arm in N-body simulations. Mon Not R Astron Soc 439:623–638. https://doi.org/10.1093/mnras/stt2483, 1310.2952

Groenewegen MAT, Udalski A, Bono G (2008) The distance to the galactic centre based on population ii cepheids and rr lyrae stars. Astron Astrophys 481:441–448. arXiv:0801.2652

Guibert J, Lequeux J, Viallefond F (1978) Star formation in interstellar gas density in our galaxy. Astron Astrophys 68:1–2

Han JL, Manchester RN, Berkhuijsen EM, Beck R (1997) Antisymmetric rotation measures in our Galaxy: evidence for an A0 dynamo. Astron Astrophys 322:98–102

Han JL, Manchester RN, Qiao GJ (1999) Pulsar rotation measures and the magnetic structure of our Galaxy. Mon Not R Astron Soc 306:371–380. arXiv:astro-ph/9903101

Han JL, Ferriere K, Manchester RN (2004) The spatial energy spectrum of magnetic fields in our Galaxy. Astrophys J 610:820–826. https://doi.org/10.1086/421760, arXiv:astro-ph/0404221

Han JL, Manchester RN, Lyne AG, Qiao GJ, van Straten W (2006) Pulsar rotation measures and the large-scale structure of the galactic magnetic field. Astrophys J 642:868–881. https://doi.org/10.1086/501444, arXiv:astro-ph/0601357

Heckman TM, Armus L, Miley GK (1990) On the nature and implications of starburst-driven galactic superwinds. Astrophys J Suppl 74:833–868. https://doi.org/10.1086/191522

Heiles C (1996) The local direction and curvature of the galactic magnetic field derived from starlight polarization. Astrophys J 462:316. https://doi.org/10.1086/177153

Heiles C, Troland TH (2005) The millennium arecibo 21 centimeter absorption-line survey. IV. Statistics of magnetic field, column density, and turbulence. Astrophys J 624:773–793. https://doi.org/10.1086/428896, arXiv:astro-ph/0501482

Holmberg J, Flynn C (2004) The local surface density of disc matter mapped by Hipparcos. Mon Not R Astron Soc 352:440–446. arXiv:astro-ph/0405155

Holmberg J, Nordström B, Andersen J (2007) The Geneva-Copenhagen survey of the Solar neighbourhood II. New uvby calibrations and rediscussion of stellar ages, the G dwarf problem, age-metallicity diagram, and heating mechanisms of the disk. Astron Astrophys 475:519–537. https://doi.org/10.1051/0004-6361:20077221, 0707.1891

Inoue M, Tabara H (1981) Structure of the galactic magnetic field in the solar neighborhood. Publ Astron Soc Jpn 33:603

Iwamoto K, Brachwitz F, Nomoto K, Kishimoto N, Umeda H, Hix WR, Thielemann FK (1999) Nucleosynthesis in Chandrasekhar mass models for type ia supernovae and constraints on progenitor systems and burning-front propagation. Astrophys J Suppl 125:439–462. arXiv: astro-ph/0002337

Jappsen AK, Klessen RS, Larson RB, Li Y, Mac Low MM (2005) The stellar mass spectrum from non-isothermal gravoturbulent fragmentation. Astron Astrophys 435:611–623. https://doi.org/10.1051/0004-6361:20042178, astro-ph/0410351

Jones A (2009) The role of dust in the interstellar medium: dust sources and dust evolution. In: Pagani L, Gerin M (eds) EAS publications series, vol 34, pp 107–118. https://doi.org/10.1051/eas:0934008

Jurić M, Ivezić Ž, Brooks A, Lupton RH, Schlegel D, Finkbeiner D, Padmanabhan N, Bond N, Sesar B, Rockosi CM, Knapp GR, Gunn JE, Sumi T, Schneider DP, Barentine JC, Brewington HJ, Brinkmann J, Fukugita M, Harvanek M, Kleinman SJ, Krzesinski J, Long D, Neilsen EH Jr, Nitta A, Snedden SA, York DG (2008) The milky way tomography with sdss. i. Stellar number density distribution. Astrophys J 673:864–914

Kalberla PMW, Dedes L (2008) Global properties of the HI distribution in the outer Milky Way. ArXiv e-prints 804. 0804.4831

Kazantzidis S, Kravtsov AV, Zentner AR, Allgood B, Nagai D, Moore B (2004) The effect of gas cooling on the shapes of dark matter halos. Astrophys J 611:L73–L76. https://doi.org/10.1086/423992, astro-ph/0405189

Kennicutt RC Jr (1998) Star formation in Galaxies along the hubble sequence. Annu Rev Astron Astrophys 36:189–232. https://doi.org/10.1146/annurev.astro.36.1.189, arXiv:astro-ph/9807187

Kroupa P (2002) The initial mass function of stars: evidence for uniformity in variable systems. Science 295:82–91. https://doi.org/10.1126/science.1067524, arXiv:astro-ph/0201098

Kroupa P, Weidner C, Pflamm-Altenburg J, Thies I, Dabringhausen J, Marks M, Maschberger T (2013) The stellar and sub-stellar initial mass function of simple and composite populations, p 115. https://doi.org/10.1007/978-94-007-5612-0

Krumholz MR (2014) The big problems in star formation: the star formation rate, stellar clustering, and the initial mass function. Phys Rep 539:49–134. https://doi.org/10.1016/j.physrep.2014.02.001, 1402.0867

Kubryk M, Prantzos N, Athanassoula E (2013) Radial migration in a bar-dominated disc galaxy - I. Impact on chemical evolution. Mon Not R Astron Soc 436:1479–1491. https://doi.org/10.1093/mnras/stt1667

Kubryk M, Prantzos N, Athanassoula E (2015a) Evolution of the Milky Way with radial motions of stars and gas. I. The solar neighbourhood and the thin and thick disks. Astron Astrophys 580:A126. https://doi.org/10.1051/0004-6361/201424171, 1412.0585

Kubryk M, Prantzos N, Athanassoula E (2015b) Evolution of the Milky Way with radial motions of stars and gas. II. The evolution of abundance profiles from H to Ni. Astron Astrophys 580:A127. https://doi.org/10.1051/0004-6361/201424599, 1412.4859

Kulkarni SR, Heiles C (1987) The atomic component. In: Hollenbach DJ, Thronson HA Jr (eds) Interstellar processes. Astrophysics and space science library, vol 134, pp 87–122

Launhardt R, Zylka R, Mezger PG (2002) The nuclear bulge of the galaxy. iii. Large-scale physical characteristics of stars and interstellar matter. Astron Astrophys 384:112–139. arXiv:astro-ph/0201294

Limongi M, Chieffi A (2009) Presupernova evolution and explosion of massive stars: the role of mass loss during the Wolf-Rayet stage. Mem Soc Ast Ital 80:151

Liszt HS, Burton WB (1980) The gas distribution in the central region of the Galaxy. III - A barlike model of the inner-Galaxy gas based on improved HI data. Astrophys J 236:779–797. https://doi.org/10.1086/157803

Lodders K (2003) Solar system abundances and condensation temperatures of the elements. Astrophys J 591:1220–1247. https://doi.org/10.1086/375492

Loebman SR, Roškar R, Debattista VP, Ivezić Ž, Quinn TR, Wadsley J (2011) The genesis of the Milky Way's thick disk via stellar migration. Astrophys J 737:8. https://doi.org/10.1088/0004-637X/737/1/8, 1009.5997

Lyne AG, Manchester RN, Taylor JH (1985) The galactic population of pulsars. Mon Not R Astron Soc 213:613–639

Maness H, Martins F, Trippe S, Genzel R, Graham JR, Sheehy C, Salaris M, Gillessen S, Alexander T, Paumard T, Ott T, Abuter R, Eisenhauer F (2007) Evidence for a long-standing top-heavy initial mass function in the central parsec of the galaxy. Astrophys J 669:1024–1041. arXiv: 0707.2382

Mannucci F, Della Valle M, Panagia N, Cappellaro E, Cresci G, Maiolino R, Petrosian A, Turatto M (2005) The supernova rate per unit mass. Astron Astrophys 433:807–814. https://doi.org/10.1051/0004-6361:20041411, astro-ph/0411450

Maoz D, Graur O (2017) Star formation, supernovae, iron, and α: consistent cosmic and galactic histories. Astrophys J 848:25. https://doi.org/10.3847/1538-4357/aa8b6e, 1703.04540

Matteucci F, Francois P (1989) Galactic chemical evolution - abundance gradients of individual elements. Mon Not R Astron Soc 239:885–904. https://doi.org/10.1093/mnras/239.3.885

McKee CF, Ostriker EC (2007) Theory of star formation. Annu Rev Astron Astrophys 45:565–687. https://doi.org/10.1146/annurev.astro.45.051806.110602, 0707.3514

McKee CF, Williams JP (1997) The luminosity function of ob associations in the galaxy. Astrophys J 476:144

Men H, Ferriere K, Han JL (2008) Observational constraints on models for the interstellar magnetic field in the Galactic disk. ArXiv e-prints 805. 0805.3454

Minchev I, Chiappini C, Martig M (2013) Chemodynamical evolution of the Milky Way disk. I. The solar vicinity. Astron Astrophys 558:A9. https://doi.org/10.1051/0004-6361/201220189, 1208.1506

Minter AH, Spangler SR (1996) Observation of turbulent fluctuations in the interstellar plasma density and magnetic field on spatial scales of 0.01 to 100 parsecs. Astrophys J 458:194. https://doi.org/10.1086/176803

Nakanishi H, Sofue Y (2003) Three-dimensional distribution of the ISM in the Milky Way Galaxy: I. The H I disk. Publ Astron Soc Jpn 55:191–202. arXiv:astro-ph/0304338

Nakanishi H, Sofue Y (2006) Three-dimensional distribution of the ISM in the Milky Way Galaxy: II. The molecular gas disk. Publ Astron Soc Jpn 58:847–860, arXiv:astro-ph/0610769

Navarro JF, Frenk CS, White SDM (1996) The structure of cold dark matter halos. Astrophys J 462:563. https://doi.org/10.1086/177173, astro-ph/9508025

Nomoto K, Tominaga N, Umeda H, Kobayashi C, Maeda K (2006) Nucleosynthesis yields of core-collapse supernovae and hypernovae, and galactic chemical evolution. Nucl Phys A 777:424–458. https://doi.org/10.1016/j.nuclphysa.2006.05.008, arXiv:astro-ph/0605725

Nordström B, Mayor M, Andersen J, Holmberg J, Pont F, Jørgensen BR, Olsen EH, Udry S, Mowlavi N (2004) The Geneva-Copenhagen survey of the solar neighbourhood. ages, metallicities, and kinematic properties of 14 000 f and g dwarfs. Astron Astrophys 418:989–1019, arXiv:astro-ph/0405198

Oey MS, Meurer GR, Yelda S, Furst EJ, Caballero-Nieves SM, Hanish DJ, Levesque EM, Thilker DA, Walth GL, Bland-Hawthorn J, Dopita MA, Ferguson HC, Heckman TM, Doyle MT, Drinkwater MJ, Freeman KC, Kennicutt RC Jr, Kilborn VA, Knezek PM, Koribalski B, Meyer M, Putman ME, Ryan-Weber EV, Smith RC, Staveley-Smith L, Webster RL, Werk J, Zwaan MA (2007) The survey for ionization in neutral gas Galaxies. III. Diffuse, warm ionized medium and escape of ionizing radiation. Astrophys J 661:801–814. https://doi.org/10.1086/517867, arXiv:astro-ph/0703033

Olling RP, Merrifield MR (2001) Luminous and dark matter in the Milky Way. Mon Not R Astron Soc 326:164–180. https://doi.org/10.1046/j.1365-8711.2001.04581.x, arXiv:astro-ph/0104465

Pagel BEJ (1997) Nucleosynthesis and chemical evolution of Galaxies. Cambridge University Press, Cambridge

Pejcha O, Prieto JL (2015) On the intrinsic diversity of type II-plateau supernovae. Astrophys J 806:225. https://doi.org/10.1088/0004-637X/806/2/225, 1501.06573

Pettini M (2004) Element abundances through the cosmic ages. In: Esteban C, García López R, Herrero A, Sánchez F (eds) Cosmochemistry. The melting pot of the elements, pp 257–298, astro-ph/0303272

Pfalzner S (2009) Universality of young cluster sequences. Astron Astrophys 498:L37–L40. https://doi.org/10.1051/0004-6361/200912056, 0904.0523

Pohl M, Englmaier P, Bissantz N (2008) Three-dimensional distribution of molecular gas in the barred Milky Way. Astrophys J 677:283–291. https://doi.org/10.1086/529004, arXiv:0712.4264

Prantzos N, Boissier S (2010) The SNIa/CCSN ratio as a function of metallicity. In: Progenitors and Environments of Stellar Explosions, p 77

Prantzos N, Silk J (1998) Star formation and chemical evolution in the Milky Way: cosmological implications. Astrophys J 507:229–240. https://doi.org/10.1086/306327

Prantzos N, Boehm C, Bykov AM, Diehl R, Ferrière K, Guessoum N, Jean P, Knoedlseder J, Marcowith A, Moskalenko IV, Strong A, Weidenspointner G (2011) The 511 keV emission from positron annihilation in the Galaxy. Rev Mod Phys 83:1001–1056. https://doi.org/10.1103/RevModPhys.83.1001, 1009.4620

Plüeschke S, Diehl R, Schöenfelder V, Bloemen H, Hermsen W, Bennett K, Winkler C, McConnell M, Ryan J, Oberlack U, Knöedlseder J (2001) In: Gimenez A, Reglero V, Winkler C (eds) Exploring the gamma-ray universe. ESA Special Publication, vol 459

Rand RJ, Kulkarni SR (1989) The local Galactic magnetic field. Astrophys J 343:760–772. https://doi.org/10.1086/167747

Rand RJ, Lyne AG (1994) New rotation measures of distant pulsars in the inner galaxy and magnetic field reversals. Mon Not R Astron Soc 268:497

Rattenbury NJ, Mao S, Sumi T, Smith MC (2007) Modelling the galactic bar using ogle-ii red clump giant stars. Mon Not R Astron Soc 378:1064–1078. arXiv:0704.1614

Robin AC, Reylé C, Derrière S, Picaud S (2003) A synthetic view on structure and evolution of the milky way. Astron Astrophys 409:523–540. arXiv:astro-ph/0401052

Rocha-Pinto HJ, Maciel WJ, Scalo J, Flynn C (2000) Chemical enrichment and star formation in the Milky Way disk. I. Sample description and chromospheric age-metallicity relation. Astron Astrophys 358:850–868. astro-ph/0001382

Roškar R, Debattista VP, Quinn TR, Stinson GS, Wadsley J (2008) Riding the spiral waves: implications of stellar migration for the properties of galactic disks. Astrophys J 684:L79. https://doi.org/10.1086/592231, 0808.0206

Russeil D (2003) Star-forming complexes and the spiral structure of our Galaxy. Astron Astrophys 397:133–146. https://doi.org/10.1051/0004-6361:20021504

Russeil D (2010) The disk of our Galaxy. High Astron 15:786–786. https://doi.org/10.1017/S1743921310011646

Salpeter EE (1955) The luminosity function and stellar evolution. Astrophys J 121:161. https://doi.org/10.1086/145971

Sawada T, Hasegawa T, Handa T, Cohen RJ (2004) A molecular face-on view of the galactic centre region. Mon Not R Astron Soc 349:1167–1178. https://doi.org/10.1111/j.1365-2966.2004.07603.x, arXiv:astro-ph/0401286

Scalo JM (1986) The initial mass function of massive stars in galaxies empirical evidence. In: de Loore CWH, Willis AJ, Laskarides P (eds) Luminous stars and associations in Galaxies, IAU Symposium, vol 116, pp 451–466

Schaller G, Schaerer D, Meynet G, Maeder (1992) A new grids of stellar models from 0.8 to 120 solar masses at $Z = 0.020$ and $Z = 0.001$. Astron Astrophys Suppl Ser 96:269–331

Schmidt M (1959) The rate of star formation. Astrophys J 129:243. https://doi.org/10.1086/146614

Schmidt M (1963) The rate of star formation. II. The rate of formation of stars of different mass. Astrophys J 137:758. https://doi.org/10.1086/147553

Schönrich R, Binney J (2009) Origin and structure of the Galactic disc(s). Mon Not R Astron Soc 399:1145–1156. https://doi.org/10.1111/j.1365-2966.2009.15365.x, 0907.1899

Sedlmayr E, Patzer ABC (2004) Grain formation and dynamical atmosphere. In: A Jorissen, S Goriely, M Rayet, L Siess, H Boffin (eds) EAS publications series, vol 11, pp 51–66. https://doi.org/10.1051/eas:2004003

Sellwood JA, Binney JJ (2002) Radial mixing in galactic discs. Mon Not R Astron Soc 336:785–796. https://doi.org/10.1046/j.1365-8711.2002.05806.x, astro-ph/0203510

Sofue Y, Honma M, Omodaka T (2008) Unified rotation curve of the galaxy—decomposition into de Vaucouleurs bulge, disk, dark halo, and the 9-kpc rotation dip –. ArXiv e-prints 0811.0859

Sommer-Larsen J, Dolgov A (2001) Formation of disk galaxies: warm dark matter and the angular momentum problem. Astrophys J 551:608–623. https://doi.org/10.1086/320211, astro-ph/9912166

Soubiran C, Girard P (2005) Abundance trends in kinematical groups of the Milky Way's disk. Astron Astrophys 438:139–151. https://doi.org/10.1051/0004-6361:20042390, astro-ph/0503498

Spano M, Marcelin M, Amram P, Carignan C, Epinat B, Hernandez O (2008) Ghasp: an hα kinematic survey of spiral and irregular galaxies - v. Dark matter distribution in 36 nearby spiral galaxies. Mon Not R Astron Soc 383:297–316. 0710.1345

Springel V, White SDM, Frenk CS, Navarro JF, Jenkins A, Vogelsberger M, Wang J, Ludlow A, Helmi A (2008) Prospects for detecting supersymmetric dark matter in the galactic halo. Nature 456:73–76, 0809.0894

Sukhbold T, Ertl T, Woosley SE, Brown JM, Janka HT (2016) Core-collapse Supernovae from 9 to 120 solar masses based on neutrino-powered explosions. Astrophys J 821:38. https://doi.org/10.3847/0004-637X/821/1/38, 1510.04643

Tammann GA, Loeffler W, Schroeder A (1994) The galactic supernova rate. Astrophys J Suppl 92:487–493

Thorsett SE, Chakrabarty D (1999) Neutron star mass measurements. I. Radio pulsars. Astrophys J 512:288–299. https://doi.org/10.1086/306742, astro-ph/9803260

Tinsley BM (1980) Evolution of the stars and gas in Galaxies. Fundam Cosm Phys 5:287–388

Troland TH, Heiles C (1986) Interstellar magnetic field strengths and gas densities observational and theoretical perspectives. Astrophys J 301:339–345. https://doi.org/10.1086/163904

Truran JW, Cameron AGW (1971) Evolutionary models of nucleosynthesis in the Galaxy. Astrophys Space Sci 14:179–222. https://doi.org/10.1007/BF00649203

Vallée JP (2005) Pulsar-based galactic magnetic map: a large-scale clockwise magnetic field with an anticlockwise annulus. Astrophys J 619:297–305. https://doi.org/10.1086/426182

Vallée JP (2017) A guided map to the spiral arms in the galactic disk of the Milky Way. Astron Rev 13:113–146. https://doi.org/10.1080/21672857.2017.1379459, 1711.05228

Vázquez-Semadeni E (2015) Interstellar MHD turbulence and star formation. In: Lazarian A, de Gouveia Dal Pino EM, Melioli C (eds) Magnetic fields in diffuse media, astrophysics and space science library, vol 407, p 401. https://doi.org/10.1007/978-3-662-44625-6-14, 1208.4132

Villalobos Á, Helmi A (2008) Simulations of minor mergers - I. General properties of thick discs. Mon Not R Astron Soc 391:1806–1827. https://doi.org/10.1111/j.1365-2966.2008.13979.x

Voss R, Diehl R, Hartmann DH, Cerviño M, Vink JS, Meynet G, Limongi M, Chieffi A (2009) Using population synthesis of massive stars to study the interstellar medium near OB associations. Astron Astrophys 504:531–542. https://doi.org/10.1051/0004-6361/200912260, 0907.5209

Wang B (2018) Mass-accreting white dwarfs and type Ia supernovae. ArXiv e-prints 1801.04031

Weidemann V (2000) Revision of the initial-to-final mass relation. Astron Astrophys 363:647–656

Weidner C, Kroupa P, Bonnell IAD (2010) The relation between the most-massive star and its parental star cluster mass. Mon Not R Astron Soc 401:275–293. https://doi.org/10.1111/j.1365-2966.2009.15633.x, 0909.1555

Weidner C, Pflamm-Altenburg J, Kroupa P (2011) The Galaxy-wide IMF - from Star Clusters to Galaxies. In: Treyer M, Wyder T, Neill J, Seibert M, Lee J (eds) UP2010: have observations revealed a variable upper end of the initial mass function? Astronomical Society of the Pacific conference series, vol 440, p 19. 1011.1905

Weidner C, Kroupa P, Pflamm-Altenburg J (2013) The m_{max}-M_{ecl} relation, the IMF and IGIMF: probabilistically sampled functions. Mon Not R Astron Soc 434:84–101. https://doi.org/10.1093/mnras/stt1002, 1306.1229

Woosley SE, Heger A (2007) Nucleosynthesis and remnants in massive stars of solar metallicity. Phys Rep 442:269–283. https://doi.org/10.1016/j.physrep.2007.02.009, arXiv:astro-ph/0702176

Woosley SE, Weaver TA (1995) The evolution and explosion of massive stars. II. Explosive hydrodynamics and nucleosynthesis. Astrophys J Suppl 101:181. https://doi.org/10.1086/192237

Zinnecker H, Yorke HW (2007) Toward understanding massive star formation. Annu Rev Astron Astrophys 45:481–563. https://doi.org/10.1146/annurev.astro.44.051905.092549, 0707.1279

Zoccali M, Renzini A, Ortolani S, Greggio L, Saviane I, Cassisi S, Rejkuba M, Barbuy B, Rich RM, Bica E (2003) Age and metallicity distribution of the galactic bulge from extensive optical and near-ir stellar photometry. Astron Astrophys 399:931–956. arXiv:astro-ph/0210660

Zuckerman B, Evans NJ II (1974) Models of massive molecular clouds. Astrophys J 192:L149–L152. https://doi.org/10.1086/181613

Chapter 12
Branching Points on the Path of the *Slow* Neutron-Capture Process

Maria Lugaro and Alessandro Chieffi

We supplement this book, and in particular the discussion of stellar nucleosynthesis presented in Chap. 3, with a list of the unstable isotopes at which branching points become relevant in the s-process reaction chain in AGB stars. For sake of clarity and a better understanding it is advisable to go through the list with a chart of the nuclides at hand. For each branching point a brief description of its operation and its relevance in the study of the s process in AGB stars is presented. The 21 branching points highlighted by a star symbol next to their atomic mass are those that Käppeler et al. (2011) considered as interesting candidates for future Time-Of-Flight (TOF) measurements of their neutron-capture cross sections. All listed isotopes suffer β^- decay, unless specified otherwise. It should also be noted that usually in s-process conditions nuclear energy metastable levels higher than the ground state are not populated, thus the effect of these states does not need to be included in the study of branching points, except for the special cases reported in the list (see also Ward 1977).

Branching factors for each branching point can be calculated in each case at a given temperature, density, and neutron density conditions referring to Takahashi

M. Lugaro (✉)
Monash Centre for Astrophysics, Monash University, Clayton, VIC, Australia

Konkoly Observatory, Research Centre for Astronomy and Earth Sciences, Hungarian Academy of Sciences, Budapest, Hungary
e-mail: maria.lugaro@csfk.mta.hu

A. Chieffi
Istituto Nazionale Astronomia Fisica INAF, Roma, Italy
e-mail: achieffi@iasf-roma.inaf.it

R. Diehl et al. (eds.), *Astrophysics with Radioactive Isotopes*, Astrophysics and Space Science Library 453, https://doi.org/10.1007/978-3-319-91929-4_12

and Yokoi (1987) for the β decay rates, and to Rauscher (2012) for the neutron-capture cross section, unless advised otherwise in the description below.[1]

^{35}S This branching point may lead to production of the rare neutron-rich ^{36}S, whose abundance can be observed in stars via molecular lines, and may be measured in sulphur-rich meteoritic materials (for discussion and models see Mauersberger et al. 2004).

^{36}Cl and ^{41}Ca These are both long-living nuclei produced and destroyed—mostly via (n, p) and (n, α) channels—via neutron captures in AGB stars, and discussed in detail in Sect. 3.6.4 of Chap. 3. While ^{36}Cl behaves as stable nucleus during the *s* process, the half life of ^{41}Ca against electron captures has a strong temperature and density dependence, which could make it act as a branching point and most importantly prevent its survival instellar environments as in the case, e.g., of the other long-living ^{205}Pb.

^{45}Ca This branching point may lead to production of the rare neutron-rich ^{46}Ca, which could be measured in Ca-rich meteoritic material.

^{59}Fe This important branching point leads to the production of the long-living radioactive nucleus ^{60}Fe. See Sect. 3.6.3 of Chap. 3 for a detailed description and AGB model results.

^{63*}Ni The half life of this nucleus decreases from 100 years to $\simeq$12 years at 300 MK. The associated branching point affects the production of the rare neutron-rich ^{64}Ni as well as the ^{65}Cu/^{63}Cu ratio.

^{64}Cu The half life of this nucleus is short, of the order of a few hours, however, this isotope is a branching point on the *s*-process paths as it has comparable β^+ and β^- decay rates. The branching point may affect the production of ^{64}Ni and ^{65}Cu.

^{65}Zn This nucleus suffers β^+ decay and the branching point may affect the production of ^{65}Cu.

^{71}Ge This nucleus suffers β^+ decay and the branching point may affect the production of ^{71}Ga.

^{79*}Se This branching point may lead to production of the long-living radioactive isotope ^{81*}Kr. This production occurs when the temperature increases in the thermal pulse, and the half life of ^{79}Se decreases from the terrestrial half life of 65,000 years to roughly 4 years at 300 MK due to population of the shorter-living isomeric state (Klay and Käppeler 1988, and see Sect. 3.6.5 of Chap. 3 for model results). Operation of this branching point also affects the ^{81}Br/^{79}Br ratio.

^{80}Br The half life of this nucleus is short, of the order of minutes, however, it is a branching point on the *s*-process paths as it can decay both β^+ and β^-, with the β^- roughly ten times faster than the β^+ channel. It can affect the production of ^{81}Kr.

^{81}Kr This nucleus is too long living ($T_{1/2}$ = 0.23 Myr, down to 2300 years at temperature 300 MK) to act as a branching point during the *s* process and rather

[1]Maria Lugaro thanks Roberto Gallino and Franz Kaeppeler for communicating the passion for branching points and for help with this section.

behaves as a stable nucleus. Its production during the *s*-process is discussed in detail in Sect. 3.5 of Chap. 3. Its radiogenic decay leads to ^{81}Br.

^{85*}Kr The relatively long half life of ^{85}Kr (11 years) allows this branching point to activate already at low neutron densities, $> 5\times10^8$ n/cm^3. The actual operation of this branching point is complicated by the fact that roughly 40% of ^{84}Kr(n,γ) reactions during the *s* process result in the production of the isomeric state of ^{85}Kr. Approximately 80% of these nuclei quickly decay into ^{85}Rb, with a half live of 4.5 h, while the remaining 20% relax into the ground state. The production of ^{87}Rb, a very long-living isotope with half live of 48 Gyr and a magic number of neutrons N = 50, has traditionally been attributed to the activation of the branching point at ^{85}Kr (Lambert et al. 1995; Abia et al. 2001). However, van Raai et al. (2012) showed that the activation of the branching point at ^{85}Kr mostly results in the production of ^{86}Kr, a nucleus with a magic number of neutrons N=50 and a very small neutron capture cross section of only $\simeq$3.4 mbarn. ^{86}Kr is thus more likely to accumulate than to capture the further neutron that would allow the production of ^{87}Rb. The importance of the production of ^{86}Kr in meteoritic SiC grains and the *s*-process is discussed in Sect. 3.5.5 of Chap. 3.

^{86}Rb The branching point at ^{86}Rb is activated at relatively high neutron densities, above 10^{10} n/cm^3, being the half life of this nucleus 18.7 days, and it leads directly to the production of ^{87}Rb. The importance of ^{87}Rb in *s*-process observations and models is discussed in Sect. 3.5.4 of Chap. 3.

^{89,90}Sr and ^{91}Y The branching point at ^{89}Sr may produce the unstable ^{90}Sr, also a branching point producing ^{91}Sr, which quickly decays into unstable ^{91}Y. This is also a branching point, producing ^{92}Y, which quickly decays into stable ^{92}Zr. The final result of the operation of this chain of branching points is to decrease the production of ^{90}Zr and ^{91}Zr, with respect to that of ^{92}Zr. This point is discussed by Lugaro et al. (2003), in relevance to the Zr isotopis ratios measured in meteoritic silicon carbide (SiC) grains from AGB stars.

^{93}Zr This nucleus is too long-living ($T_{1/2}$ = 1.5 Myr) to act as a branching point during the *s* process and rather behaves as a stable nucleus (see Sect. 3.5.2 and Fig. 3.10 of Chap. 3), with an experimentally determined neutron-capture cross section (Macklin 1985b). Its production during the *s*-process is discussed in detail in Sect. 3.6.5 of Chap. 3. Its radiogenic decay produces most of the solar abundance of ^{93}Nb. (A small fraction of the ^{93}Nb is also contributed by the radiogenic decay of ^{93}Mo ($T_{1/2}$ = 3500 years) which is not on the main *s*-process path but can be produced by neutron-capture on the relatively abundant *p*-only ^{92}Mo, 15% of solar Mo.)

^{95*}Zr This important branching point can lead to production by the *s* process of ^{96}Zr if $N_n > 5\times10^8$ n/cm^3 (Toukan and Kaeppeler 1990; Lugaro et al. 2014; Yan et al. 2017). Zr isotopic ratios have been estimated in MS and S stars via molecular lines and measured in meteoritic SiC grains, providing constraints on the neutron density in the thermal pulses. This point is further discussed in Sects. 3.4.1 and 3.4.4.

^{94,95}Nb The half life of ^{94}Nb decreases from terrestrial 20,000 years to $\simeq$0.5 years at 100 MK and $\simeq$9 days at 300 MK. This branching point can produce the unstable ^{95}Nb, which is also a branching point producing the unstable ^{96}Nb, which quickly decays into stable ^{96}Mo. Via the operation of the ^{94}Nb branching point the ^{94}Mo nucleus is skipped during the *s*-process chain, this nucleus is in fact classified among *p*-only nuclei.

^{99}Tc The half life of ^{99}Tc is 0.21 Myr, and decreases to 0.11 Myr at 100 MK and to 4.5 years at 300 MK. Thus, the neutron-capture path of the branching point is mostly open, producing ^{100}Tc, which quickly decays into ^{100}Ru, thus skipping ^{99}Ru (Fig. 3.10). Then, radiogenic decay of ^{99}Tc produces ^{99}Ru. The production of ^{99}Tc is discussed in detail in Sect. 3.5.5, and mentioned in Sect. 3.5.6 of Chap. 3 in relation to ^{99}Ru in meteoritic SiC grains.

^{107}Pd This nucleus is too long-living ($T_{1/2}$ = 6.5 Myr, down to $\simeq$700 years at 300 MK) to act as a branching point during the *s* process and rather behaves as a stable nucleus, with an experimentally determined neutron-capture cross section (Macklin 1985a). Its production during the *s*-process is discussed in detail in Sect. 3.6.5 of Chap. 3. Its radiogenic decay is responsible for production of ^{107}Ag.

^{128}I The decay half life of this nucleus is too short to allow for neutron captures, however, there is a marginal branching point here due to the fact that ^{128}I has both β^+ and β^- decay channels. The β^+ channel has significant temperature and density dependence and represents only a few percent of the decay rate. Nevertheless, this branching point has been investigated in detail because it affects the precise determination of relative abundances of the two *s*-only isotopes ^{128}Xe and ^{130}Xe, and because the timescale for its activation of the order of 25 min is comparable to that of the convective turn-over timescale of the material inside AGB thermal pulses of hours (Reifarth et al. 2004).

^{133}Xe May lead to production of the ^{134}Xe. Of interest in relation to the Xe-S component from SiC grains in primitive meteorites, as discussed in Sect. 3.5.4 of Chap. 3.

^{134*,135*,136,137}Cs The chain of branching points at the Cs isotopes is of particular interest because it affects the isotopic composition of the *s*-process element Ba and in particular the relative abundances of the two *s*-only nuclei ^{134}Ba and ^{136}Ba, as it is discussed in Sect. 3.5.5 in relation to Ba data from meteoritic SiC grains. The branching point at ^{134}Cs allows production of the long-living isotope ^{135}Cs (see Sect. 3.6.5 of Chap. 3 for model results). The half lives of both ^{134}Cs and ^{135}Cs have a strong theoretical temperature dependence, decreasing by orders of magnitude in stellar conditions. Specifically for the long-living ^{135}Cs, $T_{1/2}$ varies from terrestrial of 2 Myr down to $\simeq$200 years at 300 MK, while its neutron-capture cross section has been experimentally determined (Patronis et al. 2004). The branching point at ^{135}Cs can produce the unstable ^{136}Cs, which is also a branching point producing the unstable ^{137}Cs. With a constant half life of $\simeq$30 years, this is also a branching point producing the unstable ^{138}Cs, which quickly decays into stable ^{138}Ba.

^{141}**Ce** this branching point may lead to production of the neutron-rich ^{142}Ce, thus skipping the *s*-only ^{142}Nd and affecting the Nd isotopic ratios, which are measured in SiC stardust grains (Gallino et al. 1997).

^{142,143}**Pr** The branching point at ^{142}Pr is affected by the temperature dependence of the β^- half life of ^{142}Pr, which increases to $\simeq$4 days at 300 MK from the terrestrial 19 h. The neutron-capture branch may produce the unstable ^{143}Pr, which is also a branching point producing the unstable ^{144}Pr, which quickly decays into ^{144}Nd. The operation of this chain of branching points may affect the isotopic composition of Nd because ^{142}Nd and ^{143}Nd are skipped by the neutron-capture flux and their abundances are decreased.

^{147*}**Nd** This branching point may lead to the production of the neutron-rich "r-only" ^{148}Nd, which is of interest in relation to stellar SiC grain Nd data (Gallino et al. 1997).

^{147*,148*}**Pm** The branching point at ^{147}Pm is affected by the strong temperature dependence of the β^- decay of this nucleus, where the half life decreases from the terrestrial value of 2.6 years down to $\simeq$1 years at 300 MK. The neutron-capture cross section of this nucleus is experimentally determined (Reifarth et al. 2003). When the branching is open, it produces the unstable ^{148}Pm, a branching point that may lead to production of ^{149}Pm, which quickly decays into ^{149}Sm. The operation of this chain of branching points affects the isotopic composition of Sm, by skipping ^{147}Sm and the *s*-only ^{148}Sm. This is of interest in relation to stellar SiC grain Sm data (Gallino et al. 1997).

^{151*}**Sm** The operation of this branching point is affected by the temperature dependence of the β^- decay rate of ^{151}Sm, where the half life of this nucleus decreases from 93 years to $\simeq$3 years at 300 MK. Its operation changes the ^{153}Eu/^{151}Eu ratio, which can be measured in stars (Sect. 3.5.4 of Chap. 3) and in SiC stardust grains (Sect. 3.5.5 of Chap. 3). Note that ^{151}Sm is one of few radioactive nuclei acting as branching points on the *s*-process path for which an experimental determination of the neutron capture cross section is available (Abbondanno et al. 2004; Wisshak et al. 2006), however, some uncertainty is due to the contribution of excited states, which could be significant (Ávila et al. 2013; Rauscher 2012).

^{153}**Sm** This branching point can produce the neutron-rich ^{154}Sm and affect the ^{153}Eu/^{151}Eu ratio.

^{152}**Eu** This nucleus suffers both β^- and β^+ decays, with rates showing a strong temperature dependence covering several orders of magnitude variation in stellar conditions. The β^+ decay rate also has a strong dependence on density. The operation of this branching point, in combination with that at ^{151}Sm, makes possible the production of the rare *p*-only isotope ^{152}Gd by the *s* process.

^{154*,155*}**Eu** The decay rate of ^{154}Eu has a strong temperature dependence, with its half life decreasing from 8.8 years down to $\simeq$11 days at 300 MK. If activated, it leads to production of the unstable ^{155}Eu, a branching point also with a temperature dependence, and an experimentally determined neutron-capture cross section (Jaag and Käppeler 1995), which may produce ^{156}Eu, which quickly decays into ^{156}Gd. The operation of this chain of branching points affects the isotopic composition of Gd, which is a refractory element present in stellar SiC grains (Yin et al. 2006).

^{153*}**Gd** This nucleus suffers β^+ decay with a temperature dependence, where the terrestrial half life of 239 days increases with increasing the temperature by up to an order of magnitude in AGB stars conditions. The operation of this branching point may affect the ^{153}Eu/^{151}Eu ratio.

^{163}**Dy** and ^{163*,164}**Ho** The nucleus ^{163}Dy is stable in terrestrial conditions, but it can become unstable inside stars: at 300 MK the half life of this isotope becomes $\simeq$18 days. Thus, a branching can open on the s-process path, leading to the production of the unstable ^{163}Ho via β^- decay of ^{163}Dy. In this conditions, the β^+ half life of ^{163}Ho (which also has a strong temperature and density dependence) is $\simeq$12 years, so another branching can open on the s-process neutron capture path. Neutron captures on ^{163}Ho lead to production of the unstable ^{164}Ho, which has fast β^- and β^+ channels, both temperature dependent. The β^- channel can eventually lead to the production of ^{164}Er, a p-only nucleus, which may thus have a s-process component in its cosmic abundance.

^{169}**Er** This branching point may lead to the production of the neutron-rich ^{170}Er.

^{170*,171*}**Tm** The branching point at ^{170}Tm may produce the unstable ^{171}Tm, which is also a branching point (with a temperature dependence) producing the unstable ^{172}Tm, which quickly decays into ^{172}Yb. By skipping ^{171,172}Yb during the s-process flux, these branching points affect the isotopic composition of Yb, which is a refractory element present in meteoritic stellar SiC grains (Yin et al. 2006).

^{176}**Lu** A branching point at ^{176}Lu is activated because of the production of the short-living (half life of $\simeq$4 h) isomeric state of ^{176}Lu via neutron captures on ^{175}Lu. The situation is further complicated because, at around 300 MK, the isomeric and the ground state of ^{176}Lu are connected via the thermal population of nuclear states that can act as mediators between the two. Hence, the half life of the ^{176}Lu system can decrease at such temperatures by orders of magnitude. This branching point is of importance for the production of the very long-living ground state of ^{176}Lu (half life of 380 Gyr) and of the stable ^{176}Hf, which are both s-only isotopes, shielded by ^{176}Yb against r-process production. Hence, the relative solar abundances of these two isotopes need to be matched by s-process in AGB stars. For details and models see Heil et al. (2008) and Mohr et al. (2009).

^{177}**Lu** This branching point may lead to production of the unstable ^{178}Lu, which quickly decays into ^{178}Hf, thus decreasing the abundance of ^{177}Hf.

^{179}**Hf**, ^{179*,180}**Ta** A branching point at ^{179}Hf may be activated on the s-process path because this stable nucleus becomes unstable in stellar conditions (as in the case of ^{163}Dy) with a β^- half life of $\simeq$40 years at 300 MK. This may allow the production of the unstable ^{179}Ta, which is also a branching point with a temperature-dependent β^+ decay rate, which may lead to the production of ^{180}Ta, the least abundant nucleus in the solar system (Käppeler et al. 2004) as a few percent of neutron captures on ^{179}Ta lead to production of the very long-living isomeric state of ^{180}Ta, instead of the ground state, which suffers fast β^+ and β^- decays. As in the case of ^{176}Lu, the ground and the isomeric states of ^{180}Ta can be connected via the thermal population of nuclear states that act as mediators between the two. It is still unclear if the cosmic abundance of ^{180}Ta is to be ascribed to the *s* process or to nucleosynthetic processes in supernovae connected to neutrino fluxes.

^{181}**Hf** This branching point leads to production of the long-living radioactive nucleus ^{182}Hf (one of the few radioactive isotopes with an experimentally determined neutron-capture cross section available, Vockenhuber et al. 2007) whose decay into ^{182}W is of extreme importance for early solar system datation. The half life of ^{181}Hf is relatively long (42 days) allowing production of ^{182}Hf in AGB stars when the ^{22}Ne(α,n)^{25}Mg reaction is activated (see also Sect. 3.5 of Chap. 3).

^{182,183}**Ta** The branching point at ^{182}Ta is temperature dependent and may produce the unstable ^{183}Ta, also a branching point, producing ^{184}Ta, which quickly decays into the stable ^{184}W. These branching points may affect the isotopic composition of W, which is a refractory element that is present in stellar SiC grains.

^{185*}**W** This branching point may produce ^{186}W, and affect the isotopic composition of W as well as the ^{186}Os/^{188}Os ratio. Its signature may be seen in data from stellar SiC grains for W and Os (Humayun and Brandon 2007; Ávila et al. 2012). Note that ^{185}W is one of few radioactive nuclei acting as branching points on the *s*-process path for which an experimental determination of the neutron capture cross section is available, even thought only via indirect (γ, n) studies, which have rather large uncertainties of about 30% (Sonnabend et al. 2003; Mohr et al. 2004).

^{186}**Re** This isotope decays in $\simeq$89 h, and has both β^- and β^+ decay channel. The β^- decay channel is faster by one to two orders of magnitude depending on the temperature, which affects the β^+ decay rate. This branching point can affect the production of ^{186}Os, ^{186}W, and the very long-living ^{187}Re, whose slow decay into ^{187}Os is used as a cosmological clock (see discussion in Chap. 2).

^{191}**Os** This branching point has a mild temperature dependence whereby the half life of ^{191}Os decreases with the temperature from the terrestrial 15 days to $\simeq$8 days at 300 MK. If activated, the neutron-capture branch can decrease the *s*-process abundances of ^{191}Ir and ^{192}Pt and lead to production of ^{192}Os, thus affecting the isotopic composition of Os, which is measured in meteoritic materials (Brandon et al. 2005), and ^{193}Ir.

^{192}**Ir** This branching point can produce ^{193}Ir, and affect the *s*-process production of the rare proton-rich ^{192}Pt. A few percent of the decay rate of ^{192}Ir is made by β^+ decays.

^{193}**Pt** This isotope decays β^+ with a half life of $\simeq$50 years, which may affect the production of ^{193}Ir.

^{204*}**Tl** This branching point has a strong temperature dependence with its half life decreasing from the terrestrial value of $\simeq$3.8 years to $\simeq$7 days at 300 MK, leading to production of the *s*-only ^{204}Pb.

^{205}**Pb** This nucleus is long-living in terrestrial conditions ($T_{1/2} = 15$ Myr), but its half life against electron captures has a strong temperature and density dependence, which affects its survival in stellar environments, as in the case of ^{41}Ca. Its production during the *s*-process is discussed in detail in Sect. 3.5 of Chap. 3. Its radiogenic decay is responsible for production of ^{105}Tl.

^{210}**Bi** This temperature-dependent branching point may lead to production of the unstable ^{211}Bi, which α decays into ^{207}Tl, which quickly decays β^+ into ^{207}Pb.

^{210}**Po** May produce ^{211}Po, which quickly α decays into ^{207}Pb. The α decay of ^{210}Po, and ^{211}Bi above, represent the chain of reactions that terminates the *s* process (Clayton and Rassbach 1967; Ratzel et al. 2004).

To complete the picture we list nuclei that could be classified as potential *s*-process branching points, given that their terrestrial half life is greater than a few days, however, they do not open during the *s* process because their half life decreases with the temperature to below a few days. These are: ^{103}Ru, ^{123}Sn, ^{124}Sb, ^{156}Eu, ^{160*,161}Tb, ^{175}Yb, ^{198}Au, and ^{205}Hg. Finally, we point out the special case of ^{157}Gd, a stable nucleus which becomes unstable at stellar temperatures, but not enough to open a branching point on the *s*-process path in AGB stars.

References

Abbondanno U, Aerts G, Alvarez-Velarde F, Álvarez-Pol H, Andriamonje S, Andrzejewski J, Badurek G, Baumann P, Bečvář F, Benlliure J, Berthoumieux E, Calviño F, Cano-Ott D, Capote R, Cennini P, Chepel V, Chiaveri E, Colonna N, Cortes G, Cortina D, Couture A, Cox J, Dababneh S, Dahlfors M, David S, Dolfini R, Domingo-Pardo C, Duran I, Embid-Segura M, Ferrant L, Ferrari A, Ferreira-Marques R, Frais-Koelbl H, Furman W, Goncalves I, Gallino R, Gonzalez-Romero E, Goverdovski A, Gramegna F, Griesmayer E, Gunsing F, Haas B, Haight R, Heil M, Herrera-Martinez A, Isaev S, Jericha E, Käppeler F, Kadi Y, Karadimos D, Kerveno M, Ketlerov V, Koehler P, Konovalov V, Krtička M, Lamboudis C, Leeb H, Lindote A, Lopes I, Lozano M, Lukic S, Marganiec J, Marrone S, Martinez-Val J, Mastinu P, Mengoni A, Milazzo PM, Molina-Coballes A, Moreau C, Mosconi M, Neves F, Oberhummer H, O'Brien S, Pancin J, Papaevangelou T, Paradela C, Pavlik A, Pavlopoulos P, Perlado JM, Perrot L, Pignatari M, Plag R, Plompen A, Plukis A, Poch A, Policarpo A, Pretel C, Quesada J, Raman S, Rapp W, Rauscher T, Reifarth R, Rosetti M, Rubbia C, Rudolf G, Rullhusen P, Salgado J, Soares JC, Stephan C, Tagliente G, Tain J, Tassan-Got L, Tavora L, Terlizzi R, Vannini G, Vaz P, Ventura A, Villamarin D, Vincente MC, Vlachoudis V, Voss F, Wendler H, Wiescher M, Wisshak K (2004) Neutron capture cross section measurement of ^{151}Sm at the CERN neutron time of flight facility (n_TOF). Phys Rev Lett 93(16):161103. https://doi.org/10.1103/PhysRevLett.93.161103

Abia C, Busso M, Gallino R, Domínguez I, Straniero O, Isern J (2001) The ^{85}Kr s-process branching and the mass of carbon stars. Astrophys J 559:1117–1134. https://doi.org/10.1086/322383, arXiv:astro-ph/0105486

Ávila JN, Lugaro M, Ireland TR, Gyngard F, Zinner E, Cristallo S, Holden P, Buntain J, Amari S, Karakas A (2012) Tungsten isotopic compositions in stardust SiC grains from the Murchison meteorite: constraints on the s-process in the Hf-Ta-W-Re-Os Region. Astrophys J 744:49. https://doi.org/10.1088/0004-637X/744/1/49, 1110.4763

Ávila JN, Ireland TR, Lugaro M, Gyngard F, Zinner E, Cristallo S, Holden P, Rauscher T (2013) Europium s-process signature at close-to-solar metallicity in stardust SiC grains from asymptotic giant branch stars. Astrophys J 768:L18. https://doi.org/10.1088/2041-8205/768/1/L18, 1303.5932

Brandon AD, Humayun M, Puchtel IS, Leya I, Zolensky M (2005) Osmium isotope evidence for an s-Process carrier in primitive chondrites. Science 309:1233–1236. https://doi.org/10.1126/science.1115053

Clayton DD, Rassbach ME (1967) Termination of the s-PROCESS. Astrophys J 148:69. https://doi.org/10.1086/149128

Gallino R, Busso M, Lugaro M (1997) Neutron capture nucleosynthesis in AGB stars. In: Bernatowicz TJ, Zinner E (eds) American Institute of Physics conference series, vol 402, pp 115–153. https://doi.org/10.1063/1.53327

Heil M, Winckler N, Dababneh S, Käppeler F, Wisshak K, Bisterzo S, Gallino R, Davis AM, Rauscher T (2008) ^{176}Lu/^{176}Hf: a sensitive test of s-Process temperature and neutron density in AGB stars. Astrophys J 673:434–444. https://doi.org/10.1086/523892

Humayun M, Brandon AD (2007) s-Process implications from osmium isotope anomalies in chondrites. Astrophys J 664:L59–L62. https://doi.org/10.1086/520636

Jaag S, Käppeler F (1995) Stellar (n,γ) cross section of the unstable isotope ^{155}Eu. Phys Rev C 51:3465–3471. https://doi.org/10.1103/PhysRevC.51.3465

Käppeler F, Arlandini C, Heil M, Voss F, Wisshak K, Reifarth R, Straniero O, Gallino R, Masera S, Travaglio C (2004) Stellar neutron capture on ^{180}Ta^m. II. Defining the s-process contribution to nature's rarest isotope. Phys Rev C 69(5):055802. https://doi.org/10.1103/PhysRevC.69.055802

Käppeler F, Gallino R, Bisterzo S, Aoki W (2011) The s process: nuclear physics, stellar models, and observations. Rev Mod Phys 83:157–194. https://doi.org/10.1103/RevModPhys.83.157, 1012.5218

Klay N, Käppeler F (1988) β-decay rate of ^{79m}Se and its consequences for the s-process temperature. Phys Rev C 38:295–306. https://doi.org/10.1103/PhysRevC.38.295

Lambert DL, Smith VV, Busso M, Gallino R, Straniero O (1995) The chemical composition of red giants. IV. The neutron density at the s-Process site. Astrophys J 450:302. https://doi.org/10.1086/176141

Lugaro M, Davis AM, Gallino R, Pellin MJ, Straniero O, Käppeler F (2003) Isotopic compositions of strontium, zirconium, molybdenum, and barium in single Presolar SiC grains and asymptotic giant branch stars. Astrophys J 593:486–508. https://doi.org/10.1086/376442

Lugaro M, Tagliente G, Karakas AI, Milazzo PM, Käppeler F, Davis AM, Savina MR (2014) The impact of updated Zr neutron-capture cross sections and new asymptotic giant branch models on our understanding of the S process and the origin of stardust. Astrophys J 780:95. https://doi.org/10.1088/0004-637X/780/1/95, 1311.2660

Macklin RL (1985a) Neutron capture measurements on fission product^{107}Pd. Nucl Sci Eng 89:79

Macklin RL (1985b) Neutron capture measurements on radioactive Zr-93. Astrophys Space Sci 115:71–83. https://doi.org/10.1007/BF00653828

Mauersberger R, Ott U, Henkel C, Cernicharo J, Gallino R (2004) The abundance of ^{36}S in IRC+10216 and its production in the Galaxy. Astron Astrophys 426:219–227. https://doi.org/10.1051/0004-6361:20040451, arXiv:astro-ph/0407003

Mohr P, Shizuma T, Ueda H, Goko S, Makinaga A, Hara KY, Hayakawa T, Lui Y, Ohgaki H, Utsunomiya H (2004) s-process branching at ^{185}W revised. Phys Rev C 69(3):032801. https://doi.org/10.1103/PhysRevC.69.032801, arXiv:astro-ph/0409642

Mohr P, Bisterzo S, Gallino R, Käppeler F, Kneissl U, Winckler N (2009) Properties of the 5^- state at 839 keV in Lu176 and the s-process branching at A=176. Phys Rev C 79(4):045804. https://doi.org/10.1103/PhysRevC.79.045804, 0903.3897

Patronis N, Dababneh S, Assimakopoulos PA, Gallino R, Heil M, Käppeler F, Karamanis D, Koehler PE, Mengoni A, Plag R (2004) Neutron capture studies on unstable ^{135}Cs for nucleosynthesis and transmutation. Phys Rev C 69(2):025803. https://doi.org/10.1103/PhysRevC.69.025803

Ratzel U, Arlandini C, Käppeler F, Couture A, Wiescher M, Reifarth R, Gallino R, Mengoni A, Travaglio C (2004) Nucleosynthesis at the termination point of the s process. Phys Rev C 70(6):065803. https://doi.org/10.1103/PhysRevC.70.065803

Rauscher T (2012) Formalism for inclusion of measured reaction cross sections in stellar rates including uncertainties and its application to neutron capture in the s-process. Astrophys J 755:L10. https://doi.org/10.1088/2041-8205/755/1/L10, 1207.1664

Reifarth R, Arlandini C, Heil M, Käppeler F, Sedyshev PV, Mengoni A, Herman M, Rauscher T, Gallino R, Travaglio C (2003) Stellar neutron capture on promethium: implications for the s-Process neutron density. Astrophys J 582:1251–1262. https://doi.org/10.1086/344718

Reifarth R, Käppeler F, Voss F, Wisshak K, Gallino R, Pignatari M, Straniero O (2004) ^{128}Xe and ^{130}Xe: testing He-Shell burning in asymptotic giant branch stars. Astrophys J 614:363–370. https://doi.org/10.1086/422206, arXiv:astro-ph/0405065

Sonnabend K, Mengoni A, Mohr P, Rauscher T, Vogt K, Zilges A (2003) Determination of the (n, γ) reaction rate of unstable ^{185}W in the astrophysical s-process via its inverse reaction. Nucl Phys A 718:533–535. https://doi.org/10.1016/S0375-9474(03)00835-2

Takahashi K, Yokoi K (1987) Beta-decay rates of highly ionized heavy atoms in stellar interiors. Atom Data Nucl Data Tables 36:375. https://doi.org/10.1016/0092-640X(87)90010-6

Toukan KA, Kaeppeler F (1990) The stellar neutron capture cross sections of Zr-94 and Zr-96. Astrophys J 348:357–362. https://doi.org/10.1086/168243

van Raai MA, Lugaro M, Karakas AI, García-Hernández DA, Yong D (2012) Rubidium, zirconium, and lithium production in intermediate-mass asymptotic giant branch stars. Astron Astrophys 540:A44. https://doi.org/10.1051/0004-6361/201117896, 1202.2620

Vockenhuber C, Dillmann I, Heil M, Käppeler F, Winckler N, Kutschera W, Wallner A, Bichler M, Dababneh S, Bisterzo S, Gallino R (2007) Stellar (n,γ) cross sections of Hf174 and radioactive Hf182. Phys Rev C 75(1):015804. https://doi.org/10.1103/PhysRevC.75.015804

Ward RA (1977) The importance of long-lived isomeric states in s-Process branching. Astrophys J 216:540–547

Wisshak K, Voss F, Käppeler F, Krtička M, Raman S, Mengoni A, Gallino R (2006) Stellar neutron capture cross section of the unstable s-process branching point ^{151}Sm. Phys Rev C 73(1):015802. https://doi.org/10.1103/PhysRevC.73.015802

Yan SQ, Li ZH, Wang YB, Nishio K, Lugaro M, Karakas AI, Makii H, Mohr P, Su J, Li YJ, Nishinaka I, Hirose K, Han YL, Orlandi R, Shen YP, Guo B, Zeng S, Lian G, Chen YS, Liu WP (2017) The ^{95}Zr(n, γ)^{96}Zr cross section from the surrogate ratio method and its effect on s-process nucleosynthesis. Astrophys J 848:98. https://doi.org/10.3847/1538-4357/aa8c74, 1709.04635

Yin QZ, Lee CTA, Ott U (2006) Signatures of the s-Process in presolar silicon carbide grains: barium through hafnium. Astrophys J 647:676–684. https://doi.org/10.1086/505188

Part V
Epilogue

The previous chapters presented in detail how astrophysical studies of cosmic radioactivities proceed via measurements and theory development. Now we take a moment to offer an outlook and perspective of this field. As the book approaches its end, we append a few items, such as a glossary of a few key terms and a history of selected milestones that marked progress (from the Editors point of view), which the student may find useful when reviewing the material gathered in this book.

Chapter 13
Perspectives

Roland Diehl, Dieter H. Hartmann, and Nikos Prantzos

It is more than a century now since the phenomenon of radioactivity has been discovered. Since then, astronomical observations unravelled a fascinating trace of compositional evolution of cosmic matter, as seen in stellar atmospheres, interstellar gas, meteorites, and in the composition of matter here on Earth. Astrophysical studies established models for sources of new atomic nuclei, such as stars and their explosions, where nuclear reactions can occur. Isotopes are the fundamental sources of information about the cosmic compositional evolution, and radioactive isotopes add a natural clock. A basic understanding has been set up about the cosmic cycle of matter and nucleosynthesis in cosmic places and on cosmic time scales in the later half of the past century. Now, multiple disciplines of astronomies and astrophysics need to play together and help, so that the striking deficits we still have can be worked on. This will neither be cheap nor easy, but rewarding. Nuclear astrophysics and astronomy will have to be part of this.

Since then, the phenomenon of radioactivity evolved to a major astrophysical tool, allowing astronomers, among many other things, to

- determine the ages of stars and of our Galaxy
- probe the physical processes occurring deep inside supernova envelopes
- infer the current rate of nucleosynthesis in the Milky Way

R. Diehl (✉)
Max Planck Institut für extraterrestrische Physik, Garching, Germany
e-mail: rod@mpe.mpg.de

D. H. Hartmann
Clemson University, Clemson, SC, USA
e-mail: hdieter@g.clemson.edu

N. Prantzos
Institut d'Astrophysique, Paris, France
e-mail: prantzos@iap.fr

R. Diehl et al. (eds.), *Astrophysics with Radioactive Isotopes*, Astrophysics and Space Science Library 453, https://doi.org/10.1007/978-3-319-91929-4_13

- understand (or, at least, constrain) the timescales of the acceleration of Galactic cosmic rays and of their propagation through the interstellar medium
- infer the physical conditions inside the stars producing some of the extinct radioactivities found in meteoritic dust grains
- constrain the environment of the proto-solar nebula and the activity of the young Sun

Radioactivity is the decay of unstable nuclei, therefore it is intimately related to nuclear astrophysics and stellar nucleosynthesis, i.e. the production of such nuclei in various cosmic environments. Obviously, the steady progress in our understanding of stellar structure, evolution and explosions, and of nuclear reactions in stellar conditions, was necessary in order to turn radioactivity into a powerful astrophysical tool.

The current state of the art in modelling nucleosynthesis in single stars of low and high mass, as well as in stellar explosions occurring in binary systems, is presented in Chaps. 3, 4 and 5 of this book, respectively. Present day models are much more sophisticated (including, e.g., mass loss, stellar rotation, neutrino interactions for all flavors, general-relativistic treatment) than their earlier counterparts. Then we believe that their results are, presumably, closer to reality. Despite their sophistication, however, such models can hardly be considered as *realistic*, since key ingredients, such as convection, mass loss and stellar rotation, neutrino directional interactions, and magnetic field effects, can only be treated in a parametrised way at present; this is also the case for models of both core collapse and thermonuclear supernovae, and even more so in rare transients such as the collisions of neutron stars or even black holes. The exploration of geometrical (3D) effects has only recently been established for pioneering studies, and will without doubt bring more surprises; dealing with the large variety of possible initial conditions and sub-grid model approximations will challenge computing power, and leave a role for simpler model implementations. Interesting results have also been reported on the effects of the stellar magnetic field. As one example, at least, for positron escape from explosions such as SNIa, or from relativistic jets such as microquasars, such effects are known to play a critical role.

The book's Chap. 6 briefly describes the exciting aspects of *our* Sun and the solar system, which have been connected to measurements and insights on radioactive isotopes all along. The Sun has taught us to retain modesty and critical views: Neutrino measurements were in striking conflict with solar nuclear reaction models, and could only be reconciled through the newly-established phenomena of neutrino flavors and neutrinos with non-zero mass. Refined (3D) models of the solar photosphere have significantly reduced *metallicity* in our cosmic standard for elemental abundances. Chemical evolution descriptions have explored numerically the chances for a special nearby nucleosynthesis event making our Solar System special, with a tendency to favor a broader cosmic variance rather than very special conditions for the Sun. These recent insights tell us that it is prudent scientific endeavor to question and consolidate seemingly-obvious details, before taking giant leaps and declare physics and astrophysics *"understood"*. The early history of the solar system is now again debated with great enthusiasm, making use of more precise observations and theoretical models of radioactivity signatures in our solar system bodies, the only place where we can study cosmic materials directly. We open the curtain to cosmic chemical evolution with measurements of ^{60}Fe and ^{244}Pu on Earth, and with ^{26}Al from nearby stars helping us to understand the local interstellar environment. Detail matters.

This book's Chap. 7 then is a mixture of complex astrophysical problems and observational constraints which come into play once single objects are put into a greater context of a galaxy. The interstellar medium connects stellar sources across cosmic time intervals, in its transport properties for kinetic energy and matter. Our models for these processes are first-order and simple, still. They help us to explain the coarse aspects of galaxy evolution. But again, more detailed astrophysical understanding will be required for a *realistic* model of how stars and supernovae feed back their newly-produced isotopes and their violence from stellar winds and explosions into next generations of stars to form, and to carry on the cosmic cycle of matter. For this reason, a new Chap. 11 now addresses our contextual knowledge of our own Galaxy, and our description of chemical evolution within a galaxy and across cosmic times, as an educational chapter (see below). All great themes of current astrophysics necessarily are involved here: Nucleosynthesis yields for entire populations of source types, their occurrence rates over a galaxy's evolution, the feedback, mixing, and transport processes across the many degrees of freedom for energy. Observing electromagnetic radiation from cosmic objects is a great tool to study distant physical processes. Its limitations become obvious in view of the complexities of a galaxy's evolution.

The book's Chaps. 8–12 provide a glimpse of the variety of *tools* which are involved in today's astrophysical work; we concentrate here on tools which most-directly relate to cosmic radioactivities. Nuclear physics made great strides to establish the concepts of nature as atomic nuclei are held together and interact; in recent years, much attention turned towards nuclear-structure details which matter in cosmic environments, and involves experiments with radioactive beams or targets. This field evolves, and returns again its focus more towards fundamental science, away from the power and weapons applications of earlier years, as demonstrated in Chap. 9. Supernovae still challenge even most-advanced modern computing facilities, from their huge dynamic ranges in space and time domains. Chapter 8 gives a snapshot of how physics and mathematics ideas are employed to obtain computer simulations which are approaching reality, within such technical limitations; this is a prominent example of complex-systems study. The instruments to obtain measurements from cosmic radioactivities then are highlighted in Chap. 10. We face very different categories of equipment, from space telescopes for penetrating gamma-rays through cosmic-ray probes and to sophisticated laboratory mass spectroscopy of minute samples of cosmic material extracted from meteorites. Each of these experimental fields is in the hands of small groups of experts, only a handful of laboratories world-wide working in each of these fields. Advances are very sensitive to specific conditions under which such small laboratories operate in their countries. Neither large science communities nor large industrial applications support technological evolutions here, rather the passion of experimental physicists is the driver of progress. Then, we add a chapter on chemical evolution, describing how best we can formulate an evolution with time of the composition of cosmic matter, putting into this all the complex models described in Part II of the book, and linking it to the observations described in Part III.

While the theoretical insights of D.D. Clayton laid the foundations of *astronomy with radioactivities* in the 1960s, the field of γ-ray line astronomy was mainly driven by observations. The discovery of the 1.8 MeV γ-ray line of ^{26}Al came somewhat as a surprise (as discussed in Chap. 2), and the one of the 511 keV line (the first γ-ray line originating from outside the solar system that was ever detected) was also unexpected. Equally unexpected was the detection of the ^{56}Co lines from SN1987A about 6 months earlier than predicted from spherically symmetric models of the supernova explosion. It was the improved angular resolution of the COMPTEL imaging gamma-ray telescope within a large field of view and the multi-year CGRO mission that made possible the identification of massive stars as major ^{26}Al sources in the 1990s. Similarly, the improved spectral capabilities of SPI/INTREGRAL (and its energy range, including the few-hundred keV range) set the stage for a first clear measurement of the decay γ rays from the ^{56}Ni decay chain in a thermonuclear supernova in 2014, with 2014J being sufficiently nearby. INTEGRAL's extended mission also allowed astronomers to perform the first reliable all-sky map of the Galactic 511 keV emission, which revealed and consolidated its surprisingly large intensity ratio of the Galaxy's bulge to its disk—the current puzzle both for cosmic-ray propagation and for Galactic positron sources. Detecting fainter emission requires mastering an inherent large instrumental background, and building up experience over a longer mission helps to discriminate systematic effects from the statistical uncertainty that always accompany signals at an instrument's sensitivity threshold. Finally, an X-ray telescope played a major role for advancing supernova knowledge: NuSTAR's mirrors work well up to 80 keV, and so include the low-energy lines from ^{44}Ti decay. So, we obtained a unique *image* of radioactivity emission in a 350-year-old supernova remnant, Cas A. Together with tracking the evolution of SN1987A, these two objects likely will drive our understanding of the final explosions from massive-star evolution.

It is somewhat risky, though unavoidable, to rely on few but well-observed events towards understanding an astrophysical source. Variety often reveals interesting physical processes that were overlooked at first. The gravitational-wave event/gamma-ray burst GW170817 again illustrates this clearly: For the first time, a neutron star binary collision has been witnessed, and observed in great detail due to its proximity. But, still too distant for nuclear-line observations from radioactivity, assumptions need to be made on the nucleosynthesis and the ejection of isotopes that we have not had a chance to measure at all, far from the nuclear stability regime. Such events will remain rare, and a lively discussion of what that means for the enhancement of cosmic matter through the r-process is ongoing, and will remain with us for a while.

All these discoveries boosted an intense activity both in theoretical astrophysics and in nuclear laboratory experiments in the past four decades, generating hundreds of papers on various aspects of cosmic radioactivities. Also the cosmic environment of the Sun and the important diagnostics of radioactivities received a boost in the past decade, from more precise measurements, and not least from the discovery of ^{60}Fe radioactivity on Earth and in cosmic rays near Earth.

Compared to other fields in astrophysics, the ratio of theoretical implications to observational data has been extremely high in the case of γ-ray line astronomy. It should be noticed that some of these implications were totally *unforseen* by the pioneers of the field. For instance, long-lived radioactivities, such as ^{26}Al, can be used not only to probe the interior physics of their stellar sources, but also the physics of the Galaxy at large: star formation rate, locations of high-mass star forming sites, distribution of supernova ejecta out of the Galactic plane, etc. Similarly, the properties of positron annihilation emission can help to probe the physics of cosmic ray propagation and the interstellar medium in a new way, or the configuration of the galactic magnetic field (see Chap. 7).

One might think then that such a prolific field can only have a bright future. Laboratory equipment for the analysis of meteoritic inclusions and their isotopic abundances proceeds to ever smaller grains and precision. However, the scarcity of the astronomical data from gamma-ray studies tells a different story. Ideas and concepts promise to dig deeper into the nuclear universe, yet by modest increments, compared to other disciplines of astronomy. It will take a considerable effort to increase the number of sources where our theories can be challenged and expanded; even an increase by a modest factor of a few will require a multi-national space mission. No mission dedicated to γ-ray line astronomy is on the horizon at present, although technically, advances in sensitivity by almost two orders of magnitude have been demonstrated in lab studies (see Chap. 10). Current excitement in the astrophysical community focuses on questions of cosmology and dark components of the universe, and on consolidations of the grounds for new astronomy through gravitational waves, neutrinos, and highest-energy γ-rays and cosmic rays. Proposed nuclear-gamma-ray telescope projects may stimulate more excitement and support in new space programs only beyond the next decade, given a 10–15 year lead time for any such space mission. Other isotopic information may become available from superb resolutions now reaching into isotopic line shifts for molecular lines in the radio regime with ALMA, and even for atomic lines in the optical. X-ray lines may help to constrain elemental abundances in hot gas of the intergalactic medium, in addition to the already fruitful explorations of hot ISM in supernova remnants, and, for very few cases, even from nuclear transitions. It is clear that a considerable effort will be required in all those fields before robust theoretical predictions become available on the yields of various radioactivities. These will remain to be verified most-directly by the intensities of the corresponding γ-ray lines from radioactive decay. Valuable new insights on cosmic radioactivities may derive, as spin-offs from other fields, such as high-resolution spectroscopy resolving isotopic information, or astro-particle advances on cosmic ray details, or solid-state research connected to materials science.

Similarly, significant effort is required in theoretical and associated fields of physics: Refined descriptions of chemical evolution accounting for various galaxy components, a better understanding of nuclear structure derived from nuclear reaction experiments and theories, and more realistic models of stellar explosions. All these elements are needed in order to improve our *view and theoretical understanding* of the cosmic sources of new isotopes which have been detected up to now. This book aims to help students of astrophysics better understand the role of cosmic radioactivities in relation to their specific interests. It will hopefully also help build up resources and excitement to further advance our understanding of the origins of the cosmic elements—one of the *big questions* agreed upon in the science community.

Appendix A
Science of Cosmic Radioactivities: Milestones

This list was organised along key publications, listing first authors; often, teams of scientists and major instrumental facilities are behind such achievements. We chose a concise listing for a better overview, being aware that this may not do justice to the many contributing individuals. Figure A.1 reflects the research activity in the field of this book through the publications in scientific journals, for a broader and a more-narrow subject term (see caption).

- **1869** Mendeleyev organizes the elements
- **1895** Röntgen discovers X-rays
- **1896** Becquerel discovers radioactive Uranium
- **1899** Rutherford describes radioactive α, β decays
- **1907** Boltwood uses U-Pb radiometric dating to evaluate the age of rocks
- **1929** Rutherford estimates the age of the elements from the ^{235}U/^{238}U ratio
- **1930** Dirac predicts the anti-electron (positron)
- **1932** Anderson discovers the positron
- **1932** Chadwick discovers the neutron
- **1934** Fermi formulates theory of beta-decay
- **1938** Bethe and Critchfield describe the p-p reaction (incl. β decays) powering the Sun for billions of years
- **1939** Bethe describes the CN cycle in stars
- **1946** Hoyle suggests iron synthesis in stars (e process)
- **1947** Suess correlates nuclear magic numbers with isotopic abundances
- **1948** Haxel, Jensen, and Suess and Mayer advance shell model of magic numbers
- **1950** Fowler demonstrates that p-capture by ^{12}C yields radioactive ^{13}N
- **1950** Borst advances radioactivity as the power of supernova luminosity
- **1952** Merrill discovers radioactive Tc in stellar atmospheres
- **1952** Salpeter introduces triple-alpha reaction through quasiequilibrium ^{8}Be
- **1953** Hoyle predicts the 7.7 MeV resonance in ^{12}C
- **1954** Hoyle advances nucleosynthesis (C to Ni) in massive stars
- **1956** Suess and Urey suggest solar abundances

R. Diehl et al. (eds.), *Astrophysics with Radioactive Isotopes*, Astrophysics and Space Science Library 453, https://doi.org/10.1007/978-3-319-91929-4

- **1956** Pankey Jr. suggests that radioactive ^{56}Ni powers supernova light
- **1957** B2FH review and systematize stellar nucleosynthesis
- **1957** Cameron "Chalk River" lectures on Stellar Evolution and Nucleogenesis
- **1958** Feynman and Gell-Mann theory of Fermi interaction for beta decay
- **1960** Reynolds discovers excess ^{129}Xe from extinct ^{129}I in meteorites
- **1960** Fowler and Hoyle introduce ^{238}U/^{232}Th cosmochronology
- **1961** Clayton formulates time-dependent s-process including β decays
- **1962** Cameron suggest supernova injection of ^{26}Al at solar birth
- **1964** Clayton's cosmoradiogenic chronologies of Re*Os and U*Pb
- **1964** Reynolds and Turner discover Xe-X excess in four heavy Xe isotopes
- **1964** Bahcall studies electron capture in stellar interiors
- **1965** Seeger, Fowler, and Clayton develop time-dependent r-process
- **1965** Clayton and Craddock consider γ-ray lines from r-process ejecta
- **1966** Colgate and White advance hydrodynamic models of SNII
- **1967** Fowler assembles experimentally guided thermonuclear reaction rates
- **1968** Clayton Si-burning quasiequilibrium process
- **1969** Colgate and McKee suggest ^{56}Ni as the power source of SNIa
- **1969** Clayton, Colgate and Fishman predict ^{56}Co and ^{44}Ti γ-lines from supernovae
- **1969** Clayton and Silk predict cosmic background from ^{56}Co
- **1969** Arnett predicts ^{26}Al yield in explosive carbon burning
- **1971** Reynolds shows that extinct ^{244}Pu fission Xe exists in meteorites
- **1971** Clayton predicts ^{60}Fe gamma-ray lines from integrated supernova history
- **1972** Black discovers ^{22}Ne-rich gas in meteorites
- **1972** Johnson et al. discover galactic 511 keV line from e^+ annihilation
- **1972** Starrfield et al. describe radiogenic luminosity of novae
- **1973** Clayton proposes ^{56}Co and ^{44}Ti as a galactic positronium source
- **1973** R. Clayton, Grossman, and Mayeda discover ^{16}O-rich Al-rich inclusions
- **1974** Clayton and Hoyle consider γ-line emission from novae
- **1974** Gray and Compston find excess ^{26}Mg in Al-rich minerals in meteorites
- **1975** Clayton predicts ^{44}Ca-rich Ca from extinct ^{44}Ti in Ti-rich presolar minerals
- **1975** Clayton predicts ^{22}Na γ-ray lines from young supernovae
- **1975** Clayton and Ward predict s-process Xe in stardust
- **1976** Clayton and Hoyle predict anomalous nova dust with extinct ^{22}Na and ^{26}Al
- **1976** Ward, Newman, and Clayton analyze radioactive s-process branching
- **1976** Clayton, Dwek, and Woosley consider proton irradiation in the solar disk
- **1977** Wasserburg shows initial ^{26}Al/^{27}Al = 5 10^{-5} in Al-rich inclusions
- **1977** Cameron and Truran supernova trigger to inject ^{26}Al and other isotopes
- **1977** Clayton chemical memory model of fossil ^{26}Mg excess from ^{26}Al
- **1978** Kelly and Wasserburg discover extinct ^{107}Pd in an iron meteorite
- **1978** Srinivasan and Anders find s-process Xe in meteorites
- **1981** Clayton predicts ^{49}Ti excess from ^{49}V in supernova dust
- **1982** Mahoney et al. discover 1.8 MeV line from Galactic ^{26}Al with HEAO-C
- **1983** Yokoi, Takahashi, and Arnould calculate speedup of ^{187}Re decay

- **1985** Clayton analytic models for mean ISM radioactivity and cosmochronology
- **1985** Launch of the Solar Maximum Mission (SMM) satellite
- **1985** Share and Leising measure integrated ^{26}Al mass in the ISM with SMM
- **1987** Butcher measures galaxy age by Th/Nd in old dwarfs
- **1987** Kamiokande and IMB find prompt neutrinos from SN1987A
- **1987** Anders isolates stardust by dissolving meteorites in acid
- **1987** Zinner studies stardust isotopically using a mass spectrometer
- **1987** Bernatowicz et al. isotopically study SiC grains from red giants
- **1988** Balloon-borne Germanium spectrometers record ^{56}Co lines in SN1987A
- **1988** Matz, Share, and Chupp detect ^{56}Co gamma rays in SN1987A with SMM
- **1989** Hudson et al. measure initial solar plutonium from ^{244}Pu
- **1990** Tueller et al. measure γ-ray line profiles in SN1987A with GRIS
- **1990** Woosley et al. study ν-process, potentially enhancing ^{26}Al yields in SNe
- **1991** Launch of the Compton Gamma Ray Observatory (CGRO)
- **1991** GALLEX and SAGE neutrino observatories detect pp neutrinos
- **1991** Prantzos predicts distribution of ^{26}Al in Galactic spiral arms
- **1992** Diehl et al. observe galactic plane ^{26}Al with COMPTEL
- **1992** Kurfess et al. detect ^{57}Co in SN1987A with OSSE
- **1992** Clayton et al. introduce delayed radioactive power for SNII
- **1994** Iyudin et al. report ^{44}Ti in Cas A from COMPTEL data
- **1995** Liu and Dalgarno explain SN1987A CO by radioactive dissociation
- **1995** Diehl et al. present COMPTEL map of galactic ^{26}Al
- **1996** Srinivasan et al. find ^{41}Ca/^{40}Ca= 1.5 10^{-8} in meteorites
- **1996** delRio et al. ^{26}Al from the Cygnus region with COMPTEL
- **1996** Hoppe et al. and Nittler et al. find extinct ^{26}Al and ^{44}Ti in SiC X-grains
- **1997** Timmes et al. use ^{26}Al to constrain the galactic star formation rate
- **1997** Dupraz et al. COMPTEL search for ^{44}Ti sources
- **1999** Clayton, Liu and Dalgarno discuss how radioactivity affects graphite dust
- **1999** Korschinek, Knie et al. find live ^{60}Fe in terrestrial ocean sediments
- **2000** Plüschke/Cerviño et al. apply population synthesis to ^{26}Al in Cygnus
- **2002** Hoppe and Besmehn discover ^{49}Ti-rich titanium owing to extinct ^{49}V decay
- **2002** Diehl et al. detect ^{26}Al emission from Orion with COMPTEL
- **2003** Diehl et al. use INTEGRAL/SPI to show narrow width of ^{26}Al line
- **2003** Knödlseder et al. INTEGRAL/SPI all-sky map of positron annihilation
- **2005** Jean; Churazov et al. show annihilation to occur in warm-ionized ISM
- **2005** Smith/Harris et al. find Galactic ^{60}Fe γ rays with RHESSI/INTEGRAL
- **2006** The et al. discuss implications of the paucity of ^{44}Ti sources
- **2006** Diehl et al. find Doppler shifts of the ^{26}Al line from Galactic rotation
- **2008** Arpesella et al. detect solar ^{7}Be neutrinos with BOREXINO
- **2010** Diehl et al. detect ^{26}Al from the Sco-Cen association with INTEGRAL/SPI
- **2013** Kretschmer et al. find hot ISM with ^{26}Al at high velocity
- **2014** Grefenstette et al. image the Cas A SNR in ^{44}Ti X-rays
- **2014** Diehl et al. detect early ^{56}Ni γ-ray lines from SN2014J
- **2014** Churazov et al. detect ^{56}Co γ-ray lines from SN2014J
- **2014** Seitenzahl et al. derive radioactivities in SN1987A from its lightcurve

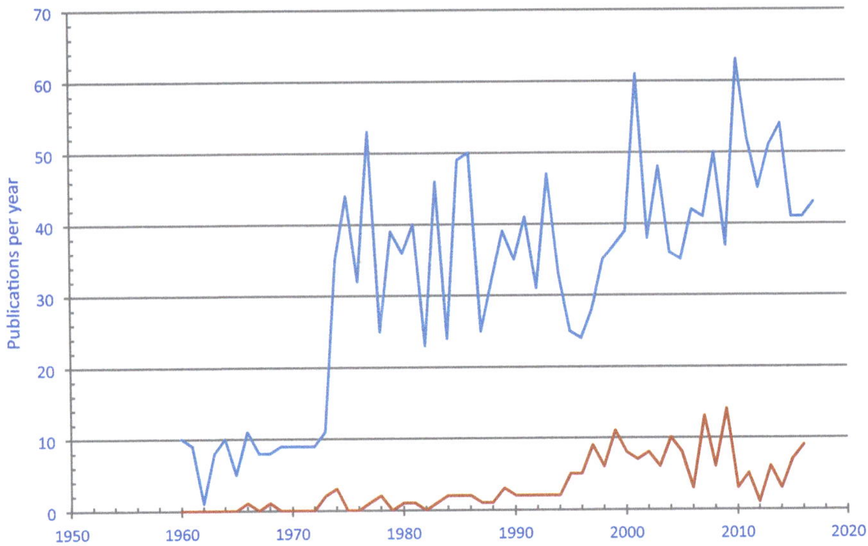

Fig. A.1 Annual publication rates for journal papers related to cosmic radioactivities (blue upper line) and their subset addressing nucleosynthesis (red lower line). Based on NASA/ADS

- **2015** Diehl et al. measure ^{56}Co γ-ray lightcurve and spectrum from SN2014J
- **2015** Boggs et al. detect ^{44}Ti in SN1987A in hard X-rays
- **2015** Siegert et al. detect e^+ annihilation from a microquasar flare (V404 Cyg)
- **2015** Izzo; Tajitsu; Molaro find nova ^{7}Be hints V1369Cen, V339Del, V5668Sgr
- **2015** Skinner; Siegert et al. find e^+ emission from a faint, thick disk component
- **2016** Wallner et al. detect ^{244}Pu in an oceancrust as evidence of rare r-process
- **2016** Fimiani et al. detect live ^{60}Fe in probes of lunar material
- **2016** Wallner et al. find live ^{60}Fe on Earth from a variety of oceanfloor probes
- **2016** Ludwig et al. find live ^{60}Fe in terrestrial microfossils
- **2016** Siegert et al. measure e^+ annihilation spectra from bulge, disk and GC
- **2016** Siegert et al. use 511 keV from dSphs to exclude dark matter origin
- **2017** Grefenstette et al. derive velocity details of Cas A ^{44}Ti clumps
- **2017** Smartt et al. and Pian et al. find kilonova emission from a neutron binary collision identified on LIGO/VIRGO gravitational wave data

Additional historic pointers in the area of gamma-ray astronomy can be found at http://heasarc.gsfc.nasa.gov/docs/history/.

Appendix B
Glossary

Often-Used Terms in Astrophysics with Radioactive Isotopes

Terminologies from very different fields are involved in the discussions of this book: Theoretical astrophysics of stars, stellar interiors, and stellar explosions is intertwined with theories of the interstellar medium and galaxy evolution. Observations from stellar-photosphere spectroscopy through radio, IR, and X- and gamma-ray spectroscopy span a wide range, additionally cosmic-ray composition measurements and laboratory mass spectroscopy of small inclusions in meteorites are discussed. Nuclear-physics experiments and theories for nuclear levels and reaction cross sections are deeply involved as well.
Here we provide a glossary of a few terms that appear often throughout the book.

- **Abundances.** The abundance of a chemical element may be measured by mass-fraction, or number-fraction. For example, the mass-fraction abundance of oxygen in water is about 90%, while the number-fraction is only 33% because 1 atom in 3 in water is an oxygen atom. On the cosmic scale, the mass-fraction of hydrogen and helium are about 74% and 23–25% respectively, while their relative abundance ratio by number is about 10:1. Abundances are often quoted relative to solar values (although not all values of the solar abundances come directly from measurements of the solar spectrum). The solar mass fraction of all elements other than hydrogen and helium (the metallicity) is slightly larger than 1%. Another notation often employed is the bracket [X/H] = $\log(X/H)/\log(X/H)_\odot$, i.e., the logarithmic (base 10) value of the number ratio of species X to hydrogen, relative to that ratio in the Sun. For example, stars in the Milky Way with the lowest metallicity exhibit [Fe/H] = −5, i.e. 10^{-5} solar iron abundance.
- **Chemical Evolution.** The change with time of relative abundances of elemental (and isotopic) species in gas and stars of galaxies. The study of this abundance evolution, by elemental or isotopic mass fraction, as a function of cosmic time

R. Diehl et al. (eds.), *Astrophysics with Radioactive Isotopes*, Astrophysics and Space Science Library 453, https://doi.org/10.1007/978-3-319-91929-4

(often expressed in redshift for the distant/early universe) in terms of its astrophysical agents (stars and gas flows) is called the science of chemical evolution (though no chemistry is involved). When treating all species collectively, one often follows the overall metallicity, instead of individual abundances. One distinguishes galactic chemical evolution (GCE) and cosmic chemical evolution (CCE) when referring to changes *with* a galaxy, or for the universe as a whole, respectively.

- **Chemodynamics.** When chemical evolution is combined with dynamic aspects, such as stellar motions or hydrodynamic feedback of massive stars on the ISM, one speaks of chemo-dynamics, thus generalizing CE to CD.
- **Core collapse (Supernova).** Once a star has consumed its nuclear fuel and cannot release nuclear binding energy in its core from nuclear reactions, it will not counteract to gravitational pressure which forces contraction. An important alternative internal energy reservoir for stellar interiors is kinetic energy of electrons confined into a small-volume stellar core, as degeneracy and the Pauli exclusion principle enforce electron energies up to the *Fermi energy*. If electrons are forced to Fermi energies reaching nuclear energies, atomic nuclei may capture such electrons in a nuclear weak transition. Such reduction of electron degeneracy pressure is the cause of core collapse in stars of masses around $10\,M_\odot$. Such core collapses, when they occur from a rotating massive star, are believed to be the origins of the *long-duration* subclass of *Gamma Ray Bursts*. Gamma-Ray Bursts are extremely bright flashes, and may arise from stellar core collapses as early as those have been formed, i.e. out to redshifts 10–20, well beyond where galaxies can be observed. Such remote light beacons allow us to study elemental abundances in the remote and young universe, from absorption line analysis (similar to the Fraunhofer lines in the Solar spectrum).
- **COMPTEL.** One of the four gamma ray telescopes aboard CGRO. The gamma-ray detectors in COMPTEL are counters of events of double Compton scattering. Their geometrical arrangement fixes the kinematics of the incident gamma ray, and thereby allows (coarse) imaging as well as spectroscopy.
- **Compton Gamma Ray Observatory (CGRO).** One of NASAs great observatories, launched by shuttle Atlantis in 1991, carrying four experiments (BATSE, COMPTEL, EGRET, and OSSE) to study gamma rays over a wide range of energies.
- **Extinct Radioactivity.** Solid objects retain a record of having once contained an abundant radioactive nucleus, which is no longer present owing to the age of the solid greatly exceeding the halflife of the radioactive nucleus. The radioactivity is therefore now extinct. The initial activity is recorded by an excess abundance of the daughter isotope of the radioactive nucleus. The effect is largest in solids having a large abundance of the parent element and a small abundance of the daughter element. Many such examples exist, and they are called extinct radioactivity.
- **Cosmic Rays.** Particles which penetrate the Galaxy and have high energies above $\sim$MeV, so that their interaction with matter causes the generation of secondary-particle avalanches. Cosmic rays mostly consist of high-energy protons, about

1% are electrons, and a small fraction are heavier atomic nuclei up to Fe. The energy spectrum reaches up to 10^{21} eV, where one (sub-atomic!) cosmic-ray particle alone has an energy comparable to a speedy tennis ball. The origins of cosmic rays is among one of nature's great mysteries and subject to a field of astrophysics and astro-particle-physics. Supernovae, pulsars, and active galaxies play a role. We distinguish Galactic cosmic rays (GCR) from extragalactic cosmic rays, from Solar Energetic Particles (SEPs), due to their (sometimes uncertain) origins, as inferred from indirect arguments.

- **Gamma Ray Lines.** A gamma-ray line is a flux of gamma-ray photons having the specific energy of a nuclear transition between two nuclear energy levels in a specific nucleus. That flux reveals the presence of those excited nuclear states. For example, the 847 keV gamma ray transition in the ^{56}Fe nucleus occurs when abundant radioactive ^{56}Co decays populate the upper ^{56}Fe level of the 847 keV transition, which then rapidly and spontaneously emits the 847 keV photon.
- **Gamov Peak.** In thermonuclear reactions inside stars energy dependent cross sections, $\sigma(E)$, must be convolved with the thermal probability density of particle velocities (or, equivalently, their relative energy, E), which is given by the characteristic Maxwell-Boltzmann, $f_{\mathrm{MB}}(E) \propto E\exp(-E/kT)$, where k is the Boltzmann constant, and the energy, E, is evaluated in the center-of-mass reference frame. Charged particle reactions rely on tunneling through the Coulomb barrier between the interacting particles, which causes the cross section to depend on energy exponentially, $\sigma(E) \propto \exp(-2\pi\eta)$, where the Sommerfeld parameter, η, depends on nuclear masses, charges, and scales with energy as $\eta \propto 1/\sqrt{E}$. Reactions rates are small at low E, as the cross section is then small, and also small at high E, as the Maxwell-Boltzmann distribution then indicates a small number of particles at such energies. The evaluation of thermonuclear reaction rates (see Chap. 9) in thermal environments (such as stellar interiors) is thus given by $< \sigma v > \propto \int_0^\infty dE\ E\ f_{\mathrm{MB}}(E)\sigma(E)$. The competition between nuclear physics ($\sigma(E)$) and thermal physics ($f_{\mathrm{MB}}(E)$) results in a product of functions in the integrand that peaks at a characteristic energy $E_0 \sim 1$ keV to 1 MeV, depending on the burning stages involved (i.e., depending on the charges of the key reactants).
- **INTEGRAL.** An ESA mission called INTErnational Gamma-Ray Astrophysics Laboratory. Launched in 2002 for a planned 3-year mission, extended to at least end 2012, due to scientific success and uniqueness of its main instruments.
- **Initial Mass Function.** A probability density function (pdf), $\Phi(m)$, describing the likelihood of forming stars in the mass range (m, m + dm) in the range from ml (the lower mass limit of stars, determined by the condition of stable core hydrogen burning) to mu (the upper mass limit, set by the limiting factor of radiation driven pulsational instabilities above which stars can not settle on a stable configuration). The commonly adopted range for these parameters is ml = 0.1 $M_\odot$, and mu = 100 $M_\odot$. A frequently adopted pdf is the Salpeter IMF, $\Phi(m) \propto m^{-\alpha}$, with $\alpha = 2.35$. As a pdf, Φ is normalized to unity when integrating over the indicated range of stellar masses. The average stellar mass is given by $<$

$m >= \int_{m_l}^{m_u} dm\Phi(m)\, m$, which evaluates to $< m >\sim 0.5\,M_\odot$. If all stars above $m_{SN} = 10\,M_\odot$ end their lives as core collapse supernovae (ccSNe), the fraction of all stars that undergo core collapse becomes $f_{SN} = \int_{m_{SN}}^{m_u} dm\Phi(m) \sim 0.001$.

- **Massive Star.** Stars may have masses from 0.1 to 100–1000 times that of the Sun. Low-mass stars are most abundant (see *Initial Mass Function*). A star massive enough so that it can ignite nuclear burnings beyond Helium burning are called *massive*. Often one draws a sharper line and uses Carbon burning as a criterion, as such a star then undergoes rapid evolution due to substantial neutrino energy losses and develops into a gravitational collapse, called a *core-collapse supernova*.
- **OSSE.** One of the four gamma ray telescopes aboard CGRO. The gamma-ray detector in OSSE was a NaI scintillator. The quantity of photo energy produced by the scintillator measures the energy of the incident gamma ray.
- **Quasiequilibrium.** During Silicon Burning, very hot ^{28}Si nuclei do not combine by nuclear fusion; rather they melt by photoejection of protons, neutrons and alpha particles. Those free particle densities assume a steady state in which they are globally recaptured by coexisting nuclei at the same rate at which they are photoejected from that distribution of nuclei. That steady state is called quasiequilibrium, and accounts for the intermediate-mass abundance distribution of nuclei between ^{44}Ca and ^{60}Ni. It provides a good description of the nucleosynthesis in that mass range. Its discovery in 1967–1968 filled the last major gap in Hoyle's theory of nucleosynthesis in supernovae. This process is of vital significance for AwR, because it provides the nucleosynthesis source for radioactive ^{44}Ti, ^{49}V, ^{52,53}Fe, and ^{56,57}Ni, whose decays are prominent for gamma-ray line astronomy and for isotopes in supernova stardust (SUNOCONs). The alpha-rich freezeout occurs when the temperature is sufficiently high to break down all ^{28}Si into primarily alpha particles, so that the reassembly during cooling also builds more nuclei in this mass range. This variant of quasiequilibrium is especially prolific in production of ^{44}Ti, ^{56}Ni and ^{57}Ni. The transition from the dominance of stable nuclei (primarily $Z = N$) below $Z = 21$ to radioactive positron emitters (also primarily $N = Z$) above $Z = 21$ occurs because abundances are restricted by the gas having low excess number of neutrons, so that $Z = N$ nuclei must dominate the abundances; however, owing to the positive Coulomb energy between nuclear protons, those nuclei having $Z>21$ have their most stable isotopes at $N = Z + 4$ rather than at $N = Z$. Therefore, the synthesis at $Z = N$ undergoes positron emission after the explosion to reach $Z = N - 4$. For example, stable ^{40}Ca ($Z = N = 20$) is the most abundant isobar at $A = 40$, whereas radioactive ^{44}Ti ($Z = 22$, $N = 22$) decays later to ^{44}Ca, the most abundant isobar at $A = 44$.
- **Population Synthesis.** Clusters of stars are prime examples of systems of stars that are believed to be coeval, or nearly so. The evolution of such a system of stars, all born at the same time, is also known as a simple stellar population (SSP). Given an initial mass function (IMF) that describes how stars are statistically distributed in their initial mass, m, the theory of stellar evolution allows an

evaluation of various quantities relevant to this SSP as a function of time. For example, the rate of supernovae of a given type can be modeled and the amount of energy returned to the ISM from these supernovae and the mass loss preceding them (e.g., STARBURST99). Such studies are relevant for the study of the properties of the ISM in galaxies as a function of time. Galaxies have a continuous star formation history, so that the delta-function burst represented by a SSP has to be convolved with an assumed or calculated SFR(t). In chapter X we discuss this in the context of radioactivities (^{26}Al, ^{60}Fe) in nearby star forming regions. Further applications are color evolution of star clusters and galaxies as a whole, and simulations of compact source populations. The latter must also include a treatment of binary stars. Population synthesis relies heavily on input from the theories of stellar evolution in single and multiple star system, but it does not address the issues of star formation itself.

- **Presolar Grain.** Inclusions in meteorites which show an clearly-unusual isotopic abundance signature (i.e. large factors beyond the spread seen among meteoritic samples) are attributed to an origin outside the solar system. This implies that they must have formed from material that is a different mix than what made the solar-system bodies. Since meteorites formed when the solar system was in its infancy, these anomalous grains must have been formed before the solar system was established. It may be more appropriate to call these grains *stardust*, as such grain formation continues to take place throughout the Galaxy.
- **Radioactivity.** Atomic nuclei which are not in their most-stable configurations of nucleons will eventually change their internal arrangement towards a more stable configuration. These nucleonic configurations are dictated by the laws of quantum physics. Transitions are made possible by the nuclear forces. Upon state transitions, secondary particles may be ejected, which causes the *harmful* effects to biological life. Radioactivity changes the type of isotope. Radioactive decay of an isotope occurs at a random moment in time, which can be estimated (though not predicted) from the characteristic decay time of each unstable isotope. Emission of secondaries from radioactive decay is thus independent of temperature or ionization state of a particular atom, therefore carries unique information not distorted by such environmental parameters. The physical unit measuring radioactivity is the *Becquerel*, counting decays per second, and named after Henri Antoine Becquerel, who discovered radioactivity in 1896.
- **Radioactive Beam Facilities.** Particle accelerators for performance of nuclear-reaction experiments, where the accelerated particles themselves are unstable. This is necessary for the study of nuclear reactions relevant for astrophysics, as in cosmic sites many reaction partners will have a composition deviant from stable nuclei as we know them.
- **Radiogenic Luminosity.** Radioactivity keeps a gas hot when it would otherwise cool, so it radiates when it would otherwise be dark. The energy input to the gas from radioactivity is first degraded to heat. Radiogenic luminosity occurs in explosive objects (supernovae and novae) because they contain sufficient radioactivity and would otherwise rapidly cool by expansion. A related phenomenon, radiogenic excitation, occurs when the non-thermal radiations from the

radioactivity cause some species to have higher excitation than it would otherwise have at the ambient temperature. Examples of the latter are (1) He+ emission lines from a gas too cool to normally contain He+ ions; (2) small abundance of the CO molecule in a gas having abundant C and O at a temperature low enough that thermal equilibrium would favor the CO molecule.

- **Solar Abundances.** See Abundances
- **Secondary-Ion Mass Spectrometry.** Called SIMS for short, the charged secondary ions liberated by sputtering are subjected to electric and magnetic fields that determine their masses accurately enough to count distinct isotopes of the common elements lighter than iron.
- **Stardust.** Stardust is a scientific word for a solid mineral grain that condensed thermally from the hot but slowly cooling gases leaving a star. Their mineral structures and high-temperature stability both attest to condensation from hot gases. Their extreme non-solar isotopic compositions are uniform throughout each grain. SiC is ^{13}C-rich and mildly ^{29,30}Si-rich compared to the sun, whereas oxides are predominantly ^{17}O-rich. Typical sizes 0.1–1 μm are not visible to the naked eye. They are solid pieces of stars. Most known examples are extracted from the meteorites, whose accumulation within the accretion disk collected also the stardust. Those extracted from meteorites are necessarily from a star that lived prior to the birth of the sun. Stardust is recognized by its extreme isotopic composition, measured by secondary-ion mass spectrometry, and contains many extinct radioactive nuclei.
- **SUNOCON.** Constructed by D.D. Clayton from the words SUperNOva CONdensate, SUNOCON is a scientific word for one specific type of stardust of exceptional importance to nucleosynthesis, supernova structure and Astrophysics with Radioactivities. Dust grains condense thermally within the hot, deep interiors of young supernovae while they are expanding explosively and cooling adiabatically. Beginning times for condensation are after a few months and end after a few years. Their mineral structures include the most refractory of high-temperature minerals, graphite (C), TiC, SiC, TiN. Their extreme isotopic compositions indicate interior supernova shells, as do their large contents of extinct radioactivities, ^{44}Ti, ^{26}Al, ^{49}V, ^{95,97}Zr and others. SUNOCONs are specifically not dust that condenses around supernovae, or as the ejecta interacts with external matter, or as H-containing supernova dust, or as dust aggregates.
- **Thermonuclear Supernova.** A compact and degenerate white dwarf star cannot expand fast enough once nuclear Carbon burning is efficiently ignited in its interior. The flame spreads so rapidly that the release of nuclear energy exceeds the gravitational binding energy of such a white dwarf, and thus the entire star is disrupted. It is unclear in detail how such Carbon ignition might occur. Thermonuclear supernovae are believed to be the cause of the observed class of *Type Ia* supernovae. These typically produce large amounts of radioactive ^{56}Ni, which makes them shine brightly so they can be seen from cosmological distances. They can be empirically calibrated in their absolute brightness, which is the basis for their role in studying the expansion history of the universe.

- **Yield.** The amount of material (usually in units of solar masses) returned to the ISM by novae, supernovae, stellar winds, etc., as a function of specific source parameters, such as the initial mass of the progenitor star. For example, the yield of ^{26}Al is sensitive function of mass, and, after weighting by the IMF, is of order $10^{-4}\,M_{\odot}$ per star massive enough to end its life as a supernova. Yields are required for the calculation of galactic or cosmic chemical evolution models.

Index

R. Diehl et al. (eds.), *Astrophysics with Radioactive Isotopes*, Astrophysics and Space Science Library 453, https://doi.org/10.1007/978-3-319-91929-4

Kolibri 360 GmbH
Ferdinand-Jühlke-Straße 7
99095 Erfurt, Deutschland
produktsicherheit@kolibri360.de